HUMANKIND EVOLVING

An Exploration of the Origins of Human Diversity

A. Roberto Frisancho

KENDALL/HUNT PUBLISHING COMPANY
4050 Westmark Drive Dubuque, Iowa 52002

Book Team

Chairman and Chief Executive Officer Mark C. Falb
Senior Vice President, College Division Thomas W. Gantz
Director of National Book Program Paul B. Carty
Editorial Development Manager Georgia Botsford
Developmental Editor Angela Willenbring
Vice President, Production and Manufacturing Alfred C. Grisanti
Assistant Vice President, Production Services Christine W. O'Brien
Prepress Editor Angela Puls
Permissions Editor Renae Horstman
Designer Suzanne Millius
Managing Editor, College Field David Tart
Associate Managing Editor, College Field Ray Wood
Acquisitions Editor, College Field David Mattaliano

ISBN 0-7575-1408-1

Printed in the United States of America
10 9 8 7 6 5 4 3 2 1

Contents

Chapter 10 Evolution of the Genus *Homo* 149

Chapter 11 Evolution of *Homo sapiens* 165

Chapter 12 The Peopling of the New World and the Origins of the Anatomically Modern *Homo sapiens* 185

Chapter 13 Diet and Brain Evolution 199

Chapter 14 Bipedalism and the Evolution of Extended Childhood 209

Section IV. Human Adaptation and Biological Diversity

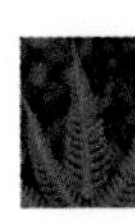

Chapter 15 General Principles for the Study of Human Adaptation 223

Chapter 16 The Human Life Cycle 235

Preface

Humans have an evolutionary history as primates stretching back nearly 60 million years, a history that was shaped by the foods on which they come to depend for their survival. Food has played a major role in the evolution of all of human biological characteristics ranging from the teeth, digestive system, brain volume to general body size. Evolution is based upon differences in reproductive efficiency, but these differences are a function of unequal access to food. This point was recognized by Darwin in 1859 using Malthus's hypothesis that all populations, including humans, grow in number at a geometric rate while food production increases only arithmetically, showing that variability in dietary resources is an important factor that influences natural selection. The purpose of this book is to examine the evolutionary foundations of human variability. For this purpose, the course will address the principles of human evolution, fossil evidence, behavior, and morphological characteristics of human and nonhuman primates. In addition, human interpopulation differences and environmental factors that account for these differences are discussed in a succinct form. To accomplish this goal the topics included in this book are divided into four sections of twenty-three interrelated chapters.

Section I. Evolution and Biological Anthropology focuses on the role of evolution from its historical roots to its contemporary use in biological anthropology and related fields, and includes three chapters:

1. Development of evolutionary theory;
2. Cellular basis of diversity; and
3. Principles of microevolution.

Section II. Macroevolution and Primate's Roots traces the primate's unique functional and behavioral traits that differentiate from other mammals. It includes five chapters:

4. Mammalian antecedents;
5. Living primates;
6. Primate behavioral ecology;
7. Fossil evidence of primate evolution; and
8. Food and evolution of digestive system.

Section III. Hominid Evolution focuses on the fossil evidence tracing the path to *Homo sapiens.* It includes six chapters:

9. Hominoid-hominid phase of evolution;
10. Evolution of the genus *Homo*;
11. Evolution of *Homo sapiens*;
12. Peopling of the new world;
13. Diet and brain evolution; and
14. Bipedalism and the evolution of extended childhood.

Section IV. Human Adaptation and Biological Diversity focuses on how humans have and are adapting to an old and new changing environment. This section includes nine chapters:

15. Principles of human adaptation;
16. Characteristics of the human life cycle;
17. Nutrition variability and the life cycle;
18. Physiological adaptive responses of contemporary populations to hot and cold climates;
19. Climate and evolution of variability in body size, body proportion, and hairlessness;
20. Human adaptation to high-altitude environments;
21. Biocultural origins of lactose intolerance and the evolution of skin color;
22. Adapting to a changing world; and
23. Globalization of obesity.

Acknowledgments

One of the greatest satisfactions of a life in science and research are the people with whom one shares ideas and stimulating discussions. Professors Paul T. Baker and Thelma Baker to whom this book is dedicated have played a pivotal role in my graduate training and thereafter. I am grateful for the friendship and the continuous collaboration of Anthony B. Way and R. Brooke Thomas with whom we conducted many aspects of the research included in this book. I am thankful for the cooperation and assistance of Guy Pawson, Michael A. Little, Jim Bindon, Daniel Brown, Tom Brutsaert, James F. Carrie, Douglas Crews, Darna Dufour, Norris Durham, Ralph Garruto, Michael Hanna, Jerry Haas, Charles Hoff, Gary James, Lynnette Leidy, Leslie Lieberman, Stephen Mcgarvey, Jay Pearson, Tom Leatherman, and Charles Weitz.

Throughout the preparation of this book I have benefited from the advice and comments of the University of Michigan professors including: C. Loring Brace, Stanley M. Garn, Frank P. Livingstone, John Mitani, Milford Wolpoff, and Rachel Caspari. Professor Mitani and Professor Wolpoff have generously supplied me with prints and photographs of extant primates and fossils that are included in this book. Professor Cynthia Beall of Case Western University has generously shared her ideas about high-altitude adaptation and supplied me with photographs of the Tibetans included here.

Many former graduate students, graduate student instructors, and undergraduate students who participated in many of the field projects have directly or indirectly contributed to the research projects we conducted in the United States, Peru, and Bolivia. I am grateful to Elizabeth T. Abrams, Rachel Albalak, Daniel Becque, William J. Babler, Veronica Barcelona, Steve Bailey, Ulana Bereza, Paul Bohensky (who also made several drawings acknowledged in the respective figures), Gary A. Borkan, Alejandro Camino, Anthony Corridore, Carmen Kudyba, Janet Dunn, Maria Estigarribia, Jeromy DeSilva, Lilian K. Gleiberman, Lorna Grindly-Moore, Scott D. Grosse, Nicole Guilding, Abigail Hardie, John Haws, Russell Herrold, Laurie Hoffman-Goetz, Erin Holmes, Virginia Hutton, Patricia C. Juliao, Elizabeth Kapp, Mary Kelaita, Jane E. Klayman, Mathew Kroot, Marquesa LaVelle, San-Hee Lee, William J. Leonard, William Lowe, Gene Mahaney, Stacy McGrath, Charles R. Marks, Elizabeth M. Miller, Christina Miranda, Pablo Nepomnaschy, Robin Nelson, Khouri Newlander, Christopher O'leary, Maria Perez, Conrad Quintyn, Karen Rosenberg, Stanley P. Sady, Nicholas Sanchez, Teryl Schessler, Emet Schneiderman, Noriko Seguchi, Shelley Smith, Sara Stinson, Krista Swaninger, Susan Tanner, Joanna Tatomir, David P. Tracer, Cecilia Tomori, Marque LaVelle, Virginia J. Vitzthum, Jessica Westin, Jason Wilson, Lucia Yaroch, Gloria Wheatcroft, and Leta Woodill. Amy Mangieri, an undergraduate student, has diligently assisted in the various aspects of proofreading the manuscript.

I am grateful to all the undergraduate and graduate students who throughout the many years I taught at the University of Michigan have enthusiastically enrolled in the several courses of Biological Anthropology. Their questions and feedback have been essential and instrumental for motivating me to put together this work.

My special thanks and appreciation goes to the many researchers from Peru, Bolivia, and the United States who have worked directly with me in many of the research topics. I am specially grateful to the members of the "Instituto de Estudios de Altura" of the University of Cayetano Heredia of Lima, Peru, especially Drs. Carlos Monge and Roger Guerra-Garcia who provided fruitful advice about high-altitude research; members of the "Instituto de Biologia Andina" of the University of San Marcos, Lima, Peru, especially Drs. Tulio Velásquez and Emilio Picon-Reategui who participated directly in the studies of high-altitude adaptation. Drs. Jorge Sanchez and Danilo Pallardel of the Department of Anthropology of the University of Cuzco, Perú, diligently participated in the studies of adaptation to high altitude and the lowlands of Peru. The faculty of the "Instituto Boliviano de Biologia de Altura" of Bolivia provided their unique facilities for the studies of high-altitude adaptation. I am extremely grateful to Drs. Enrique Vargas, Mercedes Villena, Rudy Soria, Irma Ayllon, and Hilde Spielvogel who directly participated in various aspects of research at high altitude.

Finally, my appreciation is given to my wife Hedy G. Frisancho for her continuous support and encouragement to finish this book.

Dedication

Dedicated to my former professors and mentors Paul T. Baker and Thelma Baker who have played a pivotal role in my graduate training and thereafter. Their courage and foresight made possible the development of a research program that provided the opportunity to acquire the investigative skills of a generation of biological anthropologists and give the critical impetus for the development of the field of human biology as a research strategy of biological anthropology.

ABOUT THE AUTHOR

A. Roberto Frisancho is the Arthur A. Thurnau Professor of Anthropology at the University of Michigan and a Research Professor at the Center for Human Growth and Development of the University of Michigan. He received his Ph.D. in Anthropology from Pennsylvania State University and is a distinguished biological anthropologist who has published more than 120 scientific research papers and four books.

Dr. Frisancho is the leading researcher on the study of human adaptation to environmental stresses, including high altitude hypoxia, undernutrition, obesity and heat stress. He is the author of *Human Adaptation and Accommodation to Environmental Stress* (1993) and *Nutritional Anthropometric Standards* (1994).

Dr. Frisancho was the former president of the Human Biology Association and has received the greatest number of teaching awards bestowed to professors at the University of Michigan.

SECTION I

Evolution and Biological Anthropology

Evolutionary theory is the foundation of biological anthropology in all different specialties. Chapter 1 provides a brief history of the development evolutionary theory from the theological phase to the scientific phase of Darwin and Mendel and explains how the theory of evolution became incorporated into the practice of anthropology. Chapter 2 reviews the basic principles of cell biology including the principles of cellular and molecular genetics, Mendelian genetics of dominant and recessive discontinuous traits, sex-linked recessive genes, autosomal traits, and inheritance of continuous traits. Chapter 3 focuses on the application of Mendelian and population genetics to study the role of mutations, founder effect, gene flow and isolation, and its influence on the process of microevolution through natural selection learned through observation and experimental studies of contemporary variability in animals and humans.

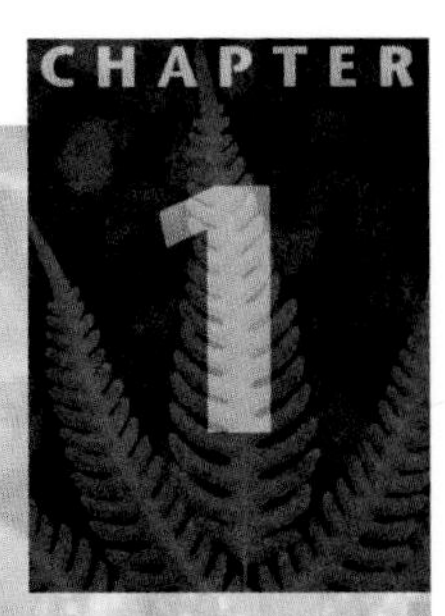

Development of Evolutionary Theory and Biological Anthropology

Our brain, dexterity, bipedal locomotion, digestive system, immunity to some diseases, susceptibility to accumulate body fat, and even spiritual perceptions all stem from the unique evolutionary history of the human species. Evolution has been called the foundation of biology, and for good reasons. The theory of evolution provides a mechanistic explanation for the amazing unity of design seen in nature and at the same time explains the incredible diversity of nature. It provides a method and a perspective for understanding how we came to be what we are today. It contributes to avenues for studying all aspects of biology, ranging from biochemistry, morphology, physiology, behavior, ecology, and paleontology. Without evolution biology becomes a disparate set of fields. Evolutionary explanations pervade all fields in biology and bring them together under one theoretical roof. Evolution unites the whole of biology and all living organisms into a coherent and comprehensive body of knowledge. Thus, no theory in biology before or since has achieved so much, and in the words of Dobzhansky[1] "nothing in biology makes sense except in the light of evolution." This chapter focuses on the development of evolutionary theory from the theological to the scientific phases. In addition, the science of biological anthropology and its subdivisions are discussed.

History of Evolutionary Theory

The existence of diverse forms of life requires explanation. The development of the evolutionary theory has its roots in the eighteenth and nineteenth centuries.[2] Explanations about the biological diversity in the world throughout history have gone through a theological phase and a scientific phase.

Theological Phase

The concept of evolution as we know it today is based on the intellectual developments in Europe in the eighteenth century. Until the middle of the eighteenth century, nature was explained almost exclusively in religious terms. It was believed that all the species that existed were created by God in their present form, and that these species were therefore fixed and unchanging. The prevailing notion was that the earth was "full" and that nothing new (such as a species) could be added. Thus, it was believed that since the Creation, no new species had appeared and none had become extinct in accord with the Church's teaching of a divine Creation. This belief that life forms could not change was known in European intellectual circles as *fixity of species*. According to religious dogma, the entire universe was created by God over a period of six days, and the earth and its inhabitants are pretty much the same now as they were when they were created. **James Ussher** and **John Ray** are the most prominent representatives of the theologian school of thought.

James Ussher (1581–1656)

The Irish archbishop **James Ussher**, based on events described in the Bible, calculated that the earth was just under 6000 years old. Part of the notion of fixity of species was also the belief that all God's creations were arranged in a hierarchy that progressed from the simplest organisms to the most complex organisms. This concept of a hierarchy of living things was first proposed in the fourth century B.C. by the Greek philosopher, Aristotle, and came to be known as the Great Chain of Being. According to the Great Chain of Being, these progressive creations had ultimately led to the creation of humans, and the unity of nature was the manifestation of a supernatural plan.

John Ray (1627–1705)

In 1691 **John Ray** published a book entitled *The Wisdom of God Manifested in the Works of Creation*, in which he postulated that God's plan in nature was the deliberate outcome of a Grand Design. Although this work represented a contribution to the theological phase of biological diversity, it also marked the beginning of the scientific era. In this work the concepts of **species** and **genus** were defined for the first time. Ray postulated that groups of plants and animals could be distinguished from other groups by their ability to reproduce with one another and produce offspring, calling such groups of reproductively isolated organisms **species**. Furthermore, Ray pointed out that many species shared similarities with other species, and these species were grouped together in a second level of classification he called **genus**. Thus, by the late 1600s, the biological definitions of species and genus were the same as they are today.

Scientific Phase

The scientific phase can be traced to Linnaeus, Buffon, Erasmus Darwin, Lamarck, Cuvier, Lyell, Malthus, and Charles Darwin.

Carolus Linnaeus (1707–1778)

Carolus Linnaeus was a Swedish naturalist. He believed in a static universe and the fixity of species. By establishing the first formal classification of all known living organisms, **Linnaeus** was instrumental in the development of the scientific approach for explaining the biological diversity of past and present forms. In 1735 he published the *Systema Naturae* (Systems of Nature), in which he standardized Ray's use of the terms genus and species, thus firmly establishing the use of binomial nomenclature for naming organisms. He also added the categories of **class** and **order**. Thus, the Linnaean classification system included four nested categories: class, order, genus, and species, which became the basis for the system of classification used today. Furthermore, Linnaeus also included humans in his classification of animals, placing them in the genus *Homo* and species *sapiens*. Placing humans in the animal kingdom was contrary to the religious dogma, which maintained that humans were created in God's image and were therefore not in the animal kingdom.

Georges-Louis Leclerc or Buffon (1707–1788)

Georges-Louis Leclerc, a French biologist who was raised to the rank of count under the name Buffon, saw nature as a dynamic system of processes instead of a static pattern of structure. Because of this dynamic relationship between environment and nature, he believed that a species could be influenced by local climatic conditions and could therefore change. However, he did not believe in the change of one species into another species. In his *Natural History*, first published in 1749, he repeatedly pointed to the importance of change in the universe and the changing nature of species.[3] Buffon was one of the first biologists who recognized the external environment as an agent of change in species. He specifically stated that when groups migrated to new areas of the world, each group would subsequently be influenced by local climatic conditions and would gradually change as a result of adaptation to the environment. Although this idea appears similar to modern principles of evolution, Buffon rejected the possibility that one species could give rise to another.

Erasmus Darwin (1731–1802)

Erasmus Darwin, the grandfather of Charles Darwin, in 1794 published his *Zoonomia*.[4] In this work, he pointed out that the similarity between organisms was related to the fact that they were derived from one parent. However, even though he commented on evolution, his ideas were much closer to the Lamarckian theory of the inheritance of acquired characteristics.

Jean-Baptiste Lamarck (1744–1829)

The French scientist **Jean-Baptiste Lamarck** in 1809 published his *Philosophie Zoologique.* Lamarck subscribed to the idea of transmutation of species but not to their extinction. He was the first to recognize and stress the importance of interactions between organisms and the environment in the evolutionary process. He proposed that organisms exist in a dynamic interaction between organic forms and the environment, such that organic forms could become altered in the face of changing environmental circumstances. In other words, evolution occurred through a natural process of organisms adjusting to their environment.[5] Thus, as the environment changed, an animal's activity patterns would also change, resulting in increased or decreased use of certain body parts. As a result of this use or disuse, body parts became altered. Because the modification would make the animal better suited to its habitat, the new trait would be passed on to offspring. This theory is known as the inheritance of acquired characteristics, or the use-disuse theory. According to Lamarck's theory acquired characteristics would pass along to offspring. In this context, the long neck of giraffes was acquired as a result of stretching to reach those leaves on upper branches of the trees, which, in turn, enabled the giraffe to obtain more food (**fig. 1.1**). The longer neck is subsequently transmitted to offspring, with the eventual result that all giraffes have longer necks than did their predecessors. In other words, a trait acquired by an animal during its lifetime can be transmitted to offspring, giving rise to new

FIGURE 1.1. Lamarckian theory of acquired inheritance and Darwin theory of natural selection. According to Lamarck the long neck of giraffes (a) is the result of continuous stretching to reach leaves on higher trees (b). This acquired trait was passed on to offspring, who were born with longer necks. In contrast, according to the Darwin-Wallace theory of natural selection among giraffes, like in other animals, there is individual variability in neck length (c). Hence, those giraffes with long necks had an advantage for feeding where trees are tall, and eventually the trait of a long neck is passed on to a greater number of offspring resulting in an increase in the length of giraffes' necks and over many generations become part of the characteristics of giraffes. (Composite made by the author from personal collection.)

forms and new species. Although Lamarck's theory became very popular, observation of animals who had lost or acquired certain characteristics but did not pass them to offspring refuted the theory of inheritance of acquired traits. For example, rats whose tails were cut off produced offspring with tails. Furthermore, the Darwin-Wallace theory of natural selection (**fig 1.1c**) later submitted that, among giraffes, like in other animals, there is individual variability in neck length. Hence, those giraffes with long necks had an advantage for feeding where trees are tall, and eventually the trait of a long neck is passed on to a greater number of offspring resulting in an increase in the length of giraffes' necks and over many generations became part of the characteristics of giraffes.

Georges Cuvier (1769–1832)

The French anatomist Georges Cuvier theorized that the earth's geological features were the result of a series of regional catastrophes that destroyed most or all of the local plant and animal life.[6] He postulated the theory of the fixity of species. Specifically, he maintained that during catastrophes all living things were destroyed, a new creation followed each catastrophe, and the newly created animals and plants survived, **fixed and unchanged**, until the next catastrophe. However, excavation and natural observation of geological strata showed the existence of fossil remnants of truly unusual creatures, such as the dinosaurs, suggesting that the world in the past was different from the present. Therefore, the concept of catastrophism could not account for the presence of different species.

Charles Lyell (1797–1875)

The British naturalist Charles Lyell in 1830 published *The Principles of Geology,*[7] in which he expounded the principle of uniformitarianism that contradicted the theory of catatrophism to explain the variety of organisms existing in the past and in the present. According to uniformitarianism, the earth's physical features were the result of natural forces operating in a uniform manner throughout time. Furthermore, he postulated that the rate of geological change was uniformly slow and steady throughout time, and the earth's crust was formed by a series of slow, gradual uniform changes over time. These changes were brought about by natural forces such as wind, water erosion, local flooding, frost, the decomposition of vegetable matter, volcanoes, earthquakes, and glacial movements, which all had contributed in the past to produce the geological landscape that exists in the present. Moreover, he noted that although various aspects of the earth's surface (e.g., climate, flora, fauna, and land surfaces) are variable through time, the underlying processes that influence them are constant. Lyell stressed the enormous time spans involved in geological changes and emphasized that the earth must indeed be far older than anyone had previously suspected. Thus, by altering perceptions of Earth's history from a few thousand to many millions of years, Lyell changed the framework within which scientists viewed the geological past.

Thomas Robert Malthus (1766–1834)

In 1798 the English economist, **Thomas Robert Malthus,** published *An Essay on the Principle of Population.*[8] In this essay, he proposed that in most species more individuals are born than can possibly survive. Based on his observation of humans he inferred that because the ability to increase food supplies artificially is limited, it poses a constraint on population growth. He specifically pointed out that if left unchecked by limited food supplies, human populations could double in size every twenty-five years (**fig. 1.2**). That is, population size increases exponentially while food supplies remain relatively stable. Thus, there is constant competition for food and other resources. The same logic could be applied to nonhuman organisms. In nature, the tendency for populations to increase is continuously checked by resource availability. If it were not for this mortality, populations would grow too large for their environments to support them. Thus, there is constant competition for food and other resources, and the ones that survive are the fittest. This concept inspired both Charles Darwin and Alfred Wallace in the development of the theory of natural selection.

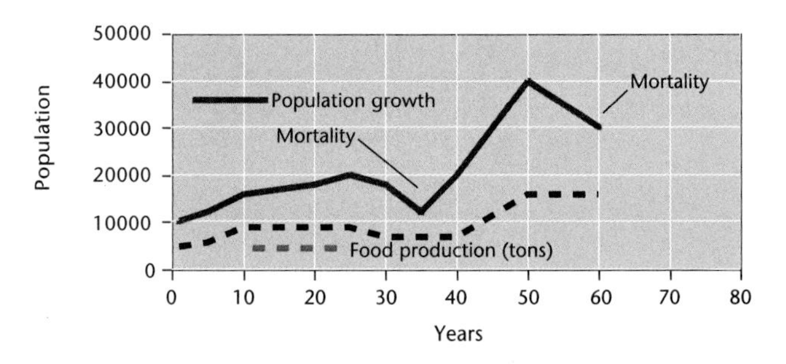

FIGURE 1.2. Malthusian population growth. According to Malthus, when a population grows faster than available food resources it causes differences in survival rates. (Composite made by the author.)

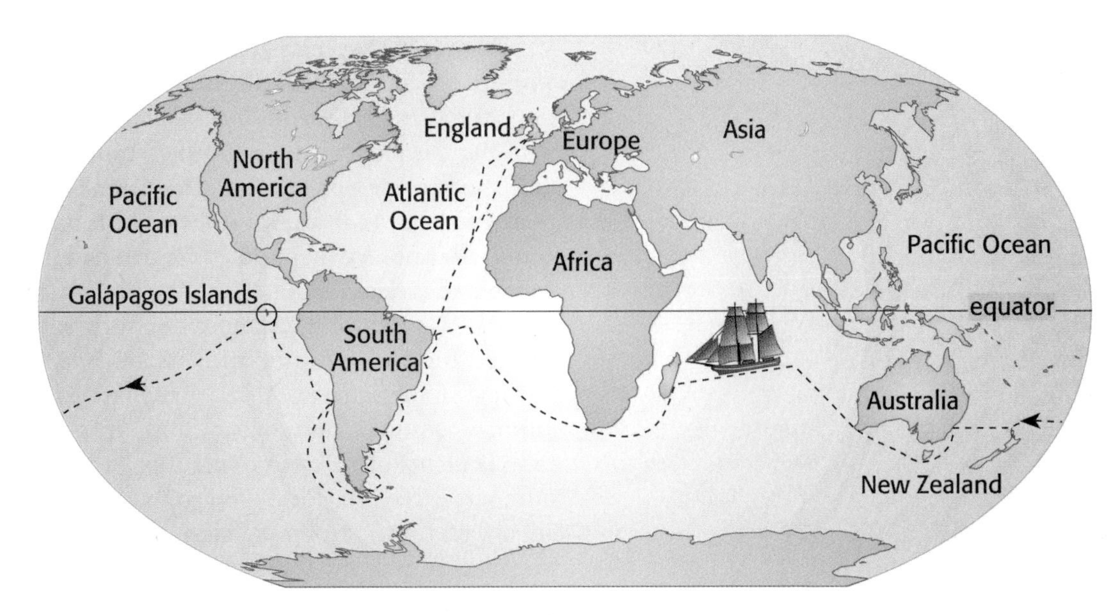

FIGURE 1.3. In 1831 Charles Darwin as naturalist with the *H.M.S. Beagle* traveled for five years around the world. The route taken included stops in South America and the Galápagos Islands off the coast of Ecuador, where he found an astounding variability in the flora and fauna of living and extinct animals such as the finches and fossils. From *BSCS Biology: An Ecological Approach, 9th Edition* by BSCS. Copyright © 2002 by BSCS. Reprinted by permission.

Charles Darwin (1809–1882) and The Origin of Species

In 1859 Charles Darwin published the treatise *The Origin of Species,*[9] a monumental work that changed how we view and interpret animal and plant diversity. During his youth, like all his contemporary naturalists, Darwin accepted the philosophy of the fixity of species. All this changed when he embarked on a five-year journey (from 1831 to 1836) around the world, collecting plant and animal specimens aboard the Royal Navy vessel *Beagle* as an unpaid naturalist.[9,10] During this trip Darwin visited the Galápagos Islands of South America and the pampas of Montevideo and Argentina and noted the striking variability in the flora and fauna of living and extinct animals such as the finches and fossils (**fig. 1.3**).

Finches' Adaptation to Variability in Food Resources

In the Galápagos Islands, Darwin observed that there were nine unique species of finches. Although they descended from a common ancestor from the South American mainland, they differed in the shape and size of their beaks. He observed that these differences were related to the food upon which they came to depend. For example, he observed that: (**fig. 1.4**)

1. Ground finches whose main foods are seeds have heavy beaks.
2. Tree finches whose main foods are leaves, buds, and fruits have thick and curved beaks.
3. Tree finches whose main foods are insects that live deep inside tree branches have long beaks.
4. Ground finches whose main foods are insects have slender beaks.

Darwin also observed the variability among Galápagos tortoises and iguana-like lizards, unique to the islands. These species differed in size and in the foods they ate: some lived on land and fed on cacti while others were marine and fed on seaweed. He was intrigued by this diversity in morphology.

Extinct Fossils

When Darwin visited the pampas of Montevideo and Argentina, he saw many bones and teeth of extinct animals. After observing fossil specimens such as large portions of the armor of a gigantic armadillo-like animal and part of the great head of a Mylodon, he con-

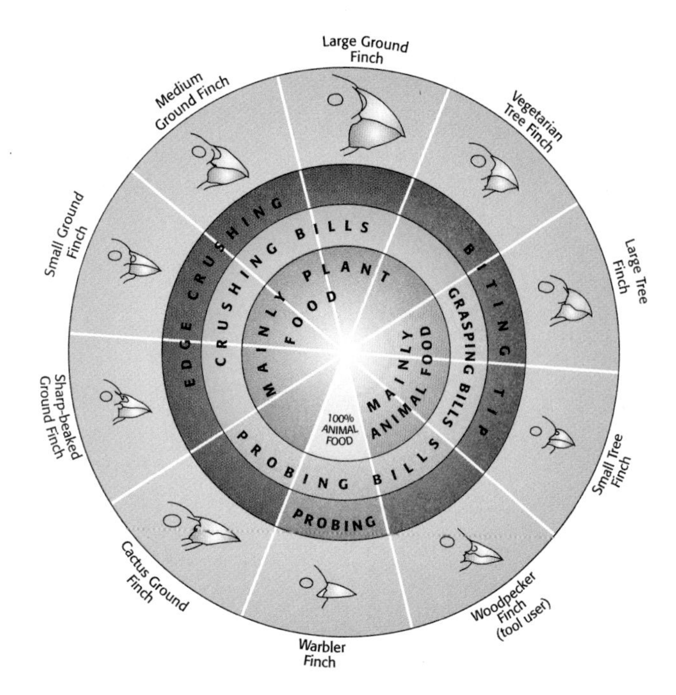

FIGURE 1.4. Finches from the Galapagos. Darwin observed that from one common ancestor from South America there had arisen nine species of finches that differed in the size and shape of their beaks, which is related to the variability of the food on which they subsist. From *BSCS Biology: An Ecological Approach, 9th Edition* by BSCS. Copyright © 2002 by BSCS. Reprinted by permission.

cluded that the whole area of the Pampas is one wide sepulchre of these extinct gigantic quadrupeds.

The Theory of Evolution and Natural Selection

By the 1830s, shortly after the *Beagle* returned to London, Darwin started formulating a mechanistic explanation for how species change through time, which he termed *natural selection*. Darwin was convinced that transmutation of species did occur—that new species did arise through natural causes. He also knew that artificial selection practiced by animal breeders could bring about this transmutation. He observed animal breeders who practiced artificial selection, whereby those animals that exhibit specific traits they hoped to emphasize in the offspring were selected as breeding stock, while those with undesirable traits are "selected against," or prevented from breeding. Nature played the role of a breeder in selecting the favorable traits or selecting out the nonadaptive characteristics (**fig. 1.5**). Thus, Darwin inferred that the environment acted to change organisms over time, a process he called natural selection.

Darwin also recognized that biological variation within a species was critically important for the operation of natural selection, but he could not come up with the mechanisms whereby such traits become established in the population. Then in 1838, he happened to read Malthus's 1798 essay on population, where after observing population growth in England, which had accelerated since about 1750, Malthus realized that as result of the competition for resources and limited food supplies organisms are in a constant "struggle for existence."[8] Malthus suggested that since animals required food to grow and reproduce, when food is plentiful animal populations grow until their numbers exceed the local food supply. Since resources are always finite, it follows that not all individuals in a population will be able to survive and reproduce. Thus, there is constant competition for food and other resources, and the ones that survive are the fittest. Darwin realized that some individuals will possess traits that enable them to survive and reproduce more successfully (producing more offspring) than others in the same environment. From this idea Darwin reasoned that if this process of differential survival continued, it would eventually result in the formation of a new species. Finally, Darwin proposed that if the advantageous traits are **inherited** by offspring, then these traits will become more common in succeeding generations. Thus, traits that confer advantages in survival and reproduction are retained in the population, and traits that are disadvantageous disappear.

Darwin summarized how species change through time as follows:

> As many more individuals are produced than can possibly survive, there must in every case be a struggle for existence, either one individual with another of the same species, or with the individuals of distinct species, or with the physical conditions of life . . . Can it, then, be thought improbable, seeing that variations useful to man have undoubtedly occurred, that other variations, useful in some way to each being in the great and complex battle of life, should sometimes occur in the course of thousands of generations? If such do occur, can we doubt (remembering that many more individuals are born than can possibly survive) that individuals having any advantage, however slight, over others, would have the best chance of surviving and of procreating their kind? On the other hand, we may feel sure that any variation in the least degree injurious would be rigidly destroyed. This preservation of favorable variations and the rejection of injurious variations, I call Natural Selection.[9]

FIGURE 1. 5. Artificial selection in chickens. Many breeds of domestic chickens were derived from the red jungle fowl by artificial selection. From *BSCS Biology: An Ecological Approach, 9th Edition* by BSCS. Copyright © 2002 by BSCS. Reprinted by permission.

From this he derived the following principles:

I. Adaptability indicates that some of the variant individuals are more adapted to live and reproduce in a given environment than are others.

II. The capability of a population to enlarge is unlimited, but the ability of any environment to support populations is always finite. Hence, because in each generation more individuals are produced than can survive, owing to limited resources, there is competition between individuals. The environmental context determines whether or not a trait is beneficial; that is, what is favorable in one setting may be a liability in another. In this way, which traits become most advantageous is the result of a natural process.

III. Radiant variation states that all forms of life, except for identical twins, vary slightly and different forms are always being produced. In addition, since organisms produce far more offspring than survive, this variation affects the capacity of individuals to survive and reproduce. Thus, those individuals who possess favorable variations or traits (e.g., speed, disease resistance, or protective coloration) have an advantage over individuals who do not possess them. By virtue of the favorable trait, these individuals are more likely to survive to produce offspring than are others. Traits are inherited and are passed on to the next generation.

IV. Natural selection refers to the tendency of certain individuals in a particular environment to be more likely to produce more offspring than other individuals. Those who reproduce the most have the highest fitness (defined as reproductive success), such that their traits are passed on to the next generation with the greatest frequency. Those characteristics that favor fitness in a certain environment will be positively selected. In other words, less favorable traits are not passed on as frequently and become less common. Those individuals who produce more offspring, compared to others, are said to have greater reproductive success.

V. Geographical isolation may also lead to the formation of new species. As populations of a species become geographically isolated from one another, for whatever reasons, they begin to adapt to different environments. Over time, as populations continue to respond to different selective pressures (i.e., different ecological circumstances), they may become distinct species, descended from a common ancestor. The nine species of Galápagos finches, presumably all descendants from a common ancestor on the South American mainland, are an example of the role of geographical isolation.[9–11]

Alfred Russel Wallace (1823–1913)

Alfred Russel Wallace, an English natural historian like Darwin, came up with essentially the same theory. In fact, in 1858 (prior to 1859 Darwin's publication *On the Origin of Species*), Wallace sent Darwin a well-researched exposition on the divergence of species to be published in the *Annals and Magazine of Natural History*.[12] In this paper, Wallace suggested that species were descended from other species and that the appearance of new species was influenced by environmental factors. Darwin, fearing that Wallace might be credited for a theory (natural selection) that he himself had formulated, quickly wrote a paper, which was read along with the paper of Wallace before the Linnean Society of London in 1858. In the end, both Wallace and Darwin concluded that natural selection is the mechanism that explains the origin and variability among species.

Historical Note

According to historical accounts Darwin waited over 20 years after his return from his trip on the *HMS Beagle* to write about his findings. Darwin, during the three years after his return, did not think of natural selection as the mechanism of evolution. He accepted the theory of Descend with Modification, which postulated that present animals and plants are descendants of similar organisms that lived a million years ago. Then, in 1859, based on his reading of several naturalists who preceded him, Darwin published *On the Origin of Species*[10] and incorporated the phrase "survival of the fittest" developed by Malthus and first used by Spencer.[13–15] Likewise, the concept of natural selection as a mechanism to explain the origin of varieties in nature was presented 65 years earlier by James Hutton. In 1794, Hutton[13–15] published a book entitled *An Investigation of the Principles of Knowledge and of the Progress of Reason, from Sense to Science and Philosophy.* In this book Hutton devoted an entire chapter to natural selection. However, Hutton specifically rejected the idea of evolution between species as a "romantic fantasy." He regarded the capacity of species to adapt to local conditions as an example of benevolent design in nature and not an evolutionary process responsible for speciation, which Darwin did. Thus, Darwin quite correctly is credited for bringing to light a concept that was previously abandoned.

The Modern or Synthetic Theory of Evolution

The modern or synthetic theory of evolution is a blend of Darwinian theory and Mendelian and population genetics. It is a multidisciplinary amalgam of well-established theories and working hypotheses, together with the observations and experiments that support accepted hypotheses, which jointly seek to explain the evolutionary process and its outcomes. It incorporates relevant knowledge regarding the role of mutations, genetic drift, and founder effect on population variability. These hypotheses, observations, and experiments originate in disciplines such as anthropology, genetics, embryology, zoology, botany, paleontology, and molecular biology.

Darwin observed that natural selection can work only if variation occurs and that variation continues to arise in a population, but he could not account for the source of variability. At that time the prevailing view was of "blending inheritance," which proposed that

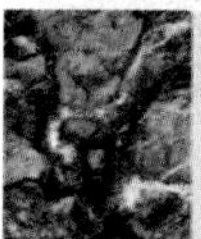

TABLE 1.1
History of the development of evolutionary theory

Name	Year	Contribution
THEOLOGICAL PHASE		
James Ussher	1600	World was 6,000 years old. Creation of the Earth began at noon on Sunday, October 23 of 4004 B.C.
John Ray	1691	World is part of God's Grand Design.
SCIENTIFIC PHASE		
Carolus Linnaeus	1758	Classification system of plants and animals
Georges L. Buffon	1749	Environment, such as climatic conditions, acts as agents of change of species.
Erasmus Darwin	1802	Advanced ideas of evolution and adaptation
Jean Baptiste Lamarck	1809	Inheritance of acquired traits
Georges Cuvier	1800	Catastrophism
Charles Lyell	1830	Uniformitarianism: The earth's crust was formed via slow uniform changes not cataclysms.
Thomas Robert Malthus	1838	Population growth faster than food supply: Struggle for existence
Alfred Russell Wallace	1858	Natural selection
Charles Darwin	1859	Natural selection and the origin of species

offspring merely struck an average between the characteristics of their parents. In such a scenario, advantageous and negative variations would be equally represented in the offspring, but because differences among variant offspring would be halved each generation, the original variation would be rapidly reduced to the average of the preexisting characteristics. Darwinian theory could not explain what provides the source of variability in the population for natural selection to work on.

The answer was provided by Mendelian genetics. In 1866 the Austrian monk Gregor Mendel published *The Principles of a Theory of Heredity*. In this work based upon his study of breeding pea plants, Mendel explained that biological traits are inherited from each parent through particulate factors (now called "genes"), which do not mix or blend but segregate in the formation of the sex cells, or gametes.[16] Mendel's discoveries, however, remained unknown until 1900, when they were simultaneously rediscovered by several scientists. The rediscovery in 1900 of Mendel's theory of heredity led to an emphasis on the role of heredity in evolution (see chapter 2).

In 1909, Hugo de Vries published *The Mutation Theory*.[17] According to de Vries, mutations (defined as novel genetic change in the genotype beyond that achieved by genetic recombination) appear randomly without influence of the environment and could produce new species. However, in 1927, Muller[19] demonstrated that mutations were affected by environmental conditions, and mutations per se do not produce new species. Then, population geneticists,[19–22] using Mendel's laws of inheritance and mathematical arguments, demonstrated that continuous variation in biological traits such as size and number of eggs laid is the result of variability that occurs during the formation of gametes and is transferred from the parents to the offspring. Then, natural selection acting cumulatively on these small variations yields major evolutionary changes in form and function. Based on this evidence, Dobzhansky formulated synthetic theory of evolution.[1,23]

As stated by Dobzhansky[1,23] in the synthetic theory of evolution, natural selection, mutation, gene flow, genetic drift, founder effect, and dispersal between populations all play roles in the evolutionary process. Each of these factors plays an important role in providing and maintaining variability on which natural selection can act. At the core of modern synthetic theory is the idea that the environment selects certain phenotypes, which are determined to a large extent by the genotype of an individual. New mutations and gene combinations continually produce some individuals that are more fit than others. Individuals with numerous viable progeny contribute the most genes to the next generation when their offspring mature and reproduce. As a result, their genes become the most common in the gene pool of the population, whereas the less fit types are selected out.[1,23] In this context, natural selection fashions individuals of succeeding generations to be ever better adapted to their circumstances. The characteristics of a population or species represent the sum of the actions of natural selection on similar individuals.

Examples of Natural Selection in Action

Moths and Pollution

There are several examples of the evidence of natural selection. One excellent example of how natural selection works is the change in the coloration of the peppered moth *(Biston betularia)* in Manchester, England over the last few centuries (**fig. 1.6**). Before 1848 and prior to the industrialization in England, 90 percent of the moths were light-colored and only 10 percent were dark-colored, but after 1898, 95 percent of the moths were dark-colored. According to the studies of Kettlewell[27,28] the soot and acid rain brought by pollution darkened the trees by first killing the lichens that festooned them and then blackening the naked trunks. The light-colored moths previously camouflaged on lichens became conspicuous and were heavily preyed upon by birds, while the less visible dark-colored moths enjoyed protection and increased in number. As a result the dark moths, now better camouflaged, survived in greater proportion (90 percent) and passed their dark color to the next generation. In other words, during the period of pollution the dark moths were selected for and the light moths were selected against.[28] After the passage of the Clean Air Acts in the 1950s, the trees regained their former appearance (became lighter in color), reversing the selective advantage of the morphs (that is, once again light moths survived better, and were selected for, whereas dark moths were selected against). As illustrated in **figure 1.7**, as the result of legislation in Britain atmospheric pollution has been significantly reduced, and concurrent with this change the proportion of dark-colored moths is declining.[29–30]

FIGURE 1.6. Adaptation of the peppered moth to industrial pollution. Selection by birds against light-colored moths. In Manchester, England, before 1848 more than 90 percent of the moths were light-colored, but after 1898, 95 percent of the moths were dark-colored. From *BSCS Biology: An Ecological Approach, 9th Edition* by BSCS. Copyright © 2002 by BSCS. Reprinted by permission of BSCS and L. M. Cook.

Finches and Food Resources

Another example of natural selection is found among the finches of the Galápagos Islands. As described by Darwin, there are nine species of finches, all of which originated from one species from the South American mainland. Each of these species of finches differs in the size and shape of the beak according to the food on which they subsisted. Studies of these birds indicate that the size of the beak is inherited, therefore large-beaked birds have large-beaked offspring and small-beaked birds have small-beaked offspring.[35,36]

The health of these bird populations depends on seed production, and seed production, in turn, depends on the arrival of the wet season. In 1977 there was a drought on the Galápagos Islands. Rainfall was well below normal, and fewer seeds were produced. As the season progressed, a population of finches (called *Geospiza fortis*) depleted the supply of small seeds. Eventually, only larger seeds remained. Most of the finches starved; the population plummeted from about 1,200 birds to less than 200. However, not all the finches starved, and some even thrived. Within the population the finches with relatively larger beaks were better able to crack the larger seeds that were more common under drought conditions. As a result, the large-beaked birds fared better than smaller-beaked ones and had offspring with correspondingly large beaks (**fig. 1.8**). Thus, there was an increase in the proportion of large-beaked birds in the population's next generation.[35,36]

African Wild Dogs and Impalas

Studies of the hunting behavior of African wild dogs show that they selectively predate those individuals within a population of impalas that are in poorer nutritional condition.[37] Fit impala do indeed have the better chance of survival.

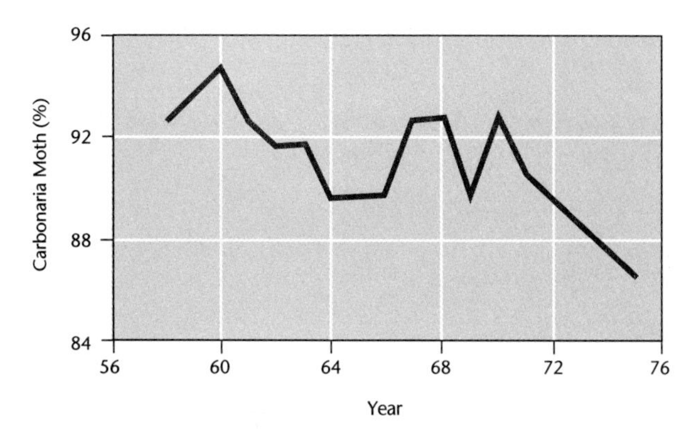

FIGURE 1.7. Decline in the frequency of the dark carbonaria moths near Liverpool (England) in response to reduced pollution levels from 1959 to 1975. Source: [Data derived from Bishop, J. A. and Cook, L. M. 1979. A century of industrial melanism. *Antenna* 3:125–128.]

In summary, these examples illustrate that natural selection allows populations to adapt to their current environments. As the environment changes, new traits may be selected. In other words, natural selection uses its genetic variability to produce a new genotype that fits the current environment.

Evolution and Religion

Religion is based upon faith and a set of principles and guidelines that help humans live within the boundaries of humanity. Religion makes statements about human morality. It sets the limits for what is right and what is wrong. Religion, by reducing fear and anxiety and providing social support, motivational enhancement, a sense of security, self-esteem, and a sense of belonging has enhanced humans' well-being. In contrast, science has nothing to say about right and wrong. Evolutionary theory is a science that is based on testable hypothesis that is applicable in all cultures, while religion and its interpretation are culture specific. Since nobody can claim that their "god" and religion is better than our "god" and religion, putting religion and evolution at odds with each other does both a disservice. Religion and science both represent ways of looking at the world and, though they work on different levels, they are not contradictory. Most informed religious people see no reason for biological facts and theories to interfere with their religious beliefs. In fact, Pope John Paul II in his 1996 message to the Pontifical Academy of Sciences stated that "my predecessor Pius XII had already stated that there was no opposition between evolution and the doctrine of the faith about man and his vocation."(38) Thus, one can be religious and believe in God and still accept the fact of evolution and evolutionary theory.(39)

FIGURE 1.8. Increased frequency of finches with large beak size in response to change in supply of seeds. The frequency of finches with large beak size increased, because the finches with larger beaks were better able to crack the larger seeds that were more common under drought conditions. (Modified from figure 2.8 of Price, P.W. 1996. *Biological Evolution.* Orlando, FL: Harcourt Brace).

Biological Anthropology and Its Subdisciplines

Anthropology is traditionally subdivided into four major subdisciplines:

- Cultural Anthropology
- Anthropological Linguistics
- Archaeology
- Biological (or Physical) Anthropology

The conceptual framework of biological anthropology, also known as physical anthropology, is that evolution is an ongoing process, in that our species and all of its constituent populations are continually being shaped by evolutionary forces. As such, biological anthropology is, and always has been, a discipline that is constantly evolving. It is a science that seeks to answer questions about when, where, and why humans appeared on the earth; how and why they have changed since then; and how and why modern human populations vary in morphological, genetic traits, behavioral, and physiological features. To achieve this goal, biological anthropologists study the fossil evidence of mammals, hominoids, and hominids that goes back several million years, the molecular composition of genes, physiological traits of extinct and extant individuals and populations, the behavioral characteristics of humans and nonhuman primates, and the biological and environmental factors involved in our physical development. In general, the various subdisciplines of biological anthropology include paleoanthropology, primatology, evolutionary behavioral ecology, human adaptation, human genetics, human osteology, applied anthropology, and forensic anthropology.

Paleoanthropology

The field of paleontology seeks to document the physical changes in organisms and their phylogenetic relationships as they can be deduced from fossils, most of which exist in the form of chemical casts of the hard parts of an animal's body. Analysis of paleontological data can offer an opportunity to understand when, where, and in relation to what environment anatomy evolved in support of particular behaviors. Paleoanthropology, which combines the fields of paleontology and anthropology, is concerned with the study of the fossil remains of our ancestors. This science has enabled biological anthropologists to propose possible lines of descent from our ancient ancestors to our present form, *Homo sapiens.*

Primatology

Primatology is the science that addresses the study of the nature and origin of the similarities and differences in behavioral and biological traits between nonhuman and human primates. Depending on the nature of study, primatology is subdivided into two interrelated specialties: comparative primate anatomy and primate ethology. Primate anatomy emphasizes the study of the nature and the origin of the similarities and differences in anatomical and morphological traits and function between nonhuman primates and humans. In contrast, primate ethology is concerned with the study of behavioral traits of primates in their natural habitat, as well as controlled laboratory conditions.

Evolutionary Behavioral Ecology

The field of evolutionary behavioral ecology, originally known as sociobiology, incorporates evolutionary theory to explain behavioral variability. According to evolutionary behavioral ecology, a given behavior has evolved through the operation of natural selection, and certain behaviors or behavioral patterns have been selected for (like any other biological trait) because they increase the reproductive fitness in individuals. In this context, an individual whose genotypes influences behaviors that lead to higher reproductive success than those who do not are selected for and become part of the population behavioral repertoire.

Human Adaptation

The science of human adaptation is concerned with understanding the biological and cultural mechanisms whereby populations survive under extreme environmental conditions. For this purpose human biologists in the field or laboratory collect such data as physiological responses to the stressful conditions of cold, heat, humidity, altitude, under- and over-nutrition, noise, pollution, and susceptibility and immunity to disease during growth and adulthood.

A basic premise of research in human adaptation is that population differences in biological traits are the result of adaptations to different environmental conditions. Humans have been able to adapt successfully to many different habitats in the world today. Humans exposed to different environmental conditions respond either through short-term reversible responses or through long-term responses that can occur during development and adulthood. Such changes and responses are part of the universal human biological adaptability, which can include responses at biochemical, physiological, and behavioral levels. How and under what conditions adaptability is expressed in growth patterns, physiology, or anatomical traits is an area of ongoing research in biological anthropology. If the biological responses become long-term adaptations and continue for several generations, they can become part of the milieu of genetic adaptation. As such, untangling the process of long-term adaptation of humans is a focus of evolutionary studies as it is dealt with in this book.

Human Genetics

It is true that children look more like their parents and their relatives than nonrelatives. This similarity is a function of the genes that parents and relatives have in common. The mechanisms through which parents transmit their biological traits to the offspring are the subject of the science of genetics. Humans, like most organisms, are the result of the interaction of genes and environment, the product of which is referred to as a **phenotype**. The phenotypic resemblance is a function of the genes, which, depending on the environmental conditions, can be expressed or suppressed. Geneticists, by determining the degree of resemblance and differences in biological traits between parents and offspring or between relatives, and based on the laws of inheritance discovered by Gregor Mendel, can quantify the environmental and genetic components that account for similarities and differences in biological traits. Using the same approach geneticists can also quantify the environmental and genetic components that account for interpopulation similarities and differences in biological traits.

In recent years, it has become evident that the function of the genes is the result of the action of the molecule called deoxyribonucleic acid or DNA. The DNA molecule provides the instruction for the building of biological structures. In this context, the expression of genetic traits is seen as the transmission of the DNA information. From this follows the premise that evolution is the transfer of DNA information from one generation to the next, along with the possibility that this code will change. The study of molecular composition and its transmission is called molecular genetics.

Human Osteology

Because archaeological remains include mostly bones and teeth (and in rare occasions dried soft tissue), biological identification has been based upon measurements of skeletal components of extant and past populations. Since skeletal remains are often found with tools, and because there is an association between types of tools and types of culture, the study of the fossil record brings about a collaboration of biological anthropologists, ethnologists, and archaeologists that enriches our understanding of the past. Similarly, biological anthropologists work in the area of paleopathology, where they inspect the skeletal material for indications of such diseases as tuberculosis, osteoarthritis and rheumatoid arthritis, dental caries, rickets, osteoporosis and osteopoenia, etc.

Applied Anthropology

Biological anthropologists, because of their unique background in anatomy, physiology, and population biology, have been involved in issues where body dimensions are involved. For example, biological anthropologists have been extensively involved in the design of protective clothing, including gas masks, oxygen masks, aviation goggles, diving goggles, and similar equipment; location of ear phones in tank helmets; in airframe design; in the design of equipment to be used at high altitude and under high G forces; and in the design of prostheses (including artificial joints) and other body parts. Similarly, biological anthropologists have been involved in the design and development of anthropometric charts for evaluation of size, growth, nutritional status, physical fitness, etc.

Forensic Anthropology

Because of their expertise in skeletal anatomy, biological anthropologists also work in "forensics," where they assist in the identification of skeletal remains of contemporary populations. Such a specialty is referred to as forensic anthropology. Forensic anthropologists usually work in conjunction with a coroner. Forensic anthropologists have been involved in human identification, including mass disaster victims, commingled remains, and the identification of isolated skeletons and single bones. With their specialized background, and through use of extensive tables of dimensions and ratios, forensic anthropologists have proven useful to coroners and in the identification of war dead.

Summary

The development of evolutionary theory has its roots in the eighteenth and nineteenth centuries. In the eighteenth century, the concept of evolution found moderate acceptance in academic circles, but the mechanistic explanation of evolution did not emerge until Darwin disclosed his mechanistic theory of evolution in 1859. Thus, the theory of evolution by natural selection was the culmination of the intellectual development of such naturalists as Linnaeus, Lamarck, Buffon, Lyell, and Malthus that took place in western Europe over the last 300 years. As knowledge about geology and fossils increased, scientists modified the concepts about the age of Earth and of the origin of life from a biblical perception of the creation of the world that was in a dynamic rather than a fixed state. Although Linnaeus believed in the fixity of species, his taxonomic approach provided the first scientific means for studying variability and diversity in the plant and animal kingdom. Malthus had noted that in most species more individuals are born than have available food resources, increasing competition for resources resulting in differences in survival. This observation provided Darwin with the mechanism through which nature acted as a selective agent and produced biological diversity.

In the modern or synthetic theory of evolution, natural selection, genetic drift, gene flow, and dispersal between populations all play roles in the evolutionary process. Thus, the characteristics of a population or species represent the sum of the actions of all these factors acting on individuals in that population. Evolution is an ongoing process, in that our species and all of its constituent populations are continually being shaped by evolutionary forces, and natural selection is the path through which evolution works. Natural selection is the force that makes evolutionary change possible. It reduces the frequency of alleles of maladaptive traits within a population and increases the frequency of alleles for adaptive ones. The change in the frequency of the size of the beak of finches of the Galapagos Islands along with the change in the food resources is an excellent illustration of the dynamic interaction between the populations' genetic traits and the environment. Thus, evolution is an opportunistic adaptation that fits the needs of the species but changes when the environment on which it evolved also changes. In other words, evolution is an ongoing and dynamic process and what is good today may not necessarily be good in the future. This flexibility to adapt and change maintains the unity of the species.

The conceptual framework of biological anthropology, also known as physical anthropology, is that evolution is an ongoing process, in that our species and all of its constituent populations are continually being shaped by evolutionary forces. It is a science that seeks to answer questions about when, where, and why humans appeared on the earth; how and why they have changed since then; and how and why modern human populations vary in morphological, genetic traits, behavioral, and physiological features. To achieve this goal, biological anthropologists study the fossil evidence of mammals, hominoids, and hominids that goes back several million years; the molecular composition of genes; physiological traits of extinct and extant individuals and populations; and the biological and environmental factors involved in our physical development, and the behavioral characteristics of humans and nonhuman primates.

Suggested Readings

Gould, S. J. 2002. *I Have Landed: The End of a Beginning in Natural History.* New York: Harmony Books.

Grant, P. R. 1999. *Ecology and Evolution of Darwin's Finches.* Princeton, NJ: Princeton University Press.

Keynes, R. 2002. *Fossils, Finches and Fuegians: Charles Darwin's Adventures and Discoveries on the Beagle.* London: HarperCollins.

Mayr, E. 1991. *One Long Argument: Charles Darwin and the Genesis of Modern Evolutionary Thought.* Cambridge, MA: Harvard University Press.

Literature Cited

1. Dobzhansky, T. 1973. *Genetic Diversity and Human Equality.* New York: Basic Books.
2. Eiseley, L. 1958. *Darwin's Century.* Garden City, NY: Doubleday.
3. Buffon, G. L. L. (Compte de). 1749–1804. *Histoire Naturelle.* 44 vols. L'Imprimerie Royale, Paris. Translated from the French by Smellie, W. 1781–1812. Natural History. London: Strahan. [Cited in Price, P. W. 1996. *Biological Evolution.* Orlando, FL: Harcourt Brace.]
4. Darwin, E. 1794. *Zoonomia.* London: Johnson. [Cited in Price, P. W. 1996. *Biological Evolution.* Orlando, FL: Harcourt Brace.]
5. Lamarck, J. B. 1809, 1984. *Zoological Philosophy.* Chicago: Univ. of Chicago Press. Translated from the French by Hugh Elliot.
6. Cuvier, G. 1837. *Edinburgh Rev.* 65:23–24.
7. Lyell, C. 1842. *Eight Lectures on Geology.* New York: Greeley and McElrath.
8. Malthus, T. R. 1798. *An Essay on the Principle of Population as It Affects the Future Improvement of Society.* London: Johnson.
9. Darwin, C. 1859. *On the Origin of Species by Means of Natural Selection, or The Preservation of Favoured Races in the Struggle for Life.* London: Murray.
10. Darwin, C. 1869. *On the Origin of Species by Means of Natural Selection,* 5th ed. London: John Murray.
11. Darwin, C. 1871. *The Descent of Man, and Selection in Relation to Sex.* London: Murray.
12. Wallace, A. R. 1858. Part III: On the tendency of varieties to depart indefinitely from the original type. *J. Proc. Linnaean Soc.* 3:45–62. [Reprinted in M. Bates, and P.S. Humphrey. 1956. *The Darwin Reader.* New York: Scribner's.]
13. Pearson. 2003. In retrospect. *Nature* 425, 665 review of "James Hutton 1794. An Investigation of the Principles of Knowledge and of the Progress of Reason, from Sense to Science and Philosopy. Thoemmes."
14. Gould, S. J. 2002. *Structure of Evolutionary Theory.* Cambridge, MA: Harvard Univ. Press.
15. Spencer, H. 1852. The development hypothesis. *The Leader.* 20 March.
16. Mendel, G. 1866. Verh. Naturforsch. Vereines Abhandlungen BrŶnn 4:3–47. [Cited in Price, P. W. 1996. *Biological Evolution.* Orlando, FL: Harcourt Brace.]
17. De Vries, H. 1900. Sur la loi de disjonction des hybrides. *Comptes Rendus de l'Academie des Sciences* (Paris) 130:845–847. Translated by Hannah, A. 1950. Concerning the law of segregation of hybrids. *Genetics* (Suppl.) 35 (5, part 2):30–32. [Cited in Price, P. W. 1996. *Biological Evolution.* Orlando, FL: Harcourt Brace.]
18. Muller, H. J. 1927. Artificial transmutation of the gene. *Science* 66:84–87.
19. Provine, W. G. 1971. *The Origins of Theoretical Population Genetics.* Chicago: Univ. of Chicago Press.
20. Fisher, R. A. 1930. *The Genetical Theory of Natural Selection.* Oxford: Clarendon Press.

21. Haldane, J. B. S. 1932. *The Causes of Evolution.* New York: Harper.
22. Wright, S. 1931. Evolution in Mendelian populations. *Genetics* 6:97–159.
23. Dobzhansky, T. 1962. *Mankind Evolving.* New Haven, CT: Yale Univ. Press.
24. Mayr, E. 1942. Systematics and the Origin of Species. New York: Columbia Univ. Press.
25. Huxley, J. S. 1942. *Evolution: The Modern Synthesis.* New York: Harper.
26. Simpson, G. G. 1944. *Tempo and Mode in Evolution.* New York: Columbia Univ. Press.
27. Kettlewell, H. B. D. 1956. Further selection experiments on industrial melanism in the Lepidoptera. *Heredity* 10:287–301.
28. Kettlewell, H. B. D. 1973. *The Evolution of Melanism: The Study of a Recurring Necessity.* Oxford: Clarendon Press.
29. Bishop, J. A. and L. M. Cook. 1979. A century of industrial melanism. *Antenna* 3:125–128.
30. Bishop, J. A. and L. M. Cook. 1980. Industrial melanism and the urban environment. *Adv Ecol Res* 11:373–404.
31. Bishop, J. A. and L. M. Cook. 1975. Moths, melanism and clean air. *Sci Amer.* 232(l):90–99.
32. Cook, L. M. and B. S Grant. 2000. Frequency of insularia during the decline in melanics in the peppered moth *Biston betularia* in Britain. *Heredity* 85:580–585.
33. Cook, L. M., G. S. Mani, and M. E. Varley. 1986. Postindustrial melanism in the peppered moth. *Science* 231:611–613.
34. Majerus, M. E. N. 1998. *Melanism: Evolution in Action.* Oxford, England: Oxford University Press.
35. Schluter, D. 1983. Diets, Distributions and Morphology of Galapagos Ground Finches: The Importance of Food Supply and Interspecific Competition. Dissertation (Ph.D.)—The University of Michigan.
36. Grant, P. R. 1999. *Ecology and Evolution of Darwin's Finches.* Princeton, NJ: Princeton Univ. Press.
37. Pole, A., I. J. Gordon, and M. L. Gorman. 2003. African wild dogs test the 'survival of the fittest' paradigm. *Proc R Soc Lond B Biol Sci.* 270:S57.
38. http://www.ncseweb.org/resources/articles/4396_message_from_the_pope_1996_1_3_2001.asp).
39. Ayala, F. J. and W. M. Fitch. 1997. Genetics and the origin of species: An introduction. *PNAS* 94:7691–7697.

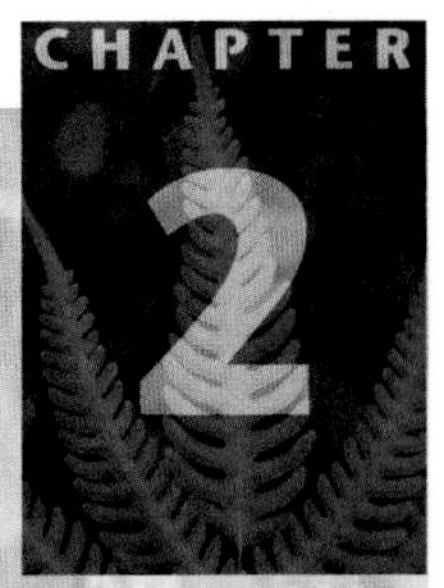

Cellular Basis of Variability and Evolution

The production of gametes (sperm and eggs) is one of the major sources of variability accounting for individual differences in biological features. It is through cell duplication, independent and random assortment of the sex cells, and the transmission of heritable traits that variability occurs. The processes of gamete formation and general cell duplication are controlled and governed by the genes and DNA genetic code. Hence, the study of cell duplication and of genetics is of paramount importance for examining the process of evolution. The science of genetics incorporates a number of different areas ranging from the molecular level, with the focus on what genes are and how they act to produce biological structures at the individual and population level. This chapter presents an introduction to the general principles of cytology and Mendelian genetics and mutations.

The Cell: Chromosomes and Genes

Cells are the basic units of life in all living organisms. The cell consists of a nucleus and the cytoplasm. The cell is enclosed by a cell membrane, within which are a variety of structures called organelles such as the nucleus, mitochondria, and cell cytoplasm (**fig. 2.1**).

Nucleus

Within the nucleus of the cell are the chromosomes. The chromosomes carry genes, which contain deoxyribonucleic acid (DNA) and ribonucleic acid (RNA) that contain the genetic information that governs the functions of the cell.

The human body consists of two kinds of cells: **somatic cells** and **sex cells**. In the somatic cells there

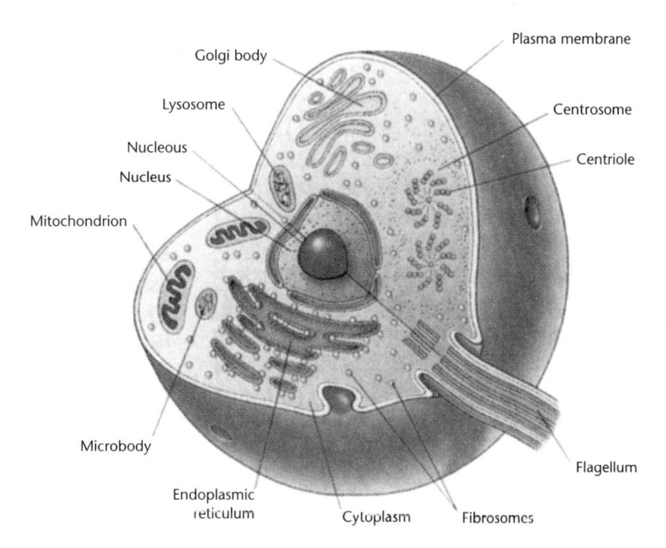

FIGURE 2.1. Schematic diagram of the cell and its major components. From *Microbiology, 3rd Edition* by Daniel Lim. Copyright © 2003 by Kendall/Hunt Publishing Company. Reprinted by permission.

are 46 chromosomes, set into 23 different pairs, one half from each parent. The first 22 pairs are called **autosomes**, and the twenty-third pair is the sex chromosome and is known alphabetically as XX or XY (**fig.** 2.2). The autosome chromosomes always occur in pairs and carry genetic information that governs the expression of all physical traits. The sex chromosomes carry the genetic information for sex determination. In general, the X sex chromosome is larger than the Y chromosome. In mammals, the Y chromosome carries genetic information involved in the development of males. Thus, all genetically normal females have two X chromosomes (XX) and have no Y chromosome. On the other hand, all genetically normal males have one X and one Y chromosome (XY). The genes carried in the autosome chromosomes and their expression are referred as autosomal inheritance, while genes carried in the sex chromosomes and their expression are referred as sex-linked inheritance.

The set of chromosomes with the basic set of genes dictates the characteristics of a species. Every species has a specific number of chromosomes. Chimpanzees and gorillas have 48 chromosomes. However, some species such as dogs and chickens have the same number of chromosomes.[7,8] This similarity does not mean that both species have the same genetic composition. The DNA that makes the gene is packed and expressed differently in each species.

Cytoplasm

The nucleus is surrounded by the cytoplasm, which contains the mitochondria and ribosomes. The mitochondria are responsible for converting energy into a form that can be used by the cell and hence they are considered the "engines" that drive the cell. The mitochondria are enclosed within a folded membrane and they contain their own DNA, called mitochondrial DNA (mtDNA). Mitochondrial DNA has the same

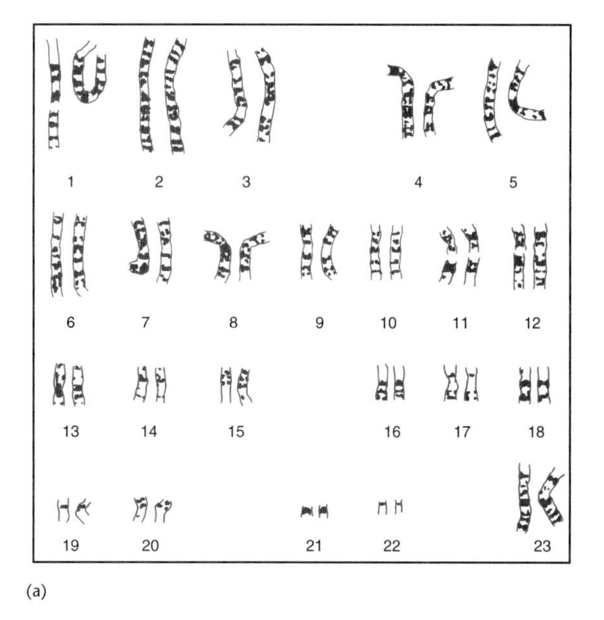

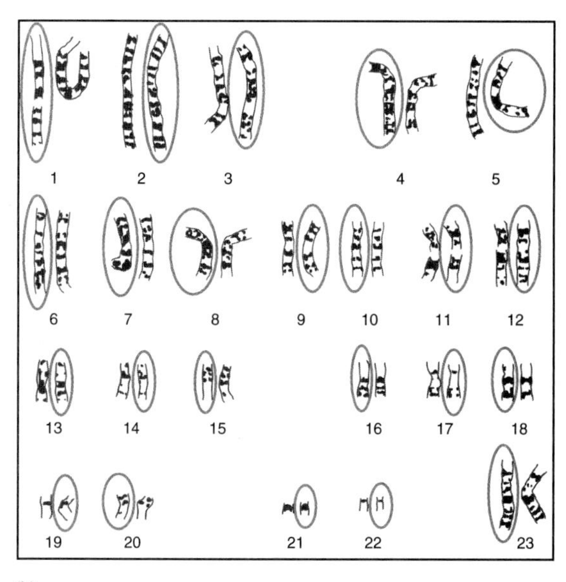

FIGURE 2.2. Human chromosomes from one somatic cell (a) and sex cells (b). Gametes or sex cells have half the number of chromosomes that the body cells have. Hence, gametes or sex cells are called haploid cells. In contrast, the body cells have both chromosomes of each pair and are called diploid.

molecular structure and function as the DNA found in the nucleus and is subject to mutations just like nuclear DNA, but differs in its form of transmission from nuclear DNA. Mitochondrial DNA is transmitted to offspring only from the mother, because at the time of fertilization egg cells retain their cytoplasm while sperm cells lose theirs. Hence, it is the mother who transmits all mitochondrial traits to her offspring of both sexes. In recent years, mtDNA has been the focus of studies of evolutionary processes (see chapter 12). Ribosomes are made up partly of RNA and are essential to the synthesis of proteins.

Mitosis and Body Cells

Every organism is in a constant dynamic state in which cell duplication and cell death occur simultaneously. Cell duplication is necessary to repair or literally replace the damaged cells of the body. Cell duplication starts with a single cell and ends with a large assemblage of genetically identical cells. For this to happen, cells have to divide in such a way that the number of chromosomes remains constant. This is achieved by a process of nuclear division called mitosis (**fig. 2.3** and **fig. 2.4**), which goes through a series of stages. In the first stage of mitosis, called prophase (see **fig. 2.3b, c**), the chromosome strands coil and become shorter and thicker. During the next stage of mitosis, metaphase (**fig. 2.3d, e**), the chromosomes move to the center of the cell and each chromosome duplicates itself, resulting in 46 pairs of chromosomes (as compared to the previous 23 double-stranded pairs). Then, during anaphase (**fig. 2.3f, g**), the chromosome strands sepa-

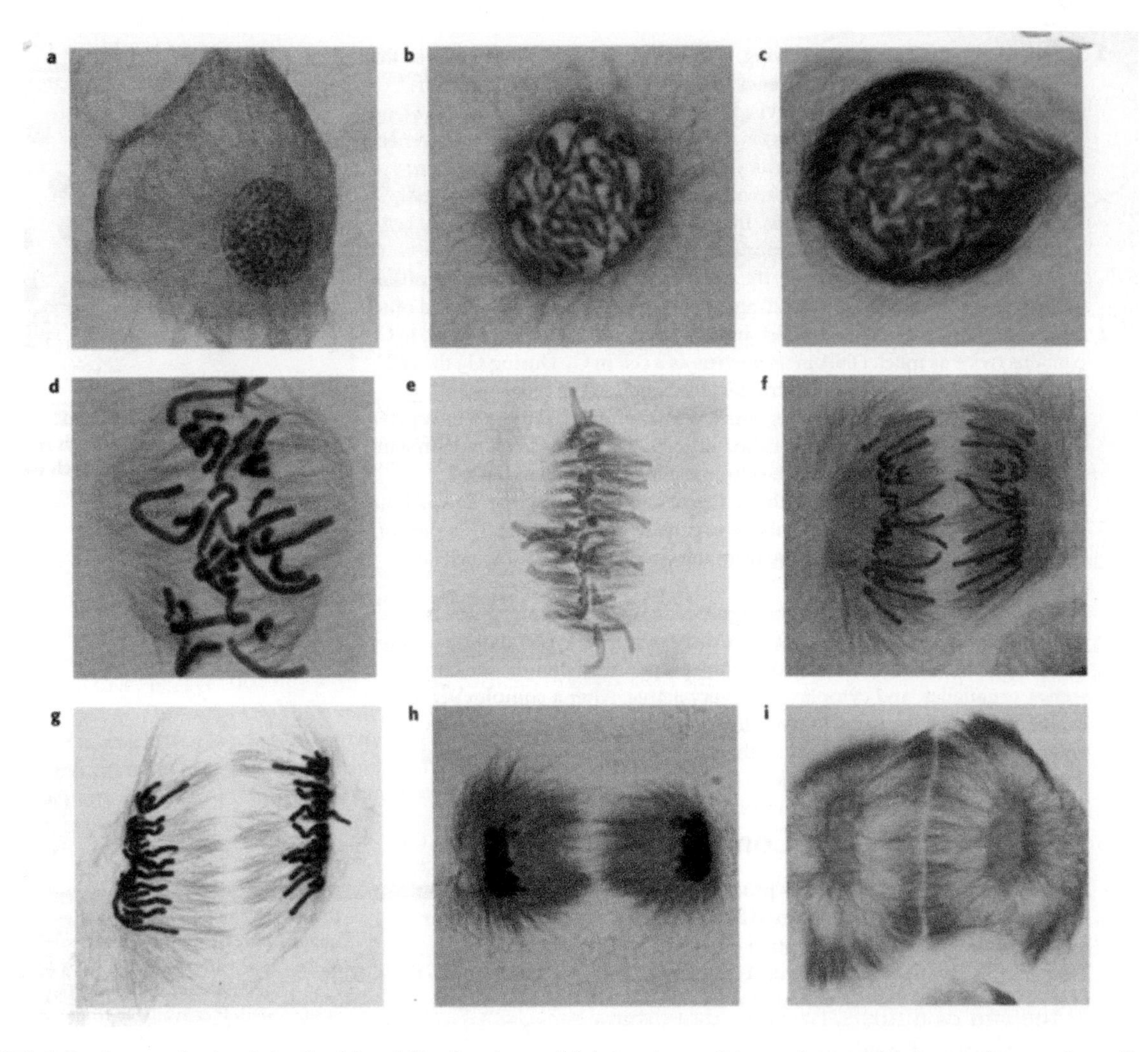

FIGURE 2.3. Stage of mitosis in the blood lily, *Scadoxus:* (a) interphase; (b) prophase; (c) late prophase, where the nuclear membrane disintegrates; (d) early metaphase; (e) metaphase; (f) early anaphase; (g) late anaphase; (h) early telophase; (i) late telophase and formation of the new two cells. Courtesy of Andrew S. Bajer, DSc.

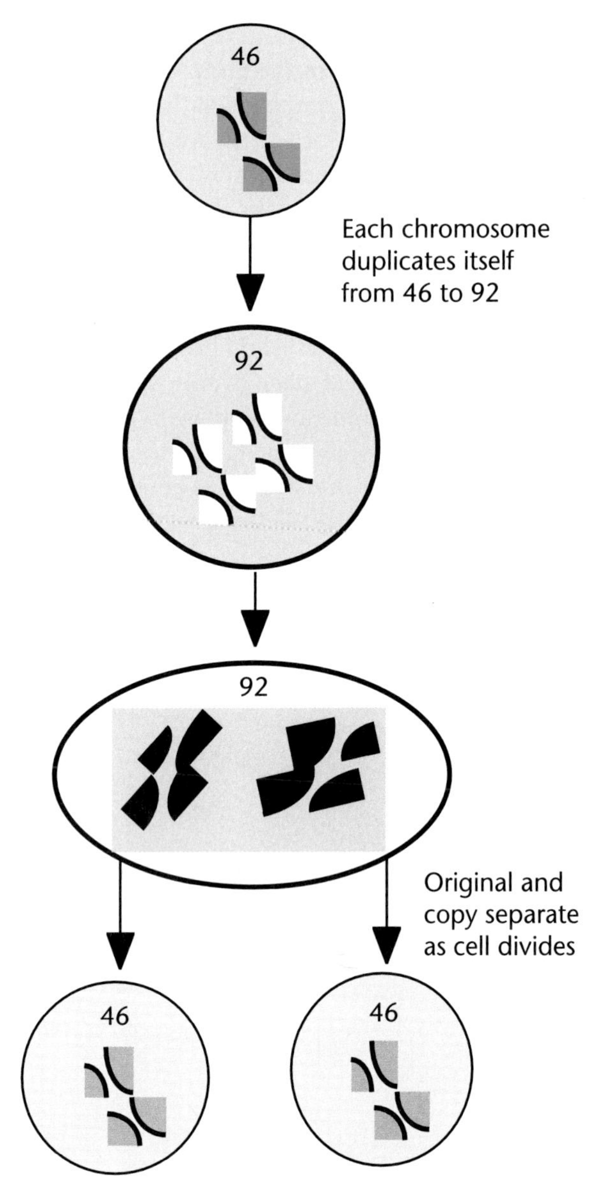

FIGURE 2.4. The process of **mitosis** and meiosis in the formation of body cells. (Composite figure made by the author.)

rate and each replicated chromosome moves along spindle fibers in opposite directions toward the two poles of the dividing cell. In the final stage, called telophase (**fig. 2.3h, i**), the centromere splits and each strand separates to one side of the cell and a new nuclear membrane forms around each group of chromosomes. At the end of mitosis, **two** new daughter cells (**fig. 2.4**) are produced, each identical to the parent and each containing one copy of each of the 46 pairs of chromosomes and DNA.[1–4]

Meiosis and Sex Cells

Gametes contain only half the number of chromosomes that body cells have. Hence, gametes are said to be haploid cells (**fig. 2.5**). The production of sex cells, sperm in males and eggs in females, is carried by the process of meiosis (**fig. 2.6**). Although the process of meiosis begins in the same way as mitosis by making a copy of each chromosome, it includes an additional cell division, which in turn is followed by another cell division without an intervening round of chromosomal replication. That is, meiosis consists of one replication followed by two divisions of the cell. Instead of ending up with two identical daughter cells, meiosis results in **four** gametes (sex cells), each with 23 single chromosomes rather than 23 pairs of chromosomes. In sperm, meiosis produces four sex cells from the initial one cell containing a set of 23 pairs of chromosomes. The process of meiosis also produces four eggs, but only one is functional.

The replication of chromosomes takes place before meiosis starts (**fig. 2.6a**). At this stage each chromosome is a double structure made of two chromatids

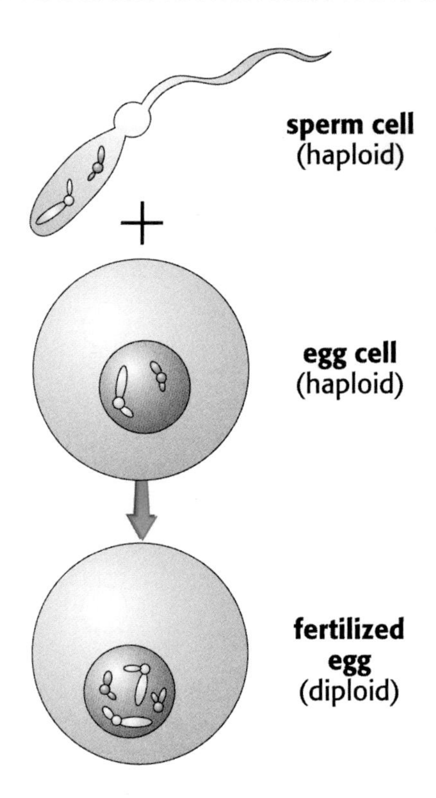

FIGURE 2.5. Gametes unite in the process of fertilization, restoring the diploid number of chromosomes. What number of chromosomes in the new individual come from each parent? Half the total number of chromosomes come from each parent. From *BSCS Biology: An Ecological Approach, 9th Edition* by BSCS. Copyright © 2002 by BSCS. Reprinted by permission.

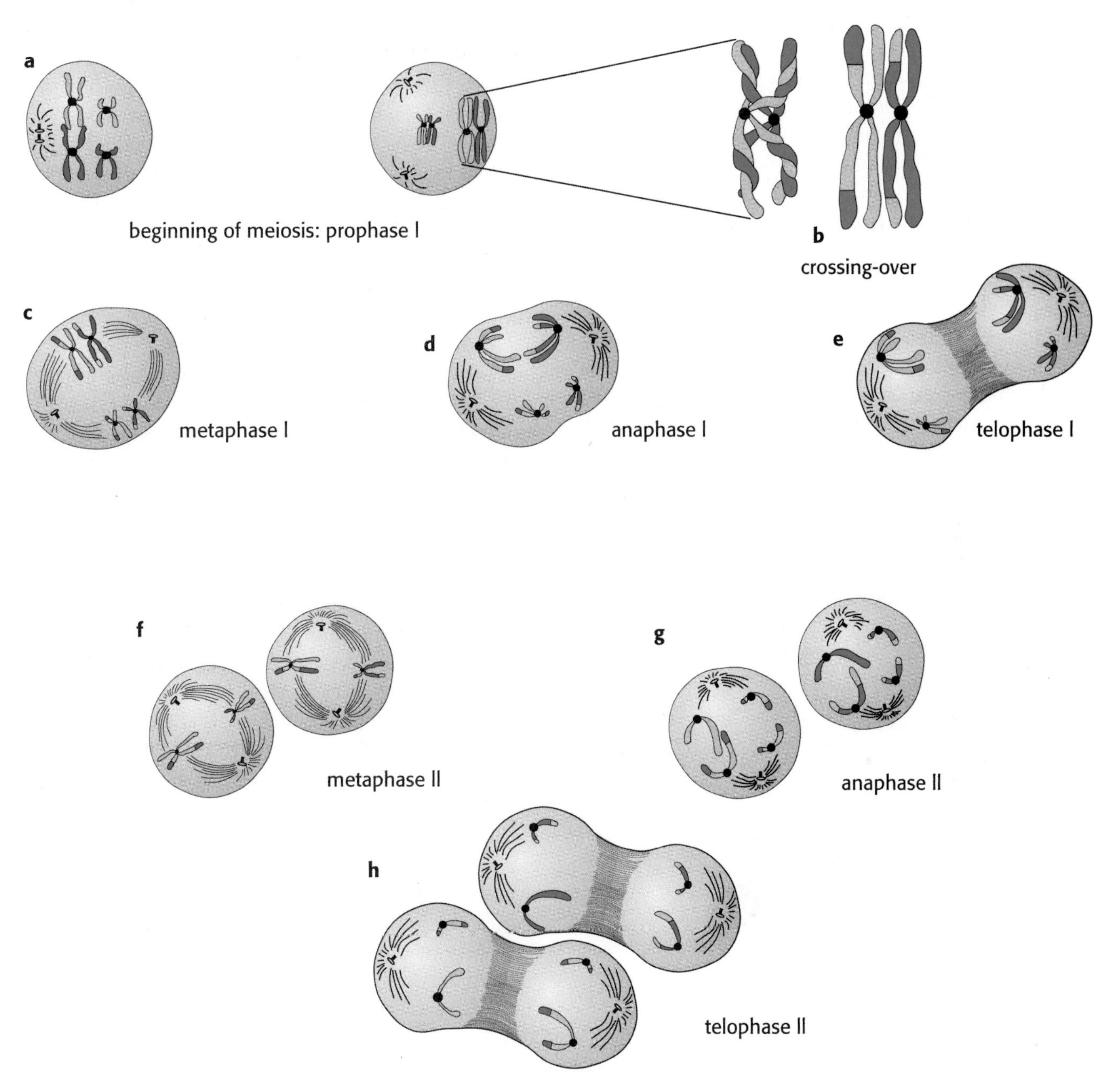

FIGURE 2.6. Meiosis. The first meiotic division is called the reduction division because each of the two resulting cells has half the number of chromosomes of the original cell. In this stage the chromosomes are still in their duplicated state. During the second meiotic division, sister chromatids are separated from each other. Each of the four resulting cells is haploid. From *BSCS Biology: An Ecological Approach, 9th Edition* by BSCS. Copyright © 2002 by BSCS. Reprinted by permission.

joined at the centromere. The two chromatids contain identical DNA. Thereafter, the cell goes into meiosis. The first meiotic division is called I; the second is called II. As meiosis begins, in prophase I, the matching pairs of homologous chromosomes come close together and twist around each other (**fig. 2.6b**). Then, parts of a chromatid of one chromosome break off and exchange places with the identical parts of a chromatid of the other matching chromosome. In this process, which is called crossing-over, chromatid segments are exchanged between the two chromosomes of a pair, which results in new combinations of genes. After crossing-over and during metaphase I (**fig. 2.6c**), the chromosome pairs move to the equator of the cell. Thereafter, during anaphase I (**fig. 2.6d**), the matching chromosomes of each pair separate and begin to move toward opposite poles of the spindle. Then the matching chromosomes, one from each pair, continue to move in opposite directions. As they approach the poles in telophase I and metaphase II (**fig. 2.6e, f**) the cell splits, resulting in two cells. Each new cell has only half as many chromosomes as a body cell and, therefore, contains half the parent cell's total genetic information. In anaphase II (**fig. 2.6g**), the centromere divides, and the two chromatids of each chromosome separate and move toward opposite poles. In telophase II (**fig. 2.6h**), the chromosomes are enclosed by new nuclear membranes and the cell divides again, resulting in four haploid cells (**fig. 2.7**), each containing 23 single chromosomes. If the production of sex cells would not include the second reduction and division, the number of chromosomes would be greater than 23 chromosomes. For example, if both chromosomes in each pair were passed on, children would have 23 pairs from both parents, for a total of 46 pairs or a total of 92 chromosomes. Thus, meiosis is the mechanism that maintains the number of chromosomes within the biological limit set by evolution that defines the human species.[1–4]

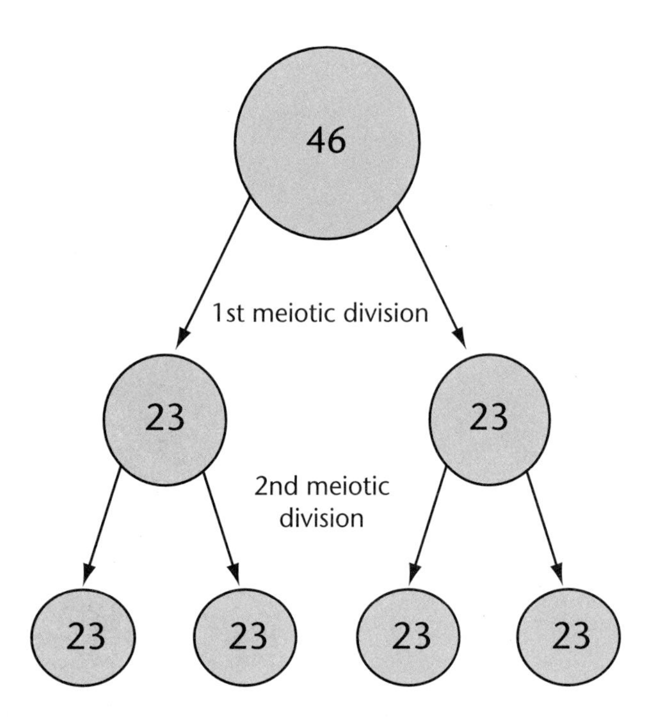

FIGURE 2.7. The process of **meiosis** and the formation of four sex cells, each having 23 chromosomes. (Composite figure made by the author.)

Meiosis as Source of Variability and Genetic Problems

Meiosis is one of the most important mechanisms and sources of variability in all sexually reproducing organisms. The nature of this variability is rooted in the fact that meiosis involves two divisions during which there is a random arrangement of chromosome pairs that can produce an almost infinite number of combinations for any human mating. The evolutionary significance of meiosis is that it ensures that chromosome sets and the genes they contain are rearranged from one generation to the next and provides the source of variability upon which natural selection can work. Meiosis is an important component of the evolutionary process, but it is also a major source of abnormalities. If for some reason the process of meiosis does not occur in the right sequence, genetic "mistakes" can occur. Many of them are associated with spontaneous abortion (miscarriages). It has been estimated that 70 percent of the spontaneous abortions (miscarriages) are caused by an improper number of chromosomes.[5,6] Failure of the chromosomes to separate during either of the two meiotic divisions can result in nondisjunction (**table 2.1**). The result of nondisjunction is that one of the daughter cells can receive two copies of the affected chromosome while the other daughter cell does not receive the particular chromosome. In that case, the affected gamete unites with a normal gamete containing 23 chromosomes, causing the zygote to have either 45 or 47 chromosomes. When nondisjunction occurs in sex chromosomes, it can produce individuals with XXY (47 chromosomes), XO (45 chromosomes), XXX (47 chromosomes), or XYY (47 chromosomes). In some conditions it may result in death or cognitive disability and/or sterility. Many nondisjunction of sex chromosomes are lethal. For example, it is possible to survive without a Y chromosome, but it is not possible to live without an X chromosome. Missing a chromosome is referred to as monosomy and having three copies of a particular chromosome (rather than two) is called trisomy. For example, Down syndrome is the result of having three copies of the twenty-first chromosome and, hence, is referred to as trisomy 21. It is estimated that trisomy 21 occurs in approximately 1 out of every 1,000 live births, and the frequency increases with maternal age. This increase is related to the fact that, in females, the numbers of gametes is set at birth. Hence, the greater the time before the formation of a zygote, the greater the possibilities that environmental factors can interfere with meiosis.

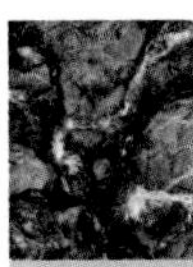

TABLE 2.1
Examples of nondisjunction in sex chromosomes

Chromosomal Complement	Estimated Condition	Incidence	Manifestations
XXX	Trisomy X	1 per 1,000	Affected women are usually clinically normal, but there is a slight increase in sterility and mental retardation compared to the general population. In cases with more than three X chromosomes, mental retardation can be severe.
XYY	XYY syndrome	1 per 1,000 male births	Affected males are fertile and tend to be taller than average.
XO	Turner syndrome	1 per 10,000 female births	Affected females are of short stature, have broad chests and webbed necks, and are sterile. Difficulties with spatial relationships, but there is usually no mental retardation. Most (95 and 99%) of the affected fetuses die before birth.
XXY	Klinefelter syndrome	1 per 1,000 male births	Symptoms are noticeable by puberty: reduced testicular development, reduced facial and body hair, some breast development, and reduced fertility or sterility.

Genes, DNA, and Genetic Code

Within the nucleus of each cell in the body are long strands of protein called chromosomes. Within each chromosome there are **genes**, and each gene contains deoxyribonucleic acid (DNA) and other organelles such as mitochondria. The cell mitochondria also contain DNA that is usually referred to as mitochondrial DNA (mtDNA). Both the nuclear DNA and the mtDNA molecules provide the code for biological structures and the means to translate this code. DNA provides the instructions and the information for building, operating, and repairing organisms. Hereditary information is transmitted through information chemically coded in DNA.[7–12] The term **genotype** refers to the genetic composition of an organism. As the result of its genotype and its interaction with the environment, an organism expresses certain observable characteristics called **phenotype.**

The physical appearance of the DNA molecule resembles a twisted ladder connected with multiple steps, or rungs, of nitrogen-containing chemicals called bases (**fig. 2.8**). Each rung is made up of four **nuclear bases** called adenine (A), thymine (T), cytosine (C), and guanine (G). The unique feature of DNA is that A always binds with T, and C always binds with G. The sequence of these combinations or spelling constitutes the genetic code (also referred to as the DNA sequence). The DNA sequence specifies the exact genetic instructions required to create a particular organism with its own unique traits.

The Genetic Code

The genetic code is the chemical equation by which hereditary information is translated from genes into structural and regulatory proteins such as hemoglobin, insulin, pepsin, enzymes, and hormones. A subunit of DNA consisting of three nitrogenous bases such as adenine, guanine, thymine, or adenine, guanine, and cytosine is called a **nucleotide**. That is, a nucleotide is made up of a 5-carbon sugar such as ribose, or deoxyribose, a nitrogen base, and a phosphate group. Within a gene, each block of three nucleotides is called a **codon** (**fig. 2.9**). Each codon designates an amino acid. A DNA molecule is formed by linking many nucleotides into a long chain. To transmit and specify all the information needed for the synthesis of

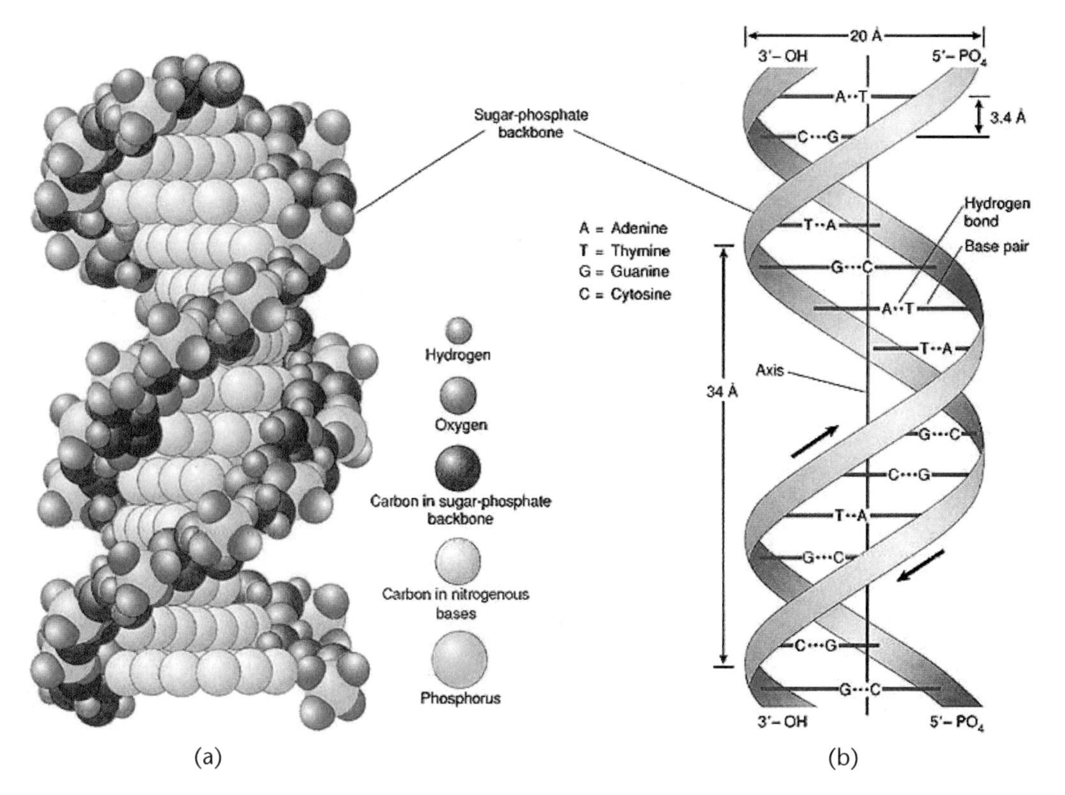

FIGURE 2.8. The DNA molecule (a) is shaped like a twisted ladder; (b) is a template for copying the whole. From *Microbiology, 3rd Edition* by Daniel Lim. Copyright © 2003 by Kendall/Hunt Publishing Company. Reprinted by permission.

proteins, the DNA molecule uses at a time only three out of four different bases. Each amino acid is coded by three of the four chemical bases. For example, the base sequence CGA provides the code for the amino acid alanine, and the base sequence TTT codes for the amino acid lysine. There are 64 possible codes that can be specified, using some combination of three bases (**table 2.2**). The ability of the DNA molecule to use the different amino acid code lies in a simple property of the chemical bases. For example, UGG codes for the production of the amino acid tryptophan. The codon AUG signals the start of translation. UAA and UGA signal a stop to translation.

However, while all genes consist of DNA, even within genes not all the DNA sequence actively codes genetic information, because the sequence of DNA can contain sections that code for amino acids that make up proteins (called exons) and also large sections that do not code for amino acids (called introns).

Replication of DNA

The complete set of instructions for making an organism is called its **genome**. The genome contains the master blueprint for all cellular structures and activities for the lifetime of the cell or organism. Each time a cell divides into two daughter cells, its full genome is duplicated. This duplication occurs in the nucleus of the cell. During cell division the DNA molecule unwinds, and the bonds between the base pairs break, allowing the strands to separate. Each strand directs the synthesis of a complementary new strand, with free nucleotides matching up with their complementary bases on each of the separated strands. Strict base-pairing rules are followed so that adenine will pair only with thymine (an A-T pair) and cytosine with guanine (a C-G pair). That is, when one side of the ladder contains the GGTCTG, the other side of the ladder automatically duplicates a matching pair of CCAGAC. Each daughter cell receives one old and one

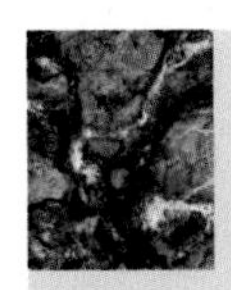

TABLE 2.2

The DNA code and the 64 codons that code for the production of some amino acid. From *BSCS Biology: An Ecological Approach, 9th Edition* by BSCS. Copyright © 2002 by BSCS. Reprinted by permission.

First Letter	Second Letter				Third Letter
	U	C	A	G	
U	phenylalanine	serine	tyrosine	cycsteine	U
	phenylalanine	serine	tyrosine	cycsteine	C
	leucine	serine	stop	stop	A
	leucine	serine	stop	tryptophan	G
C	leucine	proline	histidine	arginine	U
	leucine	proline	histidine	arginine	C
	leucine	proline	glutamine	arginine	A
	leucine	proline	glutamine	arginine	G
A	isoleucine	threonine	asparagine	serine	U
	isoleucine	threonine	asparagine	serine	C
	isoleucine	threonine	lysine	arginine	A
	(start) methionine	threonine	lysine	arginine	G
G	valine	alanine	aspartate	glycine	U
	valine	alanine	aspartate	glycine	C
	valine	alanine	glutamate	glycine	A
	valine	alanine	glutamate	glycine	G

new DNA strand. The cell's adherence to these base-pairing rules warrants that the new strand is an identical copy of the old one (**fig. 2.10**). This minimizes the incidence of errors (mutations) that may greatly affect the resulting organism or its offspring.

DNA and Protein Synthesis

The synthesis of protein follows the principle of **central dogma** (**fig. 2.11**). According to this principle protein synthesis follows two steps. First, the genetic information contained within DNA is copied into messenger RNA (mRNA) in a process called **transcription.** Second, the mRNA, acting as a messenger sends information to synthesize protein molecules. The synthesis of protein from an RNA messenger is known as **translation.**

DNA Techniques and Applications

Polymerase Chain Reaction

DNA analysis uses a technique called polymerase chain reaction (PCR). This technique produces multiple copies of DNA, making it possible to analyze segments of DNA as small as one molecule. In PCR, the two strands of a DNA sample are separated, and an enzyme synthesizes complementary strands on the ex-

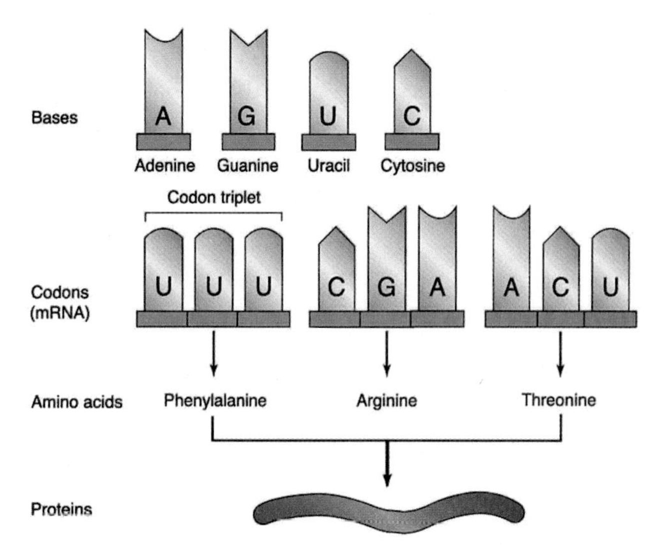

FIGURE 2.9. The genetic code and codons: The genetic alphabet consists of four bases (adenine, uracil, guanine, and cytosine), which are arranged into codon triplets in mRNA. These codons make up the genetic code of the cell. From *Microbiology, 3rd Edition* by Daniel Lim. Copyright © 2003 by Kendall/Hunt Publishing Company. Reprinted by permission.

posed bases producing copies of the original DNA material. This technique is highly useful in cases where the samples are too small for a reliable analysis of nucleotide sequences, which often happens in fossil remains and crime scenes.

Fingerprinting

Another approach for identifying individual traits is called the DNA fingerprint. DNA fingerprints are produced by placing the DNA on an electrically charged gel. This process breaks up the DNA into fragments, which in turn separate into bands that vary in thickness. Using this technique, researchers can identify the specific DNA sequence of each individual, referred to as the DNA fingerprint. For example, one individual might have a segment of eight bases such as ATTCTATA repeated three times and another might have the same sequence repeated five times. These unique banding patterns comprise the DNA fingerprint (**fig. 2.12**). DNA fingerprinting has become useful for tracing and identifying the guilty and exonerating the innocent who were wrongly convicted and imprisoned for crimes committed years previously. Likewise, DNA fingerprinting has also been used to trace the genealogy and descendants of scores of historic peoples.

Mendelian Genetics

In 1865 the Austrian monk Gregor Mendel reported his experiments on breeding pea plants to the Natural

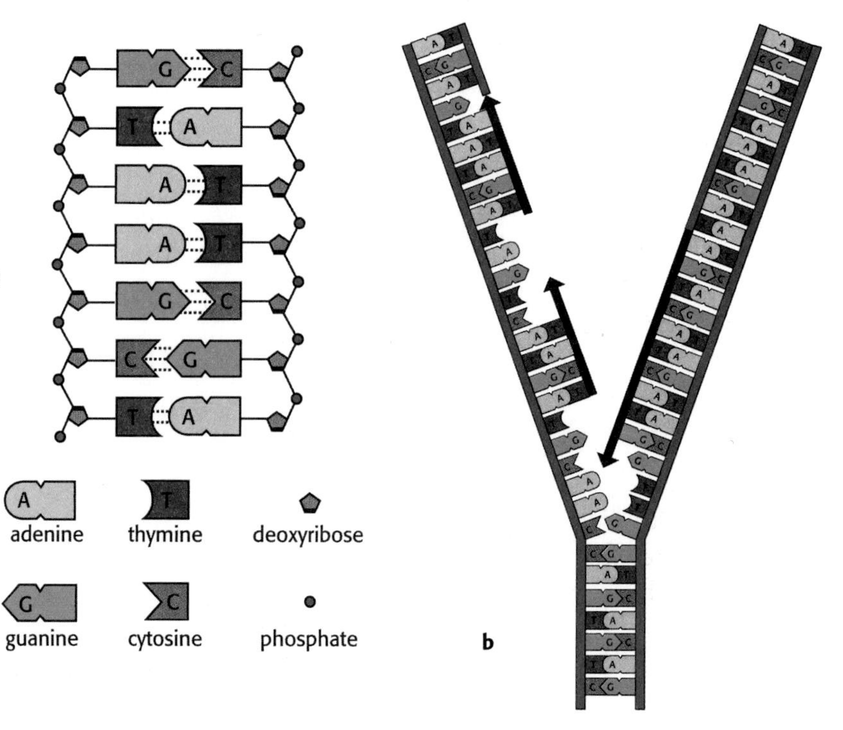

FIGURE 2.10. Replication of the DNA molecule. Once the section of initial DNA molecule (a) separates into two strands, (b) each strand attracts free-floating complementary bases such as A binds with T, C binds with G, resulting in two identical DNA molecules. From *BSCS Biology: An Ecological Approach, 9th Edition* by BSCS. Copyright © 2002 by BSCS. Reprinted by permission.

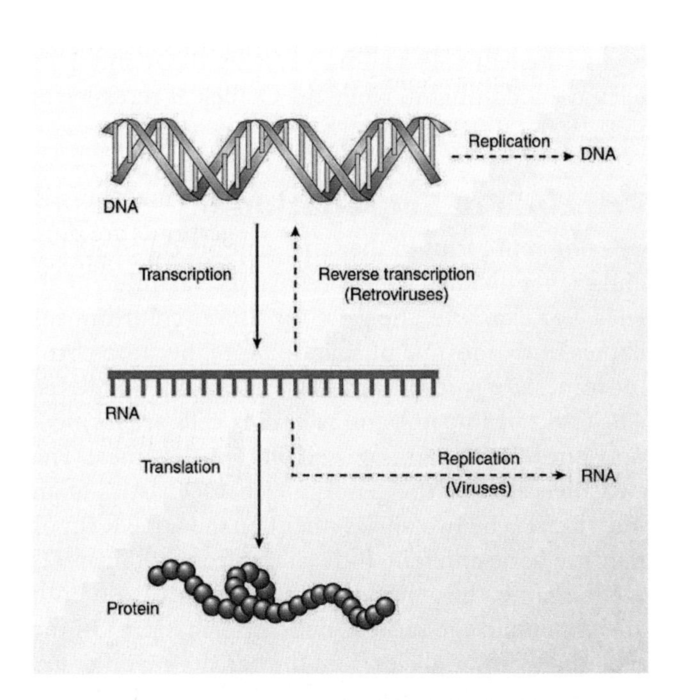

FIGURE 2.11. The central dogma of protein synthesis. The central dogma describes the process whereby messenger RNA and transfer RNA participate in the flow of genetic information in the cell for the production of a chain of proteins. From *Microbiology, 3rd Edition* by Daniel Lim. Copyright © 2003 by Kendall/Hunt Publishing Company. Reprinted by permission.

Science Society in Brünn.[13–15] Mendel discovered that biological inheritance was not an irreversible blending of parental traits. Rather, individual units of hereditary information, later called "genes," were passed from parent to offspring as discrete particles according to certain regular patterns. In one individual, a gene's effect might be blended with the effects of other genes, or even suppressed altogether. But the gene itself remained unchanged, ready to be passed on to the next generation where it might express itself and thus be available for natural selection. Over eight years (1856–1863), Mendel grew over 10,000 pea plants and discovered the fundamental laws of inheritance. Mendel selected genetically true, or true-breeding pea plants that were distinct and contrasted in the forms of only one trait. For example, plants were either tall or short, or produced either round or wrinkled, or green or yellow seeds (**fig. 2.13**). He deduced that genetic information is inherited in discrete units, which later were named genes. Mendel recognized the mathematical patterns of inheritance from one generation to the next, from which he derived the principles of segregation and independent assortment. In recognition of Mendel's accomplishments, the science of genetic inheritance is called Mendelian genetics.

Dominant and Recessive Alleles

An allele that masks the effect of the other allele is called dominant; the opposite of a dominant allele is a recessive allele, whose effect may be masked. A recessive gene will express only in the absence of the dominant or when present on the X-chromosome (see table 2.3). When Mendel crossed pea plants whose seeds were yellow with pea plants whose seeds were green, all the offspring plants had yellow seeds. This discovery suggested that somehow one trait (yellow seed color) dominated in its effects. When Mendel crossed the plants of the new generation, he found that some of their offspring had yellow seeds and some had green seeds. It became evident that the genetic information for green seeds had been hidden for a generation and then appeared again. The yellow and green traits became expressed in a 3:1 ratio so that for every three yellow seeds there was one green seed. Thus, the trait for yellow color is dominant while the trait for

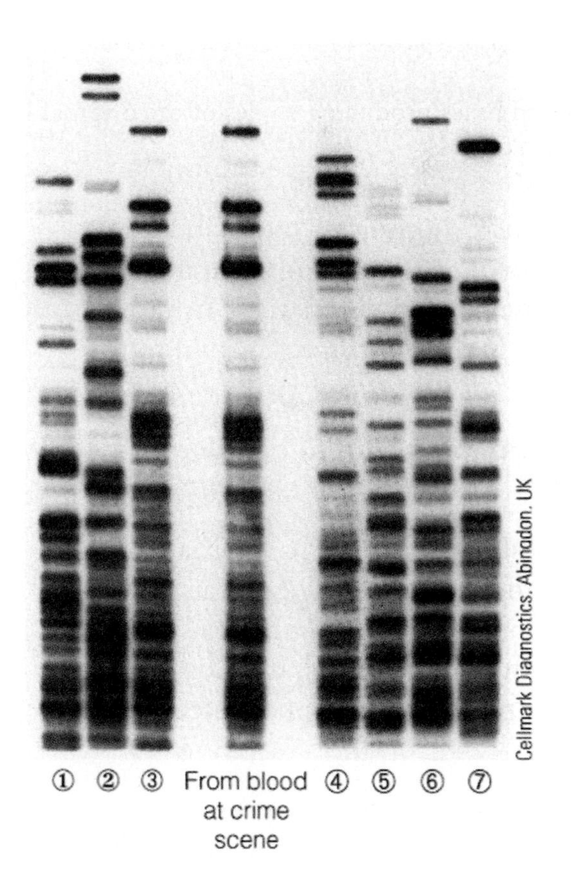

FIGURE 2.12. A DNA fingerprint derived from a blood sample left at an actual crime scene compared with seven other DNA fingerprints from suspects. Courtesy of Orchid Cellmark, Inc., Germantown, Maryland.

seed shape	seed color	seed-coat color	pod shape	pod color	flower position	stem length
round	yellow	colored	inflated	green	axial	long
wrinkled	green	white	constricted	yellow	terminal	short

FIGURE 2.13. The seven traits of peas studied by Mendel. From *BSCS Biology: An Ecological Approach, 9th Edition* by BSCS. Copyright © 2002 by BSCS. Reprinted by permission.

green color is recessive. This is a demonstration of what Mendel called the principle of dominance.

Codominant Alleles

When two different alleles are present in a genotype and both are equally expressed, they are referred to as codominant. That is, neither allele is dominant or recessive. For example, the flower color of morning glories (**fig. 2.14**) is a codominant trait. In the human blood type, MN, the M and N alleles are codominant and in the ABO blood group system, the A and B alleles are codominant, while the O allele is recessive to both A and B.

Principle of Segregation

The different forms of the same gene are called **alleles**, and each individual gene can have several different alleles. For example, the gene for flower color in peas comes in two forms: purple or white. Because chromosomes are paired, each individual has two alleles. Alleles occur in pairs, and when sex cells are formed, only one of each pair is passed on to each gamete. The two alleles define the genotypic characteristics of an individual. The two alleles might be the same form or they might be different. If the alleles from both parents are the same, the genotype is **homozygous**, and both alleles contain the same genetic information. On the other hand, if the alleles from the parents are different, the genotype is **heterozygous** and each contains different genetic information. Therefore, heterozygosity refers to the presence of different alleles at one or more loci on homologous chromosomes. The actual observable trait is known as the phenotype. The specific position of a gene on a chromosome is called a locus (plural loci).

Mendel, by crossing purple-flowered peas with white-flowered peas, (**table 2.3**) demonstrated the principle of segregation of genes. In the parental generation (*P*) each parent has identical alleles (i.e., are homozygous) for the flower-color trait represented by either *PP* for a dominant purple-flowered parent or *pp*

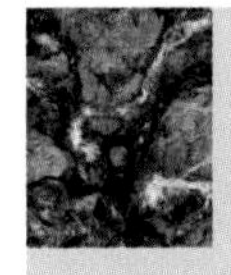

TABLE 2.3

Schematic diagram illustrating Mendel's principle of segregation of traits, where one parent has identical alleles (i.e., are homozygous) for the dominant purple flower represented by *PP* and the other parent has also identical alleles (i.e., are homozygous) for the recessive white flower represented by *pp*.

		Parental Phenotype Dominant purple flower (PP)	
Parental Phenotype Recessive white flower (*pp*)		*P*	*P*
		Offspring in First Generation	
	p	*Pp*	*Pp*
	p	*Pp*	*Pp*
		Parental Phenotype Heterozygous purple flower (*Pp*)	
Parental Phenotype Heterozygous purple flower (*Pp*)		*P*	*p*
		Offspring in Second Generation	
	P	*PP*	*Pp*
	p	*Pp*	*pp*

Note. In the first generation all the offspring have purple flowers, but in the second generation, 75 percent of plants had purple flowers and 25 percent of plants had white flowers.

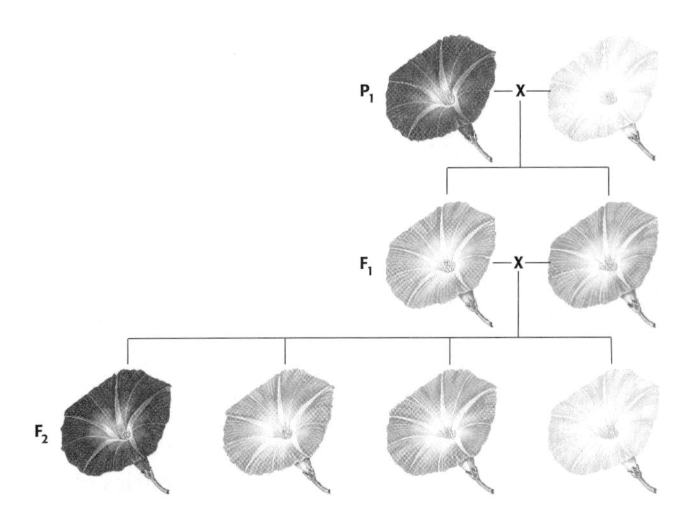

FIGURE 2.14. Codominance and inheritance of flower color of morning glories. The *P* plants are homozygous. The *F* plants are heterozygous. Their phenotype is a blend of parents. From *BSCS Biology: An Ecological Approach, 9th Edition* by BSCS. Copyright © 2002 by BSCS. Reprinted by permission.

for a recessive white-flowered parent. In the first generation (F1) all the offspring have purple flowers but are carrying the gene for white-colored flower. Therefore, when the first generation of plants mature and are crossed with each other in the second generation (F2), for every three plants with purple flowers there is one plant with white flowers. That is, 75 percent of plants had purple flowers, and 25 percent of plants had white flowers. Furthermore, one offspring will be homozygous (*PP*) for purple color, two will be heterozygous for (*Pp* genotype) purple, and one will be homozygous for white color (*pp*) (**table 2.3**). Thus, the white allele was not lost in the F2 generation, but persisted even though it was unexpressed in the visual appearance, or phenotype, of the plant.

Principle of Independent Assortment

According to Mendel's principle of independent assortment (**fig. 2.15**), genes that code for different traits assort independently of each other during gamete formation. Independent assortment occurs when the genes controlling different characteristics are located on different chromosomes. Mendel crossed a homozygous (has the same alleles) red-flowered, tall plant (*RRTT*) with a homozygous white-flowered, dwarf plant (*r r t t*). Here, each parental gamete carries two traits, one allele for each trait: one for flower color (either red, *R*, or white, *r*) and one for plant height (either tall, *T*, or dwarf, *t*). The first generation (F1) consists entirely of heterozygous (has different alleles) genotypes with the red, tall phenotype. These heterozygous parents then can produce four types of gametes. The characteristics for flower color and plant height behave independently so that all possible combinations of alleles can be found in the population. The resulting second generation (F2) is therefore composed of the following phenotype ratio: 9 red and tall, 3 white and tall, 3 red and dwarf, and 1 white and

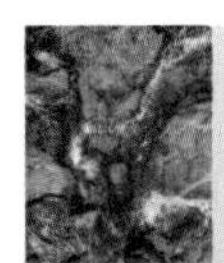

TABLE 2.4
Schematic diagram illustrating Mendel's principle of independent assortment

Heterozygous Phenotype of Parent 1 *R r T t* for Red Tall	Heterozygous Phenotype of Parent 1 *R r T t* for Red Tall			
	R T	*R t*	*r T*	*r t*
R T	***R R T T***	***R R T t***	***R r T T***	***R r T t***
R t	***R R T t***	R R t t	***R r T t***	R r t t
R T	***R r T T***	***R R T t***	R r T t	R r T t
r t	***R r T t***	R r t t	r r T t	**r r t t**

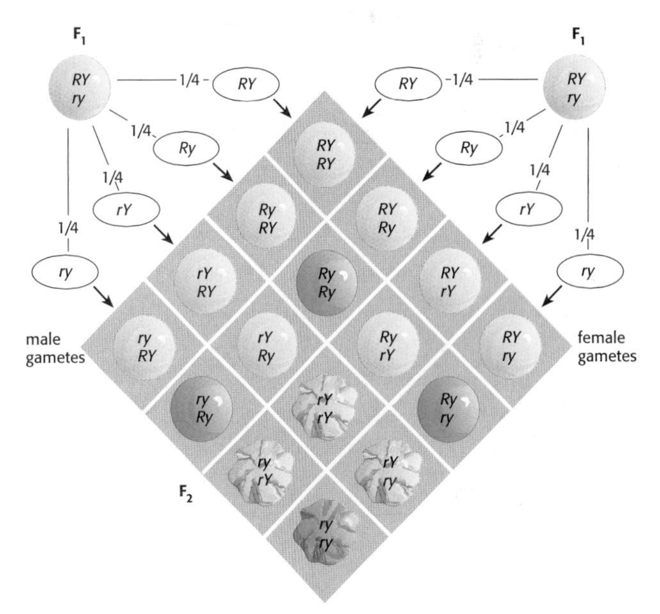

FIGURE 2.15. Principle of independent assortment of unit traits derived by Mendel. In the second generation (F2) the traits for seed shape (*R* = round, *r* = wrinkled) and seed color (*Y* = yellow, *y* = green) assorted separately. Note there are 12 round to every 4 wrinkled seeds, and 12 yellow to 4 green seeds. That is, each trait independently shows the 3:1 ratio of typical dominant and recessive traits. From *BSCS Biology: An Ecological Approach, 9th Edition* by BSCS. Copyright © 2002 by BSCS. Reprinted by permission.

dwarf (*r r t t*). In other words, alleles from separate gene pairs are inherited independently. Using the laws of probability and assuming that genes recombine at random, the genotypes and phenotypes in a plant (the progeny) can be accurately predicted.

A practical way of determining the expected frequency of trait is easily done using the Punnett square (**table 2.4**). For example, when crossing two heterozygous purple flowers in the second generation, the ratio will be three purple for one white, or 75 percent purple and 25 percent white. The genotypes appear in the ratio 1:2:1, or 25 percent homozygous purple, 50 percent heterozygous purple, and 25 percent homozygous white.

Linkage

An exception to Mendel's law of independent assortment is the process known as linkage. When alleles are on the same chromosome they are inherited together. Therefore, linked alleles are not inherited independently because they are, by definition, on the same chromosome.

Crossing-Over and Recombination of Genes

An exception to the rule of linkage is crossing-over. In many instances, during the process of meiosis chromosome pairs exchange pieces, a process known as crossing-over. For example, genes *A* and *B* are linked. *A* should always assort with *B*, however, the segment of a chromosome containing the ***a*** allele could switch with the segment of a chromosome containing the *A* allele on the other chromosome. Therefore, the gamete could have a gene with ***a*** and *B* or a gene with *A* and ***b*** (**fig. 2.16**). The result of crossing-over is referred to as recombination. Thus, crossing-over provides yet another mechanism for increasing genetic variation by providing new combinations of alleles.

In summary, Mendel demonstrated that genetic traits do not blend and are maintained in a breeding population, assuming no mortality before successful reproduction. Given the numerous genes inherited by each organism and the random nature of independent assortment, every individual in a sexually reproducing population is likely to be unique.

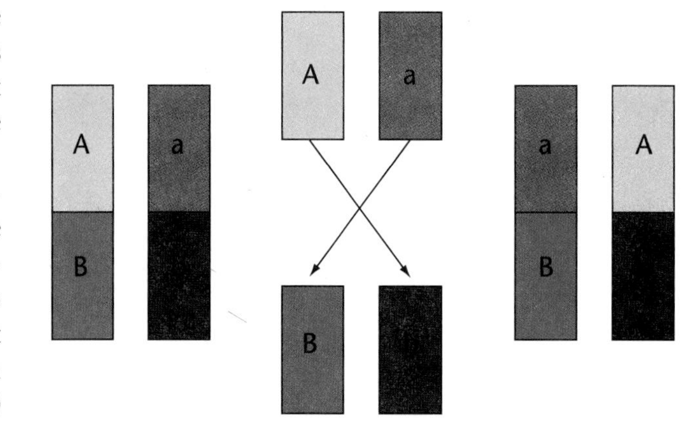

FIGURE 2.16. Crossing-over and recombination in chromosomes. During meiosis when two genes are linked chromosome pairs exchange pieces. When crossing-over occurs the genes located in the chromosomes go along; therefore, new combinations of genes take place. (Composite figure made by the author.)

Inheritance of Mendelian Discontinuous Traits

Characteristics that result from the action of a single locus with a clear-cut mode of inheritance are referred to as "simple," or discontinuous genetic traits. They are also considered discrete, meaning that they produce a finite number of phenotypes. These traits are usually inherited following Mendelian principles. According to recent summaries, there are about 4,500 human traits that are inherited according to simple Mendelian principles.[14] Some examples are given here.

PTC Tasting

Taste sensitivity, as measured by the ability to distinguish phenylthiocarbamide (PTC), is genetically determined.[16–18] Family studies indicate that sensitivity to phenylthiocarbamide (PTC) is inherited in a Mendelian form. The ability to taste is controlled by a dominant allele (*T*) and nontasters are individuals with two recessive alleles (*tt*). Subsequent evidence indicates that the taster group might be composed of two subgroups: homozygotes (*TT*) who are highly sensitive to these compounds referred to as "supertaster" and heterozygotes (*Tt*) who are moderately sensitive referred to as "medium taster." At the population level about 70 percent are tasters and 30 percent are nontasters. Given these two alleles, applying the principles of dominance and recessiveness, and depending on whether the parents are homozygous or heterozygous for the trait, the expected genotype distribution of PTC tasting in the children can be within one of the following possibilities:

a. If both parents are homozygous tasters (*TT*), 100 percent of the children are expected to be homozygous tasters (*TT*).

b. If one parent is a homozygous taster (*TT*) and the other is a homozygous nontaster (*tt*), 100 percent of the children are expected to be heterozygous tasters (*Tt*).
c. If both parents are homozygous nontasters (*tt*), 100 percent of the children are expected to be homozygous nontasters (*tt*).
d. If both parents are heterozygous tasters (*Tt*), 25 percent are expected to be homozygous tasters (*TT*), 50 percent of the children are expected to be heterozygous tasters (*Tt*), 25 percent are expected to be homozygous nontasters.

ABO Blood Groups

The ABO blood group system was discovered in the 1900s because of blood transfusions.[19] The blood of different persons usually has different antigenic and immune properties such that antibodies in the plasma of one blood type react with antigens on the surfaces of the red cells of another. An *antigen* is a substance that can induce the production of a corresponding antibody. An *antibody* is a substance capable of reacting specifically with particular antigens. Blood is divided into different *groups* and *types* in accordance with the types of antigens present in the cells. The most well known are the ABO and the Rh systems. In addition to the O-A-B and Rh blood group systems there are several other systems such as the M, N, S, s, P, Kell, Lewis, Duffy, Kidd, Diego, and Lutheran blood group systems. Here we will focus only on the ABO, Rh, and MN blood groups.

Two related antigens, named type A and type B, occur on the surfaces of the red blood cells in a large proportion of the population. Antibodies react specifically to each antigen. Blood type A has antibodies that react against the antigens of blood type B. Likewise, blood type B has antibodies that react against the antigens of blood type A. Therefore, if type B blood is injected into persons with type A blood, the red cells in the injected blood are destroyed (cause agglutination of the red cells) by the antibodies in the recipient's blood. Conversely, type A red cells are destroyed by anti-A antibodies in type B blood. On the other hand, type O blood can be injected into persons with type A, B, or O blood unless there is incompatibility with respect to some other blood group system (i.e., Rh^+, Rh^-, etc.). Persons with type AB blood can receive blood from persons with type A, B, or O.

To determine the blood type the red blood cells are first diluted with saline. One portion is then mixed with anti-A agglutinin while another portion is mixed with anti-B agglutinin. If the red blood cells have become clumped—that is, "agglutinated"—one knows that an antibody-antigen reaction has resulted. Then, on the basis of the presence of antibodies (also referred to as agglutinins) and antigens (also referred to as agglutinogens) blood is classified into four major groups: O, A, B, and/or AB as follows (**table 2.5**):

a. Blood group O does not have either A or B agglutinogen. Therefore, it does not react (coagulate) with either the anti-A or the anti-B serum.
b. Blood group A has only type A agglutinogens and therefore agglutinates (coagulates) with anti-A serum (present in type B blood).
c. Blood group B has only type B agglutinogen and therefore agglutinates (coagulates) with anti-B serum (present in type A blood).
d. Blood groups A and B have both A and B agglutinogens and therefore agglutinate (coagulate) with both types of serum.

Because the A and B alleles are codominant and the O allele is recessive, there are six possible combinations of genes: OO, OA, OB, AA, BB, and AB (**table 2.5**). From these data several facts are evident. First, a person with genotype OO produces no agglutinogens at all and therefore, does not react with either anti-A or anti-B serum. Hence, blood type O is considered a "universal donor." But an individual with blood type O can only receive blood from fellow O donors. A person with either genotype OA or AA produces type A agglutinogens and, therefore, has blood group A. A person with either genotype OB or BB produces type B agglutinogens and therefore, has blood group B. Genotype AB gives group AB blood. Second, group O blood, though containing no agglutinogens, does contain both *anti-A* and *anti-B agglutinins.* Meanwhile group A blood contains type A agglutinogens and *anti-B agglutinins,* and group B blood contains type B agglutinogens and *anti-A agglutinins.* Finally, group

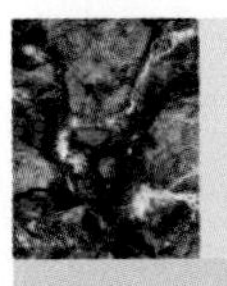

TABLE 2.5
ABO genotypes and associated phenotypes and antibodies

		Antibodies in serum that react to	
Phenotypes	Genotypes	Reaction	Cell antigens
Group O	OO	No	No
Group A	AA and AO	Yes	B
Group B	BB and BO	Yes	A
Group AB	AB	Yes	A and B

AB blood contains both A and B agglutinogens but no agglutinins at all.

World Distribution

Worldwide there is great variability in the frequency of different blood groups.[19–21] The blood group O allele is the most common in almost all human populations, reaching a frequency of 100 percent in Amerindians of South and Central America. Type B is high in Asia especially in Northern India; it is low in Europe and Africa and absent among native American and Australian Aborigines. However, type A is present among the Blackfoot and Blood Indians and surrounding tribes in Alberta and Montana (see chapter 12).

It has been suggested that natural selection may be operating on ABO blood groups through incompatibility between mother and fetus. ABO incompatibility occurs when a couple produces a fetus with an antigen not present in the mother, or when the mother has antibodies against such an antigen and she is able to damage the fetus.[22] Fetal loss and hemolytic disease of the newborn may result from such a situation.[23] An extensive study of nearly 6,000 mothers and their newborn babies from the population of Sassari (Sardinia) indicates that a woman with the anti-B antibody has a relatively higher reproductive efficiency than mothers with the anti-A antibody.[22] Thus, ABO incompatible pregnancies, in which the mother has blood type A, seem to be at an advantage. It has been suggested that maternal-fetal differences in ABO membrane protein structure could be involved in the maternal-fetal biological competition by mechanisms different than those implicated in classical immunological phenomena.[22] These findings suggest that the differential fitness associated with maternal blood types might have an important role in the maintenance of ABO polymorphism in human populations.

Rh Blood Group System

The Rh system is named after the rhesus monkeys that initially provided the blood cells with which to make the antiserum.[14,23] There are six common types of Rh antigens, each of which is called an Rh factor. These types are designated *C, D, E, c, d,* and *e* (**table 2.6**). The type *D* antigen is more antigenic than the other Rh antigens. Individuals whose blood agglutinates when mixed with the Rh serum are said to be *Rh positive* (Rh^+), while those persons who do not have type D antigen are said to be *Rh negative* (Rh^-). The antisera Rh is detected by the human antigen *D*. This gene has two alleles, *D* and *d*, where *D* is dominant. Individuals with genotypes *DD* or *Dd* are called Rh positive, and *dd* are Rh negative.

Rh Incompatibility

Rhesus incompatibility refers to a condition in which a pregnant woman and her fetus have incompatible rhesus blood groups. Rhesus incompatibility occurs when an Rh negative mother has an Rh positive fetus (**table 2.7**). Rhesus incompatibility causes *erythroblastosis fetalis*,[23] which is a disease of the fetus and newborn infant characterized by anemia and immature red blood cells. This occurs when the fetal blood carrying *D* molecules (Rh^+) enters the mother's bloodstream that has dd molecules (Rh^-). When these molecules enter the mother's bloodstream during delivery, the mother's immune system produces anti-*D* antibodies to destroy the red blood cells carrying these "invaders." Once produced, in a second pregnancy, the antibodies diffuse very slowly through the placental membrane into the

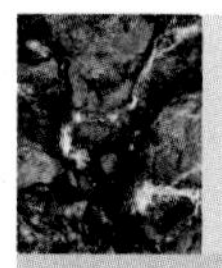

TABLE 2.6
Rh blood genotypes and phenotypes of offspring

Genotype	Phenotype	Father Genotype	Mother Phenotype	Offspring Genotype	Incompatibility with mother
DD	Rh+	*DD*	Rh+	Rh+	No
Dd	Rh+	*DD*	Rh+	Rh+	No
Dd	Rh–	*DD*	Rh+	Rh+	No
DD	Rh+	*Dd*	Rh+	Rh+	No
Dd	Rh+	*Dd*	Rh+	Rh+	No
Dd	Rh–	*Dd*	Rh+	Rh+	No
DD	**Rh+**	*dd*	**Rh–**	Rh+	Yes, 100%
Dd	Rh+	*dd*	Rh–	Rh+	Yes, 50%
Dd	Rh–	*dd*	Rh–	Rh–	No

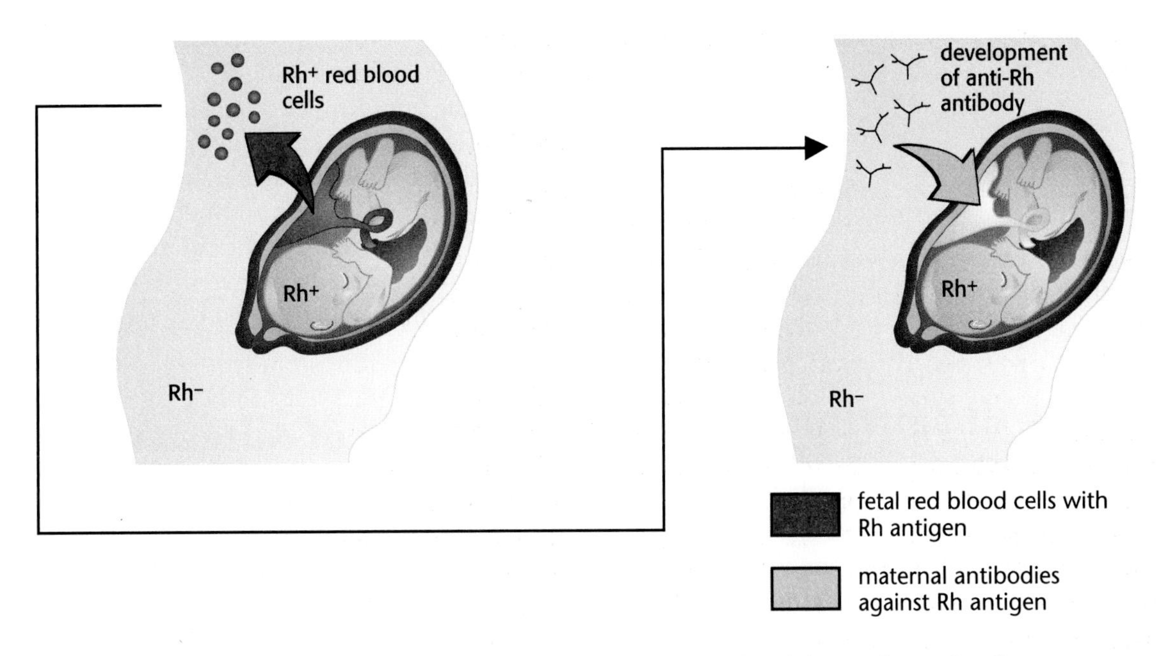

FIGURE 2.17. Rh incompatibility and anemia (called *erythroblastosis fetalis*) of the newborn after first pregnancy. If the mother is Rh negative (Rh^-) and the father is Rh positive (Rh^+), the child is Rh positive (Rh^+). Rh^+ enters into the mother's bloodstream during childbirth and stimulates the production of Rh positive (Rh^+) antibodies in the mother's blood. These antibodies, in the second and subsequent pregnancies, can enter into the fetal blood and attempt to destroy its blood cells resulting in anemia called *erythroblastosis fetalis*. From *BSCS Biology: An Ecological Approach, 9th Edition* by BSCS. Copyright © 2002 by BSCS. Reprinted by permission.

fetus's blood. There they cause slow agglutination of the fetus's blood. The agglutinated red blood cells gradually break down (hemolyze), releasing hemoglobin into the blood. The macrophages then convert the hemoglobin into bilirubin (bile pigment), which causes yellowness (jaundice) of the skin and anemia in the fetus. To compensate for the anemia the hemopoietic tissues (such as the red bone marrow and the liver) of the baby attempt to replace the red blood cells that have been destroyed. Because of the very rapid production of cells, many early and immature (blastic) forms are emptied into the circulatory system leading to the condition called *erythroblastosis fetalis*. An Rh negative mother having her first Rh positive child usually does not develop sufficient anti-Rh agglutinins to cause any harm.[24] However, an Rh negative mother having her second Rh positive child often will have become "sensitized." This is because the blood of the first child induces the production of antibodies so that in the second pregnancy she is sensitized and therefore will often develop anti-Rh agglutinins. Therefore, in the second pregnancy her anti-Rh agglutinins will attack the blood of the child (**fig. 2.17**). Fortunately, now there are various pharmacological agents that decrease the anti-Rh agglutinins of the mother.

Theoretically, the Rh negative (*dd*) should have been selected against. However, the world distribution of Rh^- shows that the *d* allele has not been eliminated. In fact, it is relatively high in a number of human populations. For example, among European populations the frequency of Rh^- (*d*) is high, averaging 40 percent. Among the Basques of Spain the frequency of Rh^- (*d*) is 53 percent. On the other hand, Australians and Native Americans are 100 percent Rh^+ (*D*) and do not have the Rh^-. It is not known why this allele persists. It is quite possible that there might be some sort of balancing selection acting to keep the d allele from disappearing.

MN Blood Group

The MN blood group system has two alleles: M and N. M allele codes for the production of M molecules, and N allele codes for the production of N molecules.[19–21] Therefore, there are three possible genotypes: MM, MN, and NN. Individuals with genotype MM have two alleles coding for the production of M molecules and will have the M molecule phenotype. Likewise, individuals with the genotype NN will have two N alleles and will have the N molecule phenotype. The M and N alleles are codominant, so when both are present (genotype MN), both are expressed as heterozygote. Therefore, an individual with genotype MN will produce both M and N molecules. Their phenotype is MN, indicating the presence of both molecules (**table 2.6**).

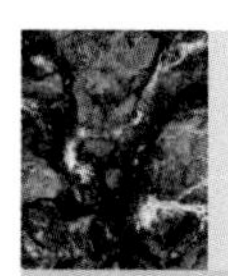

TABLE 2.6
Genotypes and phenotypes of the MN blood group system

Genotype	Phenotype
MM	M molecules
MN	M and N molecules
NN	N molecules

Inheritance of Sex-Linked Recessive Genes

Hemophilia

Mendel's law of independent assortment occurs only when genes are located on different chromosomes, but when the genes are on the same chromosome, they are inherited together or linked. Sex-linked traits are controlled by genes located on the sex chromosomes. They can be either on the X or the Y chromosome. If the gene is located on the X chromosome, males (XY) will manifest a trait, because their Y chromosome lacks the corresponding gene.

The most well-known example of a sex-linked trait is hemophilia. Hemophilia results from the lack of a clotting factor; individuals with this genetic disorder suffer bleeding episodes, which can lead to hemorrhage and death. The gene for hemophilia is recessive and is found on the segment of the X chromosome that has no corresponding portion on the Y chromosome. Therefore, even though the gene for hemophilia is recessive, it can be expressed in males with only with one allele.[25,26] As a result, hemophilia is more common in males than in females. For females to be hemophiliac, they must inherit two copies of this allele, one from each parent. Females carry the trait and pass it to their sons. A male whose mother is heterozygous (a carrier of the gene) for hemophilia has a 50 percent probability of inheriting the trait. The most famous pedigree documenting this malady is Queen Victoria and her descendants. The females did not express the trait but 50 percent of the males related to Queen Victoria were hemophiliacs.

Color Blindness

The inability to distinguish red and green is another well-known trait that is X-linked.[27] The gene for color blindness is located on the X chromosome; therefore females carry the trait and males have a 50 percent chance of getting the trait from one of the X chromosomes of their mother. Affected people tend to confuse the red and green hues (**fig. 2.18**). There are two physiological different forms of color blindness: protan and deutan. The most severe, called protanopia, cannot distinguish red color and deuteranopia cannot distinguish green color. The corresponding mild forms are known as protanomaly and deuteranomaly. While they can distinguish red and green, they cannot distinguish colors that are in between red and green on the spectrum.

Inheritance of Autosomal Traits

Dominant Traits

Some disorders are inherited either as dominant or recessive autosomal traits (**table 2.7**). Autosomal traits are controlled by genes located on any of the 22 pairs of autosome chromosomes (i.e., any chromosomes except X or Y). Autosomal traits can also be inherited as a dominant trait. For example, **brachydactyly** is an autosomal dominant trait characterized by malformed hands and shortened fingers.[29] In general, individuals who are homozygous dominant for brachydactyly are quite rare, because they usually die in the fetal stage. Therefore, the most common form of brachydactyly is caused by a dominant allele in heterozygous condition. Brachydactyly occurs in both sexes. Following Mendel's principle of segregation, approximately half the offspring of affected parents are expected to be affected. Achondroplastic dwarfism (small body size and abnormal body proportions) is also caused by a dominant allele. In human populations it occurs at very low frequencies (0.0002 to 0.0005), affecting from 2 to 4.7 per 10,000 individuals.[30,31] Although achondroplastic dwarfism is caused by a dominant allele, because most of the children with achondroplastic dwarfism have two normal parents, it is assumed that a mutation is involved. If the parent had the allele, he or she would also be a dwarf. Therefore, when both parents of a dwarf are not dwarves, we know the offspring's dwarfism is the result of a mutation in the child. In cases where two dwarves mate, the offspring can be homozygous for the disease. Such offspring generally die before or shortly after birth. In other words, offspring of homozygous dwarves, because they have an increased risk of having children with two copies of the achondroplastic allele, generally die early in life. Therefore, the low frequency of achondroplastic dwarves is the result of natural selection acting to remove the harmful allele from the population. Thus,

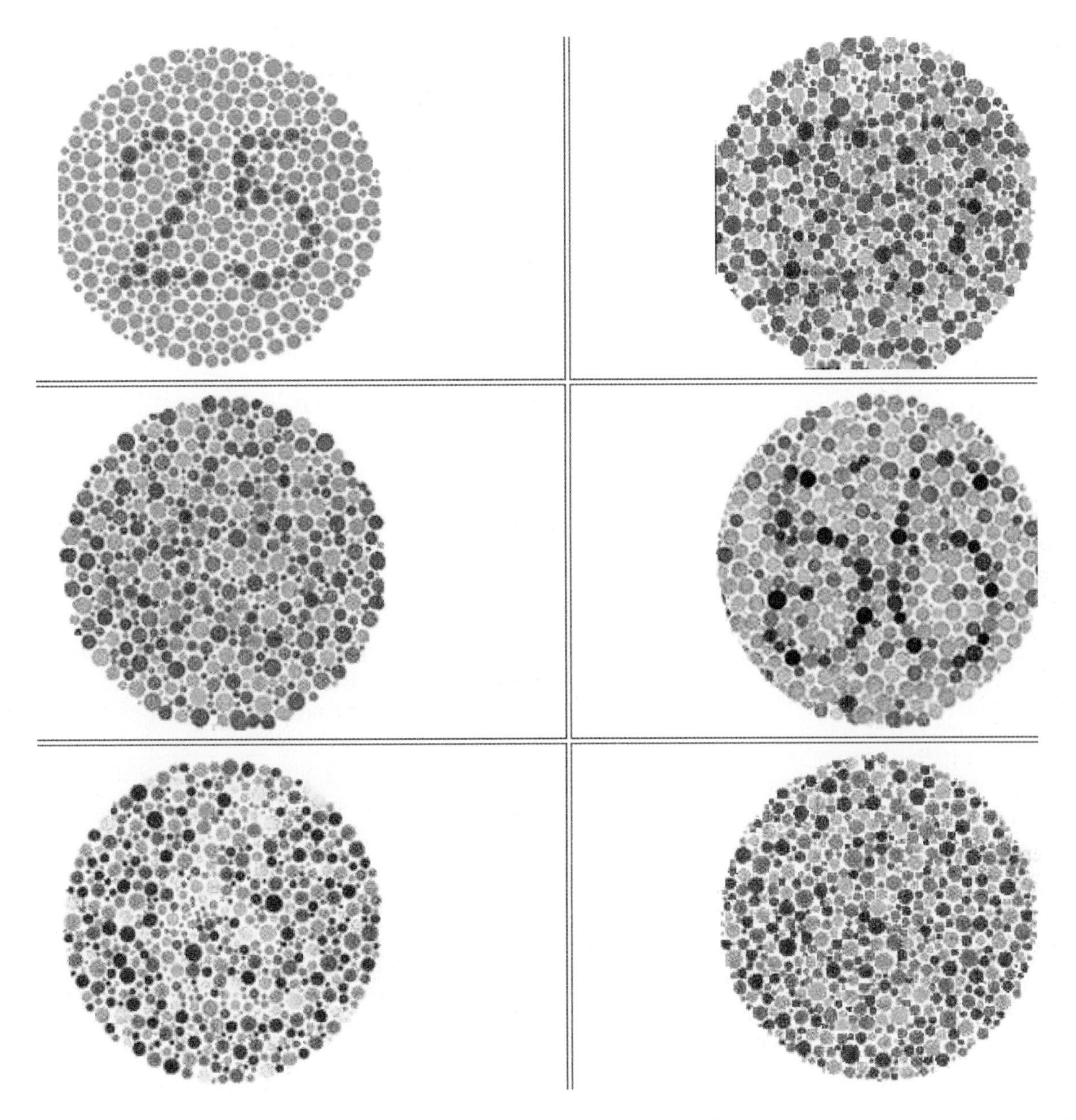

FIGURE 2.18. Color blindness. Most people can identify the numbers 25, 45, 6, 29, 56, and 8. However, 10 to 12 percent of males can identify only the number 25 and 56 because they cannot distinguish red and green colors. (Adapted from: http://www.toledo-bend.com/colorblind/Ishihara.htm.)

both differences in mortality and fertility can affect the degree of selection against an allele.

Recessive Traits

Tay-Sachs Disease

Tay-Sachs disease (TSD) is a fatal genetic disorder characterized by the accumulation of harmful quantities of a fatty substance called ganglioside GM2 in the nerve cells in the brain. The disease is accentuated after the first few months of life, when nerve cells become distended with fatty material causing a deterioration of mental and physical abilities leading to blindness, deafness, and deterioration of digestive and respiratory muscles. The condition is caused by insufficient activity of an enzyme called hexosaminidase A, which catalyzes the biodegradation of acidic fatty materials known as gangliosides,[32–35] Gangliosides are made and biodegraded rapidly in early life as the brain develops.

TSD is inherited as a recessive homozygote. Therefore, both parents must be carriers in order to have an affected child. When both parents are found to carry a genetic mutation in hexosaminidase A, there is a 25 percent chance with each pregnancy that the child will be affected with Tay-Sachs disease. The carriers of the allele do not show any major biological impairment.

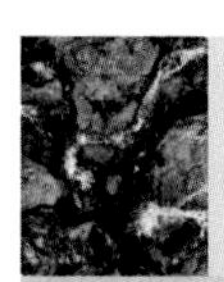

TABLE 2.7
Some Mendelian disorders in humans that are inherited as dominant and recessive traits*

Trait	Description
Dominant Traits	
Brachydactyly	Shortened fingers and toes.
Familial hyper-cholesterolemia	Elevated cholesterol levels and deposition of cholesterol plaque. Hypercholesterolemia is a leading cause of heart disease with death frequently occurring by middle age.
Neurofibromatosis	Symptoms range from the appearance of abnormal skin pigmentation to large tumors resulting in gross deformities; this so-called Elephant Man disease can, in extreme cases, lead to paralysis, blindness, and death.
Marfan syndrome	Symptoms include greater than average height, long arms and legs, eye problems, and enlargement of the aorta, which sometimes ruptures. Abraham Lincoln may have had Marfan syndrome.
Recessive Traits	
Albinism	Inability to produce normal amounts of the pigment melanin. Albinism results in very light, nonpigmented skin, light blond hair, and light eyes; may also be associated with vision problems.
Cystic fibrosis	The symptoms include abnormal secretions of the exocrine glands, with pronounced involvement of the pancreas, obstructive lung disease, and a high metabolic rate.
Tay-Sachs disease	Most common among Ashkenazi Jews; degeneration of the nervous system beginning at about 6 months, lethal by age 2 or 3 years.
Phenylketonuria (PKU)	Inability to metabolize the amino acid phenylalanine; if left untreated during childhood, results in mental retardation. Treatment involves strict dietary management and some supplementation.

*Adapted from ref. # 14. McKusick, V. A. 1998. *Mendelian Inheritance in Man: A Catalog of Human Genes and Genetic Disorders.* Baltimore; London: Johns Hopkins University Press; and ref. # 48. Cummings, M. 2000. *Human Heredity. Principles and Issues* (5th Ed.). PacificGrove: Brooks/Cole.

TSD is found at relatively high frequencies in the Ashkenazi Jewish population with an incidence of 1 per 3,600 versus 1 per 360,000 in the general non-Jewish population.[32–35]

Phenylketonuria

Phenylketonuria (PKU) is an inherited error of metabolism caused by a deficiency in the enzyme phenylalanine hydroxylase.[36] The function of phenylalanine hydroxylase is to convert the amino acid phenylalanine to tyrosine and if this enzyme is inactive or less efficient, the concentration of phenylalanine in the body can build to toxic levels. Hence, lack or deficiency of phenylalanine hydroxylase results in mental retardation, organ damage, unusual posture and can, in cases of maternal PKU, severely compromise pregnancy. With careful dietary supervision, children born with PKU can lead normal lives, and mothers who have the disease can produce healthy children. PKU is inherited as an autosomal recessive disorder, caused by mutations in both alleles of the gene for phenylalanine hydroxylase (PAH), found on chromosome 12.[37]

Inheritance of Continuous Traits

Polygenic Additive Inheritance

Continuous traits such as height, weight, body composition, lung size, heart size, red blood cell count, hemoglobin concentration, blood pressure, and skin color are controlled by many alleles, each of which make an additional contribution to the phenotypic expression of the trait. For example, height is believed to be influenced by between four to six alleles located on three loci, none of which are dominant. Individuals having only alleles for tall stature are likely to be tall, and individuals having only alleles that code for short stature are very short. However, since there are three loci and six alleles, there are numerous ways in which these alleles can combine. If an individual inherits five alleles coding for tall stature and only one for short stature, he or she will be very tall. Conversely, an individual who inherits a higher proportion of alleles that code for short stature, will be short. This is because in this system, each allele that codes for height makes a contribution to increased or decreased height. Therefore, the effect of multiple alleles at several loci, each making a contribution to individual phenotypes, is to produce continuous variation from very short to very tall. For example, height in a population, because there are many genes and many alleles affecting the trait, and the effect of each allele is small, the variation in the trait is continuous ranging from short to tall (**fig. 2.19**). The most frequent distribution falls within the average and the smaller percentage correspond to very short and very tall people.

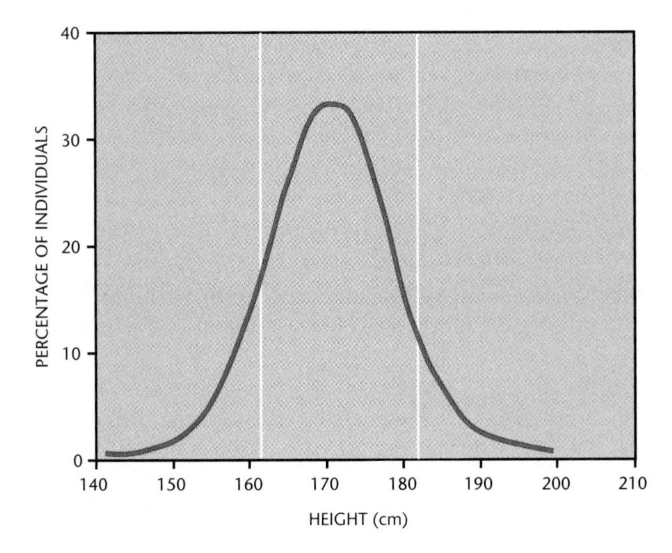

FIGURE 2.19. Histogram of a continuous polygenic trait such as height among male adults from the United States. The percentage of extremely short or tall individuals is low and the majority of people are closer to the mean, or average. (Composite made by the author.)

Environmental Influence

The phenotypic expression of continuous traits depends a great deal on the influence of the environment (see Section IV: Human Adaptation and Biological Diversity). The phenotype of an individual can depend directly on the environment in which it develops. For example, human height is influenced by prenatal and childhood nutrition as well as genetics. At present, most people in industrialized countries are on average taller now than they were a hundred years ago because of improved nutrition and medical conditions. Likewise, body weight and body composition is influenced by a variety of environmental factors. Most individuals develop in some average environment but some are exposed to more extreme environmental conditions and end up with the extreme forms of the trait. Thus, the genotype sets and limits the potential for phenotypic change. On the other hand, discrete or discontinuous genetic traits are controlled by well-defined genes that do not overlap. For example, blood types are completely distinct from one another. That is, one is either blood type A, or type B, or type AB, but one cannot intermediate between them and do not show continuous variation. Furthermore, discontinuous traits are not influenced by the environmental factors acting during the life cycle. For example, the blood type of an individual does not change once it is established during conception.

Summary

Cellular duplication is carried on by mitosis and meiosis. Mitosis is concerned with growth, repair, maintenance, and replacement of body cells. This process maintains the integrity and function of the organism until death occurs. Mitosis results in the production of two cells, each with 46 chromosomes that are exact copies of those in the parent cell. Meiosis is concerned with the production of gametes or sex cells and results in the formation of four cells with only 23 chromosomes. During fertilization the total number of 46 chromosomes is established. As demonstrated by Mendel there is no blending of genetic traits. The progeny is a product of the independent and random assortment of traits that take place during gamete formation and fertilization. Each individual receives half of his or her alleles from each parent.

The instructions for all cell duplication, whether it is mitosis or meiosis, are given by the DNA molecule.

The DNA molecule specifies the genetic code, or set of instructions, needed to produce biological structures. DNA acts along with a related molecule, RNA, to translate these instructions into proteins. The DNA is contained in long structures within the cell, called chromosomes. Chromosomes come in pairs. A segment of DNA that codes for a certain product is called a gene. The alternate forms of genes presented at a locus are called alleles. The DNA molecule has the ability to make copies of itself, allowing the transmission of genetic information from cell to cell, and from generation to generation, providing the unity of the species. However, sometimes mutations occur during the process of DNA replication causing new genes to appear. This phenomenon of mutation introduces new alleles into a population on which natural selection works—a topic that is addressed in the next chapter.

The genotype refers to an individual's entire genetic makeup and all observable traits, where the alleles together set limits and potential for growth and development. Therefore, the physical manifestation that results from interaction of the genotype and the environment is known as the phenotype. The relationship between genotype and phenotype depends on whether an allele is dominant, recessive, codominant, or polygenic. In general, discrete noncontinuous traits that follow the Mendelian pattern of inheritance are not influenced by the environment in which the organism develops. In contrast, continuous traits that are governed by several alleles that contribute to the phenotypic expression referred to as polygenic additive inheritance are influenced by the environmental factors on which the individual develops.

Suggested Readings

Cummings, M. 2000. *Human Heredity. Principles and Issues,* 5th ed. Pacific Grove: Brooks/Cole.

Ettore, E. 2002. *Reproductive Genetics, Gender and the Body.* London: New York: Routledge.

McKusick, V. A. 1998. *Mendelian Inheritance in Man: A Catalog of Human Genes and Genetic Disorders.* Baltimore; London: Johns Hopkins University Press.

Olson, S. 2002. *Mapping Human History: Discovering the Past through Our Genes.* Boston: Houghton Mifflin.

Literature Cited

1. Pollard, T. D. 2002. *Cell Biology.* Philadelphia: Saunders.
2. Diffley, J. F. and K. Labib. 2002. The chromosome replication cycle. [Review] [8 refs] *Journal of Cell Science.* 115:869–872.
3. Barry, J. M. 2002. *Molecular Embryology: How Molecules Give Birth to Animals.* New York: Taylor & Francis.
4. Mitchison, T. J. and E. D. Salmon. 2001. Mitosis: a history of division. *Nature Cell Biology.* 3:E17–21.
5. Cummings, M. 2000. *Human Heredity. Principles and Issues,* 5th ed. Pacific Grove: Brooks/Cole.
6. Sybert, V. P. 2002. Phenotypic effects of mosaicism for a 47,XXX cell line in Turner syndrome. *Journal of Medical Genetics* 39: 217–220.
7. Watson, J. D. 2003. *DNA: The Secret of Life.* New York: Alfred A. Knopf.
8. Calzada, A. and A. Bueno. 2002. Genes involved in the initiation of DNA replication in yeast. *International Review of Cytology.* 212:133–207.
9. Terry, L. and S. H. Cedar. 2001. An overview of education and "new genetics". *Nursing Standard.* 15:38–40.
10. Mark, H. F. and R. J. Deveau. 2001. Modern medical genetics: a systems approach. *Journal of Health & Human Services Administration.* 24:54–79.
11. Harpending, H. and A. Rogers. 2000. Genetic perspectives on human origins and differentiation. *Annual Review of Genomics & Human Genetics.* 1:361–385.
12. Demain, A. L. 2001. Molecular genetics and industrial microbiology—30 years of marriage. *Journal of Industrial Microbiology & Biotechnology.* 27:352–356.
13. Mendel, G. 1866. Verh. Naturforsch. Vereines Abhandlungen BrŸnn 4, 3–47. [Cited in Snustad, D.P. 2003. *Principles of Genetics.* New York: John Wiley.]
14. McKusick, V. A. 1998. *Mendelian Inheritance in Man: A Catalog of Human Genes and Genetic Disorders.* Baltimore; London: Johns Hopkins University Press.
15. Semsarian, C. and C. E. Seidman. 2001. Molecular medicine in the 21st century. *Internal Medicine Journal.* 31:53–59.
16. Blakeslee, A. F. Genetics of sensory thresholds: taste for phenylthiocarbamide. *Proc Natl Acad Sci U S A.* 1932;18:120–126.
17. Harris, H., and H. Kalmus. 1949. The measurement of taste sensitivity to phenylthiourea of 384 sib-pairs. *Ann Eugen Lond.* 15:24–31.
18. Kalmus, H. 1971. Genetics of taste. In: Beidler, L. M., ed. *Handbook of Sensory Physiology.* Vol 4(2). New York: Springer-Verlag, 1971:165–179.
19. Mourant, A. E. 1976. *The Distribution of the Human Blood Groups and Other Polymorphisms.* New York: Oxford University Press, 1976.
20. Yip, S. P. 2002. Sequence variation at the human ABO locus. *Annals of Human Genetics.* 66:1–27.
21. Nei, M. and A. K. Roychoudhury. 1993. Evolutionary relationships of human populations on a global scale. *Molecular Biology & Evolution.* 10:927–943.
22. Bottini, N., G. F. Meloni, A. Finocchi, G. Ruggiu, A. Amante, T. Meloni, and E. Bottini. 2001. Maternal-fetal interaction in the ABO system: A comparative analysis of healthy mothers and couples with recurrent spontaneous abortion suggests a protective effect of B incompatibility. *Human Biology* 73:167–174.
23. Cohen, B. H. 1970. ABO and Rh incompatibility. Fetal and neonatal mortality with ABO and Rh incompatibility: Some new interpretations. *Am J Genet.* 22:412–440.
24. Moise, K. J. Jr. 2002. Management of rhesus alloimmunization in pregnancy. *Obstetrics & Gynecology.* 100:600–611.

25. Bowen, D. J. 2002. Haemophilia A and haemophilia B: Molecular insights. *Molecular Pathology.* 55:1–18.
26. High, K. 2002. Gene-based approaches to the treatment of hemophilia. *Annals of the New York Academy of Sciences.* 961:63–64.
27. Neitz, M. and J. Neitz. 2000. Molecular genetics of color vision and color vision defects. *Archives of Ophthalmology.* 118:691–700.
28. http://www.toledo-bend.com/colorblind/Ishihara.htm.
29. Koeppen, A. H. 2001. Clinical, radiological and pathological findings in an autosomal dominant leukodystrophy. *Journal of the Neurological Sciences.* 187:107, 109, 2001.
30. Cummings, M. 2000. *Human Heredity. Principles and Issues,* 5th Ed. Pacific Grove: Brooks/Cole.
31. Baitner, A. C., S. G. Maurer, M. B. Gruen, and P. E. Di Cesare. 2000. The genetic basis of the osteochondrodysplasias. *Journal of Pediatric Orthopedics.* 20:594–605.
32. Kaback, M. M., T. J. Nathan and S. Greenwald. 1997. Tay-Sachs disease: Heterozygote screening and prenatal diagnosis. U.S. experience and world perspective. In: Kaback, M. M., editor. *Tay-Sachs Disease, Screening and Prevention,* New York: Alan R. Liss. pp. 13–36.
33. Kaback, M. M. 2000. Population-based genetic screening for reproductive counseling: the Tay-Sachs disease model. *European Journal of Pediatrics.* 159 Suppl 3:S192–S195.
34. Bach, G., J. Tomczak, N. Risch, and J. Ekstein. 2001. Tay-Sachs screening in the Jewish Ashkenazi population: DNA testing is the preferred procedure. *American Journal of Medical Genetics.* 99:70–75.
35. Liu M. C., K. C Drury, S. Kipersztok, W. Zheng, and R. S. Williams. 2000. Primer system for single cell detection of double mutation for Tay-Sachs disease. *Journal of Assisted Reproduction & Genetics* 17:121–126.
36. Zennermann, J. B., A. Loui, A. Weber, and E. Monch. 2004. Hyperphenylalaninemia in a premature infant with heterozygosity for phenylketonuria. *J Perinat Med.* 32:383–385.
37. Scriver, C. R. 2004. Translating knowledge into practice in the "post-genome" era. *Acta Paediatr.* 93:294–300.

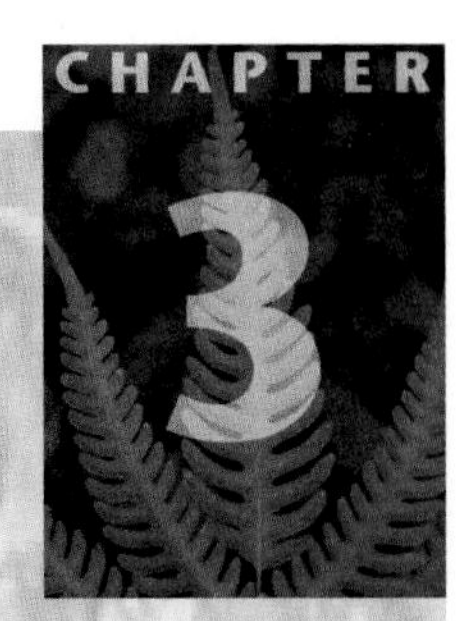

Genetics and the Process of Microevolution

Evolution is usually defined as change in the gene frequency of a population, the transfer of information from one generation to the next along with the possibility that this information will change. Thus, evolution is an ongoing process in that our species and all of its constituent populations are continually being shaped by evolutionary forces. Evolution can be divided into microevolution and macroevolution. Evolutionary changes in gene frequency from one generation to the next occurring within a population are referred to as microevolution, whereas the accumulation of genetic changes of generations over thousands or millions of generations, resulting in the differentiation of populations into different species, comprises macroevolution. This chapter focuses on the processes of microevolution.

Measuring the Occurrence of Evolution

Breeding Population

For evolution to occur there must be changes in the gene pool of the population, but the definition of a population differs according to the discipline. For sociologists, a population is defined on the basis of geographic and political boundaries and refers to all the people who live in a region or country, irrespective of whether they are able to reproduce. For biologists, population refers to a group of organisms that tends to choose mates from within the group. The importance of defining population with reference to its ability to reproduce is based upon the fact that populations, not individuals, evolve. Individuals are selected, but individual organisms do not evolve; they retain the same genes throughout their lives. When a population is evolving, the ratio of different genetic types changes. For example, as indicated in chapter 1, the increase in the frequency of peppered moths associated with industrial pollution[1,2] did not cause the moths to turn from light to dark. Therefore, in order to understand evolution, it is necessary to view populations as a collection of individuals, each harboring a different set of traits. A single organism is never typical of an entire population unless there is no variation within that population.

Hardy-Weinberg Equation to Measure Evolution

Since evolution is a change in allele frequency, we need a measure to evaluate whether there has been a change in the population. The genotype frequency is a measure of the relative proportions of different genotypes within a population. Therefore, an allele frequency is simply a measure of the relative proportion of alleles within a population. According to Mendel's principles of inheritance, when two heterozygotes reproduce, the expected ratio of dominant to recessive in the F1 generation is 3:1. But Mendelian genetics is applicable to the progeny of crosses between individuals of known genotype and is not necessarily appropriate for the treatment of large, randomly mating populations. In 1908, the English mathematician G. H. Hardy[3] and the German physician W. Weinberg[4] independently bridged the gap between Mendelian genetics and population phenomena when they developed the binomial equation referred to as the Hardy-Weinberg equilibrium. The Hardy-Weinberg equilibrium model is a mathematical statement relating allele frequencies of the expected genotype frequencies (EGF) of the offspring of heterozygotes (**table 3.1**). Let's suppose we have two alleles, A and $\boldsymbol{a}$, in which A is dominant and $\boldsymbol{a}$ is recessive. Hence, in accordance with Mendelian inheritance, the combined total frequency of $A + \boldsymbol{a} = 1$. Then, the proportions of the genotypes in the next generation are obtained by multiplying the frequencies of the different gametes of the male pool by those in the female pool. Therefore, the EGF of the offspring of this heterozygous mating would be as follows:

1. The frequency of a dominant allele produced by heterozygous mating will equal A^2.
2. The frequency of a recessive allele produced by heterozygous mating will equal $\boldsymbol{a}^2$.
3. The frequency of the dominant and recessive alleles produced by a heterozygous mating will equal $2\,A\,\boldsymbol{a}$.

The inheritance of the combined total frequency of A and $\boldsymbol{a}$ would be given by the following equation:

$$\text{EGF} = A^2 + 2A\,\boldsymbol{a} + \boldsymbol{a}^2 = 1$$

Conventionally, the frequency of a dominant allele is represented by p, and the frequency of a recessive allele is represented by q. Hence, the expected genotype frequency is given as

$$\text{EGF} = p^2 + 2\,pq + q^2 = 1$$

These frequencies should remain constant assuming that (a) the population is large, (b) the mating is random, (c) there are no mutations, and (d) the genes are not under natural selection. That is, in the absence of evolutionary pressures, the genotype frequency will remain constant generation after generation. Therefore, a difference between the expected genotype frequency and the observed genotype frequency gives an indication whether the population is in equilibrium or undergoing evolution.

Application of the Hardy-Weinberg Equation

The Hardy-Weinberg equation provides a yardstick against which we can estimate the forces of evolution that can cause deviations from the equilibrium state. This equation assumes that the mating is random, and that there are no mutations, no genetic drift and/or gene flow, and no specific selective factor. Under these conditions, the observed frequency of a given geno-

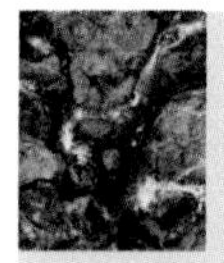

TABLE 3.1

Algebraic expression of the Hardy-Weinberg equation derived from Mendel's expected genotype frequency (EGF) among offspring of heterozygotes

	Parent 1 Genotype ($A\ a$)	
Parent 2 Genotype ($A\ a$)	A	a
A	$A\ A$	$A\ a$
a	$A\ a$	$a\ a$

	Number in Offspring	Algebraic Expression	
	$A\ A$	2	A^2
	$A\ a$	2	$2\ A\ a$
	$a\ a$	2	a^2
∴ EGF =			$A^2 + 2\ Aa + a^2$
∴ EGF =			$p^2 + 2pq + q^2$

type should equal the expected frequency. Otherwise, the population is not in equilibrium. The application of the Hardy-Weinberg equation can be illustrated by considering two examples: one in equilibrium and another not in equilibrium.

Population in Equilibrium

Consider a hypothetical population of 400 dark- and light-colored moths, where ***C*** is dominant and codes for dark color, and ***c*** is recessive and codes for light color. Of the 400 moths, 100 are homozygous dark (*CC*), 205 are heterozygous dark (*Cc*), and 95 are homozygous light (*cc*). Therefore, the genotype frequencies observed in this population would be

CC = 0.25 (100/400); *Cc* = 0.51(205/400);
cc = 0.24 (95/400).

To calculate the EGF, we use the Hardy-Weinberg equation: EGF = $p^2 + 2\,pq + q^2$. Applying this equation, we get the results shown in **table 3.2a.** It is quite evident that the observed frequency of dark- and light-colored moths does not differ by more than 0.10 (or 10 percent, if multiplied by 100) from the expected frequency. Hence, the interpretation would be that the population is in equilibrium. In other words, the EGF of dark- and light-colored moths has an equal chance of reproduction.

Population Not in Equilibrium

Consider a hypothetical population of 400 dark- and light-colored moths, where ***C*** is dominant and codes for dark color, and ***c*** is recessive and codes for light color (**table 3.2b**). Of the 400 moths, 168 are homozygous dark (*CC*), 212 are heterozygous dark (*Cc*), and 20 are homozygous light (*cc*). Therefore, the genotype frequencies observed in this population would be:

CC = 0.42 (168/400); *Cc* = 0.53 (212/400);
cc = 00.06 (20/400)

To calculate the EGF, we use the Hardy-Weinberg equation:[2] EGF = $p^2 + 2\,pq + q^2$. Applying this equation, we get the results shown in **table 3.2b.** When comparing these results, it is quite evident that the observed frequency of dark- and light-colored moths differs by more than 10 percent from the expected frequency. Hence, the interpretation would be that the population is **not** in equilibrium. In other words, the frequency of dark exceeds the frequency of light-colored moths.

The difference between the observed and expected genotype frequency may be due to the effect of mutation, natural selection, genetic drift, and/or gene flow. First, the difference between observed and expected genotype frequencies could not be due to mutation because the amount of change in a single generation

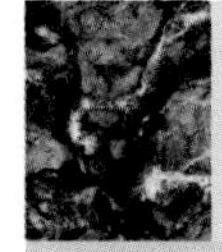

TABLE 3.2a

Calculation of observed and expected allele frequencies to determine whether the frequency of dark- and light-colored moths is in equilibrium*

Genotype	Number of Moths	Frequency	Number of C alleles	Number of c alleles	Total alleles
CC (Dark)	100	0.25 (100/400)	200	0	200
Cc (Dark)	205	0.51 (205/400)	205	205	410
Cc (Light)	95	0.24 (95/400)	0	190	190
Total	**400**	1.0	405	395	**800**
Total number of alleles = 405 + 395 = 800					
Proportion of C alleles:		405/800	= 0.51 = p		
Proportion of c alleles:		395/800	= 0.49 = q		
Total			= 1.00		

Genotype		Expected Frequency	Observed Frequency	Frequency Difference
CC	(0.51) × (0.51)	= 0.26	0.25	0.01
Cc	2 (0.51 × 0.49)	= 0.50	0.51	0.01
cc	(0.49) × (0.49)	= 0.24	0.24	0.00

*Data from ref. #2. Kettlewell, H. B. D. 1973. *The Evolution of Melanism: The Study of a Recurring Necessity.* Clarendon, Oxford.

TABLE 3.2b

Calculation of observed and expected allele frequencies to determine whether the frequency of dark- and light-colored moths is in equilibrium*

Genotype	Number of Moths	Frequency	Number of *C* alleles		Number of *c* alleles	Total alleles
CC (Dark)	168	0.42 (168/400)	336		0	336
Cc (Dark)	212	0.53 (212/400)	212		212	424
cc (Light)	20	0.06 (20/400)	0		40	40
Total	**400**	1.0	548		252	**800**
Total number of alleles = 548 + 252 = 800						
Proportion of *C* alleles:		548/800	= 0.70	= *p*		
Proportion of *c* alleles:		252/800	= 0.30	= *q*		
Total			= 1.00			

Genotype		Expected Frequency	Observed Frequency	Frequency Difference
CC	(0.70) × (0.70)	= 0.49	0.42	0.07
Cc	2 (0.70 × 0.30)	= 0.42	0.53	**0.11**
cc	(0.30) × (0.30)	= 0.09	0.06	0.03

*Data from ref. #2. Kettlewell, H. B. D. 1973. *The Evolution of Melanism: The Study of a Recurring Necessity.* Clarendon, Oxford.

exceeds the usual known rate. Mutations usually cause much lower amounts of change in a single generation. Second, we assume the mating is random. Hence, one could conclude that the observed change was due to natural selection, genetic drift, and/or gene flow.

Fitness and Selection Coefficient

Evaluation of fitness and selection coefficients is necessary to determine whether a population is at equilibrium or undergoing change. In population genetics, fitness is calculated as the ratio of survival before and after selection, while the selection coefficient is the difference between 1 and the relative fitness. The calculation of relative fitness and selection coefficient is illustrated in **table 3.3**, which is a presentation of the genotypic frequencies before and after pollution (selection) of the moths (*Biston bestularia*) studied by Kettlewell.[1,2] From this data, it is quite evident that the selection coefficient against the light-colored moths was 0.54. Using this selective coefficient (0.54), it has been estimated that it would take only nine generations to go back to the original frequency of 0.29 if the pollution was removed.[1,2]

In summary, the application of statistical methods such as those provided by the Hardy-Weinberg equation to the hereditary principles derived from Mendel was a major breakthrough that set the stage for measuring the influence of evolutionary forces. The increased frequency of dark-colored moths is an example of natural selection through predators, which in this case were the birds. Furthermore, because selection tended to prefer the dark ones, it is an example of directional selection. Hence, the mean and the proportion changed, and they will remain that way as long as the factor that caused the initial change (in this case, pollution that darkened the trees) is not removed.

Kinds of Natural Selection

It is evident that natural selection can change gene frequencies rapidly. This molding force of the evolutionary process can be one of three kinds (**fig. 3.1**): stabilizing selection, directional selection, and disruptive selection. Stabilizing selection, as the name implies, occurs when the mean of the distribution is kept in a stable condition. This happens because selection prefers individuals in the center of the normal distribution of phenotypes, in which case a population becomes more strongly represented by these phenotypes. For example, the mean birth weight in the United States is about 3,400 ± 500 grams (7.5 ± 1 lb.), and newborns who weigh less than 2,500 grams (5.5 lb.) are less likely to survive than infants born with an average weight. Conversely, infants that are born with a weight of more than 5,000 grams (11 lb.) have higher mortality and complications after birth than infants born with an average weight. As a result of stabilizing selection, the mean weight generation after generation

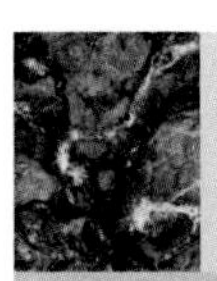

TABLE 3.3
Calculations of selection coefficients for the dark- and light-colored moths (*Biston bestularia*) genotypes*

	Carbonaria Dark (*CC*)	Carbonaria Dark (*Cc*)	Typical Light (*cc*)	Total
Number before selection	77	77	65	218
Number after selection	41	41	16	98
Frequency before selection	.35 (77/218)	.35 (77/218)	.29 (65/218)	1
Frequency after selection	.42 (41/98)	.42 (41/98)	.16 (16/98)	1
Relative survival value	.42/.35 = **1.2**	.42/.35 = 1.2	.16/.29 = .55	
Relative fitness	1.2/1.2 = 1.0	1.2./1.2 = 1.0	.55/1.2 = .46	
Selection coefficient	1 – 1 = 0	1 – 1 = 0	1.0 – 0.46 = **0.54**	

*Data from ref. #2. Kettlewell, H. B. D. 1973. *The Evolution of Melanism: The Study of a Recurring Necessity.* Clarendon, Oxford.

changes very little, and the optimum birth weight is close to the population average.

Directional selection refers to selection against either high or low values. When this happens, the mean value can change over time in one direction. The average value for a trait moves in one direction or the other. For example, changes, once they are established, tend to remain as long as the species continues to exist. For instance, over the past 2 million years, the human brain changed from about 500 grams to more than 1200 grams (see chapter 13, Diet and Brain Evolution).

Disruptive selection refers to selection against the average, resulting in two different population means. Disruptive selection can occur in a population with short and fast reproduction that is exposed to a new environment.

Additional Sources of Genetic Diversity

For natural selection to work, there must be variability in the genetic composition of the population. As discussed in chapter 2 recombination during the meiosis phase of gamete production rearranges the raw materials upon which natural selection works. In addition, genetic drift, founder effect, gene flow, and mutations are important factors that increase the genetic variation of the gene pool.

Gene Flow

Gene flow refers to the movement of alleles from one population to another brought about through mating and reproduction. The effects of gene flow depend on the size of the recipient population and the rate of gene flow. If gene flow is high and occurs between populations with rather different gene pools, it provides new alleles to the population and thus contributes new sources of variation. In some ways, it functions as mutation in accelerating change in gene frequencies and the evolution of the population. For example, if gene flow is associated with 20 individuals who enter and reproduce into a population of 20,000 adult individuals, the change in the gene frequency of the recipient population is very small. However, if the

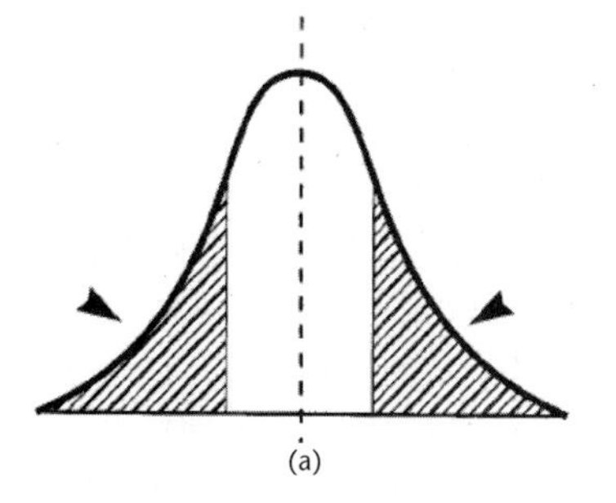

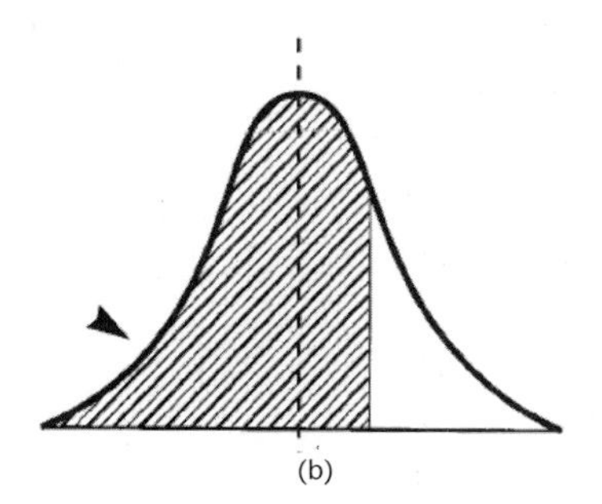

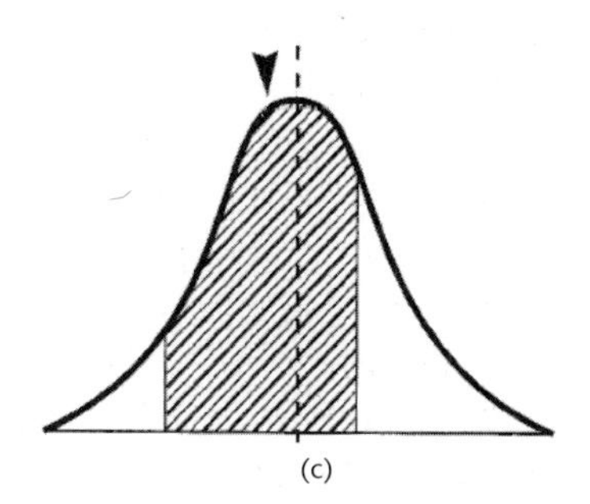

FIGURE 3.1. Types of selection. (a) Stabilizing selection, (b) directional selection, and (c) disruptive selection. (Composite illustration made by the author.)

recipient population is 300 rather than 20,000, the change in the gene frequency of the recipient population is very high. In other words, for small recipient populations, even a few immigrant individuals may represent a significant change in the gene frequency of the population. Conversely, if a large number of people move into a smaller population, the effects on the gene pool of the recipient population will be proportional to the number of immigrants. Thus, the sizes of the immigrant and recipient populations are important factors that determine the contribution of gene flow to the genetic diversity of the population.

Genetic Drift

Genetic drift is the alteration of gene frequencies through sampling error due to chance alone that can occur in small populations. The effect of genetic drift, like gene flow, on the gene frequency of population depends on the size of the breeding population.[5] The smaller the population, the more chance that drift will occur from one generation to the next. For example, a population of 10 individuals may produce 100 offspring, of which only 10 percent ($n = 10$) survive to reproduce. If the allele frequencies at a given locus are 50:50, the probability that this frequency will be represented in the 10 survivors is relatively low. On the other hand, if the population size includes 200 individuals, produces 4000 progeny, and 10 percent survive ($n = 400$) with the same allele frequency, the probability that this frequency will be represented in the 400 survivors is relatively high. Thus, the influence of genetic drift on the population pool depends on how small the population is and how continuous the drift is.[5]

Founder Effect and Genetic Drift

Founder effect refers to the disproportionate contribution of few individuals to the gene pool of the population. It occurs when a few reproductively competent individuals produce disproportionately more offspring than others. The influence of founder effect is significant if the recipient population is small—usually less than 100 reproductive individuals—and is isolated from other populations. Because of random chance, the founder is not likely to be an exact genetic representation of the original. That is, the influence of founder effect on the gene pool of the population depends on the degree of isolation of the recipient population from their parent populations.

Founder Effect and Increased Frequency of MN Blood Type

Several examples of founder effect on human genetic traits have been documented. A well-known example is the population known as the Dunkers. In the 1700s, the German religious sect group known as the Dunkers, composed of a few hundred people, immigrated to the United States and settled in Pennsylvania. Evaluation of the MN blood group system among the population of Dunkers in the United States found that the allele frequency for M = 0.655 and for N = 0.345.[6] In contrast, the German population from which the Dunkers came showed a 10 percent lower frequency (i.e., M = 0.55; N = 0.45). Furthermore, as shown in **table 3.4**, the greater frequency of the Dunkers differs by more than 10 percent from the expected theoretical frequency. Based on these data, Glass concluded that the genetics of the Dunker population were shaped to a large extent by genetic drift over two centuries.[6]

Founder Effect and the Genetic Traits of Tristan da Cunha Inhabitants

Another example is the population of Tristan da Cunha. The inhabitants of Tristan da Cunha, a remote island in the South Atlantic, number about 297

TABLE 3.4

Comparison of the allele frequency for the MN blood group system among the Dunkers of Pennsylvania

Genotype			Dunkers Observed Frequency	Germany Expected Frequency	D
MM	(0.655 × 0.655)	=	0.429	0.302	0.127
MN	2 (0.655 × 0.345)	=	0.452	0.495	0.043
NN	(0.345 × 0.345)	=	0.119	0.202	0.083
Total		=	1.000	1.000	

Source: Glass, H.B. 1953. The Genetics of the Dunkers. *Scientific American* 189: 76–81.

people, direct descendants of a small number of individuals who settled there in the first half of the nineteenth century. In 1816, the British established a garrison on Tristan to prevent the French from rescuing Napoleon Bonaparte, who was in exile on St. Helena, some 1,350 miles to the north. When the British left, they left one man, his wife, and two children. Additionally, two other men remained and were joined later by a handful of other settlers. Altogether, the current gene pool has been traced to only twenty-two ancestors, which includes seven females and fifteen males.[7–11]

Analysis of blood groups indicates that the current population has a high frequency of blood type A1 and the complete absence of A2 of the ABO blood group system. Similarly, in the Rh blood group system, they have a high frequency of Rh (cDe) and the presence of three haplotypes in the Gm system. They also have a high allele frequency for retinitis pigmentosa. All these studies indicate that the unusual gene frequencies are related to the genetic drift and founder effect.

Founder Effect, Genetic Drift, and the Gene from Limone sul Garda

Among the inhabitants of the island of Limone sul Garda of Milan, Italy, there is a high frequency of carriers of a naturally occurring variant of apolipoprotein A-I, known as ApoA-I Milano.[12] Individuals with ApoA-I Milano are characterized by very low levels (10–30 mg/dL [0.25–0.78 mmol/L]) of high-density lipoprotein cholesterol (HDL-C) and with increased triglyceridemia. In spite of the low HDL-cholesterol, carriers of ApoA-I Milano do not generally show clinical signs of atherosclerosis.[13,14] A total of 33 living carriers were identified who inherited this gene (of the ApoA-I Milano gene) as an autosomal dominant trait. (The carriers are heterozygous for the apolipoprotein variant.) The origin of the variant gene has been traced to one man who arrived in 1780 to the island of Limone, which had about 1,000 adults and a large number of children.[12] These findings indicate that this mutation was maintained by the isolation of the island, small population size, and the founder effect producing beneficial effects on the carriers of the gene. Recent experimental studies indicate that injections intravenously for 5 doses at weekly intervals of the recombinant ApoA-I Milano/phospholipid complex produced significant regression of coronary obstruction.[14] In other words, it appears that this gene indeed protects against the build-up of cholesterol on the arteries.

Mutations

Mutations are defined as the inheritable change in the sequence of DNA. Mutations may be either spontaneous or induced. Spontaneous mutations occur naturally, as result of errors in DNA replication, while induced mutations occur as result of chemical or radiation effects that increase the chances of spontaneous mutation. An organism that has undergone mutation is referred to as a mutant. Mutation introduces new alleles into a population. This occurs when DNA replication makes mistakes.[15] These mistakes alter the sequence of a gene. There are several kinds of mutations. A point mutation is a mutation in which one "letter" of the genetic code is changed to another. Lengths of DNA can also be deleted or inserted in a gene; these are also mutations. Finally, genes or parts of genes can become inverted or duplicated. Most mutations are thought to be neutral with regard to fitness, but they differ in their phenotypic expression. Most mutations have deleterious (harmful) effects and a few have good effects. Mutations that result in amino acid substitutions can change the shape of a protein, potentially changing or eliminating its function. This can lead to inadequacies in biochemical pathways or interfere with the process of development and reproduction.

Mutations and Drug Resistance

When populations are large and generations are short as with flies and mosquitoes, a mutation can be established quickly. For example, the widespread use of insecticides has amounted to a large-scale "experiment" in the natural selection of insects. In response to the insecticides, different insect species have evolved.[16] For example, in the mosquito *Culex pipiens* a gene that was involved with breaking down organophosphates (found in insecticides) became duplicated. Progeny of this mosquito with this mutation rapidly replaced the worldwide mosquito population because they were not affected by insecticides. Another example is in the increasing frequency of new types of tuberculosis. Tuberculosis is caused by the *Mycobacterium tuberculosis.* In the United States and worldwide, the frequency of tuberculosis has been declining. However, between 1985 and 1991 the frequency had increased, especially in developing nations. This increase has been attributed to the high frequency of multidrug-resistant strains, which have evolved in response to the use of isoniazid and rifampin drugs widely used in the developing nations.[16,17] Similarly, new strains of infectious diseases have appeared as result of the fact that some strains have been able to survive the effects of antibiotics and thus they have been able to multiply.[17]

Mutation, Founder Effect, and Genetic Drift: Huntington's Disease

In general, deleterious mutants are selected against but remain at low frequency in the gene pool. Some mutations can be maintained in a population due to the effect of isolation and founder effect. For example, Huntington's Disease (HD) is an autosomal dominant condition.[18] The frequency of HD varies from 3 to 10 per 100,000 individuals among populations of European descent.[18] HD is a late onset, incurable, autosomal dominantly inherited, progressive neuro-psychiatric disease. It is characterized by chorea, changes in personality, mood, and behavior, and dementia. In general, deleterious mutants are lost from the gene pool, but in the case of HD this deleterious mutation exists at a high frequency because it is expressed late in life when the genetic transmission has already taken place. In some cases, like the population of Maracaibo the trait originated in one woman who went to live in a small population.[18] Hence, the gene's high frequency has been maintained due to founder effect associated with a small population.

Mutation, Founder Effect, Genetic Drift, and Cultural Factors: Albinism

Albinism is a mutation that is inherited as an autosomal recessive trait.[19,20] Albinism is caused by a metabolic disorder characterized by deficient production of a pigment called melanin. Albinism includes two forms: oculocutaneous albinism and ocular albinism. The most common form is the oculocutaneous albinism that is characterized by a lack of pigmentation in the skin, hair, and eyes. In ocular albinism the lack of pigmentation is limited to the eyes and skin pigmentation is normal. Both types of albinism are associated with visual problems due to the low pigmentation of the iris, choroid, and retina. The frequency of the most common form of albinism among American whites varies between 1 in 10,000 and 1 in 20,000.[19] The existence of this deleterious allele or mutation load is probably the result of recurrent mutations. This allele is also probably maintained in the population due to the fact that its selection cannot be seen when it is masked by a dominant allele. People who are carriers do not suffer the negative effects of the allele. Unless they mate with another carrier, the allele may simply continue to be passed on. Furthermore, it can also be maintained by founder effect and cultural factors as shown by the example of the population of Brandywine, Maryland. In this population, the frequency of albinism is 1.2 percent, which is the highest in the United States. The origin of albinos in this group has been traced to a single pair of ancestors who were the founders of the population.[20]

Similarly, among the Hopi and Zuni Indians from the southwestern United States, the frequency of albinism is 1 in 100[21–23] and among the Cuna Indians of the San Blas islands of Panama the frequency is as high as 14 per 200.[24,25] The high frequency of albinism among the Hopi and Cuna Indians has been explained by the cultural protection that albinos receive in these populations.[24,25] Since albino males are kept indoors or stay at home during daylight hours because of their aversion to the sun, while the non-albino males go to the field to work, they have ample opportunity to maximize their reproductive fitness and thereby propagate the albino gene. Thus, a deleterious allele, once having entered in the population as result of founder effect, can remain in the population at a high frequency due to a balance between cultural factors and selection.

Natural Selection and the Sickle-Cell Hemoglobin

The best-known example of natural selection operating on a discrete genetic trait is the relationship of sickle-cell hemoglobin to malaria. Hemoglobin is the major protein component of red blood cells. Its main function is to transport oxygen from the lungs to the cells for its use and transport carbon monoxide from the blood to be exhaled by the lungs. A normal hemoglobin molecule (HbA) in an adult human is made of four polypeptide chains: two α chains with 141 amino acid residues each and two ß chains with 146 amino acid residues each. Each chain winds around a heme containing iron that is responsible for picking up oxygen in the lungs and transporting it to cells in the remainder of the body. The four polypeptide chains then interact with one another to form the hemoglobin molecule.

There are many mutations in the hemoglobin molecule, the most famous being sickle cell. In sickle cell a small mutation in the sixth codon of the chain from CTC (cytocine, tymine, cytocine) to CAC (cytocine, adenine, cytocine) causes translation of the mRNA to **valine** instead of **glutamine** (**fig. 3.2**). This mutation changes the shape of the hemoglobin molecule from round to sickle-shaped.

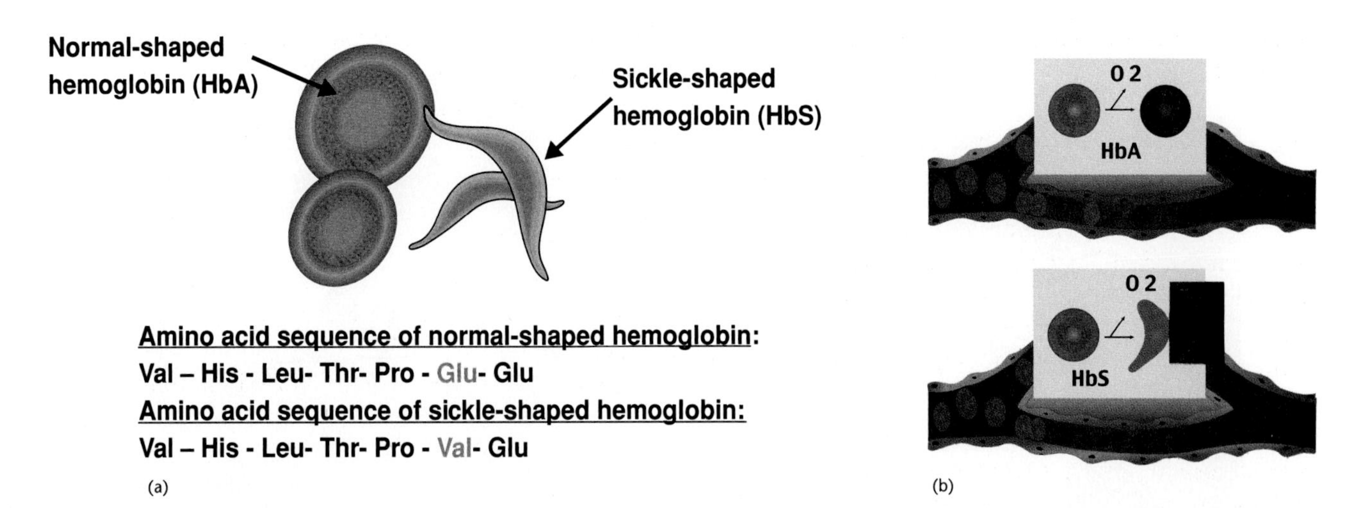

FIGURE 3.2. Sickle-cell hemoglobin. Compared to the normal hemoglobin, the sickle-cell hemoglobin is twisted in the shape of a sickle. (Modified from: Kenneth R. Bridges. 2002. Information Center for Sickle Cell and Thalassemic Disorders. Public Access: http://sickle.bwh.harvard.edu/index.html.)

Traditionally, the mutant form of hemoglobin is identified by the symbol HbS, and the normal hemoglobin is referred to as HbA. These two alleles, HbA and HbS, are at the same locus and behave as Mendelian codominant traits. Hence, there are three possible phenotypes that have different properties:

a. Homozygote (HbA HbA) normal hemoglobin produces no sickle-type hemoglobin.
b. Heterozygote (HbA HbS) produces about 60 percent normal and 40 percent sickle-type hemoglobin. Heterozygotes have a preponderance of normal hemoglobin and suffer no obvious deleterious effects unless they are exposed to high altitudes. In the United States, sickle-cell disease primarily affects African Americans, but it also occurs in some individuals of Mediterranean, Middle Eastern, and Asian Indian descent. An estimated 2.5 million people with the sickle-cell trait are carriers of the HbS gene.
c. Homozygote (HbS HbS) produces 100 percent of the mutant hemoglobin. Bearers of the mutant homozygote suffer the effects of sickle-cell anemia. When the sickle cell unloads oxygen, it forms a gel of polymers that causes the red blood cells to become stiff and distorted into sickle-like shapes. In response, the body destroys damaged cells, creating a shortage of red blood cells, which results in anemia. The sickled cells clog up blood vessels. As a result, the blood circulation is impaired, leading to tissue damage and dysfunction of internal organs. Painful attacks, severe fatigue, headache, muscle cramps, and irritability are the most common symptoms. Infections and lung damage are the leading cause of death usually occurring early in life.

Frequency of HbS Alleles in Natural Populations

Because HbS has a negative and lethal effect, selection against the mutant is stringent. Yet, in large areas of Africa, the HbS mutant is very common. Some native African populations have a rate of 40 percent of individuals heterozygous with the HbS trait, and frequencies of 5 to 19 percent are widespread. Since HbS is inherited in a Mendelian form, in matings between two carriers (HbA HbS) the probability is one in four that a child will be HbS HbS and die early. Therefore, there must be other factors maintaining the deleterious HbS allele in such a high frequency in a population. The HbA HbS genotype must confer a strong selective advantage to the individual—so strong that it more than compensates for the harmful effects in the mutant HbS.

Adaptive Advantage of the Heterozygote

Haldane solved the puzzle of the high frequency of the sickle-cell trait in some African populations.[26] Haldane proposed the concept of the "malaria hypothesis." He stated that "The corpuscles of the anaemic heterozygote are smaller than normal, and more resistant to hypotonic solutions. It is at least conceivable

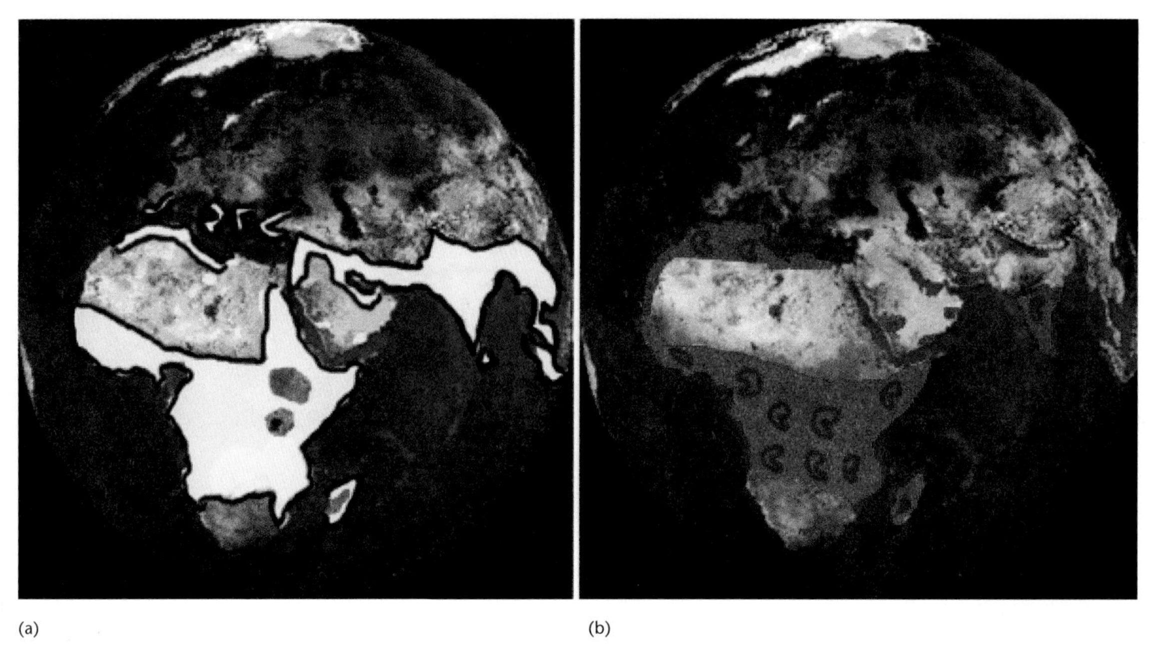

FIGURE 3.3. Geographical distribution of malaria (a) and its relationship to the distribution of sickle-cell hemoglobin (b) in Africa and Europe. Note that the areas where the frequency of malaria is high are also associated with high frequency of sickle-cell hemoglobin. (Composite made by the author.)

that they are also more resistant to attacks by the sporozoa which cause malaria." Ten years later, Allison noted that the incidence of HbS coincided with the distribution of malignant tertian malaria.[27,28] Malaria is an infectious disease caused by the parasite *Plasmodium falciparum,* which is spread through the bites of anopheles mosquitoes. Across Africa and in some parts of Europe (**fig. 3.3**), where malaria is endemic, children are exposed to malaria almost all year long.

As shown in **table 3.5**, Allison found that children with the sickle-cell trait (HbA HbS) had a considerably higher probability of surviving malaria than those who were homozygous for normal hemoglobin (HbA HbA); the heterozygote had a 17 percent advantage where malaria was most severe.[27,28] Clearly, the relationship among hemoglobin, sickle-cell anemia, and malaria supports the hypothesis of a selective advantage for the HbA HbS heterozygote.

More positive evidence comes from experimental research conducted by Allison.[27,28] In 1954 Allison injected *Plasmodium* parasites to thirty volunteers from the Luo population of East Africa. After a short time he discovered that out of the fifteen subjects with normal hemoglobin (HbA HbA), fourteen had malaria parasites, but out of the fifteen subjects who were carriers of the sickle cell (HbA HbS), only two had malaria parasites. Thus, it has been estimated that in Africa where malaria is endemic, for every 100 people with HbA HbS who survive to adulthood, 12 percent of individuals with the genotype HbA HbA die due to malaria and only fourteen of those with genotype HbS HbS survive.[29]

How Sickle Cell Kills the *Plasmodium* Parasite

Three interrelated factors are associated with the resistance of sickling hemoglobin and malaria. First, the structure of an HbS pair shows how the amino acid substitution (**fig. 3.2**) results in a sticky patch that causes the hemoglobin to polymerize into spikes. Second, the sickling of hemoglobin leaks potassium and kills the malaria-causing *Plasmodium* parasite.[30] Third, the malaria-causing *Plasmodium* parasite requires a good supply of oxygen, which the sickling does not have. Thus, in areas with malignant tertian malaria, the selective advantage of HbS in the heterozygous form counteracted the selective disadvantage of HbS in the homozygous form. Alleles for the normal and abnormal hemoglobin form a balanced polymorphism maintained by heterozygote advantage. A prospective study also showed that HbS provides significant protection against severe malarial anemia, and high-density parasitaemia especially among infants.[31]

Cultural Origin of the Advantage of Sickle Cell

The interaction of sickle cell and malaria has served as the paradigm in our understanding of how natural selection molds the genetic profile of humans. Malaria

TABLE 3.5

Genotypic frequency of HbA HbA (normal), HbA HbS (heterozygous carrier of sickling allele), and HbS HbS (sickling) hemoglobin found in an East African population*

Genotype	Number of People	Frequency	Number of HbA alleles		Number of HbS alleles
HbA HbA	8	.08 (8/100)	16		0
HbA HbS	84	.84 (84/100)	84		84
HbS HbS	8	.08 (8/100)	0		16
Total	100		100		100
Total number of alleles =	100 + 100 = 200				
Proportion of HbA alleles:		100/200	= 0.50	= *p*	
Proportion of HbS alleles:		100/200	= 0.50	= *q*	
Total			= 1.00		
Genotype		**Expected Frequency**	**Observed Frequency**		**Frequency Difference**
HbA HbA	(.50) × (.50)	= .25	.08		.17
HbA HbS	2 (.50 × .50)	= .50	.84		**.34**
HbS HbS	(.50) × (.50)	= .25	.08		.17

*Data from ref. #28. Allison, A.C. 1956. Sickle-cells and evolution. *Scientific American* 195:87–94.

transmission requires mosquitoes and the proliferation of mosquitoes requires ample sunlight and pools of stagnant water. The extensive foliage of an uncut African tropical forest prevented much of the sunlight from reaching the floor of the forest. In addition, the forest environment was very absorbent, so water did not tend to accumulate in pools. In other words, an unmodified environment was less conducive to large populations of mosquitoes. Consequently, the malaria parasite did not have a hospitable environment. According to Livingstone,[32,33] this situation changed several thousand years ago when prehistoric African populations brought horticulture into southeastern Africa. With horticulture, the land was cleared for crops, and without the many trees, it was easier for sunlight to reach the land surface. In addition, the continued use of the land changed the soil chemistry, allowing pools of water to accumulate. Both changes led to an environment conducive to the propagation and spread of large mosquito populations and the spread of the malaria parasite. The increased human population density also provided more hosts for the mosquitoes to feed on, thus increasing the spread of malaria.[32,33] In this context, before the development of horticulture in Africa, the frequency of the sickle-cell allele was probably low, as it is in non-malarial environments today; but when the incidence of malaria increased, it became evolutionarily advantageous to have the heterozygote HbA HbS genotype. Those who had it would have greater resistance to malaria without suffering the effects of sickle-cell anemia. The innovation of horticulture in human evolution occurred about 14,000 to 15,000 years ago. These findings suggest that a change in gene frequency, and hence, evolution, can occur in less than 16,000 years.

Cultural-Nutritional Practices and the Survival of Sickle Cell

Various studies have shown that the clinical symptoms of sickle-cell anemia are decreased with high consumption of cyanate- and thiocyanate-containing foods such as cassava (manioc), yams, sorghum, millet grains, sugar-cane, and dark varieties of lima beans.[34,35] According to food analysis cassava flour contains about 70 to 80 mg/100 g of thiocyanate and yam flour between 50 to 60 mg/100 g of thiocyanate.[36] Laboratory studies indicate that when cyanogen derivatives enter into the blood they can combine with hemoglobin S and alter its structure. One consequence of this alteration is the inhibition of the sickling of hemoglobin S and thus almost causing it to function like a normal hemoglobin A.[37] As a consequence the viability of individuals with homozygous hemoglobin S can increase. These findings would suggest that the persistence of high frequency of sickle cells in areas where malaria is endemic is also the result of cultural-nutritional practices.

Change in the Frequency of Sickle Cell

As native Africans migrated to other parts of the world and became exposed to new environments where there was no malaria, the selective advantage of the HbA HbS genotype decreased. With only negative selective pressure on HbS, it has declined throughout the world.[33,38] It has been estimated that the slaves from Africa included 20 to 25 percent carriers of the sickle-cell trait (HbA HbS). When the slaves arrived in the New World, there was no longer the same evolutionary advantage for high frequencies of HbS because there was less malaria. As a result, the frequency of HbS has been reduced to about 3 percent. Thus, the evolutionary trend has been reversed in populations unexposed to malaria.

Mortality and Fertility of Heterozygous Sickle Cell

It has been suggested that the high frequency of HbA HbS is also related to increased fertility of women with genotype HbA HbS.[39] Hoff et al.[39] found that in a sample of 10,000 African American women in Mobile, Alabama, those with the HbA HbS genotype had a higher number of live births, suggesting the possibility that fertility does influence selection for the S allele in some populations. However, Madrigal[40] studied the reproductive histories of black women in Limon, Costa Rica, and found no difference between genotypes HbA HbA and HbA HbS women in a number of measures of differential fertility (family size, number of pregnancies, number of live births, and number of spontaneous abortions). Therefore, it appears that the difference in the frequency of the sickle-cell allele is related to mortality rather than fertility differences. Recent studies by Abrams and associates[41,42] indicate that pregnant women with sickle-cell hemoglobin and with malaria are able to produce viable offspring in the second pregnancy due to a placental sequestration of malaria antigens. Thus, it would appear that the sickle-cell trait and malaria continue to elicit biological adaptations that enable women to reproduce despite the negative environmental conditions on which these two diseases coexist.

Summary

The population is the theater in which all the evolutionary forces such as mutation, gene recombination, gene flow, genetic drift, and founder effect create genetic variability upon which natural selection works. Depending on how each one of the factors plays its respective roles, the genetic and phenotypic constitution of the population changes, making its members adapt to any given environment, where all the forces of evolution operate. The application of statistical methods to the hereditary principles derived from Mendel and applied by the Hardy-Weinberg equation was a major breakthrough that provided a yardstick for measuring the strength and the direction of the forces of evolution.

The dynamic interaction between the frequency of sickle-cell hemoglobin and malaria is the best-known example of natural selection operating on a discrete genetic trait. It shows how in a malarial environment a lethal and deleterious allele proved to be the avenue through which the population responded and survived in an otherwise unfit environment. However, as the stress of malaria decreases, so does the selective advantage for the sickle cell change. Therefore, the genetic profile of the population also changes. Thus, evolution is an opportunistic adaptation that fits the needs of the population and changes when the environment on which it evolves also changes. In other words, evolution is an ongoing and dynamic process whereby what is good today may not necessarily be good in the future. The flexibility to adapt and change maintains the unity of the human species.

Suggested Readings

Cavalli-Sforza, L. L., P. Menozzi, and A. Piazza. 1994. *The History and Geography of Human Genes.* Princeton: Princeton University Press.

Dobzhansky, T. 1962. *Mankind Evolving.* New Haven, CT: Yale University Press.

Mayr, E. 2001. *Evolution Explained.* London: Weidenfeld & Nicolson.

Neel, J. V. 1994. *Physician to the Gene Pool: Genetic Lessons and Other Stories.* New York: John Wiley.

Literature Cited

1. Kettlewell, H. B. D. 1956. Further selection experiments on industrial melanism in the Lepidoptera. *Heredity* 10:287–301.
2. Kettlewell, H. B. D. 1973. *The Evolution of Melanism: The Study of a Recurring Necessity.* Oxford: Clarendon.
3. Hardy, G. H. 1908. Mendelian proportions in a mixed population. *Science* 28:49–50.
4. Weinberg, W. 1908. Ober den Nachweis der Vererbung beim Menschen: jahreshefte Vereins Vaterland. *Naturkunde Wiirtemberg* 64:368–382.
5. Wilson, E. O. and W. H. Bossert. 1971. *A Primer in Population Biology.* Sunderland, MA: Sinauer.

6. Glass, H. B. 1953. The genetics of the Dunkers. *Scientific American* 189:76–81.
7. Roberts, D. F. 1968. Genetic effects of population size reduction. *Nature* 220:1084–1088.
8. Roberts, D. F. 1971. The demography of Tristan da Cunha. *Population Studies* 25:465–479.
9. Thompson, E. A. 1978. Ancestral inference. II. The founders of Tristan da Cunha. *Annals of Human Genetics* 42:239–253.
10. Jenkins, T., P. Beighton, and A. G. Steinberg. 1985. Serogenetic studies on the inhabitants of Tristan da Cunha. *Annals of Human Biology* 12:363–371.
11. Soodyall, H., T. Jenkins, A. Mukherjee, E. du Toit, D. F. Roberts, and M. Stoneking. 1997. The founding mitochondrial DNA lineages of Tristan da Cunha Islanders. *American Journal of Physical Anthropology* 104:157–166.
12. Gualandri, V., G. Franceschini, and S. R. Sirtori. 1985. Identification of the complete kindred and evidence of a dominant genetic transmission. *Am J Hum Genet* 37:1083–1097.
13. Sirtori, C. R., L. Calabresi, G. Franceschini, D. Baldassarre, M. Amato, J. Johansson, M. Salvetti, et al. 2001. Related articles, links cardiovascular status of carriers of the apolipoprotein A-I (Milano) mutant: the Limone sul Garda study. *Circulation* 103(15):1949–1954.
14. Nissen, S. E., T. Tsunoda, E. M. Tuzcu, P. Schoenhagen, C. J. Cooper, M. Yasin, G. M. Eaton, et al. 2003. Effect of recombinant Apo A-I Milano on coronary atherosclerosis in patients with acute coronary syndromes: a randomized controlled trial. *JAMA* 290(17):2292–2300.
15. Fedier, A., and D. Fink. 2004. Mutations in DNA mismatch repair genes: implications for DNA damage signaling and drug sensitivity (review). *Int J Oncol* 24:1039–1047.
16. Pittendrigh, B., R. Reenan, R. H. French-Constant, and B. Ganetzky. 1997. Point mutations in the *Drosophila* sodium channel gene para associated with resistance to DDT and pyrethroid insecticides. *Mol Gen Genet* 256: 602–610.
17. Cohn, D. L., F. Bustreo, and M. C. Raviglione. 1997. Drug-resistant tuberculosis: Review of the worldwide situation and the WHO/IUATLD global surveillance project. *Clinical Infectious Disease* 24(Suppl 1): 121–130.
18. Haines, J. L., J. A. Trofatter, R. E. Tanzi, P. Watkins, N. S. Wexler, P. M. Conneally, and J. F. Gusella. Chromosome 21 genetic linkage data set based on the Venezuelan reference pedigree. 1992. *Cytogenetics and Cell Genetics* 59:88–89.
19. McKusick, V. A. 1998. *Mendelian Inheritance in Man: A Catalog of Human Genes and Genetic Disorders.* Baltimore; London: Johns Hopkins University Press.
20. Witkop, C. J., Jr. 1989. Albinism. *Clin Dermatol* 72:80–91.
21. Carey, M. C., and B. Paigen. 2002. Epidemiology of the American Indians' burden and its likely genetic origins. *Hepatology* 36:781–791.
22. Hedrick, P. W. 2003. Hopi Indians,"cultural" selection, and albinism. *Am J Phys Anthropol* 121:151–156.
23. Woolf, C. M. and F. C. Dukepoo. 1969. Hopi Indians, inbreeding, and albinism. *Science* 164:30–37.
24. Keeler, C. 1970. Cuna Moon-child albinism, 1950–1970. *J Hered* 61:273–278.
25. Durham, W. 1991. *Co-evolution.* Stanford, CA: Stanford University Press.
26. Haldane, J. 1949. The rate of mutation of human genes. *Proc VIIIth Int Congr Genet* 1949: 267–272.
27. Allison, A. C. 1954. Protection afforded by sickle-cell trait against subtertian malarial infection. *Br Med J* 1:290–294.
28. Allison, A. C. 1956. Sickle-cells and evolution. *Sci Amer* 195:87–94.
29. Bodmer, W. F. and L. L. Cavalli-Sforza. 1976. *Genetics, evolution, and man.* San Francisco: W. H. Freeman.
30. Cerami, A., and C. M. Peterson.1975. Cyanate and sickle-cell disease. *Sci Am* 232:44–50.
31. Aidoo, M., D. J. Terlouw, M. S. Kolczak, P. D. McElroy, F. O. ter Kuile, S. Kariuki, B. L. Nahlen, A. A. Lal, and V. Udhayakumar. 2002. Protective effects of the sickle cell gene against malaria morbidity and mortality. *Lancet* 359:1311–1312.
32. Livingstone, F. B. 1958. Anthropological implications of sickle cell gene distribution in West Africa. *American Anthropologist* 60:533–562.
33. Livingstone, F. B. 1967. *Abnormal hemoglobins in human populations: A summary and interpretation.* Chicago: Aldine.
34. Serjeant, G. R. 2004. The dilemma of defining clinical severity in homozygous sickle cell disease. *Curr Hematol Rep* 3:307–309.
35. Houston, R. G. 1973. Sickle cell anemia and dietary precursors of cyanate. *Am J Clin Nutr* 26:1261–1264.
36. Oke, O. L. 1969. The role of hydrocyanic acid in nutrition. *World Rev Nutr Diet* 11:170–198.
37. Jackson, F. L. 1990. Two evolutionary models for the interaction of dietary organic cyamogens, hemoglobins, and falciparum malaria. *A J Human Biology* 2:521–532.
38. Kan, Y. W., and A. M. Dozy. 1980. Evolution of the hemoglobin S and C genes in world populations. *Science* 209:388–391.
39. Hoff, C., I. Thorneycroft, F. Wilson, and M. Williams-Murphy. 2001. Protection afforded by sickle-cell trait (Hb AS): What happens when malarial selection pressures are alleviated? *Human Biology* 73:583–586.
40. Madrigal, L. 1989. Hemoglobin genotype, fertility, and the malaria hypothesis. *Human Biology* 61:311–325.
41. Abrams, E. T. 2004. Malaria during pregnancy: Physiological and evolutionary perspectives on poor birth outcomes. Ph.D. dissertation in anthropology. Ann Arbor: The University of Michigan.
42. Abrams, E. T., D. A. Milner Jr., J. Kwiek, V. Mwapasa, D. D. Kamwendo, D. Zeng, E. Tadesse, V. M. Lema, M. E. Molyneux, S. J. Rogerson, and S. R. Meshnick. 2004. Risk factors and mechanisms of preterm delivery in Malawi. *Am J Reprod Immunol* 52:174–183.

Macroevolution and Primate's Roots

In this section we look at the mammalian roots. The emphasis is on primates, which is the group of mammals to which humans belong. Chapter 4 starts with the roots of mammalian evolution starting with the successful evolutionary adaptation of mammals, which gave rise to the primates. Chapters 5 and 6 look more closely at the variation and ecological adaptations of living primates. Chapter 7 reviews the fossil evidence of the evolution of primates from their initial appearance to presently living primates. Finally, chapter 8 considers the role of food in the evolution of the primate omnivore digestive system.

Mammalian Antecedents

All life on earth is part of an organic continuum. Although as a species we are unique, we are indisputably connected by an unbroken series of ancestors extending back into the history of the other species on earth. This chapter begins with the roots of mammalian evolution, starting with fish, amphibians, and reptiles. Thereafter, the successful evolutionary adaptation of subsets of mammals that gave rise to primates will be discussed.

Age and Origin of Life on Earth

Age of the Universe

How did the evolution of the first life occur? Scientists have tried to develop a sequence of events that might have led to the first life, but in general there are more questions about the origin of life than there are answers. It is postulated that the sun condensed from the solar system's early gas cloud. These clouds had trailing arms of gases, some of which may have condensed to form the planets. The dominant theory is that the universe originated after an explosion of stupendous proportions, or the "big bang," which occurred about 15 billion years ago. Three minutes after the big bang, protons and neutrons joined to make atomic nuclei, and 100,000 years later the electrons joined the nuclei to form atoms.

Age of Earth and First Life

About 12 billion years after the big bang, galaxies and our solar system were formed around a medium-sized star in the Milky Way Galaxy. Afterward, some 4.5 billion years ago, the earth was formed out of swirling gases, which formed the atmosphere (**fig. 4.1**). According to geological evidence this atmosphere was composed of volcanic gases such as methane (CH_4) and carbon dioxide (CO_2), ammonia (NH_3), hydrogen (H_2) and water vapor (H_2O), but free oxygen was not present. With the energy of heat and lightning, the gases of the early atmosphere combined to form substances such as amino acids forming an "organic soup" (**fig. 4.2**). These simple compounds combined to form

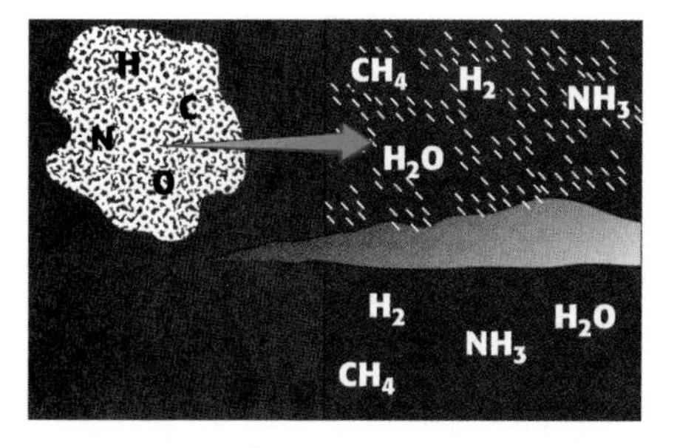

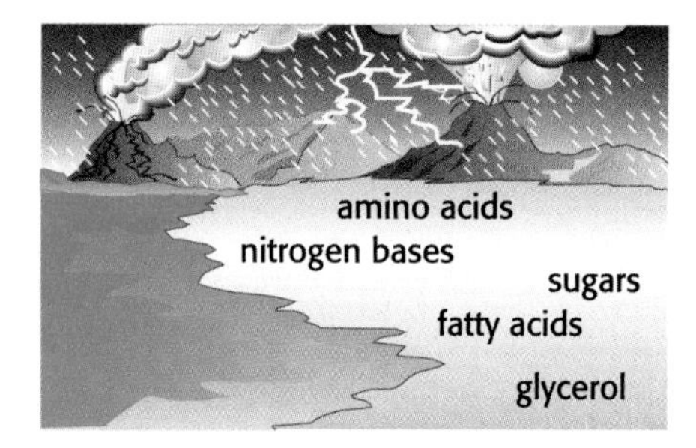

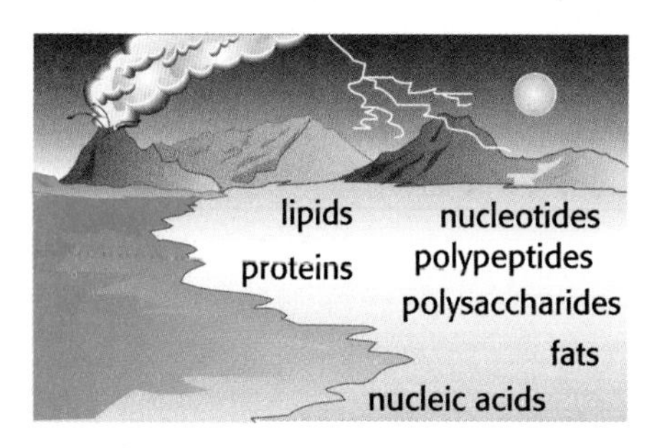

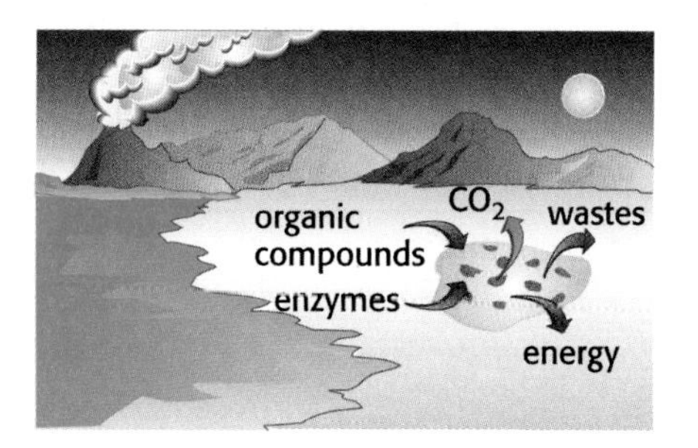

FIGURE 4.1. A sequence of events that might have led to the first life. The beginnings of the earliest life forms on earth are said to be related to those molecules that have the ability to pair up in double helices like DNA. From *BSCS Biology: An Ecological Approach, 9th Edition* by BSCS. Copyright © 2002 by BSCS. Reprinted by permission.

more complex molecules and became nucleic acid polymers that had the ability to pair up in double helices such as DNA, which are found in all organisms today.

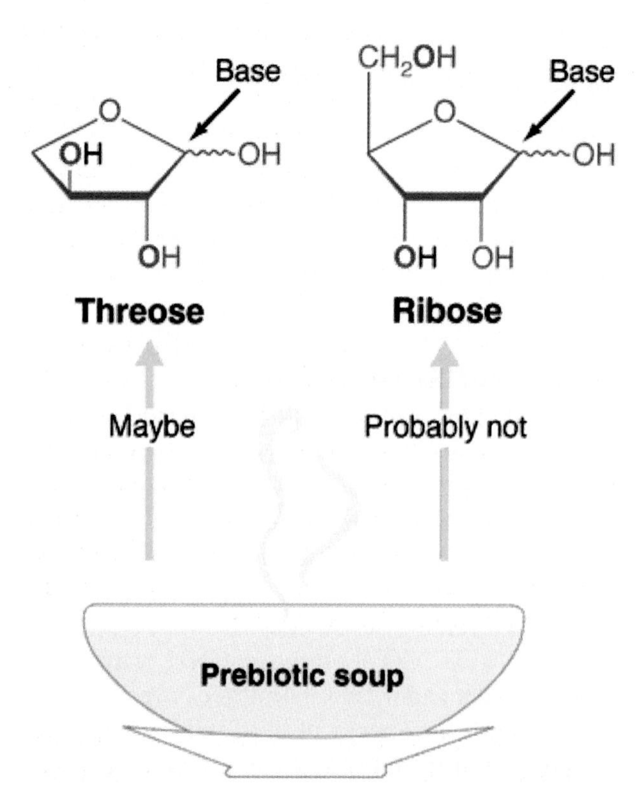

FIGURE 4.2. The beginnings of the earliest life forms on earth are said to be related to those molecules that have the ability to pair up in double helices like DNA. Source: Adapted from fig. 1 of Orgel.[1,2]

Change from Unicellular to Multicellular Organisms

Living organisms are usually separated into two groups according to their cellular structure: **prokaryotes** and **eukaryotes** (**fig. 4.3**). About **3.5 to 2.8 billion** years ago the prokaryotic organism appeared on earth. The prokaryotes do not have a nucleus or mitochondria. The genetic material of prokaryotes consists of a single thread of DNA. All these early single-celled organisms like bacteria reproduced asexually by splitting and making copies of themselves. Their major variability was only through mutation. These organisms about 1.5 billion years ago gave rise to eukaryotic organisms that have cells with true nuclei that divide by mitosis and meiosis. Then, during the Cambrian era and between 570 and 510 million years ago, the eukayrotic organisms gave rise to multicellular organisms, from which insects and eventually all the vertebrate animals evolved (**table 4.1**). These multicellular organisms reproduced sexually and now, in addition to mutation, fertilization also provided genetic and phenotypic variation. Evolutionary change quickly started to gain speed.[1–13]

Geological Time Scale

Geologists have established the geological time scale (**table 4.1**) in which very large time spans are organized into eras and periods. Periods, in turn, can be broken down into epochs. For the time span encompassing vertebrate evolution, there are three eras: Paleozoic, Meso-

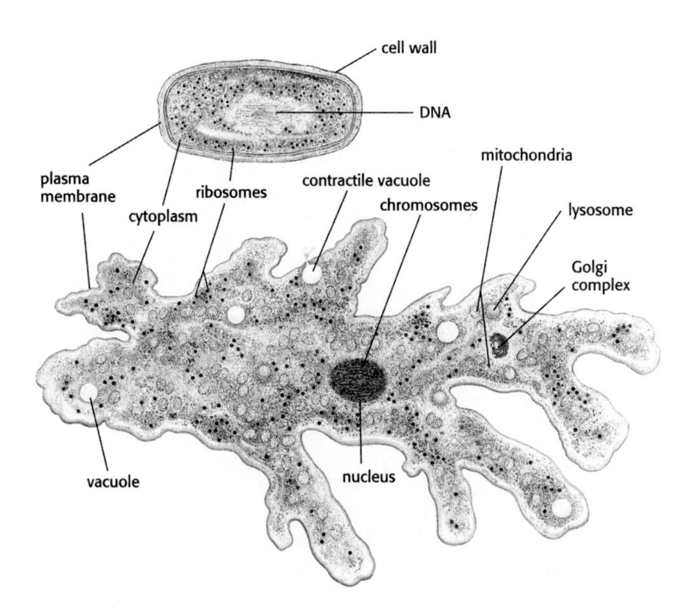

FIGURE 4.3. Comparison of prokaryote (top) and eukaryote (bottom) cell. A prokaryote cell is very simple and does not have a nucleus, whereas the eukaryote is complex and has a nucleus and many organelles. From *BSCS Biology: An Ecological Approach, 9th Edition* by BSCS. Copyright © 2002 by BSCS. Reprinted by permission.

zoic, and Cenozoic. These eras have been established based upon measurements of the age of fossils. In general, dating methods can be absolute and relative.

Relative Techniques

Currently there are two methods of relative dating, stratigraphy and fluorine analysis.

Stratigraphy

Stratigraphic dating indicates whether a given specimen is older or younger than another one, depending upon the location of the specimen. In general, materials located at the bottom of an undisturbed geological deposit are older than the ones on the top. Stratigraphy is based upon the law of superposition, which states that the layer (strata) of material situated at the bottom is older than higher stratum (**fig. 4.4**). This is because all organic and inorganic materials such as soil and animals always settle into the bottom of any location where there is water and become a part of the sedimentary deposit. The next layer of materials is deposited on top of the previous one. Over time these layers dry up, leaving a series of strata or sedimentary beds. Throughout time this layer has laid down much of the earth's crust after a layer of sedimentary rock. Paleontologists, through investigation and analysis of the composition of the layers of the earth and their relationship to one another, can determine the relative age of a given fossil.

Fluorine Analysis

Fluorine dating applies only to bone. It is based upon the fact that when bones and teeth are in contact with the ground they absorb fluorine and other minerals dissolved in the ground water, while the nitrogen content of the bones decreases as the material ages. The ratio between the amount of minerals absorbed and the amount of nitrogen lost is used to calculate the relative chronology of the material in a site.

Absolute or Chronometrical Dating

Chronometrical dating refers to specific points in time using as reference specific calendar systems, which are based upon naturally recurring units of time, such as the revolutions of the earth around the sun or the appearance of the new moon. Current systems include the Gregorian calendar, the Hebrew calendar, and the Muslim calendar. The Gregorian calendar refers to the revolutions of the earth around the sun since the traditional date of the birth of Christ; the Hebrew calendar is based on the biblical origin of the world; and the Muslim calendar is based on the date of the flight of Mohammed. Usually a chronometric date is given as years ago or years BC in which BC stands for "before Christ" or years BP in which BP stands for "before present." Obviously, the designation of a date depends and changes according to the year in which the date was determined. For example, if a date were determined to be 11,100 BP in 1960, it would have to be changed to 11,140 BP in the year 2000. Since any estimate is subject to error, most dates are given as mean plus or minus (±) a standard deviation. For example, if the mean age is 11,100 ± 100, the probability of the real date falling between 11,200 and 11,000 years is 67 percent, and the probability of the real date falling between 11,300 and 10,900 is 95 percent. There are also several methods for determining the age of fossils that include uranium-238, potassium-40, fission-track, thermoluminescence, and carbon-14.

Uranium-238 Decay

Uranium-238 decay dating is based upon the fact that one-half of the original amount of uranium-238 is lost in 4.5 billion years and through various processes becomes lead. Eventually, in 9 billion years all uranium decays and is converted to lead. Therefore, depending on how much uranium-238 is present in rock, chronometrists can determine the age of the rock. For example, if a sample of rock contains one-half of the uranium-238, the age of that rock is 4.5 billion years. If there is 25 percent of the original amount of uranium-238, then the age of the rock is 9 billion years and so on. Evaluations of uranium-238 decay have proved a useful tool in dating the age of the formation of the earth.

Potassium-Argon

Potassium-argon dating is based upon the radioactive decay of potassium-40 (^{40}K), which has a half-life of 1,250 million years and produces argon-40 (^{40}Ar). The potassium-argon method is based upon two factors. First, after volcanic activity and under very high temperatures that accompany volcanic activity, the argon gas is expelled and as it cools and solidifies, it contains a certain amount of potassium-40 but no argon-40. Second, over time the amount of ^{40}K decreases while the amount of ^{40}Ar increases. Therefore, the ratio of potassium decay and increase of argon provides information of the relative age of a specimen. To obtain the date of the rock, it is reheated, the escaping argon gas is extracted, and the amount of ^{40}Ar is measured. The technique has been used to determine the age of hominoid and hominid fossils.

Fission-Track

Fission-track dating evaluates the minerals in a deposit. This method is based upon the fact that when a heavy isotope such as uranium-238 breaks apart, its decay particles rip holes in the orderly three-dimensional crystal structure of the mineral. In so doing, the particles leave tracks in these crystals. These tracks, after being chemically treated to make them larger, can be seen under magnification and counted. If the concentration of the isotope under study is known, then the number of tracks left will indicate the age of the mineral being dated. This method can provide dates ranging from twenty years after the specimen formed to as far back as the beginning of the solar system.

Thermoluminescence

Thermoluminescence dating is based upon the fact that when crystals are heated they give off a flash of hot light, a phenomenon known as thermoluminescence. The minerals in the ground are constantly exposed to radiation from naturally occurring radioactive elements. As a result of this influence some electrons separate from the atoms. When the material is exposed to heat, the trapped electrons are expelled and, in the process, give off characteristic wavelengths of light. The amount of light is proportional to the age of the material.

Thermoluminescence has been used to date archaeological finds such as ceramics, burnt stones, and burnt flint tools, as well as volcanic debris. This method can provide dates ranging from 50 years to as far back as 300,000 years ago.

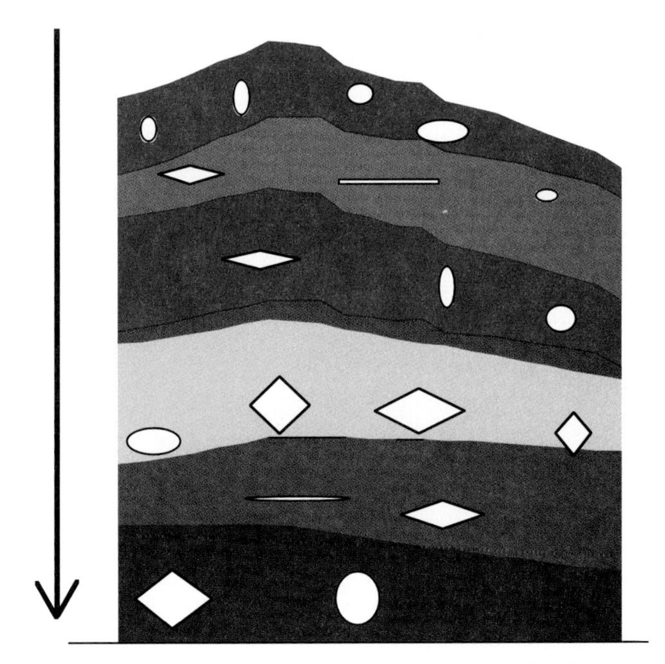

Superposition of sedimentary beds

FIGURE 4.4. Theoretical representation of stratigraphic layers. Note the lower the strata, the older the location and the specimens found in such location. (Composite made by the author.)

Carbon-14

Carbon-14 dating is based upon the principle that one-half of the original amount of carbon present in any material is lost in 5,730 years. Therefore, depending on how much carbon-14 is present in a specimen, chronometrists can determine the age of the specimen. Carbon-14 has been used to date material from less than 1,000 years to as old as 45,000 years.

The Paleozoic and Mesozoic: From Fish to Amphibians to Reptiles

During the Paleozoic era (**table 4.1**), which lasted from 570 to 280 million years ago, several varieties of fish (including the ancestors of modern sharks and bony fish), amphibians, and reptiles appeared. Thereafter, during the Mesozoic era, lasting from 225 to 135 mil-

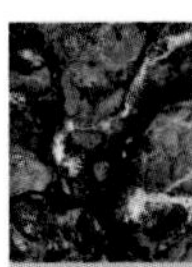

TABLE 4.1
Geological time scale

Era	Period	Epoch	(mya)	Evolution
Cenozoic	Quaternary	Holocene	0.01	Mammals
Cenozoic	Quaternary	Pleistocene	1.8	Mammals
Cenozoic	Tertiary	Pliocene	5	Mammals
Cenozoic	Tertiary	Miocene	25	Mammals
Cenozoic	Tertiary	Oligocene	35	Mammals
Cenozoic	Tertiary	Eocene	55	Mammals
Cenozoic	Tertiary	Paleocene	65	Mammals
Mesozoic	Cretaceous		135	Reptiles & Fish
Mesozoic	Jurassic		190	Reptiles & Fish
Mesozoic	Triassic		225	Reptiles & Fish
Paleozoic	Permian		280	Fish
Paleozoic	Carboniferous		345	Fish
Paleozoic	Devonian		395	Fish
Paleozoic	Silurian		430	Fish
Paleozoic	Ordovician		500	Fish
Paleozoic	Cambrian		570	Fish-
Paleozoic	Cambrian		590	Multicellular
Pre-Paleozoic	Pre-Cambrian		1,500	Bacteria-Single cell
Pre-Paleozoic	Pre-Cambrian		3,500	Bacteria like
Life on Earth			4,500	

mya = million years ago

lion years ago, amphibians that descended from lobe-finned fishes (crossopterygian) became the first animals to make the transition to land. These lobe-finned fish or amphibians possessed tiny fingerlike appendages, which eventually evolved into the digits of terrestrial vertebrates. The lobe-finned fish resembled the modern lungfish. It has a primitive simple lung that allows it to breathe air when surface water disappears and movement across short land distances in search of water is necessary. Hence, during the Miocene epoch these amphibians that belonged to the order of therapsids became the dominant land vertebrates. They were intense predators and, with few exceptions, were vertebrate competitors. These terrestrial amphibians retained important features from their aquatic ancestors. Reproduction was very much an aquatic affair. Mating and fertilization of eggs probably took place in the water, and immature stages remained aquatic.

On the other hand, the reptiles made a complete adjustment to terrestrial life. The reptilian lineage emerged some 340 million years ago. The key to continuous life on land for the reptiles was an increase in complexity of the egg that included the amniotic egg, yolk sack, allantois, chorion, and an outer shell. Like in the egg of a chicken, the amnion kept the developing embryo in an aquatic environment while the yolk sac supplied the food for development to an advanced stage, the allantois functioned as a waste excretory organ, and the chorion and outer shell prevented the egg from getting dry. All these changes in the egg design increased the chances of survival without dependence on large quantities of water. All they needed to complete their life cycle was small sources of water, such as raindrops, dew, juicy insects or plants, and small vertebrates.

The major groups of terrestrial reptiles today include snakes, lizards, and crocodiles. The most famous of these highly successful Mesozoic reptiles was the dinosaur. A key to the success of dinosaurs was the evolution of an upright stance. Amphibians and reptiles evolved a splayed out stance while dinosaurs evolved an erect stance similar to the upright stance of mammals. Animals with a splayed posture cannot sustain continued locomotion because they cannot breathe while they move; also, their walk is inefficient because their limbs are modified from fins. In contrast, an erect stance allows for continual locomotion. Yet despite their advantageous posture, the dinosaurs, during the Cenozoic era, mysteriously lumbered into oblivion.

At present it is not clear what caused the extinction of the dinosaurs. Some believe that the extinction of dinosaurs was related to the collision of asteroids or comets to earth. According to this theory, about 65 million years ago, numerous comets or asteroids collided with the earth, igniting huge fires that burned for years, pouring smoke and ash into the air and creating an enormous dust cloud of global proportions. The smoke-filled air blocked sunlight, which in turn caused plants not to photosynthesize. Because solar energy reaching the earth's surface was partially blocked, the planet also cooled and many animals, such as dinosaurs, may have been wiped out by these climatic changes. As a result of the climatic changes, most of the herbivorous animals disappeared and the food on which carnivorous dinosaurs came to depend also vanished.

The Cenozoic Era: Emergence of Mammals

The extinction of dinosaurs 65 million years ago opened up opportunities for mammals that enabled them to expand into new environments. Although the successful evolution of mammals occurred during the Cenozoic (**table 4.1**), the roots of the mammal class go back to the Triassic period. Long before the first dinosaurs evolved about 225–190 million years ago, a mammal-like reptile member of the subclass Synapsida evolved. The synapsids flourished and coexisted with the dinosaurs for nearly 100 million years, but during the Triassic period (225–195 million years ago), their skeletons were altered from the cumbersome reptilian mold to a more agile design, which included limbs that enabled them to maintain mammal-like posture. Among the synapsids, the order Therapsida prevailed and over the millennia they metamorphosed into true mammals. These first mammals were small, 5 cm (2 in) long, nocturnal, and probably partly arboreal. Their unobtrusive scuttling gave little evidence of what was to become the most exciting radiation in vertebrate history. They developed an expanded temporal skull opening, a secondary palate that formed a horizontal partition in the roof of the mouth, and a dentition with a variety of tooth shapes and sizes that were adapted to grasping, cutting, and crushing or chopping food. These changes led to an improvement of the respiratory system. For example, the false palate forms a bypass for air from the nostrils to the back of the mouth, facilitating simultaneous eating and breathing. The body morphology also changed. The ribs were no longer attached to the cervical and lumbar vertebrae, but only to the thoracic ones; the chest and abdomen were separated by the diaphragm; the pectoral and pelvic girdles were streamlined, and the limbs were aligned beneath, rather than to the side of, the body. Thus, mammals, during their coexistence with the dinosaurs, continued evolving to attain an amazing structural and functional plasticity.[14–16]

So rapid was the mammalian adaptive radiation that by the end of a 10 million-year period, more than 4,000 species, 1,000 genera, 135 families, 18 orders, and 2 subclasses of mammals appeared. The success of mammals is related to the structural changes that provided a functional flexibility that allowed them to exploit the nutritional resources of a wide range of environments and reproduce more efficiently than their predecessors had.

Central Nervous System

Mammals have larger brains than those typically found in reptiles. In particular, the cerebrum is generally enlarged, especially the outer covering, the neocortex, which controls higher brain functions (**fig. 4.5**). In some mammals, the cerebrum expanded so much that it came to comprise the majority of brain volume; moreover, the number of surface convolutions increased, creating more surface area and thus providing space for even more nerve cells (neurons). As a result of the increased complexity of the brain, mammals were able to process more information and, hence, became more adaptable than reptiles.

Reproductive Efficiency

A unique characteristic of mammals was the increased reproductive efficiency attained through extended prenatal development. While reptiles, birds, and most fish incubate their young externally by laying eggs (i.e., they are oviparous), mammals, with very few exceptions, give birth to live young and are thus said to be viviparous. Mammals are able to have live offspring because a large portion of the development occurs in utero. While other animals, such as fish or reptiles, rely on sheer numbers to ensure that at least a few offspring will survive, mammals do so with fewer offspring. In other words, in reproductive terms mammals are reproductively more efficient than reptiles or fish.

Dentition

Reptiles have several teeth that vary slightly in size, but they all have the same basic cone-shaped form (**fig.

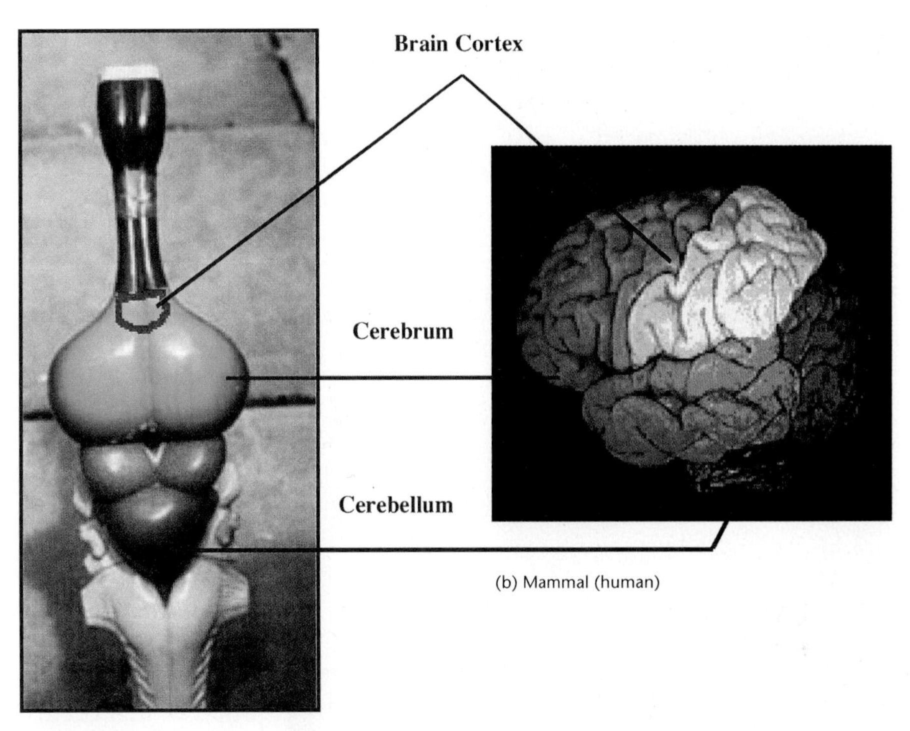

FIGURE 4.5. Enlargement and increased complexity of the mammalian brain (b) when compared to the brain of reptiles (a). (Composite picture made by the author.)

4.6). In other words, in reptiles and fish the teeth are shaped in one single form called a homodont dentition. Mammals, in contrast, are characterized by a heterodontic dentition (**fig.** 4.7). They have teeth that vary greatly in size and shape from flat, multicusped molar teeth to the sharp cone-shaped canines. Incisors are specialized for cutting, canines for grasping and piercing, and premolars and molars for crushing and grinding food. In fact, mammals are the only vertebrates that masticate their food, which is made possible by the variety of teeth. It is this trait that allows mammals to process and utilize a wider diversity of foods than reptiles could process. Thus, much of the adaptive success of mammals is related to the diversity of their teeth that allows them to eat both plant and animal foods.

FIGURE 4.6. Homodontic dentition of reptiles. Reptiles such as caiman alligators have several teeth that differ only in size but not in shape or function. © Kevin Schafer/ Tom Stack & Associates.

Dental Formula

A didactic way of describing the number of teeth uses the dental formula. In the dental formula one counts the number of each type of tooth in half of one jaw, upper or lower. Only one-half of the jaw is used because both right and left sides of the jaw contain the same number of teeth. Each tooth is represented by its initial first letters preceded by the number present. For example, the dental formula for humans is: 2I, 1C, 2PM, 3M, which refers to 2 incisors, 1 canine, 2 premolars, and 3 molars and which means that one-half of either jaw contains 8 teeth. Hence, the total number of teeth in humans is 32 ($8 \times 4 = 32$). Sometimes, the dental formula is expressed using only the number of teeth in the expected order. For example, the human dental formula is expressed as 2-1-2-3, which implies that in one-half of either jaw there are 2 incisors, 1 canine, 2 premolars, and 3 molars.

Temperature Regulation

A unique feature that distinguishes contemporary mammals from reptiles and fish is the maintenance of a constant internal body temperature irrespective of the changes in ambient temperature. The most important distinction between mammals and reptiles is how the energy to maintain homeothermy is produced and channeled. In mammals, the energy is generated inter-

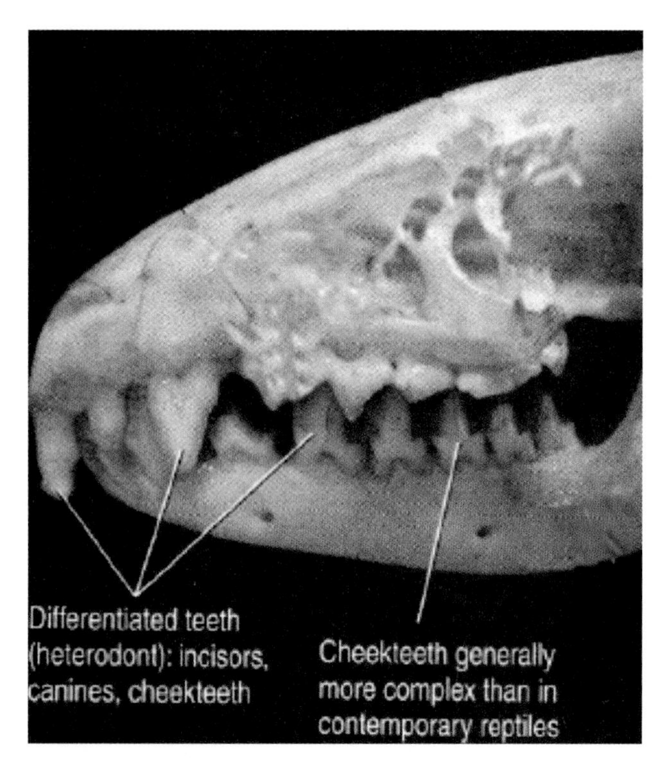

FIGURE 4.7. Heterodontic dentition of mammals. Carnivores such as wolves and herbivores such as sheep have teeth that vary greatly in size and shape from flat, multicusped molar teeth to the sharp cone-shaped canines. (Composite picture made by the author from casts.)

nally through metabolic activity by processing food or by muscle action, whereas in reptiles and fish, this energy is obtained directly from exposure (externally) to the sun. Mammals are hence referred to as endothermic and reptiles and fish are ectothermic. Because they are endothermic, mammals require a greater energy supply than ectothermic animals such as fish and reptiles. Thus, one of the major advances in evolution was the advent of animals capable of deriving their nutritional requirements not only from animal foods but also from plants. As mammals continued their evolution, some remained herbivores, while others, such as primates, became omnivores and in the process changed their digestive systems. However, recent findings of dinosaurs with feathers suggest that during the late stage of the Mesozoic they were evolving and becoming homeothermic like contemporary birds.

Continental Drift and Mammalian Evolution

About 200 million years ago, the earth was also changing as the continents altered their shape and position: a phenomenon known as **continental drift**. Continental drift refers to the split and drift of landmasses that occurred as a result of collision, subduction, erosion, and deposition of the plate tectonics of the earth. As a result of continental drift, all the continents had drifted together to form a huge supercontinent called **Pangea,** literally "all lands" (**fig. 4.8**), but during the early Mesozoic, the southern continents (South America, Africa, Antarctica, Australia, and India) began to split off from Pangea, forming a large southern continent called Gondwanaland. Similarly, the northern continents (North America, Greenland, Europe, and Asia) were consolidated into a northern landmass called Laurasia. Then, during the Mesozoic, Gondwanaland and Laurasia continued to drift apart and to break up into smaller segments. By the end of the Mesozoic (about 65 million

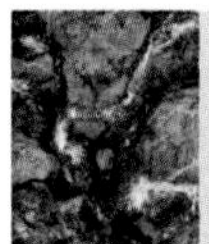

TABLE 4.2
Type and number of teeth in reptiles, fish, and mammals

	Incisors (I)	Canines (C)	Premolars (P)	Molars (M)	Total
Reptile					
Upper	4	1	3	5	
Lower	4	1	3	5	
Total	8	2	6	10	26 × 2 = 52
Mammals					
Upper	3	1	4	3	
Lower	3	1	4	3	
Total	6	2	8	6	22 × 2 = 44
Humans					
Upper	2	1	2	3	
Lower	2	1	2	3	
Total	4	2	4	6	16 × 2 = 32

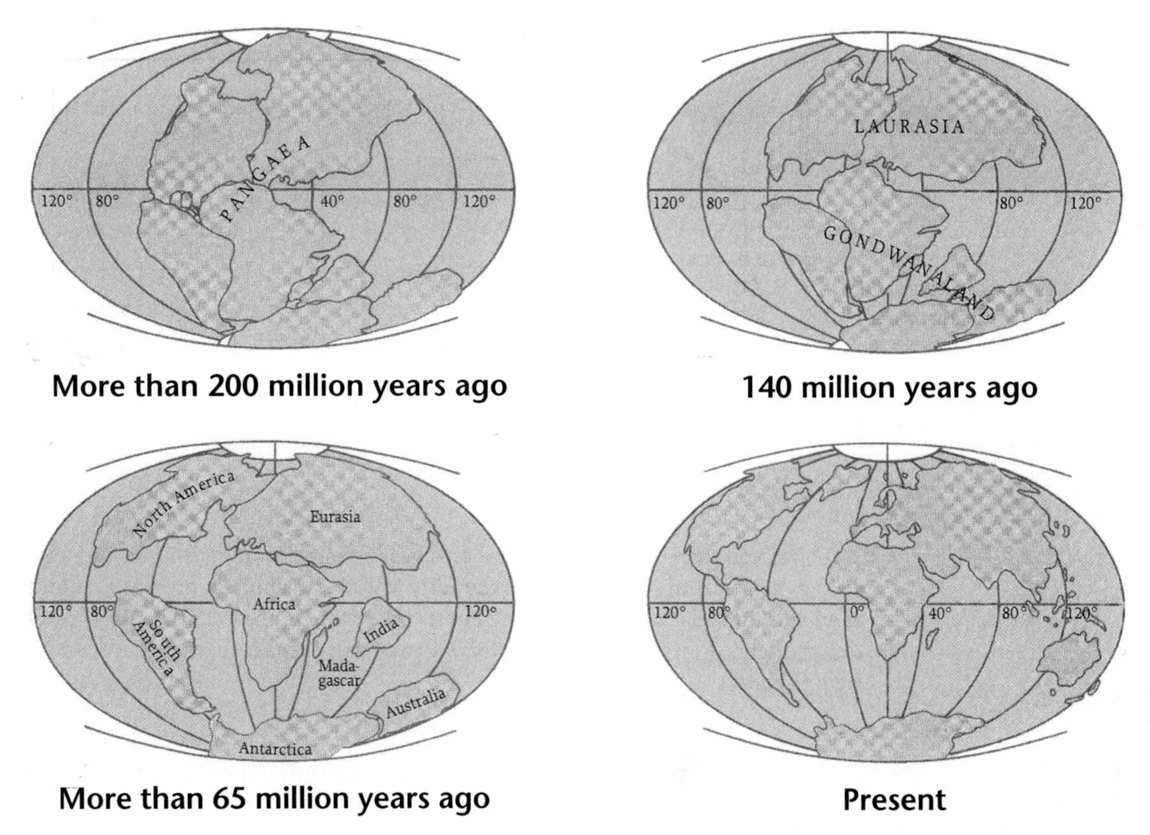

FIGURE 4.8. Pangea and continental drift. About 225 to 200 million years ago the earth had only a single landmass but, thereafter, the movement of plate tectonics caused the landmass to split into various continents. The animals that once were together became differentiated as the continents separated from each other. (Composite figure made by the author.)

years ago), the continents were beginning to assume their current positions. This is why, for example, we find fossils of the same type of dinosaur in what are now such widely separated places as China, North America, South America, and Antarctica.

The evolutionary implications of this long-term continental drift on the distribution and diversity of mammals were profound. It is estimated that as the continents were drifting together to form the supercontinent of Pangea, over 95 percent of all species of marine and terrestrial organisms suddenly became extinct. In general, groups of land animals became effectively isolated from each other by large water boundaries, and the distribution of reptiles and mammals was significantly influenced by continental movements, which, in turn, determined the fate and direction of evolution of the animals that lived in a given ecological niche. For example, before the continental split of Gondwanaland, the primates that lived in Africa, South America, Australia, India, and Antarctica had similar evolutionary histories, but after the split, the New World and Old World became completely separated and the primates that lived in each followed different evolutionary paths.

Major Mammalian Groups

A unique example of the effect of continental drift was the development of the islands of Australia and Madagascar and the continents of North and South America, which in turn contributed to the development of three major subgroups of living mammals. Based on their reproductive systems, mammals are classified into three basic grades: the egg-laying mammals, or monotremes; marsupials, or pouched mammals; and the placental mammals.

Monotremes

The monotremes lay eggs as birds do. These primitive mammals are characterized by having only one aperture for the urinary, genital, and intestinal canals. Although these animals used to be much more widespread, today there are only a few representatives of this suborder. These are the spine anteater and the duck-billed platypus of Australia (**fig. 4.9**). They are considered more "primitive" because, although they possess most of the other characteristics of mammals, they lay external eggs from which their young hatch.

FIGURE 4.9. Monotreme: A duck-billed platypus of Australia. From *BSCS Biology: An Ecological Approach, 9th Edition* by BSCS. Copyright © 2002 by BSCS. Reprinted by permission.

Marsupials

Marsupials (**fig. 4.10**) include a variety of mammals that survived in Australia and Southeast Asia (kangaroos, rabbit bandicoot, flying squirrel, wombat, marsupial mouse, and even the Tasmanian wolf) and, South and North America (opossums). They are considered in some ways to be intermediate in their biological organization and complexity between monotremes and placental mammals. Marsupials are animals that give birth to live young, born right at the end of the embryonic period. The prenatal period of mammals is usually classified into the embryonic and fetal periods. During the embryonic period, basic structures and organs are formed, and this period is referred to as the period of organogenesis. The fetal period is concerned with growth of preexisting cells and, as such, is referred to as the period of growth. In marsupials, the young are born right at the end of the embryonic period and must continue the fetal period, during which growth and maturation is completed, in an external pouch. Australia was isolated from South America before the great placental mammalian radiation of the Cenozoic (only a few were able to invade Australia via island hopping from Southeast Asia). Some of the marsupials resembled a wolf (Tasmanian wolf), a mouse (Marsupial mouse), a cat (flying squirrel), and a groundhog (wombat). The marsupials survived because they were free from the competition of the more efficient placental mammals. Furthermore, even though the reproductive pattern of giving birth to an immature offspring is an inefficient process, marsupials compensated by producing a very nutrient-rich milk that increased the survival of the offspring. In addition, the fact that marsupials can producc offspring in several stages of development that are kept in the maternal pouch ready for delivery when needed increased their reproductive efficiency.

Placental Mammals

Placental mammals represent the most derived trait of mammals. In placental mammals, the young are born after they have completed both the embryonic and fetal period (**fig. 4.11**). The process whereby the offspring is fed and nurtured is made possible by the placenta that connects the circulation of the mother with that of the offspring. It acts as a kind of filter that selectively lets nutrients reach the fetus and at the same time screens out toxins. The placenta nurtures the gestating organism in an efficient way, allowing the completion of both the embryonic and the fetal period and a more complete development of the central nervous system.

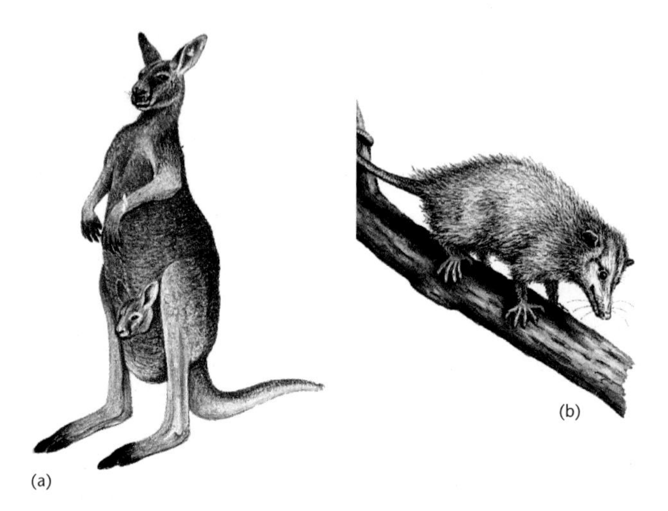

FIGURE 4.10. Marsupials. (a) Great red kangaroo of Australia. (b) Opossum of North and South America. From *BSCS Biology: An Ecological Approach, 9th Edition* by BSCS. Copyright © 2002 by BSCS. Reprinted by permission.

FIGURE 4.11. The placenta is the important trait that enhances survival that contributed to the success of mammals. (Composite illustration made by the author.)

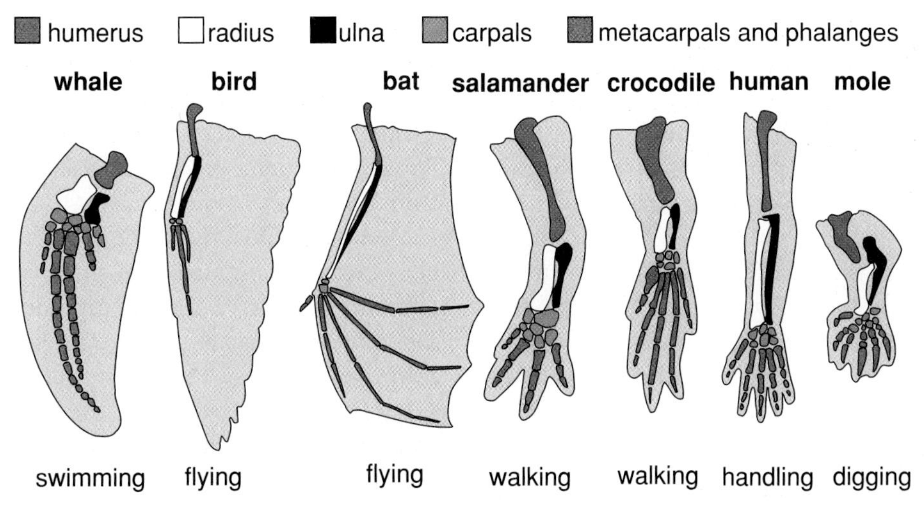

FIGURE 4.12. Homologous structures such as those of the primitive fish (Eusthenopteron), primitive amphibian (Trematops), gorilla, and human indicate a common ancestor. From *BSCS Biology: An Ecological Approach, 9th Edition* by BSCS. Copyright © 2002 by BSCS. Reprinted by permission.

Criteria for Tracing the Path of Evolution

Homologous and Analogous

Tracing the path of evolution requires classification of organisms into categories and groups based on physical similarities or differences, usually following the taxonomic procedure devised by Linnaeus. The criteria used in taxonomy is based upon basic concepts of homology and analogy. **Homology** indicates that organisms with similar anatomical structures must have been derived from a common ancestor, even if the anatomical structures have different functions. Homology can be tested by tracing origins, both through the fossil record and through the embryological history of structures in modern species. Thus, humans' front limbs are homologous to the front wings of modern birds. Their underlying structures are parallel, although their current appearance and function are quite different (**fig. 4.12**). Homologous limbs reflect different modifications of the ancestral limb plan that best fit different needs. For example, the horse hoof and the human fingernail have a similar structure but they have a different function and, as such, must be derived from structures found in the common ancestor.

On the other hand, **analogous** structures have similar functions but are without genetic affinity and are derived from different ancestors. For example, the wing of a butterfly and the wing of a bird perform a similar function but are without genetic affinity (**fig. 4.13**). Their similar functions evolved as a result of similar adaptive responses to similar environmental pressures rather than from inheritance from a common ancestor. Thus, the bird and butterfly wings evolved independently in response to similar environmental pressures.

Another approach for determining whether organisms share a common ancestry is derived from the study of embryological sequence. According to Haeckel,[17] ontogeny recapitulates phylogeny so that the embryonic stages of an individual's development

FIGURE 4.13. The wing of a bird (a) and the wing of a butterfly (b) perform a similar function. These features are considered **analogous** because they were acquired through similar function but not common ancestors. From *BSCS Biology: An Ecological Approach, 9th Edition* by BSCS. Copyright © 2002 by BSCS. Reprinted by permission.

(a) mourning dove

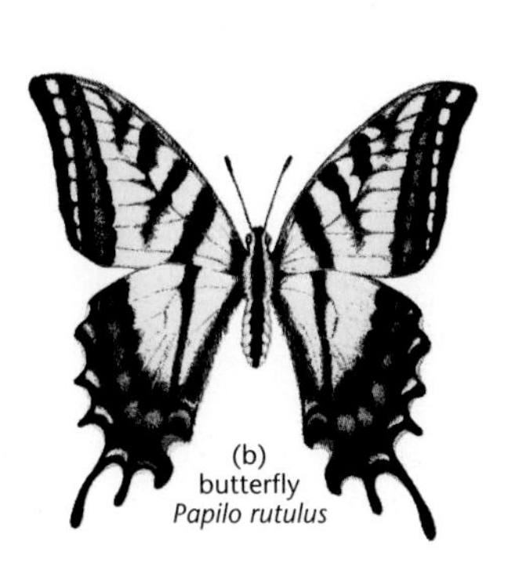

(b) butterfly *Papilo rutulus*

reflect the similar evolutionary steps through which the species passed. For example, during an early phase in embryological development, dogs, pigs, and humans resemble a fish in form and shape, which indicates that their common ancestry goes back to the fish. According to this theory, new species start out with the developmental sequence that evolved in their ancestors and use this as the basis from which their novel adaptations develop. Therefore, the study of embryological sequence has been used as a time view into the past. In this manner, by studying embryology, comparative anatomy, and paleontology, researchers have pieced together the major stages of morphological evolution.

Modes of Evolution

The process causing two living species that do not share a recent ancestor to have similar characteristics is referred to as **convergent evolution.** In convergent evolution, different groups have developed the same adaptations because they live in the same type of environment. For example, an African euphorbia has a similar body form to the Mexican cactus, but their flower structures are very different because they come from different families (**fig. 4.14**).

On the other hand, the acquisition of different characteristics among organisms that have a common ancestor is referred to as **divergent evolution.** Their different structures and functions evolved as a result of adaptive responses to different environmental pressures. In divergent evolution, the organisms became so different that they can no longer interbreed. However, even though they come from the same ancestor, they still can evolve in the same direction. When this occurs, it is referred to as **parallel evolution.** In **parallel evolution**, related organisms that live in similar environments can acquire similar characteristics but still become different species (**fig. 4.15**). For example, the African ostrich and the South American rhea are de-

FIGURE 4.14. Example of convergent evolution. The Mexican cactus (top) shows similar spines and does not have photosynthetic leaves as the African euphorbia (bottom), but differs in the structure of the flowers. (a) ©John D. Cunningham/Visuals Unlimited. (b) Courtesy of Carlye Calvin. (c) Courtesy of Carlye Calvin. (d) Courtesy of Gordon Uno.

(a)

(b)

FIGURE 4.15. Parallel evolution of (a) African ostrich and (b) South American rhea. They are different species that evolved similar adaptive traits in response to a similar environment. © Walt Anderson/Visuals Unlimited.

scended from a common ancestor and have become different species. The two species were separated when Pangea broke into a number of continents (including South America and Africa) but they evolved in similar (parallel) ways because they were subjected to the same environmental conditions on both continents.

Rates of Evolution

The development of species, as shown by the fossil record, was a gradual and slow process that changed very little for millions of years. This form of evolution, following Darwin, is referred to as **gradualism** (**fig. 4.16a**). In gradual evolution, new species arise by the transformation of an ancestral population into its modified descendants. However, occasionally after long periods of stability, or no change in the form of the species that can be seen in the fossil record, new organisms may appear without continuity. The process of alternating periods of stability and change is called **punctuated equilibrium** (**fig. 4.16b**). In punctuated equilibrium, a species remains unchanged for long stretches of time and then is quickly replaced by a new species.[14] (Interpretation about whether evolution occurred gradually or was punctuated by periods of accelerated and slow static changes requires defining the group to which the organism belongs.)

Genus and Species

In general, species are defined as members of a population that can reproduce and produce viable and fertile offspring. A genus is a group of species composed of members more closely related to each other than they are to species from any other genus. Moreover, members of a genus often share similar environments, patterns of adaptation, and physical structures. Differences among individual members of the same species are called **variations.** Variations may be small or major. For example, humans do not all look alike and, in fact, except for identical twins, are quite different from each other.

The process of forming new species is called **speciation**. Speciation usually occurs through **reproductive isolation** brought about by differences in behavior or geographic differences preventing one species from mating with another. For example, Alaskan brown bears that live in the forest and polar bears that live on snow fields and ice floes have become separate species because they never encounter one another in nature. However, in zoos, Alaskan brown bears and polar bears have mated successfully, but their offspring are sterile. Speciation can also occur through **adaptive radiation**. Adaptive radiation is when a population disperses into new areas, where it then becomes adapted

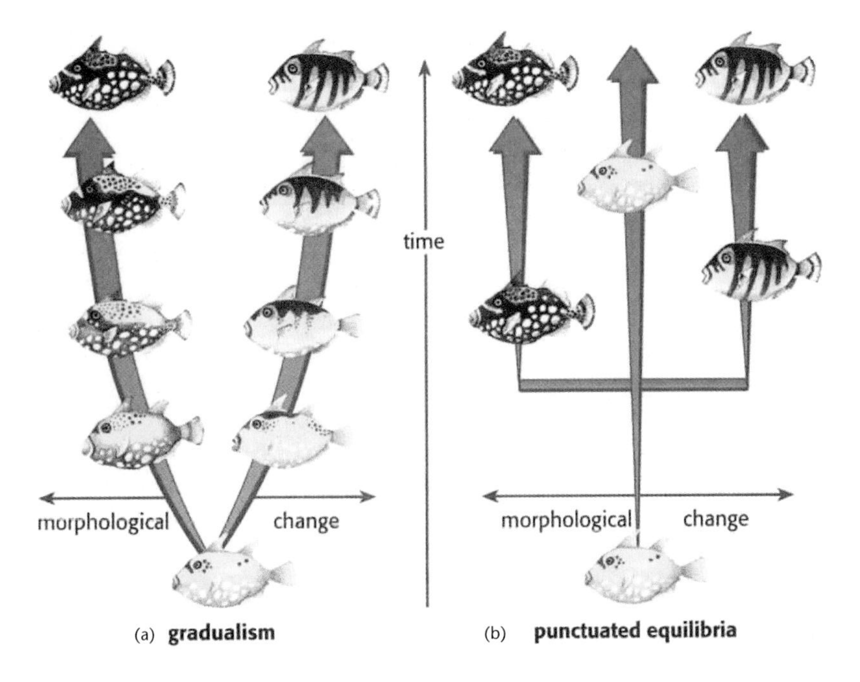

FIGURE 4.16. Speciation through (a) gradualism and (b) punctuated equilibrium. In gradualism, the species descended from a common ancestor, slowly becoming unalike as they acquired new adaptations. In punctuated equilibrium, the new species changes drastically when it branches off of the parent species and changes little afterwards. Source: From *BSCS Biology: An Ecological Approach, 9th Edition* by BSCS. Copyright © 2002 by BSCS. Reprinted by permission.

to the particular environment. In such conditions, the offspring population moves into new areas that offer opportunities and different challenges that require specific adaptations. In time, these adaptations make the populations so genetically different that they do not interbreed even when they live in the same area, and thus they become separate species. For example, the nine species of finches in the Galápagos Islands are descendants of one single species that underwent adaptive radiation.

Primitive and Derived Traits

Structures that are shared by all or most related species are referred to as primitive because they persist fairly unchanged. For example, the similar number of bones in the forelimb of frogs and lizards was inherited from a common ancestor and, as such, is referred to as a primitive, or shared, trait. On the other hand, the characteristics that are unique to a species and are changed from the ancestral form are termed derived traits. These traits are not shared with any other species because they are relatively new modifications.

Derived traits can be either generalized or specialized. A generalized trait is a structure that is adapted for many functions, like the human hand is generalized in that we have five fairly flexible digits adapted for many possible functions. On the other hand, a specialized structure is a trait that is limited to a narrow set of ecological functions. For example, the human foot is specialized in that it has been structurally modified to deal with the rigors of an upright posture and bipedal movement.

Summary

The evolution of human beings, just as of any other species that has existed on earth, ultimately began with the earliest appearance of life. As inferred from measurements of light reaching the earth from the stars, the galaxies and our solar system were formed about 12 billion years ago as a result of an explosion of stupendous proportions referred to as the big bang. Then, the earth was formed about 4.5 billion years ago. Eventually, the basic components of all of earth's organisms such as carbon, nitrogen, and oxygen appeared 4 billion years after the big bang. Thereafter, the course of life went very fast so that the first single-celled organisms appeared about 700 million years ago and the first multicellular organisms appeared between 570 and 510 million years ago. With multiple sources of variability, evolutionary change started to gain speed so that about 425 million years ago, during the Paleozoic, several varieties of fish, amphibians, and reptiles appeared. During the Paleozoic, the reptiles, such as the dinosaurs, dominated the land until the mammals displaced them 65 million years ago. Following the extinction of the dinosaurs, some of the fish returned to the sea and redeveloped adaptations for swimming while others found new opportunities for diversification into the ecological niches left vacant by the dinosaurs and differentiated into various herbivorous and carnivorous terrestrial forms. Mammals, along with birds, replaced reptiles as the dominant terrestrial vertebrates. Their evolution coincided with major changes in climate so that the initial habitat for all mammals at the start of their radiation was a type of "tropical" forest.

The success of mammals is related to their versatility, which was made possible by the evolution of an advanced central nervous system, heterodontic teeth, and improvement in reproductive efficiency and regulation of body temperature. The development of the placenta allowed mammals to become more reproductively efficient than other organisms. These general traits enabled mammals to adapt better than the specialized dinosaurs to the diverse and fluctuating environments. The mammalian adaptive radiation was very fast, so that by the end of a 10 million-year period, more than 4,000 species, 1,000 genera, 135 families, 18 orders, and 2 subclasses of mammals had appeared.

Suggested Readings

Bowen, G. J., W. C. Clyde, P. L. Koch, S. Ting, J. Alroy, T. Tsubamoto, T. Wang, and Y. Wang. 2002. Mammalian dispersal at the Paleocene/Eocene boundary. *Science* 295:2062–2065.

Carroll, R. L. 1988. *Vertebrate Paleontology and Evolution.* New York: Freeman and Company.

Macdonald, D., ed. 1985. *The Encyclopedia of Mammals.* New York: Facts of Life, Inc.

Literature Cited

1. Orgel, L. E. 2003. The origin of life—how long did it take? *Orig. Life Evol. Biosph.* 28:91–96.
2. Orgel, L. 2000. Origin of life. A simpler nucleic acid. *Science* 290:1306–1307.
3. Atsatt, P. 1988. Are vascular plants "inside-out" lichens? *Ecology* 69:17–23.
4. Awramik, S. M., and J. P. Vanyo. 1986. Heliotropism in modern stromatolites. *Science* 231:1279–1281.
5. Banin, A., and J. Navrot. 1975. Origin of life: Clues from relations between chemical compositions of living organisms and natural environments. *Science* 189:550–551.
6. Brock, T. D. 1985. Life at high temperatures. *Science* 230:132–138.
7. Cairns-Smith, A. G. 1985. *Seven Clues to the Origin of Life.* Cambridge, England: Cambridge University Press.
8. Cohen, Y., E. Padan, and M. Shilo. 1975. Facultative anoxygenic photosynthesis in the cyanobacterium, *Oscillatoria lininetica. J. Bacteriol.* 123:855–861.
9. Fox, S. W., and K. Dose. 1972. *Molecular Evolution and the Origin of Life.* San Francisco: W. H. Freeman.
10. Loomis, W. F. 1988. *Four Billion Years: An Essay on the Evolution of Genes and Organisms.* Sunderland, MA: Sinauer.
11. Margulis, L., and R. Fester., ed. 2000. *Symbiosis as a Source of Evolutionary Innovation: Speciation and Morphogenesis.* Cambridge, MA: MIT Press.
12. Oparin, A. I. 1976. Evolution of the concepts of the origin of life, 1924–1974. *Origins of Life* 7:3–8.
13. Chela-Flores, J., A. G. Lemarchand, and J. Oró., ed. 1999. Astrobiology: Origins from the big-bang to civilisation: proceedings of the Iberoamerican School of Astrobiology, Caracas, Venezuela.
14. Gould, S. J. 2002. *I Have Landed: The End of a Beginning in Natural History.* New York: Harmony Books.
15. Macdonald, D. ed. 1985. *The Encyclopedia of Mammals.* New York: Facts of Life, Inc.
16. Carroll, R. L. 1988. *Vertebrate Paleontology and Evolution.* New York: Freeman and Company.
17. Haeckel, E. H. 1897. *The Evolution of Man.* New York: D. Appleton and Company.

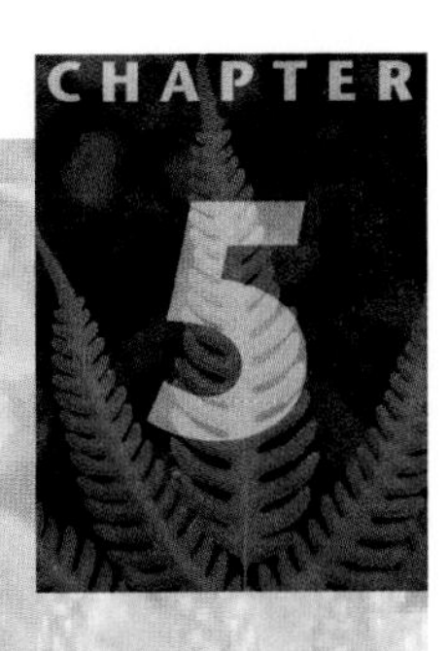

Living Primates

Humans share a common evolutionary history with other primates and, as a result, resemble other primates in many aspects of morphology, physiology, and development. Primate heritage is evident in most aspects of human anatomy, ranging from dependence on vision and decreased reliance on olfaction, to flexible limbs and grasping hands, to an extended period of juvenile development, to larger brains in relation to body size. Therefore, the study of primate biological features provides useful models for unraveling the traits that distinguish humans from nonhuman primates and for what makes us distinctively human. This chapter will focus on the unique features of primates.

Primate Taxonomy

In 1758, the anatomical similarities among monkeys, apes, and humans led the Swedish naturalist Carolus Linnaeus to place humans in the order Primates, the genus *Homo,* and the species *sapiens.* An order is divided into suborders and infraorders, which in turn, are divided into superfamily, family, genus, and finally, species. Members of the same species are defined as those animals that can interbreed and produce fertile offspring. A genus is defined as a group of species that share common morphological and behavioral features. The members of a genus are more closely related to one another than they are to species from any other genus. Traditionally, the order Primates is classified into two suborders: Prosimii and Anthropoidea[1–3] (**table 5.1**).

Prosimian

The Prosimians, or proto-monkeys, are the most ancient primates, whose descendants possess a mixture of primitive traits as well as derived or advanced features. The Prosimians include twenty-three genera found in the Old World (**table 5.1**). The Prosimians traditionally include three superfamilies: Lemuroidea, Lorisoidea, and Tarsioidea (**figs. 5.1 to 5.3**). Each of these superfamilies includes a diversity of species.

Lemurs

The superfamily Lemuroidea includes three families: Daubentoniidae, Indriidae, and Lemuridae. The Daubentoniidae is represented by one single species,

FIGURE 5.1. Prosimians: Ring-tailed lemurs. (Photo courtesy of Professor John Mitani, University of Michigan.)

the Aye-aye. The Indriidae includes several species including the Indri, sufaka, and woolly lemur. The Lemuridae is the most diverse of the prosimian families and is represented by several species ranging in size from 100 grams to 10 kilograms (such as the ruffed lemur, bamboo lemur, ring-tailed lemur (**fig. 5.1**), and other different varieties of lemurs). All lemurs are found on the African island of Madagascar.

Lemurs are arboreal but spend considerable time on the ground. Their mode of locomotion corresponds to "vertical clingers and leapers," meaning they move through the trees both by running along branches and by leaping from vertical stems or trunks. Some lemurs are frugivorous (a high proportion of their diet comes from fruits), others are folivores, specialized to feed on the leaves of bamboo, but all include some insects in their diets.

The color of the fur of lemurs is quite diverse. Some are brightly colored with contrasting patches of black, white, brown, or gray. Lemurs have short ears with tufts of fur on the tips. They have long, heavily furred tails, and some have circular stripes, such as the ring-tailed lemurs. Lemurs possess a special set of flattened comblike lower incisors that are used for extracting gum from trees and for grooming. They have 2I, 1C, 3PM, and 3M for a total of 36 teeth. The thumbs are enlarged but are opposable. The second digits of the hind feet of most lemurs have enlarged claws rather than nails, which are used in grooming. Lemurs use scents derived from urine and arm and anal glands to mark their territories and each other in a manner not seen in many other primates. Lemurs have relatively small eyes. The retina of the eyes is covered by a thin membrane called **tapetum** that reflects light and enhances night vision. Some lemurs have limited color vision. The part of the brain concerned with olfaction is relatively larger than in other primates and their rostrum is also relatively long.

Loris

The superfamily Lorisoidea includes the species called the bushbaby, galago, potto, and angwatibo, found in Africa, and the slow loris and slender loris also found in Asia. Lorises and pottos are small (85 g–1.5 kg), arboreal primates that are insectivores and frugivores. The mode of locomotion of the loris is usually slow and deliberate, but they are capable of moving rapidly if necessary. Their tails are very short and even absent in some species. The fur is thick and woolly and ranges in color from yellow to brown. Their eyes are large and directed forward (**fig. 5.2**). Their dental formula is like that of the lemurs (2I, 1C, 3PM, 3M) for a total of 36 teeth; they do have a dental comb. They have grasping

FIGURE 5.2. Prosimians: *Loris tardigradus* from Southeast Asia. Source: © Kevin Schafer/Corbis.

hands and feet. The forelimbs and hind limbs are approximately equal in length. Like the lemurs, the second digits of the hind foot have claws rather than nails.

Tarsiers

The superfamily Tarsioidea contains one genus and five species. Tarsiers are found in the islands of southeastern Asia, including Borneo and Sumatra, and some Philippine islands. Males and females are the same size and weigh 80 to 150 g. Their fur, silky and buff, ranges from light brown to grayish brown, or dark brown on the back and grayish or buff on the underside, in color. They have very distinct, forward-looking, large eyes adapted to nocturnal life (**fig. 5.3**). Their heads are round with ears that are hairless and seemingly membranous in texture. They have hairs on their nose pads. Their forelimbs are short and their hind limbs are elongated. They are unique among mammals in that the elongation of their hind limbs is the result of lengthening of the tarsals (especially the calcaneum and navicular) rather than the metatarsals. The digits are long and tipped with soft, rounded toe pads. Their second and third hind toes have claws rather than nails used for extracting insects and grooming. The tail has hair only on the tip. Tarsiers have only 34 teeth rather than the 36 teeth other prosimians have.

The dental formula for their upper teeth is 2I, 1C, 3 PM, 3M and for their lower teeth is 1I, 2C, 3PM, 3M, for a total of 34 teeth. They are nocturnal and are highly arboreal. They are insectivores that supplement with small vertebrates, which they capture by leaping on their prey and grabbing it with their hands.

In summary, some prosimians retained many traits of their nonprimate ancestors. For example, some have a moist, fleshy pad (rhinarium) at the end of their nose and a relatively long snout, both of which are functions of their pronounced reliance on olfaction (sense of smell). Some of the prosimians, such as the Tarsiers, have a claw on the second toe and a dental specialization characterized by forward-curved incisors and canines known as the "dental comb." Most are nocturnal and all have a layer on the retina of the eye called the tapetum that serves to reflect light. In some prosimians (except in the tarsiers), the eyes are somewhat more laterally placed than in anthropoids and hence their vision tends to be less than binocular and stereoscopic. Similarly, their brain size and manipulative capabilities are less well developed than in the other primates. Prosimians have 3 premolars rather than 2, giving them a dental formula of 2I, 1C, 3PM, 3M for a total of 36 (except the Tarsiers) instead of 34 teeth. They rely heavily on olfactory clues reflected in their large nasal cavities and scent glands. Also, the prosimians have nails and claws and their fingers are less dextrous than in the other primates. Therefore, as a group, the prosimians retain more an-

FIGURE 5.3. Prosimian: Tarsier. Tarsiers inhabit several islands in the western Pacific. Source: © Kevin Schafer/Corbis.

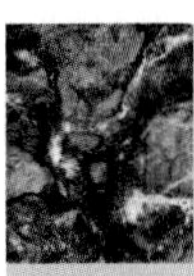

TABLE 5.1
General Taxonomy of the Order of Primates

Suborder	Infra-Order	Superfamily	Family	Common Term
PROSIMII				
	Lemuriformes			Lemur
	Lorisiformes			Loris
	Tarsiiformes			Tarsier
ANTHROPOIDEA				
	Platyrrhini	Ceboidea	Callitrichidae Cebidae	New World Monkeys
	Catarrhini	Cercopithecoidea	Cercopithecoidae	Old World Monkeys
		Hominoidea	Hylobatidae	Gibbons
		Hominoidea		Siamangs
		Hominoidea	Pongidae	Orangutans
		Hominoidea		Gorillas
		Hominoidea		Chimpanzees
		Hominoidea		Bonobos
		Hominoidea	Hominidae	Humans

cestral traits than the other primates. However, because the prosimians have prehensile hands and opposable thumbs, along with the presence of nails rather than claws on most of their fingers, they are closer to the primates than to other groups of mammals.

Anthropoid

The Anthropoidea is a suborder that includes all monkeys, apes, and humans. The anthropoids are more closely related to one another than to the prosimians. The anthropoids are subdivided into two distinct groups: platyrrhines, or New World monkeys, and catarrhines, which include the Old World monkeys, apes, and humans. These names refer to the shape of their noses. The New World monkeys have nostrils that are wide open and far apart, while in the Old World monkeys (also referred to as Cercopithecoid) the nostrils are narrow and close together. These groups also differ in the use of the tail. In the New World monkeys, the tail is prehensile and used for grasping, while in the Old World monkeys, the tail is used for balance. The New World monkeys are arboreal while the Old World monkeys are both arboreal and terrestrial.

New World Monkeys

New World Monkeys are found in the tropical forest region of the Americas. New World Monkeys have traditionally been divided into two families: Callitrichidae (marmosets and tamarins) (**fig. 5.4**) and Cebidae (all others) (**figs. 5.5** and **5.6**). The marmosets and tamarins (Callitrichidae) have retained some primitive ancestral features such as claws instead of nails and give birth to twins instead of one infant. On the other hand, most of the New World monkeys, such as spider monkeys, muriquis, howler monkeys and woolly monkeys, possess powerful prehensile tails that are used not only in locomotion, but also for suspension under branches while feeding on leaves and fruit. Most cebids are found either in groups of both sexes and all age categories, or in monogamous pairs with subadult offspring. Most of the New World monkeys have 3 premolars rather than 2, giving them a dental formula of 2I, 1C, 3PM, 3M for a total of 36 instead of 32 teeth. Most New World monkeys have relatively large brains, but among them there is large variation in brain size that is associated with their dietary preferences. As discussed in chapter 8, the howler monkeys, whose diets are mostly folivorous, have smaller relative brain sizes than the capuchin, squirrel monkeys, and spider monkeys, whose diets are insectivorous or frugivorous.

(a) (b)

FIGURE 5.4. New World monkeys: (a) Goeldis marmoset and (b) golden tamarin from South America. Source: © Kevin Schafer/Corbis.

(a) (b)

FIGURE 5.5. New World monkeys. (a) Howler monkey and (b) squirrel monkey from South America. Source: © Theo Allofs/Corbis (b) © Kevin Schafer/Corbis.

Old World Monkeys

The Old World monkeys belong to the family called Cercopithecidae. Members of the Cercopithecidae family are found only in Africa, Asia, and Europe. Compared to the New World monkeys, the Old World monkeys (and apes as well) have narrow, downward-facing nostrils (referred to as catarrhini) that are close together. They are generally larger in body size than the New World monkeys and are more terrestrial. The Old World monkeys have traditionally been divided into two subfamilies: Colobinae and Cercopithecinae.

Colobinae: Africa and Asia

The **Colobinae** subfamily includes eight genera, two found in equatorial Africa and the other six found in India and Southeast Asia.

The African colobines include the black and white *colobus* and red *colobus* monkeys. They live in multifemale, multimale groups or in one-male, multifemale groups. They range in size from 4.5 to 12 kg. The red *colobus* monkeys include five species that have varying amounts of red fur. The black and white *colobus* monkeys include five species that are quite distinctive looking with long flowing capes of white hair and long fluffy tails accenting their black bodies. The infants are born all white with pink faces. The leaf-eating **colobine monkeys** (**fig. 5.7a**) have a unique large sacculated (many-chambered) stomach, which houses bacteria that is specialized to digest their highly folivorous diet (see chapter 7). Their teeth are also adapted to a leaf-eating diet by having sharp crests that shear the leaves into small pieces. The infants are born with a different fur color than the adults, which changes to the adult shade sometime between 3 and 6 months of age.

FIGURE 5.6. New World monkeys. Capuchin monkey. Source: © Kennan Ward/Corbis.

The Asian colobines include the hanuman langur monkeys and are classified in the genus *Presbytis.* The hanuman langur monkeys (**fig. 5.7b**) are found in India, Pakistan, Bangladesh, Sri Lanka, Burma, China, Malaysia, and the Sunda Islands. They include eleven species that live in a wide variety of habitats, ranging from the tropical rain forest to urban areas to high altitudes (10,000 feet). The hanuman langur is the most terrestrial of the colobines. The fur color ranges from white, gray, and black to orange-colored adult coats. Some, like the spectacled langur monkey, have rings of white skin around each eye and around the mouth, which stands out on their otherwise dark faces. Other species have tufted crests of hair on top of their heads and prominent tufts of hair that resemble sideburns along their cheeks. The douc langurs found in Vietnam and Cambodia are brilliantly marked, with patches of different colors and shades of blue or orange on their faces. Langurs live in one-male or multimale, multifemale social groups ranging from 6 to 40 individuals. The smallest species weighs 5 kg and the largest weighs 11 kg. There is very little sexual dimorphism. (The average difference in body size between males and females.)

There are eight species of *Presbytis* that are found in the Southeast Asian islands of Borneo, Sumatra, and the Mentawai Islands, as well as in Java and on the Malay Peninsula. These species range in size from 4 to 9 kg, with little to no sexual dimorphism. The proboscis monkey, from the genus *Nasalis,* also belongs to the Colobinae subfamily. The male of this species of monkey is characterized by an extremely large and pendulous nose used to make a honking vocalization. They are found on the Southeast Asian island of Borneo. They live in multimale, multifemale groups, and males form all-male bachelor bands. Proboscis are excellent swimmers, and they can swim underwater for up to 20 meters.

(a)

(b)

FIGURE 5.7. Old World monkeys: (a) *Colobus guereza* from Central Africa and (b) langur monkeys from India. Source: (a) © Kevin Fleming/Corbis (b) © Mary Ann McDonald/McDonald Wildlife Photography.

Cercopithecinae: Africa, Europe, and Asia

The **Cercopithecinae** is composed of nine genera. The most extensive genus of the group is *Macaca*. The macaques include nineteen species, which together form the most widely distributed nonhuman primate genus. They are found on Gibraltar, across North Africa, throughout the Near East, India, Japan, and throughout Southeast Asia (**fig. 5.8**). Adult macaques range in size from 3 to 16 kg, with some species showing more sexual dimorphism than others. Their diet is varied, including flowers, seeds, bird eggs, leaves, bark, and invertebrates. The age when they reach sexual maturity averages between 3 and 4 years. Most are found in multimale, multifemale social groups, with females forming matrilineal hierarchies.

The baboons belong to the genus *Papio* of the subfamily Cercopithecinae and include five species that are widely distributed throughout Africa. Baboons are large, stocky quadrupeds with a considerable amount of sexual dimorphism in both size and weight (**fig. 5.9**). Male body weight may be twice that of females. In addition, the males have very large canine teeth, which they use in threat displays. Baboon social groups are generally large, some containing up to 200 individuals. Baboon females have conspicuous pink sexual swellings across their rumps, clearly indicating when they are in estrus or fertile. Gestation lasts 6 to 7 months. Sexual maturity is reached between 3.5 and 6 years of age, later for males than for females.

The gelada baboons and hamadryas baboons are found in the mountains of Ethiopia. On the other hand, the *Mandrillus*, which includes the mandrill and the drill species, are forest dwellers found in Africa, Cameroon, Gabon, southeast Nigeria, and the Congo. Male mandrills have brilliantly colored blue and red faces, whereas the drill's face is black. Both species have brilliantly colored perineums. Like the baboons, their sexual dimorphism is quite great. The hamadryas baboon groups of Ethiopia are multimale and multifemale in composition. These hamadryas baboons gather at night in large sleeping groups.

In summary, the Old World monkeys show a great deal of diversity in terms of body size, habitat, and social organization. The colobines tend to live in small groups, with only one or a few adult males. In contrast, cercopithecines, such as baboons and most macaque species, are found in large social units comprised of several adults of both sexes and offspring of all ages. Among cercopithecines, sexual dimorphism is particularly pronounced. In some species, such as baboons and patas monkeys, male body weight may be twice that of females. The Old World monkeys are also characterized by the development of ischial callosities—hard sitting pads on the lower side of the buttocks. Old World monkeys have been adapted to both arboreal and terrestrial lifestyles. Although they are quadrupedal, these monkeys spend a good deal of time sleeping, feeding, and grooming while sitting with their upper bodies held erect.

FIGURE 5.8. Old World monkeys: Japanese macaque. (Photo courtesy of Professor John Mitani, University of Michigan.)

FIGURE 5.9. Old World monkeys: Baboons of East Africa. (Photo courtesy of Professor John Mitani, University of Michigan.)

(a)

(b)

FIGURE 5.10. Apes: Gibbons from Borneo. Source: (a) Photo courtesy of Professor John Mitani, University of Michigan; (b) © William Manning/Corbis.

Hominoids

The hominoids include the lesser apes such as the gibbons and siamangs, the greater apes that include the orangutans, gorilla, bonobos, and chimpanzees (**figs. 5.10** to **5.13**), and humans. Two unique features of all the hominoids are that they lack any kind of tail and have shortened trunks. The habitual mode of locomotion is expanded to include terrestrial quadrupedalism and brachiation. Arm hanging and hand-over-hand progression beneath the branches rather than quadrupedal running on top of branches becomes more common because of their larger bodies. Apes are also characterized by more complex behavior and larger brains than monkeys. They also take longer to mature than do monkeys, meaning the period of infant development and dependency is greater than that of monkeys. In terms of social behavior, they range from solitary monogamous groups, such as the gibbons and siamangs, to the gregarious chimpanzees that live in fission-fusion communities.

Gibbons

The gibbons and siamangs include nine species belonging to the genus *Hylobates*. They live in the southeastern tropical areas of Asia (**fig. 5.10**) including the islands of Java, Sumatra, and Borneo. They are also found in northern Thailand, Vietnam, Laos, and China. They live in very small, exclusive groups focused on a male-female pair who bond for extended periods. However, the bonds of monogamous pairs do not necessarily imply monogamous mating patterns. They are exclusively arboreal and their primary mode of locomotion is brachiation, whereby they swing from their arms below the branches of a tree. They have a high degree of mobility. They can leap, jumping bipedally over the tops of branches as they travel from one tree to another. Like all apes, gibbons have relatively long arms, upright body posture, and no tail. However, they differ from other apes in that they are fully arboreal and are not sexually dimorphic. Males and females are the same size, both in terms of weight and length of the canine teeth. Adults weigh only between 5 and 12 kilograms.

Orangutans

The orangutans currently are restricted to the forested areas of the Indonesian islands of Borneo and Sumatra. The word *orangutan* in Malaysian and Indonesian languages means "person of the forest." The orangutans are traditionally divided into two populations: *Pongo pygmaeus* from Borneo (**fig. 5.11**) and *Pongo pygmaeus abelii* from Sumatra. Both populations are quite similar behaviorally and morphologically. They are highly dimorphic in body size. Adult males in the wild weigh about 86.3 kg and females weigh 38.5 kg,[8] resulting in extensive sexual dimorphism. Adult males also have very large cheek "pads," which are composed of fibrous fatty tissue.[9] They have relatively long arms and legs and flexible joints that enable them to brachiate with all four limbs. When they brachiate, they use their body weight to bend and sway branches toward them and propel themselves across the trees. Orangutans are fully arboreal but sometimes descend to the ground.

The orangutans tend to be solitary and do not form stable pair bonds. The basic social organization

FIGURE 5.11. Apes: Orangutans from Borneo. © Royalty-free/Corbis.

form consists of immature offspring and their mothers, and isolated females, who engage in casual mating with males whose foraging areas overlap the ranges of a number of solitary females. Orangutans differ from baboons in the absence of estrus swellings and no visual indicators of ovulation. Their menstrual cycle is similar to humans and lasts about 28 days.[8,10] Females in captivity can reach sexual maturity as early as 6.5 years[11] and the transition from the onset of maturity to adulthood is set to occur between 7 and 10 years.[12]

Gorillas

The gorillas (*Gorilla gorilla*) include three populations: the mountain gorilla (*Gorilla gorilla berengei*), the eastern lowland gorilla (*Gorilla gorilla graueri*) and the western lowland gorilla (*Gorilla gorilla gorilla*). The best-known population is the mountain gorillas that live in the Virunga Volcano region along the borders of the Democratic Republic of the Congo (Zaire), Rwanda, and Uganda. They live at an altitude of 3,650 meters (12,045 feet). Other populations of gorillas live in the Bwindi National Park of Uganda at an altitude of 1,400 to 2,300 meters (4,620 to 7,590 feet). They are by far the largest of all primates. Adult males weigh as much as 350 pounds and adult females weigh between 200 and 250 pounds. They exhibit the greatest amount of sexual dimorphism among all primates, at about 30 to 40 percent. They are both arboreal and terrestrial. Their form of locomotion on the ground is knuckle walking (**fig. 5.12**).

Gorillas live in stable, harem-type social groups with a single male (**fig. 5.12**), multiple females, and young males. The dominant male (silver back) has exclusive relationships with a group of females, and excludes rival males. However, there is variability in the group composition of gorillas that live in the lowland areas, which is related to ecological factors (see chapter 6, Primate Behavioral Ecology).

Chimpanzees and Bonobos

The chimpanzee includes two species: the "common" chimpanzee (*Pan troglodytes*) and the bonobo (*Pan paniscus*), or pygmy chimpanzee. They are found in

FIGURE 5.12. Apes: Mountain male gorilla (silver back) and gorilla infants. (Photo courtesy of Professor John Mitani, University of Michigan.)

FIGURE 5.13. Ape: Chimpanzees grooming each other. (Photo courtesy of Professor John Mitani, University of Michigan.)

the remaining patches of forest in the equatorial area of Africa, stretching through a broad belt from the Atlantic Ocean in the west to Lake Tanganyika in the east (**fig. 5.13**).

Chimpanzees

Chimpanzees live in large, complex social groupings focused on group bonded males. Chimpanzee social grouping is characterized by the so-called fission-fusion pattern, in which group size and composition fluctuate but all are part of a larger community. Males cohabit and mate with a number of female group members. Chimpanzee social behavior is complex, and individuals form lifelong attachments with friends and relatives. Chimpanzees practice group hunting, food sharing, and rudimentary tool making.

Bonobos

The bonobos (*Pan paniscus*), or pygmy chimpanzees, are found only in a small area south of the Congo River in the Democratic Republic of the Congo. Like the chimpanzee, they exhibit very complex social organization and behavior. In addition, in bonobos, sexual activity is practiced in a more extensive and generalized form than in chimpanzees. Bonobos females practice genito-genetial rubbing, absent in chimpanzees. This homoerotic behavior is said to be a tension-reducing activity, along with heterosexual behavior.[13,14]

FIGURE 5.14. Apes: Bonobos infants from Central Africa. (Photo courtesy of Professor John Mitani, University of Michigan.)

Primates' Unique Structural and Behavioral Traits

The traits that distinguish primates from other mammals include four major categories of limbs, locomotion, dentition, and brain structure and function. The combination of each of these categories of traits has permitted the primates, throughout their evolutionary history, to successfully exploit the arboreal and terrestrial environments. This adaptation has been both reciprocal and symbiotic in that some features facilitated adaptation to the arboreal habit, while other features were a consequence of life out of the trees. For example, the visual system that characterizes primates is best suited for surviving in an arboreal habitat. Hence, selection pressures may have been responsible for the evolution of primate neural and sensory systems.

Grasping Hands and Feet

Most primates have a flexible, generalized limb structure, which permits them to engage in a number of locomotor behaviors. One important aspect of this generality is the ability to rotate the forearm, a feature that had been lost in some of the more specialized mammals. Similarly, primates have retained movement of the clavicle (collarbone), which had been lost in many other quadrupedal mammals. Primates have also retained various aspects of hip and shoulder morphology that provide them with the wide range of limb movement and function found throughout the Primate order. The primate limbs show many critical adaptations. Most primates have five digits in both the forelimbs and hind limbs while in some mammals have been reduced to one or two. Similarly, while other mammals have claws, primates (except in some prosimians) have nails with sensitive pads at the tips of their digits, giving them prehensile grasping ability (**fig. 5.15**). Likewise, primates have an opposable thumb and, in most species, a divergent and partially opposable big toe. This unique combination of pad and nail provides the animal with an excellent prehensile (grasping) device for use when moving from tree to tree. The flexible primate hand has made the utilization of tools possible and paved the way for many unique abilities, such as the manufacture of tools (**fig. 5.15**).

It has been hypothesized that the order in which primate traits appeared tells us about past environments. For example, it is assumed that the grasping extremities and nails on the digits evolved for eating fruit on terminal branches of angiosperms.[15] It is also assumed that the grasping extremities, nails on the digits, and leaping adaptations of primates were favored by natural selection and evolved together for grasp-leaping locomotion during climbing and landing after leaps.[16] In this context, grasping evolved first to exploit angiosperm food products in the terminal branches, and visual convergence evolved later for visually directed predation of insects.[17] As shown by the studies of Bloch and Boyer,[18] the grasping hands evolved in primates before visual specialization and leaping.

Locomotion and Body Morphology

Flexibility in limb structure allows many primates to employ more than one form of locomotion. Even though most primates are arboreal quadrupeds, their form of locomotion can be grouped into three forms of locomotion: vertical clinging and leaping, semibrachiation, and brachiation.[2,19] **Vertical clingers** and **leapers** support themselves vertically by grasping onto trunks of trees while their knees and ankles are tightly flexed (**figs. 5.16a and b**). This mode of locomotion is used primarily by prosimians. **Semibrachiation** involves a combination of leaping with some arm swinging and the use of a prehensile tail, which in effect serves as a marvelously effective grasping fifth "hand." This mode of locomotion is used by New World monkeys only (**fig. 5.16c**). In contrast, **brachiation** is locomotion where the body is alternatively supported under either forelimb. It is used by orangutans, gibbons, and siamangs of Southeast Asia (**fig. 5.16d**).

All primates, whether they are arboreal or terrestrial, have a tendency toward erectness while sitting, leaping, standing, and walking, and have a mode of quadrupedal locomotion (**fig. 5.16**). Furthermore, primates spend much more of their waking hours in a vertical, upright position. As a result of the tendency to be erect, the organs and skeletal structures of primates are organized perpendicular (**orthograde**) to the ground while in other mammals the body is organized parallel to the ground in a plantigrade manner. In addition, in

(a)

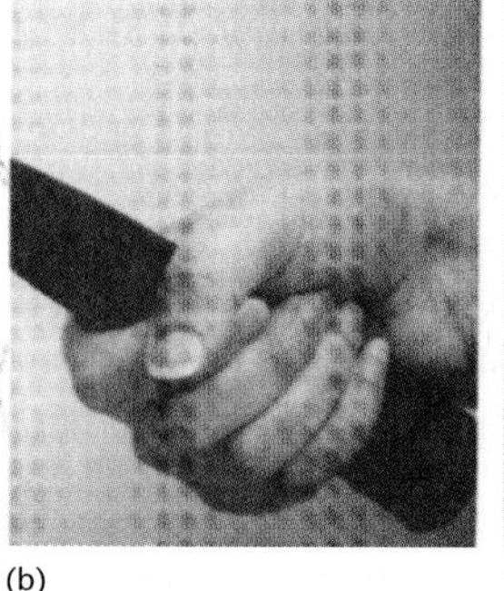

(b)

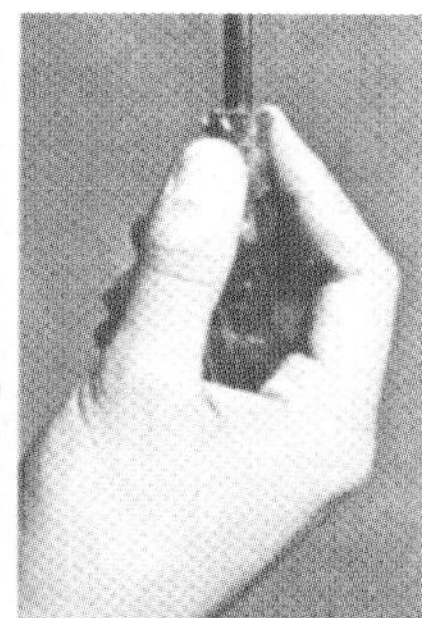

(c)

FIGURE 5.15. (a) Howler monkey. (b and c) Human showing flexible digits and an opposable thumb. (Composite illustration made by the author.)

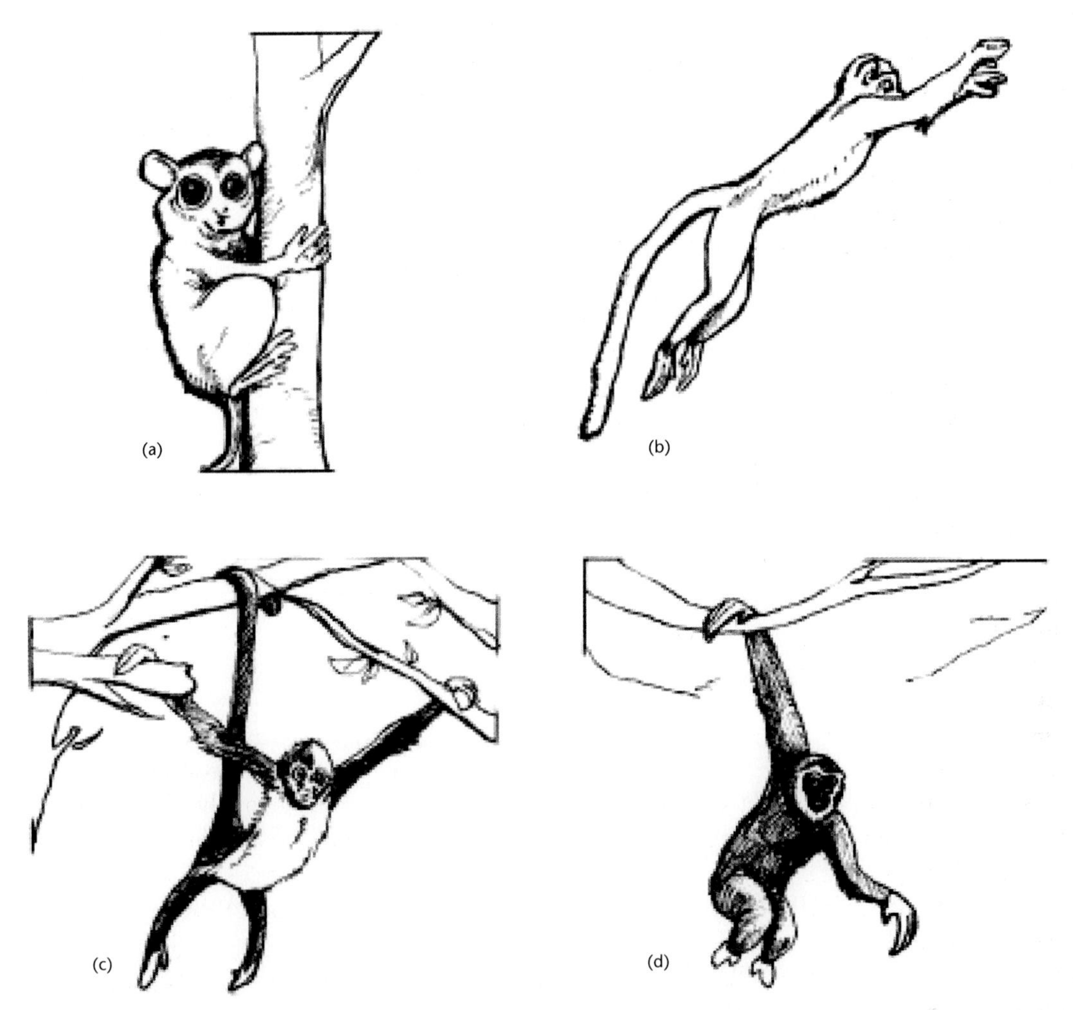

FIGURE 5.16. Forms of arboreal primate locomotion: (a) vertical, (b) leaper, (c) semibrachiation, and (d) brachiation. (Composite illustration made for the author by Paul Bohinsky, former graduate student in Biological Anthropology, University of Michigan.)

arboreal quadrupeds such as primates, the forelimbs are relatively shorter and may be only 70 to 80 percent as long as hind limbs. In contrast, the limbs of terrestrial quadrupeds are of equal length approximately, with forelimbs being 90 percent (or more) as long as hind limbs. Primates are also characterized by a relatively long and flexible lumbar spine, which enhances their ability to propel the animal forward.[19]

Generalized Dentition Adapted to Grinding and Chewing

Like the majority of other mammals, most primates have four kinds of teeth: incisors (I), canines (C) for biting and cutting, premolars (PM) and molars (M) for chewing and grinding food (**fig. 5.18**). However, in primates, there has been a general evolutionary trend toward reduction of the number of teeth when compared to other mammal groups.[2,20–22] For example, while in generalized placental mammals the dental formula is 3.1.4.3 (three incisors, one canine, four premolars, and three molars), in Old World monkeys the anthropoid dental formula is 2.1.2.3 and in some of New World monkeys, the dental formula is 2.1.3.3. (That is, they have two incisors, one canine, three premolars, and three molars.) In general, the size and shape of the teeth of primates is adapted more to chewing and grinding than to biting or tearing. For example, most primates possess premolars and molars that have low, rounded cusps, a molar morphology that enables them to process most types of foods, while carnivores typically have premolars and molars with high pointed cusps adapted for tearing meat.

Large and Complex Brain

One of the biological traits that contributed to the success of primates in exploiting the arboreal and terrestrial environments is the relatively large and complex brain.

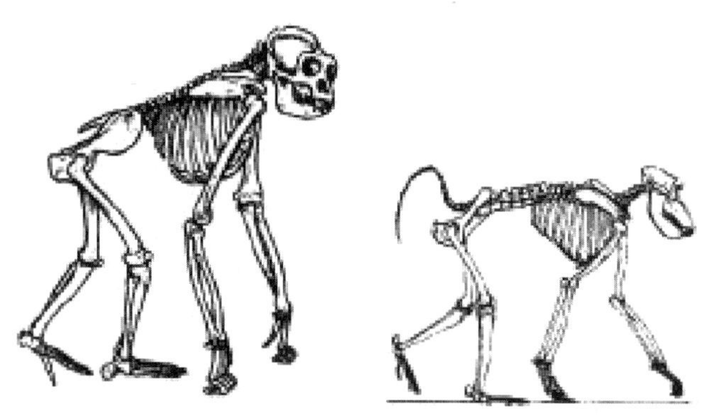

FIGURE 5.17. Quadrupedal locomotion. To an extent, all primates have a mode of quadrupedal locomotion. Source: Modified from Richmond, B. G., D. D. S. Begun, and D. S. Strait. 2001. Origin of human bipedalism: The knuckle-walking hypothesis revisited. *Yrbk. Phys. Anthropol.* 44:70–105.

Large and Expanded Neocortex of the Brain

Compared to other mammals, primates exhibit a large and complex brain.[24] Almost all parts of the brain, and especially the cerebrum and brainstem, are enlarged. This expansion is most evident in the neocortex and cerebrum that represent the visual and association areas of the neocortex (portions of the brain where information from different sensory modalities is integrated). Expansion in regions involved with the hands (both sensory and motor) is seen in many species, particularly in humans. As a result, the cerebrum in all primates and humans expanded so much that it came to comprise the majority of brain volume, and the convoluted neocortex provided more surface area and space for even more nerve cells (neurons) than would be possible otherwise. In humans, the cortex of the brain expanded in size by a factor of two and it is clear from the organization of the cortex that the area devoted to the hands is much larger than equivalent areas in monkeys and apes. As illustrated in **figure 5.19**, the primate brain (b and c) compared to that of a dog exhibit: (1) a drastic reduction of the olfactory bulbs; (2) expansion of the prefrontal cortex onto the lateral surface of the brain; (3) expansion of the visual and auditory areas of the temporal lobe; and (4) expansion of the occipital lobe.

Stereoscopic Vision

The visual system of primates is the most complex processing system in nature that has ever evolved. Nearly a third of the cerebral cortex in most primates is directly involved in visual processing. The primary visual cortex of primates is also unique among mammals because of its high density of cells and the complexity of its cellular architecture, making the primary visual cortex of primates unique. This unique feature allows for stereoscopic vision, or the ability to perceive objects in three dimensions. This characteristic is made possible by reorganization of the brain's visual

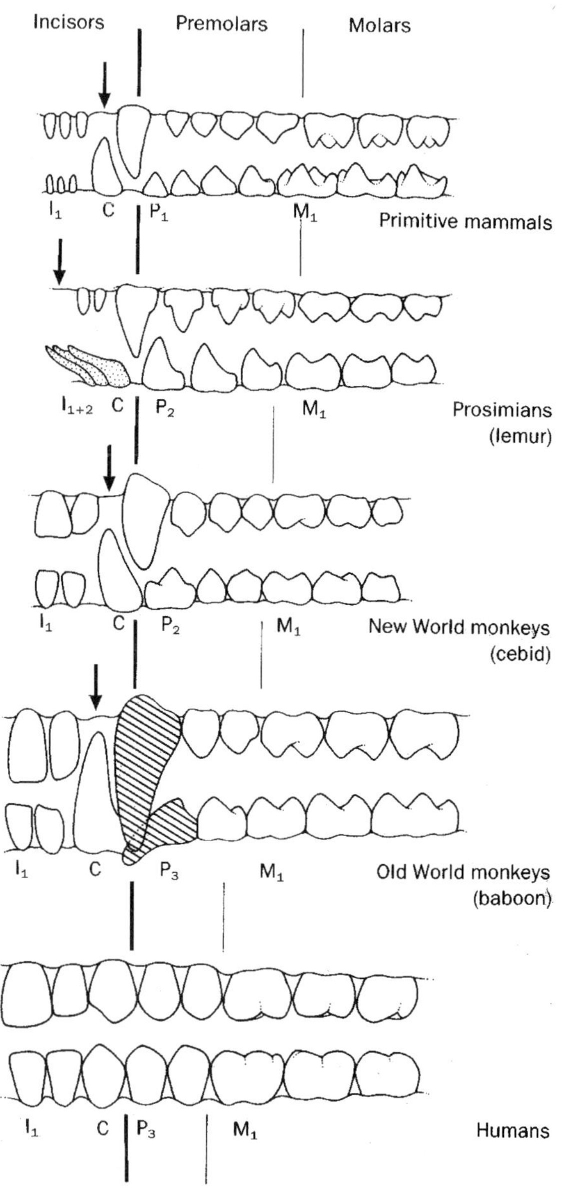

FIGURE 5.18. Evolution of primate teeth. Primates' teeth changed from numerous sharp teeth adapted to the soft diet of fruits or insects that characterized the prosimians, to fewer teeth albeit bigger teeth with larger surface area adapted to chewing and grinding the hard plant foods. The teeth came to form a larger part of the diets of anthropoids. Source: Jones, Martin, and Pilbeam, fig. 2.[23]

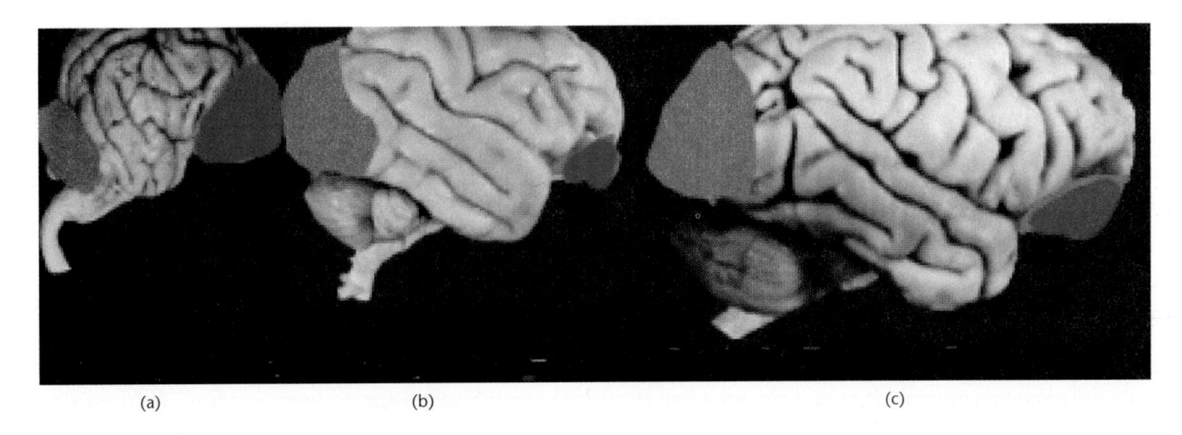

FIGURE 5.19. Comparison of brain of dog (a), spider monkey (b), and chimpanzee (c). In primates the brain is enlarged and the areas concerned with vision (occipital lobes) and touch (parietal lobes) are dramatically expanded, while in mammals such as dogs, the olfactory bulb is expanded. (Composite made by the author.)

structures so that the eyes are positioned toward the front of the face (not to the sides), creating overlapping visual fields, or binocular vision.

The primate brain is organized into two distinct but interconnected right and left hemispheres (**fig. 5.20**). Although the same functions are represented in both hemispheres, one hemisphere is dominant with regard to certain functions, so that certain functions appear to be more common in one hemisphere than in the other. In other words, the brain function is not symmetric. The capacity for stereoscopic vision is dependent on each hemisphere of the brain having received visual information from both eyes and from overlapping visual fields. In nonprimate mammals, most optic nerve fibers cross to the opposite hemisphere through a structure at the base of the brain. In contrast, in primates, about 40 percent of the fibers remain on the same side. The communication between the hemispheres is made primarily through the *corpus callosum* (a thick bundle of fibers). The way in which the two hemispheres are integrated is made possible by a system of contralateral control, in which the right hemisphere is "in charge" of the left side of the body and vice versa. Each hemisphere has association areas on the cortex that specialize in particular functions. For example, the left visual field is composed of visual input from both eyes, but information is processed more thoroughly on the right. The information from both fields of vision crosses at the optic chiasma, where the right and left optic nerves cross on the way to their respective control hemispheres.

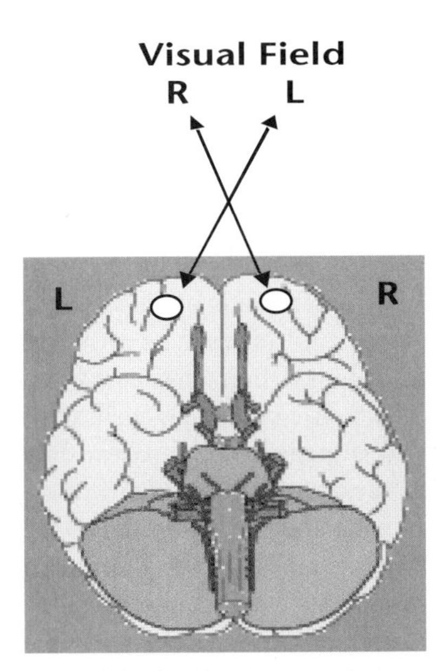

FIGURE 5.20. Stereoscopic vision characteristic of primates. Stereoscopic vision results from the visual overlap and message crossing from the right and left hemispheres. (Composite illustration made by the author.)

Night Vision

All diurnal primates, except some prossimians and nocturnal primates such as the nocturnal owl monkey of South America, have color vision, a feature that is rare in other mammals, but is common in birds, reptiles, and even fish. The evolution of color perception in primates required specialization of the retina and of the cerebral cortex. The primate secondary visual cortex has a distinct area specialized for calculating the relative color of objects. Color vision probably evolved at the same time that primates began moving towards the day, perhaps to accompany a shift to feeding on fruits that advertise ripeness by changing color. Furthermore, color vision may also have evolved as part of social communication. For example, some Old World monkeys (*Cercopithecus*) have bright blue underbellies and bright red genitals, indicating reproductive maturity, while male mandrills (*Papio* or *Mondrillus sphinx*) have bright red, white, and blue skin on their faces, indicating dominance.

Hearing

The areas of the cortex specialized for auditory perception in primates are no more or less specialized than those of other mammals. Many primates use the ears as sound amplifying and directional systems. Among nocturnal prosimians, both ears can be aimed towards a sound to amplify and locate it. In diurnal monkeys and apes, external ears are small and comparatively immobile.

Sense of Smell

Primates have a decreased reliance on the sense of smell (olfaction). This trend is seen in an overall reduction in the size of olfactory structures in the brain. Corresponding reduction of the entire olfactory apparatus has also resulted in decreased size of the snout. Apes, monkeys, and tarsiers have a smaller olfactory system and have lost the damp nasal skin (rhinarium) common to lemurs, lorises, and other mammals. Humans have the most reduced olfactory apparatus of all mammals and primates; only dolphins and whales have more reduced olfaction. (In some species, such as baboons, the large muzzle is not related to olfaction, but to the presence of large teeth, especially the canines).

Touch and Movement

In primates, the parts of the brain cortex and spinal tracts associated with touch and movement are particularly well developed. As shown by spatial maps of sensory surfaces and muscle systems in the cerebral cortex in primates, the hand occupies a uniquely large part of these maps. This corresponds to primate specialization of the hands for grasping and manipulation and the face for communicating by gestures. The relationship between the size of the cortical area and use appears to be evident even in other appendages such as the prehensile tail of some South American monkeys. The sensory and motor cortex is much larger than in the corresponding area in other monkeys that have tails but do not have prehensility. Similarly, the muscles of the mouth and throat are well represented in the primate motor cortex.

Communication

Primates use four main modes of communication: tactile, visual, olfactory, and vocal. These can be used singly or in combination, depending on the circumstances.

Tactile Communication

Tactile communication is the most common among primates and ranges from gentle bites on the neck to violent physical contacts. Most primates groom each other with their hands, teeth, and mouth and give their friends reassuring touches or hugs when they meet up with one another. Reassuring physical contact including copulation is said to be used to avoid aggression and initiate reconciliation interactions following conflicts between opponents.[5] When conflicts heighten and are unavoidable, primates do resort to violent physical attacks that can lead to wounds and infections.

Visual Communication

One of the distinguishing features of primates is their high visual acuity characterized by three-dimensional sight and color vision. Reflecting this high visual acuity, primates exhibit diverse visual cues that range from the color of the fur, skin, and the genitals (e.g., blue testicles of vervet monkeys,[5] the red swelling evident in many anthropoids) to facial expressions, yawns exposing teeth, eyelid flashes or lip slaps, throwing vegetation, chest beating, yawning, ground slapping, branch shaking, head bobbing and bouncing, strutting and scratching. These cues serve to communicate emotional signals such as fear, threat, greeting, danger, pain, hunger, courtship, and many other messages.

Olfactory Communication

Olfactory communication is practiced by all primates including humans and varies in its form of expression. Some primates mark their territory with scent and urine. This practice is extremely well defined among the ring-tailed langurs of Madagascar. Furthermore, squirrel monkeys, and owl monkeys coat their hand and tail with urine and then rub their soaked appendage against vegetation. Ring-tailed lemurs also practice ritualized stink-fights when competing for mates during breeding season. These cues provide olfactory or pheromonal cues about their ovulatory condition.

Vocal Communication

Vocal communication varies from species to species. It ranges from the loud sounds emitted by the howler monkeys to the "humming" of mountain gorillas. Field experimental studies of gibbons from Indonesia indicate that each species produces vocal sounds that are species specific. The specificity of vocal sounds enables gibbons to distinguish members of their own kind from others.[26] The selective advantage of having species-specific vocal sounds is related to the fact that it enables the identification of mates of the same species and hence reproduction. Long-term field studies of baboons in Botswana's Okavango Delta conducted by Bergman et al.[27] indicate that adult female baboons are capable of evaluating the status of another individual in terms of both rank and kinship by identifying threat vocalizations and submission screams from different female baboons. The specificity of vocal sounds suggests that baboons have developed a rudimentary form of communication that enables them to establish their status and that of their own kind.

Origin of the Unique Primate Traits

Four complementary hypotheses have been advanced to explain the evolutionary origin of the primate traits: arboreal adaptation, visual predation, emphasis on energy-rich foods, and complex social groups and life span.

Arboreal Adaptation

Most researchers relate the primate features, such as stereoscopic vision, excellent grasping ability, single births, and omnivorous diet, to an arboreal tree-living adaptation. According to this hypothesis, the arboreal habitat presented different stresses on its inhabitants than the terrestrial niche that other mammals had filled.[15] The arboreal habitat, which is characterized by low light and an abundance of predators, required the development of a high degree of visual acuity with binocular and frequently color vision, as well as auditory acuity. Selective pressures in the open arboreal habitat fostered enhancement of the visual system (including depth perception, sharpened acuity, and color vision), which enabled primates to travel rapidly through the three-dimensional space of the forest canopy and easily discern the presence of ripe fruits and young leaves. The continuous pressure of remembering where the right foods are located in a complex environment is probably an important factor for the development of the large brain that characterizes primates, especially in anthropoids. Thus, the primary visual centers, located in the occipital lobe, and those for auditory stimuli, located in the temporal lobe, are expanded in anthropoids and even more in humans.

Visual Predation

According to the "visual-predation hypothesis," orbital convergence, grasping extremities, and nails on the digits evolved together for visually directed predation of insects on terminal (slender) branches. According to this hypothesis,[28] the primate ancestors were first adapted to living in the bushy forest undergrowth and the lowest branches of the forest canopy. Early primates relied heavily on vision to exploit the fruits and insects occupying the same habitat. Once in the high arboreal habitat, the visual system adapted for predation became even more valuable. Thus, the close-set eyes allowed the development of binocular vision and depth perception, which permitted primates to judge accurately the distance from their prey without moving their heads. Similarly, the five fingers that were initially adapted for the pursuit of insects in the forest undergrowth became even more adaptive, and the grasping hand enabled primates to select foods and climb with ease in their arboreal habitat. Thus, according to the proponents of the visual-predation hypothesis, the primate characteristics are a continuation of preadaptations acquired by primate ancestors who were adapted to living in the bushy forest undergrowth and the lowest branches of the forest canopy.

Emphasis on Energy-rich Foods

This hypothesis maintains that the large and complex brain of primates was made possible by a shift to a high-quality diet (derived either from animal foods or plants).[29,30] A high-quality diet per unit of weight includes a higher concentration of nutrients and energy than a low-quality diet. For example, fruit and animal foods have a higher energy content than shoots and leaves. From **table 5.2** it is evident that the primarily folivorous howler monkey, for its body weight, has a smaller brain size (<1% of body weight) than the frugivores capuchin monkey (2 to 3% of body weight). Similarly, the folivorous Indriidae shows a lower comparative brain size (<1% of body weight) than the more frugivorous or omnivorous Lemurinae (about 1% of body weight). In the leaf-eating Colobinae, the brain represents less than 1% of body weight, while in the more frugivorous Cercopithecinae such as *Cercocebus* the brain accounts for about 1.5% of body

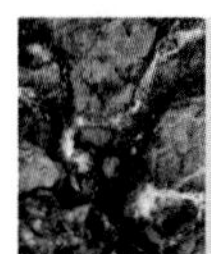

TABLE 5.2

Brain Size in Primates as a Percent of Body Weight by Dietary Pattern

Species	Dietary Pattern	Brain (%)
Howler Monkey	Folivore	< 1
Capuchin Monkey	Frugivore	2.5
Indriidae	Folivore	< 1
Lemurinae	Frugi-Omnivore	1.5
Colobinae	Folivore	< 1
Cercopithecinae	Frugivore	1.5
Gorilla	Folivore	<1
Hylobatidae	Frugivore	2.0

Source: Allman, McLaughlin, and Hakeem.[31]

weight. The highly frugivorous Hylobatidae and Pongidae have greater relative brain size (2% of body weight) than the folivorous gorillas (<1% of body weight). Furthermore, as illustrated in **fig. 5.21** the cerebral cortex of the frugivore spider monkey is more convoluted than the brain of the leaf-eating howler monkey. Additionally, the large and complex mixture of different types of trees of the tropical rain forest presents the forager with complex problems for harvesting fruit. The successful harvesting of ripe, digestible fruit requires relatively complicated strategies for extracting high-quality foodstuffs. In contrast, leaves are present everywhere and can be easily harvested with little competition. It is in response to this environment that frugivores developed a larger brain than folivores. The existence of larger brains in fruit-eaters than in leaf-eaters supports the hypothesis that the brain helps the animal to cope with environmental variation.

In summary, primates who selected plant foods of high nutritional quality had more energy available and larger brains than those who depended on low-quality foods. Therefore, it is quite possible that the evolution of the large and complex brain of primates was associated with a shift to a high-quality diet.

Complex Social Groups and Life Span

This hypothesis proposes that the development of social expertise was a key factor in the evolution of the brain in primates.[24,33] All primates, except for some nocturnal primates, have a high level of sociality, tend to associate with others, at least with offspring, and have a tendency to live in social groups. They also maintain a permanent association of adults, which is uncommon in other mammals. Comparative studies indicate that the neocortex increases across the range of primates[31–35] (**fig. 5.22**). For example, in insectivores the neocortex is approximately the same size as the medulla, in prosimians it is about 10 times bigger than the medulla, it is 20 to 50 times larger in the anthropoids, and more than 100 times larger in humans. Social-group size has been used as a proxy measure of the complexity of social life of primates. The assumption is that the complexity of social interactions in different species increases in direct proportion to the size of the group and the neocortex plays a major role in such interactions. Analysis of the relationship of social-group size for many species found a significant relationship between group size and the ratio of neocortex volume to the volume of the rest of the brain[35] (**fig. 5.23**). According to Allman,[32] a social-group size predicted about 45 percent of the variation in the neocortex ratio among primate species. That is, the larger the size of the neocortex relative to the rest of the brain, the larger the social-group size in primates.

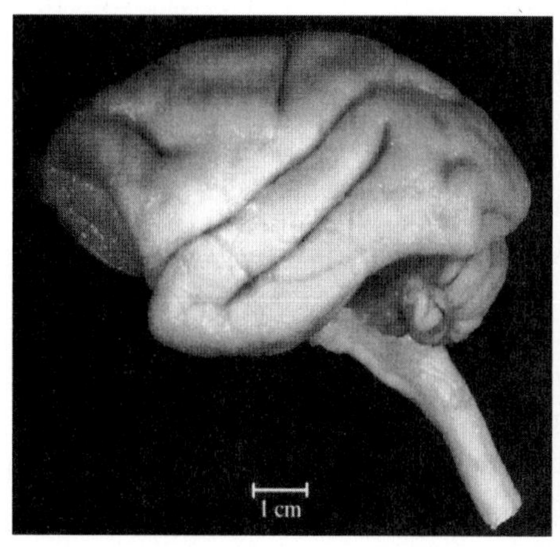

(a)

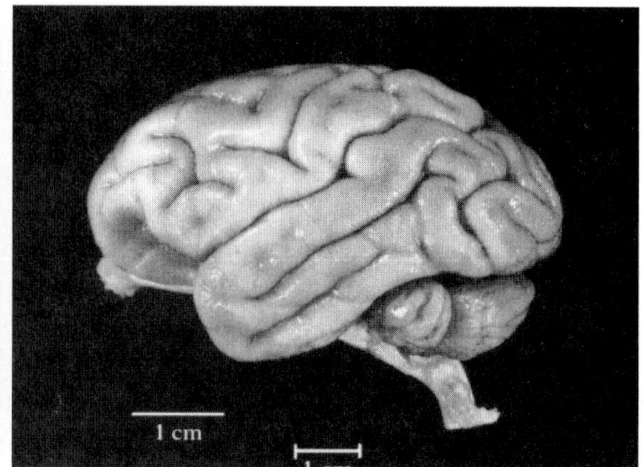

(b)

FIGURE 5.21. Brain size comparison of howler and spider monkeys. The leaf-eater howler monkey (a) has a less convoluted and less complex cortex than the fruit-eater spider monkey (b) of the same body weight. Source: Adapted from Allman p. 166.[24]

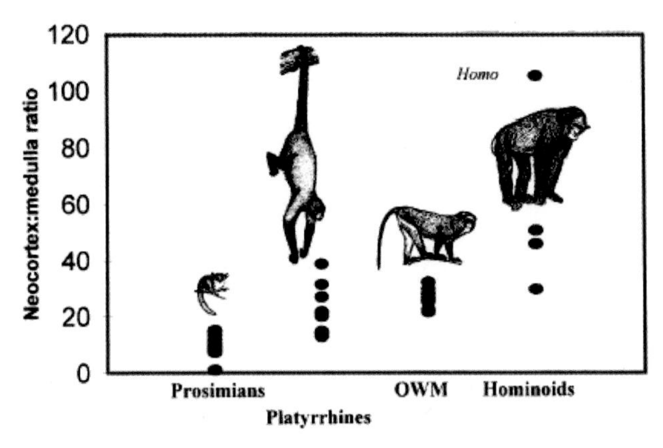

FIGURE 5.22. Ratio of neocortex volume to medulla volume in prosimians, platyrrhines (New World monkeys), Old World monkeys (OWM), hominoids, and humans. Source: Passingham, fig. 13.[32]

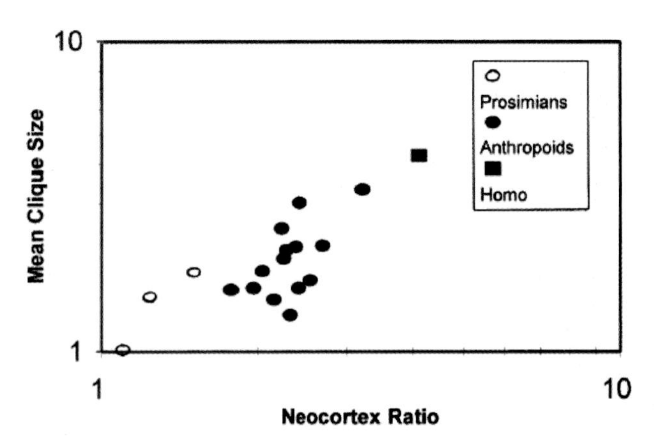

FIGURE 5.23. Neocortex ratio and clique size. Source: Adapted from Dunbar, fig. 3.[34]

The complexity of social groups in primates is also associated with extension of life span.[24,31] In general, animals that have longer life spans are likely to experience more extreme environmental fluctuations and thus are exposed during their longer lives to more severe crises than are animals with shorter life spans.[31,32] Studies by Allman and colleagues[24,31] have shown that, except for prosimians, among primates relative brain size is linked to longevity, so that the greater the longevity the greater the size of the brain. Furthermore, the structures involved in memory and strategies to cope with environmental variation such as the neocortex, amygdala, hypothalamus, and cerebellum are also related to longevity.

In summary, it is quite possible that the evolution of a large brain in all primates was driven primarily by the complexities of social life. However, it is also feasible that having a large brain enabled the attainment of complex social relations.

Summary

The order Primates is divided into suborders: Prosimii and Anthropoidea. Primates structurally are not differentiated from the rest of the mammals by any single feature. As a group, the primates are by far the most common of living mammals. They possess numerous characteristics they share in common with other placental mammals, and they can be distinguished from other mammals by the configuration created by the combination of a series of generalized or primitive traits and derived or specialized traits.

The increased brain size and increased complexity of the brain enabled primates to succeed in the arboreal environment where much more accurate judgment of distances is required, which was facilitated by the development of stereoscopic vision and color vision than life on the ground. The forward-facing eyes of primates are adaptations that facilitated this type of visual precision. In such an environment, manual dexterity is more important than sense of smell. Hence, the areas of the brain concerned with motor function and especially those that allow manual dexterity are enlarged, while those of olfaction are reduced.

Suggested Readings

Napier, J. R., and P. H. Napier. 1967. *A Handbook of the Living Primates: Morphology, Ecology, and Behavior of Nonhuman Primates.* London: Academic Press.

Roe, N. 1999. *The Pictorial Guide to the Living Primates.* Charlestown, RI: Pogonia Press.

Literature Cited

1. Groves, C. 2001. *Primate Taxonomy.* Washington, DC: Smithsonian Institution Press.
2. Fleagle, J. G. 1999. *Primate Adaptation and Evolution.* New York: Academic Press.
3. Dolhinow, P., and A. Fuentes. 1999. *The Nonhuman Primates.* Mountain View, CA: Mayfield Publishing Company.
4. Primate Center Library, Wisconsin Primate Research Center. Internet resource: http://www.primate.wisc.edu
5. Strier, K. B. 2000. *Primate Behavioral Ecology.* Boston: Allyn and Bacon.
6. Oregon Zoo. Internet resource: http://www.zooregon.org
7. McDonald Wildlife Photography, Inc. :http://www.webshots .com.
8. Markham, R., and C. P. Groves. 1990. Weights of wild orangutans. *American Journal of Physical Anthropology* 81:1–3.

9. Winkler, L. 1989. Morphology and relationships of the orangutan fatty cheek pads. *American Journal of Primatology* 17:305–319.
10. Nadler, R. D. 1988. Sexual and reproductive behavior. In *Orangutan Biology,* ed. T.H. Schwartz, 31–51. Oxford: Oxford University Press.
11. Kingsley, S. 1982. Causes or non-breeding and the development of the secondary sexual characteristics in the male orangutan: A hormonal study. In *The Orangutan: Its Biology and Conservation,* ed. L. E. M. de Boec, 215–219. The Hague: DrWJunk.
12. Graham, C. E., and R. D. Nadler. 1990. Socioendocrine interactions in great ape reproduction. In *Socioendocrinology of Primate Reproduction,* ed. T. E. Ziegler and F. B. Bercovitch, 33–58. New York: Wiley-Liss.
13. Wrangham, R. W. 1993. The evolution of sexuality in chimpanzees and bonobos. *Human Nature* 447–479.
14. de Waal, F. B. M., and F. Lanting. 1998. *Bonobo: The Forgotten Ape.* Berkeley: University of California Press.
15. Sussman, R. W. 1991. Primate origins and the evolution of angiosperms. *Am. J. Primatol.* 23:209–223.
16. Szalay, F. S., and M. Dagosto. 1988. Evolution of hallucial grasping in the primates. *J. Hum. Evol.* 17:1–33.
17. Rasmussen, D. T., and K. A. Nekaris. 1998. Evolutionary history of lorisiform primates. *Folia Primatol.* (Basel). 69 (Suppl 1):250–285.
18. Bloch, J. L., and D. M. Boyer. 2002. Grasping primate origins. *Science* 298:1606–1610.
19. Morikawa, T. 1985. *Primate Morphophysiology: Locomotor Analyses and Human Bipedalism.* Tokyo: University of Tokyo Press.
20. Napier, J. R., and P. H. Napier. 1967. *A Handbook of the Living Primates: Morphology, Ecology, and Behavior of Nonhuman Primates.* London: Academic Press.
21. Simons, E. L. 1972. *Primate Evolution: An Introduction to Man's Place in Nature.* New York: Macmillan.
22. Oxnard, C. H. E. 1990. *Animal Lifestyles and Anatomies: The Case of the Prosimian Primates.* Seattle: University of Washington Press.
23. Jones, S., R. Martin, and D. Pilbeam eds. *Encyclopedia of Human Evolution.* New York: Cambridge University Press.
24. Allman, J. M. 1999. *Evolving Brains.* New York: Scientific American Library.
25. University of Wisconsin, Michigan State Comparative Mammalian Brain Collections, and the National Museum of Health and Medicine. Internet resource: http://www.brainmuseum.org
26. Mitani, J. C., and J. Gross-Louis. 1995. Species and sex differences in the screams of chimpanzees and bonobos. *International Journal of Primatology* 16:393–411.
27. Bergman, T. J., J. C. Beehner, D. L. Cheney, and R. M. Seyfarth. 2003. Hierarchical classification by rank and kinship in baboons. *Science* 302:1234–1236.
28. Cartmill, M. 1975. *Primate Origins.* Minneapolis: Burgess Pub. Co.
29. Milton, K., and M. W. Demment. 1988. Digestion and passage kinetics of chimpanzees fed high and low fiber diets and comparison with human data. *American Institute of Nutrition* 118:1082–1088.
30. Milton, K. 1993. Diet and primate evolution. *Sci. Am.* 8:86–93.
31. Allman, J., T. McLaughlin, and A. Hakeem. 1993. Brain weight and life-span in primate species. *Proc. Natl. Acad. Sci.* 90:118–122.
32. Passingham, R. E. 1982. *The Human Primate.* San Francisco: Freeman.
33. Dunbar, R. I. M., and M. Spoor. 1995. Social networks, support cliques and kinship. *Hum. Nature* 6:273–290.
34. Dunbar, R. 1998. The social-brain hypothesis. *Evol. Anthropol.* 6:178–190.
35. Barton, R. A., and R. L. M. Dunbar. 1997. Evolution of the social brain. In *Machiavellian Intelligence,* Vol. 2. B. R. Whiten A, ed. Cambridge, MA: Harvard University Press.

Primate Behavioral Ecology

During the last two decades, the study of nonhuman primate behavior changed from a descriptive and classificatory phase into an analytic science that incorporates the principles of evolutionary theory. The use of evolutionary theory has enabled researchers to understand the origins of diversity in behavior among primates. This approach is currently referred to as behavioral ecology (previously called sociobiology). This chapter focuses on the principles and behavioral traits of primates that have been interpreted within the framework of behavioral ecology.

Evolution and Social Behavior

Premise

The major premise of behavioral ecology is that genetic and behavioral traits are two distinct expressions of a single evolutionary process.[1–6] In behavioral ecology, behaviors are treated like any other biological trait and are potentially subject to natural selection. The processes involved in behavioral evolution are equivalent to those in genetic evolution as far as the forces affecting the frequency of a trait obey to the influence of natural selection transmitted from parent to offspring through differential fertility and mortality. In evolutionary perspective, biological structures have been custom tailored to motivate behaviors that are likely to enhance individual fitness. Therefore, behavioral variants with a high fitness have been favored and these perpetuate the evolutionary origin of fitness-enhancing biological traits. It follows then that behavioral traits that enhance fitness also accentuate biological fitness. In other words, a change occurring in one system entails a change in the fitness governing evolution in the other system. Therefore, both genetic and behavioral selection tend to favor those existing variants whose net effect is to increase the average fitness of the individual and population to the prevailing conditions.

Primates use a diversity of behaviors that increase the likelihood of gaining access to mates and guarantee the survival of their offspring, which in turn assures the passing of their traits to the next generation.[2–6] In this context, behavioral actions that lead to a higher reproductive success will become adaptive and the genes associated with such behavior will be transferred to the next generation faster than those that are less fit. Therefore, the differences in **fitness** between individuals and populations will determine the behavioral pattern of a given primate group. In other words, a specific behavioral **strategy** that contributes to the survival and **reproductive fitness** of the individual—and eventually the population—becomes part of the genetic milieu of the species. In summary, primates behave in ways that promote their survival and reproduction.

Kin Selection, Altruism, and Inclusive Fitness

In primates the first individuals who detect a predator give an alarm call that alerts others in the group to take cover while possibly attracting the predator's attention to the alarm caller. Similarly, primates are known to come to the aid of one another in a fight against a third party, potentially risking injury or retaliation aimed at themselves. This type of behavior is referred to as **altruistic behavior**. An altruistic behavior benefits the fitness of the recipient at a cost to the fitness of the giver. At the individual level, altruistic behavior has a negative fitness when compared to more selfish ways that enhance fitness. A primate who keeps quiet when it sees a predator or does not get involved when a member of its troop is attacked is more likely to survive than one who altruistically alerts its neighbors or participates in a fight. If the individual is the unit of selection, altruists should be selected against and selfish individuals favored. Then, the important question is how such altruistic behavior evolved. According to behavioral ecology, altruistic behavior evolved within the context of relatives with whom the altruists shared some proportion of their genes through common genealogical descent. In this view, altruism actually benefits the altruists when the recipient of the favor is a relative. Although giving an alarm call may increase the altruist's risk of being eaten, it helps the relatives to survive and hence the fitness of the altruist and its relatives is enhanced. In other words, the negative costs of altruism are offset by the benefits gained through the survival of the genes of the relatives. This type of altruism results in enhanced fitness referred to as **inclusive fitness**. As defined by Hamilton,[7] **inclusive fitness** is the personal fitness an individual actually expresses in its production of adult offspring. Therefore, inclusive fitness includes the individual's own fitness plus the fitness of each of his or her relatives, weighted in proportion of the genes shared between them. Further, the closer the biological relationship among relatives when one helps one's relative, the greater the potential benefits of increasing one's own fitness.[7]

Viewed in this context, assistance in rearing offspring given by a childless individual to a relative increases the inclusive fitness; much as altruistic behavior evolved to benefit the individual and the group. The effectiveness of kin selection also increases when the relatives live in close proximity. Studies of Japanese macaques, vervet monkeys, and chimpanzees indicate that relatives that live in close proximity are likely to form alliances with one another more frequently than with distant relatives. (They form alliances to either resist takeover attempts by other males or attack unrelated communities.)

Reciprocal Altruism

The fitness of an individual and the group at large also depends on the altruistic behavior of nonrelatives. When an altruist gives an alarm call, it benefits not only his or her relatives, but also other unrelated members of the group because a primate troop does not include only relatives. Thus, altruism can be selected if these nonrelated individuals can be counted on to reciprocate the favor when the need arises. A recipient of an altruistic behavior who fails to reciprocate is a cheater.[2] A cheater may gain in the short run by receiving aid without any costs to their own fitness. However, reciprocity is necessary for future support in the long run because the cheater's fitness is lower when compared to the individual who reciprocated. In view of the fact that primates constantly need to protect themselves from neighboring communities and predators, one can assume that reciprocal altruism must have been selected for because it enhanced their fitness, not because the animals are conscious of their motives or the reproductive consequences of their behavior.

Sociality and Reproductive Fitness

Grooming plays an important role in the life of most primates. It is a social lubricant because it eases the interaction between the male/female and higher/lower ranks, where there is a possibility of friction. Grooming is often reciprocal; with roles interchanged, the groomer may become the groomee, a process resulting in the establishment of friendly social relations among the animals in the group (**fig. 6.1**). Thus, grooming is not an indiscriminate practice, who grooms whom requires understanding and an evaluation of the status of another individual in terms of both rank and kinship. It is also hygienic since it removes ectoparasites,

FIGURE 6.1. Grooming in baboons. (Photo courtesy of Professor John Mitani, University of Michigan.)

dead skin, and debris from the skin and fur. As inferred from long-term studies of baboon monkeys (going back more than 15 years), sociality among females has a direct impact on their reproductive fitness.[8] Silk et al.[8] describe how female baboons in Amboseli, Kenya, who are significantly more social (defined by the amount of time that others spend grooming them), have more than the average number of infants surviving to 12 months of age. Thus, it would appear that sociality is positively associated with overall fitness.

Mother-Infant Relationship

The mother-infant pair is the basic social group for many primates. In some primates, the mother-infant closeness continues after infancy. All primates differ from other mammals in the high degree of attachment of mother and infants. Unlike other animals, in which the newborn are left in a nest or den, the primate infant develops a closeness with the mother that does not end with weaning (**fig. 6.2**). The mother-child bond plays an important role in the development and transmission of various skills. The maternal influence is evident in the case of the Japanese macaques: one individual accidentally learned that the sweet potatoes they were being fed tasted better if the sand was washed off (this is not a normal food for these animals). This behavior has now spread through the whole group and is being passed on to infants. It is now part of their culture. Similarly, chimpanzee's tool-use abilities used in termite fishing are also passed from mother to offspring.

The importance of the mother-child bond was clearly demonstrated by the studies of Harlow and Zimmermann,[9] who raised rhesus monkeys in captivity. In this study, the infants were separated at birth; some of them were suckled by surrogate mothers fashioned from cloth that resembled their mothers, while others suckled from the milk bottles attached to surrogate mothers that did not resemble their mothers. When these infants matured, the ones raised by the surrogate mothers that resembled their mothers were more adept at caring for their offspring and attained more normal sexual behavior than infants that were left to suckle milk from their wire-attached bottles. The results of this study have been interpreted to indicate that the experience and the attachment of mother and infant is necessary for the expression of the maternal instinct to care for and survival of the infant. Thus, the mother-infant relationship plays an important role in the social and reproductive development of primates.

In cases where the mother is either deceased or unable to take care of the offspring, adult female members of a troupe contribute to the care of young infants.[10] For example, female adult baboons will protect children of members of a troupe (**fig. 6.3**). Similarly, among langur monkeys, as many as eight females may hold the infant during its first day of life. Sometimes adult males, as in the case of hamadryas baboons, "adopt" an infant and provide nourishment and protection. The importance of alloparental care (i.e., any type of parenting behavior shown toward an

FIGURE 6.2. Parent-offspring bond in Hanuman langur monkey. © Kevin Schafer/Corbis.

FIGURE 6.3. Female adult macaque monkeys protecting children of members of the troupe. Courtesy of Professor Meredith F. Small, Cornell University, Ithaca, NY.

infant by an individual other than the infant's biological mother or father) is that it increases the survival of the infant in the event that the mother dies and binds the adults of the group together.

Kin Recognition and K-Selection

Kin Recognition

Primates are able to recognize one another individually and to retain recognition over long periods. Among chimps from Gombe, related males form close, cooperative groups. When male rhesus monkeys emigrate to join other groups, related males sometimes leave together, joining the same new group; males also tend to transfer to new groups into which a male relative has previously immigrated.[11] Kinship may be an important factor in male alliance formation as well. Thus, it would appear that primates recognize some kin and behave differentially (usually favorably) toward relatives.

K-Selection

In comparison with other mammals of similar body weights, all primates (except marmosets) have reduced litter size to a single individual. Although primates have a longer period of gestation for their body weight, their offspring are born less mature than other mammals. Therefore, in primates, a much greater portion of postnatal development takes place outside the womb. Juvenile primates develop more slowly in comparison with most other animals of similar body size. Because of delayed maturation, parental investment (i.e., the behavior or the contribution that parents make to the fitness of their offspring) in each offspring is increased. Thus, although fewer offspring are born, they receive more intense and efficient rearing. Another feature of a prolonged childhood is that play lasts longer in primates than in other mammals, which represents the rehearsal of adult roles. In other words, primates produce only a few young immature offspring, each requiring larger amounts of parental and kinship care. This reproductive strategy, which is referred to as K-selection, enables primates to maintain a high reproductive fitness despite their low birth rate.

Sexual Selection

Sexual Dimorphism and Parental Investment

Sexual dimorphism (defined as the difference between genders in body size and body morphology) in baboons and gorillas, like in many mammals, is quite high. According to Darwin,[12] sexual dimorphism across primates and other animals is related to intrasexual competition driven by a process of sexual selection. Sexual selection as proposed by Darwin and expanded upon by behavioral ecologists[13] involves two processes: (a) competition over access to mates, and (b) choice of mates.

Competition over access to mates characterizes males. It is related to morphological features such as brightly colored faces, bright fur, and larger bodies or larger canines. These features are thought to provide benefits in their competition with other males for mates, or in enhancing their attractiveness to females.[14] These features are emphasized in one-male–multifemale mating groups (**fig. 6.4**).

On the other hand, selection for choice of mates is within the realm of females. The difference in these reproductive strategies is said to be related to the differences in the reproductive potential between males and females. In females, the reproductive potential is more limited than in males. Females are born with a finite number of sex cells, which at maturity are released from the ovaries during ovulation once a month. Hence, females can conceive when an ovum

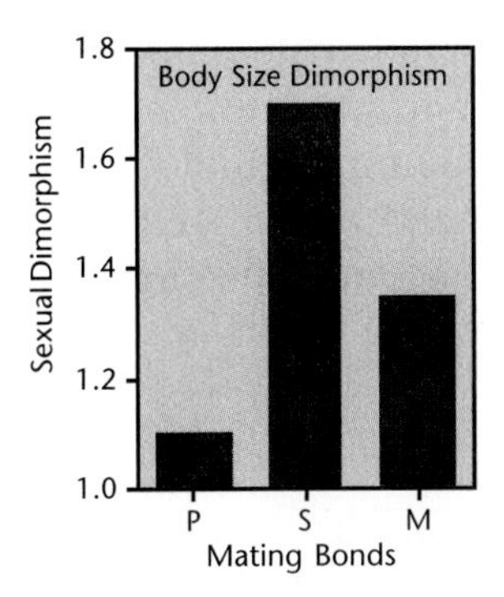

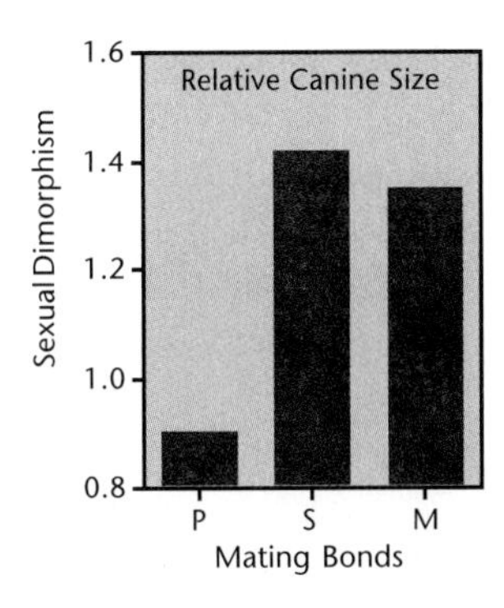

FIGURE 6.4. Body size dimorphism and relative canine size for primate genera belonging to different breeding systems. P = monogamous pair mating bonds; S = Single male–multifemale mating bonds; M = multimale–multifemale mating bonds.Source: Based upon data from Harvey, P. H. and A. H. Harcourt. 1984. Sperm competition, testes size, and breeding systems in primates. In ed. R. L. Smith, *Sperm Competition and the Evolution of Animal Breeding Systems*, 589–600.

reaches maturity and is released from the ovaries during ovulation. Once fertilization has occurred, females endure the time and energy costs of gestation and lactation. Because hormonal feedback mechanisms inhibit ovulation during pregnancy and lactation, a female cannot conceive during this time. In contrast, male sex cells, or sperm, are more numerous and are constantly being replenished during spermatogenesis throughout adulthood. Hence, males are biologically capable of fertilization whenever they have produced enough viable sperm. Therefore, the biology of internal reproduction and lactation limits mothers, but not fathers, from pursuing new reproductive opportunities throughout the life cycle. As a result of these differences, ovulating females represent a limiting resource for male reproduction. The lack of ovulating females leads to high male-male competition for access to ovulating females. Similarly, the energy costs of abortion and stillbirths are more costly for females than they are for males. A female who spontaneously aborts a fetus or loses a live infant afterward is the parent who has irrevocably lost time and energy that instead could have been invested in a more successful reproductive effort.

In other words, the female investment (that is, the contributions that parents make to the fitness of their offspring) is much higher than that of the male. Given these evolutionary constraints, females have to be selective about with whom they mate. In this context, evolution and sexual selection favor females who are selective about their mates. Conversely, the intensity of male competition is directly affected by the physiological status of females, as well as by what other males are doing. One way of increasing the competitive advantage of males is to be bigger. The advantage of being bigger in body size or canine size is that big males can bully or coerce—through aggression or the threat of aggression—other males, as well as females, into compliant behaviors. For example, among the polygamous hamadryas baboons and mountain gorillas, the opportunities for males to impose their own reproductive strategies are determined mostly by their large body size.

Sexual Signals

For mating, most male primates rely on cues given by females about ovulation and estrus. Some females send signals about their ovulatory cycles through their urine, which they distribute by rubbing on their fur and on branches. In female baboons or chimpanzees, estrus may last 28 to 30 days. The reproductive cycle is signaled by estrus and by patches of sex skin on the rumps that inflate and deflate like a balloon, brighten and swell up. The sex skin reaches its maximum peak size and color for two to three days near the day of ovulation. Following conception, the sex skin will return to its anovulatory size and color.[2] During female estrus, male aggressive competition is most intense.

According to behavioral ecologists, the sexual signals in cercopithecines, colobines, and great apes have evolved to select the best mates. Inciting competition among males is advantageous for females because the outcome of male competition can help females choose their mates. Males who win contests against other males at these critical times are good bets for good genes. Giving in to the sexual advances of the male who has managed to beat his competitors represents an excellent strategy for females in selecting strong males. However, as discussed later (see Mating Patterns), befriending a female can offset the advantage of the successful and strong male.

Group Membership

Depending on whether more males or females leave their natal groups, the group membership of primates can be classified in two groups: patrilineal and matrilineal.[2,3]

Matrilineal Societies and Male Dispersal

In most multimale primate species (common baboons, rhesus monkeys, vervet monkeys, and Japanese macaques), it is the males who disperse at adolescence. Females of these species remain in their natal group and live in subunits of female groups that contain mothers with their young; in some cases these units extend over several generations, producing "matrilines." In the matrilineal societies, a female's position in the hierarchy is inherited from her mother. Daughters of high-ranking females become high ranking themselves, while daughters of low-ranking females remain subordinate. High-ranking females have significantly higher infant survival, faster maturing daughters, more rapid production of young, and better access to good foraging areas than lower-ranking females. However, the rank or position of a male in the male group is not known how it is established.

Patrilineal Societies and Female Dispersal

Among chimpanzees, the females of one group tend to migrate out at adolescence and mate with males of other groups.[2,3] In contrast, males nearly always stay together in their natal community for life, maintaining close kin connections among themselves. Since these males associate closely with one another and cooperate to defend a territory, as well as access to females of their group, they establish a kind of "patrilocal" residence pattern. Males that remain in their natal groups have the advantage that there aren't unattached, surplus males wandering around in "patriline" societies, looking for female groups to join or take over. Male kin groups form coalitions that prevent other males from gaining access to their groups containing females. Unrelated males also form coalitions in response to high levels of within-group competition. Extended patriline hierarchies differ from matrilineal ones because males do not inherit their father's rank the way that daughters in many matrilineal societies inherit their mother's rank.

Mating Patterns

The behavioral options and reproductive opportunities reflect the demographic distribution of males and females and the ecological conditions where they live. The most common mating pattern in primates is polygamy, whereby both males and females mate with two or more different partners. In polygynous mating, such as multimale groups or multimale, multifemale groups, mating occurs with males within the group, but also with males outside the group. One unusual mating pattern involves infanticide.

Infanticide

Infanticide, or the killing of infants, is not uncommon among primates. Infanticide has been observed in redtail monkeys, red colobus, blue monkeys, savanna baboons, howler monkeys, gorillas, and chimpanzees, but occurs most frequently among langur monkeys of India.[17–19] Among langurs, the mothers and other females attempt to resist infanticide, but their efforts in the face of large, aggressive, and determined males usually fail. Interestingly, after the male succeeds in killing her infant, she is likely to turn around and mate as soon as she can possibly conceive again with the male who killed her infant. This paradoxical behavior has been interpreted as a behavioral adaptation that has been shaped by evolution.[15,16] According to Hrdy,[15,16] infanticide is related to the social organization and reproductive cycles of langur monkeys. First, langurs typically live in social groups composed of one adult male, several females, and their offspring. Other males without mates associate in bachelor groups. These peripheral males occasionally attack and take over the group. At this time, the infants fathered by the previous male are attacked and killed. Second, similar to how other nonhuman primate females become sexually accessible only during estrus, but during lactation, ovulation and estrus are inhibited. Once the infants are dead, the females stop lactating, stop the hormonal inhibition of ovulation, and enter into estrous, becoming reproductively accessible to the newly arrived male. Thus, by killing the infants, he avoids a two- to three-year wait before the females come back into estrus. On the other hand, the non-infanticidal male would have had to wait until the mothers wean their offspring and resume their ovulatory cycles. Furthermore, the females who mate with the infanticidal male also increase their chances of passing along their genes and those of the infanticidal males to the next generation.

In this view, infanticidal langurs will have higher reproductive success than the individuals behaving in another way. Consequently, their genes (including any that underlie this behavior) are differentially passed onto the next generation. Therefore, infanticide among langur monkeys has been considered a behavioral adaptation that has been shaped by evolution.[15,16] Other researchers, however, consider infanticide a pathological behavior limited largely to the crowded conditions that stimulate male takeovers of troops.[17–19]

Monogamous Pair Bonds

Monogamy refers to mating with just one partner. In gibbons and some cebids (titis and owl monkeys), the typical social group is the one male–one female pair. These pair-bond primates have been assumed to be monogamous, but recent studies indicate that not all monogamous pair-bond primates practice monogamy when it comes to mating. In both Sumatra and Thailand, female gibbons that typically live as territorial pairs mate with different males.[20,21] Similarly, male tarsiers mate with different female partners than their own pair.[21] Thus, monogamous pair bonds appear to be opportunistic arrangements governed by factors other than mating preference.

Polygyny: One Male and Multifemale Bonds

Polygyny, in which a single male monopolizes access to multiple females, is widespread across primates. Polygyny resulting from a male patrolling multiple solitary females is common among orangutans. Female orangutans are thought to avoid exclusive associations with a single male because of the greater energy and time expenditure they would incur in finding sufficient food with another adult around. A polygynous mating pattern is associated with a high degree of sexual dimorphism, as seen in gorillas and orangutans. Polygynous males invest a great deal of time and effort in preventing other neighboring males from intruding into the troupe. Paternity studies in chimpanzees indicate that the community is the primary but not exclusive unit for reproduction in wild chimpanzees, and females do not typically reproduce with outside males.[22]

Friendship-Based Bonds

As a consequence of the territorial males that control a large number of females, there is a surplus of less fortunate males in the population. Older males, who are no match in fights against younger, high-ranking males in the prime of their lives, are unable to compete for mates. These males, however, resort to the strategy of befriending a female to improve their reproductive chances. Males attempting to immigrate into the troupe often form affiliative or "friendship" relationships with females in a troupe. These "friendships" are often established by older and low-ranking males, who cuddle, nuzzle, and groom their female friends and tolerate the annoying antics of infants. Then, when the female enters into estrus she chooses her friend as a mate independent of his rank in the male hierarchy. Field observations of baboon monkeys indicate that male friends accounted for 40 percent of the copulations by estrous females.[23] Similarly, low-ranking male chimpanzees shower females with friendly attention before they go into estrus. Then, when the female enters in estrus, she frequently goes on safaris with her low-ranking friend.[23] In both chimpanzees and baboons, although males may vary in their ability to cultivate female friends, female choice is a powerful selection pressure and the strategy of befriending a female offsets the priority of access that high-ranking males have.

In summary, the different mating patterns, including infanticide, monogamous pair bonds, and polygamous bonds, represent reproductive strategies. These reproductive strategies differ between primate societies depending on the adaptations to specific ecological conditions on which the population evolved. As such, they reflect the compromise between males and females that allows them to gain access to mates, food, and protection that they and their offspring need to survive.

Food Availability, Food Distribution, and Group Size

Differences among primate social systems are to a certain extent related to food quality and the availability and distribution of food sources. The influence of these variables is evident in the variability of body size and group size.

Food Preference and Body Size

For the most part, there is an inverse association between body size and metabolic rate: the smaller the body size, the greater the nutritional requirements.[2,24,25] Likewise, there is an association between food preference and body size. Smaller primates (<10 kg) are typically frugivore-insectivores while larger primates (>10 kg) are typically frugivore-folivores (**fig. 6.5**). These associations are related to the fact that the

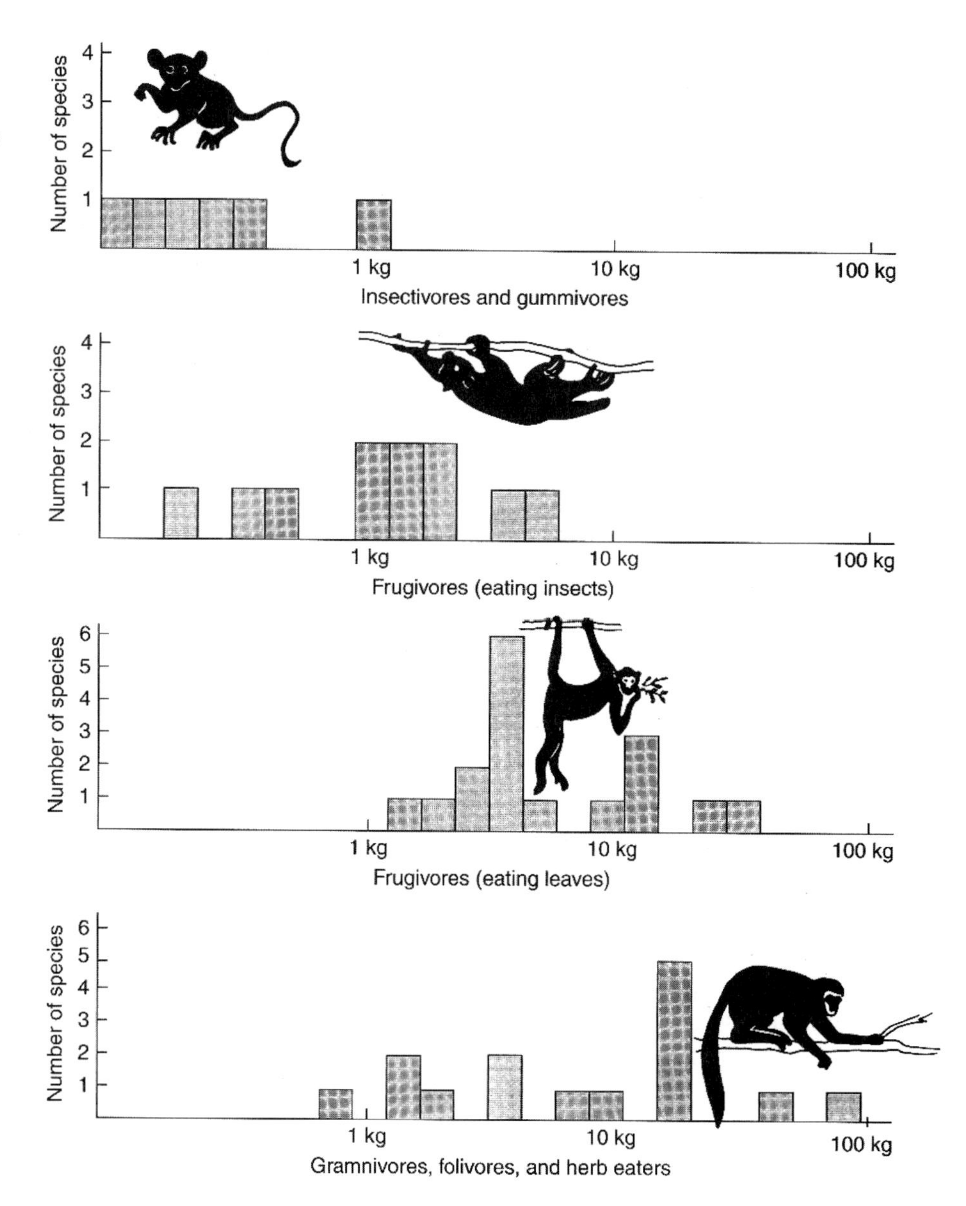

FIGURE 6.5. Range of body sizes in the major dietary categories. Note that frugivore-insectivores tend to be lighter in weight than frugivore-folivores. Source: Richards fig. 2.[25]

nutritional quality of foods per unit of weight is greater in animal foods than in plant foods, and fruits have high nutritional quality than leaves. (High-quality foods refer to diets that are rich in easily digestible energy and protein; low-quality diets are those that are poor in these nutrients.) In other words, the relatively high metabolic rate of small-bodied primates restricts them to animal foods (such as insects) that can be digested more rapidly than leaves, whereas the larger-bodied primates within each of the major taxonomic groups, with their relatively low metabolic rates, rely on the abundant supply of plant leaves as well as underground storage organs of grasses, called corms, which are rich in carbohydrates as well as high in protein and water.

Group Size and Competition for Food

In general, group size reflects the balance of costs and benefits associated with the competition within the group for food. In areas where food is plentiful, the competition for food decreases and so does the need for protection of nutritional resources. For example, folivorous (leaf-eating) primates, because of the ubiquity of plant foods, do not need to travel far each day in search of food. They can also live in relatively large groups without experiencing high levels of competition among members. Therefore, among primarily folivorous primates, variability in group size is not related to traveling distance. This is due to the primate's reliance on herbaceous vegetation that reduces feeding competition within groups. Therefore, where food is abundant, primates do not need to modify their group size.

In contrast, because the availability of fruits is influenced by season and is not evenly distributed in the environment, frugivores (fruit eaters) have to travel great distances each day in search of food. This creates the tendency to live in small social groups (e.g., monogamous gibbons and solitary orangutans) so as to decrease the levels of feeding competition. Likewise, gorilla populations that live at the lowest altitude forests are highly frugivorous, more arboreal, and travel far distances daily. In some cases, the gorilla population may live in a less cohesive, smaller social group than the mountain gorillas.[2,3,26,27] Comparative studies indicate that the mountain gorillas live in relatively stable groups of 2 to 34 members, with an average of 9.2 individuals. The western lowland gorilla also has a similar mean group size (9.5 individuals), but western lowland gorilla groups rarely exceed 16 to 18 members.[28] On the other hand, at high altitudes where there are few fruit sources, gorillas are folivorous, terrestrial, travel short distances daily, and live in cohesive, stable groups.

The size of food patches and the distribution of items within them determines the group size and the grouping patterns and patterns of cooperation and competition within and between groups. As a result, nearly all primates adjust the size of their feeding groups, or feeding parties, to the size of their food patches. Primates with fission-fusion grouping patterns, such as chimpanzees and spider monkeys, divide themselves into smaller feeding parties when their preferred fruit resources occur in small patches; solitary females are frequently sighted foraging alone. Whenever the fruit patches are large enough that they can accommodate more individuals without competition, larger feeding aggregates are formed.[29–33] Even the solitary orangutans aggregate at large patches of ripe fruit.[34]

Summary

The major premise of behavioral ecology is that genetic and behavioral traits are two distinct expressions of a single evolutionary process. Thus, behaviors are treated like any other biological trait that is potentially subject to natural selection. It assumes that a change occurring in one system entails a change in the fitness governing evolution in the other system. Therefore, both genetic and behavioral selection tend to favor those existing variants whose net effect is to increase the average fitness of the individual and population to the prevailing conditions.

There is abundant evidence indicating that primates use diverse behavioral actions that are associated with enhanced reproductive fitness. These traits, in turn, become part of the genetic repertoire of the population and lead to the evolution of present behavioral and biological traits of primates. In other words, individuals become vehicles of behavior as well as vehicles of genes, so that the behaviors influencing reproductive success become established in the population gene pool.

Suggested Readings

Strier, K. B. 2000. *Primate Behavioral Ecology.* Boston: Allyn and Bacon.

Williams, G. C. 1966. *Adaptation and Natural Selection.* Princeton: Princeton University Press.

Wilson, E. O. 1975. *Sociobiology: The New Synthesis.* Cambridge, MA: Harvard University Press.

Wynne-Edwards, V. C. 1962. *Animal Dispersion in Relation to Social Behaviour.* New York: Hafner Publishing Co.

Literature Cited

1. Trivers, R. L. 1971. The evolution of reciprocal altruism. *Q. Rev. Biol.* 46:35–57.
2. Strier, K. B. 2000. *Primate Behavioral Ecology.* Boston: Allyn and Bacon.
3. Mitani, J. C., D. P. Watts, and M. N. Muller. 2002. Recent developments in the study of wild chimpanzee behavior. *Evolutionary Anthropology* 11:9–25.
4. Silk, J. 2001. Ties that bind: The role of kinship in primate societies. In: *New Directions in Anthropological Kinship* ed., Stone, L. Lanham: Rowman & Littlefield Publishers. 71–92.
5. Cronk, L. 1991. Human behavioral ecology. *Annu. Rev. Anthropol.* 20:25–53.
6. Charnov, E. L., and D. Berrigan. 1993. Why do female primates have such long lifespans and so few babies? *or* Life in the slow lane. *Evol. Anthropol.* 1:191–194.
7. Hamilton,W. D. 1964. The genetical evolution of social behaviour, I and II. *Journal of Theoretical Biology* 7:1–52.
8. Silk, J. B., S. C. Alberts, and J. Altmann. 2003. Social bonds of female baboons enhance infant survival. *Science* 302:1231–1234.
9. Harlow, H. F., and R. R. Zimmermann. 1959. Affectional responses in the infant monkey. *Science* 130:421–432.
10. Rosenblatt, J. S., and C. T. Snowdon, ed. 1996. *Parental Care: Evolution, Mechanisms, and Adaptative Significance.* San Diego: Academic Press.

11. Gouzoules, S., and H. Gouzoules. 1987. Kinship. In *Primate Societies,* eds., B. Smut., D. L. Cheney, R. M. Seyfarth, R. W. Wrangham, and T. T. Struhsaker. Chicago: University of Chicago Press, 299–305.
12. Darwin, C. 1877. *The Descent of Man, and Selection in Relation to Sex.* London: Murray.
13. Trivers, R. L. 1972. Parental investment and sexual selection. In *Sexual Selection and the Descent of Man 1871–1971.* ed., B. Campbell. Chicago: Aldine, 136–179.
14. Harvey, P. H., and A. H. Harcourt. 1984. Sperm competition, testes size, and breeding systems in primates. In *Sperm Competition and the Evolution of Animal Breeding Systems,* ed., R. L. Smith. New York: Academic Press, 589–600.
15. Hrdy, S. B. 1974. Male-male competition and infanticide among the langurs (*Presbytis entellus*) of Abu, Rajasthan. *Folia Primatol.* (*Basel*) 22(1):19–58.
16. Hrdy, S. B. 1977. Infanticide as a primate reproductive strategy. *Am. Sci.* 65:40–49.
17. Boggess, J. 1979. Troop male membership changes and infant killing in langurs (*Presbytis entellus*). *Folia Primatol* 32: 65–107.
18. Dolhinow, P., and G. Murphy. 1983. Langur monkey mother loss: Profile analysis with multivariate analysis of variance for separation subjects and controls. *Folia Primatol.* (*Basel*) 40:181–196.
19. Harris, T. R., and S. L. Monfort. 2003. Behavioral and endocrine dynamics associated with infanticide in a black and white colobus monkey (*Colobus guereza*). *Am. J. Primatol.* 61:135–142.
20. Palombit, R. A. 1994. Extra-pair copulations in a monogamous ape. *Animal Behaviour* 47:721–723.
21. Bartlett, T. Q. 1998. Within and between group social encounters among white-handed gibbons (*Hylobates lar*) in Khao Yai National Park, Thailand. *Am. J. Phy. Anthropol.* (Suppl.) 26:68–69.
22. Vigilant, L., Hofreiter, M., Siedel, H., and C. Boesch. 2001. Paternity and relatedness in wild chimpanzee communities. *PNAS* 98:12890–12895.
23. Smuts, B. B. 1985. *Sex and Friendship in Baboons.* New York: Aldine.
24. Tutin, C. E. G. 1980. Reproductive behaviour of wild chimpanzees in the Gombe National Park, Tanzania. *J. Reprod. Fertil.* (Suppl.) 28:43–57.
25. Richards, A. F. 1985. *Primates in Nature.* New Jersey: W. H. Freeman and Company.
26. Wrangham, R. W. 1979. On the evolution of ape social systems. *Social Science Information* 18:355–368.
27. Harcourt, A. H., K. J. Stewart, and M. Hauser. 1993. Functions of wild gorilla "close" calls. I. Repertoire, context and interspecific comparison. *Behaviour* 124:89–121.
28. Goldsmith, M., 1999. Gorilla socioecology. In *The Nonhuman Primates,* ed. P. Dolhinow, and A. Fuentes. Mountain View, CA: Mayfield Publishing Company, 58–63.
29. Chapman, C. A. 1990. Ecological constraints on group size in three species of neotropical primates. *Folia Primatologica* 55:1–9.
30. Leighton, M., and D. R. Leighton. 1982. The relationship of size of feeding aggregate to size of food patch: Howler monkeys (*Alouatta palliata*) feeding in *Trichilia cipo* fruit trees on Barro Colorado Island. *Biotropica* 14:81–90.
31. Overdorff, D. J. 1996. Ecological correlates to social structure in two lemur species in Madagascar. *Am. J. Phy. Anthropol.* 100:487–506.
32. White, F. J., and R. W. Wrangham. 1988. Feeding competition and patch size in the chimpanzee species *Pan paniscus* and *Pan troglodytes. Behaviour* 105:148–164.
33. Moraes, P. L. R., O. Carvalho, Jr., and K. B. Strier. 1998. Population variation in patch and party size in muriquis (*Brachyteles arachnoides*). *International Journal of Primatology* 19:325–337.
34. van Schaik, C. P. 1999. The socioecology of fission-fusion sociality in orangutans. *Primates* 40:69–86.

Fossil Evidence for the Emergence of Primates

As mammals underwent their adaptive radiation during the Cenozoic, some became terrestrial herbivores, some became carnivores, and some went back to the sea as whales, while others exploited the arboreal ecological niche that would give rise to the primates. This chapter will focus on the fossil evidence that traces the evolution of nonhuman primates.

Earliest Primate Roots

The fossil record of primate evolution dates back nearly 60 million years ago. The earliest common ancestor for all primates evolved from a rat-sized mammal that had plesiomorphic ("primitive") characteristics. The oldest primate-like animal with a reasonable fossil record is Plesiadapis.[1–6] The Plesiadapiformes were small mammals that collectively have been referred to as "archaic primates." The Plesiadapis were characterized by a long tail, agile limbs, fingers with claws rather than nails, rodent-like jaws and teeth, eyes at the side of the head, long snout, and no postorbital bar. There are several species that have been found in North America, Europe, and Asia, such as *Carpolestes simpsoni* and *Teilhardina asiatica.*

Carpolestes simpsoni

Carpolestes simpsoni (**fig. 7.1**) is a Paleocene primate found in Wyoming and Montana, USA,[6] that belongs to the Plesiadapiformes dating back 56 million years ago. This species had long fingers and an opposable hallux (an opposable big toe) with a nail rather than a claw. However, the eye sockets were on the side of the face rather than on the front of the face, which differ from the orbital convergence to contemporary primates. Therefore, grasping hands evolved in primates before visual specializations and leaping.[6]

Teilhardina asiatica

Chinese paleontologists[7] report the finding of a fossil skull from China, dating back 55 million years ago, referred to as *Teilhardina asiatica* (**fig. 7.2**). This fossil has several characteristics of modern primates (referred to as euprimates) such as a bony ring around each eye socket (orbit), forward rotation of the orbits, and a relatively large braincase. But the skull is very small, with an overall length of about 25 mm, and the animal's body mass is estimated at 28 g—less than in any modern primate species. The dentition is relatively primitive (unlike that of euprimates), including four premolars (the maximum known for any primate). This fact and its small body size suggests that the animal was an insectivore. Analysis of the skull morphology indicates that *Teilhardina asiatica* was a diurnal primate[7] characterized by relatively smaller orbits, but other research suggests that it may have been nocturnal.[8] In any event, the discovery of the Chinese primate fossil indicates that earlier primates lived in both Asia and Europe before the landmass of Eurasia was

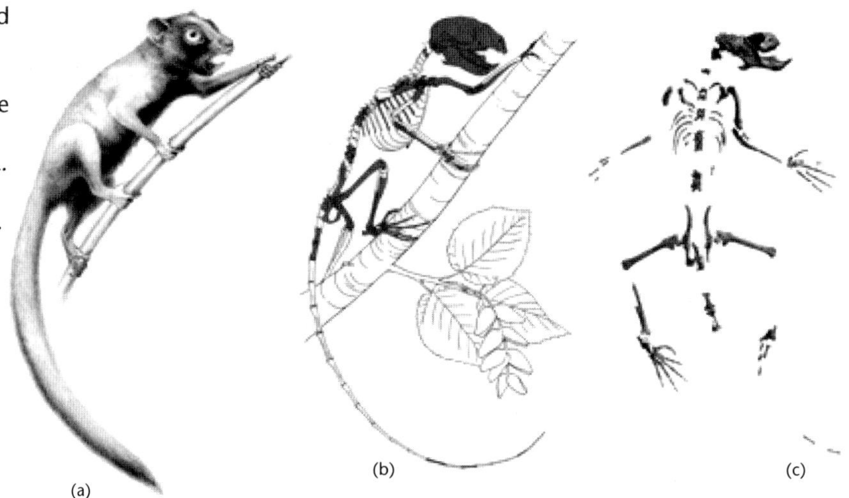

FIGURE 7.1. *Carpolestes simpsoni* found in Wyoming, USA. Reconstruction (a and b) of *C. simpsoni* (c). This specimen had an opposable big toe with a nail. Reprinted with permission from "Grasping Primate Origins" by J. L. Bloch and D. M. Boyer from *Science 298:*1606–1610. Copyright 2002 AAAS.

split. Prior to this finding, it was widely held that the only possibility for interchange of mammals between Asia and Europe was indirect migration across the Bering Straits, through North America, and across the Greenland land bridge, or vice versa, but such migration around 55 million years ago was ruled out by a transcontinental marine barrier.

Prosimians

Notharctus

The first unequivocal primates have been dated to the Eocene about 50 million years ago. Two main groups have been identified: Adapiformes, which are usually considered to be ancestral to modern lemurs, and Tarsiformes, which are (mostly) considered to be early ancestors of haplorhines or tarsiers. Several fossils have been found in North America, Western Europe, and Asia. These fossils had typical primate morphological features. **Fig. 7.3a** illustrates the craniofacial bones of *Notharctus* found in the western United States. *Notharctus*, like contemporary prosimians (**fig. 7.3b**), were characterized by a complete postorbital bar and eye sockets placed toward the front of the face, rather than the side.[9] They also had rounder braincases and relatively smaller snouts than other mammals. The jaws are made up of left and right parts joined in the middle, or at the mandibular symphysis, by strong ligaments. Like contemporary prosimians, *Notharctus* had 36 teeth with a dental formula of 2I, 1C, 3PM, 3M. They had small, unspecialized incisors, elongated digits with nails rather than claws, long hind limbs, and mobile toes.

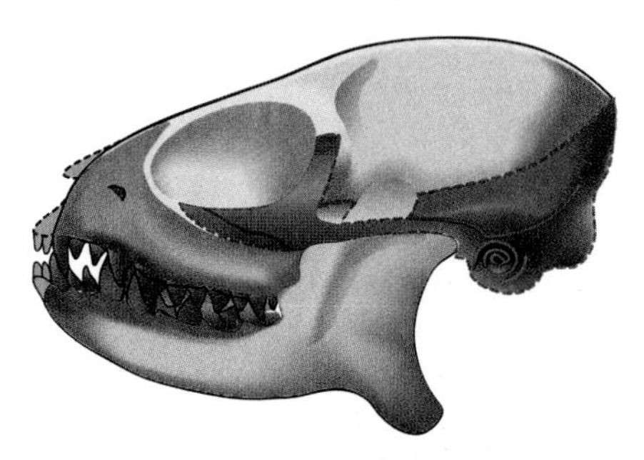

FIGURE 7.2. Primate ancestor from China. Reconstruction of the skull (grey shadow indicating the missing parts) of *Teilhardina asiatica*. Source: Adapted from Ni, Wang, Hu, and Li, fig. 1.[7]

In summary, Eocene primates such as *Notharctus* had clearly attained the prosimian grade of organization and established the fundamental adaptive trends of the primate order.[9] It was at this time that the adoption of a distinctive primate survival strategy emerged—the fine coordination of eyes and hands. For example, the placement of eye sockets toward the front of the face rather than the side suggests that these primates might have had overlapping fields of perception and binocular vision. Likewise, their relatively small snouts suggest that they had a reduced olfactory apparatus and, as such, they were beginning to depend more on eyesight to locate food, detect danger, and keep track of one another. Finally, their elongated digits with nails rather than claws, long hind limbs, and mobile toes suggest that the Eocene primates committed themselves to a life in the trees beyond that seen in the more generalized Paleocene forms.

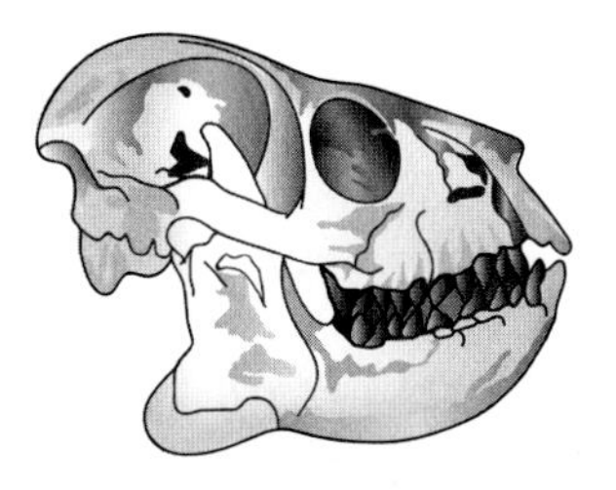
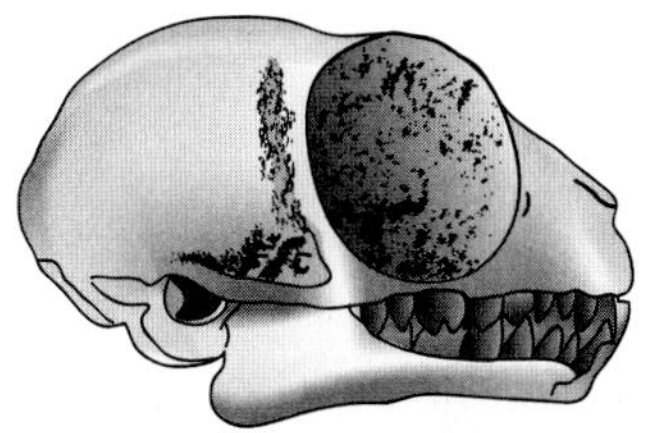

(a) *Notharctus* (b) Living *Tarsier*

FIGURE 7.3. *Notharctus*, found in the western United States, is characterized by the postorbital bar of bone. *Notharctus* (a) shows the same craniofacial form as the living prosimians (b). (a) From fig. 2 of Harwig, W.C. 1999. Primate Evolution. In *The Nonhuman Primates.* Source: P. Dolhinow, and A. Fuentes, 10–17. CA: Mayfield Publishing Company.

Anthropoids

Aegyptopithecus

The ancestry of Anthropoids, such as Old World and New World monkeys, can be traced back to the genera *Apidium* and *Aegyptopithecus*, which lived during the early **Oligocene** about 40 to 35 million years ago.[10–13] *Aegyptopithecus* was relatively heavy bodied, roughly the size of a modern howler monkey weighing about 5.9 kg to 9.1 kg (13 to 20 pounds) (**fig. 7.4**). Several fossils of *Aegyptopithecus* were found in Egypt in the locality called Fayum. These genera are characterized by a skull with a complete wall of bone around the eye called a postorbital plate (an anthropoid feature) and only two premolar teeth (a defining feature of Old World monkeys, apes, and humans). *Aegyptopithecus* differs from prosimians in having 32 teeth like modern primates. They had a dental formula of 2.1.2.3. The brain size was intermediate between that of prosimians and other anthropoids. The visual cortex was large compared to the prosimians, but the snout and nasal cavity were reduced. As the olfactory apparatus diminished, the orbits moved further forward on the face, and each eye was protected by a complete postorbital plate. Furthermore, the two halves of the mandible were firmly fused at the chin like in contemporary anthropoids. They were omnivorous and lived on fruits, leaves, and some insects as inferred from the shape of the molars and incisors.

In summary, early Oligocene primates such as *Aegyptopithecus* show characteristics that bridge the gap between the basic prosimian traits of the Eocene and the advanced monkeys and apes that first appear in the Miocene—the stock from which the Old World and New World monkeys will evolve.

New World Monkeys

Cebupithecia

Several kinds of fossils have been found at the locality of La Yenta in Colombia.[14] These primates are called *Cebupithecia* and they lived 16 to 12 million years ago during the **Miocene.** The *Cebupithecia* is supposed to be the ancestor of the living New World monkeys[14,15] (**fig. 7.5**). These had 36 teeth and, like the living New World monkeys, retained the third premolar of their early mammalian progenitors, so that they had a dental formula of 2.1.3.3. The teeth morphology suggests a frugivore-folivore dietary pattern.

Old World Monkeys

Sixteen to twelve million years ago, during the Miocene, descendants of *Aegyptopithecus* (**fig. 7.4**) gave rise to the Old World monkeys.[16,17] Fossils from this time period have been found in Africa, Europe, and Asia. The anthropoid primates of the Miocene were much larger than their Oligocene forebear, ranging from about the size of a vervet monkey, weighing about 20 lbs to the size of a large chimpanzee, weighing about 100 lbs. The few skulls show reduction of the olfactory apparatus and the braincase to be full and round, as in modern apes and monkeys. *Aegyptopithecus* had 32 teeth as in humans and dental formula of 2.1.2.3.

In summary, through the evolution of primates the number of teeth has been reduced from 36 in the prosimians, to 32 in the anthropoids. These changes are also accompanied by modifications in tooth shape

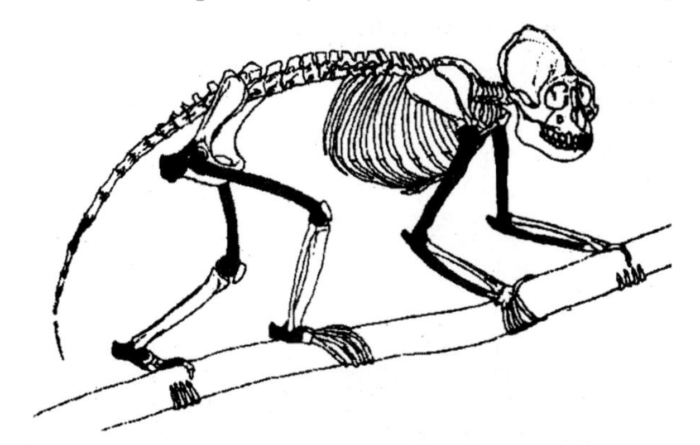

FIGURE 7.4. *Aegyptopithecus* found in Egypt. Probable ancestor of both Old World and New World monkeys. Source: Redrawn from Ankel-Simons, F., J. G. Fleagle, and P. S. Chatrath. 1998. Femoral anatomy of *Aegyptopithecus zeuxis*, an early Oligocene anthropoid. From *American Journal of Physical Anthropology,* 106(4) by F. Ankel-Simons et al. Copyright © 1998 by Wiley-Liss, Inc. Reprinted with permission of Wiley-Liss, Inc., a subsidiary of John Wiley & Sons, Inc.

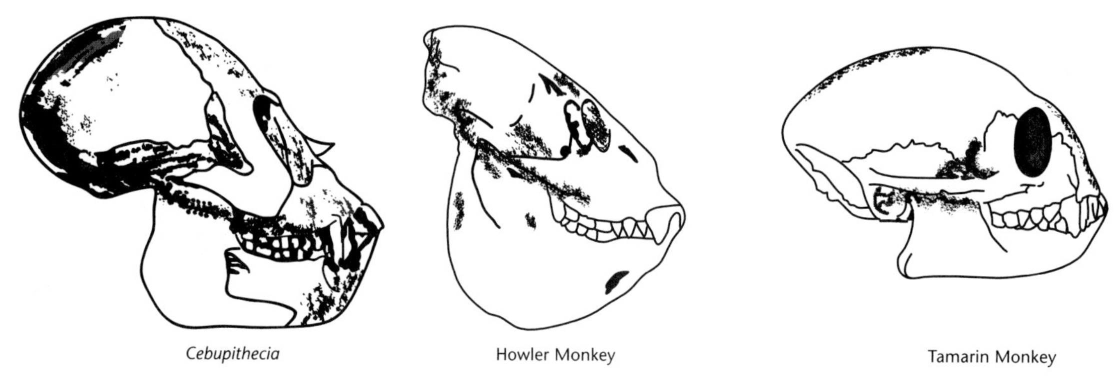

FIGURE 7.5. *Cebupithecia* lived in South America about 16 to 12 million years ago. Probable ancestor of New World monkeys. Note the similarities with the skulls of contemporary New World monkeys on the right. (a) From figure 2 of Harwig, W. C. 1999. Primate Evolution. In: Dolhinow, P., and A. Fuentes, *The Nonhuman Primates* 10–18. Mayfield Publishing Company: California.[9]

from the simple puncturing and crushing teeth of insectivorous prosimians to the powerful grinders of modern humans and their fossil hominid ancestors. Likewise the jaw morphology also evolved from a unit formed by two bones joined by strong ligaments to a single rigid bone.

Hominoids

Proconsul: Dryopithecine Pattern

The hominoid primates originated in Africa during the early **Miocene** about 18 million years ago from primates belonging to the genus *Proconsul.*[3,18–20] Several fossils of the genus *Proconsul* (**figs. 7.6, 7.7** and **7.8**) have been found in the African Rift Valley. Descendants of *Proconsul* later migrated over to the tropical latitudes of eastern Africa and the neotropical latitudes of Europe, Asia, and Southeast Asia. As such, members of the *Proconsul* genus have also been found in Turkey, Pakistan, Russia, China, and India. Tooth morphology and postcranial anatomy show *Proconsul* to be a fruit-eating, arboreal quadruped. The general anatomy shows a mosaic of features that illustrate more specific similarities with later forms. They were large bodied, weighing 20 to 30 kg, and their brain was as large as other living hominoids. *Proconsul* had eye sockets located on the front of the face, rather than on the side. The limb structure also shows a mosaic of features. The elbow, shoulder joint, and feet resemble those of a chimpanzee, the wrist resembles that of monkeys, and the lumbar vertebrae are like those of a gibbon.[20]

Like all cercopithecines, *Proconsul* had a dental formula of 2.1.2.3 for 32 teeth. However, *Proconsul* is

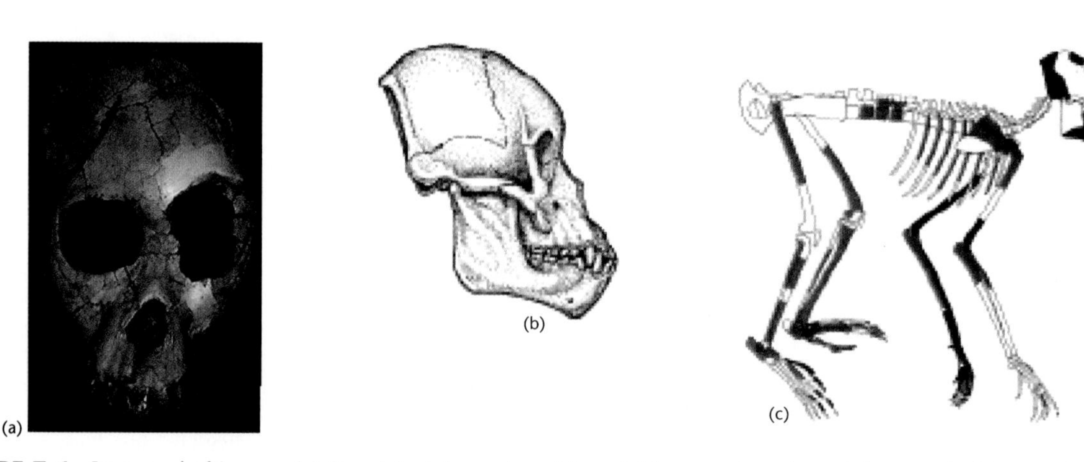

FIGURE 7.6. *Proconsul africanus.* (a) Craniofacial remains; (b) and (c) reconstruction of *Proconsul africanus.* Source: (a) Courtesy of Professor Milford Wolpoff, University of Michigan, Ann Arbor. (b) and (c) adapted from Pilbeam, fig. 3.[3]

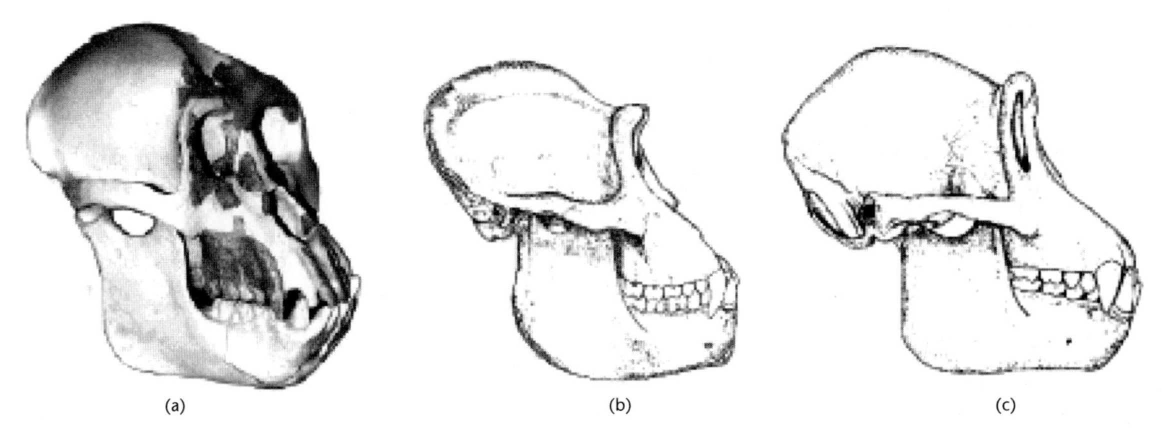

FIGURE 7.7. *Proconsul* (a) dated to about 23 to 17 million years ago found on Rusinga Island in Kenya. Probable ancestor of hominoids such as great apes (b) gorilla and (c) chimpanzee. Source: (a) From Kohler, Moya-Sola, and Alba, fig. 2.[18] From *American Journal of Physical Anthropology, 115* by M. Mohler et al. Copyright © 2001 by Wiley-Liss, Inc. Reprinted by permission of Wiley-Liss, Inc., a subsidiary of John Wiley & Sons, Inc.

the first example of a uniquely apelike molar tooth. A unique trend in hominoids is an elaboration of the cusp pattern of the molars. Throughout primate evolution, the trend in the morphology of the molars has been to increase the number of cusps from two to four or five. *Proconsul* molars have 5 cusps known as the **Y-5** pattern or **dryopithecine** pattern (**fig. 7.8**). An effect of the Y-5 pattern of the molar is that it expands the area of the tooth, which probably was an adaptation to a herbivorous diet. All specimens found during the Miocene have the dryopithecine molar pattern and as such are assumed to belong to a single related subfamily of Pongidae. According to the investigations of MacLatchy and associates,[21,22] the fossil referred to as *Morotopithecus* was the oldest ape from which the hominoids might have evolved.

Sivapithecus (**fig. 7.9**), dated to about 12 to 8 million years ago and found in Pakistan and Turkey, is considered to be the ancestor of present-day orangutans and gibbons.[23]

Summary

The earliest common ancestor for all primates evolved from a rat-sized mammal that had plesiomorphic ("primitive") characteristics nearly 60 million years ago. As shown by the recent findings from North America, Europe, and Asia, the ancestor of primates gave rise to the contemporary prosimians such as lemurs, lorises, and tarsiers about 55 million years ago. From there onwards, the prosimian level of organization and the first signs of distinctive primate charac-

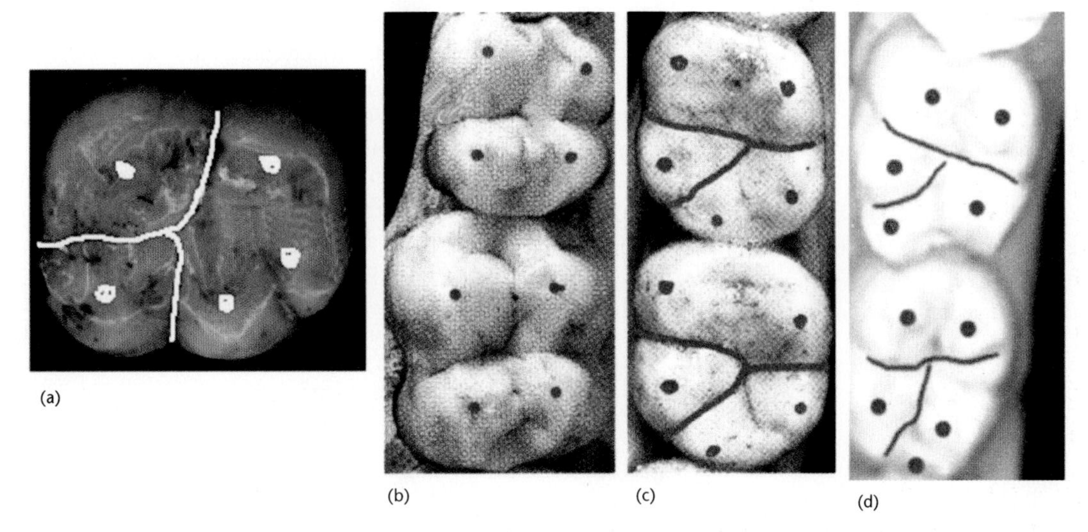

FIGURE 7.8. Dryopithecine or Y-pattern of the molar teeth. (a) *Proconsul*, (b) baboon monkey, (c) orangutan, and (d) modern humans. Both hominoids and humans have a similar Y-pattern like *Proconsul* (a) who was the ancestor of hominoids but differ from that of the baboon. (Illustration made by the author.)

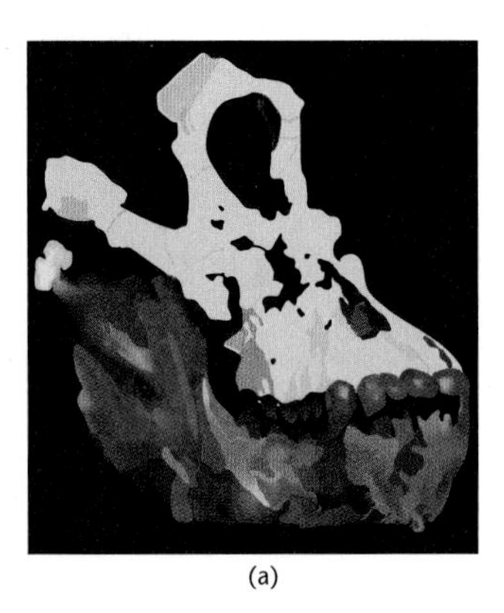

(a)

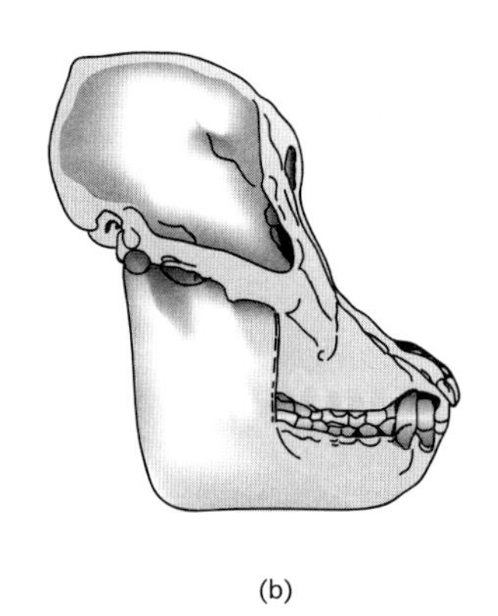

(b)

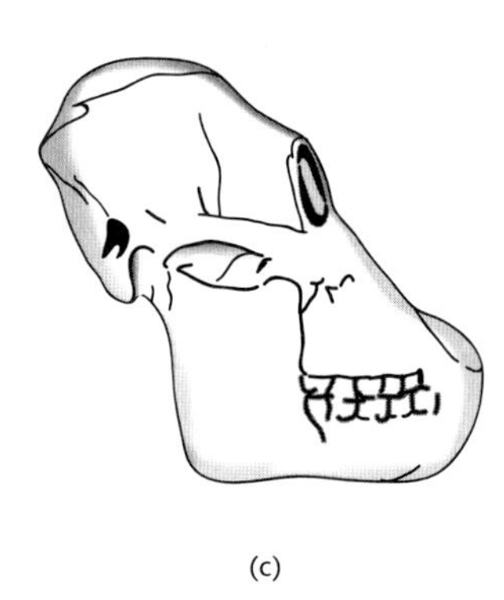

(c)

FIGURE 7.9. *Sivapithecus.* (a) Craniofacial remains and (b) reconstruction of *Sivapithecus.* The facial features of *Sivapithecus* (a) are similar to the modern orangutan (c), suggesting a close evolutionary link. Source: (a) Photo courtesy of Professor Milford Wolpoff, University of Michigan, Ann Arbor; (b) and (c) from Pilbeam, fig. 2.[3]

teristics appear, such as relatively larger brains, dependence upon eyesight, unspecialized dentition, and elongated digits for grasping with nails rather than claws. Continuing this evolutionary trend, the animals that lived during the Oligocene exhibit features that characterize anthropoids. The evidence suggests that the New and Old World (cercopithecoid) monkeys, and the hominoids might have evolved from Oligocene primates. The Oligocene primates, in turn, gave rise to *Dryopithecus,* which included the common ancestor of pongids and hominids.

Throughout evolution, acquisition of new traits without loss of versatility was a central theme of primate origins that have enabled them to evolve further than other mammal species. It is in the arboreal environment where primates acquired their unique traits, such as binocular vision, prehensile hands, opposable thumbs and fingers with nails rather than claws.

Suggested Readings

Cartmill, M. 1972. New views on primate origins. *Evolutionary Anthropology* 1:105–111.

Conroy, G. C. 1990. *Primate Evolution.* New York: W. W. Norton.

Fleagle, J. G. 1988. *Primate Adaptation and Evolution.* San Diego: Academic Press.

Grine, F. E., J. G. Fleagle, and L. B. Martin. eds. 1987. *Primate Phylogeny.* London, San Diego: Academic Press.

Literature Cited

1. Martin, R. D. 1990. *Primate Origins and Evolution: A Phylogenetic Reconstruction.* Princeton, NJ: Princeton University Press.
2. Carroll, R. L. 1988. *Vertebrate Paleontolgy and Evolution.* New York: Freeman and Company.
3. Pilbeam, D. R. 1984. The descent of hominoids and hominids. *Sci. Am.* 250:84–97.
4. Bauer, K. 1996. Primate phylogeny from a human perspective: A study based on the immunological technique of comparative determinant analysis (CDA). Stuttgart, Germany, New York: G. Fischer.
5. Szalay, F. S., M. J. Novacek, and M. C. McKenna, eds.1993. *Mammal Phylogeny.* New York: Springer-Verlag
6. Bloch, J. L., and D. M. Boyer. 2002. Grasping primate origins. *Science* 298:1606–1610.
7. Ni, X., Y. Wang, Y. Hu, and C. H. Li. 2004. A euprimate skull from the early Eocene of China. *Nature* 427:65–68.
8. Martin, R. D. 2004. Chinese lantern for early primates. *Nature* 427:22–23.
9. Harwig, W. C. 1999. Primate Evolution. In *The Nonhuman Primates.* 10–18. California: Mayfield Publishing Company.
10. Simons, E. L., and D. T. Rasmussen. 1989. Cranial morphology of *Aegyptopithecus* and *Tarsius* and the question of the tarsier-anthropoidean clade. *Am. J. Phys. Anthropol.* 79(1):1–23.
11. Simons, E. L. 1997. Discovery of the smallest Fayum Egyptian primates (Anchomomyini, Adapidae). *Proc. Natl. Acad. Sci. USA* 94(1):180-184.
12. Seiffert, E. R., and E. L. Simons. 2001. Astragalar morphology of late Eocene anthropoids from the Fayum Depression (Egypt) and the origin of catarrhine primates. *J. Hum. Evol.* 41(6):577–606.
13. Jaeger, J. J., T. Thein, M. Benammi, Y. Chaimanee, A. Naing Soe, T. Lwin, T. Tun, S. Wai, and S. Ducrocq. 1999. A new primate from the Middle Eocene of Myanmar and the Asian early origin of anthropoids. *Science* 15; 286:528–530.
14. Meldrum, D. J., and R. F. Kay. 1997. *Nuciruptor rubricae,* a new pitheciin seed predator from the Miocene of Colombia. *Am. J. Phys. Anthropol.* 102(3):407–427.
15. Kay, R. F., D. Johnson, and D. J. Meldrum. 1998. A new pitheciin primate from the Middle Miocene of Argentina. *Am. J. Primatol.* 45(4):317–336.

16. Benefit, B. R., and M. L. McCrossin. 1993. Facial anatomy of *Victoriapithecus* and its relevance to the ancestral cranial morphology of Old World monkeys and apes. *Am. J. Phys. Anthropol.* 92(3):329–370.

17. Benefit, B. R., and M. L. McCrossin. 1997. Earliest known Old World monkey skull. *Nature.* 388(6640):368–371.

18. Ward, S., B. Brown, A. Hill, J. Kelley, and W. Downs. 1999. *Equatorius*: A new hominoid genus from the Middle Miocene of Kenya. *Science* 285:1382–1386.

19. Richmond, B. G., J. G. Fleagle, J. Kappelman, and C. C. Swisher III. 1998. First hominoid from the Miocene of Ethiopia and the evolution of the catarrhine elbow. *Am. J. Phys. Anthropol.* 105:257–277.

20. Kohler, M., S. Moya-Sola, and D. M. Alba. 2001. Cranial reconstruction of *Dryopithecus* without forming a sagittal crest. *Am. J. Phys. Anthropol.* 115:284–288.

21. MacLatchy, L., D. Gebo, R. Kityo, and D. Pilbeam. 2000. Postcranial functional morphology of *Morotopithecus bishopi,* with implications for the evolution of modern ape locomotion. *J. Hum. Evol.* 39:159–183.

22. Young, N. M., and L. MacLatchy. 2004. The phylogenetic position of *Morotopithecus. J. Hum. Evol.* 46:163–184.

23. Andrews, P., and J. E. Cronin. 1982. The relationships of *Sivapithecus* and *Ramapithecus* and the evolution of the orangutan. *Nature.* 297(5867):541–546.

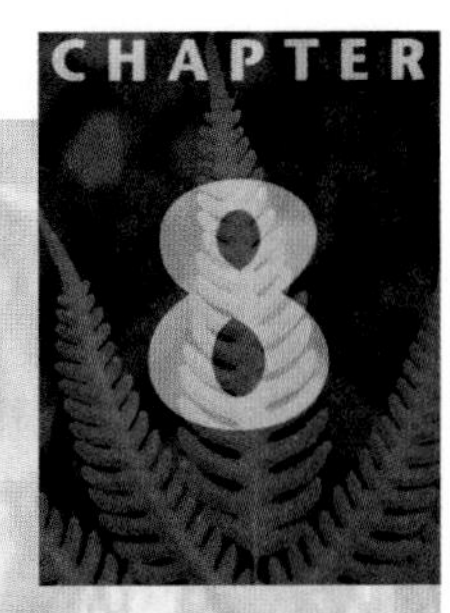

Evolution of the Digestive System of Omnivores

In general, the digestive tract of herbivores, carnivores, and omnivores is composed of the mouth, tongue, teeth, esophagus, stomach, and the small and large intestines, but the relative function and specialization of each compartment differ depending upon the digestive strategy used to extract nutrients from the foods on which they came to depend. The survival of any organism depends on its capacity to extract nutrients from the foods that it eats. Throughout evolution, mammals have developed an efficient digestive system for extracting nutrients from plant foods. As primate ancestors became omnivores, the capacity to extract nutrients from plant foods diminished. Reflecting this change, the digestive system of primates and humans were transformed both morphologically and physiologically. This chapter will focus first on the human digestive system, then the different digestive strategies and the trajectory of the digestive system from herbivores to omnivores will be discussed.

Omnivore Digestive System: Human

The digestive system of humans is similar to other primates, differing only in some aspects that can be traced to the dietary pattern on which they have come to depend. The digestive system includes the gastrointestinal tract and accessory digestive organs. The gastrointestinal tract includes the oral cavity, pharynx, esophagus, stomach, small intestine, large intestine, and rectum (**fig. 8.1**). The teeth, tongue, salivary glands, liver, gallbladder, and pancreas are the accessory digestive organs.

The Oral Cavity and Pharynx

The first chamber of the typical mammalian digestive tract is the oral cavity, which contains teeth that function in the mechanical breakdown of food by biting, shearing, and chewing. In the oral cavity, food is

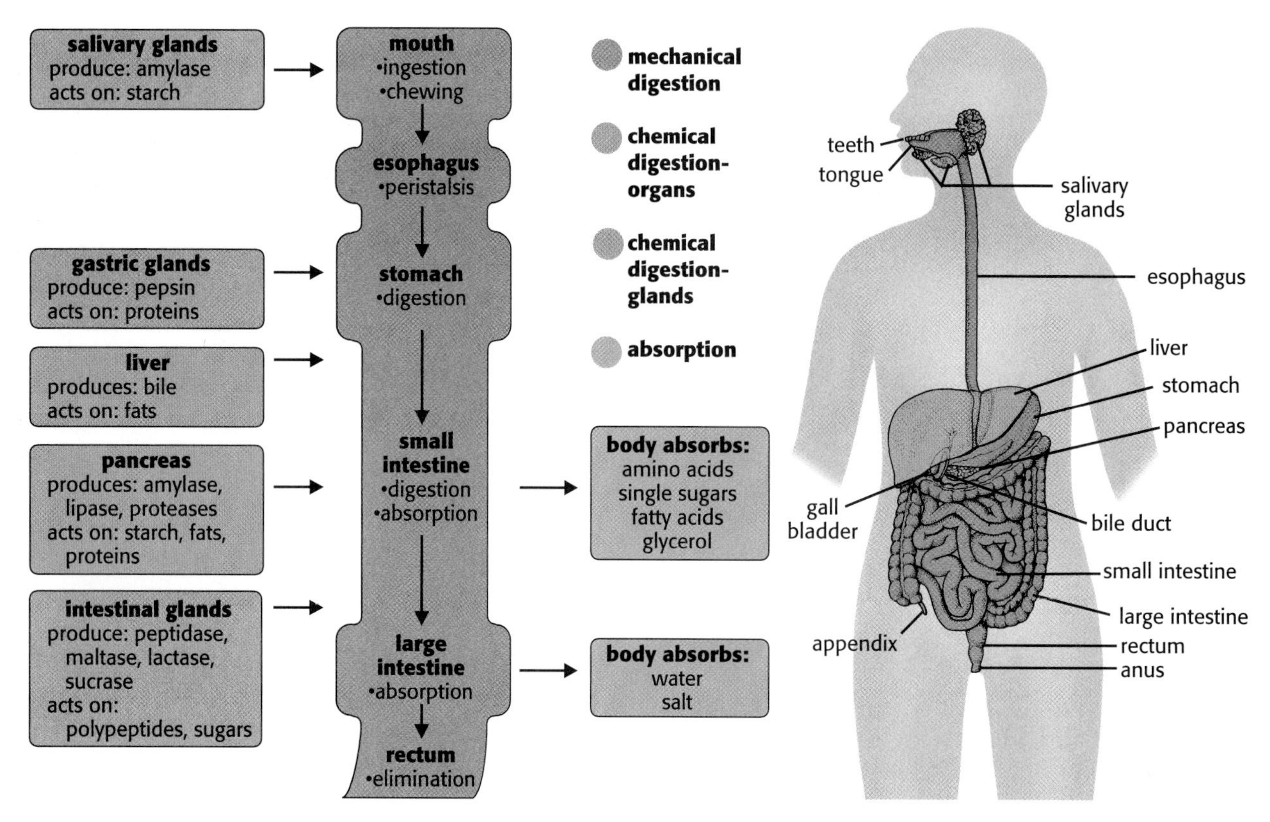

FIGURE 8.1. The human digestive system. The digestive system is a continuous tract that starts in the mouth and continues through the rectum, where digestion of food and eventual absorption of nutrients takes place. From *BSCS Biology: An Ecological Approach, 9th Edition* by BSCS. Copyright © 2002 by BSCS. Reprinted by permission.

tasted, smelled, and mixed with saliva secreted by several sets of salivary glands. The saliva dissolves some of the food and acts as a lubricant, facilitating passage through the subsequent portions of the digestive tract. The saliva contains a starch-digesting enzyme called **amylase** (ptyalin), which initiates the process of enzymatic hydrolysis; it splits starch into molecules of the double sugar maltose. Many carnivores, such as dogs and cats, have no amylase in their saliva since their natural diet contains very little starch. The tongue manipulates the food as it is chewed, forming it into a mass called a bolus in preparation for swallowing. The tongue pushes the bolus backward through the pharynx and into the esophagus. The control of swallowing is done by the epiglottis, a small flap that slips over the top of the trachea, or windpipe, and prevents food from entering the respiratory tract and lungs.

The Esophagus

The dissolved macromolecules pass through the **pharynx** and **esophagus** into the stomach. The esophagus is a tube in the upper portion of the abdominal cavity, which extends from the pharynx through the neck and chest to the stomach. The dissolved macromolecules, called a bolus, are pushed along by waves of muscular contractions in a process called peristalsis, passing through the esophagus into the stomach. The sphincter, located at the junction between the esophagus and the stomach, closes the entrance to the stomach after food enters into the stomach, thus preventing the contents of the stomach from moving back into the esophagus. However, the sphincter opens when a peristaltic contraction from the esophagus reaches it.

The Stomach: Digestion

After food passes through the esophagus, it rapidly enters the stomach, where the enzymatic digestion begins. The stomach, located at the end of the esophagus, is a saclike organ that consists of three sets of muscles: an inner mucous membrane of connective tissue and many glands, a thick middle layer of smooth muscle, and an outer layer of connective tissue. The muscle layer contains fibers running around the stomach, others running longitudinally, and still others oriented diagonally. Thus, the stomach is capable of a great variety of movements. When it contains food, the stomach is swept by powerful waves of contraction, which churn the food, mixing and breaking it. In this manner, the stomach supplements the mechanical action of the teeth. In the stomach, food is mixed with the gastric secretions to form a homogeneous mass called **chyme**.

The gastric secretions consist of (a) hydrochloric acid (secreted by parietal cells in the gastric mucosa, or lining); (b) mucus (secreted by goblet cells); (c) pepsinogen, an inactive form of the enzyme pepsin (secreted from chief cells); (d) the hormone gastrin (secreted by G cells); and (e) intrinsic factor, a protein that is required for the absorption of vitamin B_{12} in the intestine. Once the food is in the stomach, the hormone gastrin is released and, in turn, this hormone stimulates the secretion of hydrochloric acid (HCl) and **pepsin** from the stomach wall. The HCl is a powerful acid ($pH < 2.0$) that performs two main functions. First, it dilutes ingested proteins, unfolding the molecules in a way that makes their breakdown easier. Second, it converts the pepsinogen into the active enzyme pepsin. Then the pepsin goes on to break down proteins into small peptides and amino acids. The wall of the stomach is protected from the strongly acidic environment in the stomach by the secretion of mucus, which forms a protective coat over the lining, or mucosa, of the stomach. The chyme mixes in the stomach for approximately 4 to 6 hours, preparing the nutrients in food for their eventual absorption. The **chyme** contains molecular fragments of proteins and polysaccharides and droplets of fat.

The Small Intestine: Nutrient Absorption

The digestion and absorption is completed in the **small intestine**. The absorption of nutrients in the small intestine is assisted by the action of **bile** secreted by the **liver** and digestive enzymes and a fluid rich in bicarbonate ions secreted by the **pancreas**. These secretions enter the small intestine through a duct leading from the pancreas to the duodenum. The function of the pancreatic bicarbonate ions is to neutralize the high acidity of the chyme coming from the stomach. The fat droplets entering the small intestine from the stomach are dissolved by the action of bile salts secreted by the liver. Between meals, secreted bile is stored in the **gallbladder**, which is a small sac located underneath the liver. During a meal, the smooth muscles in the gallbladder wall contract, causing a concentrated bile solution to be injected via the bile duct into the duodenum.

The small intestine is a long coiled tube that is 2.4 to 3.6 meters (9 to 12 feet) long with an inner surface that has fingerlike projections, or circular folds. Although the small intestine is the longest organ in the body, it is called small because its diameter (about 8.8 cm or 1.5 inches) is less than that of the large intestine. The small intestine consists of the duodenum, jejunum, and ileum. These folds are 8 to 10 mm in height and increase the intestinal surface area to at least three times what it would be without them. On the surface of these folds and over the rest of the surface of the small intestine are small projections called villi. Villi are from 0.5 to 1.5 mm high and are tightly packed at a density of about 20 per 40 mm^2. The villi increases the inner surface area of the small intestine to about 15 times what it would be without the villi and circular projections. Each villus is, in turn, covered with even smaller projections called microvilli, which are each 1 μm (micro millimeters) high and are packed together at a density of 200,000 per mm^2. The microvilli multiply the inner surface area of the small intestine by 20 times. The combined effect of the circular folds, the villi, and the microvilli increases the total surface area of the small intestine by about 300 times the size it would be if the small intestine were simply a smooth-walled tube. The surface of each villus is made of a one-cell-thick layer of specialized absorptive cells.

In the small intestine, monosaccharides and amino acids are absorbed by specific carrier-mediated transport processes in the plasma membranes of the intestinal epithelial cells, whereas fatty acids enter these cells by diffusion. Most mineral ions are actively absorbed, and water diffuses passively down osmotic gradients through the small intestine. In addition to the digestion and absorption of nutrients, the small intestine participates in the protection of the host through a strong defense against aggressions from the external environment. The small intestine defends the organism through the action of the microflora, mucosal barrier, and local immune system. The gut microflora plays a major role against exogenous bacteria through colonization resistance. The intestinal mucosa provides a cellular barrier against foreign substances and exogenous microorganisms. The intestine's immune system associated with the lymphoid tissue enables the organism to tolerate dietary antigens and rejects enteropathogenic microorganisms that may challenge the body's defenses.

The Large Intestine: Excretion and Reabsorption

The undigested material is passed to the **large intestine** (or **colon**). The large intestine also consists of three parts: the cecum, the colon, and the rectum. The undigested material is passed to the large intestine (or colon). The **large intestine** is 10-feet long and 5 inches diameter. As the mass enters the large intestine, water is absorbed into the body, resulting in a progressively more solid mass that eventually reaches the anus as semisolid fecal matter, or feces. From there, undigested matter passes to the **rectum** and the associated sphincter muscles for its **defecation** in the form of **feces**. The feces consist almost entirely of bacteria and ingested material such as dietary fiber that was not digested and absorbed. On average, the feces contain about 10 to 20 percent of undigested and nonabsorbed food. The feces may also contain nutrients that are part of the various digestive secretions, which are not reabsorbed. For example, the cells lining the intestinal tract are completely replaced every 3 to 4 days, with the old cells and any nutrients they contain excreted in the feces. Excretion of some nutrients occurs through the skin, in perspiration; within epithelial cells, which are constantly dying and being shed from the skin; or within hair and nails. Dietary fiber increases the bulk of the stools and facilitates their passage through the colon. Lack of fiber in the diet produces dry, hard stools that are passed with difficulty. This results in constipation and sometimes contributes to diverticulitis of the large intestine.

Small amounts of certain metabolic products are excreted via the gastrointestinal tract, but the elimination of most waste products from the internal environment is achieved by the lungs and kidneys. Almost all of the carbon dioxide waste and 10 percent of the excreted water is excreted from the lungs. The kidneys selectively filter waste products of metabolism from the blood and excrete them within urine. The kidneys also allow excess nutrients, such as water-soluble vitamins, to leave the body in urine. However, as the blood is filtered through the kidneys, they reabsorb almost all of the glucose and protein present. With some other nutrients, such as sodium, the kidneys reabsorb only the amounts needed to maintain normal blood and tissue levels and excrete any excess in urine. The kidneys are the central organs that maintain the organism's homeostasis.

In summary, the gastrointestinal tract is a highly complex ecosystem whose major function is to digest food and absorb nutrients. In addition, the gastrointestinal tract protects the organism through the action of three lines of defense: the intestinal microflora, intestinal mucosa, and intestinal immune system. To understand the difference between the human digestive system and that of primates we need to trace the trajectory of the digestive system from herbivores to omnivores.

Digestive Strategies of Herbivores

Nutrients in plant foods such as fruits, seeds, leaves, and flowers are encased in cellulose, yet vertebrate animals do not produce the enzymes necessary to break down this abundant material. However, many microbes secrete enzymes that break the cellulose down, allowing them to utilize plant food and other plant wall materials as dietary sources. These microbes, whose functions are to ferment and break down the cellulose, inhabit the digestive tract of all animals. Their amount and location depend on the digestive strategies that the animals use. Among herbivores two distinct strategies to facilitate fermentation have evolved.[1,2] These include (a) monogastric (hindgut) fermentation, and (b) multigastric fermentation (or rumination).

Monogastric Fermentation (Hindgut Fermentation)

Nonruminant herbivores are usually referred to as monogastric or hindgut fermenters. Hindgut fermenters (**fig. 8.2**), such as elephants, horses, rabbits, and other herbivores, utilize cellulose and other fermentable substrates, but, lacking fore-stomachs, perform fermentation in their large intestines. In all monogastric, nonruminant herbivores the large intestine is massive and anatomically complex. The colon itself is huge in size and is divided into three segments: ascending colon (by far the largest), transverse colon, and descending colon (small colon). In monogastric animals, dietary protein is digested and absorbed as amino acids and much of the soluble carbohydrate is hydrolyzed (diluted with water) and absorbed as monosaccharides in the small colon. Cellulose and related molecules pass through the small colon intact and then some enter into the cecum and some into the large intestine. In the large intestine the cellulose and hemicellulose, along with starch and other soluble carbohydrates that escape digestion in the small intestine, are subjected to fermentation.

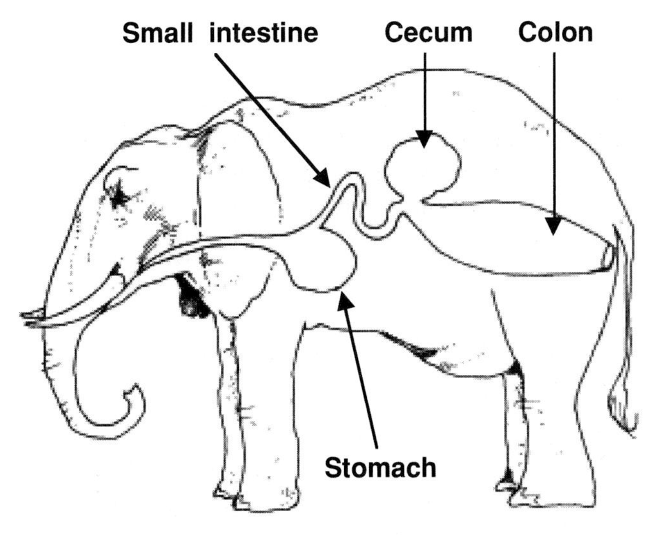

FIGURE 8.2. Monogastric herbivores, also referred to as hindgut fermenters, include most grazing animals that digest their food in the large intestine and cecum. The stomach of an elephant works as temporary storage and preliminary fermentation of the food. (Drawing made by Paul Bohensky.)

The process of fermentation that occurs in the large intestine hindgut is so efficient that the animals are able to extract all the essential amino acids. They do this because they are able to produce large quantities of bacteria capable of breaking the cellulose that encases the protein of plant cells down. However, the digestive strategy of monogastric herbivores is less efficient than that of ruminants. Compared to ruminants, monogastric herbivores waste a large quantity of microbial protein generated in the equine large gut because there is no opportunity in the large intestine hindgut for significant absorption of amino acids. Hence, nonruminant herbivores digest only about 45 percent of the ingested cellulose.[1,2] To compensate for this deficiency, monogastric herbivores (hindgut fermenters) need to consume large quantities of food per day. Furthermore, monogastric herbivores, because they do not recycle urea, need to drink daily to produce sufficient water to balance the urea in the urine.

Multigastric Fermentation (Rumination)

Anatomy

Ruminants are herbivorous animals that ferment foodstuffs prior to their entry into the glandular stomach (so-called foregut fermentation) (**fig. 8.3**). Ruminants regurgitate and masticate their food after swallowing. Like in monogastric herbivores, the digestive tract of ruminants is composed of the mouth, tongue, teeth, esophagus, stomach, and the small and large intestines. The difference is that the stomach is divided into either three or four compartments: **rumen**, **reticulum**, **omasum**, and **abomasum**. In ruminants such as cows, sheep, goats, antelope, deer, gazelles, and giraffes, the stomach is divided into four chambers, while ruminants such as the chevrotains, camel, dromedary, llama, alpaca, guanaco, and vicuña have a stomach divided into three chambers (rumen, omasum, and abomasum).

Rumen. The rumen, by far the largest of the four stomachs, is itself sacculated by muscular pillars called the dorsal, ventral, caudodorsal, and caudoventral sacs. The rumen is the main fermentation vat where billions of microorganisms attack and break down the relatively indigestible food components of the ruminant's diet. The rumen is a fermentation vat that provides an anaerobic environment, constant temperature and pH, and good mixing. The rumen is the most intricate and sensitive part of the ruminant's digestive system, providing a complex environment composed of billions of bacteria, protozoa, and fungi. These microbes, or lysozymes, work together to attack and digest the consumed food. There are two main types of microbes: fiber- (forage) digesting and starch (feed grains) digesting bacteria. The fiber-digesting bacteria attack forages and other fiberlike materials. These lysozymes are most active at a low pH (around 6.7). Once food enters into the rumen, huge quantities

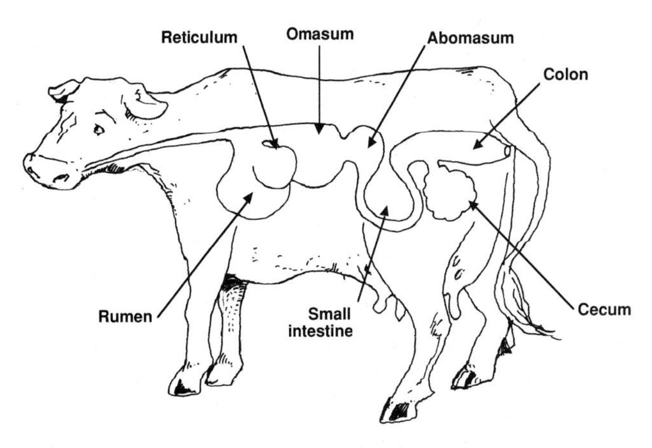

FIGURE 8.3. Multigastric fermentation, also referred to as foregut fermenters or ruminants, include cows, sheep, chevrotains, and cameloids that digest their food in multicompartmentalized digestive chambers: rumen, reticulum, omasum, and abomasum. In ruminants, food is chewed several times. It is fermented and digested by bacteria and microorganisms in the rumen and the reticulum. (Drawing made by Paul Bohensky.)

of fermentative bacteria flow into the stomach. Depending on the feeding, these bacteria can double in number within 1 to 10 hours. The starch-digesting bacteria grow best at an acidic pH (<7.4). It is said that the evolution of foregut fermentation was accompanied by recruitment of lysozymes as lytic digestive enzymes, and that the selection of this enzyme to act in stomach fluid has driven its molecular evolution.

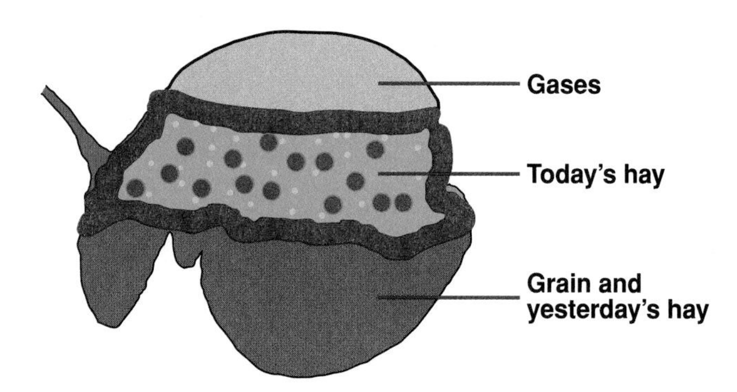

FIGURE 8.4. Functional partition of the rumen into three primary zones based on their specific gravity. Gas rises to fill the upper regions, grain and fluid-saturated roughage ("yesterday's hay") sink to the bottom, and newly arrived roughage floats in a middle layer. (Composite made by the author.)

Reticulum. The reticulum functions as an intermediary connection between the rumen and the omasum. It secretes hydrochloric acid to be used by the rumen in fermentation. The contents of the reticulum and rumen intermix freely because the wall of the reticulum is honeycomb in structure.

Omasum. The omasum's function is to absorb residual volatile fatty acids, water, minerals, nitrogen, and bicarbonates. The omasum acts as a filter pump to sort liquid from fine food particles. Periodic contractions of the omasum break and separate flakes of material out of the leaves for passage into the abomasum.

Abomasum. The abomasum is a true stomach and functions very similarly to the stomach of a monogastric animal. The abomasum secretes hydrochloric acid, pepsin, rennin, and the enzyme lysozyme in order to efficiently break down bacterial cell walls.

Process of Digestion

Physical Process

Initially the undigested foods, such as grass, hay, water and saliva, are delivered to the rumen through the esophageal orifice. Next the food passes into the rumen, where it is fermented by microorganisms. Ruminants produce large quantities of saliva, which provides fluid for the fermentation vat, large amounts of alkaline saliva is rich in bicarbonate, which buffers the large quantity of acid produced in the rumen. Once the food enters the rumen, it is partitioned into three primary zones based on its specific gravity (see **fig. 8.4**). Gas rises to fill the upper regions, grain and fluid-saturated roughage (yesterday's hay) sink to the bottom, and newly arrived roughage floats in a middle layer. The food is regurgitated into the mouth and completely masticated, then swallowed again and passed to the reticulum, omasum, and abomasum. The regurgitation and chewing in the mouth is called rumination (usually referred to as chewing the cud). Large amounts of bacteria and protozoa live in the rumen and reticulum. When food enters these chambers, the microbes begin to digest and ferment it, breaking down protein, starch, fats, and cellulose. The larger, coarser material is periodically regurgitated as the cud; after further chewing, the cud is reswallowed. Slowly the products of microbial action, and some of the microbes themselves, move into the ruminant's true stomach and intestine, where further digestion and absorption take place. The contents of the plant cells are thus released for digestion.

Biochemical Process of Digestion

Ruminants, like other mammals, because they do not have cellulose-digesting enzymes of their own, rely upon the digestive activity of symbiotic microbes present in their digestive tract. These symbiotic microorganisms produce cellulase enzymes, which break down the cell walls of plant materials to release fatty acids into the gut. These fatty acids are absorbed by the ruminant and constitute a significant portion of the overall energy budget input. In the rumen, any ingested protein is degraded into fatty acids and byproduct ammonia. The ammonia and other simple nitrogen-containing substances are used by the microorganisms for their own cell-protein synthesis. These organisms are ultimately digested in the abomasum and small intestine, thus providing the ruminant with protein. Protein and other nutrients then pass back to the second chamber (reticulum), bypassing the rumen. From there, food and the accompanying bacteria enter the third (omasum) and fourth (abomasum) stomachs. Digestion is completed in the abomasum and nutrients are absorbed in the small intestine. Later, when the food passes into the small intestine, absorption of amino acids, lipids, carbohydrates, etc., takes place as in other animals.

Once the food material goes through the "double digestion," it enters the small intestine. Here, in the upper portions of the intestine, secretion of enzymes, pancreatic juice, and bile induces further breakdown. The end products of the digestion process, such as proteins and fats, are absorbed in the lower section of the small intestine. Then all residues of the ingested food are deposited in the large intestine. In the large intestine there are absorption sites for water, minerals, and nitrogen. This "double digestion" system of ruminants allows them to use a lot of the cellulose in their herbivorous diet, which otherwise would have no nutritional value. Therefore, this cellulose is available to the ruminant animal but denied to other herbivores. Thus, the presence of symbiotic microorganisms in the forestomach turns a ruminant into a mobile fermentation plant for cellulose that results in the utilization of **60 percent** of the available cellulose and **90 percent** of the nutrient content encased in cellulose.[1,2]

Similarly, the products of protein digestion are absorbed at a very high rate among ruminants. This efficiency of protein absorption is attained through a complex process for recycling urea, the nitrogen-rich waste product that is normally excreted in the urine. In ruminants, the level of urea is controlled, in part, by microorganisms in the gut that use the urea for sustenance. Owing to this symbiotic relationship, ruminants such as buffalo, camels, and giraffes can survive by drinking water occasionally because their bodies do not deplete internal water reserves like other animals that need to flush the toxic urea from their systems by combining it with water to form urine.

It is quite evident that the evolutionary success of the ruminants, as a group, is based on the efficiency of their digestive systems to extract nutrition from low-quality food. Fermentation in the rumen generates enormous quantities of gas. In adult cattle, gas production may range from 30 to 50 liters per hour and about 5 liters per hour in a sheep or goat. Eructation, or belching, is how ruminants continually get rid of fermentation gases. Anything that interferes with eructation is life threatening to the ruminant because the expanding rumen rapidly interferes with breathing. Animals suffering from ruminal tympany (bloat) die from asphyxiation.

In summary, the digestive activity of symbiotic microbes present in the digestive tract play a major role in the digestion process and the supply of nutrients to the ruminant. These microbes attack food particles and supply a major portion of the protein and energy to the ruminant. Thus, the ruminants obtain nutrients and energy from the digested microbes and from the volatile fatty acids produced as a by-product of the fermentation and digestion process.

Digestive Strategies of Omnivores: Nonhuman Primates

From the start, the earliest primate ancestors were partially omnivorous, supplementing their vegetarian diet with insects and bugs. Eventually the primate order became fully omnivorous. Adaptation to an omnivore diet has resulted in changes in the digestive strategies and the evolution of certain morphological and behavioral traits that distinguish primates from other mammals. These morphological and physiological adaptations represent a continuation of the adaptations of herbivore mammals and are oriented at increasing the efficiency of nutrient extraction and absorption in an environment that offers little energy per unit of food.

Sacculated Stomach

Among the Old World monkeys, the *Colobus* monkeys of Africa are the most folivorous. Their diet consists primarily of young leaves but they also consume some termites, clay, fruits, and flowers. There are several species of *Colobus* monkeys that differ in fur color, including red hair, and white and black hair. As illustrated in **fig. 8.5**, the stomach of the *Colobus* monkey is characterized by a sacculated or compartmentalized stomach instead of the single chamber found in the other primates, such as the vervets. Furthermore, anatomical studies indicate that the upper region of the stomach is expanded and separated from its lower acidic region.[3–6] This anatomic specialization has two critical functions: (a) it creates a chamber for the fermentation of foliage by anaerobic bacteria, and (b) it provides a buffer fluid between the two regions of the stomach to detoxify plant toxins with bacteria. The end result is that the colobines can digest and break down the cellulose and potentially toxic substances by gut bacteria more efficiently than any of the other primates.

Once the chewed leaves flow from the esophagus into the upper region of the stomach (one of the two stomach compartments in colobines), the cellulose and hemicellulose in plant cell walls are broken down by cellulytic bacteria through the process of fermentation. This results in the production of gases, which

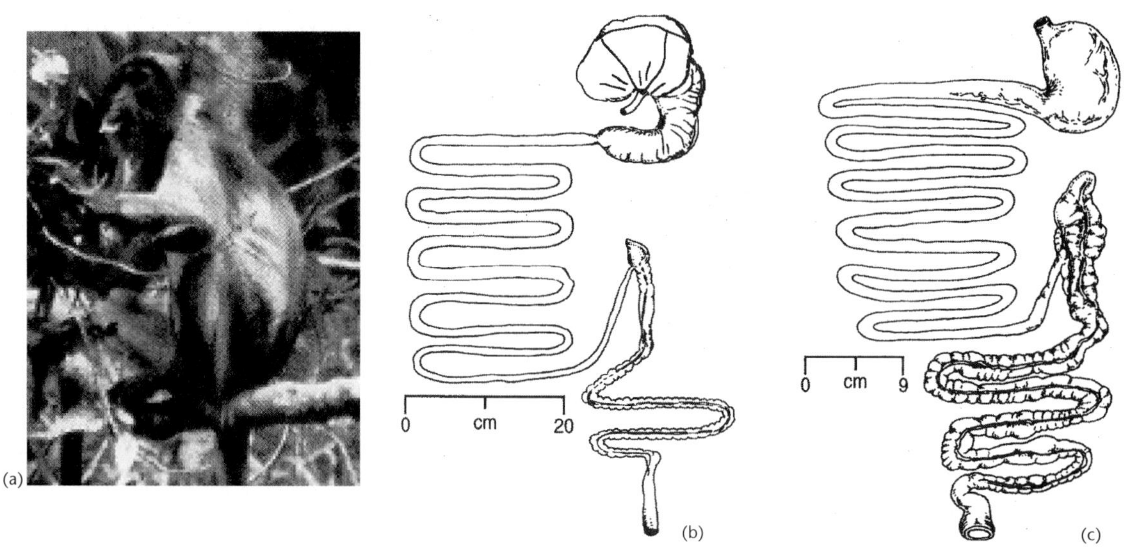

FIGURE 8.5. A leaf-eating *Colobus* monkey (a) has a larger and more compartmentalized stomach (b) than that of the vervet monkey, whose diet is based primarily on fruits (c). Source: (a) Courtesy of Professor John Mitani, University of Michigan, Ann Arbor; (b) and (c) from Milton, figs. 3 and 4.

pass through the stomach wall into the bloodstream, where they provide energy for body tissues or are delivered to the liver for conversion into glucose. The undigested portion of the fiber passes out of the forestomach and into the second compartment of the stomach, where special enzymes (lysozymes) cleave the bacterial cell walls and release protein and other nutritious materials of which the cellulytic bacteria are composed.

In summary, the sacculated stomach allows the colobine to eat items that would otherwise be poisonous, since plant defense compounds are found in all trees. Primates without such bacteria cannot break down the cellulose and obtain energy from it. This multichambered stomach and the lysozymes of the colobines must have evolved independently like the ruminants in response to a herbivorous diet.

Enlargement of the Colon and Small Intestine

Primates whose diets are not exclusively folivorous, such as the New World monkeys, have adapted by enlarging the digestive area of the large intestine. For example, the dietary differences between the spider monkey and the howler monkey are reflected in differences in the size of colon. The spider monkey's diet is based on ripe fruits (**fig. 8.6a**), while the howler monkey's diet is based on leaves (**fig. 8.6b**). The howler monkeys, compared to spider monkeys, have considerably wider and longer colons. This difference is related to the fact that leaves require more space and need to stay longer than fruits to ferment and extract energy from the fiber-bound leaves. In fact, experimental studies[5] indicate that howler monkeys, when fed fruits and leaves, digest food about 16 hours more slowly than spider monkeys, and howler monkeys also retain more of the nutrients. Thus, by having a large colon, howlers are able to obtain as much as 31 percent of their required daily energy from volatile fatty acids produced during fermentation. Despite the fact that howler monkeys feed very selectively, primarily on tender young leaves and sugary fruits and flowers that can be rapidly fermented, they are not very efficient at obtaining nutrients and energy from leaves. On the other hand, spider monkeys, by passing food more quickly through their shorter and narrower colons, are more efficient than howlers at extracting energy from the fiber in their diet. By selecting fruits, which are highly digestible and rich in energy, they can attain all the necessary calories, and by supplementing their diet with young leaves, they can derive the required protein without an excess of fiber. Thus, spider monkeys, as rapid digesters, tend to maximize the efficiency of energy extraction from high-energy fruits, while howlers, as slow digesters, tend to maximize their energy intake by extracting from a high volume of hard-to-digest, low-energy food.

Another digestive strategy involves enlargement of the colon only. The enlargement of the colon is quite evident when one compares the digestive system of the

(a)

(b)

FIGURE 8.6. New World monkeys. Spider monkey eating fruit and a Howler monkey chewing leaves adapted from pp. 166 and 167 from *Evolving Brains* by John Morgan Allman. Illustrations by Joyce Powzyk, © 1999 by Scientific American Library. Reprinted by permission of Henry Holt and Company, Inc.

orangutan to that of the chimpanzee (**fig. 8.7**). Orangutans have an enlarged colon, relating to their predominantly folivorous diet. Since the amount of energy per unit of weight is less in plant foods than in animal foods, a diet that emphasizes plant foods requires a large colon where energy can be efficiently extracted. Accordingly, in mountain gorillas, orangutans, and siamangs, who are highly folivorous, a large proportion of the gut is represented by the colon. Likewise, the chimpanzees, who are more folivorous than humans, have a proportionally larger colon than humans (**fig. 8.8**) do. As illustrated in **figure 8.9**, the small intestine in humans accounts for about 60 percent of total gut volume, while in orangutans and chimps it accounts for only 24 percent. In contrast, the colon represents only 20 percent of total gut volume in humans, while it is more than 50 percent in orangutans and chimps.

In summary, it is quite evident that whether a primate is a folivore, foli-frugivore or omnivore is reflected in the morphology of the digestive system. The general pattern is clear: humans have "large" intestines, while chimps and orangutans have "large" colons. The enlargement of the colon among folivorous primates represents an adaptation that facilitates the extraction of nutrients by increasing the amount of time for fermentation and breakdown of the cellulose. This adaptation is necessary because plant matter must be held in the gut for some time to be fermented and broken down by cellulolytic bacteria present in the colon. This inference is supported by the finding that capuchin monkeys (*Cebus* species) and savanna baboons (*Papio*), who have a high-quality diet of sweet fruits, oily seeds, and animal foods—invertebrates (insects) and small vertebrates, have a small colon and large small intestine.[3–6] Thus, it would appear that the enlargement of the small intestine in humans represents an adaptive response to diets made up of high-quality dietary items that are capable of being digested and absorbed primarily in the small intestine.

Primate Taxonomy by Dietary Preference

Although the majority of primate species tend to emphasize some food items over others, most will eat a combination of fruit, leaves, other plant materials, insects, and animal foods. Many obtain animal protein from birds and amphibians as well. Some (baboons

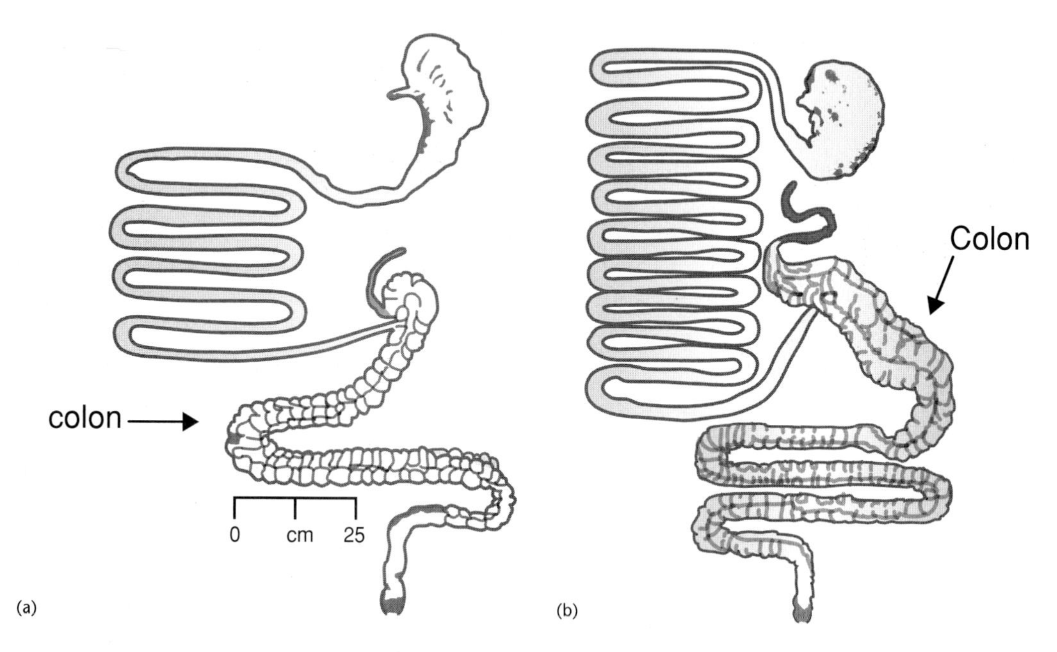

FIGURE 8.7. Comparison of the digestive system of the chimpanzees (a) and orangutans (b). The small intestine in the omnivorous chimpanzee accounts for a greater proportion of the gut volume than in the orangutans. Source: Milton, fig. 4.[4]

and especially chimpanzees) occasionally kill and eat small mammals, including other primates. Others, such as African *Colobus* monkeys and the leaf-eating monkeys (langurs) of India and Southeast Asia, have become more specialized and subsist primarily on leaves.

For didactic purposes, a dietary index that incorporates the three main primate dietary bases has been constructed.[8–10] Based on such an index, primates are classified into three groups: faunivore, frugivore, and folivore, depending upon which foods contribute more than 50 percent of the diet. For example, folivores such as *Colobus* monkeys routinely consume significant amounts of insects (animal matter), but leaves account for more than 50 percent of their daily diet. In contrast, prosimians eat plant foods, but animal matter contributes the largest proportion of their daily diet. Hence, they would be considered more in-

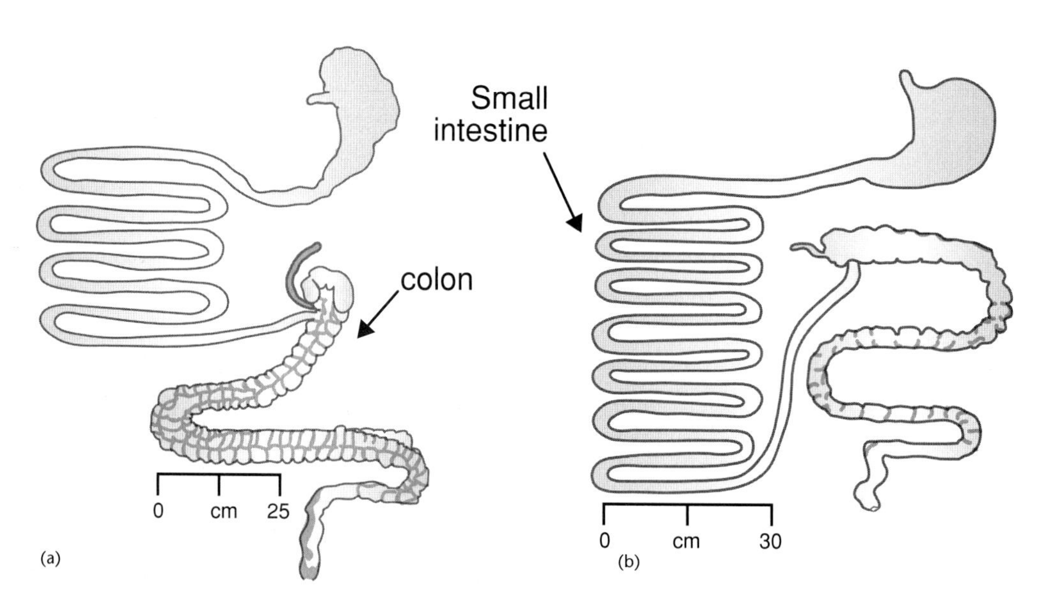

FIGURE 8.8. Comparison of the digestive system of the chimpanzees (a) and humans (b). The small intestine in humans accounts for a greater proportion of the gut volume than in chimpanzees. Source: Milton, fig. 5.[4]

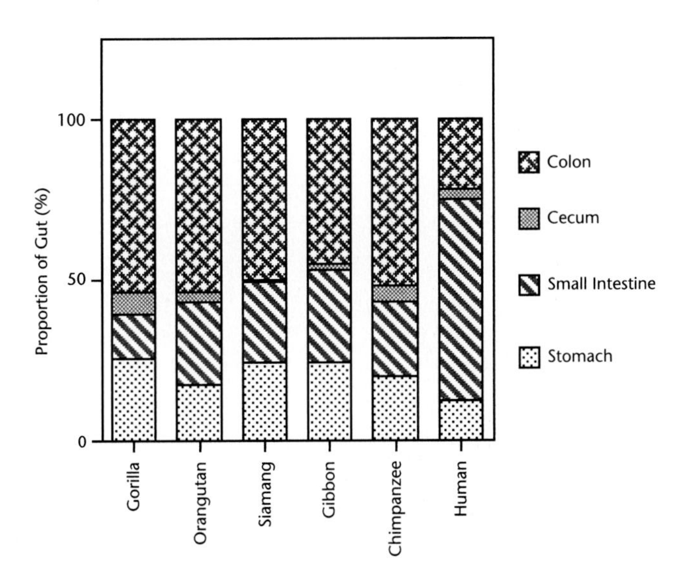

FIGURE 8.9. Differences in the proportion of the colon with respect to the gut in hominoids. Source: Data derived from Milton.[4]

sectivorous than folivorous. Likewise, frugivores normally eat some leaves and/or animal matter in addition to their primary food of fruit. Thus, depending on the preference and the proportion of specific foods consumed, primates can be classified into five broad categories of diet (**table 8.1**): (a) Insectivore–frugi-folivore, (b) folivore, (c) frugi-folivore, (d) foli-frugivore, and (e) omnivore.

Insectivore–Frugi-Folivore

Insectivore–frugi-folivores derive a major part of their protein requirements from insects, and some from fruits and leaves. Prosimians are diverse in dietary preference. The tarsiers, for the most part, are insectivores, while the lemurs and lorises are folivores and frugivores. In prosimians, foods such as sugary fruit or gum are usually rapidly digestible and are often supplemented largely by animal matter consisting typically of insects.

Frugi-Folivore

Despite the fact that animal matter represents a significant component of the diet of all the New World monkeys, some can be considered to be frugi-folivores. This is because they get a major part of their protein requirements from fruits and some from leaves. Some New World monkeys are frugi-folivores and, to a lesser extent, faunivores. The marmosets and tamarinds, owl monkeys, titi monkeys, squirrel monkeys, and capuchin monkeys are frugivores. The uakaris and sakis are fruit-seed eaters that use leaves as the main supplement of their diet. The spider, woolly, and howler monkeys are frugi-folivores.

Foli-Frugivore

Foli-frugivores get a major part of their protein requirements from leaves and some from fruits. The Old World colobine monkeys, with one exception, have leaves as a major component of their diet. The general preference is for young rather than mature leaves. Many species of *Colobus* monkeys (**fig. 8.5**) have been observed to eat repulsive and foul-smelling, latex-bearing plants, such as the ak (*Calotropis*), which are avoided by most animals. A small proportion of the diet of *Colobus* monkeys includes insect galls, fungi, lichen, pith, dead wood, roots, gum, and sap. It has been postulated that the ancestors of *Colobus* monkeys were able to survive in open woodland and savanna areas by preferring leaves and other plant parts that are less susceptible than fruits to seasonal fluctuations in availability. The apes such as the orangutans and siamangs are foli-frugi-faunivores, in that both eat notable amounts of leaves, shoots, stems, and/or bark as well as fruit and some insect matter. Gibbons are predominately frugivores. Their diet consists primarily of fruit with some foliage and animal matter.

Omnivore

Omnivores get a major part of their protein requirements from plant and animal foods. Cercopithecine Old World monkeys are omnivores. Among cercopithecine monkeys, such as baboons and macaques, a large proportion of the diet is derived from animal matter, including small vertebrates. Baboons will often prey on small mammalian herbivores and occasionally hunt vervet monkeys, the kids of gazelles, and hares. In some cases, the preference for animal foods sometimes overrides the plant component of their diet. For example, chacma baboons may feed for days on an insect outbreak.[11] Species such as Japanese macaques that live near the seashore eat seaweed while savanna baboons on the coast of South Africa take shellfish off the rocks. Talapoins will eat fish, grasshoppers, and caterpillars as well as leaves. It has been reported that in Botswana, when the numbers of insects on certain kinds of trees increased, baboons switched from their usual diet, consisting largely of plants, to capturing insects. Thus, cercopithecines seem to be quite opportunistic so long as they achieve a long-term balance in the intake of primary nutrients.

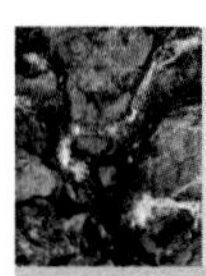

TABLE 8.1

Dietary Patterns of Extant Primates

Taxonomic Name	Common Name	Dietary Pattern	Dietary Source	Dietary Supplement
PROSIMII				
Lorisids	Slow loris	Frugivore	Fruits & gums	Insects
Tarsiers	Spectral	Carnivore	Arthropods, birds	Bats & snakes
Lemurs	Aye-aye	Insectivore	Bamboo shoots	Insects
	Western mouse	Vegetarian	& leaves	Insects
N.W. MONKEYS				
Pithecine	Aotus	Folivore-frugivore	Leaves, flowers, & fruit	Insects
	Callicebus	Folivore-frugivore	Leaves, flowers, & fruit	Insects
	Chiropotes & Cacajao	Folivore-frugivore	Leaves, flowers, fruits, & seeds	Insects
Alouatta	Howlers	Folivore-frugivore	Leaves, flowers, fruits, & seeds	Insects
Cebids	Capuchin	Frugivore	Fruits	Insects
	Squirrel monkey	Frugivore	Fruits	
O.W. MONKEYS				
Papio	Hamadryas	Folivore	Leaves & flowers	Insects
	Gelada	Folivore	Leaves, flowers, seeds, roots, & bulbs	Insects
Papio	Mandrills	Folivore	Fruits & shoots	Insects
HOMINOIDEA				
Hylobatidae	Gibbon	Frugivore	Fruits & young leaves	Insects
Pongid	Gorilla	Folivore	Young leaves, fruits, stems, vines, & shrubs	Insects
Pongid	Orangutan	Frugivore-folivore	Fruits & leaves, bark, seeds, & acorns	Ants & termites
Pongid	Chimp	Omnivore	Fruits, bark, leaves, seeds, & animal foods	Ants & termites
Pongid	Bonobo	Omnivore	Fibrous plants, herbs, & shoots	Earthworms

*Source: Chivers.[9]

Among the hominoids, orangutans and gorillas are the least omnivorous. Their diet includes mature foliage, bark, and unripe fruits, but they do consume invertebrates such as worms and insects. Although a heavy reliance on invertebrates as a protein source may not provide sufficient protein or calories for most larger-bodied anthropoids, such foods do provide these primates with particular essential nutrients that are inadequately supplied by their particular plant-based diets.

On the other hand, chimpanzees, including the pygmy chimpanzee (bonobo), are the most omnivorous of all the hominoids. Their diet includes mostly herbs, shoots, pith, and stems of ground plants supplemented with animal foods that include a wide range of invertebrates, including earthworms and millipedes.[7,12] Current research indicates that the diet of the chimpanzee includes a great deal more meat than previously thought. Studies conducted in Zaire and Gombe National Park indicate that chimpanzees prey on red *Colobus* monkeys and at least 25 different species of mammals, including wild pigs and small antelopes,[13–16] making 5 to 10 percent of their daily calories derived from hunted meat.[14] Thus, it is evident that the chimpanzees are more omnivorous than folivorous because they supplement their plant foods with some animal products such as small game, animals, and insects deliberately obtained. However, it should be noted that chimpanzees differ from humans in that when given a choice, they will avoid eating hard-to-chew nuts and seeds and prefer leaves or fruits.

Why Primates Search for Animal Food

It is quite evident that all primates, irrespective of their dietary emphasis, supplement their diet with varying amounts of animal foods. Various studies indicate that all primates supplement their diet by eating termites, ants, and other insects that are easily accessible to both males and females. In addition, anthropoids, such as baboons, will often prey on small mammalian herbivores and occasionally hunt vervet monkeys, the kids of gazelles, and hares. Adult chimpanzee males, assisted by adult and adolescent females, actively engage in hunting *Colobus* monkeys.(19) In addition, females obtain meat through theft or from males who share with them.(19) Sometimes chimpanzee females do not hunt to avoid taking risks that would jeopardize their offspring's survival or because they can get food more efficiently and with less risk than by hunting. An important question is, why do primates search for animal food? Milton(17) proposed that meat consumption is a way of circumventing the constraints imposed by the gut anatomy and digestive kinetics that characterize hominoids. Within this context, I propose that the primate's digestive system cannot extract all the essential amino acids and purified fatty acids from plant foods.

Variability in Capacity of the Digestive System to Extract Amino Acids

There are 20 amino acids that are concentrated in most foods (**table 8.2**). Of these, nine are considered essential because the body cannot synthesize them in quantities to meet its need.(18) Hence, the kind and relative proportion of amino acids contained in the food determine the quality of dietary protein. To be properly utilized, proteins must have the minimum nine essential amino acids. Animal protein, except gelatin (which has limited amounts of tryptophan and lysine), contains all of the nine essential amino acids in proportions capable of promoting growth. As such, animal proteins are considered good-quality proteins (also known as complete proteins). This means that if animal foods are used as the sole source of protein in sufficient amounts to meet total protein needs, they

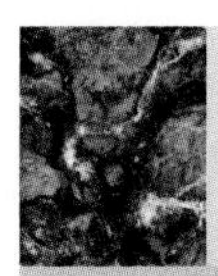

TABLE 8.2
Naturally Occurring Amino Acids in Food and Body Tissues

Essential amino acids	Nonessential amino acids
Histidine	Alanine
Isoleucine	Arginine
Leucine	Asparagine
Lysine	Aspartic acid
Methionine	Cysteine
Phenylalanine	Glutamic acid
Threonine	Glutamine
Tryptophan	Glycine
Valine	Proline
	Serine
	Tyrosine

Source: Guthrie and Picciano.(18)

usually provide enough of all the essential amino acids. Any excess amount of the essential amino acids can be used to synthesize nonessential amino acids and maintain growth. On the other hand, proteins that lack or have limited amounts of one or more essential amino acids are considered poor-quality proteins (also known as incomplete proteins). In general, vegetable protein (with the exception of quinoa, peanuts, and grasses) lacks one or two essential amino acids. Using incomplete proteins as the sole source of dietary protein will eventually result in malnutrition, along with impaired growth and reproduction.

Faunivores do not have any cellulytic enzymes capable of breaking down the cellulose that encases the plant cells. Consistent with this trait, their gut structure consists of a simple globular stomach, a tortuous small intestine, a short conical cecum, and a simple smooth-walled colon. Hence, faunivores have no capacity to extract all the amino acids encased in the cells of plant foods. This is not surprising since the animal foods provide all the necessary amino acids and when carnivores consume plant foods, they are used only as roughage or a laxative and not as a nutrient source. In contrast, herbivores are very efficient at extracting all the amino acids from plant foods such as grass (that contains all the essential amino acids). They are able to extract all the amino acids because their digestive system, compartmentalized into three or four chambers, contains high concentrations of cellulytic enzymes, which are very effective at breaking down and dissolving the cellulose that encases the plant cells, within which are located the amino acids.

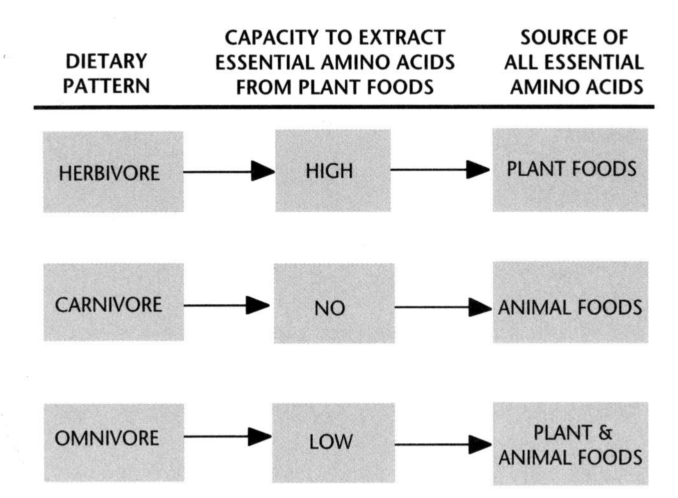

FIGURE 8.10 Development of the low capacity to extract all the essential amino acids from plant foods in omnivores.

On the other hand, the digestive system of omnivores has a low concentration of cellulytic enzymes. Therefore, omnivores are only able to extract some of the amino acids from plant foods. For this reason, omnivores such as humans have a limited ability to utilize plant foods and, as such, cannot subsist solely on plant foods unless they consume a mixture of plant foods that complements or balances all essential amino acids. For example, wheat has ample methionine but lacks lysine, and soybeans have ample lysine but are limited in methionine, so a combination of wheat and soybeans provides a mixture of amino acids capable of promoting growth. Obviously, nutrition complementation requires a thorough knowledge about the amino acid composition of all plant foods.[18]

Need for Polyunsaturated Fatty Acid

The polyunsaturated fatty acids (PUFA) are considered essential nutrients,[18] which means that they cannot be synthesized and must come from the diet. PUFA include two types, Ω 3 and Ω 6. Fatty acids are referred to by numerical nomenclature. For example, linoleic acid has 18 carbon atoms and 2 methylene-interrupted double bonds that are located on the sixth position of the first double-bond carbon. Hence, the chemical nomenclature of linoleic acid is 18:2 ω 6. Herbivores can synthesize sufficient quantities of PUFA by converting the 18-carbon from plants to arachidonic acid (AA) and docosahexaenoic acid (DHA).[21–23] In contrast, omnivores, such as rabbits, pigs, primates, and humans, cannot synthesize sufficient quantities of PUFA due to a combination of high rates of PUFA oxidation for energy, inefficient enzymatic conversion, and substrate inhibition.[24,25]

In summary, the preference for animal foods among primates is related to evolutionary changes in the digestive system that resulted in a low capacity to extract all the essential amino acids and purified fatty acids from plant foods. Viewed in this context, the primates' search for animal foods, which some primates do through hunting, scavenging, and termite and insect fishing, is an adaptation to supplement the incompleteness of a purely plant-based diet (**fig. 8.10**).

Summary

Herbivores have developed two distinct strategies to extract the nutrients encased in plants in protective cellulose. These include monogastric fermentation and multigastric fermentation (or rumination). Each of these adaptations required modifications of the digestive system oriented at favoring the proliferation of large amounts of microbes, bacteria, and lyzosomes capable of breaking down the protective cellulose and digesting the nutrients of plants.

As schematized in **figure 8.11**, primates have several different digestive systems, each adapted to increase the efficiency of nutrient extraction and absorption of nutrients from foods that are difficult to digest.

First, primates who are primarily folivores, such as the colobine monkeys, have adapted by retaining the primitive sacculated stomach, which is similar to the compartmentalized stomach of the ruminants that evolved among the seed-eaters of the Miocene epoch. They also have adapted by developing an enlarged colon. Together, these adaptations provide an expanded compartment for bacterial fermentation of the cellulose cell walls in leaves and nutrient absorption of plant foods.

Second, primates who are frugi-folivores (because fruits are easily digestible but leaves are not) have developed adaptations for facilitating nutrient absorption and providing greater room for the digestion of plant foods. This is evident in the lengthening and widening of the cecum and enlargement of the colon, which provides an expanded compartment for the bacterial fermentation of the cellulose cell walls in leaves and fruits. This is also the case in leaf-eating primates such as mountain gorillas, lemurs, and howler monkeys.

Third, humans, as a result of the reliance on a diet containing less fiber, have attained a small colon (large

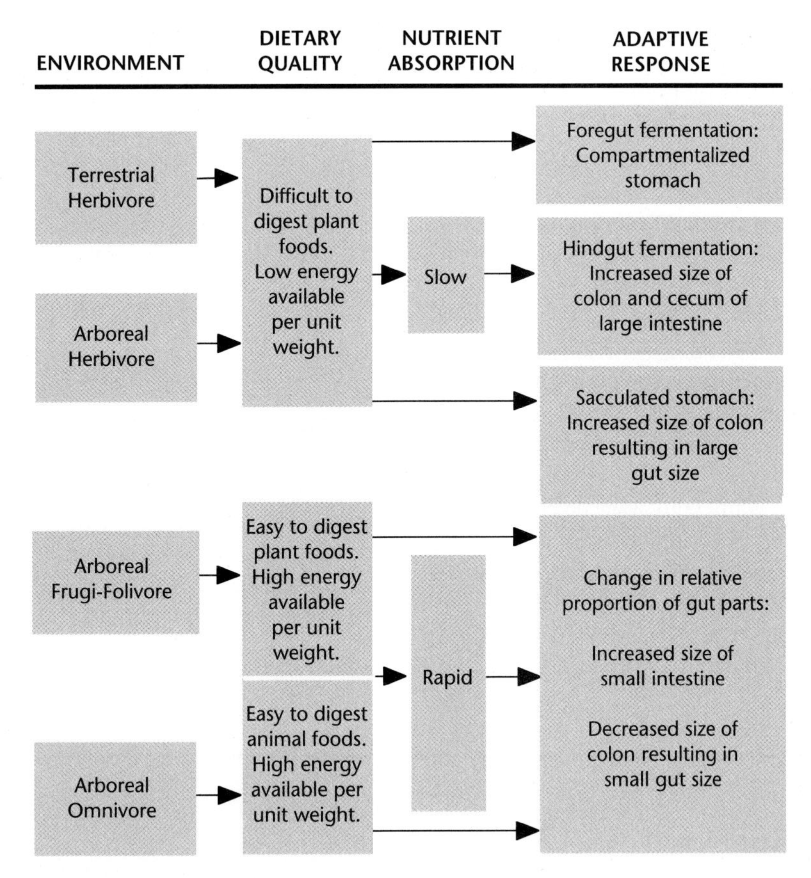

FIGURE 8.11. Schematization of the evolution of the hominid digestive system.

intestine) and relatively large small intestine. Together, these changes decrease food transit time and increase the rate of nutrient absorption.

Fourth, primates, in general, hunt for meat to compensate for their inability to extract all the necessary amino acids from plant foods.

Suggested Readings

Church, D. C., ed. 1988. *The Ruminant Animal: Digestive Physiology and Nutrition.* Englewood Cliffs, NJ: Prentice-Hall.

Dolhinow, P., and A. Fuentes. 1999. *The Nonhuman Primates.* CA: Mayfield Publishing Company.

Literature Cited

1. Church, D. C., ed. 1988. *The Ruminant Animal: Digestive Physiology and Nutrition.* Englewood Cliffs, NJ: Prentice-Hall.
2. Wyburn, R. S. 1979. Digestive physiology and metabolism in ruminants. In *Digestive Physiology and Metabolism in Ruminants: Proceedings of the 5th International Symposium on Ruminant Physiology,* ed. Y. Ruckebusch and P. Thivend. Lancanter, UK: MTP Press Ltd.
3. Milton, K. 1981. Food choice and digestive strategies of two sympatric primate species. *American Naturalist* 117:496–505.
4. Milton, K. 1987. Primate diets and gut morphology: Implications for hominid evolution. In *Food and Evolution: Toward a Theory of Food Habits.* ed. M. Harris and E. B. Ross, 93–115. Philadelphia: Temple University Press.
5. Milton, K. 1993. Diet and primate evolution. *Scientific American* 269: 86–93.
6. Milton K., and M. W. Demment. 1988. Digestion and passage kinetics of chimpanzees fed high and low fiber diets and comparison with human data. *Journal of Nutrition* 118:1082–1088.
7. Dolhinow, P., and A. Fuentes. 1999. *The Nonhuman Primates.* Mountain View, California: Mayfield Publishing Company.
8. Chapman, C. A., and L. J. Chapman. 1990. Dietary variability in primate populations. *Primates* 31:121–128.
9. Chivers, D. J. 1992. Diets and guts. In *The Cambridge Encyclopedia of Human Evolution,* ed. S. Jones, R. Martin, and D. Pilbeam, 60–64. New York: Cambridge University Press.
10. Chivers, D. J., and C. M. Hladik. 1984. Diet and gut morphology in primates. In *Food Acquisition and Processing in Primates.* ed. D. J. Chivers, B. A. Wood, and A. Bilsborough, 213–230. New York: Plenum Press.
11. Hamilton, W.J. III, R. E. Buskirk, and W. H. Buskirk. 1978. Omnivory and utilization of food resources by chacma baboons. *American Naturalist* 112:911–924.
12. Cichon, R. L., and R. A. Nisbett, eds. 1993. *The Primate Anthology.* Upper Saddle River, NJ: Simon & Schuster.

13. Teleki, G. 1981. The omnivorous diet and feeding habits of chimpanzees in Gombe National Park, Tanzania. In *Omnivorous Primates.* ed. R. S. O. Harding and G. Teleki, 303–343: New York: Columbia University Press.
14. Stanford, C. 1999. *The Hunting Apes.* Princeton, NJ: Princeton University Press.
15. Stanford, C. 2001. The ape's gift: Meateating, meatsharing, and human evolution. In *The Tree of Origin.* ed. F. deWaal, 95–117. Cambridge: Harvard University Press.
16. Wrangham, R., W. McGrew, F. deWaal, and P. Heltne. 1994. *Chimpanzee Cultures.* Cambridge: Harvard University Press.
17. Milton, K. 1999. A hypothesis to explain the role of meat-eating in human evolution. *Evol. Anthrop* 8:11–21.
18. Guthrie, H. A., and M. F. Picciano. 1995. *Human Nutrition.* St. Louis, MO: Mosby.
19. Mitani, J. C., and D. P. Watts. 2001. Why do chimpanzees hunt and share meat? *Animal Behavior* 61:915–924.
20. McGrew, W. C. 1974. Tool use by chimpanzees in feeding upon driver ants. *J. Human Evolution* 1:501–508.
21. Kelly, M. L., J. R. Berry, D. A. Dwyer, J. M. Griinari, P. Y. Chouinard, M. E. Van Amburgh, and D. E. Bauman. 1998. Dietary fatty acid sources affect conjugated linoleic acid concentrations in milk from lactating dairy cows. *Journal of Nutrition* 128:881–885.
22. Jenkins, T. C. 1998. Fatty acid composition of milk from Holstein cows fed oleamide or canola oil. *Journal of Dairy Science* 81:794–800.
23. Dhiman, T. R., L. D. Satter, M.W. Pariza, M. P. Galli, K. Albright, and M. X. Tolosa. 2000. Conjugated linoleic acid (CLA) content of milk from cows offered diets rich in linoleic and linolenic acid. *Journal of Dairy Science* 83: 1016–1027.
24. Broadhurst, C. L., Y. Wang, M. A. Crawford, S. C. Cunnane, J. E. Parkington, and W. Schmidt. 2002. Brain-specific lipids from marine, lacustrine or terrestrial food resources; Potential impact on early African *Homo sapiens. Comp. Biochem. Physiol.* Part B 131:653–673.
25. Bazinet, R. P., E. G. McMillan, R. Seebaransingh, A. M. Hayes, and S. C. Cunnane. 2003. Whole-body beta-oxidation of 18:2Ω6 and 18:3Ω3 in the pig varies markedly with weaning strategy and dietary 18:3Ω3. *J. Lipid Res.* 44:31–49.

Hominid Evolution

Current opinions concerning the interpretation of the fossil record of hominid evolution are about as divergent as they can possibly be. Some consider that the hominid origins can be traced to 6 to 8.5 million years ago while others traced them to 2.5 million years ago. Researchers have been very free in naming some of the fossil findings and identifying them as species in the direct line of the hominid ancestry. The preference for one as opposed to the other interpretation would seem to be as subjective as the recognition of what counts as specific distinction. What is clear is that the transition from the earliest hominid fossils to modern humans follows well-defined trends that include bipedal locomotion, large brain size, increased body size, decreased tooth size, and decreased skeletal robustness.

Structures that are shared by all or most related species are referred to as primitive because they persist fairly unchanged. For example, the similar number and morphology of molar teeth among hominoids, a relatively small brain, a large sagittal crest, a U-shaped dental arcade, and large canines are referred as primitive, or shared, traits because they were inherited from a common ancestor and remain unchanged. On the other hand, the characteristics that are unique to a species and are changed from the ancestral form are considered to be derived traits. For example, a centrally placed foramen magnum, a wide and large pelvis, and straight lower limb bones (associated with bipedal locomotion) are considered derived traits because they are relatively new modifications. A derived trait that is adapted for many functions, like the human limb, is considered a generalized structure, whereas a feature such as the human foot, because it has been structurally modified to deal with the rigors of an upright posture and bipedal movement, is considered to be a specialized derived trait. The archaeological record shows that several species of hominoids and hominids coexisted for many millions of years and the early hominid ancestors had a mixture of primitive and derived traits. Based on the relative frequency of primitive and derived traits, to be considered in the hominid line, a species must have a preponderance of derived traits relative to primitive ones. Conversely, to be considered in the hominoid line, a species must have a preponderance of primitive traits relative to derived ones. Following this model, the evolution of the early primates into the modern hominids can be grouped into various **phases** depending on the preponderance of primitive or derived traits. In this framework the term **phase** implies comparable levels of adaptive integration where similar traits allow the pursuit of similar ways.

Hence, here I propose that the path to the modern hominid line has gone through four phases: hominoid-hominid, early hominids, true hominids, and *Homo sapiens.*

A. Hominoid-hominid, which includes species such as *Sahelanthropus tchadensis, Orrorin tugenensis*, and *Ardipithecus ramidus*, is characterized by possessing a high frequency of primitive traits and a low frequency of derived (advanced) traits. By all definitions they are mi-bipedal.
B. Early hominids embody species with a high frequency of derived characteristics and a low incidence of primitive traits. This category includes the several species of *Australopithecus robust, Australopithecus gracile,* and the early hominids such as *Homo habilis.*
C. Hominids are species with all or mostly derived traits, including *Homo erectus* and archaic *Homo sapiens.*
D. *Homo sapiens* have the biological and behavioral traits that characterize modern humans.

This section includes six chapters. Chapters 9 to 11 focus on the evolution from the earliest hominoid-hominids to the anatomically modern *Homo sapiens.* In addition, the peopling of the New World and hypothesis about the origin of the anatomically modern *Homo sapiens* is discussed in chapter 12. Thereafter, chapter 13 discusses the role of food on the evolution of human brain size and chapter 14 focuses on the effects of bipedalism on childbirth and altricial development of humans.

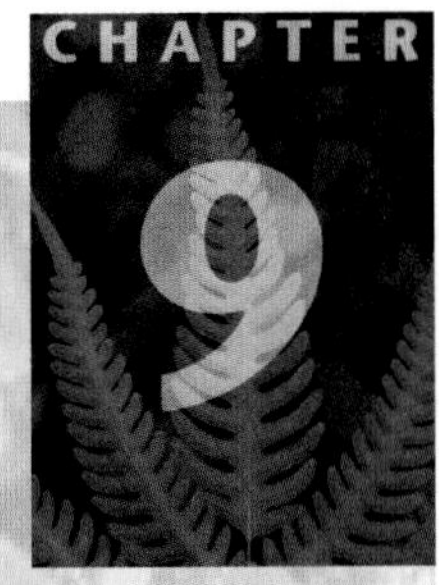

The Hominoid-Hominid and Early Hominid Phase of Evolution

Hominoid-Hominid Phase

Estimates based on molecular genetics place the separation between human and the African great apes as having occurred between 8 and 6 million years ago.[1] Measurements of potassium-argon and magnetostratigraphy indicate that the hominid fossil record begins in Africa approximately 7 to 6 million years ago. This age range encompasses the fossils that are considered in the hominoid-hominid phase such as members of the genus *Sahelanthropus tchadensis, Orrorin tugenensis,* and the species *Ardipithecus ramidus.* This phase includes those species associated with a high frequency of primitive traits such as small brain size, quadrupedal locomotion, large canine teeth, U-shape dental arcade and a low frequency of derived features such as large brain size, small canines, and skeletal adaptations to upright posture and terrestrial bipedal locomotion.

Sahelanthropus tchadensis

Sahelanthropus tchadensis was found in Chad, Central Africa, 2,500 km from the East African Rift Valley in the southern Sahara desert.[2] The associated fauna suggests that the fossils are between 7 and 6 million years old. The fossils include a nearly complete cranium and fragmentary lower jaws (**fig. 9.1**). This species is considered in the hominoid-hominid phase because it has more primitive traits than advanced traits. First, *Sahelanthropus tchadensis* has a small brain (between 320 and 380 cc), and a relatively long, flat nuchal plane with a large external occipital crest similar to large living and known fossil hominoid genera. Second, the canine breadth is similar to the chimpanzee mean, being within the range of both chimpanzee and gorilla females and of chimpanzee males.[3] Third, the dental arch is small, narrow, and U-shaped like in earlier hominoids. Fourth, the enamel thickness of the post-canine teeth is intermediate between hominoids and *Australopithecus.*

On the other hand, *Sahelanthropus tchadensis* has advanced hominid features.[2] First, the face is more orthognathic (vertical) than in apes, and the orbits are separated by a very wide interorbital pillar and crowned with a large, thick, and continuous raised thick brow ridge. Second, the cheek teeth (upper

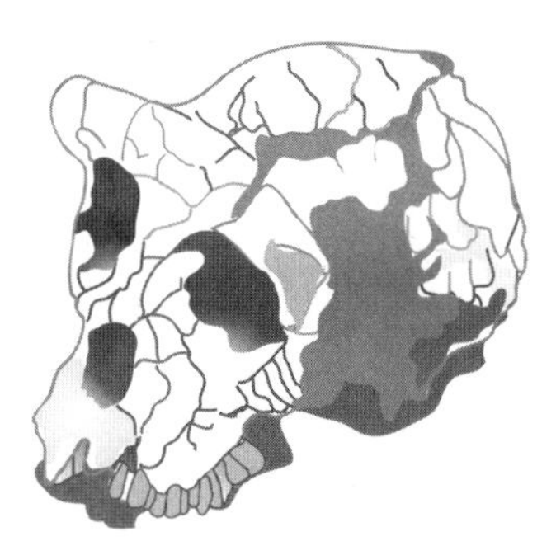

FIGURE 9.1. *Sahelanthropus tchadensis.* It has heavy brow ridges and a vertical face as seen in later hominids, but it is not known whether it had bipedal locomotion. (Picture courtesy of Bone Clones, Inc. 2146 Chase St. # 1; Canoga Park, CA.)

premolars and molars) are smaller than in living hominoid genera and within the size range of early hominids. Finally, the basicranium has small occipital condyles associated with an apparently large foramen magnum. However, because the base of the skull is missing, it is not known whether the foramen magnum was located under or behind the skull.

In summary, *Sahelanthropus tchadensis* displays a unique mosaic of primitive and derived characteristics, and constitutes the beginning of the trend toward the hominid pattern of morphology. The fact that *S. tchadensis* is from the Chad desert suggests that the beginnings of hominid evolution were spread throughout East and North Africa.

Orrorin tugenensis

Orrorin tugenensis lived about 6 million years ago in East Africa. French and Kenyan researchers discovered the remains of 13 fossils including fragmentary thigh and arm bones as well as several teeth. The fossils belong to at least five individuals found at four localities in the Lukeino Formation and at Tugen Hills, Kenya[4] (**fig. 9.2**).

This species is considered within the hominoid-hominid phase because it has more primitive than advanced traits. First, the upper central incisors were large and robust like those of apes. Second, the upper canines were also large and retained a narrow and shallow anterior groove, like in apes. Third, the premolars are relatively small and apelike. Fourth, the molars are relatively small, with thick enamel, like in apes.

On the other hand, the postcranial remains of *Orrorin tugenensis* exhibit both primitive and advanced traits. First, the humerus, represented by a distal shaft, shows a straight lateral crest, onto which the brachial muscle was attached like in modern chimpanzees. Second, the digits (phalanx) of the hands are curved like in arboreal primates. From the upper limb morphology, it appears that *Orrorin* possessed some arboreal adaptations.

In contrast, size and morphology of the lower limbs exhibit advanced derived traits. First, the femur is large and thick. Second, the head of the femur (trochanter) has a well-developed fossa trochanterica situated just above the trochanter minor like in hominids. Third, the femur has deep grooves where thigh muscles are attached and is characteristically associated with frequent bipedalism.

Therefore, the discoverers of the Lukeino hominid postulate that *Orrorin tugenensis* was already adapted to habitual or perhaps even obligate bipedalism when on the ground, while its humerus and manual phalanx show that it possessed some arboreal adaptations.[4]

Ardipithecus ramidus

Ardipithecus ramidus lived about 5.8 to 4.4 million years ago in East Africa. Several fragments of teeth (**figs. 8.3** and **8.4**), skulls, and parts of the upper arms of 17 individuals along with a skeleton that is about 45 percent complete were found at the Aramis site in the Middle Awash Basin of Ethiopia.[5,6] The discoverers[5,6] named

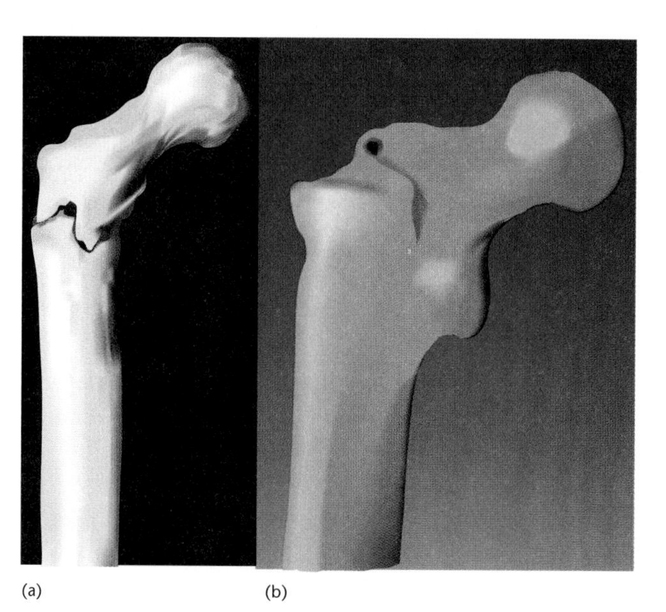

FIGURE 9.2. Femur of *Orrorin tugenensis.* (a) Proximal portion of the right femur and (b) 3-D model reconstructed from CT sections. Both pieces show that the femurs were long and shaped like those of hominids indicating the *O. tugenensis* was bipedal. (Composite illustration made from figure 1 & 2 Galik K., B. Senut, M. Pickford, D. Gommery, J. Treil, A.J. Kuperavage, and R.B. Eckhardt. 2004. External and Internal Morphology of the BAR 1002'00 *Orrorin tugenensis* Femur. *Science.* 305:1450–1453.)

the species *Ardipithecus ramidus*, from the words *ardi*, meaning "ground or floor," and *ramis*, meaning "root" in the local Afar language. This species is considered within the hominoid-hominid phase because it has more primitive than advanced traits.

First, the teeth show a mosaic of primitive and derived morphological features. For example, the enamel thickness is comparable to that of the hominids. On the other hand, the third molar has four distinct cusps like most Miocene apes. Second, the curvature of the phalanx of the upper limbs is similar to that of early hominids such as *Africanus afarensis*, but the ulnar shaft is more curved than in hominids. These traits indicate that *Ardipithecus ramidus* might have retained arboreal adaptations. On the other hand, the anatomical features of the lower limbs suggest that *A. ramidus* might have been adapted to bipedal locomotion. For example, the lower limb bones are thick with a big trochanter suggesting that *Ardipithecus ramidus* was possibly bipedal. The persistence of primitive dental and postcranial characteristics indicates that *Ardipithecus* was phylogenetically close to the common ancestor of chimpanzees and humans,[6] which would correspond to the hominoid-hominid phase that I propose here.

In summary, it is quite evident that the craniofacial morphology and dental features resemble the hominoids adapted to an arboreal environment more than the hominids. Yet, the postcranial skeletal morphology shows that they were partially adapted to the terrestrial environment and therefore, when they were on the ground, they probably had bipedal locomotion. The preponderance of primitive traits relative to the derived ones justifies considering all these specimens in the hominoid-hominid phase of evolution.

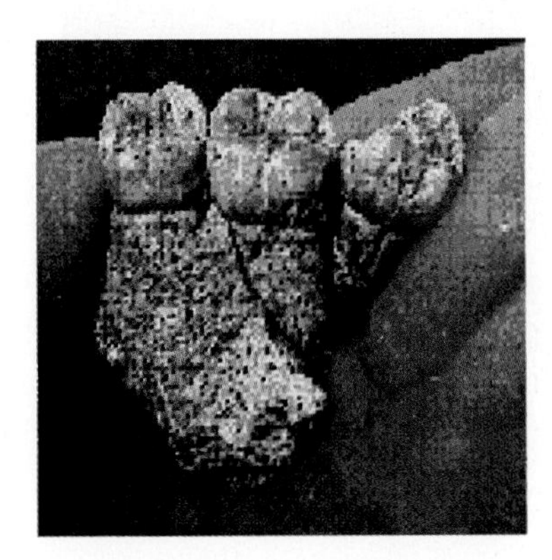

FIGURE 9.3. *Ardipithecus ramidus.* Premolars. Source: White, Suwa, and Asfaw.[5] Permission given to author by Tim White.

Early Hominid Phase

Early hominids include species with a high frequency of derived characteristics and low incidence of primitive traits. That is, the early hominids have traits that are uniquely associated with hominids rather than with hominoids. This group includes the species that are within the genus *Australopithecus* (**tables 9.1, 9.2,** and **9.3**). Collectively, the australopithecines are characterized by reduced size of the canines, and bipedal locomotion, which is associated with a wide pelvis. The australopithecines are divided into two distinct forms: robust and gracile australopithecines, hereafter referred to as robust and gracile hominids, respectively.

Robust Hominids

The robust australopithecines existed between 4.2 and 2.5 million years ago and occupied both East and South Africa. These specimens are known by a number of taxonomic names that include (**table 9.2**) *Australopithecus anamensis, Australopithecus aethiopicus, Australopithecus robustus*, and *Australopithecus boisei.*

Common features of the species referred to as *Australopithecus robustus* are a prominent sagittal crest (a thin plate of bone extending down the center of the top of the brain case), enlarged and flaring zygomatic arches (cheek bones), and an enlarged mandible, associated with large premolars and molars. These craniofacial characteristics of the robust hominids reflect their masticatory adaptations for the processing of fibrous (vegetable) material such as nuts, tough seeds, and chutes (see below). There is a general consensus among anthropologists that the robust australopithecines are an evolutionary dead end.

Australopithecus anamensis

Australopithecus anamensis existed between 4.2 and 3.9 million years ago and has been found in Turkana in Kenya (East Africa).[7,8] The fossil record suggests that *Australopithecus anamensis* is the ancestor of *A. aethiopicus*, which in turn gave rise to the species of robustus: *A. robustus* and *A. boisei.* The remains of *Australopithecus anamensis* include parts of the upper and lower jaw, parts of a tibia, a humerus, and numerous teeth of 21 individuals. *A. anamensis* had large molars with a thick enamel, and reduced canines (**fig. 9.4**). The lower jaw had an apelike receding chin and the dental arcade is more like the U-shape seen in chimpanzees and gorillas. A partial tibia (the larger of

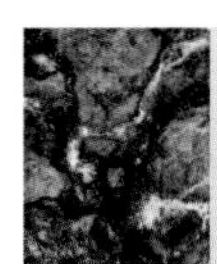

TABLE 9.1

List of the Skeletal Remains of Early Hominid Fossils That Lived Million Years Ago (mya) in East and South Africa

Site Name (location)	Age (mya)	Fossil Sample
Lothagam (Kenya)	5.8–5.6	1 specimen (partial mandible)
Mabaget (Kenya, Chemeron Beds)	5.1–5.0	1 specimen (Partial sub-adult humerus)
Tabarin (Kenya, Chemeron Beds)	5.0	1 specimen (partial mandible)
Aramis (Ethiopia)	4.4	More than 50 individuals (*Ardipithecus ramidus*)
Kanapoi (Kenya)	4.2–3.9	9 specimens (*Australopithecus anamensis*)
Allia Bay (Kenya, East Turkana)	3.9	13 specimens (*Australopithecus anamensis*)
Belohdelie (Ethiopia)	3.9–3.8	5 pieces of cranium (*Australopithecus afarensis*)
Olduvai (N. Tanzania)	1.85 -1.0	48 specimens of early *Homo*
Turkana (N. Kenya, Eastern Side of Lake Turkana)	1.9–1.3	More than 150 specimens; many australopithecines
Turkana (N. Kenya, West Side of Lake Turkana)	2.5–1.6	1 cranium, 1 nearly complete skeleton (australopithecine)
Middle Awash (N.E. Ethiopia, Bouri)	2.5	5 hominids; 1 cranium; Parts of limb skeleton
Hadar (N.E. Ethiopia)	3.9–3.0	40 to 65 individuals
Laetoli (N. Tanzania)	3.7–3.5	24 hominids: early australopithecines (*A. afarensis*)

Source: Ciochon, R.L., and J. Felagle. 1993. *The Human Evolution Source Book.* Englewood Cliffs, NJ: Prentice-Hall.

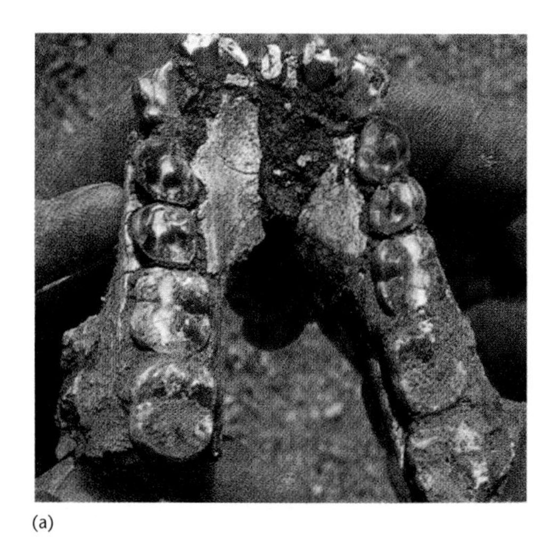

(a)

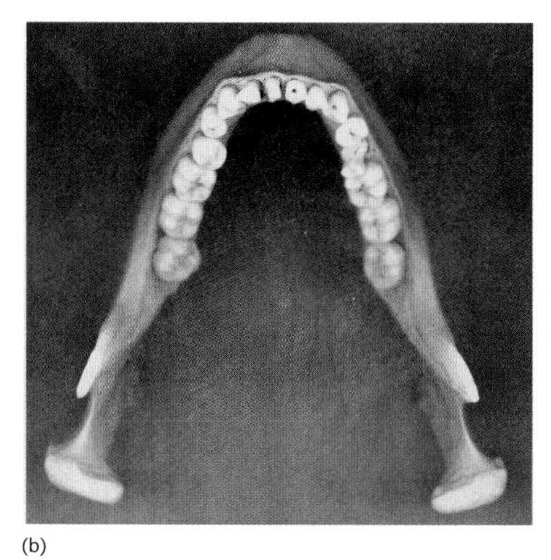

(b)

FIGURE 9.4. *Australopithecus anamensis* (a) and modern human (b). The dental arcade of *A. anamensis* is U-shaped unlike that of human. (Source: (a) courtesy of Professor Carol Ward, Department of Anthropology, University of Missouri (b), picture taken by author.)

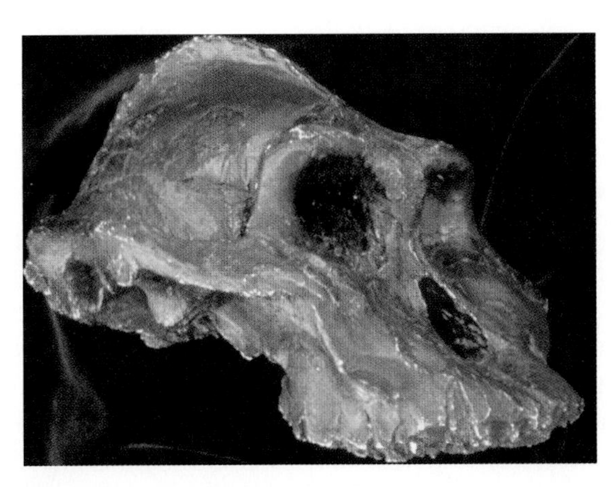

FIGURE 9.5. *Australopithecus robustus: aethiopicus* (Black Skull).

the two lower leg bones) shows strong evidence of weight bearing, suggesting bipedality, and the lower humerus (the upper arm bone) is extremely human-like. In other words, the *anamensis* had a **mixture** of primitive and derived features that were intermediate between those of *Ardipithecus ramidus* and the later *Australopithecus afarensis*.

Australopithecus aethiopicus

Australopithecus aethiopicus existed between 2.6 and 2.3 million years ago. This species is known from one major specimen, the Black Skull (KNM-WT 17000) discovered in 1985 near West Turkana in Kenya. It has a mixture of primitive and derived traits.[9, 10] The brain size is very small at 410 cc. The massiveness of the face and jaws and the large sagittal crest are similar to those of *A. boisei*.

Australopithecus robustus

There are several fossils of australopithecines that have been grouped in the species named *Australopithecus robustus* (TM 1517, SK 48, DNH 7, "Eurydice"). *A. robustus* were found in 1950 at Swartkrans in South Africa[11–14] and they existed between 2 and 1.5 million years ago. *A. robustus*, like all the australopithecines, were characterized by a massive and robust skull and large teeth. The massive face is flat or dished with no forehead and large brow ridges. It has relatively small front teeth but massive grinding teeth in a large lower jaw. The skull has a sagittal crest, the brain size is about 530 cc, and they weighed about 48 kg.

Australopithecus boisei

Australopithecus boisei (was *Zinjanthropus boisei*) is otherwise referred to as the "Nutcracker Man." *Australopithecus boisei* includes several specimens (OH 5, "Zinjanthropus," KNM-ER 406, KNM-ER 732, KGA10-525) and was discovered in 1959 at Olduvai Gorge in Tanzania.[15–18] *A. boisei* existed about 1.8 million years ago. There is an almost complete cranium, with a brain size of about 530 cc (**fig. 9.6**).

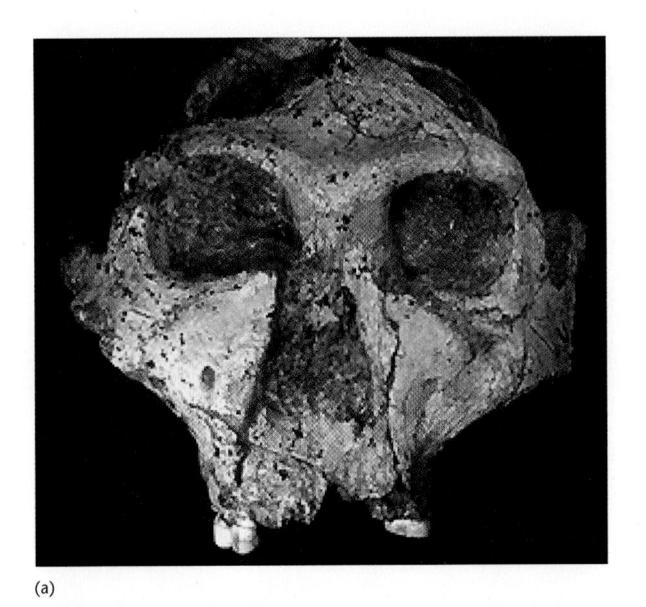

(a)

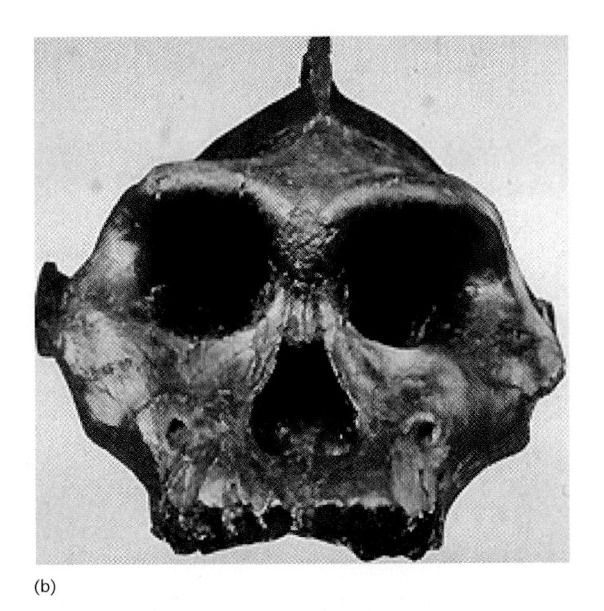

(b)

FIGURE 9.6. *Australopithecus robustus.* (a) Zinjanthropus or "Nutcracker Man" and (b) *Australopithecus boisei.* Courtesy of Professor Milford Wolpoff, Department of Anthropology, University of Michigan.

Gracile Hominids

The gracile australopithecines have been found throughout both East and South Africa and are known by a number of taxonomic names that include *Australopithecus afarensis, Australopithecus africanus,* and *Australopithecus garhi.* Gracile australopithecines also relied on bipedal locomotion but their craniofacial features were not as pronounced as those in the robust australopithecines.

Australopithecus afarensis

Australopithecus afarensis includes several specimens: AL 129-1, AL 288-1, "Lucy," AL 333, AL 444-2, and the "Laetoli" footprints. *A. afarensis* have been firmly dated to between 3.9 and 3.0 million years ago. The skeleton of the fossil referred to as "Lucy" was discovered in 1974 at Hadar in Ethiopia.[19–22] "Lucy" is the best-known example of the hominid *Australopithecus afarensis.* "Lucy" was an adult female of about 25 years. About 40 percent of her skeleton was found and her pelvis, femur (the upper leg bone), and tibia show her to have been bipedal (**fig. 9.7**). She was about 107 cm (3'6") tall (small for her species) and weighed about 28 kg (62 lbs). At a site known as Al 333, also of Hadar, the partial remains of 13 individuals of all ages were discovered. The remains of "Lucy" and other members of *Australopithecus afarensis* are about 3 million years old.

Australopithecus africanus

Australopithecus africanus consists of several remains including the "Taung Child" and "Mrs. Ples" (TM 512) (**figs. 9.8** and **9.9**). The "Taung Child" was discovered in 1924 at Taung in South Africa.[23] The find consisted of a full face, teeth and jaws, and a partial skull. "Taung Child" lived between 2 and 3 million years ago. The teeth of this skull showed it to be from a child of about 3 or 4 years. The brain size was 410 cc, and would have been around 440 cc as an adult. The large rounded skull, canine teeth that were small and not apelike, and the position of the foramen magnum, below and the middle of the skull, indicate that the "Taung Child," or *Australopithecus africanus* (African southern ape), was a biped. Another specimen (TM 512) of *Australopithecus africanus,* known as "Mrs. Ples," includes remains found at Sterkfontein in South Africa[24, 25] that consisted of parts of the face, upper jaw, and cranium of an adult female, which was named "Mrs. Ples." Remains of a nearly complete vertebral column, a pelvis, some rib fragments, and part of a femur of a very small adult were also found. It is

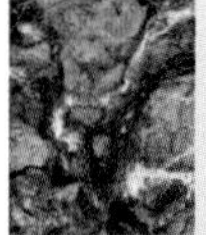

TABLE 9.2
Distinguishing Characteristics of *Australopithecus robustus* Found in Eastern and Southern Africa

Traits	*Australopithecus boisei*	*Australopithecus robustus*
Brain size (g)	410–530	530
Height (m)	1.2–1.4	1.1–1.3
Weight (kg)	40–80	40–80
Sexual dimorphism	High > 20%	High > 20%
Physique	Very heavy build	Heavy build
Limb proportions	Relatively long arms	Relatively long arms
Skull form	Prominent sagittal crests	Prominent sagittal crests
Face	Broad, flat face Strong facial buttressing	Broad, semi-flat face Moderate facial buttressing
Teeth	Small incisors and canines Large premolars Large molars	Small incisors and canines Large premolars Large molars
Jaw	Thick and large	Thick and large
Distribution	Eastern Africa	Southern Africa

Source: Ciochon, R.L., and J. Fleagle. 1993. *The Human Evolution Source Book.* Englewood Cliffs, NJ: Prentice-Hall.

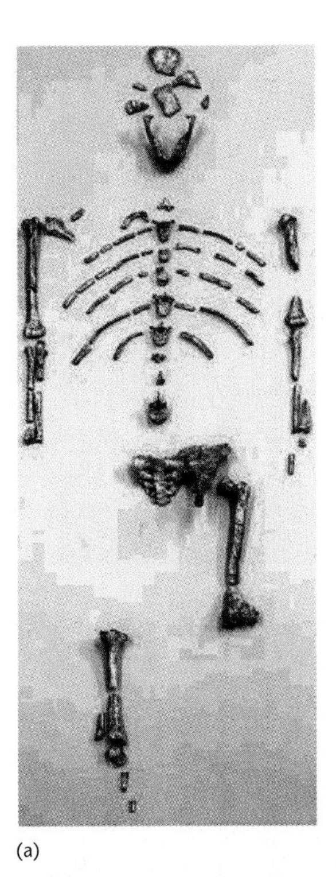

(a)

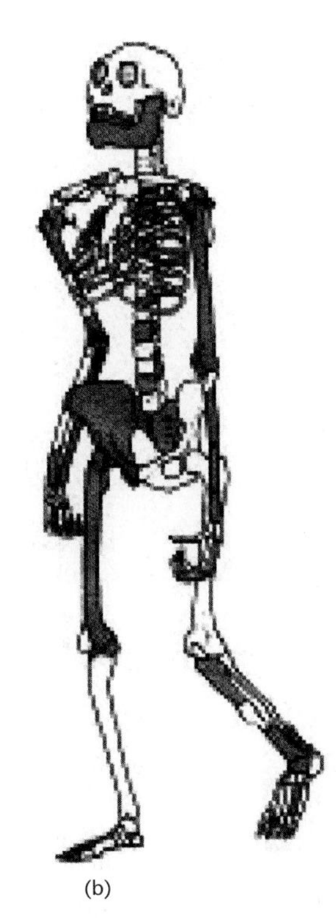

(b)

FIGURE 9.7. The skeletal remains of *Australopithecus afarensis:* "Lucy" (AL 288-I). This specimen (a) is the most complete of the australopithecines thus far discovered. (b) Forensic reconstruction of this specimen. (Source: (a) Picture from the cast collection of the Department of Anthropology, University of Michigan; (b) Modified from Ref. #50. Wolpoff, M. H. 1999. *Paleoanthropology*. Boston, Massachusetts: McGraw-Hill.)

estimated that this specimen lived about 2.5 million years ago and had a brain size of about 485 cc.

Australopithecus garhi

The species *Australopithecus garhi* is represented by a partial skull, which includes an upper jaw with teeth, and was found at Bouri in Ethiopia.[26, 27] It lived about 2.5 million years ago. This species had small brains (approximately 450 cc) with large canines, premolars, and molars. Observation of the archaeological remains suggests that *A. garhi* made and used stone tools to dismember animal carcasses, cut meat from bones, and smash open bones to extract the marrow. According to the fossil evidence, *Australopithecus garhi* was the possible descendant of the *gracile australopithecines* and, as such, is a candidate ancestor for early *Homo.*[26, 27] The postcranial remains feature a derived humanlike humeral/femoral ratio and an apelike upper arm–to–lower arm ratio.

Brain Size

As shown in **table 9.3** and **figures 9.8** and **9.9**, the brain size (using as proxy the volume within a skull) of *austrolopithecine hominids* averaged about 450 *grams*, which (corrected for body size) falls either within or just above the upper range of modern chimpanzees (350 grams). For example, the 3- to 4-year-old "Taung child" (*Australopithecus africanus*) had a brain size of about 405 grams and if he had reached adulthood would have had a brain volume of 440 grams, which falls below that of humans.

Bipedal Locomotion

Several lines of evidence conclusively show that australopithecines had a bipedal locomotion. The foramen magnum (**fig. 9.10**) (through which the spinal cord enters into the brain) is located below the middle base of the skull, as in modern humans, indicating that the australopithecines maintained an erect posture. Similarly, the anatomy of the pelvis and the placement of the femoral neck with the relationship to the pelvis indicate that the australopithecines had a mode of bipedal locomotion (**fig. 9.11**). The pelvis of the *Australopithecus afarensis* is bucket-shaped, similar to that of humans and different from the long and narrow pelvis of the chimpanzee. The most dramatic confirmation that *Australopithecus* walked bipedally is found in the "footprints" discovered in 1978 at Laetoli in Tanzania (**fig. 9.12**).[28] These footprints were made nearly 3.7 million years ago by two or three individuals that walked across newly fallen volcanic ash. Evaluations of the size and shape of the footprints, the linear distance between where the heels struck, and the amounts of "toe out" indicate that they are effectively identical to those of modern humans.[29] Based on the size and stride length, it is estimated that they were about 140 cm (4'8") and 120 cm (4'0") tall.[29]

In summary, all the gracile australopithecines (including *A. garhi*) collectively had bipedal locomotion and nonprojecting, small canine teeth. These features, despite the fact that they have a small brain size, clearly justify considering the australopithecines as hominids rather than as hominoids.

The development of bipedal locomotion represents a major shift in the evolutionary path in that by freeing the hands, bipedal locomotion expanded the possibilities for using and making tools and exploring the environment on the ground. As shown by the relatively long arms and curved digits, it appears that the early hominids retained the ability to climb the trees. This dual retention of bipedality and quadrupedalism

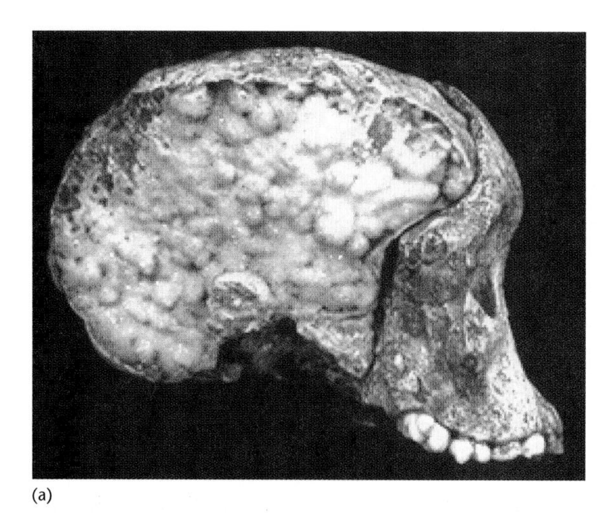

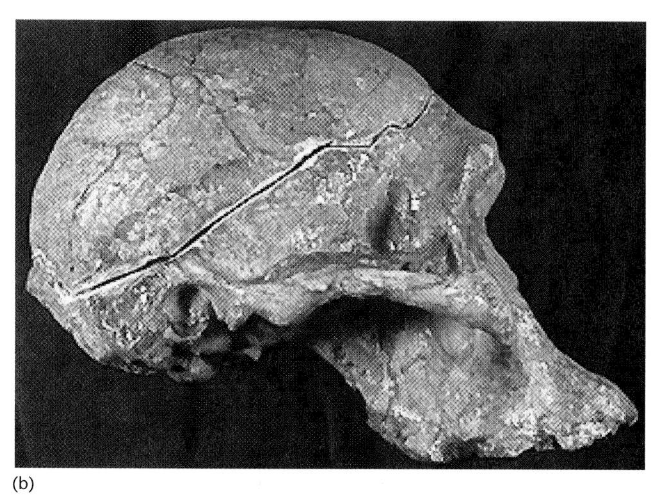

FIGURE 9.8. Gracile *Australopithecus africanus*: (a) the "Taung Child" and (b) "Mrs. Ples" (TM512). Courtesy of Professor Milford Wolpoff, University of Michigan, Ann Arbor.

probably enabled the early hominids to explore the arboreal and terrestrial habitat more than could be possible before.[30]

Body Size

The consensus among paleoanthropologists is that the body size of the multiple species of *Australopithecus* was small, under 5 feet. Current estimates indicate that the average weight of *A. afarensis* for males was 44.6 kg ± 18.5 and 29.3 kg ± 15.7 for females.[31]

Sexual Dimorphism

Sexual dimorphism refers to the percent difference between males and females in measurements of body size. In humans and chimpanzees today, males are bigger than females by about 10 to 15 percent, whereas among the hominoids such as gorillas and orangutans, males exceed females by 25 to 50 percent. The consensus among paleoanthropologists has been that the sexual dimorphism (male australopithecines were 25 to 35 percent bigger than females) of australopithecines was similar to the gorilla and orangutan. However, close scrutiny of the fossil record suggests that this hypothesis is built on a data set replete with limitations. For example, the sample used to estimate dimorphism is very small (fewer than six individuals for *A. afarensis*). Furthermore, males and females did not come from the same sites. Therefore, the observed difference in body size might reflect the

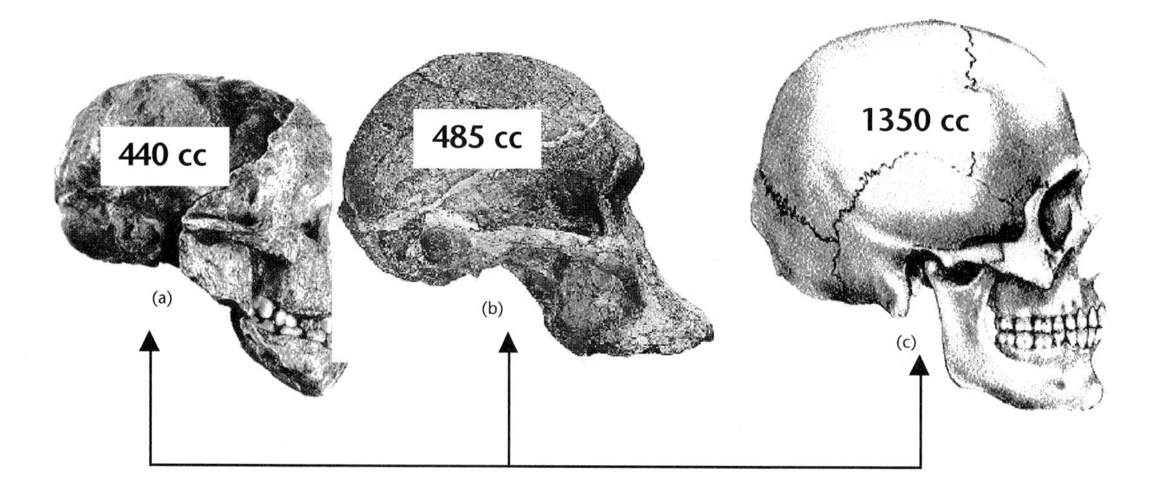

FIGURE 9.9. Gracile *Australopithecus africanus*: (a) the "Taung Child," (b) "Mrs. Ples," and (c) modern human. The central location of the foramen magnum indicates that the head of *Australopithecus africanus* was balanced above the spine similar to that of modern human, indicating erect posture. (Composite illustrations made by the author using the Paleoanthropology picture archive of Professor Milford Wolpoff of the Department of Anthropology, University of Michigan.)

effects of temporal and geographic variation as well as the limitations of small sample size. Reno et al.[32] applied a new statistical modeling method of simulating dimorphism to an assemblage of *A. afarensis* representing the remains of individuals who likely died simultaneously in a single catastrophic event some 3.2 million years ago at site AL 333, Hadar, Ethiopia. Using the 40 percent complete skeleton ("Lucy") from site AL 288 as a morphometric base line, they calculated femoral head diameters from measurements for the postcranial elements from AL 333 and other *A. afarensis* remains. This analysis revealed the sexual dimorphism in body size in *A. afarensis* was like that of contemporary *Homo sapiens* and chimpanzees.[32]

These findings have implications to the understanding of the evolution of social behavior and organization in early hominids. Larsen[33] suggests that the relatively low amount of dimorphism of *A. afarensis* is more consistent with paired social behavior. Studies among baboons of Africa indicate that where dimorphism is high, male–male competition is high; conversely, where dimorphism is low, competition among males is less frequent.[34] On the other hand, among chimpanzees, where dimorphism is moderate, although adult males express aggressive behavior toward one another, males tolerate each other, live in multimale kin groups, and engage in cooperative, coalitionary behavior.[34]

Although we will never know what the social organization and mating systems were for early hominids given the similarity of sexual dimorphism of australopithecines and chimpanzees, it would appear that the social organization of *A. afarensis* might be best characterized as multimale, cooperating kin groups.

Diet and Dental Evolution

The evolution and success of mammals is related to the adaptation in the number, shape, and structure of the teeth, which is reflected in their morphology. Teeth, because they are protected by enamel, are more resistant to destruction and as such are preserved in greater numbers than are other parts of the skeleton. Moreover, different aspects of teeth can also be associated with diet, allowing the estimation of dietary specialization in extinct species. In other words, dental characters reflect the functional demands, opportunities, and constraints afforded by their evolutionary origins. Thus, teeth provide an excellent opportunity for studying evolutionary changes and the effects of the environment throughout human evolution.

Reduction of the Size of the Incisors and Canines and Shape of the Palate

A unique feature of *Australopithecus* is that while the configuration of the face resembles that of a chimpan-

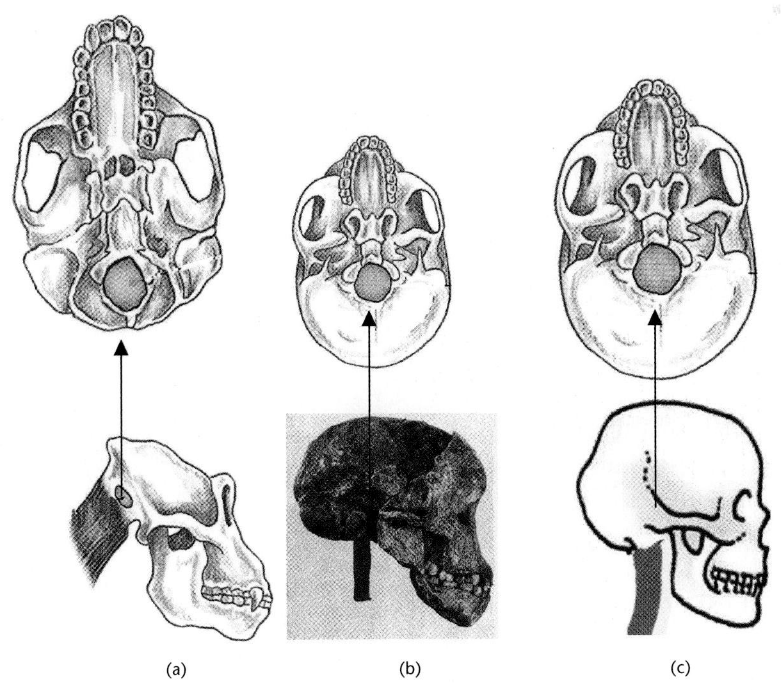

FIGURE 9.10. Position of the foramen magnum. In the gorilla (a), the foramen magnum is behind the skull while in *Australopithecus africanus* (b) and modern human (c), it is located below the skull. (Composite illustrations made by the author using the Paleoanthropology picture archive of Professor Milford Wolpoff of the Department of Anthropology, University of Michigan.)

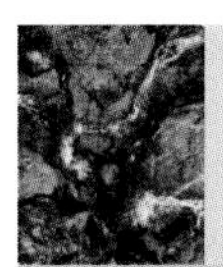

TABLE 9.3
Distinguishing Characteristics of the Gracile Australopithecines Found in Eastern and South Africa

	Australopithecus afarensis "Lucy"	*Australopithecus africanus*
Brain size (g)	400–500	400–500
Height (m)	1–1.5	1.1–1.4
Weight (kg)	30–70, Light build	30–60, Light build
Sexual dimorphism	Moderate to high, 15 to 20%	?
Shape of thorax	Broad-pyramidal	?
Finger and toes	Curved fingers and toes	?
Limb proportions	Long arms relative to legs	?
Skull form	Low, flat forehead	Higher forehead
Face shape	Projecting face	Shorter face
Brow ridges	Moderately prominent	Moderately prominent

Source: Ciochon, R. L., and J. Felagle. 1993. *The Human Evolution Source Book.* Englewood Cliffs, NJ: Prentice-Hall.

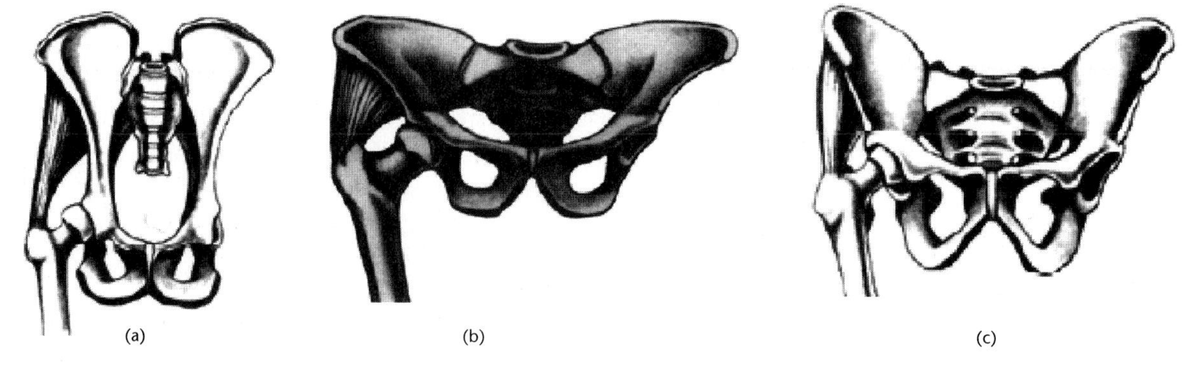

FIGURE 9.11. Pelvises and bipedal locomotion. The pelvis of the *Australopithecus afarensis* (b) is very human-like (c) and differ from the long and narrow pelvis of the chimpanzee (a). (Composite illustrations made by the author.)

zee, the size and shape of their teeth, and the palate where the teeth lie is different (**fig. 9.13**). For example, the canines in the australopithecines are reduced and do not extend much above the height of the surrounding teeth. The incisors are relatively small but the molars are large and have thick enamel. Likewise, the teeth of modern primates and of modern humans show clear differences. In humans, the teeth lie in small, arched, or V-shaped dental arcades, whereas those of apes are aligned in a U-shaped pattern with long, straight sides with the sharp-edged incisors in front, the pointed canines at the corners, and the premolars and molars on the sides. The dental arcade of the australopithecines is in between that of the hominoids and modern humans. In the course of evolution, the height of the mandible and the length of the jaw have changed, resulting in hominids with less protruding jaws and taller faces.

Large Size of the Molars

One of the hallmarks of the australopithecines is their large, relatively flat molars. The general consensus is that the large molars reflect an extreme adaptation to hard, gritty, and coarse foods (**table 9.4**). One productive way of determining the relative effectiveness of the molars in chewing hard foods is to calculate the ratio of the area of first molar (M_1) to third molar (M_3). It has been shown that this ratio is inversely related to the percentage of leaves, flowers, and shoots in the diet. [35–37] A high ratio indicates low use of the molars

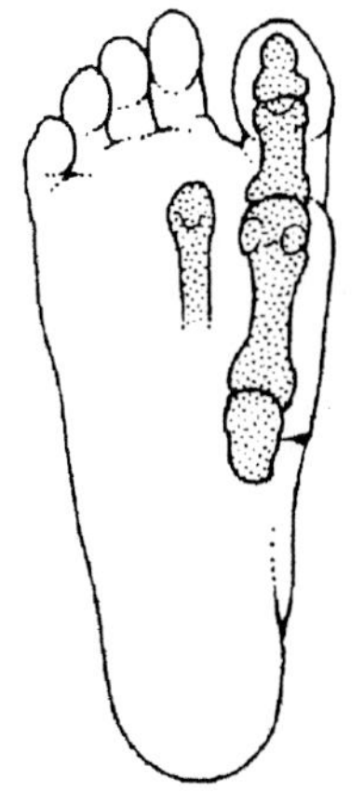

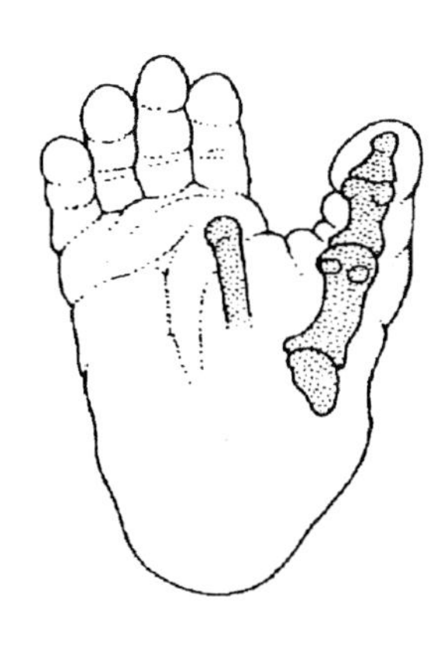

FIGURE 9.12. The footprints from Laetoli, Tanzania show that the big toes were in line with the rest of the toes like that in humans (left). The parallel position of the great toe makes possible the formation of an arch, which works as an energy absorber. In comparison, in the footprint of a chimpanzee (right) the big toe extends from the foot, and that is why a chimpanzee is able to use it like a thumb and does not have an arch. (Composite tracing made by the author using a reference the Laetoli footprints of the Museum of Natural History, New York, and presented in the Public Broadcasting Service, http://www.pbs.org/wgbh/evolution/library/07/1/1_071_03.html.).

while a low ratio indicates heavy use of the molars. That is, anthropoids with a high ratio of M_1 to M_3 area consumed more fruit than did those with a low M_1 to M_3 ratio. The australopithecines (**fig. 9.14**), when compared to the orangutan and chimpanzee, had a low M_1 to M_3 ratio, suggesting that the large molars of the australopithecines were associated with hard-to-chew foods such as seeds.

Size of the Mandible

In humans the teeth are arranged in upper and lower arches (**fig. 9.16**). The upper arches are called maxillary and the lower arches are called mandibular. The maxilla, or upper jaw, is formed by two bones which are rigidly attached to the skull. On the other hand, the mandible consists of one single horseshoe-shaped bone. The ascending branch of the mandible, which articulates with the skull by way of the temporomandibular joint, is called the mandibular ramus. Based on the overall diameter and height of the mandibular corpus, the index of robusticity of the mandible is derived by computation:

Mandibular Robusticity Index (%) =
(corpus breadth, mm/corpus height, mm) × 100.

A. afarensis, A. africanus, and *A. robustus* have relatively thick mandibles and a larger mandibular ramus than extant hominoids.[39–41] Along with the large mandible, the zygomatic facial bones, especially of *A. robustus,* are enlarged (**fig. 9.16**). As shown in **figure 9.17**, the robust australopithecines have a higher mandibular robusticity index than extant great apes and extinct gracile australopithecines. The large size and unique shape of the australopithecine mandibular corpus has been attributed to the functional demands of mastication. Strong and thick mandibles are able to

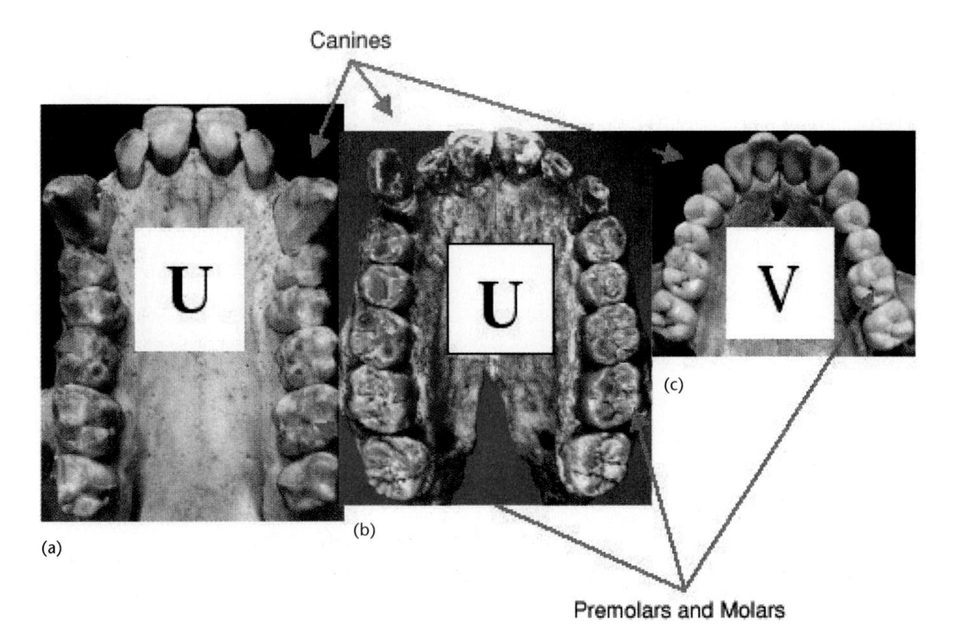

FIGURE 9.13. Teeth and dental arcade of gorilla (a), australopithecines (b), and modern humans (c). Compared to the hominoids the australopithecines have small incisors and large premolars and molars, a trend that continues in humans. The dental arches in (c) of humans are more V-shaped while in the (a) hominoids it is U-shaped and (b) australopithecines is in between that of the hominoids and modern humans. (Composite illustrations made by the author using the paleoanthropology picture archive of Professor Milford Wolpoff of the Department of Anthropology, University of Michigan.)

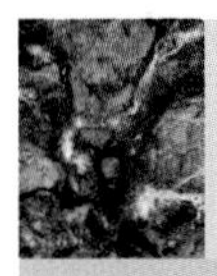

TABLE 9.4
Summed Areas of Posterior Tooth Row for Maxilla and Mandible (mm^2)

Sample	Maxilla	Mandible
Australopithecines	935	906
H. erectus	630	638
Neanderthals	543	533
Anatomically modern *H. sapiens*	490	487

Source: Brose and Wolpoff.[38]

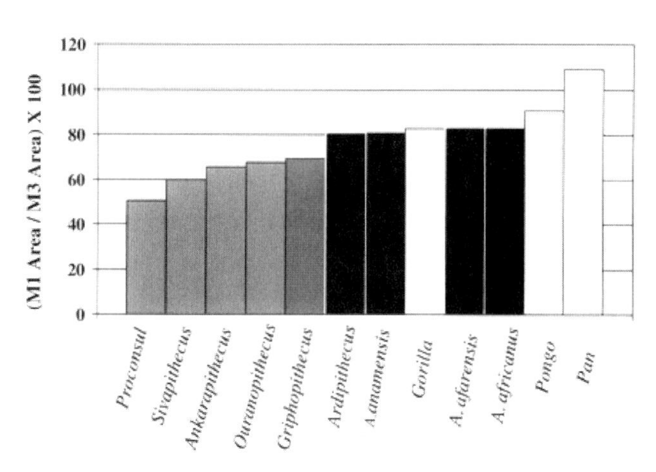

FIGURE 9.13. Large molars of the australopithecines. The australopithecines had a lower ratio of the areas of first molar (M_1) to third molar (M_3) than fossil and extant hominoids. Source: Teaford and Ungar, fig. 3.[35]

resist extreme stresses associated with chewing coarse foods. This response is called "wishboning." As such, the australopithecine mandibular morphology probably reflects the elevated stresses associated with unusual mechanical demands of the fibrous, coarse foods, which required repetitive loading.

In summary, the small incisors suggest that these hominids ate foods that did not require extensive incisal preparation, such as fruits and soft bark. Their large flat molars would have served them well for crushing. Their mandibular corpora would probably have conferred an advantage for transmitting force from the heavy zygomatic muscles to which the jaw is attached.

Dental Structure and Wear

Dental Enamel Thickness and Structure

Teeth are composed of a crown, root, cervical line, and pulp cavity (**fig. 9.18**). Teeth also consist of four tissues: enamel, dentin, pulp, and cement. Enamel is the hard substance that covers the crown of the tooth. It is a highly mineralized, totally acellular bondlike substance. In its mature state, it is 96 percent mineral, composed of hydroxy apatite, calcium phosphate, and calcium carbonate. Because enamel is the hardest substance in the body, it protects the tooth from the wear of chewing and provides strength to the teeth against abrasion and friction caused by chewing of coarse foods. Therefore, the thicker the enamel, the greater is the protection against hard objects, which commonly causes fracture of enamel. Accordingly, there is a direct correlation between the consumption of hard or abrasive food items and thick enamel.

FIGURE 9.15. Components of the upper and lower jaw. The upper arches are called maxillary (a) and the lower arch is called mandible, which consists of the corpus and ascending ramus. (Composite made by the author.)

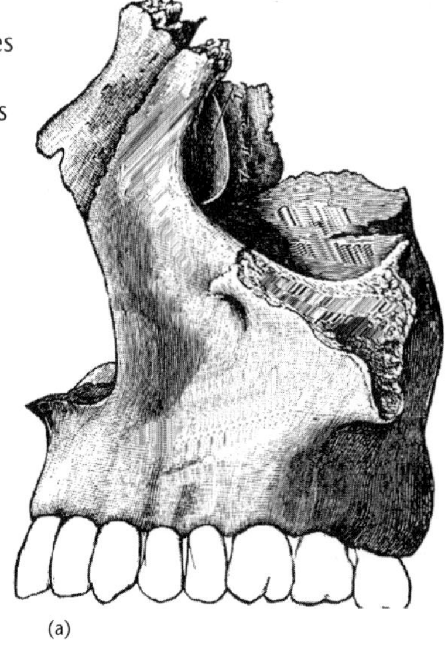

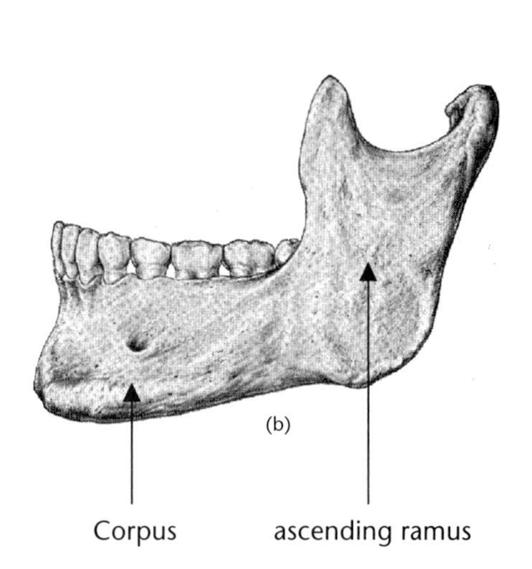

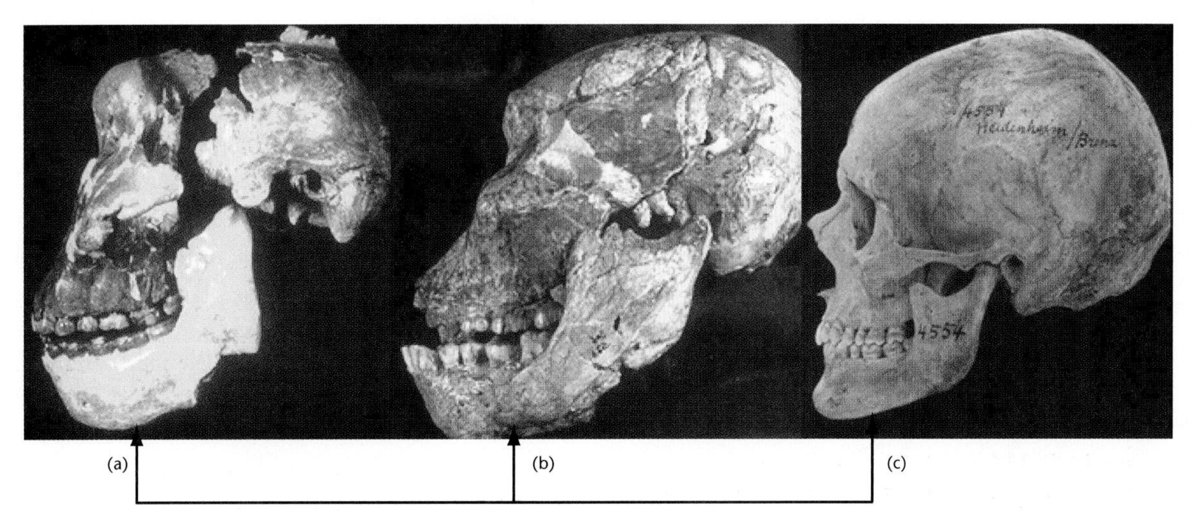

FIGURE 9.16. Mandibular robusticity in (a) *A. boisei robustus* ("Nutcracker Man"), (b) *A. africanus,* and (c) modern humans. (Source: (a) courtesy of Professor C. Loring Brace, of the Department of Anthropology, University of Michigan, (b) courtesy of Professor Milford Wolpoff of the Department of Anthropology, University of Michigan.)

A general feature of the teeth of australopithecines, like many of the Miocene apes, is that they had relatively thick enamel compared with living primates. Various studies indicate a relationship between the consumption of hard food items or abrasive food items and thick molar enamel.[37, 38, 42, 43] Therefore, it has been postulated that the relationship between thick dental enamel and abrasive food reflects an adaptation to the hard coarse diet of australopithecines. However, others have noted that *Otavipithecus*[44] and *Ardipithecus ramidus*[45] had thin enamel even though their diet included hard and coarse foods. Furthermore, the presence of interweaving of the enamel, referred to as crystallite decussation, reflects the organism's attempts to prevent the teeth from cracking.[37, 42] Based on this evidence, the presence of crystallite decussation indicates whether or not the animal eats coarse foods. Since the teeth of australopithecines did have a significant degree of prism decussation, it is quite possible that the thick enamel of the early hominids may have been a means of resisting breakage during the consumption of hard objects and an adaptation that prolonged the life of the tooth.

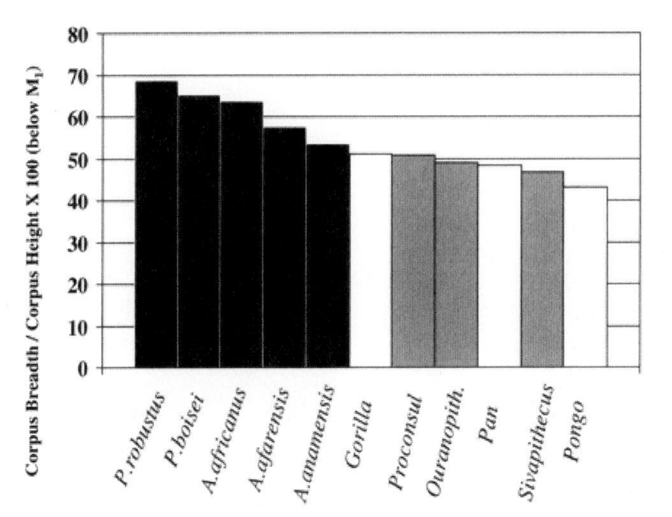

FIGURE 9.17. Index of mandibular robusticity in early hominids and hominoids. The australopithecines, and especially the *A. robustus,* had a large mandible. Source: Teaford and Ungar, fig. 4.[35]

Dental Microwear

Dental microwear refers to the microscopic scratches and pits that form on a tooth's surface as the result of its use. Dental microwear has been examined on the molar teeth and incisors of a wide variety of primates and other animals. These studies indicate that patterns of dental microwear on mammalian incisors and molar teeth reflect the material properties of food items. For example, as illustrated in **figure 9.19**, animals that have a diet consisting of harder, more brittle food items (such as hard seeds, nuts, or bone) have heavily pitted molar surfaces. In contrast, animals whose diet requires shearing tough food items (such as leaves or meat) have heavily scratched molar teeth. Intermediate patterns of pits and scratches on the molar surfaces indicate mixed diets. It may also indicate diets with in-

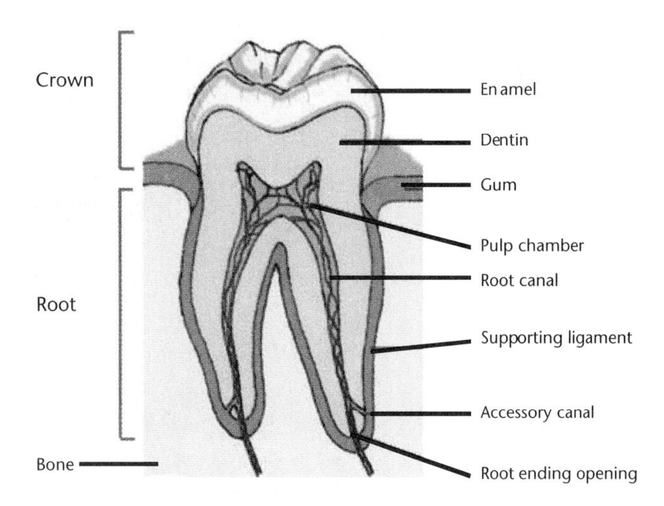

FIGURE 9.18. Basic structure of a tooth. The tooth consists of a root and a crown that is protected by the enamel. (Composite made by the author.)

termediate food properties (such as soft fruits). Likewise, patterns of dental microwear on the incisors indicate the importance of anterior or posterior teeth use in food processing.

A. africanus exhibited anterior dental microwear,[46] suggesting that these hominids ate fibrous and coarse foods. As shown in **figure 9.20**, the second molar of *Australopithecus africanus* shows a higher frequency of microwear features and scratches that are longer, narrower, and have lower incidences of pitting than *A. robustus*. This finding suggests that the gracile australopithecines may have already begun to incorporate some abrasive, terrestrial resources that required incisal stripping into their diets.

Teeth and Type of Food

Whether the australopithecines were primarily adapted for the consumption of coarse plant foods or soft animal foods can be determined by evaluating the concentration of carbon isotope in the teeth enamel.[48, 49] Animals that eat primarily plant food [i.e., fruits, leaves, and the roots of trees, bushes, and some tropical grasses (i.e., blades, seeds, and roots)] have a high concentration of carbon 13 (^{13}C) in their enamel. On the other hand, animals that primarily eat animal foods (i.e., carnivores) have high concentration of carbon 12 (^{12}C) in their tooth enamel. Hence, the ratio of ^{13}C to ^{12}C in tooth enamel can be used to provide dietary information. Stable carbon isotope analysis of A. africanus demonstrates that this early hominid consumed a mix of fruits and leaves along with large quantities of grasses and sedges.[48–50] Additionally, there is evidence to suggest they consumed varied amounts of high-quality animal foods that they were themselves eating, such as the C4 plant.[48–50]

In summary, the high frequency of microwear and large mandibular corpus of the robust australopithecines indicates that their dietary pattern was very specialized. They probably relied on hard-to-chew coarse

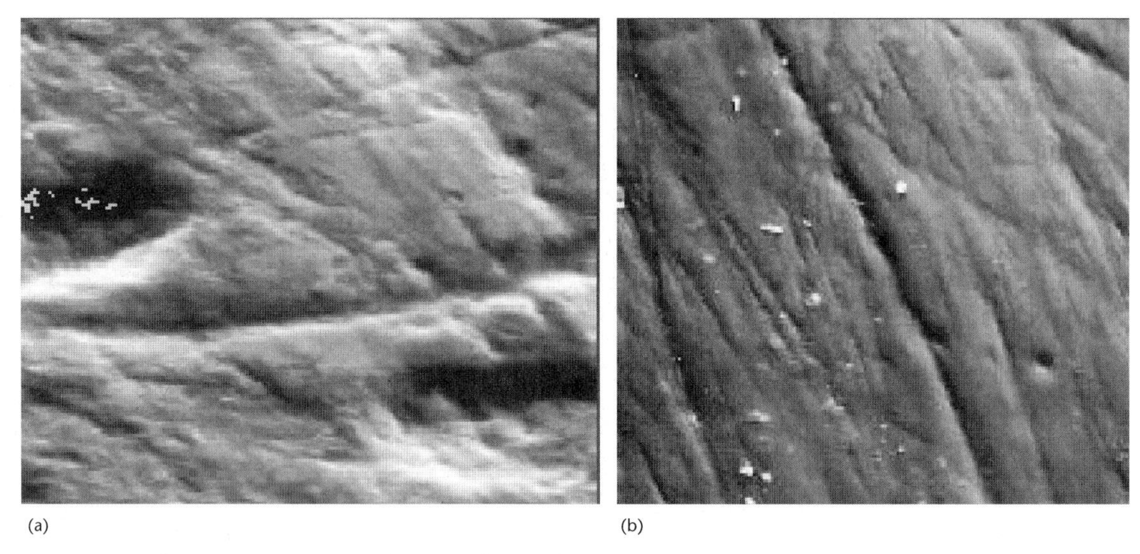

FIGURE 9.19. Dental microwear associated with different dietary patterns. The surface of the second molar of an *Ouranopithecus macedoniensis* (a) (fossil specimen from Macedonia, in northern Greece) is heavily pitted, indicating the consumption of hard food items such as nuts or tubers. In contrast, the surface on the second molar of an *Oreopithecus bambolii* (b) (fossil specimen from Tuscany in Italy) is heavily scratched, indicating a diet comprised of leaves or other soft objects. Source: Dental Microwear website.[47] (a) Courtesy of Professor C. Loring Brace, Department of Anthropology, University of Michigan. (b) Courtesy of Professor Peter S. Ungar, University of Arkansas.

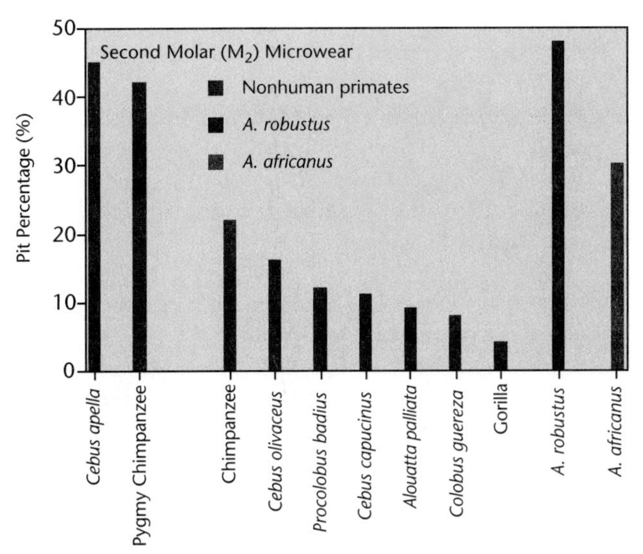

FIGURE 9.20. Dental microwear in early hominids. *Australopithecus africanus* had higher microwear features than *A. robustus*. Source: Teaford & Ingar fig. 5[35].

foods such as nuts and seeds. In contrast, the low frequency of microwear and small mandibular corpus of the gracile australopithecines suggests that their dietary pattern was less specialized. Evaluation of the concentration of carbon isotope in the teeth enamel suggests that the australopithecines exploited a wider range of food sources and had a high dietary breadth where both plant and animal foods played important roles.

Diet-Associated Dental Evolutionary Trends

The morphological features of the robust australopithecines, as shown by the microwear and large mandibular corpus, indicate that their dental anatomy was specialized to chew hard and coarse foods, such as nuts and seeds.

The generalized dental pattern of the gracile australopithecines is also associated with a broader dietary breadth than that of extant apes. In evolutionary perspective, this dietary strategy had a great adaptive significance. First, it would have allowed australopithecines to survive in a greater variety of habitats than have modern great apes. This dietary flexibility could also have increased dietary quality over that of extant apes by adding low-fiber underground storage organs and protein- and lipid-rich animal foods to australopithecine diet. Second, the increased dietary breadth could have buffered australopithecines against climatic change and habitat loss.

Viewed in the context of dietary breath, the fate of the robust australopithecines may be related of their high dietary specialization. They were eventually replaced by our *Homo* forebears who were less specialized and had regular access to higher-quality animal foods. The high dietary flexibility of early *Homo* allowed for increased access to rich sources of protein and fat, such as those derived from bone marrow of scavenged or hunted prey. Thus, where the gracile australopithecines may have eaten when they stumbled upon animal foods, the robust australopithecines might not have done so and thus become an evolutionary dead end. The gracile australopithecines might have continued through the evolutionary path, and eventually, about 2.5 million years ago, were replaced by *Homo habilis*—a toolmaker—and with that a whole new way of dealing with the world came about (see next chapter).

Summary

According to the fossil evidence, the trend toward hominid evolution might have started 7 to 6 million years ago with the appearance of *Sahelanthropus tchadensis, Orrorin tugenensis,* and *Ardipithecus ramidus.* The craniofacial morphology and dental features of these species more closely resemble hominoids adapted to an arboreal environment than that of terrestrial hominids, yet postcranial skeletal morphology shows these species were partially adapted to the terrestrial environment and therefore, when they were on the ground, they probably had a bipedal locomotion. The preponderance of primitive traits relative to the derived ones justifies considering all these specimens in the hominoid-hominid phase of evolution.

The trend toward bipedal locomotion becomes more evident with the early hominids such as the six species of australopithecines that lived between 3 and 4 million years ago in South and East Africa. The most important features that distinguish australopithecines from the species in the hominoid-hominid phase was their locomotion—the shoulder hip structure, and lower limbs tell us they were completely bipedal, but also in a forest environment retained the ability to climb trees. The similarity of sexual dimorphism in body size of australopithecines and chimpanzees suggests that the social organization of *A. afarensis* might be best characterized as multimale, cooperating kin groups.

Dental features of australopithecines provide interesting insight into hominid evolution. A unique feature of australopithecus is that while the configuration of the face resembles that of a chimpanzee, the dental palate and size and shape of the teeth are different. For example, the canines are reduced and do not extend much above the height of the surrounding teeth, the incisors are relatively small, and the molars are large, with thick enamel. Likewise, the teeth of modern primates and of modern humans do show clear differences. In humans the teeth lie in small, V-shaped dental arcades whereas those of apes are aligned in a U-shaped pattern with long, straight sides. In the course of evolution, the height of the mandible and the length of the jaws have also changed, resulting in hominids with less-protruding jaws and taller faces. According to analysis of the gene that controls growth of the jaw, the small jaw of hominids evolved approximately 2.4 million years ago.

Suggested Readings

Conroy, G. C. 1997. *Reconstructing Human Origins. A Modern Synthesis.* New York: Norton.

Delson, E., ed. 1985. *Ancestors: The Hard Evidence.* New York: Liss.

Grine, F., ed. 1988. *Evolutionary History of the Robust Australopithecines.* New York: de Gruyter.

Tearford, M. E., M. M. Smith and M. J. W. Ferguson, eds. 2000. *Development, Function and Evolution of Teeth.* New York: Cambridge University Press.

Wolpoff, M. H. 1999. *Human Evolution,* 2d ed. Boston: McGraw-Hill.

Ciochon, R. L., and J. Fleagle. 1993. *The Human Evolution Source Book.* Englewood Cliffs, NJ: Prentice-Hall.

Literature Cited

1. Goodman, M. 1999. The genomic record of humankind's evolutionary roots. *Am. J. Hum. Genet.* 64:31–39.
2. Brunet, M., F. Guy, D. Pilbeam, H. T. Mackaye, A. Likius, D. Ahounta, A. Beauvilain, C. Blondel, H. Bocherens, J. R. Boisserie, L. De Bonis, Y. Coppens, J. Dejax, C. Denys, P. Duringer, V. Eisenmann, G. Fanone, P. Fronty, D. Geraads, T. Lehmann, F. Lihoreau, A. Louchart, A. Mahamat, G. Merceron, G. Mouchelin, O. Otero, P. Pelaez Campomanes, M. Ponce De Leon, J. C. Rage, M. Sapanet, M. Schuster, J. Sudre, P. Tassy, X. Valentin, P. Vignaud, L. Viriot, A. Zazzo, and C. Zollikofer. 2002. A new hominid from the Upper Miocene of Chad, Central Africa. *Nature* 418:145–151.
3. Wolpoff, M. H., B. Senut, M. Pickford, and J. Hawks. 2002. Palaeoanthropology (communication arising): *Sahelanthropus* or *'Sahelpithecus'*? *Nature* 581–582.
4. Senut, B., M. Pickford, D. Gommery, P. Meind, K. Cheboie, and Y. Coppens. 2001. First hominid from the Miocene (Lukeino Formation, Kenya). *C.R. Acad. Sci. Paris* 332: 137–144.
5. White, T. D., G. Suwa, and B. Asfaw. 1994. *Australopithecus ramidus,* a new species of early hominid from Aramis, Ethiopia. *Nature* 371:306–312.
6. Haile-Selassie, Y. 2001. Late Miocene hominids from the Middle Awash, Ethiopia. *Nature* 412:1–181.
7. Patterson, B., and W. W. Howells. 1967. Hominid humeral fragment from early Pleistocene of northwestern Kenya. *Science* 156:64–66.
8. Leakey, M., C. S. Feibel, I. McDougall, and A. C. Walker. 1995. New four-million-year-old hominid species from Kanapoi and Allia Bay, Kenya. *Nature* 376:565–571.
9. Leakey, R. E., and R. Lewin. 1992. *Origins Reconsidered: In Search of What Makes Us Human.* New York: Doubleday.
10. Walker, A. C., and P. Shipman. 1996. *The Wisdom of the Bones.* New York: Alfred E. Knopf.
11. Broom, R. 1936. A new fossil anthropoid skull from South Africa. *Nature* 138:486–488.
12. Broom, R. 1938. The Pleistocene anthropoid apes of South Africa. *Nature* 142:377–379.
13. Johanson, D. C., and B. Edgar. 1996. *From Lucy to Language.* New York: Simon and Schuster.
14. Keyser, A. W. 2000. The Drimolen skull: The most complete australopithecine cranium and mandible to date. *South African Journal of Science* 96:189–193.
15. Leakey, L. S. B. 1959. A new fossil skull from Olduvai. *Nature* 184:491–493.
16. Lewin, R. 1987. *Bones of Contention: Controversies in the Search for Human Origins.* New York: Simon and Schuster.
17. Leakey, R. E. 1973. Evidence for an advanced Plio-Pleistocene hominid from East Rudolf, North Kenya. *Nature* 242:447–450.
18. Leakey, R. E. 1974. Further evidence of Lower Pleistocene hominids from East Rudolf, North Kenya, 1973. *Nature* 248:653–656.
19. Johanson, D. C., and M. A. Edey. 1981. *Lucy: The Beginnings of Humankind.* New York: Simon and Schuster.
20. Johanson, D. C., and M. Taieb. 1976. Plio-Pleistocene hominid discoveries in Hadar, Ethiopia. *Nature* 260:293–297.
21. Johanson, D. C., and T. D. White. 1980. On the status of *Australopithecus afarensis. Nature* 207:1104–1105.
22. Kimbel, W. H., D. C. Johanson, and Y. Rak. 1994. The first skull and other new discoveries of *Australopithecus afarensis* at Hadar, Ethiopia. *Nature* 368:449–451.
23. Dart, R. A. 1925. *Australopithecus africanus:* The man-ape of South Africa. *Nature* 115:195–199.
24. Broom, R. 1936. A new fossil anthropoid skull from South Africa. *Nature* 138:486–488.
25. Broom, R., and J. T. Robinson. 1947. Further remains of the Sterkfontein ape-man, Plesianthropus. *Nature* 160:430–431.
26. Asfaw, B., T. D. White, C. O. Lovejoy, B. Latimer, S. Simpson, and G. Suwa. 1999. *Australopithecus garhi:* A new species of early hominid from Ethiopia. *Science* 284:629–635.
27. Asfaw, B., W. H. Gilbert, Y. Beyene,. W. K. Hart, P. Renne, G. WoldeGabriel, et al. 2002. Remains of *Homo erectus* from Bouri, Middle Awash, Ethiopia. *Nature* 416:317–320.
28. Leakey, M. D. 1984. *Disclosing the Past.* New York: Doubleday.

29. Tattersall, I. 1993. *The Human Odyssey: Four Million Years of Human Evolution.* New York: Prentice Hall.
30. Brace, C. L. 1995. *The Stages of Human Evolution.* Englewood Cliffs, N.J.: Prentice Hall.
31. McHenry, H. M. 1994. Behavioral ecological implications of early hominid body size. *J. Hum. Evol.* 27:77–87.
32. Reno, P. L., R. S. Meindl, M. A. McCollum, and C. O. Lovejoy. 2003. Sexual dimorphism in Australopithecus afarensis was similar to that of modern humans. *Proc. Natl. Acad. Sci. USA* 100:9404–9409.
33. Larsen, C. S. 2003. Equality for the sexes in human evolution? Early hominid sexual dimorphism and implications for mating systems and social behavior. *Proc. Natl Acad. Sci. USA* 100:9103–9104.
34. Mitani, J. C., D. P. Watts, and M. N. Muller. 2003. Recent Developments in the study of wild chimpanzee behavior. *Evol. Anthrop.* 11:9–25.
35. Teaford, M. F., and P. S. Ungar. 2000. Diet and the evolution of the earliest human ancestors. *Proc. Natl. Acad. Sci. USA* 97:13506–13511.
36. Teaford, M. F., and P. S. Ungar. 1999. Paleontological perspective on the evolution of human diet. Proceedings of the 14th International Congress of Anthropological and Ethnological Sciences, Williamsburg, Virginia.
37. Lucas, P. W., R. T. Corlett, and D. A. Luke. 1986. Postcanine tooth size and diet in anthropoid primates. *Z. Morphol. Anthropol.* 76(3):253–257.
38. Brose, D. S., and M. H. Wolpoff. 1993. Early Upper Paleolithic and Late Middle Paleolithic Tools. In *The Human Evolution Source Book,* ed. R. L. Ciocho, and J. G. Fleagle, pp. 523–545. Englewood Cliffs, N.J.: Prentice Hall.
39. Hylander, W. L. 1975. Incisor size and diet in anthropoids with special reference to Cercopithecoidea. *Science* 189: 1095–1098.
40. Hylander, W. L. 1988. Implications of in vivo experiments for interpreting the functional significance of "robust" australopithecine jaws. In *Evolutionary History of the "Robust" Australopithecines,* ed. F. E. Grine. New York: Aldine.
41. Daegling, D. J., and F. E. Grine. 1991. Compact bone distribution and biomechanics of early hominid mandibles. *Am. J. Phys. Anthropol.* 86:321–339.
42. Kay, R. F. 1981. The Nut-Crackers—A new theory of the adaptations of the Ramapithecinae. *Am. J. Phys. Anthropol.* 55:141–151.
43. Dumont, E. R. 1995. Enamel thickness and dietary adaptation among extant primates and chiropterans. *J. Mammal.* 76:1127–1136.
44. Conroy, G. C., M. Pickford, B., Senut, J. Van Couvering, and P. Mein. 1992. Otavipithecus namibiensis, first Miocene hominoid from southern Africa. *Nature* 356:144–148.
45. White, T. D., G. Suwa, and B. Asfaw. 1994. *Australopithecus ramidus,* a new species of early hominid from Aramis, Ethiopia. *Nature* 371:306–312.
46. Ryan, A. S., and D. C. Johanson. 1989. Anterior dental microwear in *Australopithecus afarensis. J. Hum. Evol.* 18: 235–268.
47. Dental Microwear. Internet resource: http://comp.uark.edu/~pungar
48. Sponheimer, M., and J. A. Lee-Thorp.1999. Isotopic evidence for the diet of an early hominid, *Australopithecus africanus. Science* 283:368–370.
49. Sponheimer, M., and J. A. Lee-Thorp. 2003. Differential resource utilization by extant great apes and australopithecines: Towards solving the C4 conundrum. *Comparative Biochemistry and Physiology. Part A: Molecular & Integrative Physiology* 136:27–34.
50. Scott, R. S., Ungar, P. S., Bergstorm, T. S., Brown, S. A., Orine, E. E., Teaford, M. F., Walker, A. 2005. Dental microwear texture analysis shows within-species diet variability in fossil hominids. *Nature* 436: 693–695.
51. Wolpoff, M. H. 1999. *Paleoanthropology.* Boston: McGraw-Hill.

Evolution of the Genus *Homo*

The last 2.5 million years have been a unique period in the history of humankind. It was the time that marked the appearance of the earliest members of the genus *Homo*. It was the time during which hominids such as *Homo habilis* and *Homo erectus* came to occupy most continents of the earth: Africa, Asia, Australia, and Europe. The appearance of *Homo erectus* in Africa, Asia, and Europe is characterized by a marked increase in body size and energy requirements. As such, *Homo erectus* was the first hominid to display the strategies that maximize the acquisition of food, allowing them to become large in body size and capable of maintaining this large body size. The strategies were different than those of earlier hominids, but similar to the strategies used by *Homo sapiens*. This chapter focuses on the fossil evidence that documents this trajectory.

Homo habilis

Brain Size and Body Size

About 2.5 million years ago in East Africa and later, 1.6 million years ago, in Southern Africa, *Australopithecus* began to be replaced by a hominid species with bigger brains, referred to as *Homo habilis* ("handy man") (**table 10.1**). As exemplified by the KNM-ER 1470 skull (**figs. 10.1** and **10.2**), *H. habilis* shows traits that are in the intermediate stage between the more primitive characteristics of the gracile australopithecines and the more rounded skull of modern *H. sapiens*.[1–5] The braincase of *Homo habilis* was rounder and more like that of modern humans, but the face was very long, broad, flat, and with prominent australopithecine-like cheek bones. At an average of approximately 700 grams (ranging from 510 to 770 cc), the *H. habilis* brain was about 40 percent bigger than that of the gracile australopithecines.

The brain of *H. habilis* was not only larger than that of the australopithecines, but it also showed signs of structural reorganization in the direction of the modern human form.[1–5] Humans differ from apes in their possession of Broca's area on the parietal lobe of the brain. This controls the impulses that regulate the proper flow of air and the vocal cords necessary for speech (**fig. 10.3**). The endocranial cast of KNM-ER 1470 shows the presence of Broca's area, suggesting that the functional specialization of cerebral hemispheres that characterizes *H. sapiens* was already beginning in *H. habilis*. The leg bones of *H. habilis* were larger than that of australopithecines. The hip and leg anatomy was similar to the later species of *Homo erectus*. Likewise, the sexual dimorphism in body size of *H. habilis* was about 10 to 15 percent, which is similar to modern humans.

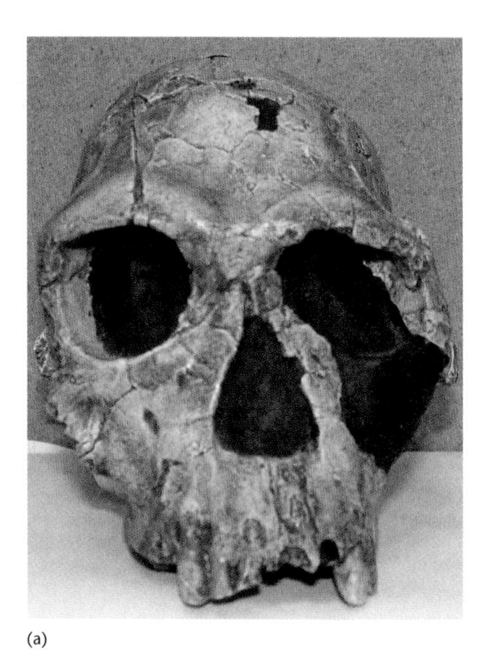

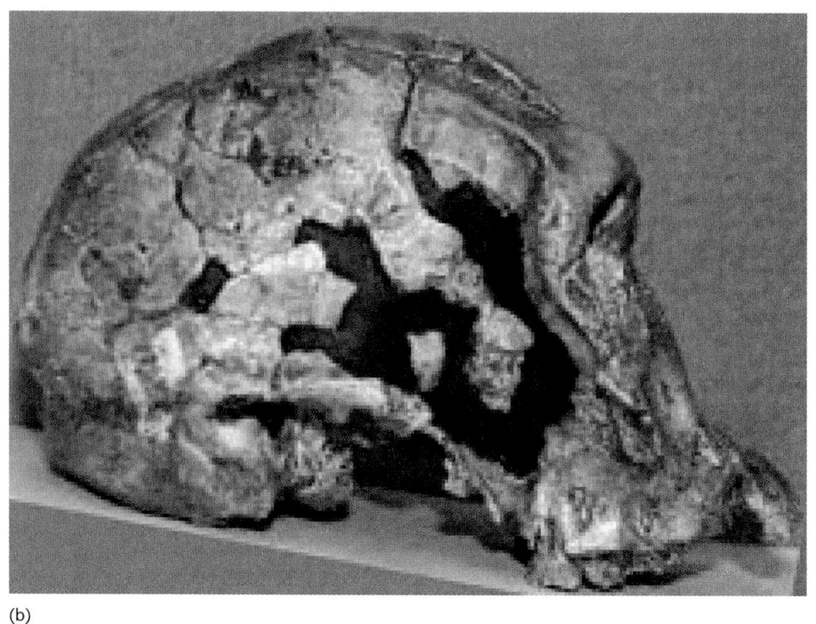

(a) (b)

FIGURE 10.1. *Homo habilis.* Represented by KNM-ER 1813 (Picture taken by the author using the Paleoanthropology cast collection of the Department of Anthropology, Ann Arbor, University of Michigan.)

Reduction in Size of the Mandible

As discussed in the previous chapter, a distinguishing difference between hominoids such as chimpanzees and gorillas is that they have powerful masticatory muscles that are associated with large jaws. This trend still persists among the australopithecines who have massive and large jaws (see **fig. 9.19**). On the other hand, hominids starting with *Homo habilis* are characterized by relative small jaws. An important question is why did the reduction in the size of the jaw occur? According to Stedman and colleagues[5] the reduction in the size of the jaw in *Homo sapiens* is associated with the inactivation (due to a mutation) of the gene that controls muscle growth in the masticatory muscles as such those found in the temporal area. These authors have identified the gene (called MYH16) that controls contraction of jaw muscles of humans and monkeys and found that, in humans, the MYH16 has mutated and as result have a low level of activity and hence the effect on the muscle fibers of masticatory muscles is decreased causing a reduction in the size of the jaw (**fig. 10.4**). In contrast, all non-human primates have an intact copy of the gene, and have a high level of MYH16 protein in their jaw muscles. The date when the jaw became small coincides with the appearance of *Homo habilis* whose dietary pattern included probably

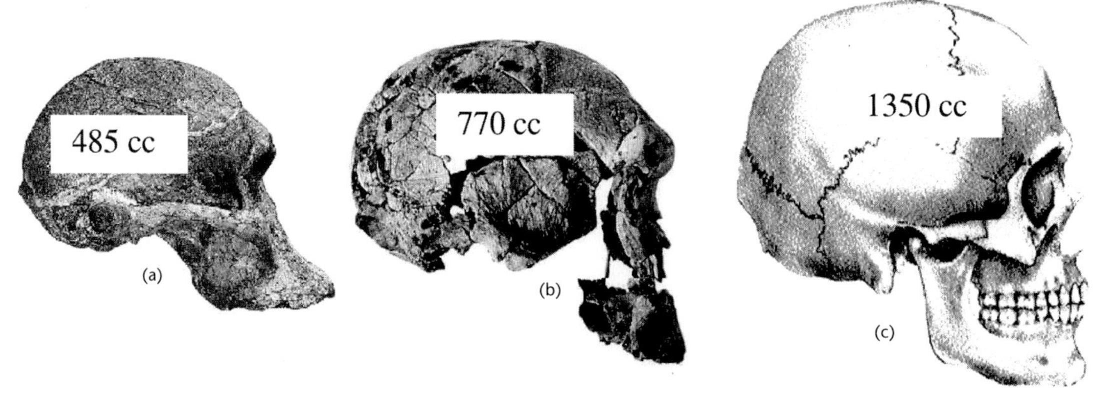

FIGURE 10.2. The skull of (a) gracile *Australopithecus africanus,* (b) *Homo habilis,* as represented by KNM-ER 1470, and (c) modern Homo sapiens. *Homo habilis* had a cranial capacity of 770 grams and lived about 1.8 million years ago. (Composite illustrations made by the author using the Paleoanthropology picture archive of Professor Milford Wolpoff of the Department of Anthropology, University of Michigan.)

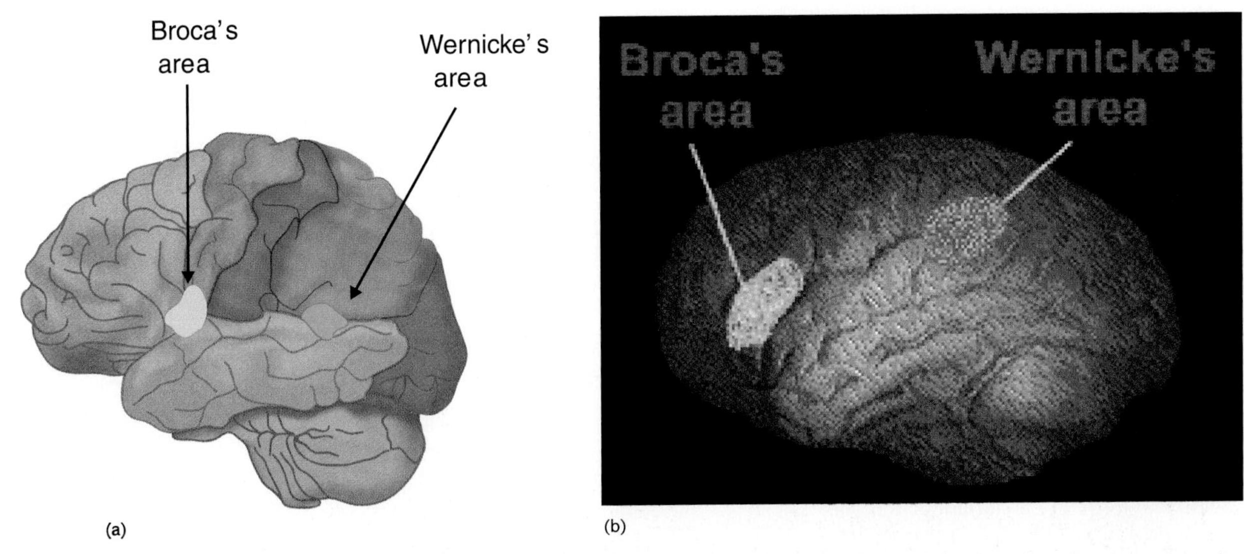

FIGURE 10.3. Functional areas of the human brain (a) and (b) brain endocast of *Homo habilis* showing the presence of the Broca's and Wernicke's area. Source: (a) Composite made by the author; (b) Walker, P. L., and E. H. Hagen. 2003. *Human Evolution. A Multimedia Guide to the Fossil Record.* Student Edition CD-ROM. W.W. Norton & Company, Inc.

a high proportion of animal foods that required less masticatory effort than a plant-based diet. Viewed in this context, the mutation of the MYH16 gene became established in the hominid line as a result of the shift in dietary preference from a plant-based diet to an animal-based diet.

Oldowan Stone Tool Tradition

The earliest direct evidence of hominid technology is the Oldowan Industrial Complex dated to 2.5 million years ago in the Ethiopian Rift Valley and associated with *Homo habilis*. The stones have been found in caves along with skeletal remains of *H. habilis* and large animals. Oldowan stone tools have been found in piles (referred to as stone caches) or in a formation of a circular scatter of stones (referred to as a stone ring). Archaeologists use patterns of how tools are abandoned to infer information about tool use and ways of life. The pattern of how these stones were placed in the caves suggests that *Homo habilis* might have established a home base where food was derived either from scavenging or hunting; it might have been used for food sharing and perhaps storage for later con-

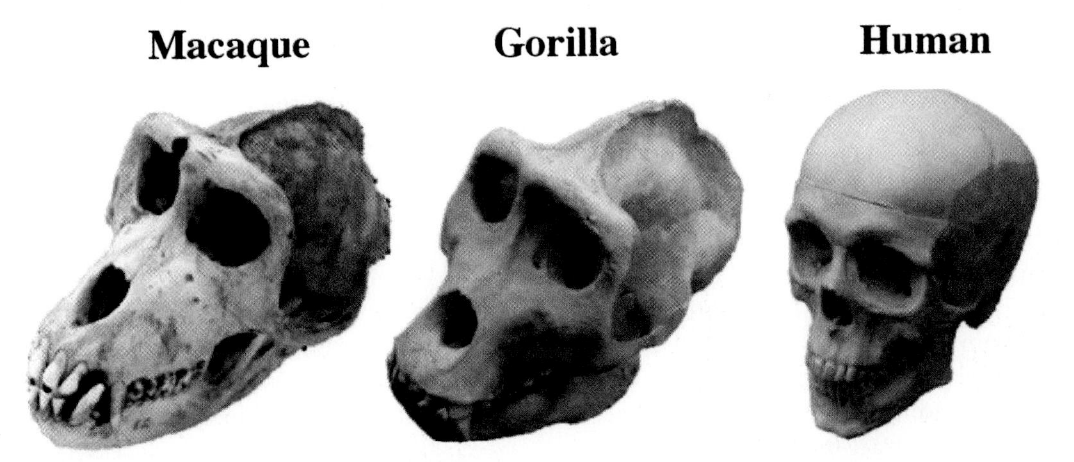

FIGURE 10.4. Reduction of jaw size in *Homo sapiens*. The reduction in the size of the jaw in *Homo sapiens* is associated with the inactivation (due to mutation) of the gene that controls muscle contraction and growth of the masticatory muscles such as those found in the temporal area shown in **orange color**. It is estimated that this mutation appeared approximately 2.4 million years ago and coincide with the evolution of *Homo habilis*. (Source: Stedman, H. H., B. W. Kozyak, A. Nelson, D. M. Thesier, L. T. Su, D. W. Low, C. R. Bridges, J. B. Shrager, N. Minugh-Purvis, and M. A. Mitchell. 2004. Myosin gene mutation correlates with anatomical changes in the human lineage. *Nature* 428:415–418.)

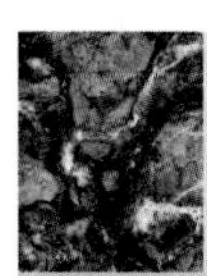

TABLE 10.1
Skeletal Remains of *Homo habilis*

Identification (Location)	mya	Skeletal Remains Features	Brain Size (cc)
OH 7 (Olduvai Gorge in Tanzania)	1.8	lower jaw and two cranial fragments of a child and a few hand bones	680
OH 8 (Olduvai Gorge in Tanzania)	1.8	set of foot bones	
OH 13 (Olduvai Gorge in Tanzania)	1.6	lower jaw and teeth, bits of the upper jaw and a cranial fragment	650
OH 16 (Olduvai Gorge in Tanzania)	1.7	teeth and fragmentary parts of the skull	640
OH 24 (Olduvai Gorge in Tanzania)	1.85	*Homo habilis* fairly complete cranium	590
KNM-ER 1470 (Koobi Fora in Kenya)	1.9	*Homo habilis* (or *Homo rudolfensis*?) skull and face without the large brow ridges	750
KNM-ER 1805 (Koobi Fora in Kenya)	1.85	heavily built cranium contains many teeth	600
KNM-ER 1813	1.9–1.8	*Homo habilis*? similar to 1470	510
(Koobi Fora in Kenya) Sts53 (Sterkfontein in South Africa)	2.0–1.5	number of cranium fragments including teeth, many stone tools	
OH 62 (Olduvai Gorge in Tanzania)	1.8	*Homo habilis*: portions of skull, relatively long arms in proportion to the legs Height: 105 cm (3′5″)	

Source: Leakey,[1] Leakey,[2] Leakey,[3] Tobias,[4] Blumenschine, Peters, and Masao et al.[31]

sumption. Thus, *Homo habilis* signals the beginnings of recognizable sapient behavior such as foresight and planning.

As shown by findings in Eastern Europe, the Oldowan tool technology has persisted through the early stages of *Homo erectus*.[6] The **Oldowan** stone tool tradition primarily consists of pieces of stone about the size of a tennis ball (pebble stones) with sides trimmed to provide a *sharp cutting edge* (**figs. 10. 5a and b**). Analysis of the process of tool manufacturing suggests that early hominids used the strategies with the least amount of effort to obtain large sharp-edged flakes from available raw materials. Individuals who produced Oldowan tools possessed an excellent empirical understanding of the mechanical properties of lithic raw materials, fracture mechanics, and geometry. Two techniques were used for producing the sharp-edged slivers. The first technique, referred to as the hand-held method, involved striking the edge of a platform and a hand-held block or cobble (a core) with a hammer stone to produce thin, sharp-edged flakes (**fig. 10.5b**). The other method, referred to as the bipolar technique, involved simply placing the core stone on an anvil and smashing it with a hammer stone. Bone tools were also probably flaked like stone or were used without intentional modification.

As shown by the Oldowan technology, the early hominids' manual skills exceeded those of chimpanzees. For example, the bonobo called Kanzi was taught the technique of stone tool flaking by direct percussion.[8, 9] After 3 years of training, he was unable to strike forcefully and accurately at the correct angle or position on the platform. This difference is related to anatomical limitations posed by the wrist and fingers of chimpanzees, which are adapted to quadrupedal and arboreal locomotion. Not only does the chimpanzee lack the forearm muscle necessary for powerful thumb flexion, it also has narrow fingertips[8–12] and a wrist with the limited rotation necessary to avoid overextension during knuckle walking. In contrast, humans have short straight fingers, a long stout thumb, and fingertips with broad fleshy pads underlain by

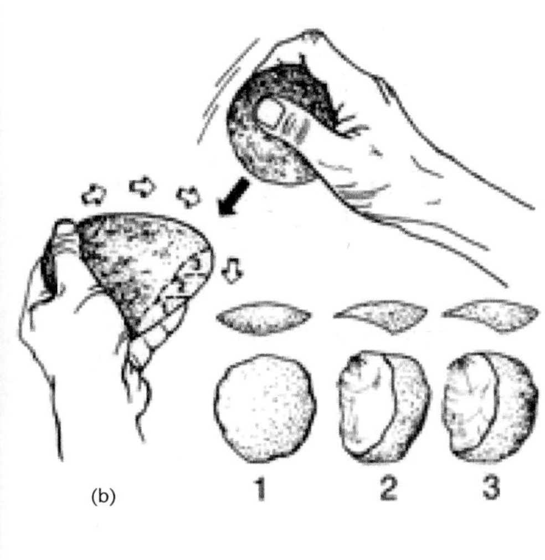

FIGURE 10.5. Oldowan stone tool technology. Dated to about 2.5 million years ago, it is associated with *Homo habilis* from East Africa. Stones with *sharp cutting edges* (a) that were developed using a hand-held technique, (b) suggesting that the early hominids had manual skills far exceeding those of chimpanzees. Reprinted with permission from "Paleolithic Technology and Human Evolution" by S. H. Ambrose, *Science* 291:1748–1753. Copyright 2001 AAAS.

wide apical tufts of bone, all of which increase stability when gripping small tools. Therefore, assuming that the behaviors of the chimpanzees were similar to those performed by the early hominids, we can infer that the manual dexterity necessary for making stone tools must have evolved about 2.5 million years ago, when *Homo habilis* appeared.

Stone Tools and Cut Marks on Bones

As shown in **figure 10.6**, the animal bone assemblages excavated from Olduvai and Koobi Fora tend to consist mainly of meat-rich bones. Many, but not all, of the cut marks found on these bones were caused by a slicing action in areas of muscle attachment. Besides these definite signs of meat acquisition, there is evidence that hominids also broke the long-shafted leg bones of animals to acquire marrow. Stone tool percussion marks have been detected on these leg bones, and the middle or shaft portions tend to be fractured all the way down to their ends. Humans use the same methods when using stone tools to obtain bone marrow.[13–15] The cut marks and signs of tool breakage of animal bones indicate that Oldowan artifacts (at least after 1.9 million years ago) were used to slice and bash bones to obtain meat and marrow. The evidence of bone modification suggests that hominids were able to gain access to the carcasses of large mammals, and possibly new food resources, by virtue of making and using sharp stone implements. Thus, it appears that Oldowan technology was an adaptive strategy that expanded the nutritional resources and quality of food of early hominids.

Scavenging and Small Game Hunting

There are two hypotheses for the ways in which early hominids obtained meat: hunting and scavenging.[16] The hypothesis that early hominids were scavengers is based on several observations. First, in order for human hunters to carry food back to a central place or camp they must systematically cut their kill into small pieces, unless the animals are small enough to be eaten on the spot. Disarticulation leaves cut marks in a predictable pattern on the major joints of the limbs: shoulder, elbow, carpal joint (wrist), hip, knee, and ankle. Second, butchering leaves marks on the shafts of bones from the upper part of the front or hind limb, since this is where the big muscle masses lie. Third, to remove skin or hide to be used for clothing, bags, thongs, and so on, human hunters must separate the skin or hide from the bones in areas where there is little flesh, such as the lower limb bones. In all such cases, cutting the skin leaves a cut mark on these bones. There are no cut marks on the midshaft on the animal bones found with the fossil bones. All the remains found with *Homo habilis* suggest that they took what they could get—be it skin, tendon, or meat, in an unsystematic manner.

Carnivore tooth marks and stone tool cut marks were often found on the same bones. Occasionally, these marks overlapped, and when they did, it was

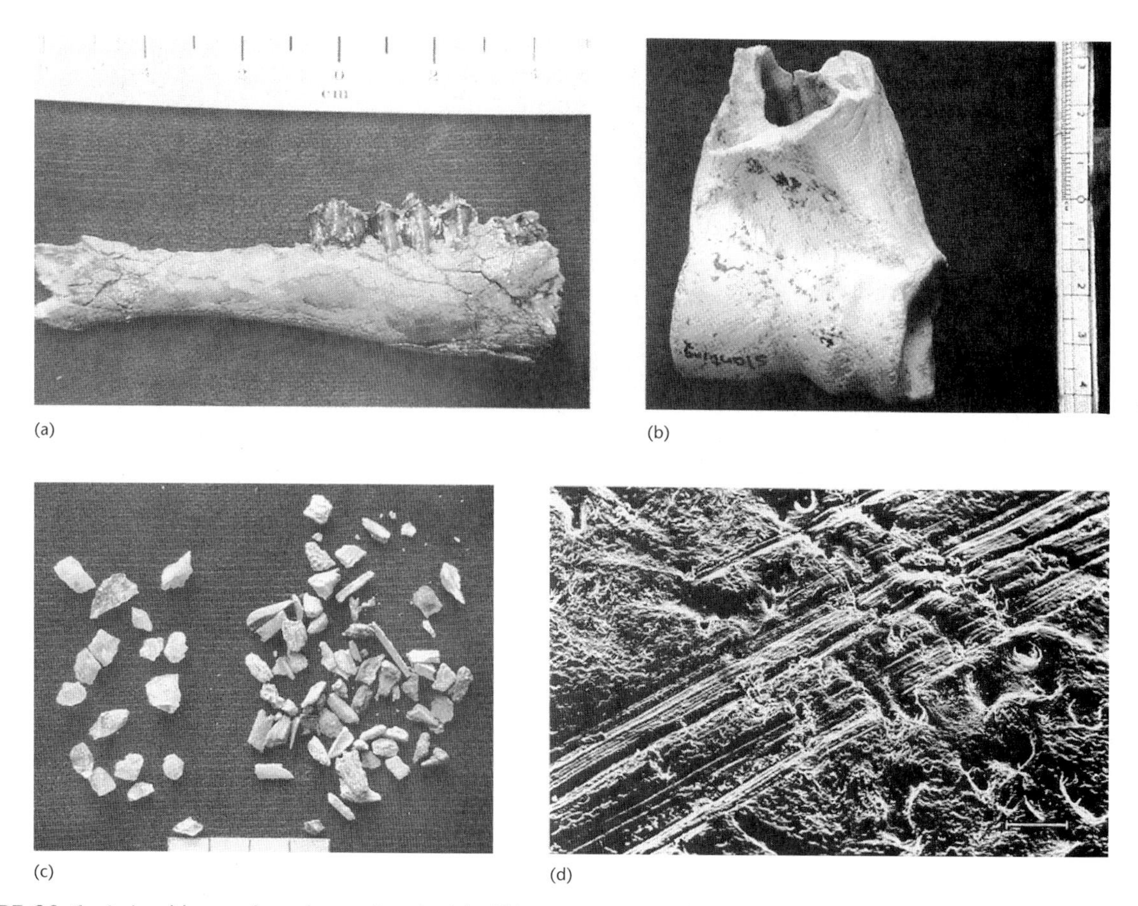

FIGURE 10.6. Animal bones found associated with Oldowan stone tool technology: (a) bovid (antelope) jaw, (b) distal humerus, (c) bone fragments. The scanning electron micrograph (d) shows that the grooves or cut marks in bone (b) were made by a stone tool while the smooth contour of grooved cut marks were made by carnivore teeth. Rasmussen, D. T., *The Origin & Evolution of Humans & Humanness,* 1993; Jones and Bartlett Publishers, Sudbury, MA. www.jbpub.com. Reprinted with permission.

possible to say which had been made first. In at least a few instances, hominids were not the first possessors of the carcass, since the cut marks were often superimposed on previous carnivore tooth (gnaw) marks. In other words, those hominids were not the sole users of animal carcasses. Given the abundant flora and fauna in the African savanna and the large numbers of migratory species in certain seasons, there were probably a lot of carcasses available, increasing the nutritional resources available for scavengers. In fact, even more recently during one week in East Africa, investigators were able to secure more than 655 pounds of meat without resorting to hunting.[17]

The hypothesis that early hominids hunted small game is derived from the observations that extant primates such as modern baboons, great apes (bonobos, *Pan paniscus*, and orangutans, *Pongo pygmaeus*) hunt and scavenge meat. Field observation has shown that chimpanzees obtain about 10 percent of the total daily calorie intake from meat of red *colobus* monkeys.[18] Chimpanzee hunters typically consume the entire carcass, including the flesh, viscera, and bone of their prey. Monkeys are often hunted using a cooperative hunting strategy. As shown in **table 10.2**, the body size of early hominids was as small as modern chimpanzees (weighing about 40 kg and standing no more than 5 feet tall).[19] Therefore, early hominids probably hunted small game like the modern chimpanzees.

Homo erectus

Homo habilis began to be replaced by a bigger kind of hominid referred to as *Homo erectus* ("erect man") in Asia and *Homo ergaster* in Africa (**fig. 10.7** and **table 10.3**). The available evidence suggests that the genus *Homo* arose in an African savanna. Subsequently, the area of habitation of *Homo erectus* was extended to ecologically comparable stretches northward into the Temperate Zone of Asia and Europe.[23-27] Archeological studies indicate that the migration route out of Africa would have been inland, through the Sinai and the Levant (the eastern Mediterranean coast). Comparative studies of the cranial morphology indicate that *Homo erectus* and *Homo ergaster* were members of the

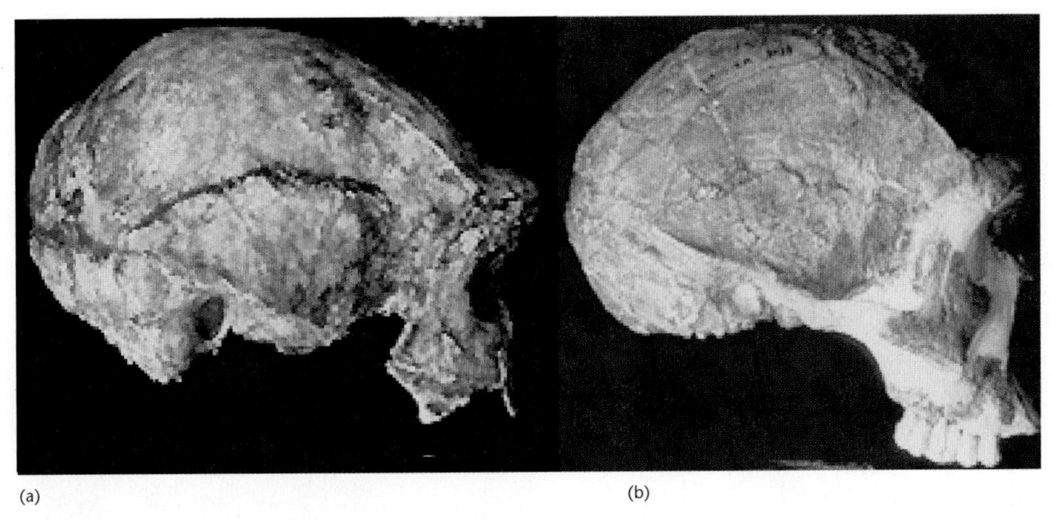

(a) (b)

FIGURE 10.7. African *Homo erectus.* (a) *H. ergaster,* ER 3883 and (b) *H. ergaster,* ER 3733. (Courtesy of Professor Milford Wolpoff, University of Michigan, Ann Arbor.)

same paleo-species.[21, 22] In other words, about 2 million years ago there was a single species rather than several different species that spread from Africa to Europe and Asia.

Body Size: The Nariokotome Boy

In 1985, the remains of a nearly complete skeleton were found in West Lake Turkana of East Africa, and were referred to as KNM-WT 15000, "Turkana Boy" or the Nariokotome fossil.[28–30] Detailed analysis indicates that the Nariokotome fossil belonged to a 14-year-old boy that was nearly 168 cm (5 ft. 6 in.) tall and, had he lived to adulthood, might have been taller than 6 ft. (**fig. 10.8**).

Craniofacial Morphology

Homo erectus,[21–30] whether in Africa (**figs. 10.7** and **10.8**) or in Asia (**figs. 10.9** and **10.10**), was characterized by heavy and thick brow ridges. The continuous bar of bone above the eyes created a conjoined brow known as the supraorbital torus, and just behind it, there was a marked depression referred to as the supraorbital sulcus. The forehead recedes instead of rising sharply as it does in modern humans. At the rear end of the skull, the occipital bone protrudes beyond the occipital, forming an extension called the nuchal torus.

In summary, the craniofacial morphology of *Homo erectus* compared to that of *Homo sapiens* (**fig. 10.11**) differs by large size of the brow ridges (or referred to as the supraorbital torus), low forehead, large occipital torus, and absence of chin.

Brain Size

The brain size of *Homo erectus* varied considerably. African and Asian *Homo erectus* had a brain size that ranged from 800 to 1,225 cc. However, the European

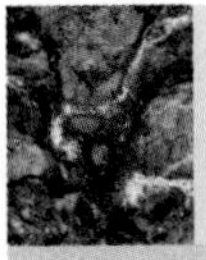

TABLE 10.2

Estimated Body Weight and Stature among the Plio-Pleistocene Hominids*

Hominid	Body Weight Male	Body Weight Female	Stature Male	Stature Female
A. afarensis	45 kg (99 lb)	29 kg (64 lb)	151 cm (59 in.)	105 cm (41 in.)
A. africanus	41 kg (90 lb)	30 kg (65 lb)	138 cm (54 in.)	115 cm (45 in.)
A. robustus	40 kg (88 lb)	32 kg (70 lb)	132 cm (52 in.)	110 cm (43 in.)
A. boisei	49 kg (108 lb)	34 kg (75 lb)	137 cm (54 in.)	124 cm (49 in.)
H. habilis	52 kg (114 lb)	32 kg (70 lb)	157 cm (62 in.)	125 cm (49 in.)

*Adapted from McHenry, H. M. 1992. How big were early hominids? *Evolutionary Anthropology* I:15–20.

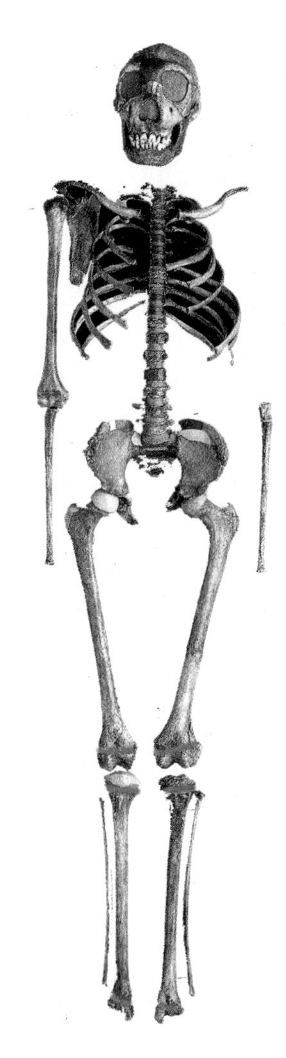

FIGURE 10.8. *Homo ergaster* from Turkana, East Africa (KNM-WT 15000, "Turkana boy" or "Nariokotome fossil"). The epiphyses (end of the bone) of the end long bones (marked in red) are not fused to the diaphyses (shaft of the long bones), which indicate that the "Turkana boy" when he died was less than 12 to 14 years old. If he had reached adulthood, he would have been over 6 ft. (Picture taken by the author using the cast collection of the Department of Anthropology of the University of Michigan.)

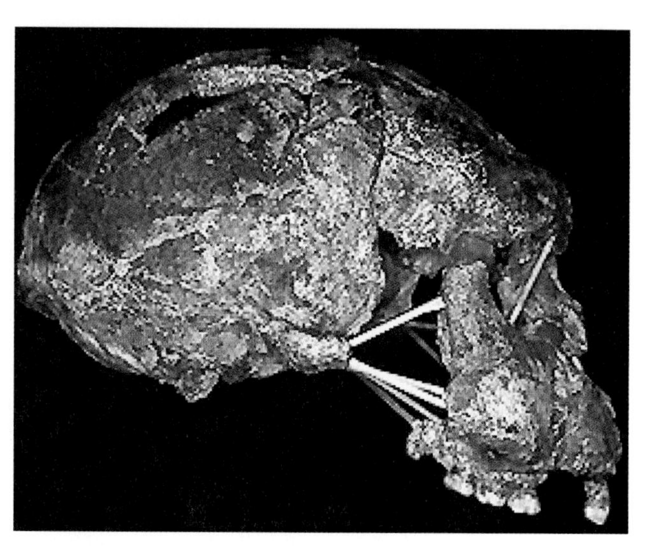

FIGURE 10.9. *Homo erectus* craniofacial features. Reconstruction of *Homo erectus* (Sangiran 17, Indonesia) shows a marked supraorbital torus, a receding inclined forehead, and a strong muscle attachment area for back muscles as indicated by the marked occipital torus. (Courtesy of Professor Milford Wolpoff, University of Michigan, Ann Arbor.)

Homo erectus found at the site of Dmanisi had a brain size that ranges from 600 to 670 cc,[26, 27] which is well below the average capacity of both *Homo erectus* and *Homo ergaster* and is much closer to the *H. habilis* mean.[27] If the Dmanisi fossils (**fig. 10.11**) were smaller than the Africans, then it is quite possible that the small brain size of Dmanisi is related to their smaller body size. In addition, the internal configuration of the skull of *Homo ergaster* showed the presence of the area of the brain (Broca's area) that controls articulate speech.[28–30] At first it was thought that *Homo erectus* could have had the ability to speak, but studies of the size of the foramen magnum indicate that it was too small for the passage of nerve fibers that control speech.

Dmanisi of Georgia

In 2000 Gabunia and Vekua[27, 28] reported the finding of *Homo erectus* remains in the site of Dmanisi in the Eurasian republic of Georgia date to have lived 2.5 million years ago (**fig. 10.12**). While the Dmanisi skulls had similar features as those of the African and Asian *Homo erectus,* they differed in several traits. First the Dmanisi had a brain size that ranges from 600 to 670 cc,[26, 27] which is well below the average capacity of both *Homo erectus* and *Homo ergaster* and is much closer to the *H. habilis* mean.[27] Second, the Dmanisi skulls are small and round instead of angled at the back similar to that of *Homo habilis.* Accordingly, it is has postulated that the Dmanisi represent an intermediate population between *Homo habilis* and *Homo erectus.* This inference suggest that Dmanisi-like hominids about 2 million years ago evolved from *Homo habilis* in Africa and then migrated to the Middle East and Georgia and eventually arrived in Asia and returned to Africa, which gave rise to *Homo ergaster.* A unique feature about Dmanisi is that some of the individuals lived to old age. One skull and jawbone was edentulous and exhibited complete resorption of the tooth sockets and extensive remodelling of the alveolar process.[28] This finding indicates that this individual's loss of teeth occurred several years before death as a result of aging and/or pathology. On these bases the authors suggest that the Dmanisi must have developed behavioral or cultural means such as help from other individuals who assisted the survival of individuals with masticatory.[28] If this is the case, it is quite possible that subsistence strategies that differentiate humans from other primates might have its roots nearly 2 million years ago.

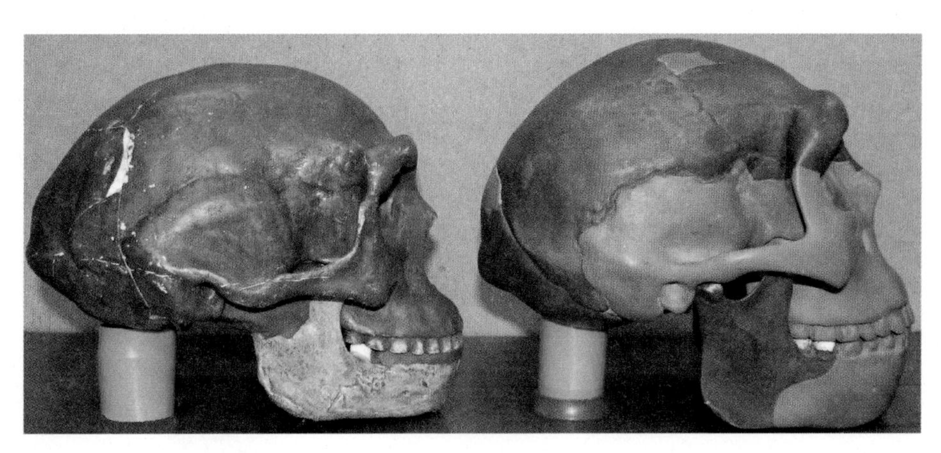

FIGURE 10.10. Asian *Homo erectus* (Zhoukadian). Reconstruction of "Peking man" from fragments of several individuals that were found in Zhoukadian (China) that probably lived about 500,000 years ago.
(a) Cast 11 reconstructed by Weidenreich and (b) cast 12 reconstruction by Weinert. (Composite illustration made by the author using the cast collection of the Department of Anthropology, University of Michigan, Ann Arbor.)

Oldowan-Acheulean Stone Tool Tradition

The stone technology of *Homo erectus* is associated with both late Oldowan tools (similar to that of *Homo habilis*) and the **Acheulean** tool tradition that appeared between 1.3 and 1.8 million years ago. This tradition became widespread throughout Africa, Europe, and across Asia (including Mongolia, South Korea, and the Middle East) and coincides with the appearance of *H. erectus.* On the other hand, the stone technology of *Homo erectus* from Eastern Europe, such as Dmanisi, was more similar to the Oldowan than the Acheulean tradition. In any event, the fact that the stone technology is more like what is found in the Oldowan technology supports the inference that the first hominid migration out of Africa might have been *H. habilis,*[27] after which *Homo erectus* followed.

The Acheulean tools consisted of large cutting tools about 10 to 17 cm long (**fig. 10.13**). The Acheulean tradition takes its name from the French village of St. Acheul, where these tools were first identified. In the Acheulean tradition, the stone tool making advanced toward a form that included stones that were bifacially worked (or flakes are taken off of both sides). The Acheulean bifacial stone tool technology represents an advance over the Oldowan tool tradition in several ways. In the Oldowan tradition, tools were

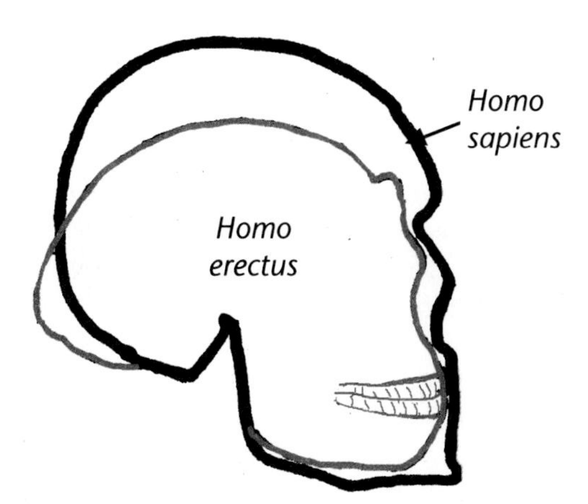

FIGURE 10.11. Comparison of the tracing of the *Homo sapiens* superimposed on the tracing of *Homo erectus.* (Composite illustration made by the author using the paleonthropology cast archive of the Department of Anthropology, University of Michigan, Ann Arbor.)

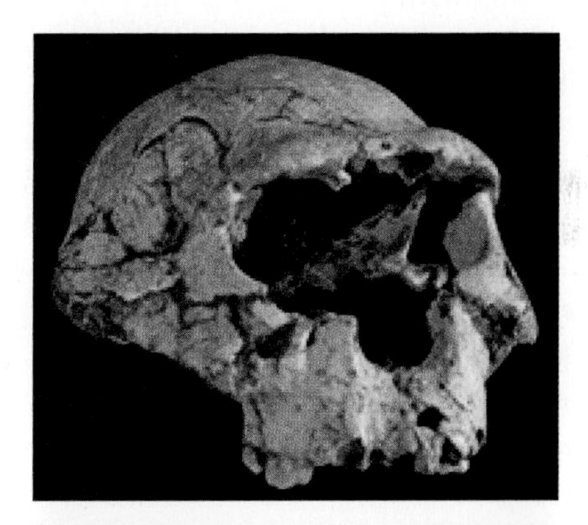

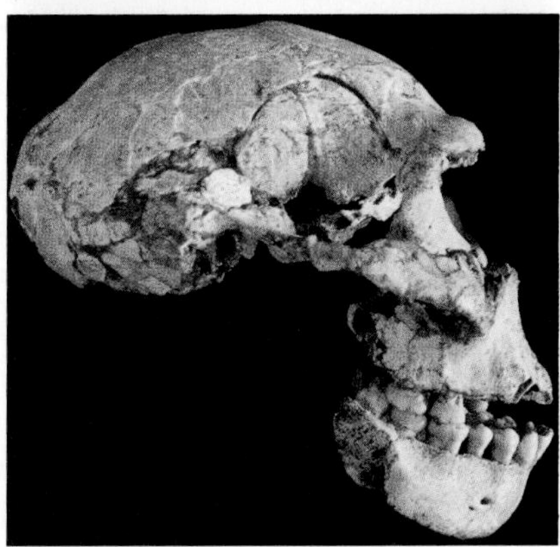

FIGURE 10.12. *Homo erectus* found in Dmanisi, Georgia of Eastern Europe. (Courtesy of Professor Milford Wolpoff of the Department of Anthropology, University of Michigan.)

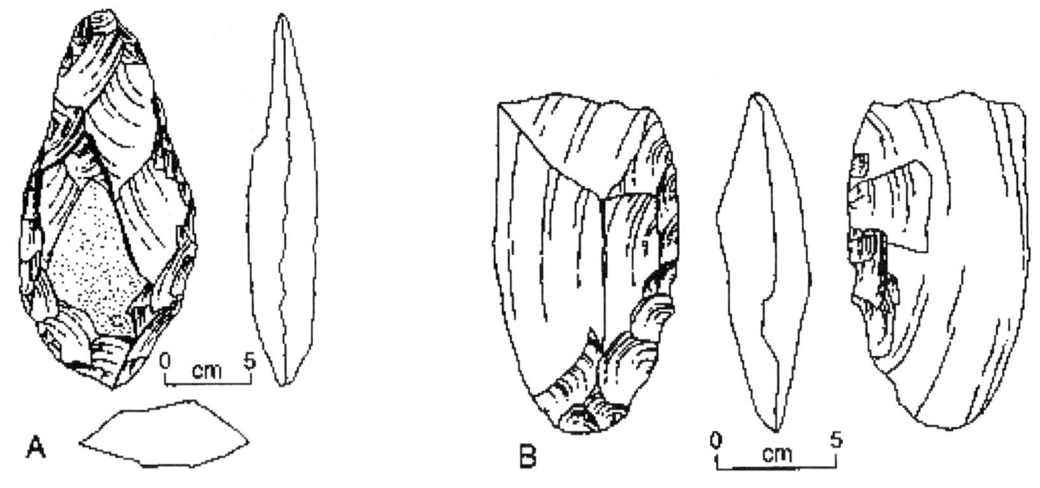

FIGURE 10.13. Acheulean stone tool tradition associated with *Homo erectus*. These techniques, associated with the emergence of *H. erectus*, date between 2.8 and 2.3 million years ago. Reprinted with permission from "Paleolithic Technology and Human Evolution" by S. H. Ambrose, *Science* 291:1748–1753. Copyright 2001 AAAS.

made by picking up pebbles the size of tennis balls, or smaller, and chipping off a few flakes from one end to form a rough and irregular edge. The Acheulean technique involved chipping the core all over rather than merely at one end. The core was converted from a round piece of rock into a flat oval hand ax about 15 cm (6 in.) long with a cutting edge far superior to that of the pebble tool. Hand axes could be used for digging edible roots and other foods from the ground. Cleavers, core tools with a straight cutting edge at one end, were probably used for heavy chopping and for hacking away at the tough sinews of larger slaughtered animals. Wooden spears and bone daggers were used to hunt animals as large as elephants. Microwear studies and experiments suggest that the Acheulean tools may have been multipurpose tools, and they may have been used to skin and cut their prey (**fig. 10.13**).

Technological and Biological Factors that Contributed to the Evolution of *Homo erectus*

Range of Migration

It is quite evident that, after its origin in East Africa, *H. erectus* quickly reached Southeast (1.8 mya) and Western Asia (1.7 mya), and Eastern Europe (1.9 mya). Expansion of *Homo erectus* was not limited to land routes. Stone tools of the Acheulean tradition associated with *Homo erectus* have been found on the Island of Flores, Indonesia, dating back to 800,000 years ago.[32] *H. erectus* probably reached the island in the hunt for stegodons a million years ago, either by building some kind of boat or by walking across a short-lived landbridge.[33]

In any event, these findings together indicate that *Homo erectus* moved throughout Africa, Eurasia, and Asia several times.[34] These migrations probably consisted of small bands of hominids that left Africa in search of new horizons. An important question is why *Homo erectus* dispersed so much throughout the world. Here, I propose that the expansion of *Homo ergaster* toward Asia and Europe is related to the ability to control fire and changes in the nutritional resources of the African environment that resulted in greater body size.

Control of Fire and Nutritional Quality of Plant and Animal Foods

The ability to control fire represents one of the most important landmarks in the saga of human evolution. The emergence of *Homo erectus* and its full commitment to control fire and to manufacture tools represents a new way of dealing with the world. According to fossil evidence, *Homo erectus* might have learned to control fire. The first signs of human control of fire come from burned bones found at the caves of Swartkrans in South Africa,[35] dated at 1 million years ago, and 1.6 million years ago at Koobi Fora, Kenya.[36, 37] For millions of years the climate of northern Europe and Asia was very cold, colder than it is at present. So cold were areas of Europe that 1 million years ago it became a tundra.[34] For *Homo erectus* to survive in such an environment would have required the use and control of fire. Thus, under these conditions, the suc-

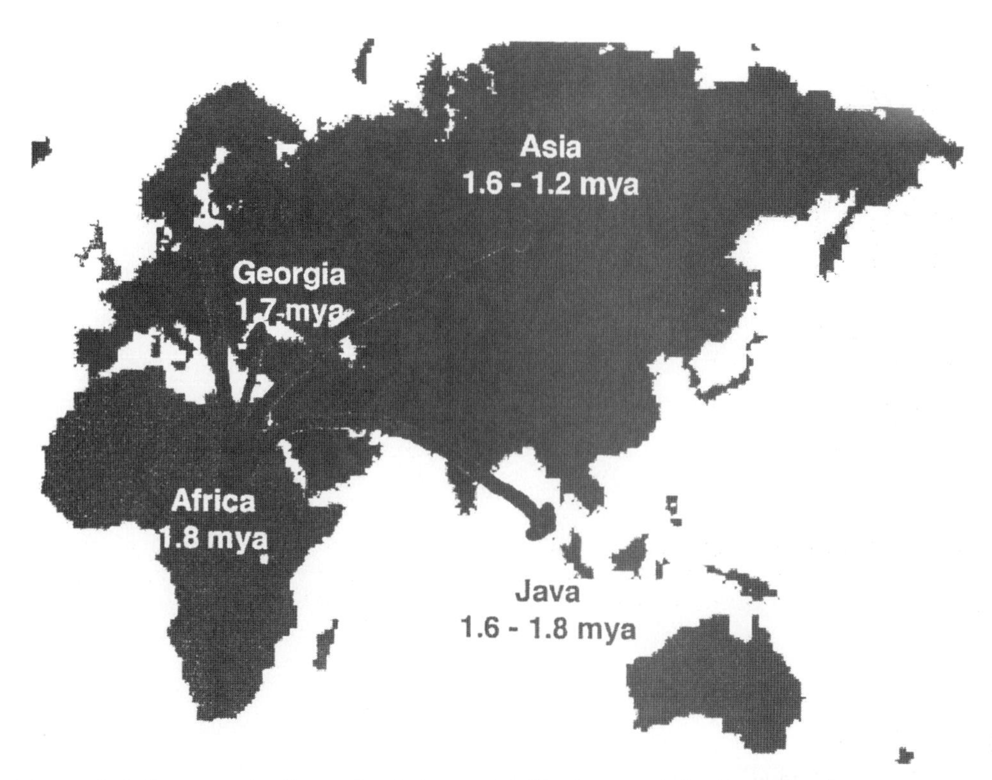

FIGURE 10.14. Dispersal of *Homo erectus* throughout the world. (Tracing made by the author.)

cess of *Homo erectus* as a species depended a great deal on their ability to master the control of fire.[38] Hence, at the beginning, fire was probably used for keeping warm. As stated by Brace,[39] "The application of heat to food, if for no other purpose than to thaw the frozen remainders of yesterday's haunch, made an important contribution to subsistence at the northern edges of human occupation." Thus, because people could not have survived winters without being able to defrost meat from kills, cooking became obligatory for hominids occupying glacial zones 250,000 years ago.[40–43]

Excavations by Goren-Inbar and associates[63] at the Acheulean open-air site of Gesher Benot Ya'aqov (GBY) in northern Israel, known for its excellent preservation of wood remains, indicate *Homo erectus* more than 790,000 years ago evolved the ability to control and use fire. These studies found that the distribution of burned wood, plant seeds, and flint artifacts were in discrete clusters within the stratigraphic layers of the site, indicating evidence of hearths in specific locations. These findings suggest that *Homo erectus* might use fire not only to keep warm and keep away predators but also for roasting plant and animal foods.

Once *Homo erectus* mastered the control of fire, they started using it to improve the nutritional quality of animal and plant foods. As summarized by Wrangham et al.,[44, 45] roasting plant foods has several positive nutritional effects. First, it softens the plant cellulose, making the cell contents more easily available for digestion and absorption. Second, it modifies the physical structure of molecules, such as proteins and starches, into forms more accessible for digestion by enzymatic degradation. Third, it reduces the chemical structure of indigestible molecules into smaller forms that can be fermented more rapidly. Fourth, it can detoxify and denature toxins,[46–48] making otherwise poisonous plant food edible.

Since some of the *Homo erectus* sites[34] show the exploitation of many individuals of a single species, such as baboons or large bovines and the remains of butchered elephants, in association with Acheulean tools and scatters of charcoal. This suggests that *Homo erectus* was able to hunt big game in an organized form, in which they may have employed fire as a tool to drive animals into a swamp or cave. In addition, *Homo erectus* may have used fire to roast animal foods. Roasting animal foods has several positive nutritional effects. First, it tenderizes and softens meat. The meat of wild animals is leaner and tougher than that of domesticated animals. Roasting softens muscle fibers and fat tissue, making animal food easier to bite and chew.

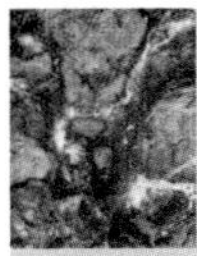

TABLE 10.4
Skeletal Remains of *Homo erectus*

Identification (Location)	mya	Skeletal Remains Features	Brain Size (cc)
"Java man," *Pithecanthropus* I (Trinil, Indonesian island of Java)	700,000	A flat, very thick skullcap and a femur	940
"Peking Man," *Sinanthropus pekinensis* (Choukoutien, China)	300,000 to 500,000	14 partial craniums, 11 lower jaws	
Skull III (Choukoutien, China)		Skull of juvenile	915
Skull II (Choukoutien, China)		Skull of adult	1,030
Skulls X, XI, and XII (Choukoutien, China)		Skull of adult man Skull of adult woman Skull of young man	1,225 1,025 1,030
Skull V (Choukoutien, China)		Cranial fragments	1,140
Sangiran 2, *Pithecanthropus* II (Indonesian island of Java)		Skull cap	815
OH 9 (Olduvai Gorge, Tanzania)	1.4	Partial braincase with massive brow ridges	1,065
OH 12 (Olduvai Gorge, Tanzania)	600,000 to 800,000	Partial braincase	750
Sangiran 17, *Pithecanthropus* VIII Sangiran (Indonesian island of Java)	1.7	Complete cranium, large face with flaring cheekbones, thick brow ridges	1,000
KNM-ER 3733 (Koobi Fora, Kenya)	1.7	Complete cranium, large face with flaring cheekbones, thick brow ridges	850
KNM-WT 15000, "Turkana boy" (Nariokotome near Lake Turkana)	1.6	Almost complete skeleton of a 14-year-old boy Robust and tall Height: 160 cm (5'3") By adulthood would have been about 185 cm (6'1")	880
D2700 (Dmanisi, Georgia)	1.7	Complete skull and lower jaw	600–670
TD6-69: *Homo* Ancestor (Atapuerca, Spain)	780,000	Partial face of a 10- to 11.5-year-old child	

Sources: Theunissen,[23] Wiedenreich,[24] Leakey and Walker,[25] Gabunia, Vekua, and Swisher et al.,[26] Vekua, Lordkipanidze, and Rightmire et al.,[27] Brown, Harris, Leakey, and Walker,[28] Walker and Leakey,[29] Walker and Shipman.[30]

Second, roasting meat kills the bacterium that causes botulism and salmonella.[48]

In summary, roasting increases the digestibility, palatability, and nutritional quality of meat. Although humans in some cases eat raw meat as a delicacy (e.g, steak tartar, caviar), they cannot subsist just eating raw meat without suffering undernutrition.[49] Hence, it is quite possible that fire might have been used by *Homo erectus* for roasting meat. By doing so, the total calorie and protein intake of *Homo erectus* would certainly have been enhanced.

Increase in Body Size and Change in Home Range

Compared to the australopithecines and early hominids, the body size of *Homo erectus* nearly doubled.[19, 50–53] The increase in body size was associated with an increase in the relative length of the lower limbs and general robusticity. Since a large body size requires a greater calorie intake than a small body size, the dietary needs of *Homo erectus* were greater than those of earlier hominids. It has been estimated that the total energy requirements of *Homo erectus* females during lactation and pregnancy were much greater than for australopithecines.[54] **Table 10.5** gives the total energy requirements for the gracile australopithecines, *Homo habilis* and *Homo erectus*. These estimates were based upon the equations used to estimate the total energy requirements for humans.[55] These data show that to support an active lifestyle, australopithecines would have required between 1,354 and 1,741 calories per day. In contrast, *Homo erectus*, to support an active lifestyle, would require between 2,133 and 2,483 calories per day. To have such a high-calorie intake, *Homo erectus* must have been eating either more food or eating food of higher quality than the australopithecines. It is very plausible then, that through the use of fire, *Homo erectus* would have improved not only the digestibility and nutritional quality of animal foods, but also would have enhanced nutritional absorption of plant and animal foods, resulting in increased calorie and protein intake. It is reasonable to assume that this increase in body size of *Homo erectus* was due to the shift in dietary quality and quantity with an emphasis on animal foods.

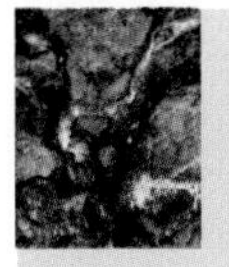

TABLE 10.5

Estimates of Daily Energy Requirements for Australopithecines and *Homo erectus* Based Upon Estimates of Body Mass

	Sex	Body mass[1] (kg)	Total Energy Expenditure[2] (calories/day)
Australopithecus afarensis	m	44.6	1,741
Australopithecus afarensis	f	29.3	1,354
Australopithecus africanus	m	40.8	1,760
Australopithecus africanus	f	30.2	1,386
Homo habilis	m	52	2,132
Homo habilis	f	32	1,452
Homo erectus	m	63.0	2,483
Homo erectus	f	52.0	2,133

[1] Source: McHenry.[51]

[2] Calculated from: Total Energy Expenditure (calories/day) = 93.3 $W^{0.792}$ given in Leonard and Robertson.[55]

Coinciding with the appearance of *Homo erectus* (during the Plio-Pleistocene), the environment in both eastern and southern Africa changed from a tropical lush forest (mesic) to a dry (xeric), sparsely forested savanna.[33, 56, 57] Under a dry or xeric savanna, the long lower limbs and more linear body form associated with increased body size of *Homo erectus* would have had a significant adaptive advantage for thermoregulation.[58, 59] On the other hand, the dry and sparse savanna implies that the nutritional resources are more dispersed and patchy than in forested environments, resulting in larger home ranges for *Homo erectus*. The game animals on which *Homo erectus* had come to depend for food also migrated in search of better pastures. There is evidence that *Homo erectus* was the first hominid group to forage daily over relatively larger ranges than seen in *Australopithecus*.[55, 60–62]

Viewed in the context of ecology and adaptation, the greater ranges of foraging done by *Homo erectus* are a response to the patchy quality and the low density of animal foods that characterize the dry and sparse savanna. Therefore, expanding the size of the home range from Africa into other parts of the Old World may be viewed as an adaptation of *Homo erectus* to overcome the dispersed food resources that occurred in Africa. Thus, the expansion of *Homo erectus* from Africa into other parts of the Old World was nutritionally motivated.

Summary

The third phase of *Homo habilis* and *Homo erectus* is characterized by the beginning of control of the environment. *Homo erectus* became the hominid species to expand its ecological niche beyond their tropical African homeland so that by about 1.2 to 1.8 million years ago, *Homo erectus* had reached the Middle East, Asia, and Europe. Throughout this process *Homo erectus* developed ways of coping with new environments.

The relatively large legs and wide pelvis of the australopithecine came to be expressed even more among *H. erectus*. The continuation of these traits suggests that the selective forces that led to the development of these traits among the australopithecines allowed them to also become part of the genetic repertoire of the genus *Homo*. Likewise, the increasing sophistication in the manufacture of stone tools of the Acheulean bifaces suggests that the selective advantage of high-quality energy extracted from foods increased even more among *H. erectus*.

In terms of food extraction during the late Pliocene and early Pleistocene, the subsistence pattern of the australopithecines was a primarily vegetarian diet combined with the scavenging of carrion. Thereafter, for *Homo habilis,* which was characterized by a relatively bigger brain than earlier hominids and the use of Oldowan stone technology (albeit rudimentary), hunting became a major component of the subsistence pattern. As inferred from skeletal remains of butchered animals found in caves, the dietary repertoire of *Homo habilis* expanded to include animal food that was derived from small game hunting and scavenging.

With the evolution of *Homo erectus,* characterized by a bigger body size and equipped with a somewhat advanced Acheulean stone technology, the quest for food switched from a passive role into an active role. During the *Homo habilis* stage, high-quality foods were probably derived mostly from scavenging carrion and less by hunting. With the appearance of *Homo erectus,* animal foods were mostly derived from hunting big game. Hunting of large game was the major cultural episode that might have catapulted *Homo erectus* into the *Homo sapiens* stage. The expansion of *Homo erectus* toward Asia and Europe is related to a dynamic interaction of increased nutritional requirements and the depletion of the nutritional resources in the African environment. Control of fire might have catapulted *Homo erectus* into the *Homo sapiens* stage. The epitome of this two-million-year saga through the world will be expressed in the success of the evolution of *Homo sapiens,* to be discussed in the next chapter.

Suggested Readings

Conroy, G. C. 1997. *Reconstructing Human Origins. A Modern Synthesis.* New York: Norton.

Delson, E., ed. 1985. *Ancestors: The Hard Evidence.* New York: Liss.

Falk, D. 1993. A good brain is hard to cool. *Natural History* 102(8):65.

Grine, F., ed. 1988. *Evolutionary History of the Robust Australopithecines.* New York: de Gruyter.

Wolpoff, M. H. 1999. *Human Evolution.* 2d ed. Boston: McGraw-Hill.

Literature Cited

1. Leakey, L. S. B., P. V. Tobias, and J. R. Napier. 1964. A new species of the genus *Homo* from Olduvai Gorge. *Nature* 202:7–10.
2. Leakey, R. E. 1973. Evidence for an advanced Plio-Pleistocene hominid from East Rudolf, Kenya. *Nature* 242: 447–450.
3. Leakey, R. E. 1974. Further evidence of Lower Pleistocene hominids from East Rudolf, North Kenya, 1973. *Nature* 248:653–656.
4. Tobias, P. V. 2003. Encore Olduvai. *Science* 299:1193–1194.
5. Stedman, H. H., B. W. Kozyak, A. Nelson, D. M. Thesier, L. T. Su, D. W. Low, C. R. Bridges, J. B. Shrager, N. Minugh-Purvis, and M. A. Mitchell. 2004. Myosin gene mutation correlates with anatomical changes in the human lineage. *Nature* 428:415–418.
6. Gabunia, L., A. Vekua, D. Lordkipanidze, C. Swisher, III, R. Ferrign, A. Justus, M. Nioradze, M. Tvalchrelidze, S. Anton, G. Bosinski, O. Jories, M. de Lumley, G. Majsuradze, and G. Mouskhelishvili. 2000. Earliest Pleistocene hominid cranial remains from Dmanisi, Republic of Georgia: Taxonomy, geological setting, and age. *Science* 288:1019–1025.
7. Ambrose, S. H. 2001. Paleolithic technology and human evolution. *Science* 291:1748–1753.
8. Toth, N., and K. D. Schick. 1993. Pan the tool-maker: Investigations into the stone tool-making and tool-using capabilities of a bonobo (*Pan paniscus*). *J. Archaeol. Sci.* 20:81–91.
9. Schick, K. D., N. Toth, G. Garufi, S. E. Savage-Rumbaugh, D. Rumbaugh, and R. Sevcik. 1999. Continuing investigations into the stone tool-making and tool-using capabilities of a bonobo (*Pan paniscus*). *J. Archaeol. Sci.* 26:821–832.
10. Susman, R. L. 1998. Hand function and tool behavior in early hominids. *J. Hum. Evol.* 35:23–46.
11. Marzke, M. W., and R. F. Marzke. 2000. Evolution of the human hand: Approaches to acquiring, analysing and interpreting the anatomical evidence. *J. Anat.* 197:121–140.
12. Richmond, B. G., and D. S. Strait. 2000. Evidence that humans evolved from a knuckle-walking ancestor. *Nature* 404:382–385.
13. Bunn, H. 1983. Evidence on the diet and subsistence patterns of Plio-Pleistocene hominids at Koobi Fora, Kenya, and at Olduvai Gorge, Tanzania. In *Animals and Archaeology*, J. Clutton-Brock, C. Grigson, ed. London: British Archaeological Reports.
14. Potts, R., and P. Shipman. 1981. Cutmarks made by stone tool on bones from Olduvai Gorge, Tanzania. *Nature* 291:577–580.
15. Blumenschine, R. 1986. What lions leave behind: Scavenging versus hunting in the human past. Paper presented to the Anthropological Society of Washington, January 18, 1986.
16. Shipman, P. L. 1986. Scavenging or hunting in early hominids: Theoretical framework and tests. *Am. Anthropol.* 88:27–43.
17. Schaller, G. B., and G. R. Lowther. 1959. The relevance of carnivore behavior to the study of early hominids. *Southwest J. Anthropol.* 25:307–341.
18. Stanford, G. 1998. The social behavior of chimpanzees and bonobos: Empirical evidence and shifting assumptions. *Current Anthropology* 39:399–420.
19. McHenry, H. M. 1992. How big were early hominids? *Evolutionary Anthropology* I:15–20.
20. Potts, R. 1993. Archeological interpretations of early hominid behavior and ecology. In *The Origin and Evolution of Humans and Humanness,* D. T. Rasmuseen, ed. Boston: Jones and Barlett Publisher.
21. Asfaw, B., W. H. Gilbert, Y. Beyene, W. K. Hart, P. R. Renne, G. WoldeGabriel, E.S. Vrba, and T. D. White. 2002. Remains of

Homo erectus from Bouri, Middle Awash, and Ethiopia. *Nature* 416:317–320.

22. Antón, S. C. 2002. Evolutionary significance of cranial variation in Asian *Homo erectus. Am. J. Phys. Anthropol.* 118:301–323.
23. Theunissen, B. 1989. *Eugene Dubois and the Ape-Man from Java.* Dordrecht, The Netherlands: Kluwer Academic Publishers.
24. Weidenreich, F. 1937. The new discovery of three skulls of *Sinanthropus pekinensis. Nature* 139:269–272.
25. Leakey, R. E., and A. C. Walker. 1976. *Australopithecus, Homo erectus* and the single species hypothesis. *Nature* 261:572–574.
26. Gabunia, L., A. Vekua, C. C. Swisher, III, R. Ferring, A. Justus, M. Nioradze, et al. 2000. Earliest Pleistocene hominid cranial remains from Dmanisi, Republic of Georgia: Taxonomy, geological setting, and age. *Science* 288:1019–1025.
27. Vekua, A., D. Lordkipanidze, G. P. Rightmire, J. Agusti, R. Ferring, G. Maisuradze, et al. 2002. A new skull of early *Homo* from Dmanisi, Georgia. *Science* 297:85–89.
28. Lordkipanidze, D., Vekua, A., Ferring, R., Rightmire, G. P., Agusti, J., Kiladze, G., Mouskhelishvili, A., Nioradze, M., Ponce de Leon, M. S., Tappen, M., Zollokofer, C. P. 2005. Anthropology: the earliest toothless hominin skull. *Nature* 434:717–718.
29. Walker, A. C., and R. E. Leakey. 1993. *The Nariokotome* Homo erectus *Skeleton.* Cambridge, MA: Harvard University Press.
30. Walker, A. C., and P. Shipman. 1996. *The Wisdom of the Bones.* New York: Alfred E. Knopf.
31. Blumenschine, R. J., C. R. Peters, F. T. Masao, R. J. Clarke, A. L. Deino, R. L. Hay, C. C. Swisher, I. G. Stanistreet, M. G. M. Gail, L. J. McHenry, N. E. Sikes, N. J. van der Merwe, J. C. Tactikos, A. E. Cushing, D. M. Deocampo, J. K. Njau, and J. I. Ebert. 2003. Late Pliocene *Homo* and hominid land use from Western Olduvai Gorge, Tanzania. *Science* 299:1217–1221.
32. Morwood, M. J., P. B. O'Sullivan, F. Aziz, and A. Raza. 1998. Fission-track ages of stone tools and fossils on the East Indonesian island of Flores. *Nature* 392:173–176.
33. Antón, S. C., W. R. Leonard, and M. L. Robertson. 2002. An ecomorphological model of the initial hominid dispersal from Africa. *J. Hum. Evol.* 43:773–785.
34. Dennell, R. 2003. Dispersal and colonisation, long and short chronologies: How continuous is the Early Pleistocene record for hominids outside East Africa? *J. Hum. Evol.* 45:421–440.
35. Brain, C. K. 1993. The occurrence of burnt bones at Swartkrans and their implications for the control of fire by early hominids. In *Swartkrans. A Cave's Chronicle of Early Man,* ed. C. K. Brain, 229–242. Transvaal Museum Monograph No. 8, Transvaal.
36. Straus, L. G. 1989. On early hominid use of fire. *Curr. Anthrop.* 30:488–491.
37. Rowlett, R. M. 2000. Fire control by *Homo erectus* in East Africa and Asia. *Acta Anthropol. Sin.* 19:198–208.
38. James, S. R. 1989. Hominid use of fire in the Lower and Middle Pleistocene: A review of the evidence. *Curr. Anthrop.* 30:1–26.
39. Brace, C. L. 1999. An anthropological perspective on "race" and intelligence: The non-clinal nature of human cognitive capabilities. *J. Anthrop. Res.* 55:245–264.
40. Brace, C. L. 1995. *The Stages of Human Evolution.* Englewood Cliffs, NJ: Prentice-Hall.
41. Brace, C. L. 2002. The raw and the cooked: A Plio/Pleistocene Just So Story, or sex, food, and the origin of the pair bond. *Social Sci. Inf.* 39:17–27.
42. Brace, C. L., K. R. Rosenberg, and K. D. Hunt. 1987. Gradual change in human tooth size in the late Pleistocene and post-Pleistocene. *Evolution* 41:705–720.
43. Ragir, S. 2000. Diet and food preparation: Rethinking early hominid behavior. *Evol. Anthropol.* 9:153–155.
44. Wrangham, R. W., J. H. Jones, G. Laden, D. Pilbeam, and N. L. Conklin-Britain. 1999. The raw and the stolen: Cooking and the ecology of human origins. *Curr. Anthrop.* 40:567–594.
45. Wrangham, R. W. and N. L. Conklin-Brittain. 2003. Cooking as a biological trait. *Comp. Biochem. Physiol. A. Mol. Integr. Physiol.* 136:35–46.
46. Bravo, L. 1999. Effect of processing on the non-starch polysaccharides and in vitro starch digestibility of legumes. *Food Sci. Technol. Int.* 5:415–423.
47. Periago, M. J., G. Ros, and J. L. Casas. 1997. Non-starch polysaccharides and in vitro starch digestibility of raw and cooked chick peas. *J. Food Sci.* 62:93–96.
48. Orta-Ramirez, A., and D. M. Smith. 2002. Thermal inactivation of pathogens and verification of adequate cooking in meat and poultry products. *Adv. Food Nutr. Res.* 44:147–194.
49. Koebnick, C., C. Strassner, I. Hoffmann, and C. Leitzmann. 1999. Consequences of a longterm raw food diet on body weight and menstruation: Results of a questionnaire survey. *Ann. Nutr. Metab.* 43:69–79.
50. McHenry, H. M., and L. R. Berger. 1998. Body proportions in *Australopithecus afarensis* and *A. africanus* and the origin of the genus *Homo. J. Hum. Evol.* 35:1–22.
51. McHenry, H. M. 1994. Tempo and mode in human. PNAS. 91:6780–6786.
52. Ruff, C., and A. Walker. 1993. Body size and body shape. In *The Nariokotome* Homo erectus *Skeleton,* ed. A. Walker, and R. Leakey. 234–265 Cambridge, MA: Harvard University Press.
53. Tardieu, C. 1998. Short adolescence in early hominids: Infantile and adolescent growth of the human femur. *Am. J. Phys. Anthrop.* 107:163–178.
54. Aiello, L. C., and C. Key. 2002. Energetic consequences of being a *Homo erectus* female. *Am. J. Human Biol.* 14:551–565.
55. Leonard, W. R., and M. L. Robertson. 1997. Comparative primate energetics and hominid evolution. *Am. J. Phys. Anthropol.* 102:265–281.
56. Vrba, E. S. 1995. The fossil record of African antelopes relative to human evolution. In *Paleoclimate and Evolution, with Emphasis on Human Origins,* ed. E. S. Vrba, G. H. Denton, T. C. Partridge, and L. H. Burkle, 385–424. New Haven: Yale University Press.
57. Owen-Smith, N. 1999. Ecological links between African savanna environments, climate change and early hominid evolution. In *African Biogeography, Climate Change, and Human Evolution,* T. G. Bromage, and F. Schrenk, ed., 138–149. New York: Oxford University Press.

58. Frisancho, A. R. 1993. *Human Adaptation and Accommodation.* Ann Arbor, MI: University of Michigan Press.

59. Wheeler, P. E. 1991. Thermoregulatory advantages of hominid bipedalism in open equatorial environments. The contribution of increased convective heat loss and cutaneous evaporative cooling. *J. Hum. Evol.* 21:107–115.

60. Potts, R. 1988. *Early Hominid Activities at Olduvai.* New York: Aldine de Gruyter.

61. Isbell, L. A., J. D. Pruetz, M. Lewis, and T. P. Young. 1998. Locomotor activity differences between sympatric patas monkeys (*Erythrocebus patas*) and vervet monkeys (*Cercopithecus aethiops*): Implications for the evolution of long hindlimb length in *Homo. Am. J. Phys. Anthropol.* 105:199–207.

62. Kramer, P. A., and G. G. Eck. 2000. Locomotor energetics and leg length in hominid bipedality. *J. Hum. Evol.* 38:651–666.

63. Goren-Inbar, N., N. Alperson, M. E. Kislev, O. Simchoni, Y. Melamed, A. Ben-Nun, and E. Werker. 2004. Evidence of Hominid control of fire at Gesher Benot Ya'aqov, Israel. *Science* 30:725–727.

CHAPTER 11

Evolution of *Homo sapiens*

Between 50,000 and 700,000 years ago, the population of *Homo erectus* in Africa, Asia, and Europe either evolved into, or were replaced by *Homo sapiens.* The *Homo sapiens* stage includes hominids with a mosaic of biological and cultural traits whose origins and adaptations go back to the earlier stages of evolution. Based on morphological evidence, *Homo sapiens* are classified into two populations: archaic *Homo sapiens* and modern *Homo sapiens.* This chapter focuses on the cultural and biological processes that define these populations.

Archaic *Homo sapiens*

Archaic *Homo sapiens* were a diverse group whose craniofacial features display a mosaic of transitional traits between *Homo erectus* and *H. sapiens.* Age estimates indicate that populations of archaic *H. sapiens* may have lived between 50,000 and 600,000 years ago.[1] Some even estimate as late as 780,000 years ago. There were several populations of archaic *H. sapiens* found in Africa, Asia (**table 11.1**), and Europe (**table 11.2**). In Europe, two of the most well-known populations of archaic *H. sapiens* were the *Neanderthals* and the *Heidelbergensis.*

Neanderthals

The name Neanderthal comes from the Neander Valley, the site where the first skeletal remains were discovered. The Neanderthal populations lived in Western, Central, and Eastern Europe, and the Middle East. Neanderthal remains have been found dating between roughly 125,000 and 35,000 years ago. The Neanderthals had a large brain volume, which averaged about 1,400 grams. Although at first it appears that Neanderthals had larger brains than modern *H. sapiens,* when adjusted for their large frame size,[2] their brain size is similar to modern *H. sapiens.*

As illustrated in **figures 11.1, 11.3, 11.4,** and **11.5** the Neanderthals differed qualitatively from modern *H. sapiens* in some aspects of craniofacial and body morphology. For example, Neanderthals had massive brow ridges that somewhat resemble those of *H. erectus,* but the brow ridge of archaic *H. sapiens* is not continuous. The forehead was low and sloping, like *Homo erectus,* rather than vertical, like *Homo sapiens.*

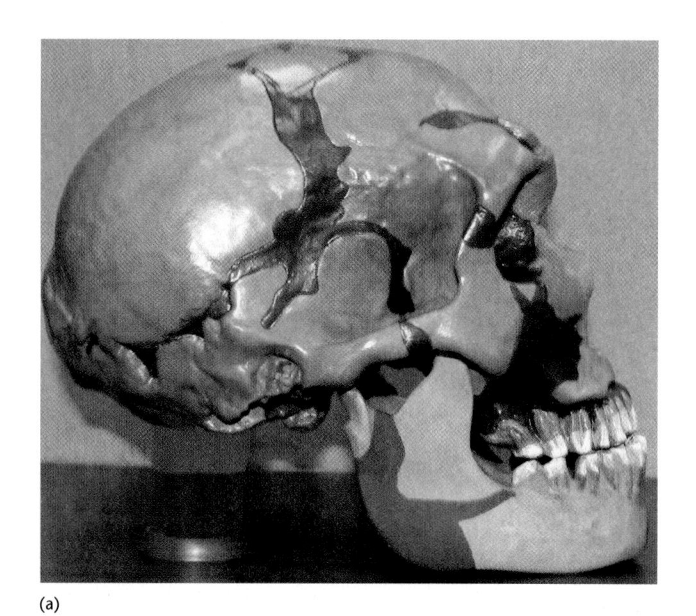

(a)

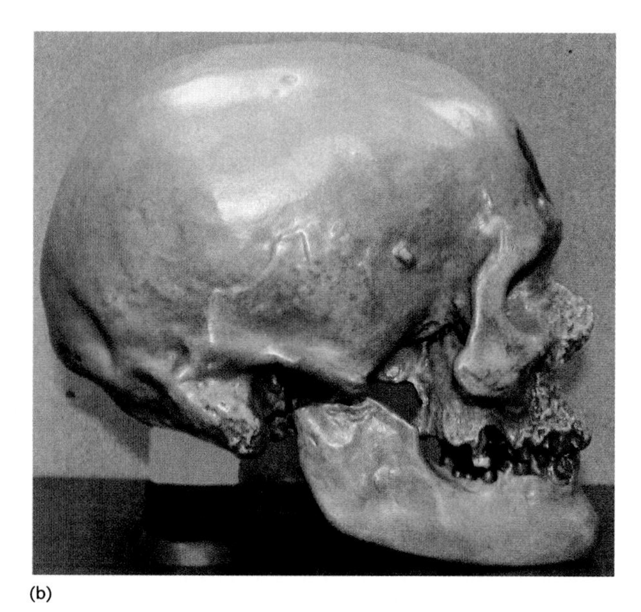

(b)

FIGURE 11.1. The craniofacial characteristics of (a) Neanderthal ("La Ferrassie") and (b) modern *Homo sapiens* (Cro-Magnon). (Picture taken by the author using the Paleoanthropology Cast Collection of the Department of Anthropology, University of Michigan, Ann Arbor.)

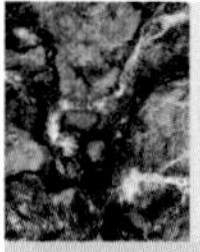

TABLE 11.1

Archaic *Homo sapiens* Found in Africa and Asia

Name	Site	Date (1,000 years ago)	Human Remains	Brain size (grams)
AFRICA				
Bodo	Awash River Valley, Ethiopia	Middle Pleistocene 600	Incomplete skull, part of braincase	1,300
Broken Hill (Kabwe)	Cave deposits near Kabwe, Zambia	Late Middle Pleistocene 130 or older	Nearly complete cranium, cranial fragments of a second individual, miscellaneous postcranial bones	1,280
CHINA				
Dali	Shaanxi Province, North China	Late Middle Pleistocene 230 to 180	Nearly complete skull	1,120–1,200
Jinniushan	Liaoning Province, Northeast China	Late Middle Pleistocene 200	Partial skeleton, including a cranium	1,260
Maba	Cave near Maba village, Guangdong Province	Early Upper Pleistocene 140 to 120	Incomplete skull of middle-aged male	Insufficient remains for measurement

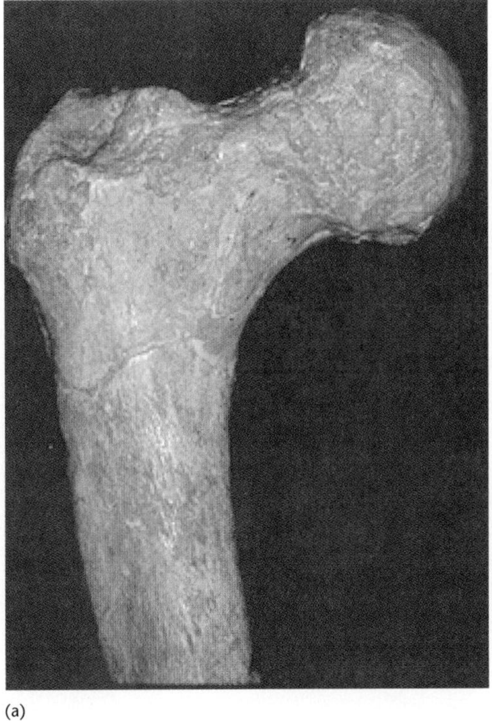

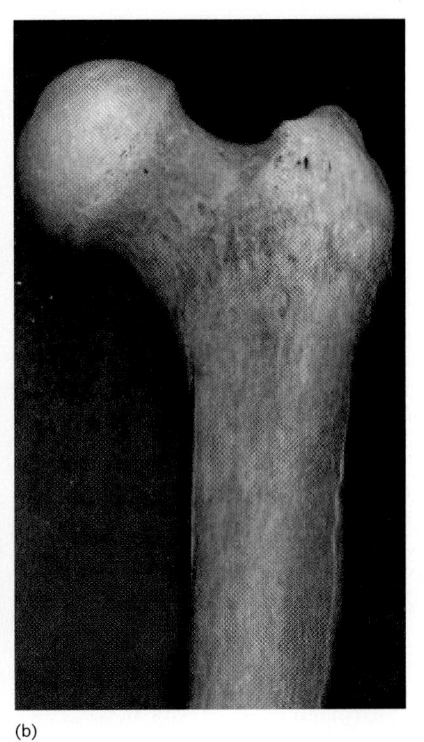

(a) (b)

FIGURE 11.2. Femur of Neanderthal. The femur of Neanderthals (a) was more robust than that of modern *Homo sapiens* (b). (Photos (a) and (b) Courtesy of Professor Milford Wolpoff, University of Michigan, Ann Arbor.)

The back of the skull had a marked bulge, or occipital bun, and the large face had a massive jaw without a chin, like *H. erectus*, and unlike modern *H. sapiens*. The face had a large nose with well-developed nasal bones and high nasal bridges. The front teeth (incisors) were large.

FIGURE 11.3. Archaic *Homo sapiens* from Kabwe, Zambia. (Courtesy of Professor Milford Wolpoff, University of Michigan, Ann Arbor.)

Overall, Neanderthals were robust and heavily muscled but not tall (**table 11.4**). The limb bones were large and were characterized by a thick and dense cortical layer around the bone marrow and large muscle attachments that were thicker and greater than in modern *Homo sapiens* (**fig. 11.3**). These morphological traits, as shown in Chapter 18, reflect adaptations to both the frigidly cold climates and variability in nutritional resources of the environment in which Neanderthals lived.[1–3, 9–11]

Heidelbergensis and Other Archaic *Homo sapien* Populations

Heidelbergensis is another representative of the population of "archaic *Homo sapiens*." In 1907, gravel pit workers near Heidelberg, in Germany, discovered the "mauer jaw." Hence, the fossil is referred to as Heidelbergensis. This find consisted of an extremely large and robust lower jaw with a receding chin like the Neanderthals. The estimated age is dated between 400,000 and 700,000 years. Recently, a large collection of fossil hominids belonging to the same species of Heidelbergensis has been found at the Sima de los Huesos site, in Atapuerca, in northeast Spain.[4–8] These specimens also display a mixture of modern and ancient features. The main noticeable features of the

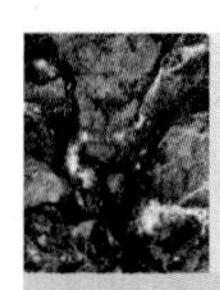

TABLE 11.2

Archaic *Homo sapiens* Found in Spain, Greece, Germany, Hungary, and England

Name	Site	Date (1,000 years ago)	Human Remains	Brain size (grams)
Arago (Tautavel)	Cave site near Tautavel, Vercouble Valley, Pyrenees, Southeastern France	400–300 date uncertain	Face; parietal perhaps from same person, many crania fragments; up to 23 individuals represented	1,150
Atapuerca	Sima cle los Huesos, Northern Spain	320–190 300	32 individuals; including some nearly complete crania	1,125–1,390
Petralona	Cave near Petralona, Khalkichiki, Northeastern Greece	300–200 date uncertain	Nearly complete skull	1,190–1,220
Bilzingsleben	Quarry at Bilzingsleben, near Erfurt, Germany	425–200 probably 280	Skull fragments and teeth	Insufficient material
Steinheim	Gravel pit at Steinheim, Germany	Mindel-Riss Interglacial 300–250	Nearly complete skull, lacking mandible	1,100
Ehringsdorf	Fossil quarry, travertine deposits, Eastern Germany	245–190	Minimum of 9 individuals, including partial cranium adult occipital bone, fragments of infant teeth	1,450
Vertesszolloss	Near village of Vertesszolloss, 30 miles west of Budapest, Hungary	210–160 date uncertain		1,115–1,434
Swanscombe	Swanscombe, Kent, England	Mindel-Riss Interglacial 300–250	Occipital and parietals	1,325

fossils are the more prominent face and nose. The mandibles display a set of Neanderthal traits, such as the position of the mental foramen (aperture located on the mandible and below the first molar) and the retromolar space (a gap between the anterior border of the ramus and the distal end of the third molar) (**fig. 11.4**). According to recent analysis, the remains from Atapuerca represent an evolutionary transition between European Middle Pleistocene hominids (*Homo erectus*) and the Neanderthals.[7,8] In other words, the Heidelbergensis may have been ancestral to Neanderthals in Europe and to *Homo sapiens* in Africa. Other investigators considered Heidelbergensis as evidence of a transition toward modern *Homo sapiens.*

In addition to the Neanderthals and Heidelbergensis there were other populations of archaic *H. sapiens* who lived in Africa, Asia, and Europe. The populations of archaic *H. sapiens* do not resemble the classic Neanderthals. For example, these populations were taller and less robust than the European Neanderthals (**table 11.2**). Furthermore, the faces of Middle Eastern and European archaic *H. sapiens* are generally shorter, protrude less, and are a bit more well-rounded than most western European Neanderthal skulls. The nasal region is small. The skull does not have the large crest of bone running from behind the ear toward the back of the skull that is present in Neanderthals. Similarly, the occipital bun is less common than that seen in Neanderthals.

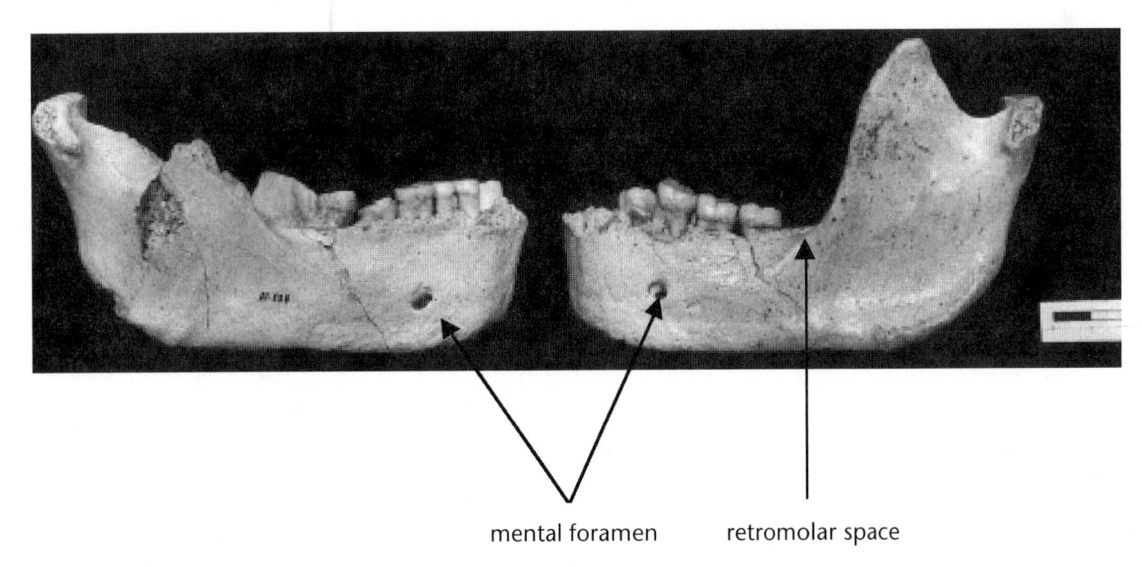

FIGURE 11.4. The jaw of Heidelbergensis (from Atapuerca, Spain). The mandible shows the Neanderthal features such as the retromolar space (a space behind the third molar) and the mental foramen located below the first molar. From *American Journal of Physical Anthropology,* 114(1) by A. Rosas. Copyright © 2001 by Wiley-Liss, Inc. Reprinted by permission of Wiley-Liss, Inc., a subsidiary of John Wiley & Sons, Inc.

Mousterian Tool Technology

Between 130,000 and 500,000 years ago, new strategies of tool manufacture and regionally distinct industries called Mousterian technology appeared.[12] The Mousterian tradition (named after the rock shelter at Le Moustier in the Dordogne region of southwestern France) became established in western Europe, North Africa, and the Middle East. The appearance often coincides with the presence of Neanderthal and other populations of archaic *Homo sapiens.* In Africa and much of Asia, industries equivalent to the Mousterian tool tradition were practiced. The Mousterian tradition differed from the preceding Acheulean primarily in the absence of hand axes and other large, bifacially worked tools (**fig. 11.6**). The Mousterian period included a tool-making technique called **Levallois.** The Levallois technique was based on Acheulean invention for predetermining flake size and shape. In the Levalloisian technique, the toolmaker first shaped the core and prepared a "striking platform" at one end. Subsequent retouching of these flakes produced special-purpose tools. The Mousterian stone tool tradition included at least fourteen categories of special-purpose tools designed for different jobs.

Choice of Large Prey

By the Middle Pleistocene, archaic *Homo sapiens,* using the Mousterian tool tradition, had expanded their nutritional resources by hunting large and dangerous prey. In East and South African caves,[13,14] a high frequency of remains of young giant and Cape buffalo, blue antelope, and extinct species of the giant gelada baboon were found. In the sites of Torralba and Ambrona, in Spain and in the Ukraine, remains of numerous elephants and the extinct cave bear were found[14–18] in a situation that suggests that the animals

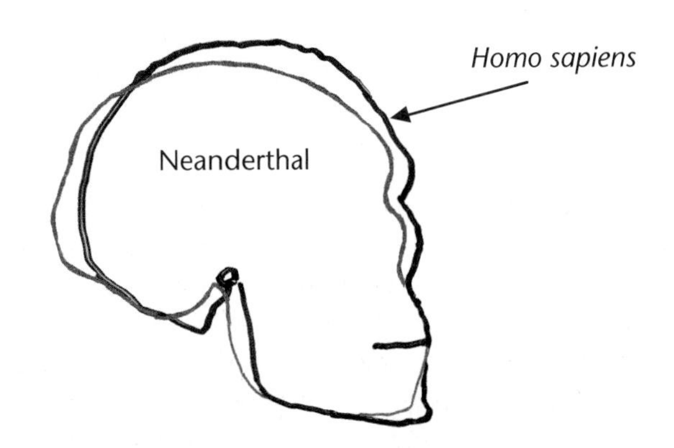

FIGURE 11.5. Comparison of the tracing of modern *Homo sapiens* superimposed on the tracing of Neanderthal. The Neanderthals retained the heavy brow ridges that characterize *H. erectus* but the brow ridges are not continuous. Compared to the anatomically modern *Homo sapiens,* the Neanderthals had a low sloping forehead, and a massive jaw without a chin, whereas anatomically modern *H. sapiens* has a high and vertical forehead and small jaw with a prominent chin.

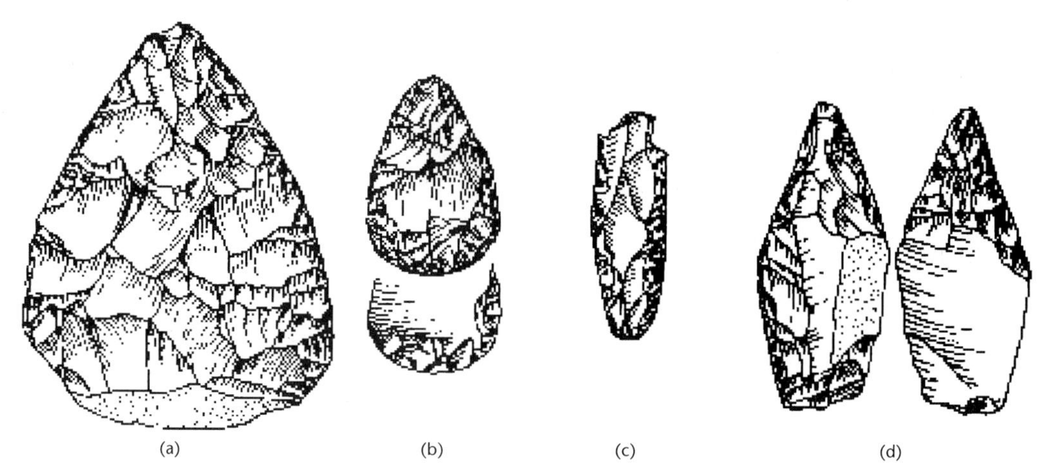

FIGURE 11.6. Mousterian tradition associated with the emergence of archaic *Homo sapiens.* The tools include (a) heart-shaped hand ax, (b) point with a thinned butt, (c) a double-sided scraper on a blade, and (d) bifacial points. The Levalloisian method required that the toolmaker first shaped the core and prepared a "striking platform" (a) from which large, sharp, and well-formed flakes were struck (b–d). Reprinted with permission from "Paleolithic Technology and Human Evolution" by S. H. Ambrose, *Science* 291:1748–1753. Copyright 2001 AAAS.

were driven and trapped in marshy valley bottoms where they were killed and butchered. These and other sites have convinced many investigators that Middle Pleistocene hominids, presumably archaic *Homo sapiens,* were able to use both fire (food technology, see Extinction of Large Herbivores) and had a complex understanding of geography and animal behavior to kill animals much larger than themselves. Eventually they were also able to kill animals in greater quantities than other predators.[19] The fact that some species were killed more frequently than others implies the development of hunting strategies tailored to individual species.

Symbolic Behavior and Language

At the Shanidar Cave site in Iraq, burials contained skeletal remains that were placed in a fetal position (**fig. 11.7a**). The position and placement of the dead has been interpreted as indicative that Neanderthals had a concept of life after death. Many of these burial sites show evidence of intentional burials, some of which had grave goods, like stone tools, animal bones, and perhaps flowers. However, it is not clear if these remains represent offerings to the dead. For example, flower pollen found in the graves might also have been brought by rodents or deposited by water or wind at the time of the burial. Nevertheless, the fact that some burials also contain the remains of elderly individuals with healed fractures, arthritis, and without teeth suggests that Neanderthals had a social system that cared for others and allowed for the sharing of food and other resources.[1] Excavations at the sites in France yielded a large assortment of personal ornaments (**fig. 11.7b**), which included different kinds of animal teeth, ivory beads, and bones decorated with sequences of regularly spaced notches, perforations, or grooves.[20, 21] It has been proposed that Neanderthals used the personal ornaments to transmit autonomous codes and express social role, thus making them cultured human beings capable of symbolic, artistic, and modern behavior.[21]

Whether Neanderthals had the capacity for language as modern humans do is a matter of some debate. According to some investigators, the anatomical, skeletal vocal structures of Neanderthals were too small with too little control possible, therefore they did not have the capacity to produce all the vowel sounds for language.[22] Other investigators indicate that the portions of the Neanderthals' brains responsible for speech control fall within the range of modern humans, thus they could make all the sounds of which we are capable.[23] On the other hand, recent studies indicate that the size of the hypoglossal canal that controls speech overlaps in living hominoids, fossil hominids, and modern humans.[24] It appears that there is no quantifiable, morphological diagnostic criteria for determining whether or not Neanderthals could speak like humans and when this capability finally emerged.

In summary, the transition from *Homo erectus* to archaic *Homo sapiens* is a mosaic of both biological and cultural traits. Some of the morphological features associated with *Homo erectus* continue to be expressed (albeit in a less prominent fashion) among archaic *H. sapiens* while others became less frequent. As shown in

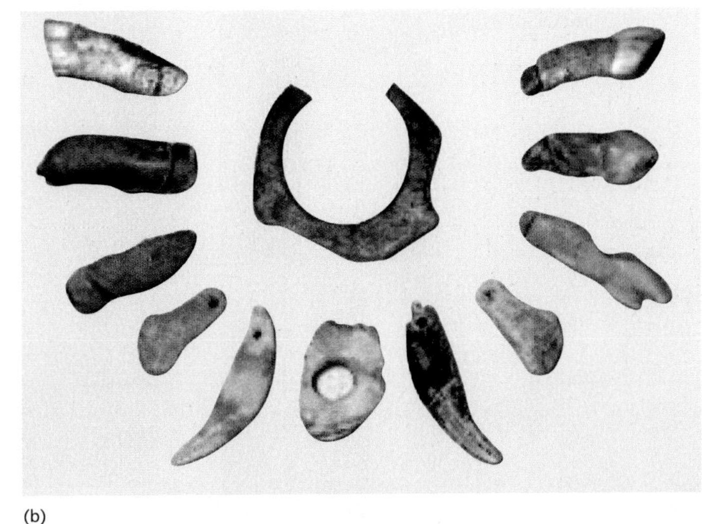

(a) (b)

FIGURE 11.7. Neanderthal burial and art. (a) Neanderthals buried their dead in a fetal position, which suggests that they believed in life after death. (b) Ornaments made of pierced teeth and ivory rings indicate that Neanderthals had the intellectual ability to develop such modern features as personal ornaments. Source: Composite derived from figures in Trinkaus and Shipman.[1] Bahn, fig. 1.[25] From *The Neanderthals* by Erik Trinkaus and Pat Shipman, copyright © 1994 by Erik Trinkaus and Pat Shipman. Used by permission of Alfred A. Knopf, a division of Random House, Inc.

Chapter 18, the body morphology of archaic *H. sapiens* such as Neanderthals reflects adaptations to both the frigidly cold climates and variability in nutritional resources on which they survived. It is evidence that Neanderthals survived and thrived in some of the harshest and least hospitable habitats ever occupied by hominids. It is this adaptability that enabled archaic *H. sapiens* to persist and evolve into modern *Homo sapiens.*

Anatomically Modern *Homo sapiens*

At the same time as the Neanderthal phase, the world was populated by a hominid that was morphologically indistinguishable from modern humans. This species, referred to as anatomically modern *H. sapiens,* has been found in Europe, Asia, Africa, the Middle East, and Australia. The date that anatomically modern *H. sapiens* lived ranges from 35,000 to 200,000 years ago (**table 11.3**). The appearance of modern *H. sapiens* coincided with drastic changes in climatic conditions, some of which were quite rapid. For example, about 20,000 years ago a climatic "pulse" caused the weather to become noticeably colder in Europe and Asia, causing the continental glaciations (called the "Wurm" in Eurasia). During times of glaciation, sea levels dropped, which extended the landmass of the continents and connected Asia and Europe to North America by land bridges. Populations of anatomically modern *H. sapiens* crossed these landmasses following game herds; first reaching southeast Asia and eventually, by at least 20,000 years ago, migrating into the New World. As they entered new areas, populations were faced with novel and diverse environments. Hence, much diversity can be expected among populations of modern *Homo sapiens.*

On the other hand, the anatomically modern *H. sapiens* from Africa, such as those from Klasies River of South Africa[35, 36] and those in Ethiopia,[37] show the classical features associated with modern *Homo sapiens* rather than Neanderthals.[47]

Fossils with Classical Morphological Features

Researchers consider the skeletal features of Cro-Magnon as representative of anatomically modern *Homo sapiens.*[26, 27] Cro-Magnon was discovered in 1868 in a rock shelter in the village of Les Eyzies, in the Dordogne region of southern France.[26, 27] It has been dated to have existed around 30,000 years ago. As illustrated in **figures 11.8 and 11.9**), the anatomically modern *Homo sapiens* is characterized by the presence of a chin, reduction in the size of the jaw, and a rounded skull with a forehead that rises vertically above the eye orbits and does not slope like archaic *H. sapiens.* Furthermore, the brow ridges are small, the face does not protrude much, and there is no occipital

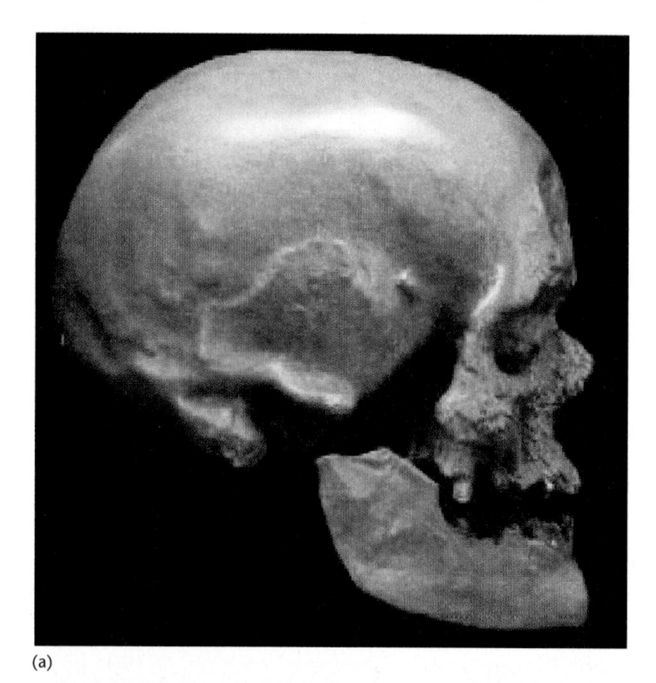

(a)

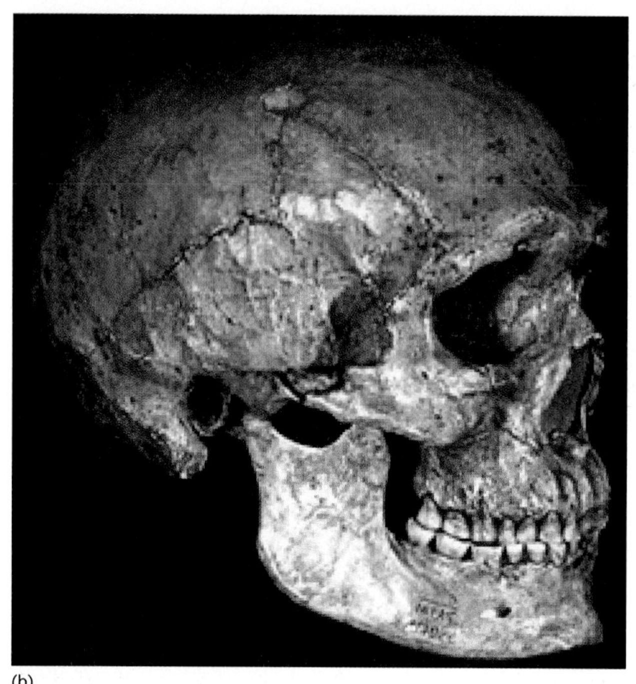

(b)

FIGURE 11.8. Anatomically modern *Homo sapiens*. (a) Cro-Magnon (France), (b) Zhoukoudian (China). (Courtesy of Professor Milford Wolpoff, University of Michigan, Ann Arbor.)

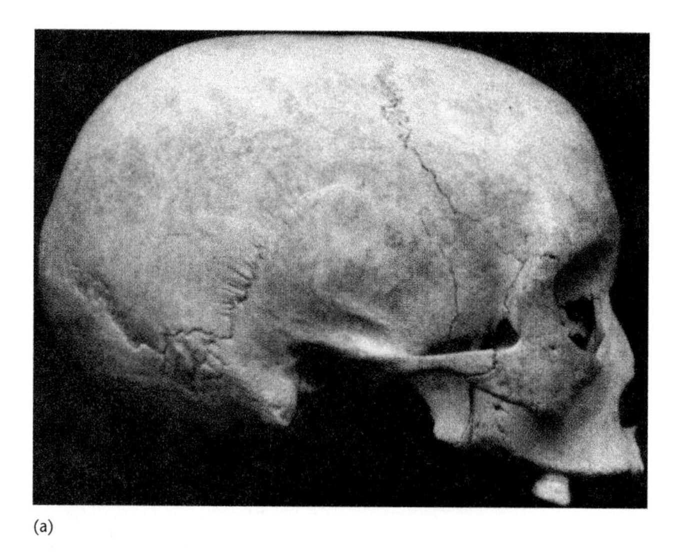

(a)

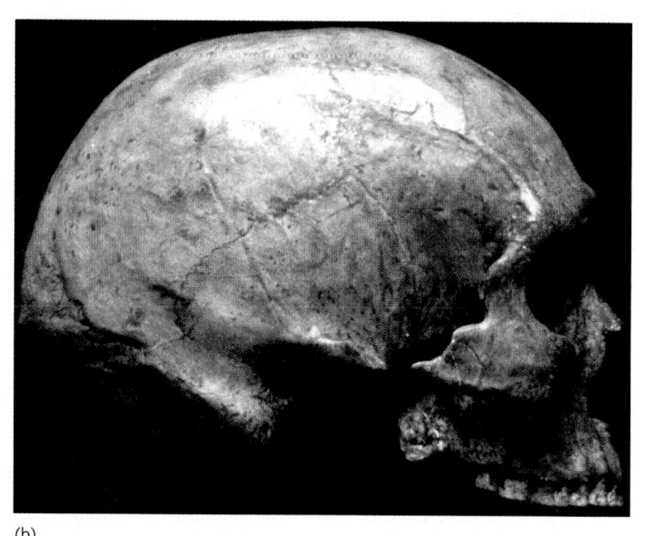

(b)

FIGURE 11.9. Anatomically modern *Homo sapiens*. (a) Bokbaai (South Africa, Cape) and (b) Upper C 101. (Courtesy of Professor Milford Wolpoff, University of Michigan, Ann Arbor.)

bun on the back of the skull like archaic *H. sapiens*. The anatomically modern *Homo sapiens*, with an average brain volume of 1,300 to 1,400 cubic centimeters (cc), do have somewhat larger brains than those of earlier *Homo* species. But the Neanderthals had a larger brain size (averaging 1,400 cc) than modern humans. However, adjusted for the Neanderthals' more robust bodies, their brains were comparable to *Homo sapiens*.

Fossils with Morphological Features Resembling Archaic *Homo sapiens*

Many fossils found in Asia[28] and Europe[26, 27] display the typical anatomical features of the Cro-Magnon. However, some fossils found in Czechoslovakia[29] (**fig. 11.11a**) and Israel (**fig. 11.11b**) had large brow ridges, and a massive face that was similar to Neanderthal morphology. Fossils found in Australia[33, 34] also exhibit features that are similar to Neanderthals.

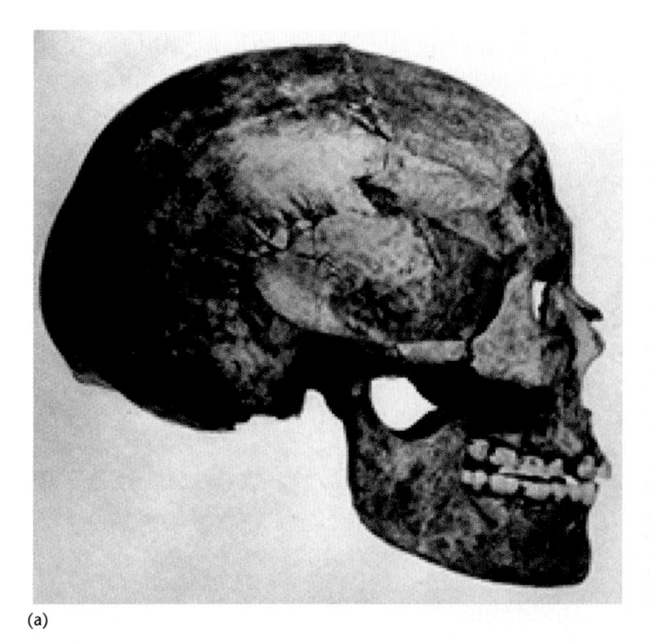
(a)

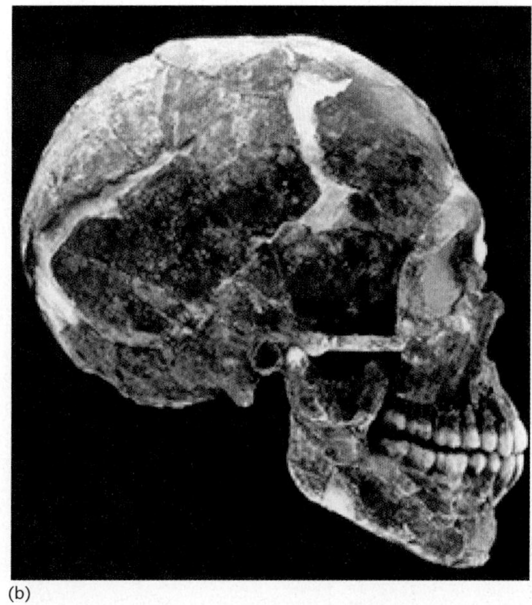
(b)

FIGURE 11.10. Anatomically modern *Homo sapiens*. (a) Predmotsi (Czech Republic) and (b) Qafzeh (Israel). (Courtesy of Professor Milford Wolpoff, University of Michigan, Ann Arbor.)

Recently, White and associates[48] reported the finding of one immature and two nearly complete adult skulls (**fig. 11.12**) recovered from the Herto Bouri Formation in the Middle Awash of Ethiopia. These new remains have been dated to about 154,000 to 160,000 years old, making them the oldest remains that can be firmly assigned to modern *Homo sapiens*. These fossils show a combination of features from archaic, early modern, and recent humans. The cranium shares with ancient African crania a wide interorbital breadth, anteriorly placed teeth, and a short occipital area. It also has a wide upper face and moderately domed forehead, as do the archaic *H. sapiens* from the Middle East. On the other hand, the low nose and globular braincase are typically modern. Overall, the Herto specimen shows a great deal of resemblance to the anatomically modern *H. sapiens* from Broken Hill (Zambia). Based on this evidence and despite the presence of some primitive features, the Herto specimen has been considered as the form from which all morphologically "modern" humans are derived.[48]

Homo sapiens in Indonesia

Recently, Brown and colleagues[52] reported the finding at the cave in Liang Bua, Island of Flores of a skeleton belonging to an adult woman who lived 18,000 years ago (**fig. 11.13**). This woman had a stature of approximately 1 meter (3.3 ft.) and a brain volume of 380 cc, which is equal to that of the australopithecines. Yet craniofacial morphology corresponds to that of modern *Homo sapiens*. In addition, as inferred from the hip bone and the legs, this individual had a bipedal pattern of locomotion. The presence of *Homo sapiens* about 18,000 years ago on this island raises several questions. First, how did this individual arrive on the island? It is quite possible that like *H. erectus, Homo floresiensis* reached the island in a hunt for stegodons a million years ago, either by building some kind of boat or by walking across a short-lived landbridge.[33] Second, this individual is the smallest of any known population of *Homo sapiens*. Although some authors have assigned this individual to the species *Homo floresiensis*,[52] others assume that its small size is related to a pathological condition such as microcephaly. Given the fact that so far nine individuals have been identified[53] and the small brain size is proportional to the small body size[54] supports the hypothesis that the overall reduction in size of *Homo floresiensis* is related to poor resource availability and long-term isolation on the comparatively small island.[52] In any event, this finding suggests that the wide diversity in morphological features of *Homo sapiens* goes back a long time ago.

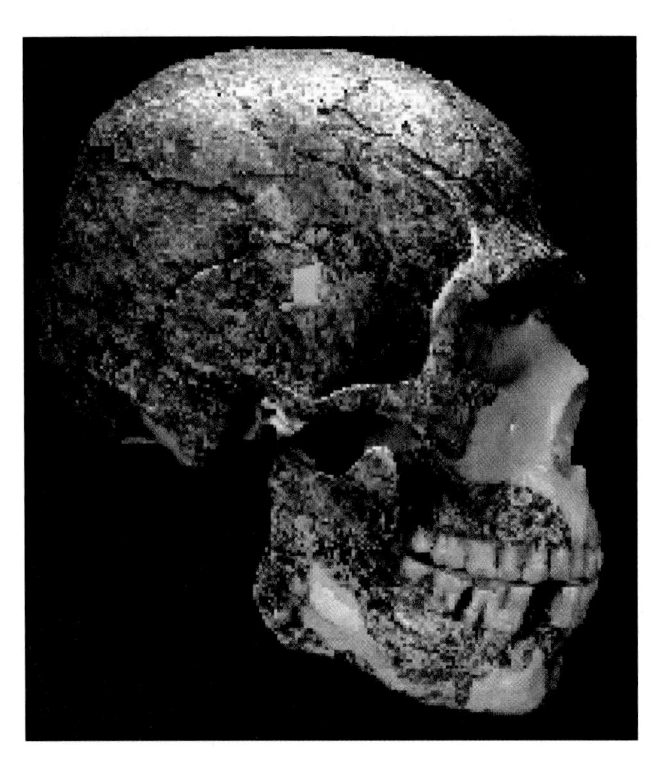

FIGURE 11.11. Anatomically modern *Homo sapiens:* Skhul 5 (Israel). (Courtesy of Professor Milford Wolpoff, University of Michigan, Ann Arbor.)

In summary, although the dating is not precise, it appears that about 40,000 to 164,000 years ago the world was populated by both the archaic *Homo sapiens* (including the Neanderthals) and the anatomically modern *Homo sapiens.* The overall characteristics of the anatomically modern *Homo sapiens* is associated with decreased robustness in both cranial and post-cranial morphology when compared to the features of archaic *Homo sapiens.* The skull became rounder with less prominent brow ridges. The face has a vertical forehead and a mandible with a chin. The effect of these changes is that modern *Homo sapiens* have become gracile and less robust than archaic *Homo sapiens.*

Technology and Demographic Changes During the Upper Paleolithic Period

Traditionally, the cultural trajectory associated with the tool technology of early humans is divided into three periods: the Paleolithic ("old Stone Age"), Mesolithic ("middle Stone Age"), and the Neolithic ("new Stone Age"). The Paleolithic encompasses all of the Pleistocene up to the end of the last Ice Age to about 12,000 years ago (**table 11.4**). The lower Paleolithic is roughly associated with the appearance of the australopithecines, *Homo habilis* and *Homo erectus*; the middle Paleolithic with Neanderthals and archaic *Homo sapiens*; and the upper Paleolithic with the anatomically modern *H. sapiens.*

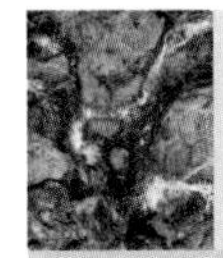

TABLE 11.3
Early Modern *Homo sapiens* Specimens

Site and Specimens	Remains	Geological Age (1,000 y)	Brain Size (cm³)
AFRICA			
Ethiopia (Omo-Kibish)	Partial skeleton	130	1,430
Sudan (Singa)	Braincase	80	1,500
S. Africa (Klasies River)	Skull, jaw & fragments	120–80	
S. Africa (Border Cave)	Skull, jaw & fragments	110–90	1,507
Morocco (Jebel Irhoud)	Crania	50	1,305, 1,450
ASIA			
Israel (Skhul)	Crania & skeleton	50–40	1,554, 1,450
Israel (Qafzeh)	Crania & skeleton	70–50	1,568, 1,523
Borneo (Niah)	Cranium	40	
EUROPE			
Bulgaria (Bacho Kiro)	Bone fragments	43	
W. Germany (Hahnofersand)	Frontal	36	
Yugoslavia (Velika Pecina)	Frontal	34	
Czechoslovakia (Predmosti)	Several specimens	26	
France (Cro-Magnon)	Several specimens	20	

Sources: Vallois,[26] Vallois and Billy,[27] Weidenreich,[28] Karavanic and Smith,[29] Seidler, Falk, and Stringer et al.[30]

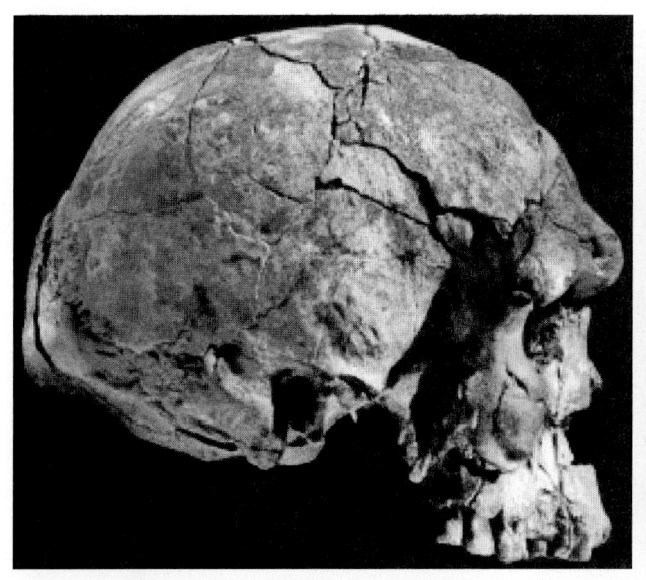
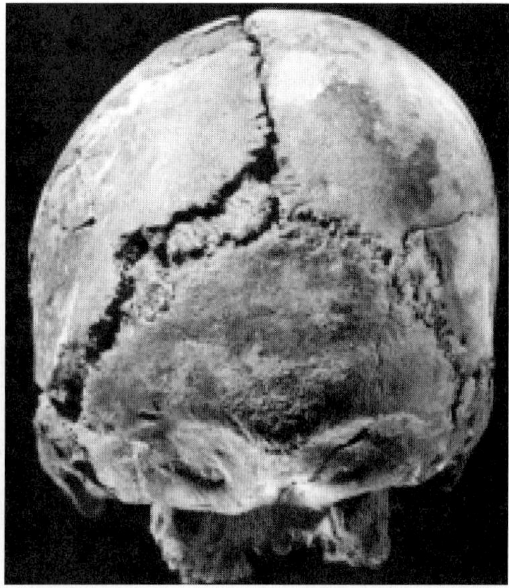

FIGURE 11.12. Anatomically Modern *Homo sapiens* from Ethiopia, Africa. Although this fossil exhibits a rounder skull and more vertical forehead, the size and shape of the brow ridges are intermediate between those shown by Neanderthals and early anatomically modern *Homo sapiens.* Source: Adapted from White, Asfaw, and DeGusta et al.[37] Reprinted by permission of Tim White.

Tool Techniques During the Upper Paleolithic

The Upper Paleolithic period is a time of cultural explosion that coincides with the appearance of the anatomically modern *H. sapiens.* Technologically it is characterized by the improvement of preexisting techniques and the invention of new tools such as the bow and arrow, spear thrower, complex bone tools, nets and snares, new methods of food storage, improved hearths, clothing, and more efficient, larger dwellings. The Upper Paleolithic period is divided into five cultural traditions: Aurignacian, Chatelperronian, Gravettian, Solutrean, and Magdalenian.

Some of these cultural traditions used the *indirect percussion* and *pressure flaking* techniques for making spear points.[12] In the technique of *indirect percussion,* narrow flakes are produced from specialized cores with a prismatic form. Each successive blade takes its form from the ridges left on the core after the removal of the previous blades. *Pressure flaking,* rather than using percussion to strike off flakes as in previous technologies, employs pressure with a bone, wood, or antler tool at the edge of the tool to remove small flakes. After shaping a core into a pyramidal or cylindrical form, the toolmaker puts a punch of antler, wood, or another hard material into position and strikes it with

FIGURE 11.13. *Homo floresiensis.* While the craniofacial features of (a) *H. floresiensis* are similar to (a) anatomically modern *H. sapiens,* it is less than one third the size of modern *Homo sapiens.* (Sources: (a) Brown et al.[52] (b) Courtesy of Professor Milford Wolpoff, Unniversity of Michigan, Ann Arbor.)

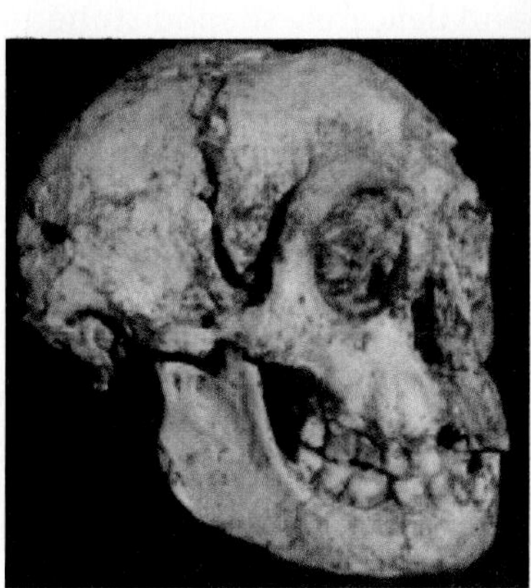
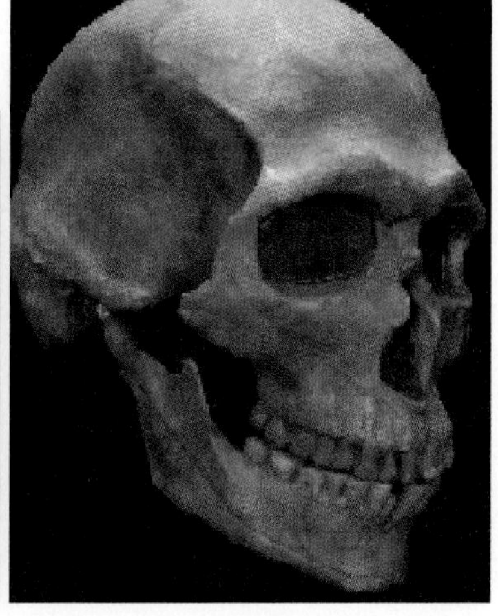

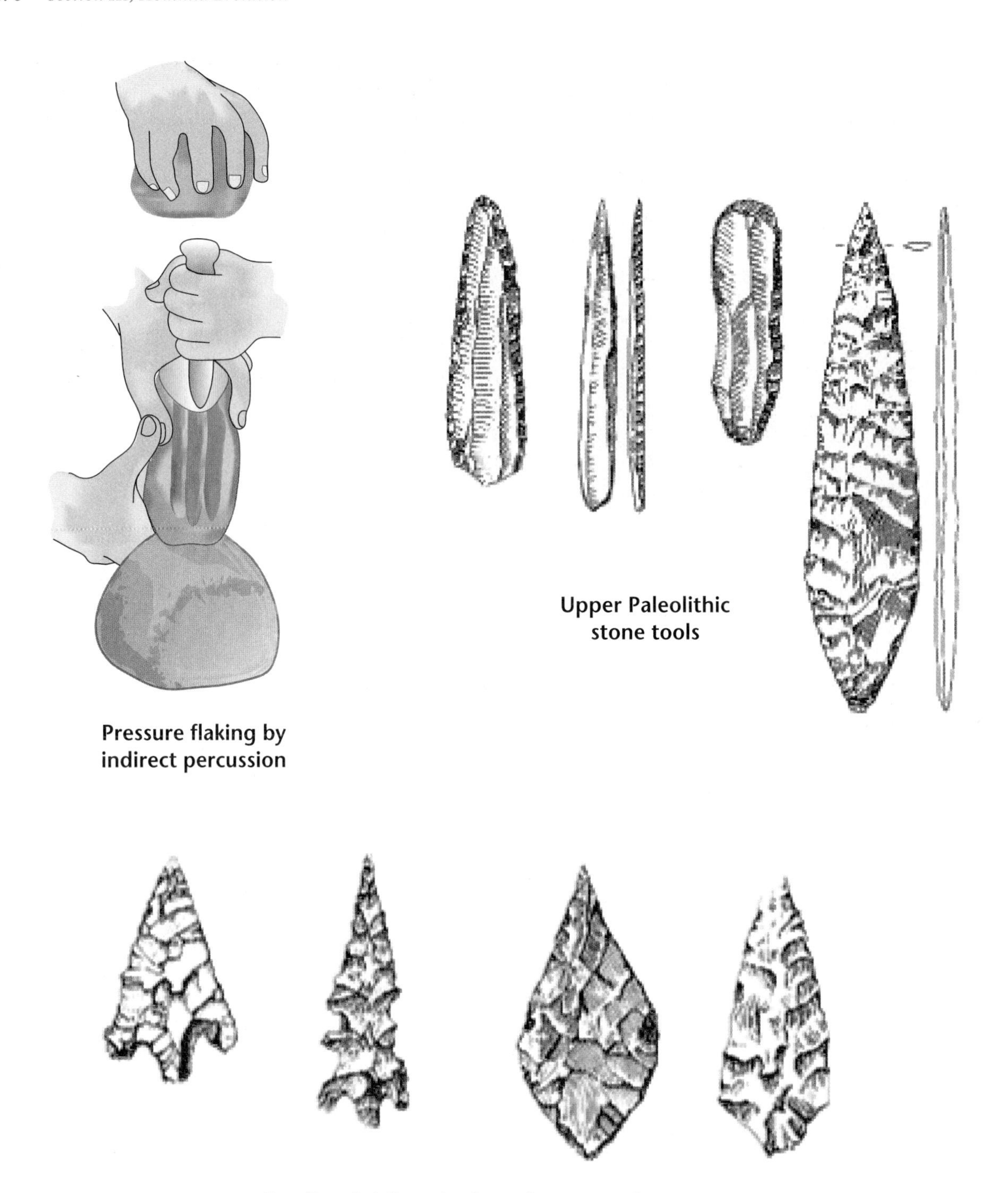

FIGURE 11.14. Pressure flaking by indirect percussion and Upper Paleolithic stone tools. Pressure flaking works by employing pressure with a bone, wood, or antler tool at the edge of the tool to remove small flakes. These small and delicately shaped stone blades were often fitted into handles, one blade at a time or several blades together, to serve as spears, axes, knives, and sickles. All of these tools were used for hunting big game. Source: Composite picture: (a) from Fagan, B. M. 1972. *In the Beginning*. Boston: Little Brown; (b) Bordes, F. 1961. Mousterian cultures in France. *Science* 803–810.

a hammer (**fig. 11.14**). Because the force is readily directed, the toolmaker is able to strike off consistently shaped blades that are more than twice as long as they are wide.

The most recent part of the Upper Paleolithic has traditionally been called the *Mesolithic*. The Mesolithic also had a characteristic tool type—the *microlith* (Greek for "small stone"), which was a small and delicately shaped stone blade. These blades were often fitted into handles, one blade at a time or several blades together, to serve as spears, adzes, knives, and sickles (**fig. 11.14**). During this period, the technological and cultural sophistication reached its highest level of the Paleolithic. Beyond the microlith, other mesolithic innovations include the development of projectile weapons such as the spear thrower made from bone, antler, and ivory; manufacture of dugout canoes; manufacture of nets, ropes, bags, and clothing woven from fiber; production of art and ornamentation; development of symbolism and ritual burial; sophisticated architecture; land use planning; resource exploitation; and strategic social alliances. All of these clearly characterize modern human behavior.

In summary, during the Upper Paleolithic, the tools of the anatomically modern *Homo sapiens* became more refined and complex. Microliths were used as fishhooks and as parts of harpoons. Throughout the territory they exploited, hunters probably used dugout canoes in fishing. The bow and arrow became an essential tool as people exploited swamps and marshes where they were able to prey on waterfowl.

Extinction of Large Herbivores

In many parts of the world except in Africa, between 100,000 and 1,000 years ago, the large game animals such as mammoths, ground sloths, and other mammal

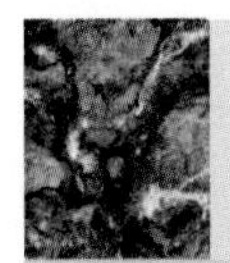

TABLE 11.4

The Subdivisions of the Pleistocene and Its Relative Association with the Paleolithic, Cultural Tool Traditions, and Hominid Stages

Geological Epoch (Years Ago)	Glacial Period	Paleolithic Stage	Cultural Tool Tradition	Hominid Stage
UPPER PLEISTOCENE				
10,000				
14,000		Neolithic	Agriculture/Herding	Modern *H. sapiens*
17,000			Magdalenian	Modern *H. sapiens*
21,000			Solutrean	Modern *H. sapiens*
27,000			Gravettian	Modern *H. sapiens*
33,000	Last glacial period	Upper Paleolithic	Aurignacian	Modern *H. sapiens*
40,000			Chatelperronian	Modern *H. sapiens*
50,000				
100,000	Last interglacial			
200,000	period	Middle Paleolithic	Mousterian:	Neanderthals &
250,000			blades made	Archaic *H. sapiens*
300,000			from specially	Archaic *H. sapiens*
			retouched stones	
MIDDLE PLEISTOCENE				
400,000	Earlier glacial	Lower Paleolithic	Acheulean:	*Homo erectus*
700,000	periods		bifacial	*Homo erectus*
900,000			chopper stones	*Homo erectus*
1,200,000			Use of fire	*Homo erectus*
LOWER PLEISTOCENE				
1,800,000			Oldowan:	*Homo habilis*
2,200,000			pebble stone	*Homo habilis*
2,400,000			size of tennis balls	*Homo habilis*
2,600,000			perishable	*Australopithecus*
3,500,000			tools	*Australopithecus*
4,500,000				*Ardipithecus*
6,000,000				*Orrorin tugenensis*
8,000,000–7,000,000				*Sahelanthropus*

"megafauna" went extinct. Because this time frame coincides with the appearance of the improved hunting technology of *Homo sapiens*, it has been hypothesized that humans were involved in the extinction. It has been postulated that the extinction of large herbivores is related to overkill and low reproductive rate.

According to the "overkill" hypothesis, the extinction was due to heavy and selective hunting of large-bodied prey by people. An extreme form of the overkill hypothesis is the "blitzkrieg" model. The "blitzkrieg" model envisages that when people invaded new lands they initially specialized in hunting large-bodied mammals, which were both abundant and easy to kill.[38, 39] This intensive exploitation of large prey supported rapid growth and geographical expansion of invading human populations, but it quickly changed from widespread hunting to extinction of the megafauna species. Support for this hypothesis comes from the findings of great numbers of skeletal remains of large herbivores. For example, at Solutré, in France, and Predmost, Czechoslovakia, remains of a large number of horses (at least 100,000) and mammoth skeletons (more than 1,000 mammoths) have been found.[40] Comparative studies of skeletal remains found in the Iberian Peninsula[41] indicate that the range of species hunted by Paleolithic hominids was much greater than that of any other predator. A computer simulation of North American end-Pleistocene human and large herbivore population dynamics indicates that for 32 out of 41 prey species, hunting alone could have produced the observed pattern of extinction within short periods of time.[42] For example, of the 35 genera of mammals that were lost toward the end of the Pleistocene in North America, only 15 species survived after the Clovis culture[43] peopled the New World about 14,000 years ago (see next chapter). However, the overkill argument is contested mainly on the grounds that the archaeological and paleontological records provide too little support for heavy hunting of large mammals that went extinct.

Others postulate that the extinction of large herbivores was due to a low reproductive rate of the species.[44] This inference is based upon the observation that within taxonomic families of living mammals the likelihood of extinction is higher for groups with lower reproductive rates, regardless of their body size.[44] Even relatively small mammals become extinct if their fecundity is low enough. Specifically this investigation demonstrates that, for all groups of animals, the threshold reproductive rate, at which the chance of extinction exceeds 50 percent, is about one offspring per female per year.

In summary, it would appear that during the late Pleistocene, the population of *Homo sapiens* increased drastically. To satisfy nutritional needs, humans practiced general overkill where species of all body sizes were hunted and, of course, those with low reproductive rates were more vulnerable to extinction than those with high fecundity.

Increased Consumption of Meat

In any event, increased hunting success translated into greater meat consumption per capita in the Middle and Upper Pleistocene. Because plants do not fossilize as commonly as bone, plant utilization is more difficult to document than that of skeletal remains. Nevertheless, there is some direct evidence that late Pleistocene hunter-gatherers complemented their meat diet with plant food by collecting and preparing certain vegetable foods. This evidence is in the form of large quantities of roasted hackberry seeds found in deposits at Zhoukoudian, China[45] and some fragmentary seeds of legumes and chenopods, pine charcoal, and broken pine cones found in Mousterian sites in Europe.[46]

Increased Consumption of Marine Foods

By the Upper Paleolithic in Europe and the Middle Stone Age in southern Africa, marine foods (including shellfish, fish, and marine mammals) had become a significant component of the diet of the anatomically modern *Homo sapiens.*[30] As a result of improved technology throughout Europe, gathering of large quantities of shellfish and mollusks appears to have intensified. The abundant remains of shellfish found in the site of Indian Knoll[32] suggest that in the North American archaic period, intensive shellfish harvesting also became an important component of the diet.

Seasonal Meat Consumption in Subarctic Regions

Beginning in the Middle Pleistocene, sites found in Africa and Europe suggest that *Homo sapiens* inhabited tropical regions as well as cold, periglacial regions. In subarctic environments, there were probably few edible plant foods available to humans. Similar to modern Arctic hunters, meat probably comprised all or most of the winter diet of *Homo sapiens.* This inference is based on observations from occupation sites in the Ukraine[45] that appear to have been primarily winter encampments inhabited during very cold periods.

FIGURE 11.15. Cave painting in Lascaux (Southern France). Painted by anatomically modern *Homo sapiens* who lived about 17,000 years ago. Source: Artists Rights Society (ARS), New York/DACS, London, 2002.

The inhabitants of these encampments derived their subsistence from intensive hunting of reindeer and horse. These findings suggest that during the climatic fluctuations that characterized the interglacial periods, human populations moved into more diverse habitats, seasons, and diets. Thus, modern *Homo sapiens,* with their increased sophistication and improvement of food procurement technology, were able to exploit both land and marine sources while occupying diverse habitats that ranged from tropical to arctic climates.

Art in the Upper Paleolithic Period

During the Upper Paleolithic Period 40,000 to 50,000 years ago, art and technological achievements reached their maximum expression. This fundamental behavioral expression is the result of gradual change that started during the African Middle Stone Age about 250,000 to 300,000 years ago.[50] Many of the components of the Upper Paleolithic art and technology such as blade and microlithic technology, bone tools, increased geographic range, specialized hunting, the use of aquatic resources, long-distance trade, systematic processing and use of pigment, and art and decoration are found in the African Middle Stone Age.[50]

Cave Art

The high intellectual capacity and appreciation of art of early modern *Homo sapiens* are evident by the paintings that adorned the walls of the more than 150 caves found in northern Spain, southwestern France,[51] and China. A common theme of the cave paintings is the depiction of Ice Age mammals along with horses and bovids (**fig. 11.15**). These paintings were made 15,000 to 30,000 years ago. Rock art of similar antiquity has also appeared on other continents, such as Africa and South America. In Africa (Namibia), a site containing such art is dated between 19,000 and 28,000 years ago. Australia also has several caves with ancient finger markings and deeply engraved circles and grids. Some of the finger markings probably date to between 15,000 and 24,000 years ago. Even in the Americas, the rock shelters of Pedra Furada, and Perna in the Piaui region of Brazil have red-painted human figures dating to at least 10,000 to 12,000 years ago.

Several explanations for the function of the ancient art have been given. One of the early explanations emphasized the relationship of the paintings to hunting. Hunting rituals may be viewed as a kind of imitative magic that would increase prey animal populations or help hunters successfully find and kill their quarry. A recent explanation suggests that the caves were meeting places for local bands of people and group activities. In this context, the paintings and engravings served as "encoded information" that could be passed on across generations.

Sculpture

In addition to cave art, there are numerous examples of small sculptures frequently termed "portable art,"

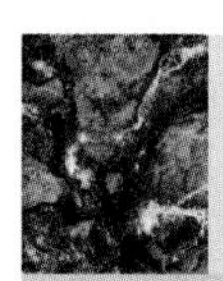

TABLE 11.5

Number of Old, Young, and Ratio of Old to Young among Hominid Fossil Samples

	Old	Young	Total	Old to Young Ratio
Australopithecines	37	316	353	0.12
Early *Homo*	42	166	208	0.25
Neanderthals	37	96	113	0.39
Early Upper Paleolithic	50	24	74	**2.08**
All groups combined	166	602	768	0.28

Source: Adapted from Caspari and Lee.[49]

excavated from sites in Africa and Europe (**fig. 11.16**). These include elaborate engravings on tools, and tool handles made of bone and ivory.

The most famous sculptures in stone and bone, representing females and emphasizing reproductive features ("Venus" figure), have been found in Europe and Siberia. Some of these figures were realistically carved. The faces of these sculptures appear to be modeled after actual women. Another stylistic innovation was the partial sculpting of a rock face. In one rock cave in southwest France, several animals (including mountain goats, bison, reindeer, and horses) and one human figure were depicted in bas-relief.

FIGURE 11.16. Upper Paleolithic Artifacts. From utilitarian tools, such as harpoons, which are artistically carved with figures representing animals, to the Venus figurine (lower right), who is said to have been a fertility symbol. (Photo by Kirschner. Courtesy of Department of Library Services, American Museum of Natural History.)

Longevity and the Upper Paleolithic Period

Humans differ from nonhuman primates by having an increased longevity. An important question is when in the path of hominid evolution did changes in longevity occur. To answer this question, Caspari and Lee[49] examined the available fossil record, which represents the last 3 million years of human evolution. The study included evaluations of dental age of 768 individuals that belonged to four separate groups of hominids, from australopithecines to more recent societies that replaced Neanderthals in the Late Pleistocene. A salient finding is that longevity increased between all groups. The ratio of old to young adults increases gradually from 0.12 in australopithecines to 0.40 among the Neanderthals. There is a particularly sharp increase in the last part of the human evolutionary record, the Upper Paleolithic, when they found a fourfold increase in the number of adults old enough to be grandparents (**table 11.5**). The importance of having more older adults in a population is that older age increases the bank of experience, information, and knowledge that could be transferred to the next generation, which in turn might allow for the development of culture and complex societies that characterize humans. Consequently it is quite possible that the demographic expansion that gradually started in the African Middle Stone Age[50] and culminated with the "creative explosion" of the Upper Paleolithic period was the result of increased longevity.

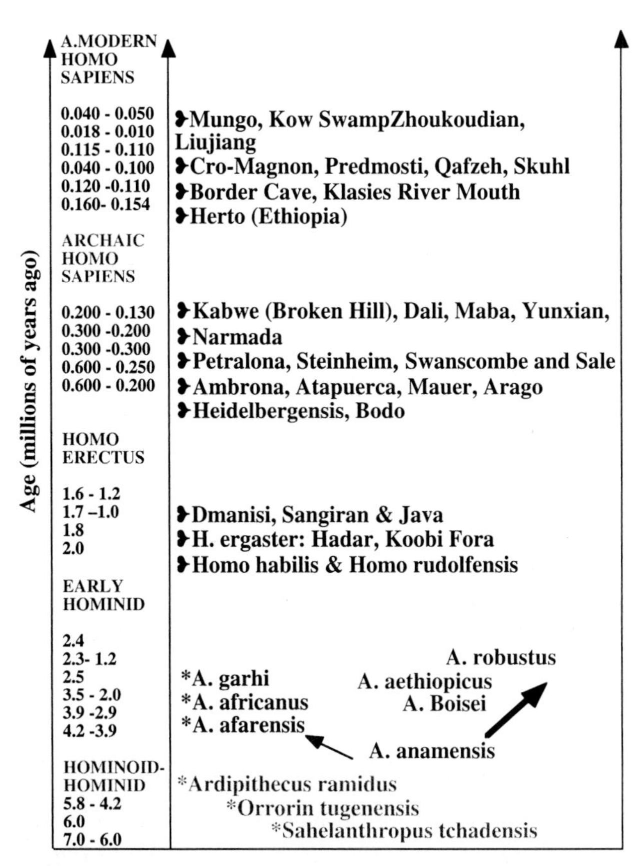

FIGURE 11.17. Schematization of the phylogenetic view about the origins of *Homo sapiens.*

Summary

There are several interpretations about the phylogenetic relationship between the early hominids and anatomically modern *Homo sapiens.* Paleoanthropologists usually base their phylogenetic interpretations on either the cladistic or phenetic models. The **cladistic** model assumes that all morphological differences whether they were shared or derived reflect through species differences. Using the cladistic model between 5 to 20 hominid species are identified. In contrast, the **phenetic** model assumes that population and individuals are continuously evolving through time, sometimes as a group and sometimes splitting to form new lineages depending on the needs to adapt to a specific environmental condition. Accordingly, in the **phenetic** perspective few species are identified.

In this book, we assume that biological and behavioral variability is the necessary landmark of survival in contemporary populations and in the past there must have been the same degree of variability. Accordingly, it is inferred morphologically that small differences between fossils reflect natural population differences rather than species differences. As illustrated in **figure 11.17** we have placed the path to *Homo sapiens* into four grades: Hominoid-Hominid, Early Hominid, *Homo erectus,* and *Homo sapiens.* The Hominoid-Hominid grade includes in chronological order *Sahelanthropus tchadensis,* Orrorin *tugenensis,* and *Ardipithecus radimus.* In the early Hominid grade are included the australopithecines. In the *Homo erectus* grade are included *Homo habilis* and the several populations of *Homo erectus.* In the *Homo* ***sapiens*** grade are included populations of Neanderthals and anatomically modern *Homo* ***sapiens.*** In this view, the archaic *H. sapiens* and Neanderthals were all part of one potentially interbreeding and, evolving species. On the other hand, in the splitters' perspective the origin of modern *Homo sapiens* can be traced to *Homo heidelbergensis* (such as Petralona and Sima de los Huesos in Europe and Bodo and Kabwe in Africa) from which both Neandertals and the anatomically modern *H. sapiens* evolved. In both the phenetic and cladistic approach the robust australopithecines are considered as evolutionary dead ends.

Irrespective of whether one accepts the phenetic or cladistic perspective of human evolution it is clear that about 40,000 to 50,000 years ago in the period referred to as the Upper Paleolithic the cultural and technological achievements of *Homo sapiens* reached to its maximum sophistication. This trend started gradually during the African Middle Stone Age about 100,000 years ago. The achievements included dramatic technological innovations, such as the harpoons, spear throwers, and bow and arrow, all of which increased *Homo sapiens* ability to hunt big game. The archaeological record seems to explode with creative activity, including personal ornaments, elaborate ritualistic burials, and fantastic cave paintings. As inferred from recent studies of the fossil record, it appears that an important factor that contributed to the cultural and technological sophistication that characterizes modern *Homo sapiens* might have been the increased survivorship of older age individuals who became the repositories of information and knowledge that could be transferred across generations. Thus, present-day human achievements in art and technology are simply a continuation of the skills acquired and knowledge accumulated since the Middle African Stone Age reaching their maximum expression during the Upper Paleolithic period.

Suggested Readings

Brace, C., K. Rosenberg, and K. Hunt. 1987. Gradual changes in human tooth size in the Late Pleistocene and Post-Pleistocene. *Evolution* 41:705–720.

Conroy, G. C. 1997. *Reconstructing Human Origins. A Modern Synthesis.* New York: Norton.

Delson, E., ed. 1985. *Ancestors: The Hard Evidence.* New York: Liss.

Falk, D. 1993. A good brain is hard to cool. *Natural History* 102(8).

Wolpoff, M. H. 1999. *Human Evolution,* 2d ed. Boston: McGraw-Hill.

Literature Cited

1. Trinkaus, E., and P. Shipman P. 1992. *The Neandertals: Changing the Image of Mankind.* New York: Alfred E. Knopf.
2. Kappelman, J. 1996. The evolution of body mass and relative brain size in fossil hominids. *J. Hum. Evol.* 30:243–276.
3. Ruff, C. B. 1994. Morphological adaptation to climate in modern and fossil hominids. *Yearb. Phys. Anthropol.* 37: 65–107.
4. Aguirre, E., J. L. Arsuaga, J. M. Bermúdez de Castro, E. Carbonell, M. Ceballos, C. Díez, P. Enamorado, Y. Fernández-Jalvo, E. Gil, A. Gracia, A. Martín-Najera, I. Martínez, J. Morales, A. I. Ortega, A. Rosas, A. Sánchez, B. Sánchez, C. Sesé, E. Soto, and T. Torres. 1990. The Atapuerca sites and the Ibeas hominids. *Human Evol.* 5:55–73.
5. Arsuaga, J. L., I. Martínez, A. Gracia, and C. Lorenzo. 1997. The Sima de los Huesos crania (Sierra de Atapuerca). A comparative study. *J. Hum. Evol.* 33:219–281.
6. Arsuaga, J. L., I. Martínez, A. Gracia, J. M. Carretero, C. Lorenzo, and N. García. 1997. Sima de los Huesos (Sierra de Atapuerca). The site. *J. Hum. Evol.* 33:109–127.
7. Bermudez de Castro, J. M., J. L. Arsuaga, E. Carbonell, A. Rosas, I. Martínez, and M. Mosquera. 1997. A hominid from the Lower Pleistocene of Atapuerca, Spain: Possible ancestor to Neanderthals and modern humans. *Science* 276:1392–1395.
8. Rosas, A. 2001. Occurrence of Neanderthal features in mandibles from the Atapuerca-SH site. *American Journal of Physical Anthropology* 114:74–91.
9. Mellars, P. 2004. Neanderthals and the modern human colonization of Europe. *Nature* 432:461–465.
10. Helmuth, H. 1998. Body height, body mass and surface area of the Neanderthals. *Morphol. Anthropol.* 82:1–12.
11. Franciscus, R. G. 2003. Internal nasal floor configuration in *Homo* with special reference to the evolution of Neandertal facial form. *J. Hum. Evol.* 44:701–729.
12. Ambrose, S. H. 2001. Paleolithic technology and human evolution. *Science* 291:1748–1753.
13. Isaac, G. L. 1977. *Olorgesailie: Archaeological Studies of a Middle Pleistocene Lake Basin in Kenya.* Chicago: University of Chicago Press.
14. Binford, S. 1968. Early upper Pleistocene adaptations in the Levant. *Am. Anthropol.* 70:707–717.
15. Howell, F. C. 1966. Observations on the earlier phases of the European lower Paleolithic. *Am. Anthropol.* 68:88–201.
16. Howell, F. C., and L. G. Freeman. 1982. Ambrona: An early Stone Age site on Spanish Meseta. *The L.S.B. Leakey Foundation News* 22(1):11–13.
17. Klein, R. G. 1973. *Ice-Age Hunters of the Ukraine.* Chicago: University of Chicago Press.
18. de Lumley, H. 1969. A Paleolithic camp at Nice. *Sci. Am.* 220:42–50.
19. Gordon, K. D. 1987. Evolutionary perspectives on human diet. In *Nutritional Anthropology,* Johnston, F. M., ed. New York: Alan R. Liss, Inc.
20. Hublin, J. J., F. Spoor, M. Braun, F. Zonneveld, and S. Condemi. 1996. A late Neanderthal associated with upper Palaeolithic artifacts. *Nature* 381:224–226.
21. d'Errico, F., J. Zilhão, M. Julien, D. Baffier, and J. Pelegrin. 1998. Neanderthal acculturation in western Europe? A critical review of the evidence and its interpretation. *Current Anthropol.* 39:1–44.
22. Lieberman, D. E., B. M. McBratney, and G. Krovitz. 2002. The evolution and development of cranial form in *Homo sapiens. PNAS* 99:1134–1139.
23. Kay, R. F., M. Cartmill, and M. Balow. 1998. The hypoglossal canal and the origin of human vocal behavior. *PNAS* 95:5417–5419.
24. Jungers, W. L., A. A. Pokempner, R. F. Kay, and M. Cartmill. 2003. Hypoglossal canal size in living hominoids and the evolution of human speech. *Human Biology* 75:73–90.
25. Bahn, P. G. 1998. Neanderthals emancipated. *Nature* 394:719–721.
26. Vallois, H. V. 1968. La découverte des hommes de Cro-Magnon, son importance anthropologique. In *L'Homme de Cro-Magnon,* 11–22. G. Camps, and G. Olivier, ed. *Anthropologie et archéologie,* Arts et métiers graphiques édit, Paris.
27. Vallois, H. V., and G. Billy. 1965. Nouvelles recherches sur les hommes fossiles de l'abri de Cro-Magnon. *L'Anthropologie* 69:47–74.
28. Weidenreich, F. 1938/39. On the earliest representatives of modern mankind recovered on the soil of East Asia. *Peking Nat. Hist. Bull.* 13:161–174.
29. Karavanic, I., and F. H. Smith. 1998. The Middle/Upper Paleolithic interface and the relationship of Neanderthals and early modern humans in the Hrvatsko Zagorje, Croatia. *J. Hum. Evol.* 34(3):223–248.
30. Seidler, H., D. Falk, C. Stringer, H. Wilfing, G. B. Muller, D. zur Nedden, G. W. Weber, W. Reicheis, and J. L. Arsuaga. 1997. A comparative study of stereolithographically modelled skulls of Petralona and Broken Hill: Implications for future studies of Middle Pleistocene hominid evolution. *J. Hum. Evol.* 33:691–703.
31. Stringer, C. B. 1983. Some further notes on the morphology and dating of the Petralona hominid. *J. Hum. Evol.* 12: 731–742.
32. Stringer, C. B. 2002. Modern human origins: progress and prospects. *Philos. Trans. R. Soc. Lond. B. Biol. Sci.* 357:563–579.

33. Vandermeersch, B. 1978. Quelques aspects du problème de l'origine de l'homme modern. In *Les origines humaines et les époques de l'intelligence.* Bordes, F. ed. 153–157: Masson, Paris.
34. Thorne, A., R. Grun, G. Mortimer, N. A. Spooner, J. J. Simpson, M. McCulloch, L. Taylor, and D. Curnoe. 1999. Australia's oldest human remains: Age of the Lake Mungo 3 skeleton. *J. Hum. Evol.* 36:591–612.
35. Singer, R., and J. Wymer. 1982. *The Middle Stone Age at Klasies River Mouth in South Africa.* Chicago: University of Chicago Press.
36. Rightmire, G. P., and H. J. Deacon. 2001. New human teeth from Middle Stone Age deposits at Klasies River, South Africa. *J. Hum. Evol.* 41:535–544.
37. Clark, J. D., Y. Beyene, G. WoldeGabriel, W. K. Hart, P. R. Renne, H. Gilbert, A. Defleur, G. Suwa, S. Katoh, K. R. Ludwig, J. R. Boisserie, B. Asfaw, and T. D. White. 2003. Stratigraphic, chronological and behavioural contexts of Pleistocene *Homo sapiens* from Middle Awash, Ethiopia. *Nature* 423:747–752.
38. Flannery, T. F., and R. G. Roberts. 1999. Late Quaternary extinctions in Australasia: An overview. In *Extinctions in Near Time: Causes, Contexts and Consequences,* R. D. E. MacPhee, ed., 239–256. New York: Kluwer/Plenum.
39. Martin, P. S., and H. E. Wright. 1967. *Pleistocene Extinctions.* New Haven: Yale University Press.
40. Howell, F. C. 1966. Observations on the earlier phases of the European lower Paleolithic. *Am. Anthropol.* 68:88–201.
41. Freeman, L. 1981. The fat of the land: Notes on Paleolithic diet in Iberia. In *Omnivorous Primates,* R. S. Harding, and G. Teleki, ed., 104–165. New York: Columbia University Press.
42. Alroy, J. 2001. A multispecies overkill simulation of the end-Pleistocene megafaunal mass extinction. *Science* 292: 1893–1896.
43. Grayson, D. K., J. Alroy, R. Slaughter, and J. Skulan. 2001. Did human hunting cause mass extinction? *Science* 294:1459–1462.
44. Johnson, C. N. 2002. Determinants of loss of mammal species during the Late Quaternary 'megafauna' extinctions: Life history and ecology, but not body size. *Proc. R. Soc. Lond. B. Biol. Sci.* 269:2221–2227.
45. Klein, R. G. 1973. *Ice-Age Hunters of the Ukraine.* Chicago: University of Chicago Press.
46. Wu, R., and S. Lin. 1983. Peking Man. *Sci. Am.* 226(6):86–94.
47. Dean, C., M. G. Leakey, D. Reid, F. Schrenk, G. T. Schwartz, C. Stringer, and A. Walker. 2001. Growth processes in teeth distinguish modern humans from *Homo erectus* and earlier hominids. *Nature* 414:628–631.
48. White, T. D., B. Asfaw, D. DeGusta, H. Gilbert, G. D. Richards, G. Suwa, and F. C. Howell. 2003. Pleistocene *Homo sapiens* from Middle Awash, Ethiopia. *Nature* 423:742–747.
49. Caspari, R., and S. H. Lee. 2004. Older age becomes common late in human evolution. *PNAS* 101:10895–10900.
50. McBrearty, S., and A. S Brooks. 2000. The revolution that wasn't: A new interpretation of the origin of modern human behavior. *Journal of Human Evolution* 39:453–563.
51. Clottes, J., and J. Courtin. 1996. *The Cave Beneath the Sea. Paleolithic Images at Cosquer.* New York: Harry N. Abrams, Inc.
52. Brown, P., T. Sutikna, M. J. Morwood, R. P. Soejono, J. Jatmiko, E. W. Saptomo, and R. A. Due. 2004. A new small-bodied hominin from the Late Pleistocene of Flores, Indonesia. *Nature* 431:1055–1061.
53. Morwood, M. J., Brown, P., Jatmiko, J., Sutikna, T., Wahyu Saptomo, E., Westaway, K. E., Due, R. A., Roberts, R. G., Maeda, T., Wasisto, S., and Djubiantono, T. 2005. Further evidence for small-bodied hominids from the Late Pleistocene of Flores, Indonesia. *Nature* 437:1012–1017.
54. Falk, D., Hildebolt, C., Smith, K., Morwood, M. J., Sutikna, T., Brown, P., Jatmiko, J., Saptomo, E. W., Brunsden, B., and Prior, F. 2005. The brain of LB1, *Homo floresiensis. Science* 308:242–245.

The Peopling of the New World and the Origins of the Anatomically Modern *Homo sapiens*

As Europe and Asia went through the Ice Age, the game animals continued migrating toward better climates and better pastures. Following the game animals, on which the anatomically modern *Homo sapiens* came to depend, resulted in their migration into the New World. Upon arriving on the continent, some of the small bands settled in Alaska, while others continued to migrate and finally to populate most of the regions of North and South America. An important question is who were those new settlers: were they a new species of *Homo sapiens* or a simple and evolved form of archaic *Homo sapiens*? This chapter presents the archaeological and genetic evidence about the peopling of the New World. In addition, this chapter summarizes the hypotheses postulated to explain the origins of the anatomically modern *Homo sapiens.*

The Peopling of the New World

The ancestors of present-day Native Americans arrived from Northeast Asia, South Asia, Central Asia, and possibly from the Pacific Rim in two or three successive migrations about 15,000 to 12,000 years ago (**fig. 12.1**). These migrations occurred through the Beringia land bridge and coast of the landbridge. During the late Pleistocene, the Beringia landbridge (which was about 1,300 miles long) was formed as a result of the continental glaciers that lowered the sea levels by trapping water in glaciers, causing an emergence of land that joined North America and Asia. The two or three major glaciers of North America reached their peaks at different times. During the coldest periods of the late Pleistocene, about 18,000 to 22,000 years ago, North America and the Beringia landbridge offered hunter-foragers varied resources depending on the period in which the migrants crossed. In some periods, the Beringia landbridge was so dry and barren that it did not have particularly rich resources, but at other times it had abundant animal resources, such as musk oxen and other large herbivores, along with the fish, birds, and sea mammals along the shore. Between 10,000 and 30,000 years ago, from Siberia into the New World there was a continuous back and forth migration of animals, including species of deer, bison, camels, bears, foxes, mammoths, moose, caribou, woodchucks, and mice. In the mid-Wisconsin interglacial period, humans may have had a narrow but clear ice-free corridor all the way to South America. Thus, the peopling of the Americas involved two or three migrations of populations composed of small

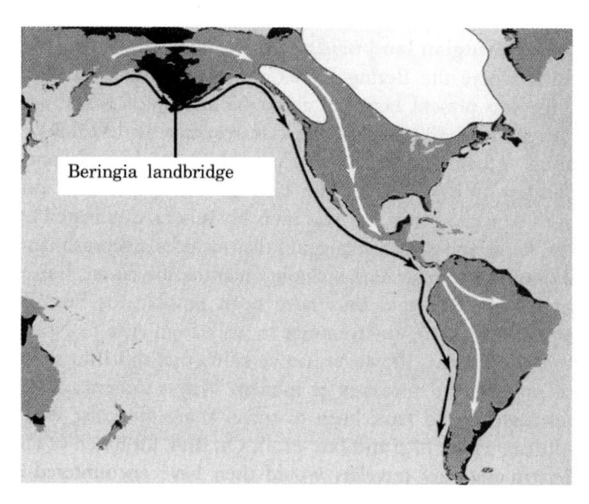

FIGURE 12.1. Possible routes of migration into the New World. The dark black area denotes the landmass exposed during the Ice Age. The first migration occurred by land and along the coast (black arrows). After crossing the Beringia landbridge they traveled along the coast (black arrows) reaching South America. The second migration, associated with the Clovis Big Game Hunter, occurred by land (white arrows) across the Beringia landbridge into Alaska and then down into North and South America. From *The Settlement of the Americas* by T. D. Dillehay. Copyright © 2000 by T. D. Dillehay. Reprinted by permission of Basic Books, a member of Perseus Books, L.L.C.

bands that came from Asia and across the Beringia landbridge in search of new horizons and food.

First Migration through the Coast of the Landbridge

The most likely course that the first Americans took would be along the coast of the landbridge, into Alaska then into the Mackenzie River Valley, and from there southward along the eastern slopes of the Rocky Mountains and on into the Dakotas and then further southward. Along the coast of the landbridge were probably abundant resources in the form of fish, birds, eggs, invertebrates, and many plant foods. Although such sites would lie deep beneath the current coastal seas, the first migration might have been by the Pacific Coast route. Excavations in the eastern caves of Chile indicate that the first migrants (**fig. 12.1**) retained some elements of the generalized hunting, fishing, and foraging economies required in the intertidal zone.

Second Migration through the Middle of the Landbridge and the Clovis Big Game Hunter

The second migration probably took place around 11,000 to 12,000 years ago. This migration also occurred using the Beringia landbridge, but then they proceeded across the middle of the landbridge. After crossing the Beringia landbridge, they moved down into what is now Alberta and, splitting into smaller groups, they developed new cultures and eventually reached the tip of South America. Except for some later-arriving groups, including today's Inuits, or Eskimos, these people would have been the ancestors of nearly all of today's Native Americans.

The "Clovis" people were named for a site near New Mexico where the first spear points were found. The Clovis spear points found in New Mexico (**Fig. 12.2**) displayed distinctive grooves or "flutes" (two long strips removed near the base and parallel to the length of the point) where they attached to wooden shaft.[1] The Clovis spear points were modified into several varieties of fluted spear point points and have been found in much of North and South America, ranging from the Arctic to the tip of South America. As inferred from the spear points found embedded in the ribs of bison and mammoth, the "Clovis" people or Paleo-Indians were exceptional hunters of large mammals. The fauna inhabiting the plains 12,000 to 14,000 years ago was abundant and included: giant moose, along with mammoths, giant ground sloth and beaver as large as a modern bear, vast herds of straight-horned bison, caribou, musk-oxen, mastodon (mammoth's cousin), rabbits, armadillos, birds, camels, peccaries, and other animals. Central and South America were also rich game preserves including deer, antelope, cameloids (such as llama, alpaca, huanaco, vicuna), and dozens of other large vertebrate species. The economy, upon which these migrations were based, was on big-game hunting. However, soon thereafter the arrival of human hunters to the New World, the mammoths, ground sloths, and other large mammals referred to as "megafauna" went extinct. This extinction is attributed to two factors: First, the extinction was due to heavy and selective hunting by human hunters of large-bodied mammals that were both abundant and easy to kill. An extreme form of the overkill theory is the "blitzkrieg" model, in which the earlier hunters were big-game hunters, who selectively and rapidly hunted the largest species to extinction as they swept through newly colonized continents.[18] This excessive hunting resulted in rapid population growth and geographical expansion of the people in the New World, but this expansion within short periods of time resulted in widespread hunting to extinction of megafauna species. Second, extinction was affected by slow reproduction of large-size animals. In general, slowly reproducing species can rapidly become extinct because the rate at which a population

FIGURE 12.2. Types of Paleo-Indian projectile points. (a) Clovis, (b) Folson, (c) Agate basin, (d) Angostura, (e) Alberta, (f) Eden, and (g) Scottsbluff. Source: Nebraska Game and Parks Commission. Modified by the author.

can replace animals killed by hunters is low.[19] These two factors along with the climate and vegetation change at the end of the last ice age might have contributed to the extinction of large mammals such as mammoths.

In summary, the evidence supports the hypothesis that once *Homo sapiens* crossed the Beringia landbridge 12,000 to 15,000 years ago, they continued south along the Pacific coastal route and reached South American sites such as Monte Verde. The second migration, about 11,000 to 12,000 years ago, also occurred through the Beringia landbridge, but they continued south on land and reached North and South American sites.

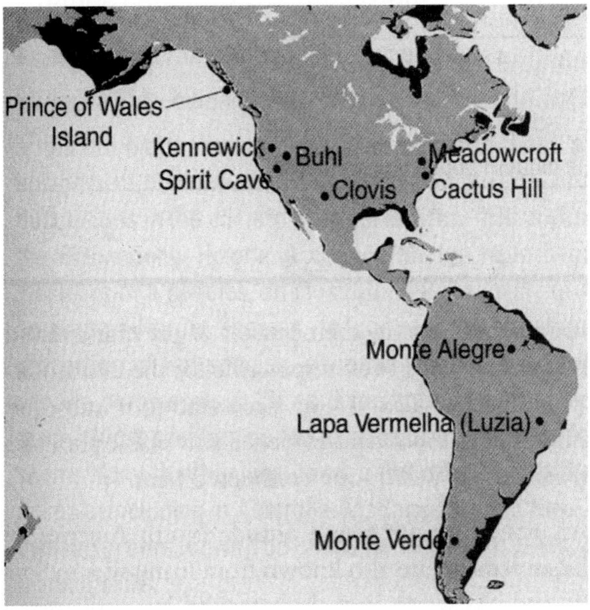

FIGURE 12.3. Locations of archaeological remains of early Native Americans. From *The Settlement of the Americas* by T. D. Dillehay. Copyright © 2000 by T. D. Dillehay. Reprinted by permission of Basic Books, a member of Perseus Books, L.L.C.

Paleo-American and Archaic Populations

Archaeologists usually refer to those ancient Americans dated to have lived around 8,000 years ago as Paleo-Indian period and those that lived after 8,000 years as archaic.[10, 11] The antiquity of the peopling of the New World is evident from the various archaeological sites found throughout the Americas (**fig. 12.2**). Radiocarbon dates of hearths, numerous stone tools and animal bones, and charcoal from sites near Pedra Furada, in eastern Brazil, indicate that the Paleo-Indians might live here between 17,000 to 22,000 years ago. Similar dating methods of crude stone tools and animal bones found in the Pikimachay Cave in Peru indicate that people might have lived here between 8,000 to 12,000 years ago.[6] Likewise, excavations at the Meadowcroft Rock Shelter near Avella, Pennsylvania, found interspersed in the many layers of alluvial sediments clear indications of human occupation, including tools such as prismatic flint blades, baskets, bones, and hearths. A bit of charred basketry from near the bottom and the charcoal were radiocarbon-dated to between 19,600 ± 2,400 and 19,150 ± 800 years ago, respectively.[7] Some archaeological sites in South America date from 12,500 years ago, which suggests that the first humans arrived at least 15,000 years ago. Hence, it is quite possible that the first colonizers arrived between 12,000 and 15,000 years ago.

Excavations at Monte Verde, in south central Chile, and radiocarbon analysis of the tools and debris found in the deep layers indicate that thirty to fifty people lived here around 13,000 years ago.[5] Monte Verde is an open-air site located in a cold, wet, forested area that was covered by a peat bog and thereby extremely well preserved (**fig. 12.3**). Bones from seven mastodons, which appear to have been killed elsewhere and carried home in large segments, were found along with a human footprint, log foundations for huts, animal skins, numerous plant remains, and many other remains.

Number of Migrations and Linguistic Groups

Based on linguistic and genetic information research, it has been postulated that the peopling of the Americas was the result of two or three migrations. Linguists have classified the Native American languages into three groupings:[2–4] Amerind, Na-Dene, and Eskimo-Aleut.

Amerind. The language spoken by the native populations of North and South America.

Na-Dene. The language spoken by people who lived on the Northwest coast of Canada and the United States.

Eskimo-Aleut. Includes languages spoken by the Eskimos and Aleutian Islanders living in Alaska, Greenland, and parts of Siberia.

It has been postulated that these three groups were the result of three distinct migrations from Siberia. According to this model, the first migration gave rise to the so-called Paleo-Indians, represented today by Amerind speakers, and two subsequent migrations originated the Na-Dene and Eskimo-Aleut.

Number of Migrations and Mitochondrial DNA

Mitochondrial DNA (mtDNA) studies show that genetic patterns cluster by geographical regions and that relationships between populations can be established from these patterns.[14–17] These studies using mitochondrial DNA conclude that all Native Americans come from four genetic lineages, labeled A through D. Significantly, Amerinds have all four lineages, Na-Dene only A, and Eskimo-Aleuts A and D—suggesting different migrations at different times.[10–15] These studies also found that the Beringian groups such as the Na-Dene, Eskimo, and Chukchi from Siberia have a close affinity and a lower mtDNA diversity (primarily haplogroup A sequences)[20, 21] than the Amerind and they have a more distant relationship to the Amerind inhabiting the rest of the continent. Hence, it has been proposed that the Amerind originated from two migrations, the first carrying the so-called Native American mtDNA haplogroups A, C, and D,[22–26] and the second mtDNA sequences from haplogroup B only. A third migration would have given rise to the Na-Dene speakers. However, these studies found that all four lineages showed up in all three language groups.[27–29] In other words, according to these studies, all the North, Central, and South America populations have all four lineages. Since it is unlikely that the same lineages and languages ended up in all these populations across two continents, their similarities were probably related to interbreeding among the four linguistic groups.[29]

Craniofacial Similarities of Asians and Native Americans

In terms of craniofacial morphology, natives of North America show ties to the Ainu of Hokkaido and Jomon from Japan. On the other hand, the Inuit (Eskimo), the Aleut, and the Na-Dene speakers show more similarities to the mainland populations of East Asia.[8] Similarly, in terms of the variability of teeth shape, such as incisor "shoveling" (refers to the curvature in the lingual surface of incisors), the Native Americans are more similar to Asians than to Europeans.[9]

Analysis of a worldwide sample of craniofacial measurements indicates that *Homo sapiens* from the Upper Cave of Zhoukoudian (UC 101 and UC 103) exhibit similarities to Paleo-American and Archaic Indian samples from North America.[12] This finding supports the hypothesis that these individuals were members of an unspecialized, pre-"Mongoloid" group that was involved in the peopling of the Americas. Recent morphometric analysis of the skeletal remains of Amerindians from early historic times, excavated from the tip of the Baja California, indicates that they show a closer affinity to the Paleo-American skulls than to modern Amerindians.[13] The Baja Amerindian and Paleo-American skulls have similar long and narrow braincases and relatively short, narrow faces, implying common ancestry with the ancient inhabitants of South Asia and the Pacific Rim. A number of early fossils have been found, of which the most prominent are the Buhl woman, Luzia, Prince of Wales Island Man, and Kennewick Man.

Buhl Woman

In 1989, the remains of a female about 17 to 21 years old were found along with an obsidian bifaced stone tool and a bone needle in a quarry near Buhl, Idaho. Dating of bone collagen gave a date of 12,000 years ago. The remains were reburied in 1991 in accordance with the Native American Graves Protection and Repatriation Act.

Luzia

In 1975, the skull of a female named "**Luzia**" was discovered in the Lapa Vermelba rock shelter in Minas Gerais, Brazil. Carbon-14 measurements gave a date of 13,500 years ago.[35, 36]

Prince of Wales Island Man

In 1996, the jaw, vertebra, and pelvis of a man were found with a stone point in a cave near the coast of Prince of Wales Island, Alaska. The remains belonged to a man who lived about 11,000 years ago. This finding provides support for the use of the coastal route for early travel to the New World.

The Kennewick Man

In 1966, the remains of a skull and an almost complete skeleton of a 40- to 45-year-old man, named Kennewick Man (**fig. 12.4a**), were discovered eroding from the banks of the Columbia River in Washington State. The radiocarbon date indicated that Kennewick Man lived about 9,300 years ago.[30–34] Because the reconstruction of the skull resembled European skeletal morphology, some argued that Kennewick Man was not ancestral to Native Americans. However, as shown in **figure 12.4b**, the forensic reconstruction of the skull does resemble living Native Americans. Besides, any difference in cranial morphology between the skeletal remains of Native Americans and living Native Americans does not necessarily point to the replacement of the oldest American populations by more recent immigrants.[37] It is a general principle that each individual and population, by adapting to the local environment, might acquire traits that differentiate them from other populations.[38] This principle is supported by the finding that the skulls from Baja California, Mexico resemble Paleo-Americans more than recent archaic samples,[13] despite the fact that they both come from the same population stock. The high de-

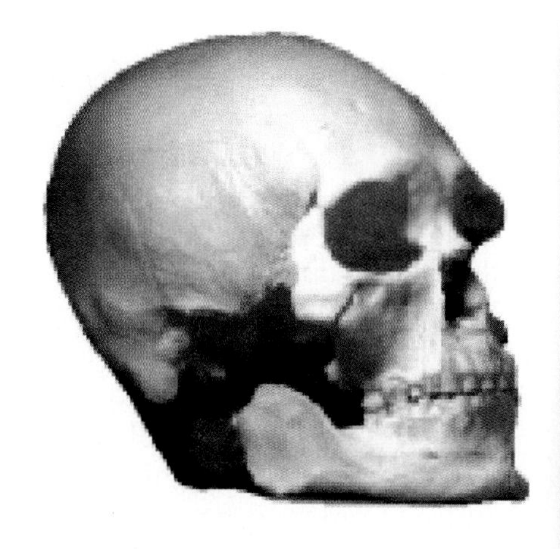

(a)

(b)

FIGURE 12.4. Paleoanthropology of Native Americans: (a) the skull of the Kennewick Man; (b) forensic reconstruction of Kennewick man; (c) Native American Sioux from Cherry Creek. Source: (a) Adapted from Dillehay ,T.D. 2003. Paleoanthropology: Tracking the first Americans. *Nature* 425:23–24; (b) Dillehay[5]. From *The Settlement of the Americas* by T. D. Dillehay. Copyright © 2000 by T. D. Dillehay. Reprinted by permission of Basic Books, a member of Perseus Books, L.L.C.

gree of morphological similarity between the Baja California samples and the Paleo-American remains has been attributed to the climatic changes during the Middle Holocene, which generated the conditions for isolation from the continent, restricting the gene flow of the original group with northern populations.[13]

In summary, the various skeletal remains of Native Americans show that there was as much interpopulation morphological variability as occurs in contemporary populations.

ABO Blood Group Difference between Asians and Native Americans

As shown in **figure 12.5**, in terms of ABO blood groups, modern native North American populations are genetically similar to Asian populations. On the other hand, native South American populations are genetically less variable than Asian populations, from which they are supposed to have descended. In general, the highest values for O found in the world are in native South American populations. Conversely, blood type B is the rarest of the three alleles, and except among the Inuit, it appears to have been completely absent in the pre-Columbian New World. In contrast, the B allele occurs in more than 20 percent of the populations of central Asia, western Siberia, central Mongolia, and the Himalayas. Likewise, the A allele occurs in more than 50 percent of the Blackfoot Indians. An important question is, why do frequencies of the ABO alleles vary so much in different populations?

The source of these differences is probably related to the influence of genetic drift, founder effect, and natural selection. Genetic drift eliminates genetic variation at a rate that depends on the size of the population. In large populations, drift removes genetic variants slowly, while in small populations drift eliminates variation more quickly. In addition, founder effect would have a high influence since the original populations that came to the New World included small bands of less than 100 people. Suppose that populations in different parts of Asia were isolated from each other they would have diverged genetically and if emigrants to the New World were drawn from one local Asian population, they would carry with them only a fraction of the total genetic variation present in Asia. As a result, the Native American descendants of this regional Asian population would be less variable than modern populations across Asia. These populations over the next 10,000 to 20,000 years continued to diverge in the New World. Natural selection would also have contributed to maintaining these preexisting differences. Studies of Asian populations indicate that the incidence of male individuals with blood type B with cardiac cancer (CC) and patients with carcinoma in the upper third esophagus (EC) was 2.3 percent and 4.7 percent higher than the corresponding controls.[39] If the increased risk of these cancers is also associated with lower reproductive fitness, it is quite possible that the maintenance of O blood type in Native American populations is indirectly related to natural selection.

Hypotheses about the Origins of the Anatomically Modern *Homo sapiens*

It is generally accepted that the human lineage evolved in Africa, but when and where this lineage became *Homo sapiens* has been the most contentious issue in anthropology over the last 20 years. The debate over the origin of *Homo sapiens* has focused on three models that are referred to as : (a) African replacement (AR), (b) multi-regional evolution (MRE), and (c) admixture of Neanderthals and anatomically modern *Homo sapiens* (ANHS). Each postulate will be addressed in this section.

The African Replacement Model

The replacement model proposes that anatomically modern *Homo sapiens* evolved in Africa about 100,000 to 200,000 years ago.[40–43] Then they migrated out into Asia and Europe replacing the several species of *archaic Homo sapiens* such as the *Neanderthals* that it encountered during its expansion (**fig. 12.5**). In other words, the Neanderthals, following the nearly 200,000 years of successful adaptation to the glacial climates of northwestern Eurasia, about 30,000 to 40,000 years ago disappeared, to be replaced by the anatomically modern *H. sapiens.* This model proposes that all current regional morphologies, especially those outside Africa, developed within the last 100,000 years. This model is based upon data derived from mitochondrial DNA (mtDNA) and fossil remains.

Mitochondrial DNA

This hypothesis is based on studies of the DNA structure of present-day human populations in different areas of the world and residual traces of "ancient" DNA extracted from a number of Neanderthal and

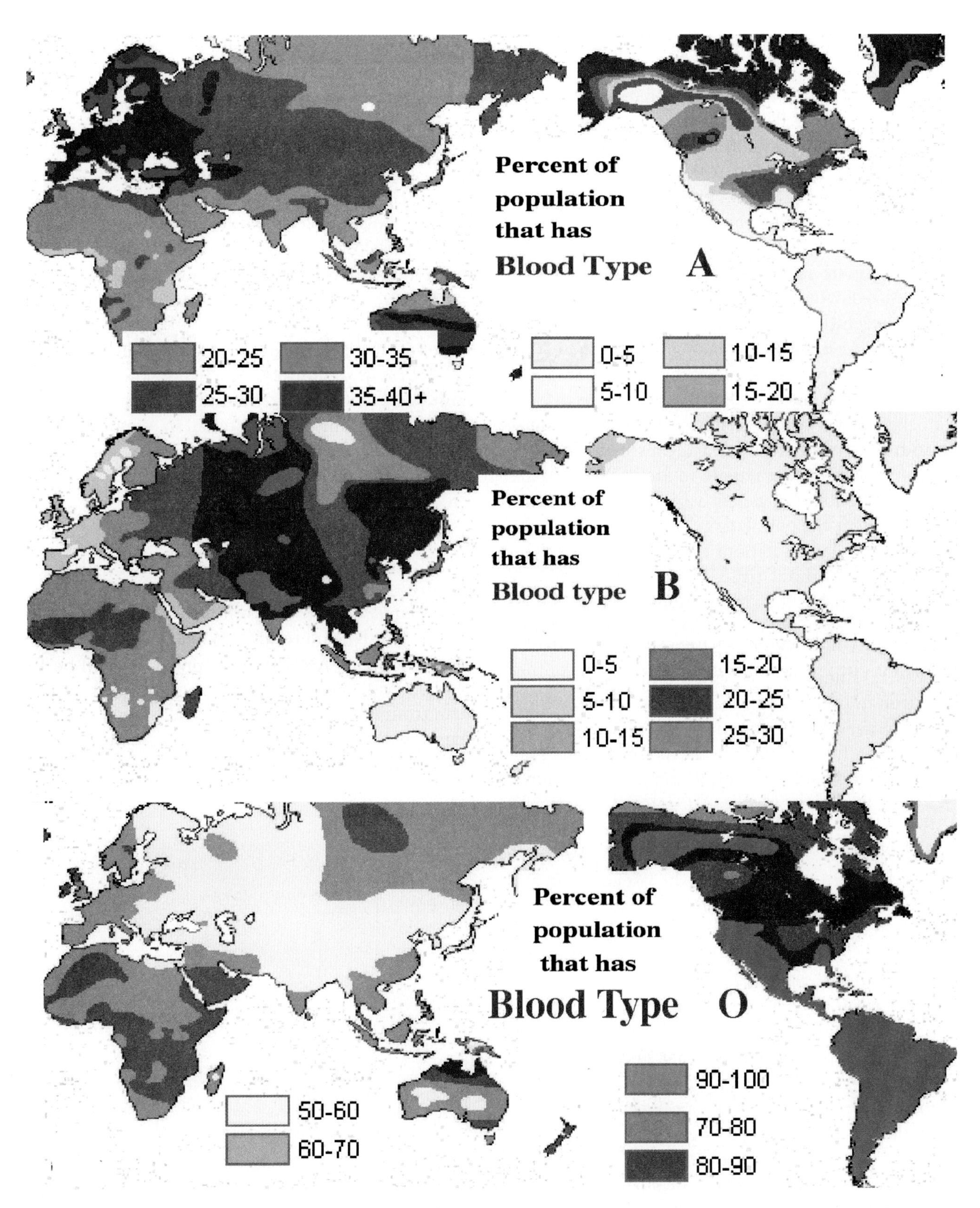

FIGURE 12.5. ABO blood group distribution. Note that the ABO blood type and especially the frequency of O blood type among North and South American Indigenous populations is markedly different from their Asian ancestors. (Composite illustration made by the author.)

early anatomically modern human remains. All mammalian cells contain small organelles called mitochondria (see chapter 2), which in addition to being the energy center of the cell also carry genetic information. On average, the mitochondria contain about 0.05 percent of the DNA present in all the chromosomes. This DNA is referred to as mitochondrial DNA (mtDNA). The mtDNA has two important unique properties. First, mtDNA is inherited maternally, because the mtDNA present in the sperm cells do not enter into the fertilized ovum. Therefore, there is no recombination between the father's and mother's mtDNA, so that a child carries the same mitochondria genes as the mother. Furthermore, within each cell there are hundreds of mitochondria, each with its own copy of each mitochondrial gene. Hence, it is easier to recover mitochondrial genes than nuclear genes from the tiny amounts of DNA that remain in fossils. Second, the greater the amount of genetic variation in mitochondria, the older the population. Third, mitochondrial genes accumulate mutations that are supposed to be constant throughout time, and as such provide a means of establishing a mitochondrial genetic clock for dating of events in the last few hundred thousand years.

Several studies indicate that the amount of genetic variation in mitochondrial and nuclear DNA within contemporary African populations is greater than the variation within European or Asian populations.[(40–43)] Therefore, the population explosion of *Homo sapiens* must have begun recently in an African population.[(40–43)] Based on analysis of the mtDNA of Neanderthal remains (dated to be between 29,000 and 40,000 years old), it has been postulated that the last common ancestor of Neanderthals and modern humans lived about 500,000 years ago and the divergences in mtDNA of modern humans occurred between 106,000 and 246,000 BP.[(44)] Analyses of mitochondrial DNA of seven separate Neanderthal specimens indicate that they are different from those of all known present-day populations in either Europe or other parts of the world, and they are equally different from those recovered from five early specimens of anatomically modern populations from European sites.[(45)] Based on these studies it is assumed that there was no interbreeding between the local Neanderthals and the modern populations in Europe. In this view, all of the archaic *H. sapiens* populations outside Africa, including Neanderthals, would be classified as belonging to a different species of *Homo* and, as such they would have evolutionary dead ends.[(40–42)]

In summary, studies of mtDNA do support the hypothesis that modern *Homo sapiens* evolved recently in Africa, then spread throughout Europe, Asia, and the New World. However, some geneticists are skeptical of the main conclusions derived from the genetic data.[(48–51)] For example, Ayala[(48)] concludes that the mitochondrial replacement (or Eve) hypothesis emanates from a confusion between gene genealogies and individual genealogies. Similarly, Adcok and associates[(51)] analyzed the mtDNA of ancient Australians that lived 10,000 and 60,000 years ago and found that all of them fall within the skeletal range of living Australians, but the late Pleistocene samples fall outside the skeletal range of contemporary indigenous Australians. Results of this study indicate that the anatomically modern humans were present in Australia before the complete fixation of the mtDNA lineage now found in all living people.[(51)]

Fossil Remains

The African replacement model is supported by a recent discovery of the remains of anatomically modern *Homo sapiens* in the Herto Bouri Formation in the Middle Awash of Ethiopia by White and associates.[(52)] These fossils are dated to have lived about 154,000 to 160,000 years ago and have been considered as the form from which all morphologically "modern" humans are derived.[(52)] The form and shape of the brow ridges are intermediate between those seen in Nean-

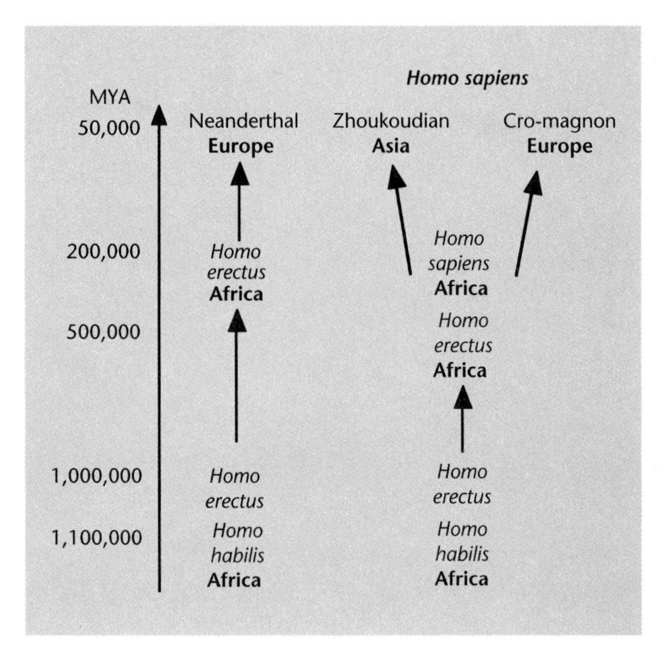

FIGURE 12.6. A simplified schematization of the African replacement model.

derthals and early modern *Homo sapiens* such as those from Broken Hill (Africa) and Skhul (Israel). In other words, the heavy brow ridges that characterized Neanderthals persists among the early modern *Homo sapiens.* Morphometric analysis of craniofacial form indicates that the anatomically modern *H. sapiens* differ in the shape of the cranial base angle, cranial fossae length and width, and facial length from archaic *Homo sapiens.*[46] Evaluations of genetic divergence and differentiation in the craniometric data between Neanderthals and Upper Paleolithic modern humans inferred that the Neanderthals are taxonomically different from modern humans as suggested by proponents of the replacement model.[47]

A unique feature of dental development is that the crown enamel and the underlying dentine is marked by periodic disruptions that cause the formation of a series of horizontal ridges known as **perikymata (fig. 12.7)**. Measurement of this perikymata can provide a record of dental development. Analysis of the frequency of perikymata from large numbers of anterior teeth (incisors and canines) of samples of *archaic Homo sapiens* including Neanderthals and Upper Paleolithic Mesolithic samples of *Homo sapiens* indicate that the dental development (measured by the much wider spacing of the perikymata in the bottom half of the crown) of Neanderthals is faster than in *Homo sapiens* and is similar to that of the early Homo species. That is, the Neanderthals might have had a different developmental trajectory of dental development than modern *Homo sapiens.*[53]

In summary, the recent discovery of modern *Homo sapiens* who lived between 154,000 to 160,000 years ago in Ethiopia, Africa[52] supports the replacement hypothesis. However, metric comparison of the Ethiopian skull and European Neanderthal according to Wolpoff and Lee[54] show that the Europeans are *more similar* to the Ethiopians over time, not less similar. Likewise, Brace,[55] superimposing the traces of the skull of Ethiopian and European Neanderthals (La Ferrassie) and modern *Homo sapiens* (**fig. 12.8**), finds that the Neanderthal requires less modification to transform into a modern *Homo sapiens* form than the Ethiopian skull. These trends suggest that the Ethiopian African fossil, even though it appeared before the modern *Homo sapiens*, already exhibited the facial morphology that characterizes Neanderthals.

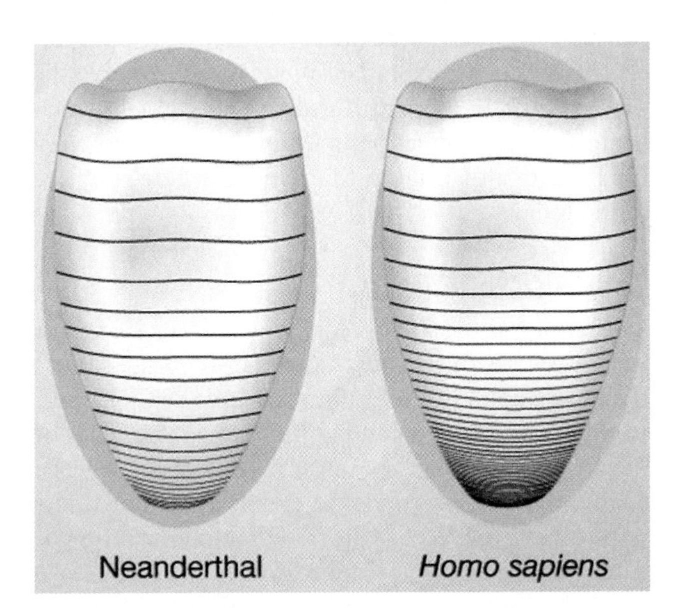

FIGURE 12.7. Representations of the horizontal ridges, or perikymata, caused by brief, periodic disruptions in enamel deposition of the incisor surface of a Neanderthal and Paleolithic *Homo sapiens.* As inferred by the wider space between horizontal ridges and less densely packed perikymata towards the base of incisors, Neanderthals' dental development and crown formation was more rapid, and the overall duration of crown formation shorter, than in *H. sapiens.* Source: Adapted from Ramirez Rozzi and Bermudez de Castro, fig. 1.[77]

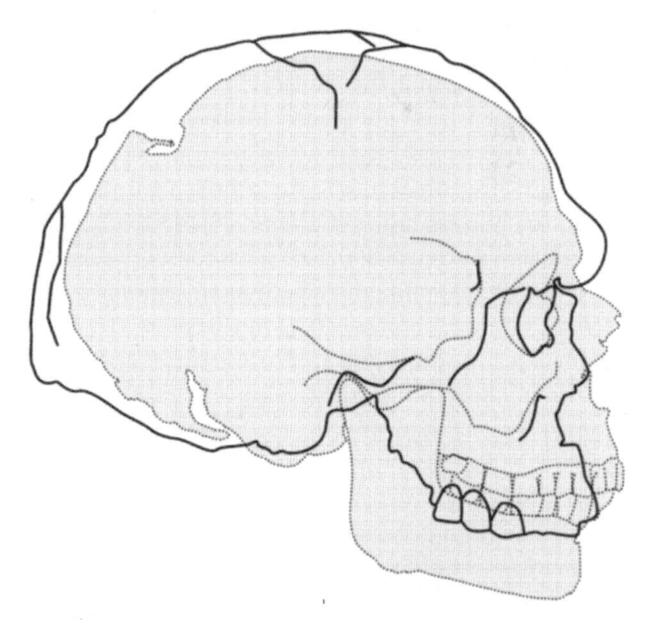

FIGURE 12.8. Comparison of the tracing of the Ethiopian Herto skull superimposed on the Neanderthal skull of La Ferrassie. Source: Brace, C. L. 2004. *"Neutral theory" and the dynamics of the evolution of "modern" human morphology* (unpublished manuscript). Courtesy of Professor C. Loring Brace, University of Michigan, Ann Arbor.

Multiregional Evolution

Fossil Evidence

The multiregional evolution (MRE) hypothesis was first put forward by Weidenreich[56, 57] and Brace.[58–62] According to Weidenreich,[56, 57] multiple lineages were evolving toward modern *Homo sapiens* in several regions of the world. These regional lineages preserved regional continuity in terms of morphology but were interconnected to prevent speciation. In other words, the descendants of nonmodern fossils of any geographical region has evolved into the modern local populations. This inference was based on the realization that the ways of life of prehistoric hominid contemporaries were essentially identical from one portion of the world to another, and that adjacent groups were almost certainly related to each other throughout the entire extent of occupation. Furthermore, according to Brace[58–62] the fact that the inhabitants of Asia were pursuing the same kind of subsistence strategy reflected in similar stone technology and they adopted technological innovations that had initially arisen elsewhere indicates that they were not a separate species.[54–58] Therefore, following the initial occupation by *Homo erectus* in Africa, Asia, and Europe there was a "regional" or "in situ continuity" through time in each of the major regions of the Old World.[58–62]

Wolpoff and associates,[63–67] like Weidenreich and Brace, propose that the transformation to anatomically modern humans occurred in situ in different parts of the world (**fig. 12.9**). That is, there was no single geographical origin for modern humans but that, after the radiation of *Homo erectus* from Africa into Europe and Asia 800,000 to 1.8 million years ago, there were independent transitions in regional populations from *H. erectus* to *Homo sapiens*. In other words, local populations in Europe, Asia, and Africa continued their indigenous evolutionary development from archaic *H. sapiens* into anatomically modern humans.[63–67] The unity of the species was maintained by some gene flow (migration) between archaic populations. Through gene flow and local selection, local populations would not have evolved totally independently from one another, and such mixing would have "prevented speciation between the regional lineages and thus maintained human beings as a single, although obviously polytypic species throughout the Pleistocene."[67] Archaeological studies in Europe and the Far East show considerable overlap in morphological and technological traits among the various archaic forms linking the fossil and modern populations of any geographical region. This model is supported

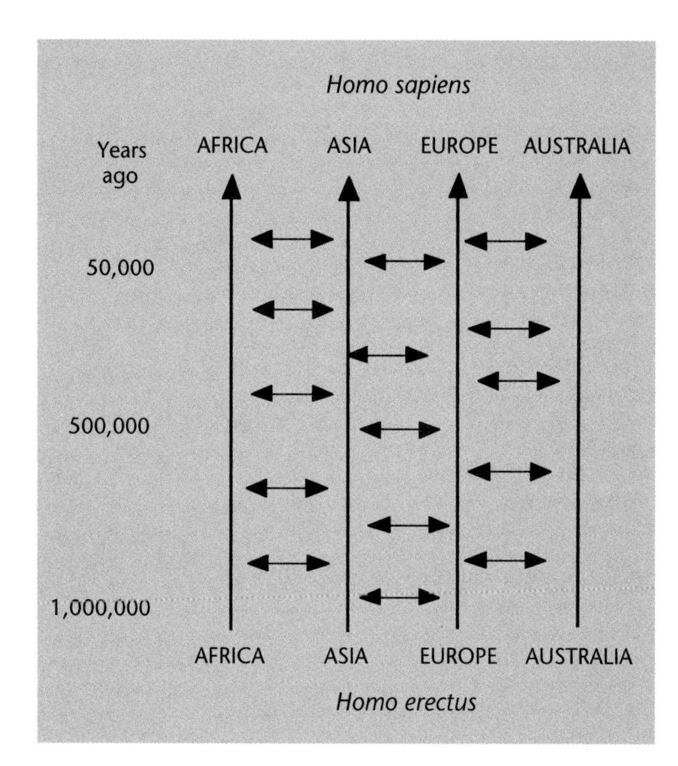

FIGURE 12.9. A simplified schematization of multiregional evolution model.

primarily by the continuity of certain morphological traits in the fossil record. For example, the robust cheekbones observed in *H. erectus* fossils from Southeast Asia and in modern Australian aborigines, which indicate that modern populations evolved over very long periods of time in the regions where they are found today.

Tool Technology

Evaluations of the stone technology indicate that there was extensive cultural exchange between the Mousterian technology of Neanderthals and the Chatelperronian technology of the upper Paleolithic tools of modern humans.[68–70] Specifically, a number of modern features of technology such as simple bone tools and a number of grooved or perforated animal-tooth pendants have been found among some of the final Neanderthal communities of central and western Chatelperronian levels in Central France. The appearance of these technologies coincides closely with the appearance of early Aurignacian *Homo sapiens* populations in the nearby regions of central Europe. As a consequence of this exchange, new "transitional" tools were produced. Naturally then, one would expect that along with the cultural exchange, there was also gene flow between Neanderthals and the anatomically modern *Homo sapiens.* That is, early and archaic humans in Europe and the Middle East contributed some of their genes to modern *Homo sapiens* populations. Reflecting this exchange contemporary humans do exhibit, albeit at low frequency, some of the morphological traits that characterized Neanderthals, such as heavy brow ridges and shovel-shaped incisors.[65–67] In other words, the transition from archaic to modern humans was a gradual and complex process of hybridization and continuity of some morphological traits.[71]

DNA

Recently Templeton[72] analyzed the sequences of DNA from ten genetic regions on regular chromosomes, sex chromosomes, and mitochondria. Based upon the observation of the variation in genes from different places and the pattern of particular mutations, he inferred that there were not changes in the mtDNA and Y-DNA distributions before, during, or after the last expansions (i.e., 100,000 years ago). Therefore, the occasional major movements out of Africa and out of Asia acted to extend the geographical range of the human species and to establish additional areas linked by enhanced gene interchange through interbreeding.[72] Furthermore, the DNA sequences in all the genetic systems studied among African and Eurasian populations are quite similar and overlap considerably, which suggests that over the last half million years they were linked by recurrent gene flow and not replacement.[72]

Admixture of Neanderthals and Anatomically Modern *Homo sapiens*

The admixture (ANHS) model proposes that Africa was the original source of anatomically modern *Homo sapiens*, but there was interbreeding with archaic *Homo sapiens* or Neanderthals.[64] Evaluations of the stone technology indicate that there was extensive cultural exchange between the Mousterian technology of Neanderthals and the Chatelperronian technology of the Upper Paleolithic tools of modern humans.[68–70] Specifically, a number of modern features of technology such as simple bone tools and a number of grooved or perforated animal-tooth pendants have been found among some of the final Neanderthal communities of central and western Chatelperronian levels in Central France. The appearance of these technologies coincide closely with the appearance of early Aurignacian *Homo sapiens* populations in the nearby regions of central Europe. As a consequence of this exchange new "transitional" tools were produced. Naturally then, one would expect that, along with the cultural exchange, there was also interbreeding or gene flow between Neanderthals and the anatomically modern *Homo sapiens.*[71] That is, early and archaic humans in Europe and the Middle East contributed some of their genes to modern *Homo sapiens* populations. Analysis of mtDNA lineages indicate that due to admixture, some of our genes could have come through multiple archaic populations and not just down a single lineage in Africa.[73] According to carbon-14 dating of fossils and artifacts, Neanderthals and modern humans about 28,000 years ago coexisted for several thousand years in western Europe. If Neanderthals coexisted and traded cultural items, they may have also mated with the anatomically modern *Homo sapiens.* Reflecting this exchange contemporary humans do exhibit, albeit at low frequency, some of the morphological traits that characterized Neanderthals such as heavy brow ridges and shovel-shaped incisors.[65–67] In other words, the anatomically modern *Homo sapiens* is the product of a complex process of admixture of archaic and modern biological traits.

Summary

The evidence supports the hypothesis that the peopling of the Americas involved two or three migrations of populations that came from Asia and across the Beringia landbridge 8,000 to 15,000 years ago. These migrations probably included small bands of people, who, after crossing the Beringia landbridge, continued traveling inland, while others continued south through the coast area, both reaching North and South America. As they populated the continent, these small bands of people had different lifestyles and technologies. Some populations not only hunted big game but also exploited a wide range of plant and animal life.

Based upon information from fossils, archaeological remains, the study of modern human genetic variations, and the analysis of ancient DNA, two hypotheses have been advanced: the replacement model (or recent out of Africa) hypothesis and the multiregional evolution hypothesis. The replacement model maintains that modern humans originated in Africa about 100,000 to 200,000 years ago, with populations subsequently spreading globally, replacing other species of archaic *Homo* that it encountered during its expansion. In contrast, the multiregional hypothesis postulates that the transition from *Homo erectus* to *Homo sapiens* occurred in a number of places in the Old World, with diverse modern human traits arising at different times and in different places. The hypothesis that there was admixture between Neanderthals and the anatomically modern *Homo sapiens* is supported by archeological and genetic information. If that is the case, some of the Neanderthal genes are among us.[74]

Given the limited amount of information and the diversity of interpretations of the data, the debate about whether the origin of modern *Homo sapiens* is attributed to a recent migration out of Africa or the result of gradual evolution occurring in Africa, Asia, and Europe will continue for many years. What is clear is that humans belong to one single species that is continuously evolving and adapting to the changing environment. That topic is addressed in the next chapter.

Suggested Readings

Crawford, M. H. 1998. *The Origins of Native Americans: Evidence from Anthropological Genetics.* New York: Cambridge University Press.

Dillehay, T. D. 2001. *The Settlement of the Americas: A New Prehistory.* New York: Basic Books.

Templeton, A. R. 2002. Out of Africa again and again. *Nature* 416: 45–51.

Wolpoff, M. H., and R. Caspari. 1997. *Race and Human Evolution: A Fatal Attraction.* New York: Simon and Schuster.

Literature Cited

1. Figgins, J. D. 1927. The antiquity of man in America. *Natural History* 27:229–239.
2. Greenberg, J. H., C. G. Turner, II, and S. L. Zegura. 1986. The settlement of the Americas: A comparison of the linguistic, dental, and genetic evidence. *Current Anthropology* 27:477–496.
3. Nichols, J. 1990. Linguistic diversity and the first settlement of the New World. *Language* 66:475–521.
4. Petit, C. 1998. Rediscovering America: The New World may be 20,000 years older than experts thought. *U.S. News and World Report* (October 12):56–64.
5. Dillehay, T. D. 2001. *The Settlement of the Americas: A New Prehistory.* New York: Basic Books.
6. MacNeish, R. S. *The Science of Archaeology.* North Scituate, MA: Duxbury Press.
7. Adovasio, J. M., J. Donahue, and R. Stuckenrath. 1990. The Meadowcroft Rockshelter radiocarbon chronology, 1975–1990. *American Antiquity* 55:348–354.
8. Brace, C. L., A. R. Nelson, N. Seguchi, H. Oe, L. Sering, P. Qifeng, L. Yongyi, and D. Tumen. 2001. Old World sources of the first New World human inhabitants: A comparative craniofacial view. *Proc. Natl. Acad. Sci. USA* 98:10017–10022.
9. Turner, C. G., II. 1987. Tell-tale teeth. *Natural History* 96(l):6–10.
10. Bamforth, D. B. 1988. *Ecology and Human Organization on the Great Plains.* New York: Plenum Press.
11. Fagan, B. M. 1995. *Ancient North America: The Archaeology of a Continent.* London: Thames & Hudson.
12. Cunningham, D. L., and R. L. Jantz. 2003. The morphometric relationship of Upper Cave 101 and 103 to modern *Homo sapiens. J. Hum. Evol.* 45:741–758.
13. Gonzalez-José R. I., A. Gonzalez-Martın, M. Hernandez, H. M. Pucciarelli, M. Sardi, A. Rosales, and S. Van der Molen. 2003. Craniometric evidence for Palaeoamerican survival in Baja California. *Nature* 425:62–65.
14. Barbujani, G., and A. Pilastro. 1993. Genetic evidence on origin and dispersal of human populations speaking languages of the Nostratic macrofamily. *Proc. Natl. Acad. Sci. USA* 90:4670–4673.
15. Crawford, M. H., J. McComb, M. S. Schanfield, et al. 2002. Genetic structure of Siberian populations: The Evenki of Central Siberia and the Kizhi of Gorno Altai. In *Human Biology of Pastoralist Populations,* W. R. Leonard, and M. H. Crawford, ed., 10–49. Cambridge, UK: Cambridge University Press,.
16. Ruhlen, M. 1998. The origin of the Na-Dene. *Proc. Natl. Acad. Sci. USA* 95:13994–13996.
17. Rubicz, R., K. L. Melvin, and M. H. Crawford. 2002. Genetic evidence for the phylogenetic relationship between Na-Dene and Yeniseian speakers. *Human Biology* 74.6:743–760.
18. Cavalli-Sforza, L. L., and M. W. Feldman. 1981. *Cultural Transmission and Evolution.* Princeton, NJ: Princeton University Press.
19. Chen, J., R. R. Sokal, and M. Ruhle. 1995. Worldwide analysis of genetic and linguistic relationships of human populations. *Hum. Biol.* 67:595–612.
20. Howell, F. C. 1966. Observations on the earlier phases of the European lower Paleolithic. *Am. Anthropol.* 68:88–201.
21. Shields, G. F., A. M. Schmiechen, B. L. Frazier, A. Redd, M. Voevoda, J. K. Reed, and R. H. Ward. 1993. mtDNA sequences suggest a recent evolutionary divergence for Beringian and northern North American populations. *Am. J. Hum. Genet.* 50:758–765.
22. Szathmary, E. J. 1993. mtDNA and the peopling of the Americas. *Am. J. Hum. Genet.* 53:793–799.
23. Torroni, A., T. G. Schurr, C. C. Yang, E. J. Szathmary, R. C. Williams, M. S. Schanfield, G. A. Troup, W. C. Knowler, D. N. Lawrence, K. M. Weiss, and D. C. Wallace. 1992. Native American mitochondrial DNA analysis indicates that the Amerind and the Nadene populations were founded by two independent migrations. *Genetics* 130:153–162.
24. Torroni, A., J. V. Neel, R. Barrante, T. G. Schurr, and D. C. Wallace. 1994. Mitochondrial DNA "clock" for the Amerinds and its implications for timing their entry into North America. *Proc. Natl. Acad. Sci. USA* 91:1158–1162.
25. Bonatto, S. L., and F. M. Salzano. 1997. A single and early migration for the peopling of the Americas supported by mitochondrial DNA sequence data. *Proc. Natl. Acad. Sci. USA* 94:1866–1871.

26. Szathmary, E. J. 1993. mtDNA and the peopling of the Americas. *Am. J. Hum. Genet.* 53:793–799.
27. Merriwether, D., and R. E. Ferrell. 1996b. The four founding lineage hypothesis for the New World: A critical reevaluation. *Molecular Phylogenetics and Evolution* 5:241–246.
28. Merriwether, D., F. Rothhammer, and R. Ferrell. 1994. Genetic variation in the New World: Ancient teeth, bone, and tissue as sources of DNA. *Experientia* 50:592–601.
29. Merriwether, D., F. Rothhammer, and R. Ferrell. 1995. Distribution of the four founding lineage haplotypes in Native Americans suggests a single wave of migration for the New World. *Am. J. Phys. Anthropol.* 98:411–430.
30. Holden, C. 1999a. Kennewick Man gets his day in the lab. *Science* 283:1239–1240.
31. Holden, C. 1999b. Australasian roots proposed for "Luzia." *Science* 286:1467.
32. Holden, C. 2000a. More tests for K Man. *Science* 287:963.
33. Holden, C. 2000b. Bones decision rattles researchers. *Science* 289:2257.
34. Morell, V. 1998. Kennewick Man's trials continue. *Science* 280:190–192.
35. Munford, D., M. C. Zanini, and W. A. Neves. 1995. Human cranial variation in South America: Implications for the settlement of the New World. *Braz. J. Genet.* 18:673–688.
36. Neves, W. A. 2000. Luzia is not alone. *Science* 287:974–975.
37. van Vark, G. N., D. Kuizenga, and F. L. E. Williams. 2003. Kennewick and Luzia: Lessons from the European Upper Paleolithic. *Am. J. Phys. Anthropol.* 114:146–155.
38. Frisancho, A. R. 1993. *Human Adaptation and Accommodation.* Ann Arbor, MI: University of Michigan Press.
39. Su, M., S. M. Lu, D. P. Tian, H. Zhao, X. Y. Li, D. R. Li, and Z. C. Zheng. 2001. Relationship between ABO blood groups and carcinoma of esophagus and cardia in Chaoshan inhabitants of China. *World J. Gastroenterol.* 7:657–661.
40. Cann, R. L., M. Stoneking, and A. C. Wilson. 1987. Mitochondrial DNA and human evolution. *Nature* 325:31–36.
41. Stoneking, M., and H. Soodyall. 1996. Human evolution and the mitochondrial genome. *Curr. Opin. Genet. Dev.* 6:731–736.
42. Vigilant, L., M. Stoneking, H. Harpending, K. Hawkes, and A. C. Wilson. 1991. African populations and the evolution of human mitochondrial DNA. *Science* 253:1503–1507.
43. Tishkoff, S. A., and S. M. Williams. 2002. Genetic analysis of African populations: Human evolution and complex diseases. *Nature Reviews Genetics* 3:611–621.
44. Ovchinnikov, I. V., A. Götherström, G. P. Romanova, V. M. Kharitonov, K. Liden, and W. Goodwin. 2000. Molecular analysis of Neanderthal DNA from the northern Caucasus. *Nature* 404:490–493.
45. Serre, D., A. Langaney, M. Chech, M. Teschler-Nicola, M. Paunovic, P. Mennecier, M. Hofreiter, G. G. Possnert, and S. Paabo. 2004. No evidence of Neanderthal mtDNA contribution to early modern humans. *PLoS Biol* 2:313–317.
46. Lieberman, D. E., B. M. McBratney, and G. Krovitz. 2002. The evolution and development of cranial form in *Homo sapiens. Proc. Natl. Acad. Sci. USA* 99:1134–1139.
47. Schillaci, M. A., and J. W. Froehlich. 2001. Nonhuman primate hybridization and the taxonomic status of Neanderthals. *Anthropol.* 115:157–166.
48. Ayala, F. J. 1995. The myth of Eve: Molecular biology and human origins. *Science* 270:1930-1936.
49. Templeton, A. R. 1992. Human origins and analysis of mitochondrial DNA sequences. *Science* 255:737–740.
50. Relethford, J. H. 2001. *Genetics and the Search for Modern Human Origins.* New York: Wiley.
51. Adcock, G. J., E. S. Dennis, S. Easteal, G. A. Huttley, L. Jermiin, W. J. Peacock, and A. Thorne. 2001. Mitochondrial DNA sequences in ancient Australians: Implications for modern human origins. *Proc. Natl. Acad. Sci. USA* 98:537–542.
52. White, T. D., B. Asfaw, D. DeGusta, H. Gilbert, G. D. Richards, G. Suwa, and F. C. Howell. 2003. *Pleistocene Homo sapiens* from Middle Awash, Ethiopia. *Nature* 423:742–747.
53. Ramirez Rozzi, F. V., and J. M. Bermudez de Castro. 2004. Surprisingly rapid growth in Neanderthals. *Nature* 428:936–939.
54. Wolpoff, M., and S. H. Lee. 2004. Herto and the Neanderthals. (Unpublished manuscript.)
55. Brace, C. L. 2004. "Neutral Theory" and the dynamics of the evolution of "modern" human morphology. (Unpublished manuscript.)
56. Weidenreich, F. 1943. The "Neanderthal Man" and the ancestors of "*Homo sapiens.*" *Amer. Anthropol.* 47:39–48.
57. Weidenreich, F. 1946. *Apes, Giants, and Man.* Chicago: Univ. Chicago Press.
58. Brace, C. L. 1964. The fate of the "classic" Neanderthals. A consideration of hominid catastrophism. *Current Anthropol.* 5:3–43.
59. Brace, C. L. 1967. *The Stages of Human Evolution: Human and Cultural Origins.* Englewood Cliffs, NJ: Prentice-Hall.
60. Brace, C. L. 1979. Krapina, "classic" Neanderthals, and the evolution of the European face. *J. Hum. Evol.* 8:527–550.
61. Brace, C. L. 1995. *The Stages of Human Evolution.* New Jersey: Prentice-Hall.
62. Brace, C. L. 1998. An anthropological perspective on "race" and intelligence: The non-clinal nature of human cognitive capabilities. *J. Anthropol. Res.* 55:245–264.
63. Thorne, A. G., and M. H. Wolpoff. 1992. The multiregional evolution of humans. *Sci. Am.* 266:76–79.
64. Smith, F. H., A. B. Falsetti, and S. M. Donnelly. 1989. Modern human origins. *Yb. Physical Anthrop.* 32:35–68.
65. Wolpoff, M. H., J. Hawks, and R. Caspari. 2000. Multiregional, not multiple origins. *Am. J. Phys. Anthropol.* 112:129–136.
66. Hawks, J. D., and M. H. Wolpoff. 2001. The accretion model of Neanderthal evolution. *Evolution Int. J. Org. Evolution.* 55:1474–1485.
67. Wolpoff, M. H., J. Hawks, D. W. Frayer, and K. Hunley. 2001. Modern human ancestry at the peripheries: A test of the replacement theory. *Science* 291:293–297.
68. Bahn, P. G. 1998. Neanderthals emancipated. *Nature* 394:719–721.

69. Hublin, J. J., F. Spoor, M. Braun, F. Zonneveld, and S. Condemi. 1996. A late Neanderthal associated with Upper Palaeolithic artifacts. *Nature* 381:224.

70. d'Errico, F., J. Zilhão, M. Julien, D. Baffier, and J. Pelegrin. 1998. Neanderthal acculturation in Western Europe? A critical review of the evidence and its interpretation. *Current Anthropol.* 39:1–44.

71. Brauer, G. 1984. A craniofacial approach to the origin of anatomically modern *Homo sapiens* in Africa and implications for the appearance of modern humans. In *The Origins of Modern Humans*, eds. F. H. Smith, and F. Spencer, 327–410. New York: Liss.

72. Templeton, A. R. 2002. Out of Africa again and again. *Nature* 416:45–51.

73. Harding, R. M. and G. McVean. 2004. A structured ancestral population for the evolution of modern humans. *Curr. Opin. Genet. Dev.* 14:667–674.

74. Sarich, V. M. 1980, A macromolecular perspective on "The Material Basis of Evolution." *Experientia Supp.* 35: 27–31.

Diet and Brain Evolution

One of the biological traits that contributed to humans' success as a species was the relatively large and complex brain. Understanding the origins of this large and complex brain that distinguishes humans from other primates presents one of the great new challenges in biological anthropology. This chapter will focus on the development of and the hypotheses to explain the evolution of the hominid brain.

Brain, Body Size, and Energy

Brain Complexity

The human brain, like that of other primates, consists of the cerebral cortex, brain stem, and cerebellum (**fig. 13.1**). The cerebral cortex, the outermost layer of the brain, is composed of gray matter about 2–4 mm (0.080–0.16 in.) thick and contains billions of nerve cell bodies. The cerebral cortex is folded and convoluted, forming bulges called gyrus (pl. gyri) and deep grooves between folds called sulcus (pl. sulci). The cerebral cortex is divided into four sections, called "lobes": the frontal lobe, parietal lobe, occipital lobe, and temporal lobe. The cerebrum is the upper main part of the brain and constitutes the largest component of the brain. It is divided into a right and a left hemisphere. The brain stem is the lower extension of the brain where it connects to the spinal cord. It is the pathway for all fiber tracts passing up and down from peripheral nerves and spinal cord to the highest parts of the brain. Because it connects the brain with the rest of the body, all information to and from our bodies passes through the brain stem on the way to and from the brain. The cerebellum, or "little brain," is located just above the brain stem and toward the back of the brain. It is similar to the cerebrum in that it has two hemispheres and a highly folded surface, or cortex. It is involved in the coordination of voluntary motor movement, balance and equilibrium, and muscle tone.

The cerebral cortex includes three major areas involved in speech and in communication: Broca's area, Wernicke's area, and the planum temporale. Broca's area is present only in humans and is located on the frontal lobe of the cerebral cortex.

These areas have been correlated with the process of language receptiveness and in both spoken and gestural aspects of human communication. All these areas are characterized by a left-right asymmetry. That is, the left hemisphere is proportionally larger than the right. The left hemisphere usually controls right-handedness and processes involved in speech and communication.

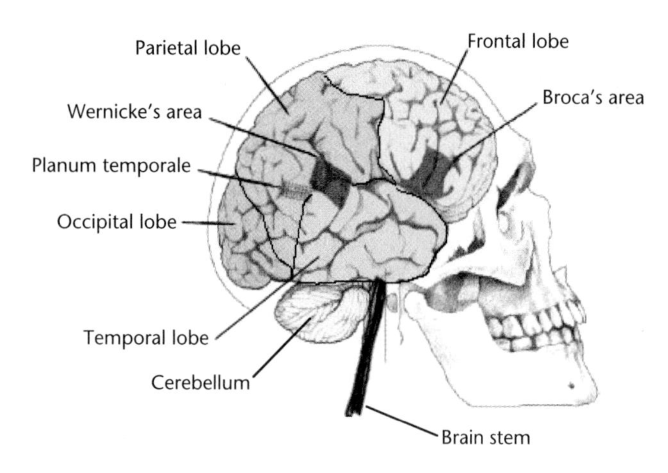

FIGURE 13.1. External lateral view of a modern human brain. (Composite made by the author.)

Allometry of Brain Size

When a given body part varies in direct proportion to body size, it is referred to as an **isometric** trait. For example, the size of the lungs or the size of the heart varies in direct proportion to body size, as does the volume of blood circulating in the body or the weight of the skeleton. However, many parts do not vary in direct proportion to body size. For example, in primates, the size of the eyes does not increase in direct proportion to body size. In other words, when the proportional relationship between a given characteristic and body size does not change progressively with increasing body size, it is referred to as an **allometric** trait. Therefore, when comparing the body parts of animals that differ in body size and different adaptations, the scaling effects of body size need to be taken into account. One productive way of taking into account the scaling effects of body size is to apply the allometric formula:

$$Y = K X^{ê}$$

Where Y is brain weight, K is the scaling constant factor, X is body weight, and ê is an exponent.

Because the values of X and Y vary a lot, it is often convenient to transform them into logarithmic scales, so that the equation becomes:

Log Y (brain weight (grams)) = ê log X (body weight (grams)) + log k.

Applying these principles several equations have been derived to calculate brain from a given body weight in mammals and primates.[1,2] The equations to calculate the expected brain size for a given body for mammals in general and for primates in particular are given in the following two equations:

Expected brain weight for mammals (1)
Log brain weight (grams) = 0.755 log body weight (kg) + 1.774

Expected brain weight for primates (2)
Log brain weight (grams) = 0.755 log body weight (kg) + 2.061

Encephalization Quotient (EQ) in Humans versus Other Mammals

The ratio between the predicted (**fig. 13.2**) and actual value of brain weight for a given individual is called the encephalization quotient (EQ).[2] For example, the observed brain weight for an individual weighing 50 kg averages 1350 grams. Applying the equation for mammals the expected brain weight (grams) for a 50 kg individual would equal 205 grams [$Y = 0.755 \times \log 50000 + 1.774 = 5.321$; e^X of 5.3217= 204.7 grams]. Similarly, using the equation for primates this individual would be expected to have a brain weight of 273 grams [$Y = 0.755 \times \log 50000 + 2.061 = 5.608$; e^X of 5.6087= 272.8 grams]. Therefore, this individual has an EQ of 6.6 (1350/205 = 6.6) and 4.9 (1350/273 = 4.9) when compared to mammals and nonhuman primates, respec-

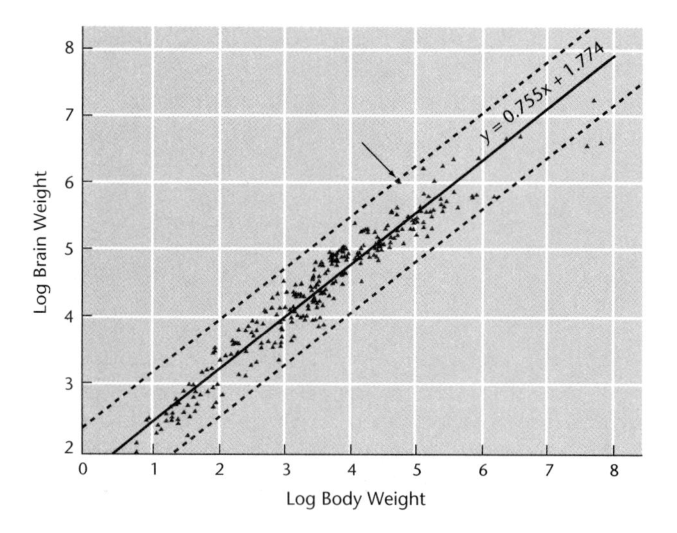

FIGURE 13.2. Logarithmic relationship of brain weight (in mg) against body weight (in grams) for 309 placental mammal species. Note that primates for their body size have larger brains than other mammals. (Adapted from figure 3, Martin, R. D. 1983. *Human Brain Evolution in an Ecological Context.* New York: American Museum of Natural History.)

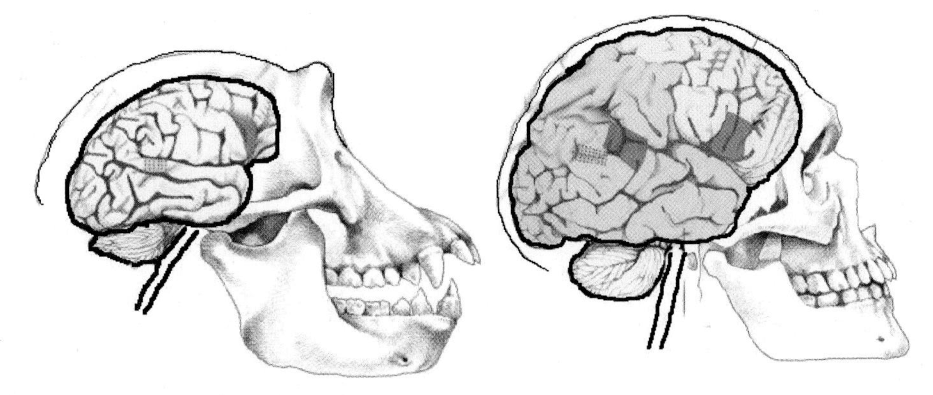

FIGURE 13.3. Human and chimpanzee brain. Although humans and chimpanzees have similar body sizes, the human brain is about three times larger than the chimpanzee. (Composite figure made by the author.)

tively. In other words, this individual for its body weight has a brain size that is 6.6 and 4.9 times larger than other mammals and nonhuman primates. This difference is illustrated in **figure 13.3**, which shows that humans' brain size is about three times larger than that of chimpanzees with similar body weight.

Resting Metabolic Rate

An individual's total daily energy needs include the energy to maintain basal metabolism, the energy required to carry on with physical activity, as well as the energy released when consuming foods. The basal metabolic rate (BMR) represents the largest portion of daily energy expenditure (60 to 75 percent). It is the minimum level of energy required to sustain the body's vital functions. These needs include the energy to maintain the nervous system's activity; ventilate the lungs and keep the heart pumping to circulate the blood; maintain minimal levels of protein synthesis, hormone production, glandular secretion, nutrient uptake, and waste excretion; maintain muscle tonus; and generally keep the internal environment of the body, including all its cells, in a functioning biochemical condition. The BMR is usually measured under fasting conditions. When the metabolic rate is measured under resting conditions, it is referred to as resting metabolic rate (RMR) or resting energy expenditure (REE). The RMR is usually measured three or four hours after a meal and under resting conditions of thermal neutrality. In general, the RMR is basically a combination of basal energy needs, plus the thermic effect of food, plus a small amount of energy needed to perform the most basic sedentary activities, such as sitting quietly. However, for most purposes the terms basal metabolic rate and resting metabolic rate are used interchangeably.

Human Brain and Its Metabolic Requirements

In general, the larger the animal, the more energy it needs to maintain its bodily functions; so that the bigger the mass, the higher the energy demands. In general, the basal metabolic rate accounts for the largest proportion of energy needs. Traditionally, the relationship between body mass and basal metabolic requirements has been described by the Kleiber's equation[5]:

$$\text{BMR (Klcal/day)} = 70\ \text{Wt}^{0.75}$$

That is, in mammals the basal metabolic energy (BMR) varies in relation to body mass raised to a power of 0.75. As illustrated by **figure 13.4**, humans and primates conform to the Kleiber relationship of body mass and metabolic rate. On the other hand, the relationship of resting metabolic rate (RMR) and brain size in humans is above average and departs substantially from the expected relationship of RMR and brain size. This means that humans expend a relatively larger proportion of resting metabolic rate on brain metabolism than other primates do.

It is estimated that during the fetal stage of prenatal development and by birth, brain metabolism accounts for upwards of 70 percent of RMR, and during infancy (body weight 10 kg), brain metabolism accounts for 50 percent of RMR. At adulthood (weight 70 kg), brain metabolism represents 20 to 25 percent of RMR.[4] In contrast, in other primates the energy expenditure of the brain accounts for less than 10 percent of the RMR.[4] **(See table 13.1.)**

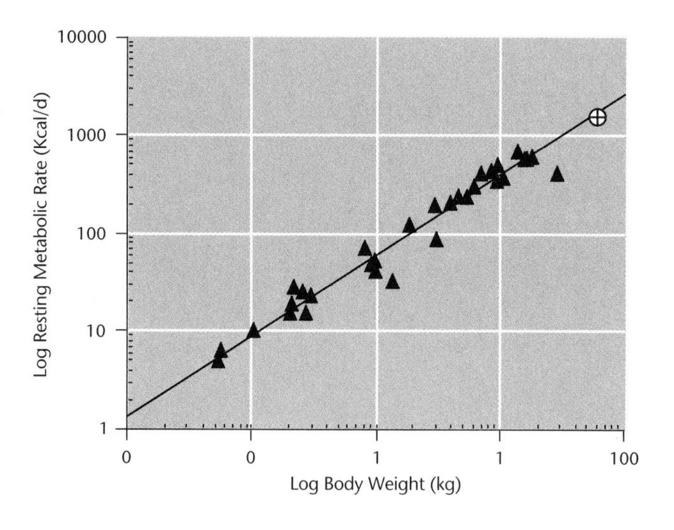

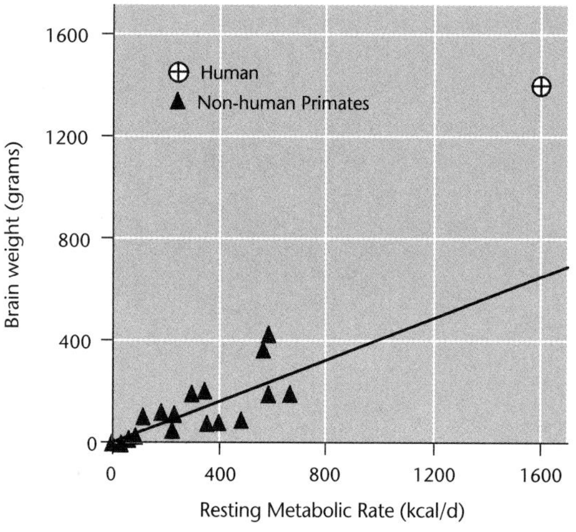

FIGURE 13.4. Relationship of (a) body weight and resting metabolic rate, and (b) resting metabolic rate and brain weight in nonhuman primates and humans. (Based upon data from Leonard, W. R., and Robertson, M. L. 1997. Comparative primate energetics and hominid evolution. *Am. J. Phys. Anthropol.* 102: 265–281.)

Evolution of Brain Size and Fossil Evidence

The increase in brain size in hominids did not occur at the same relative rate through time. The increase in brain size over 4 to 5 million years was static at times, faster in some intervals, and reversed slightly more recently. About 3 to 4 million years ago when the australopithecines lived in South Africa, the brain size, corrected for body size, averaged approximately 400 to 500 grams, which is similar to that of extant great apes (**fig. 13.5**). The first clear increase in hominid brain size occurred about 2.5 million years ago with the appearance of *Homo habilis,* who had a brain size that ranged between 650 to 750 grams. This is an evolutionarily significant change that is not accounted for change in body size since *Homo habilis* was as small as the australopithecines (under 5 ft). With the appearance of *H. erectus* about 1.75 million years ago, except for the European *Homo erectus* from Dmanisi who had a brain size that ranged from 600 to 670 cc, brain encephalization continued, reaching to 1100 grams. Thereafter, and with the appearance of archaic *Homo sapiens,* brain size continued to increase, attaining the present levels of modern *Homo sapiens* (ranging from 1200 to 1500 grams); among the Neanderthals it reached even higher (1600 grams). Thereafter, and through the Neolithic there appears to be a decrease in average absolute brain size, which is paralleled by a corresponding decrease in average body size.

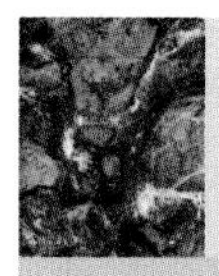

TABLE 13.1

Energy Requirements of the Human Brain from Birth to Adulthood[6]

Body Weight (kg)	Age Group	Brain Weight (grams)	Brain's Energy Consumption (kcal/day)	Body's[8] Total Energy Consumption (kcal/day)	Energy to Brain (% of) Body
3.5	newborn	400	118	248	74
5.5	4–6 months	650	192	545	64
11	1–2 years	1045	311	943	53
19	5–6 years	1235	367	1516	44
31	10–11 years	1350	400	2464	34
50	14–15 years	1360	403	3039	27
70	Adult	1250	414	2800	23

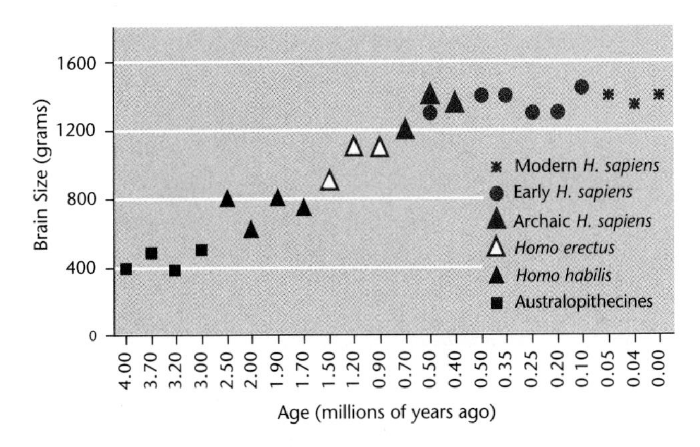

FIGURE 13.5. Fossil evidence and evolutionary trend in brain size. The trend toward large brain size started with the appearance of *Homo habilis* 2.5 million years ago whose members attained a brain weight that exceeded 700 grams. Thereafter, the positive encephalization continued to reach the present modern levels of 1100 to 1600 grams. (Source: Data from references 1, 2, 4, and 6–8.)

Hypotheses to Explain the Human Brain Expansion

Obviously, the attainment and maintenance of a large and complex brain requires an abundant supply of energy. The important question is then during evolution, where did the energy come from to fuel the enlargement of the brain? Several hypotheses have been forwarded to explain the development of a large brain in humans: (1) a shift to a high-quality diet (derived either from animal foods or technological induced change in dietary quality); (2) reduction in gut size; (3) slow growth, late maturation, and extension of the period of growth in body size; and (4) shift to land-based and marine foods.

Selection of Energy-Rich Diet

Several researchers postulate that a shift to a high-quality diet (derived either from animal foods or plants) contributed to increase the energy allocation to the brain of humans.[6, 9] A high-quality diet contains more fat and meat and less plant material. In general, meat or fish, at an equivalent weight (except for nuts), has a higher concentration of nutrients and energy than plant foods such as shoots, leaves, or fruit. The assumption is that early hominids such as *Homo habilis* may have been forced to expand their dietary repertoire from plant to animal foods by ecological changes that took place in Africa. About 2.5 million years ago, the environment was continuing to become drier, creating more arid grasslands.[10, 12] The climatic change made animal foods more abundant and thus, an increasingly attractive food resource,[11, 12] which might have provided *H. habilis* with an energy-rich diet.

Expensive Tissue: Reduction of Gut Size

This proposition, referred to as the expensive tissue hypothesis, maintains that the metabolic requirements of relatively large brains were made possible by a corresponding reduction in size of the gut.[7] The premise of this hypothesis is that the splanchnic organs, such as the liver and gastrointestinal tract, are as metabolically expensive as brains. Additionally, a small gut is compatible with high-quality, easy-to-digest food. Support for this hypothesis is shown by the inverse relationship between relative gut size and relative brain size.[7] That is, primates with relatively large guts have relatively small brains, while primates with relatively small guts have relatively large brains. In other words, the average primate, with a larger relative brain size than the average mammal, also has a smaller relative gut size than the average mammal.[7] In general, a small gut is compatible with high-quality (such as animal foods), easy-to-digest food, while poor-quality diet (such as plant foods) is associated with a large gut (see Chapter 6). Based on this finding it has been proposed[7] that the concomitant reduction in energy associated the reduction in gut size has been used for developing and maintaining a large brain.

Slow Growth Rate and Extended Period of Growth in Body Size

This hypothesis maintains that the metabolic requirements of relatively large brains were made possible by having a slow growth rate and an extended period of growth.[13] The basis of this hypothesis is that a rapid growth rate (faster changes in mass per unit of time) requires more energy than slow rate (small changes in mass per unit of time), and the saving of energy associated with a slow growth rate is used to grow a large brain.

At birth, humans have a brain weight on the average of about 400 grams, which is equivalent to 32 percent of its adult value. Likewise, chimpanzees at birth have a brain weight on the average of about 140 grams, which is equivalent to 31 percent of its adult value. That is, at birth humans and chimpanzees have similar relative brain size, yet, by adulthood in humans the brain reaches more than 1250 grams (average ranges from 1100 to 1500 grams) while in chimps it does not exceed 500 grams. This difference is attained through differences in pattern of brain growth. As illustrated in **figure 13.6**, the human brain continues growing until about the age of eight years, while in chimpanzees it stops by four years. Thus, by extending the period of growth into the childhood stage, humans are able to have a large brain, which otherwise would not have been possible. From the behavioral perspective, the additional and/or relatively delayed time course of human brain development has several important benefits. The additional time allows the developing and susceptible human brain to gain more experience and knowledge that otherwise would not be possible. In this context, the trend toward prolongation of the period of childhood may be considered a component of the evolutionary trend associated with human evolution[13–16] (see next chapter).

In absolute terms, humans' sexual maturity and growth in body size occurs at a much later age than in other primates (**fig. 13.7**). For example, chimpanzees reach sexual maturity by the age of 8 years and finish growth in body size (measured by adult height) by the age of 12 years, while for humans, the onset of sexual maturity (measured by age at menarche) occurs at around the age of 12 years; adulthood is attained by about the age of 20 years in males and 18 years in females. In other words, humans compared to other primates have a slow pattern of growth and late maturation. This slow pattern of growth is also associated with lower daily energy requirements. The total energy requirements for children under 18 years of age averages about 1512 Kcal/day (table 13.3). For example, assuming a length of 50 cm (20 inches) at birth to attain an adult height of 176 cm, by the age of 18 years one needs to grow about 7 cm/year for a total growth of 126 cm in 18 years. Hence, the total energy requirements per year equals 551,880 Kcal (1512 Kcal/day × 365 = 551,880) or 78,840 Kcal/year per centimeter of height (551,880 Kcal/year/7 cm = 78,840). In contrast, to attain the same adult height of 176 cm in 12 years rather than 18 years one needs to grow about 10.5 cm/year (126 cm/12 years = 10.5 cm/year). To attain this growth rate of 10.5 cm/year the total energy requirements per year equals 827,820 Kcal (78,840/cm × 10.5 cm/year = 827,820 Kcal/year) or 2268 Kcal/ day (827,820 Kcal/year/365 = 2268 Kcal/ day). Therefore, in humans, extending the period of growth in height to 18 years rather than 12 years as in chimpanzees results in a saving of 756 Kcal/day (2268 Kcal/day – 1512 Kcal/day = 756 Kcal/day).

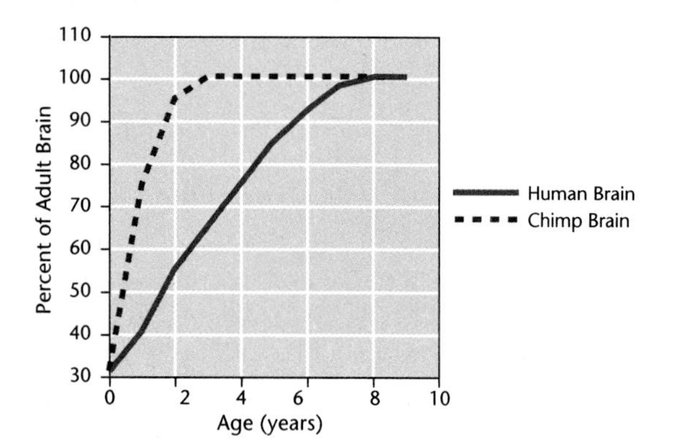

FIGURE 13.6. Relative growth in brain size in humans and chimpanzees.

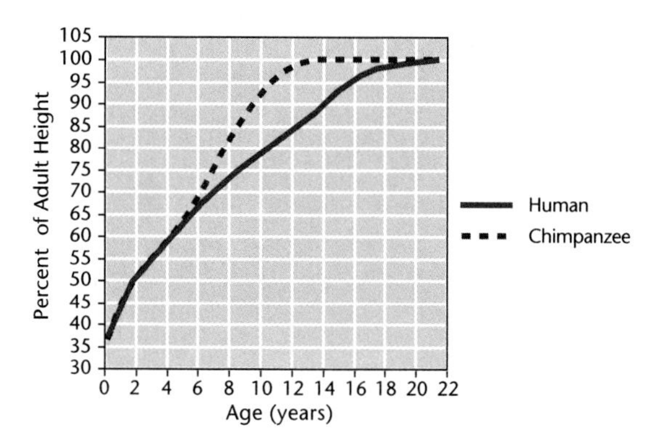

FIGURE 13.7. Relative growth in body size in humans and chimpanzees. Apes are characterized by rapid growth in body size so that by the age of 12 years they have attained their adult height. Humans grow at slower rate and for a greater period of time, attaining their adult size between ages 18 and 20.

Viewed in this context, the slow pattern of growth and late maturation might have evolved as a compensatory mechanism to spare energy for the development of a large brain. That is, a slow rate in growth of body size may offset the risks of developing metabolically expensive structures such as the human brain. In addition, slow growth rates, because of their lower metabolic costs per unit of time and because feeding competition influences growing individuals more drastically than adults, might have been selectively favored. Hence, when the availability of food is low, juveniles whose growth is slow may be more favored than those with high growth rates. Analyses of tooth eruption suggest that the juvenile period in hominids is greater than that of chimpanzees and australopithecines.[18] Thus, it is quite possible that in evolutionary perspective slow postnatal growth and postponement of the age of sexual maturity could have been an important factor that may have contributed to spare energy for the development and maintenance of a large brain.

Newborn Fat

This hypothesis proposes that the metabolic requirements of relatively large brains were made possible by stored energy in the form of fat.[18] About 90 percent of the human fetal weight gain during the last trimester of pregnancy is accounted by fat deposition.[18] Comparative studies of body composition indicate that relative to other primates, humans are born with relatively more fat than other mammals.[19] On the average human newborns have 15 percent fat while other mammals have only between 1 and 5 percent body fat (**fig. 13.8**). Furthermore, humans continue gaining fat during the first two years of infancy and lose this fat between the age of 2 and 8 years.[20–23] As illustrated in **figure 13.9**, the decline in body fat coincides with the increase of brain size. The high amount of newborn body fat represents a large fuel storage from which the brain can derive energy to satisfy the high energy requirement after birth. Additionally, newborn fat provides a source of ketone bodies (ketones: hydroxybutyrate, acetoacetate, and acetone). Human babies are endowed with proportionally by far the largest ketone reserve (body fat) of any mammalian infants.[22] Ketones can be oxidized and used as fuel for the brain. In human fetuses at mid-gestation, ketones supply as much as 30 percent of the energy requirement of the brain.[23] Likewise, ketones are a key source of carbon for the brain to synthesize lipid, cholesterol, and fatty acids necessary for brain development.

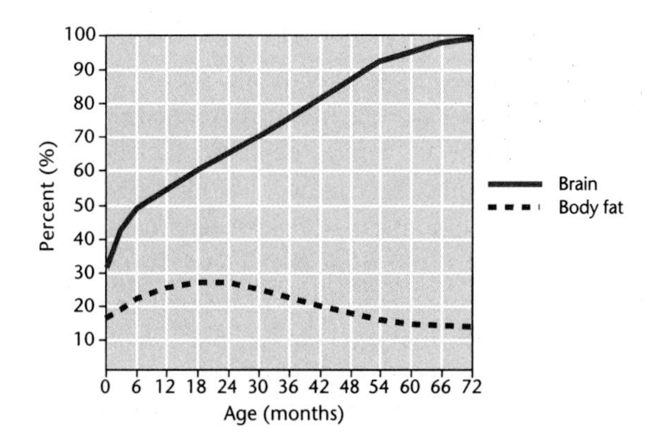

FIGURE 13.9. Relative changes in brain size and body fat during infancy and childhood. The high amount of body fat of the newborn provides energy storage that can be used by the brain when energy availability is low. (Data derived from Kuzawa CW. 1998. Adipose tissue in human infancy and childhood: an evolutionary perspective. Yearbook Physical Anthropology 41:177–209.)

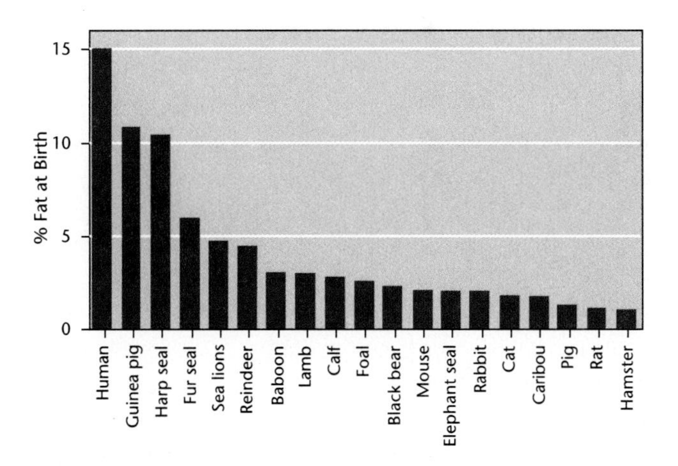

FIGURE 13.8. Percent body fat at birth of fifteen mammalian species. (Source: Data from Kuzawa.[21])

In summary, newborn fat provides a reliable source of brain energy that can be used in between feedings, and they provide a major proportion of the lipid building blocks for developing brain cells. The high amount of energy storage is suitably matched to the high energy and structural demands of the developing infant brain.

Emphasis on Shore-Based Diet

This hypothesis proposes that the initial impetus for brain evolution of early hominids depended upon the exploitation of shore-based foods rather than land-based animal foods.[24–28] Specifically, it postulates that along the shores of fresh or salt water, mollusks, crustaceans, birds' eggs, spawning fish, frogs, turtles, and a variety of plants would have provided an extensive selection of nutrient- and energy-rich, highly accessible foods.[24–28] The easy availability of high-quality shore-based foods could be gathered by anyone irrespective of gender, social position, physical stature, or degree of specialized hunting skills.[29] Additionally, the composition of the human central nervous system is almost wholly composed of two long-chain polyunsaturated fatty acids (LC-PUFA), docosahexaenoic acid (DHA), and arachidonic acid (AA), respectively known as omega-3 and omega-6 fatty acids. The concentration of these essential nutrients is greater in marine animals than in land-based game.[27] Dietary levels of

DHA are 2.5- to 100-fold higher for equivalent weights of marine fish or shellfish vs. lean or fat terrestrial meats. The littoral marine and lacustrine food chains consistently provide larger amounts of preformed LC-PUFA than the terrestrial food chain. Only the internal organs and brains of game have a concentrated sources of AA and especially DHA. But to obtain this source of nutrients, hominids would have to compete with carnivores and primary scavengers who typically consume the first parts of a carcass. Brain and bone marrow tissue are reasonable options for hominids with tool technology.[30–31] However, even a very large herbivore such as the rhino has but 350 g brain tissue for one ton body weight. A carcass would need to be scavenged quickly, as brain, organ, and marrow tissues spoil very quickly. In addition, the fat content of game is exceedingly low (1–7 percent) compared to typical modern domestic meats.[32, 35)]

The South African Cape sites such as Klassies river dated from 100,000 to 180,000 years ago, Rift Valley lakes of the Nile Corridor, and the Middle East, Katanda, Zaire contain large shell middens and fish. The author proposes that given the limitations of obtaining foods rich in the essential lipids, shore-based foods provided the initial impetus for the rise of brain evolution. Then, as skills improved gathering of shore-based foods would have been gradually supplemented by fishing, hunting, and experimenting with cooking less nutritious roots, tubers, etc.

Summary

It is evident that the human brain is not a product of simple shifts in growth relationships, but of multiple, independent, and superimposed modifications of existing structures and developmental pathways, rather than the invention of new features.[1] From the studies discussed above, three key points emerge that have a bearing on attempts to reconstruct the evolutionary events that underlie the origin and modification of the human brain (**fig. 13.10**).

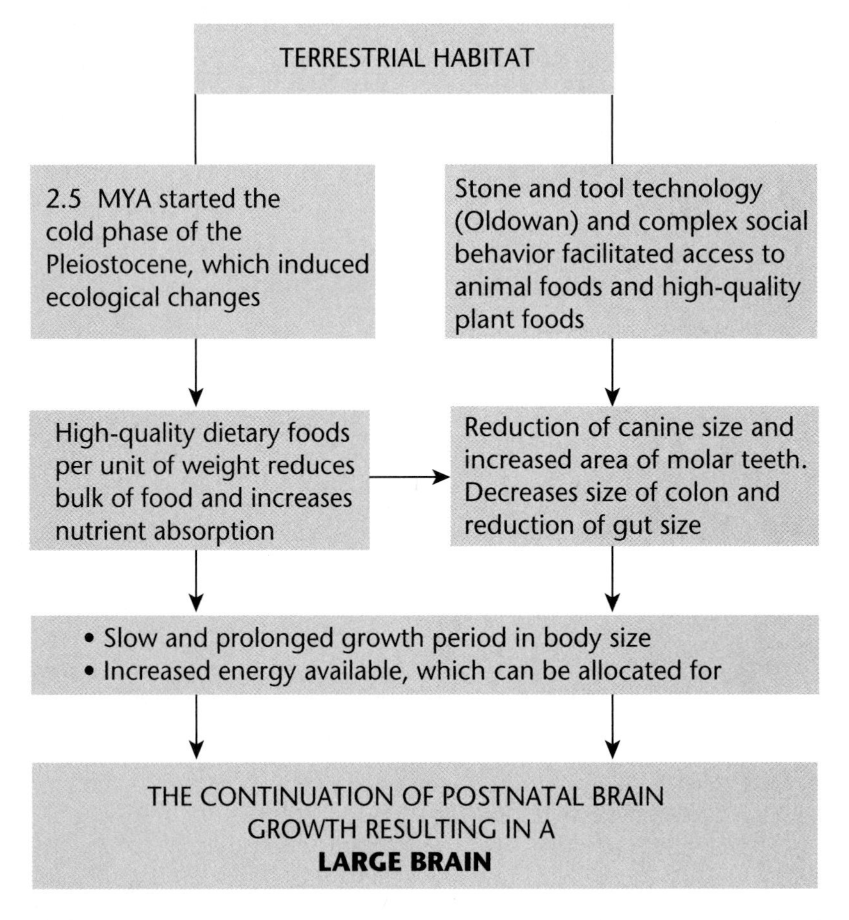

FIGURE 13.10. Schematization about the evolutionary factors that are postulated to account for the emergence of human brain size. The shift to a diet of high nutritional quality made possible by the ecological changes and stone tool technology practiced by *Homo habilis* resulted in increased energy availability, which, in turn, could be allocated for the postnatal continuation of fast brain growth resulting in a large brain.

First, as primates expanded into the terrestrial habitat, the omnivorous dietary predilections of the early ancestors were enhanced by climatic changes and by complex social behaviors. These changes resulted in an increased availability of animal foods and plant foods of high nutritional quality, and concomitant with this change, brain sizes of primates have increased throughout the phylogenetic order.

Second, the brain size during hominid evolution underwent explosive growth about 2.5 million years ago with the appearance of *Homo habilis* that attained a brain size of 750 grams. This increase is associated with the Oldowan tool technology, and the adoption of this technology unquestionably expanded the abilities of early hominids to modify wood, bone, and other materials, with which they gained greater access to high-quality food resources, including the meat and marrow of very large thick-skinned mammals.[36, 37] The fact that the dentition of *Homo habilis* was smaller than that of australopithecines suggests that technology had begun to intervene in the dietary behavior procurement of the early member of our genus.[37] In other words, stone tools have facilitated access to formerly unavailable foods, both plant and animal, or upgraded existing food quality.

Third, concomitant with such a shift in dietary preference, gut size was reduced, which in turn resulted in an energy sparing that could be used to continue brain growth.

Finally, the unique features associated with speech such **Broca's area** and the left-right brain asymmetry that characterize *Homo sapiens* have its precursors among nonhuman primates such as chimpanzee brains. Hence, the neuroanatomical substrate of left-right dominance in speech and right-handedness occurred before the origin of hominids.[38] These features continue its path as shown by the fact that the early hominids such *H. habilis* and *H. erectus* had the left-right hemisphere asymmetry like in modern humans.

Suggested Readings

Allman, J. M. 1999. *Evolving Brains.* New York: Scientific American Library.

Falk, D., and K. R. Gibson, eds. 2001. *Evolutionary Anatomy of the Primate Cerebral Cortex.* New York: Cambridge University Press.

Martin, R. D. 1990. *Primate Origins and Evolution: A Phylogenic Reconstruction.* Princeton: Princeton University Press.

Literature Cited

1. Martin, R. D. 1983. *Human Brain Evolution in an Ecological Context.* New York: American Museum of Natural History.
2. Jerison, H. J. 1973. *Evolution of the Brain and Behavior.* New York: Academic press.
3. Kleiber, M. 1961. *The Fire of Life.* New York: Wiley.
4. Holliday, M. A. 1986. Body composition and energy needs during growth. In: Falkner, F., and J. M. Tanner, eds., *Human Growth: A Comprehensive Treatise.* New York: Plenum, 101–117.
5. Butte, N. F. 2000. Fat intake of children in relation to energy requirements. *Am. J. Clin. Nutr.* 72(suppl):1246-1252.
6. Leonard, W. R., M. L. Robertson. 1997. Comparative primate energetics and hominid evolution. *Am. J. Phys. Anthropol.* 102: 265–281.
7. Aiello, L. C. and P. Wheeler. 1995. The expensive-tissue hypothesis: The brain and digestive system in human and primate evolution. *Curr. Anthropol.* 36:199–221.
8. Ruff, C. B., E. Trinkaus, and T. W. Holliday. 1997. Body mass and encephalization in Pleistocene *Homo. Nature* 387: 173–176.
9. Milton, K. 1993. Diet and primate evolution. *Sci. Am.* 8:86–93.
10. Vrba, E. S. 1995. The fossil record of African antelopes relative to human evolution. In: E. S. Vrba, G. H. Denton, T. C. Partridge, and L. H. Burkle, eds., *Paleoclimate and Evolution, with Emphasis on Human Origins.* New Haven: Yale University Press, 385–424.
11. Owen-Smith, N. 1999. Ecological links between African savanna environments, climate change and early hominid evolution. In: Bromage, T. G., and F. Schrenk, eds., *African Biogeography, Climate Change, and Human Evolution.* New York: Oxford University Press, 138–149.
12. Behrensmeyer, K., N. E. Todd, R. Potts, G. E. McBrinn. 1997. Late Pliocene faunal turnover in the Turkana basin, Kenya and Ethiopia. *Science* 278:1589–1594.
13. Leonard, W. R. and M. L. Robertson. 1994. Evolutionary perspectives on human nutrition: The influence of brain and body size on diet and metabolism. *Am. J. Hum. Biol*. 6:77–88.
14. Bogin, B. 1990. *Patterns of Human Growth.* Cambridge: Cambridge University Press.
15. Allman, J. M. 1999. *Evolving Brains.* New York: Scientific American Library.
16. Quartz, S. and T. Sejnowski. 1997. The neural basis of cognitive development. *Behavioral and Brain Sciences* 20:537–596.
17. Smith, B. H. 1992. Life history and the evolution of human maturation. *Evol. Anthropol.* 4:134–142.
18. Cunnane, S. C. and M. A. Crawford. 2003. Survival of the fattest: fat babies were the key to evolution of the large human brain. *Comp. Biochem. Physiol. A. Mol. Integr. Physiol.* 136:17–26.
19. Battaglia, F. C., and G. Meschia. 1973. *Fetal and Neonatal Physiology.* Cambridge (UK): Cambridge University Press, 383–397.
20. Kuzawa, C. W. 1998. Adipose tissue in human infancy and childhood: An evolutionary perspective. *Yrbk. Phys. Anthropol.* 41:177–209.

21. Dewey, K. G., M. J. Heinig, L. A. Nommsen, J. M. Peerson, and B. Lonnerdal 1993. Breast-fed infants are leaner than formula-fed infants at 1 y of age: The darling study. *Am. J. Clin. Nutr.* 52:140–145.
22. Fomon, S. J., F. Haschle, E. F. Ziegler, and S. E. Nelson. 1982. Body composition of reference children from birth to 10 years. *Am. J. Clin. Nutr.* 35:1169–1175.
23. Widdowson, E. M. 1974. Changes in body proportion and composition during growth. In: J.A. Davies, J. Dobbing, eds., *Scientific Foundations of Pediatrics.* London: Heinemann, 153–163.
24. Adam, P. A. J., N. Raiha, E. L. Rahial, and M. Kekomaki. 1975. Oxidation of glucose and D-b -hydroxybuyrate by the early human fetal brain. *Acta. Paediatr. Scand.* 64:17–24.
25. Crawford, M. A., and D. Marsh. 1989. *The Driving Force.* London: Heinemann.
26. Crawford, M. A., M. Bloom, C. L. Broadhurst, W. Schmidt, S. C. Cunnane, and C. Galli, et al. 1999. Evidence for the unique function of docosahexaenoic acid during the evolution of the modern hominid brain. *Lipids* 34:S39–S47.
27. Broadhurst, C. L., S. C. Cunnane, and M. A. Crawford. 1998. Rift Valley lake fish and shellfish provided brain-specific nutrition for early *Homo. Br. J. Nutr.* 79:3–21.
28. Broadhurst, L. C., Y. Wangc, M. A. Crawford, S. C. Cunnane, J. E. Parkingtone, and W. F. Schmidt. 2002. Brain-specific lipids from marine, lacustrine, or terrestrial food resources: Potential impact on early African *Homo sapiens. Comparative Biochemistry and Physiology,* Part B: Biochemistry and Molecular Biology. 131:653–673.
29. Cunnane, S. C., L. S. Harbige and M. A. Crawford. 1993. The importance of energy and nutrient supply in human brain evolution. *Nutr. Health* 9:219–235.
30. Morgan, E. 1994. *The Scars of Evolution.* New York: Oxford University Press.
31. Eaton, S. B., A. J. Sinclair, L. Cordain, and N. J. Mann. 1998. Dietary intake of long-chain polyunsaturated fatty acids during the Paleolithic. *World Rev. Nutr. Diet* 83:12–23.
32. Cordain, L., J. Brand Miller, S. B. Eaton, N. Mann, S. H. A. Holt, and J. D. Spet. 2000. Plant-animal subsistence ratios and macronutrient energy estimations in worldwide hunter-gatherer diets. *Am. J. Clin. Nutr.* 71:682–692.
33. Cordain, L., B. A. Watkins, G. L. Florent, M. Kehler, L. Rogers, and Y. Li. 2002. Fatty acid analysis of wild ruminant tissues: evolutionary implications for reducing diet related chronic disease. *Eur. J. Clin. Nutr.* 56:1–11.
34. Speth, J. D. 1989. Early hominid hunting and scavenging: The role of meat as an energy source. *J. Hum. Evol.* 18:329–349.
35. O'Dea, K. 1991. Traditional diet and food preferences of Australian Aboriginal hunter-gatherers. *Philos. Trans. R. Soc. Lond. B* 334:233–241.
36. Blumenschine, R. J. and J. A. Cavallo. 1992. Scavenging and human evolution. *Sci. Am.* (Oct): 90–96.
37. Potts, R. 1988. *Early Hominid Activities at Olduvai.* New York: Aldine.
38. Carroll, S. B. 2003. Genetics and the making of *Homo sapiens. Nature* 422: 849–857.

Bipedalism and the Evolution of Extended Childhood

Bipedal locomotion brought about changes in pelvic morphology that reduced the birth canal and led to the altricial birth of humans. The altricial birth, in turn, resulted in an extended period of childhood that is unique among primates. Humans have developed various biocultural adaptations that assist in the survival of the immature and dependent offspring and, by extension, the species in general. The purpose of this chapter is to describe the process of how bipedalism might have evolved, its consequences on postnatal development, and the biocultural responses that increased the survival of the immature human infant.

Forms of Locomotion Preceding Bipedalism

Primate locomotion is usually divided into leapers, which describes some prosimians; vertical clingers, which describes some New World monkeys; brachiation, practiced by orangutans and gibbons; and quadrupeds, such as the gorillas, chimps, and baboons. Hence, there is great interest in determining the forms of locomotion that preceded bipedalism and their transition to bipedality. These hypotheses include the arboreal quadruped, terrestrial quadruped, arboreal climbing, and knuckle-walking.

Arboreal Quadruped

According to the arboreal quadruped hypothesis, bipedalism evolved from an ancestor adapted to above-branch, pronograde (relatively level trunk) quadrupedalism, much like that observed in most living anthropoids.

Terrestrial Quadruped

This hypothesis posits that bipedalism evolved from an ancestor adapted to some form of terrestrial quadrupedal locomotion.[1-4] The latter includes digitigrady, palmigrady, fist-walking, and knuckle-walking. Digitigrady primates support their weight on their fingers and metacarpal heads, while palmigrade primates use the entire hand, including the digits and palm. Fist-walking involves weight support in the backs of the proximal phalanges, and knuckle-walking involves weight support on the backs of the middle phalanges.

Arboreal Climbing

The arboreal climbing hypothesis maintains that the common primate manner in which hominids climb trees, by always keeping the head above the trunk, involves considerable fore- and hind-limb mobility, suspensory postures, and use of multiple supports, constitutes, a suitable preadaptation for bipedal locomotion on the ground.[2, 3] The climbing hypothesis explicitly argues that there was no significant terrestrial component, including knuckle-walking, to the locomotor repertoire of the ancestor of the earliest hominid bipeds. The fact that the australopithecines have very curved digits, which are much more similar to chimpanzees than modern humans, supports the assumption that early hominids depended on life in the trees and on the ground. In other words, bipedalism is an extension of the preadaptation acquired during arboreal life.

Knuckle-Walking

The knuckle-walking hypothesis postulates that knuckle-walking was a significant component of the locomotor repertoire of the ancestor of the first bipeds. During knuckle-walking, African apes bear their weight on the backs of their middle phalanges (middle segments of their fingers), which involves strongly flexed proximal interphalangeal joints, and extended metacarpo-phalangeal joints. The retention of some vestiges of former knuckle-walking adaptations in hominids, especially in the earliest bipeds, provides support for this hypothesis.[4]

Hypotheses to Explain the Origins of Bipedalism

Several hypotheses and scenarios have been put forward regarding the selective agents underlying the origin of bipedalism. There is general consensus that the change from an arboreal to a terrestrial habitat brought about new environmental stresses such as high heat stress and dispersion of food resources. In this scenario the early hominids developed biological adaptations that increased their survival. The various hypotheses that have been postulated to explain these adaptations involve changes in temperature regulation, food provisioning, and environmental exploration.

Temperature Regulation

At present, evidence indicates that the evolution of bipedalism was associated with change in the climate of East Africa. During the Miocene period (25 to 6 million years ago) there was a worldwide decrease in rainfall that resulted in the shrinking of the tropical forests that were the home of early hominoids and the expansion of the woodland-savanna zones. In these new habitats, the ex-forest dwellers hominoid primates, who were essentially quadrupeds, underwent an adaptive radiation where bipedalism became an adaptive advantage. The adaptive significance of bipedalism includes reduction of heat absorption and an increase in heat dissipation.[5] In comparison with quadrupedal animals, true bipeds have much less of their body surface area exposed to the direct rays of the sun, and thus a decrease in heat absorption and an increase in the efficiency of heat dissipation (**fig. 14.1**). It is estimated that in bipedal posture when the sun is directly overhead, the heat load is only about 40 percent than that of quadrupedal posture.[5] An extension of the

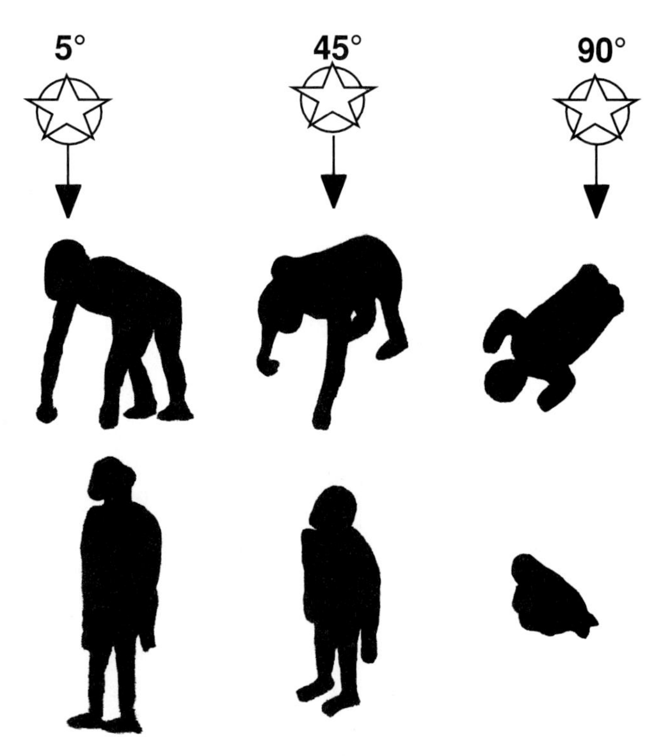

FIGURE 14.1. Bipedalism and heat adaptation. In quadrupedal posture (above) more body surface is exposed to the sun than in bipedal posture (below). (Adapted from Wheeler, P. E. 1991. The thermoregulatory advantages of hominid bipedalism in open equatorial environments: The contribution of increased convective heat loss and cutaneous evaporative cooling. *J. Hum. Evol.* 21: 107–115.)

temperature regulation hypothesis, referred to as the radiator theory, has been postulated.[6–7] This proposition is based upon the fact that the brain must remain cool in order to function properly. Observations of the inner casts of the brain show that the early hominids developed a special cooling system of veins in the head that led to more efficient cooling. In other words, the need to cope with physiological stress that our ancestors encountered in tropical Africa was the main selection pressure that encouraged bipedalism.

Predator Avoidance

According to the predator hypothesis, once the early hominids were forced to live on the ground, the main selective pressure was the avoidance of predators. For this, hominids developed the ability to be habitual bipeds so that they could spot predators (and perhaps prey) by seeing over tall savanna grass. This hypothesis, however, is not supported by the fact that none of the numerous African mammals of the savanna became bipeds.

Provision of Food

According to the provisioning hypothesis, the main selection pressure that encouraged bipedalism was the necessity to provision food to women and children.[8] Hominids evolved a form of bipedalism that allowed them to cover a relatively great distance in a single day, and bipedalism would also have freed the hands for tool use and for carrying food derived from scavenging or hunting. This attributes that males were the main food providers; but mothers, among contemporary hunter and gatherers, despite the burdens imposed by children, did manage to go about the countryside collecting most of the hunter and gatherer diet.

Environmental Exploration

The change from a tropical humid climate to a dry climate typical of the savanna caused the food sources to become thinly dispersed. The exploration of such an environment demanded a more energy-efficient mode of travel. In this scenario, bipedalism might have evolved as an adaptation to cover a wider environment. Although bipedalism is slower and less efficient than quadrupedalism at high speeds, it is more energy efficient when traveling a long distance. For example, chimpanzees are 50 percent less energy efficient than conventional quadrupeds when walking on the ground—whether bipedal or knuckle-walking. Contemporary hunter and gatherers walk an average of 15 km per day and use more than 200 km^2 in a single year and many cover more than 1,000 km^2 in a year. In contrast, male chimpanzees move less than 3 km per day and cover only about 10 km^2 in a lifetime.[9, 10] If the food resources were as thinly dispersed as they are today, it is quite possible that food scarcity provided the selection pressure favoring improved energetic efficiency in locomotion. Therefore, in addition to being an adaptation to heat stress or protection against predators, bipedalism facilitated an efficient terrestrial locomotion and higher daily mobility through a wide range of territory.

In addition, once bipedalism evolved, it also became advantageous in terms of carrying or making things with freed hands. Such an adaptation would certainly improve the means for exploring the environment, for extractive foraging, hunting, etc., which together would increase the acquisition of a high-quality diet. Bipedalism may also have freed the hand to carry and transport infants and food simultaneously from the place where food is obtained to the place where it can be shared with the mates and offspring.[11, 12] It is quite evident that once the early hominids acquired bipedalism, it expanded the environment and exposed the early hominids to a wide range and unpredictable food resources.

In summary, the origin of bipedalism has been attributed to several competing hypotheses. The thermoregulatory model views the increased heat loss, increased cooling, reduced heat gain and reduced water requirements conferred by a bipedal stance in a hot, tropical climate as the selective pressure leading to bipedalism. The ecological and behavioral model attributes bipedality to the social, sexual, and reproductive conduct of early hominids in the search of food in a new expanding environment where the resources are dispersed. Each of the different theories overlap and many times are mutually supportive, but what is clear is that bipedalism initiated changes that would become profound and forever change the path of human evolution.

Changes in Pelvic Size and Shape Associated with Bipedalism

Changes in Pelvic Size and Shape

The pelvic girdle consists of the two **coxal** bones, or hip bones (**fig. 14.2**). The two bones are united to each other anteriorly at the **symphysis pubis** and posteriorly at the **sacrum**. The sacrum and coccyx (which are parts of the vertebral column) and the two coxal bones

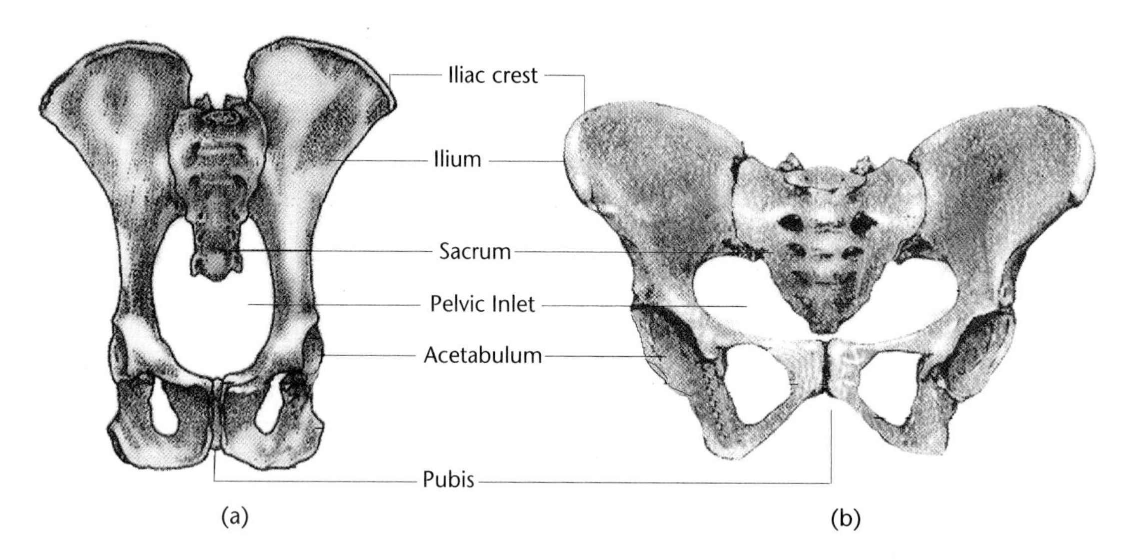

FIGURE 14.2. Anatomic landmarks of the pelvis of (a) chimpanzee and (b) human. (Composite picture made by the author.)

form the basinlike structure called the pelvis. The two coxal consist of three separate bones called: **ilium, ischium,** and **pubis.** The area where the three fuse form a lateral fossa called the acetabulum. The upper part of the coxal bone is the ilium, and the superior border of the ilium is known as the iliac crest. The inferior portions of the ilium, sacrum, coccyx, and the pubis merge, forming an opening called the **pelvic inlet,** which represents the birth canal.

Throughout evolution the pelvis changed to permit stable weight transmission from the upper body to the legs, to maintain balance through pelvic rotation, and consequently altered proportions and orientations of several key muscles. As shown in **table 14.1** and **figures 14.2** and **14.3**, the size and shape of the pelvis of quadrupedal primates, such as chimpanzees and gorillas, is different than that of humans. In quadrupeds, the coxal or iliac blades of the pelvis are positioned on

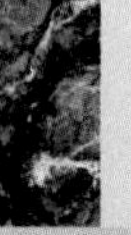

TABLE 14.1

Adult Female Pelvic Inlet Dimensions and Newborn Dimensions for Selected Primate Species

	Pelvis (1) Trans. (mm)	Newborn Head (2) A. P. (mm)	Newborn Shoulder (3) Length (mm)	Ratio 1 (3/1) (%)	Ratio 2 (3/2) (%)	(4) Width (mm)	Ratio 3 (4/1) (%)	Ratio 4 (4/2) (%)
Ateles geoffroyi	54·4	90·3	66·9	123	74	50·4	93	56
Macaca mulatta	50·9	67·7	66·3	130	98	49·2	97	73
Nasalis larvatus	51·8	71·5	64·0	124	89	59·0	114	82
Hylobates lar	55·9	78·7	64·2	115	82	51·2	92	65
Pongo pygmaeus	102·5	149·6	84·1	82	56	81·8	80	55
Gorilla gorilla	122·6	175·7	97·0	79	55	92·0	75	52
Pan troglodytes	98·0	149·5	83·0	85	55	84·9	85	57
Homo sapiens	121·6	112·9	123·8	102	110	118·3	97	105

Source: Schultz.[17]

Trans.= transverse diameter of the pelvic inlet; A.P. = anteroposterior diameter of the pelvic inlet; Length = maximum length of the newborn head from glabella to occipital; Width = shoulder breadth—breadth across the acromial points.

Ratio 1 = ratio of newborn head length to transverse diameter of pelvis.

Ratio 2 = ratio of newborn head length to anteroposterior diameter of pelvis.

Ratio 3 = ratio of newborn shoulder width to transverse diameter of pelvis.

Ratio 4 = ratio of newborn shoulder width to anteroposterior diameter of pelvis.

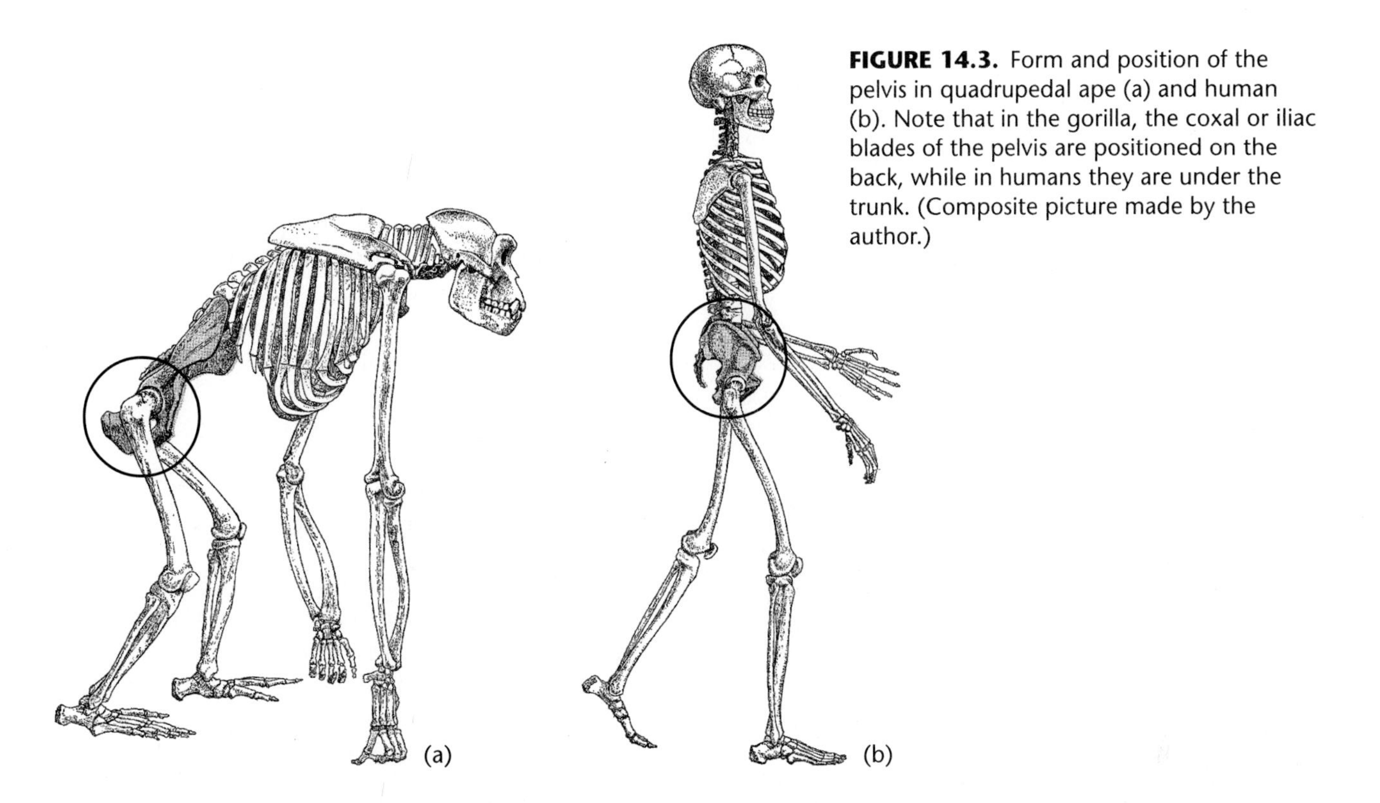

FIGURE 14.3. Form and position of the pelvis in quadrupedal ape (a) and human (b). Note that in the gorilla, the coxal or iliac blades of the pelvis are positioned on the back, while in humans they are under the trunk. (Composite picture made by the author.)

the back, while in humans they are under the trunk. In addition, the pelvis size and shape also changed with bipedalism.[4, 16] The ilium of the pelvis of chimpanzees is long and narrow, and the sacrum is also long, resulting in a long and narrow pelvic outlet. In humans, the iliac blades are wide, the iliac crest flares out, and the two coxal bones join together to form a bowl-shaped pelvis. The ischium is short, but the acetabulum that forms the socket for the head of the femur is broad and thick. The change in the orientation of the narrow iliac blades from parallel to flared out position results in the reduction of the pelvic inlet that forms the birth canal. Although the pelvic inlet of the chimpanzee is smaller than that of humans, it is much bigger with reference to the size of the newborn.

Pelvic Shape and Childbirth

As a result of the changes in pelvic shape brought about by bipedalism, the process of giving birth in early hominids became a major obstetric problem.[18–25] Studies of the remains of the australopithecines indicate that the pelvis and birth canal are proportional to the size of the brain (about 450 to 500 grams) and are comparable to that of the chimpanzee[20] (**figs. 14.4, 14.5**). However, according to the studies of Trevathan and Rosenberg,[18–23] the australopithecines had more problems with childbirth than the chimpanzees. This

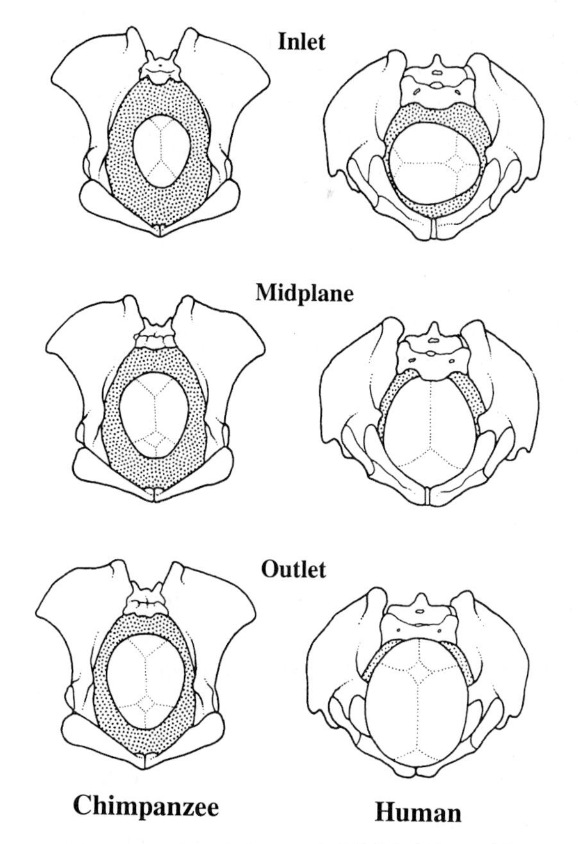

FIGURE 14.4. Pelvis shape and childbirth in a chimpanzee and human. In chimpanzee because the birth canal is longer, the newborn does not have a difficult birth. In contrast, in humans the fetal head is considerably larger than the birth canal so childbirth is a very difficult process.

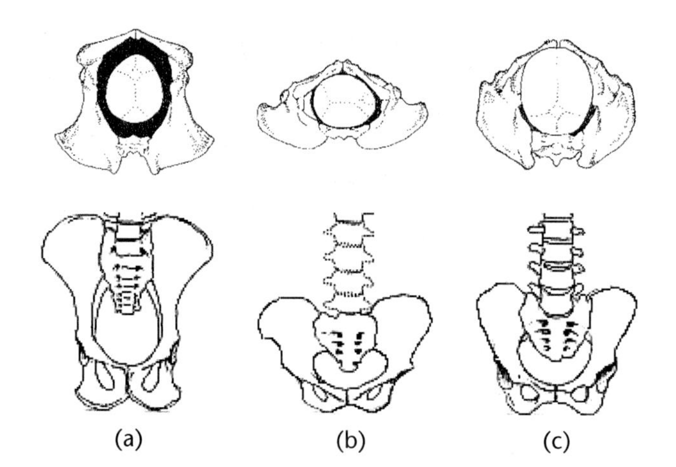

FIGURE 14.5. Pelvic inlet and newborn head size in (a) chimpanzee, (b) Australopithecus afarensis, and (c) modern humans. The size of the pelvic inlet or birth canal with reference to the size of the head of the newborn of the chimpanzee was similar to that of the australopithecines, but is much smaller in humans.

problem is related to the shape of the birth canal. The shape of the australopithecine birth canal is a flattened oval with the greatest dimension from side to side at both the entrance and exit, resulting in a symmetrical opening of unchanging shape.

Under this condition, the fetus of australopithecines could pass through the birth canal either in a face forward position or facing backward like modern human births, but a successful delivery would have required some kind of assistance.[21, 22] The problem of giving birth became accentuated with the evolution of *Homo habilis* about 2.5 million years ago, whose brain size increased from under 500 grams among the australopithecines to more than 650 grams. The increase in brain size meant that the early hominids have been subject to intense selective pressure. Based on skeletal reconstruction of the pelvis, researchers conclude that the pelvic anatomy of early *Homo* may have limited the growth of the human brain until the evolutionary point at which the birth canal expanded enough to allow a larger infant head to pass.[26] In other words, from an evolutionary perspective, bigger brains and roomier pelvises were symbiotically related, and the early hominids who displayed both characteristics were more successful at giving birth to offspring who survived to pass on the traits. As such, the trend toward a large brain size that started about 2.5 million years ago with *Homo habilis* occurred only when the pelvis had become large enough to accommodate carrying and delivering a fetus with a large head.

In summary, the problems of childbirth started with the evolution of bipedal locomotion and the appearance of the australopithecine. However, the trend toward a bigger brain size started with *Homo habilis* that lived about 2.5 million years ago. Hence, the problems of childbirth and high altriciality resulting in an extended period of childhood go back to 2.5 million years when *Homo habilis* evolved. Thereafter, the problems and risks of childbirth have continued throughout hominid evolution and became a major selective factor.

Altricial Pattern of Development and Its Consequences

The pattern of growth and development (ontogeny) of primates is different from that of other mammals. All primates are characterized by an "altricial" pattern of growth while other mammals are "precocial."[27, 28] That is, other mammals at birth are not as dependent on their parents as primates are. This pattern of altriciality is more exaggerated in humans than in other primates. As a result of the reduction of the pelvic inlet, humans are born with a relatively small brain size and in a relatively immature state, compared to other primates. Newborns of monkeys and apes have the ability to cling to the bodies of their mothers and are able to be transported quite effectively. On the other hand, in human infants the hand-grasping ability develops after the age of six months and the capacity to actually cling and be carried by the mothers is expressed after the age of four years. Furthermore, as previously discussed (see chapter 5), in humans about 25 percent of the adult brain is attained at birth, while chimpanzees are born with about 35 to 40 percent of the adult brain. In other words, the altricial pattern of development in humans is more exaggerated than in other hominoids. Researchers have used several approaches to determine the consequence of the altricial pattern of development on the attainment of size and shape of humans. These propositions can be grouped into those concerned with the role of neoteny in trajectory of morphological traits and those addressed at determining the function of hypermorphosis in the origin of the growth pattern itself.

Neoteny and the Trajectory of Morphological Traits

Neoteny refers to the slowing down of the rate of development that results in the retention of formally

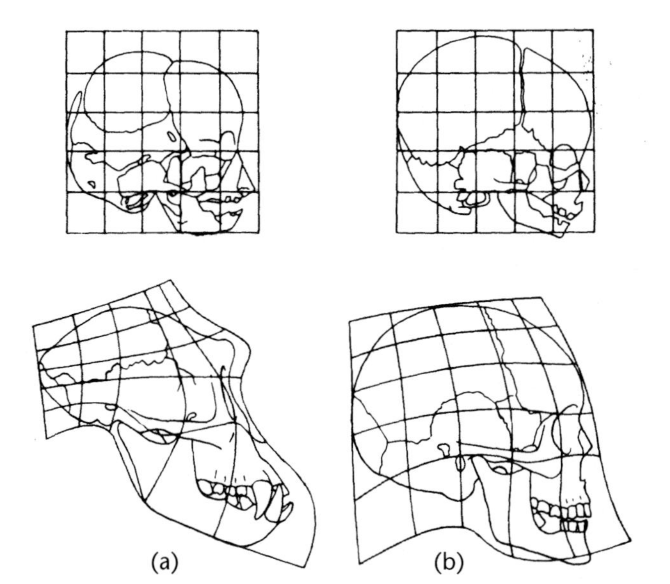

FIGURE 14.6. Neoteny and evolution of craniofacial morphology of chimpanzees (a) and humans (b). Source: Adapted from Thompson.[29]

juvenile characteristics by adult descendants. In 1942, Thompson[29] applied the principle of neoteny to explain differences in skull form between chimpanzees and humans. He postulated that the skull form of humans and chimpanzee is derived from a common neonatal form (**fig. 14.6**). Specifically, he indicated that adult humans retain the skull form present during juvenile stage (above) in the chimpanzee and that the adult differences can be accounted for by differences in growth rates. For example, adult differences in skull shape are due to different patterns of growth of the cranial bones, maxilla, and mandible. Following this school of thought, in 1977, Gould[30] postulated that adult human morphological features are the result of neoteny. In other words, the adult human morphological features reflect the primate juvenile stages of their ancestors. He based this hypothesis upon the observation that the infant chimpanzee is characterized by a large, rounded cranium, flat face, and erect posture like an adult human. In other words, the primate juvenile stages of ancestors become adult features of humans. That is, according to Gould[30] humans "in a metaphorical sense are permanent children." In this context, adult humans are retaining the characteristics of the fetal, infantile, and childhood periods of development. Recent studies of the cranial dimensions of an extensive sample of human and chimpanzee skulls at various stages of growth by Penin and colleagues[31] support the neotenic origin of the human skull. Specifically, they[31] demonstrate that humans, with respect to chimpanzee growth, are retarded in terms of both the magnitude of changes and shape alone. Similarly, at the end of growth, the adult skull in humans reaches an allometric shape equivalent to that of juvenile chimpanzees with no permanent teeth, and a size equivalent to that of adult chimpanzees. Therefore, human neoteny involves not only shape retardation (paedomorphosis), but also changes in relative growth velocity. For example, before the eruption of the first molar, human growth is accelerated and then strongly decelerated relative to the growth of the chimpanzee.[31]

Hypermorphosis and the Human Pattern of Growth

Hypermorphosis refers to the extension of the period of growth and development of the descendant beyond that of the ancestor.[32, 33] That is, hypermorphosis is a prolongation of the time for development. Application of the principle of hypermophosis to humans produces in theory a giant with disproportional features.[32] On the other hand, analysis of the profiles indicates that human and chimpanzee growth curves are the result of a combination of neoteny and sequential hypermorphosis.[33] Obviously, the validity of such inferences depends on the statistical techniques used and the sample on which the inferences are based, but the fact remains that humans and hominoids differ in the timing when various landmarks of the life cycle occur.

Developmental Stages of Humans and Hominoids

In general, the age of eruption of the first molar (M^1) marks the end of childhood, while the age at menarche, the onset of reproductive maturity of females, and the age at first estrus signals the onset of maturity in hominoids. On the average, the first molar (M^1) in humans erupts about the age of 6.5 years[34, 35] and by 4.5 years in chimpanzees. The age at menarche occurs around the age of 12 years in humans and in hominoids the age of estrus occurs between 8 and 10 years. Therefore, humans, compared to the gorilla and chimpanzee, have a longer childhood and later onset of reproductive maturity (**fig. 14.7**).

Evolution and Biocultural Adaptation to Altricial Birth and Extended Childhood

It is quite evident that bipedal locomotion brought about changes in pelvic morphology that reduced the birth canal and led to the altricial birth of humans.

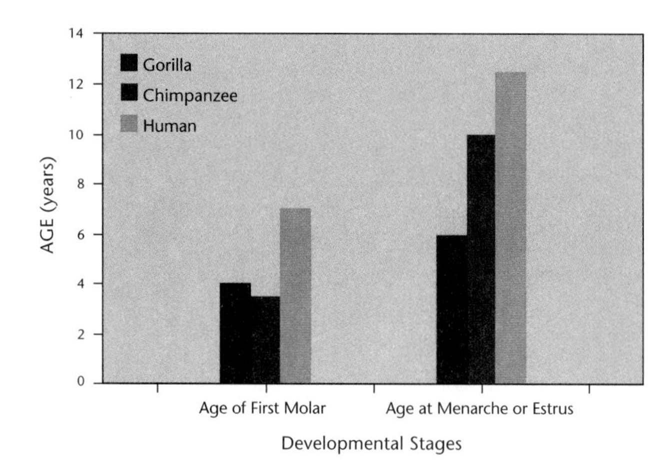

Figure 14.7. Comparison of human and hominoid developmental stages. Source: Dental data from Dimirjian[34] and Smith;[35] menarche data from Tanner[36] and Garn;[37] estrus data from Nishida et al. [38] Watts and Pusey,[39] and Bogin.[41]

The altricial birth, in turn, resulted in an extended period of childhood that is unique among primates. Therefore, the result was that the early hominids became progressively delayed.[42] In response to this unique developmental trait, humans have developed various biocultural adaptations that assist in the survival of the immature and dependent offspring and, by extension, of the species in general (**fig. 14.8**). These biocultural adaptations induce a food sharing and feeding adaptation as indirect effects derived from biological immaturity.

Food Sharing and Evolution

In nonhuman primates, only animal food is shared with their mates and other adults who help them procure the food.[43–47] In terms of reproduction, it has been shown that by sharing hunted animal food with adult females, males gain increased reproductive opportunities,[43, 44] or by sharing with adult primates, they form alliances that increase both their mating possibilities and future hunting opportunities.[45–47] The net effect of this pattern of selective food sharing is that the reproductive opportunities of the individual giver are increased, but it does not enhance the survival of the other members of the troop, such as the young and helpless children.

In contrast, humans are characterized by a pattern of general food sharing, whereby both animal and plant foods are shared regularly, willingly, and not only with other adults, but with the helpless young. Food sharing among hunter and gathering societies follows two patterns: (a) foods that are abundant and to which everybody has access, such as insect grubs and plant foods (fruit, and the starchy pith of the palm) are shared only by the immediate family—the spouse and the children of the person who acquired them; (b) foods that are not abundant and are risky and unpredictable to acquire, such as animal foods, are shared among all members of the band, irrespective of family ties so that Ache women and children are no more likely to eat meat from animals killed by their husbands and fathers than by other group members.[48, 49] In this manner, men, by sharing meat within a network of other hunters, guarantee themselves that their families on a regular basis are provided with a very valuable resource. According to these studies, division of labor and sharing results in an 80 percent increase in nutritional status and nearly a threefold increase in the predictability and regularity of daily food intake.

The transition to general food sharing probably capitalized upon and intensified the pattern of selective food sharing that was present in the hominoid social repertoire. This new order required heightened cooperation between males, females, and children. With selective food sharing primates, associations were probably either: (a) male like chimps and some multi-male and multi-female monkeys; or (b) all-female-with-young as among multi-male groups with matrilines. Then, after the shift to general food sharing, early human females probably became interested in strengthening alliances with males because they wanted a regular supply of meat for themselves and their children. Demographic studies indicate that the infant mortality rate of foragers is about 1 percent per year, whereas the rate for chimpanzees is about 3.5 percent per year.[9] It is quite possible that despite the low level of altriciality of chimpanzees, the higher infant mortality is related to the general lack of food sharing between adult and children. Life-history models postulate that parental investment during infancy and childhood are compensated by higher productivity during adulthood.[9] While it is true that mother-child food sharing occurs among many primates,[45–47] only in humans do both mothers and fathers provide a substantial fraction of their plant and animal food to their children. In this manner parents share resources that they themselves can gather at high rates but that their children cannot. These findings together suggest that food sharing in humans evolved as a strategy to assist the young and the help-

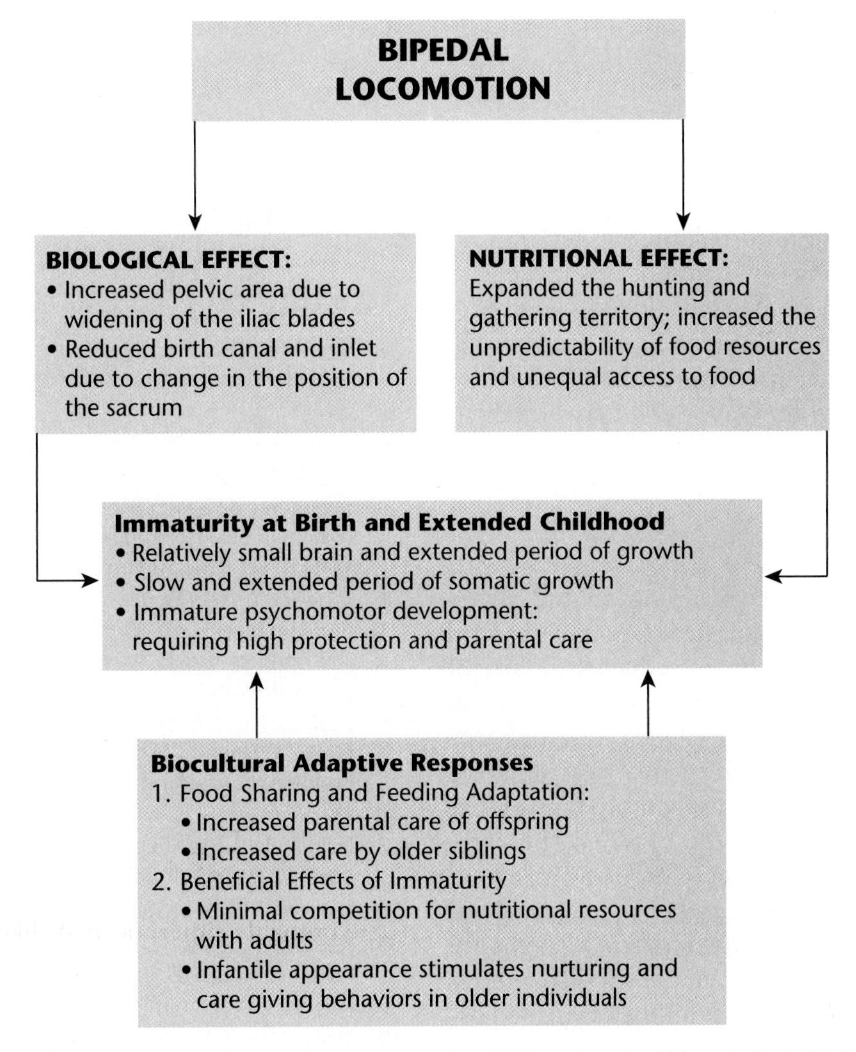

FIGURE 14.8. Schematization of the evolution of altricial development and extended period of childhood brought about by bipedalism and change in size and shape of pelvis. (Source: Frisancho, A. R.)

less obtain food without which survival would not have been possible. Thus, the commitment to caring for and feeding vulnerable young renders food sharing as a fundamental feature of human adaptation to the extended period of childhood.

Beneficial Effects of Immaturity

The allometric pattern of human growth, characterized by slow growth in body size and accelerated growth in brain size, results in the maintenance of an infantile appearance. This infantile appearance (characterized by large head size, small face, small body size, and little sexual development) stimulates nurturing and caregiving behaviors in older individuals.[40, 41] Since adults are compelled to feed infants and young dependent children, feeding competition with these younger offspring is minimized so as not to bring the interests of juveniles into conflict with the interests of adults. Furthermore, in many traditional societies juveniles share the responsibilities of caretaking of their young siblings and practically become surrogate parents. The economic, social, and reproductive value of juveniles, as "babysitters," gives delayed sexual maturation added selective advantages.[40, 41] Likewise, a small body size prior to adolescence, coupled with delayed maturation is advantageous because it reduces some social and sexual competition between adults and young during the years when technological, social, and cultural learning occur.[40, 41] On the other hand, the relatively early neurological maturity versus late sexual maturity allows juveniles to provide much of their own care and also provide care for young children. This frees up adults for subsistence activity, adult social behaviors, and further childbearing.

In summary, in an unpredictable environment, an immature and helpless infant without parental care would not be able to obtain food. In this context, the food sharing and feeding adaptation would have evolved in response to the altricial birth and extended childhood of humans. In other words, once altricial development occurred, the survival and reproductive fitness of early hominids was made possible by the uniquely human biocultural adaptive responses such as increased parental care of offspring, food sharing, etc. In addition, the altriciality and biological immaturity of humans would have been beneficial to the species.

Summary

Bipedalism had two effects: biological and nutritional (**fig. 14.8**). In the biological aspect, it changed the size and shape of the pelvis, reducing the size of the birth canal, which in turn produced the altricial pattern of growth characterized by a slow and prolonged childhood that requires a great deal of parental care. In the nutritional aspect, bipedalism expanded the environment and exposed the early hominids to a wide range and to unpredictable food resources, in which the young and helpless could not survive without the assistance of parents. Then once altricial development occurred, the survival and reproductive fitness of early hominids was made possible by the uniquely human biocultural adaptive responses such as food sharing and feeding adaptation associated with increased parental care of offspring, which together increased the reproductive fitness of our species. Thus, food sharing and feeding adaptation, at some point in human evolution, changed from being simply a nutritional supplement to a social instrument that enabled the young and helpless to survive in an unpredictable environment. In this scenario, the development of cultural responses is at the root of human evolution and is the behavioral trait that enabled hominids to survive and attain its present status despite its biological immaturity.

Suggested Readings

Bogin, B. 2001. *The Growth of Humanity.* New York: Wiley-Liss.

Falk, D. 2001. *Evolutionary Anatomy of the Primate Cerebral Cortex.* Cambridge, UK, and New York: Cambridge University Press.

Gould, S. J. 1977. *Ontogeny and Phylogeny.* Cambridge: Harvard University Press.

Rosenberg, K. R. 1988. The functional significance of Neanderthal pubic length. *Curr. Anthrop.* 29:595–617.

Thompson, D. W. 1942. *On Form and Growth.* Cambridge: Cambridge University Press.

Literature Cited

1. Hunt, K. D. 1996. The postural feeding hypothesis: An ecological model for the evolution of bipedalism. *S. Afr. J. Sci.* 92:77–90.
2. Sarmiento, E. E. 1988. Anatomy of the hominoid wrist joint: Its evolutionary and functional implications. *Int. J. Primatol.* 9:281–345.
3. Sarmiento, E. E. 1994. Terrestrial traits in the hands and feet of gorillas. *Am. Mus. Nat. Hist. Novitates* 3091:1–56.
4. Richmond, B. G, D. D. S. Begun, and D. S. Strait. 2001. Origin of human bipedalism: The knuckle-walking hypothesis revisited. *Yrbk. Phys. Anthropol.* 44:70–105.
5. Wheeler, P. E. 1991. The thermoregulatory advantages of hominid bipedalism in open equatorial environments: The contribution of increased convective heat loss and cutaneous evaporative cooling. *J. Hum. Evol.* 21:107–115.
6. Falk, D., J. C. J. Redmond, J. Guyer, C. Conroy, W. Recheis, G. W. Weber, and H. Seidler. 2000. Early hominid brain evolution: A new look at old endocasts. *J. Hum. Evol.* 38(5): 695–717.
7. Falk, D. 2001. *Evolutionary Anatomy of the Primate Cerebral Cortex.* New York: Cambridge University Press.
8. Lovejoy, C. O. 1981. The origins of man. *Science* 211:341–348.
9. Kaplan, H., K. Hill, J. Lancaster, and M. Hurtado. 2000. A theory of human life history evolution: diet, intelligence, and longevity. *Evol. Anthrop.* 9:156–185.
10. Wrangham, R. W. and B. Smuts. 1980. Sex differences in behavioral ecology of chimpanzees in Gombe National Park, Tanzania. *J. Reprod. Fertil.* (Suppl) 28.
11. Lancaster, J. B. 1978. Carrying and sharing in human evolution. *Hum. Nat.* 1:82–89.
12. Lancaster, J. B. 1997. The evolutionary history of human parental investment in relation to population growth and social stratification. In *Feminism and Evolutionary Biology,* P.A. Gowaty, ed. New York: Chapman and Hall, 466–489.
13. LaVelle, M. 1995. Natural selection and developmental sexual variation in the human pelvis. *Amer. J. Phys. Anthrop.* 98:59–72.
14. Berge, C. 1998. Heterochronic processes in human evolution: an ontogenetic analysis of the hominid pelvis. *Am. J. Phys. Anthropol.* 105:441–459.
15. Novotny, R., J. Davis, R. Wasnich, I. Biernacke, and A. Onaka. 2000. Maternal pelvic size, measured by dual energy x-ray absorptiometry, predicts infant birthweight. *Am. J. Hum. Biol.* 12:552–557.
16. Videan, E. N. and W. C. McGrew. 2002. Bipedality in chimpanzee (*Pan troglodytes*) and bonobo (*Pan paniscus*): Testing hypotheses on the evolution of bipedalism. *Am. J. Phys. Anthropol.* 118:184–190.
17. Schultz, A. 1949. Sex differences in the pelvis of primates. *Am. J. Phys. Anthrop.* 7:401–424.

18. Tague, R. G. and C. O. Lovejoy. 1986. The obstetric pelvis of A. L. 288-1 (Lucy). *J. Hum. Evol.* 15:237–255.

19. Trevathan, W. and K. R. Rosenberg. 2000. The shoulders follow the head: Postcranial constraints on human childbirth. *J. Hum. Evol.* 39:583–586.

20. Leutenegger, W. 1987. Neonatal brain size and neurocranial dimensions in Pliocene hominids: Implications for obstetrics. *J. Hum. Evol.* 16:291–296.

21. Rosenberg, K. R. 1988. The functional significance of Neanderthal pubic length. *Curr. Anthropol.* 29:595–617.

22. Rosenberg, K. R. 1992. The evolution of modern human childbirth. *Yearb. Phys. Anthropol.* 35:89–124.

23. Rosenberg, K. R. and W. T. Trevathan. 2001. The evolution of human birth. *Sci. Am.* (Nov):77–78.

24. Trevathan, W. R. 1988. Fetal emergence pattern in evolutionary perspective. *Am. Anthropol.* 90:674–681.

25. Tague, R. G. 1994. Maternal mortality or prolonged growth: age at death and pelvic size in three prehistoric amerindian populations. *Am. J. Phys. Anthropol.* 95:27–40.

26. Ruff, C. B. 1995. Biomechanics of the hip and birth in early Homo. *Am. J. Phys. Anthropol.* 98:527–574.

27. Leigh, S. 2001. Evolution of human growth. *Evol. Anthropol.* 10:223–236.

28. Martin, R. D. and A. M. MacLarnon. 1990. Reproductive patterns in primates and other mammals. The dichotomy between altricial and precocial offspring. In *Primate Life History and Evolution,* ed. C. J. DeRoussea. New York: Wiley-Liss, 47–88.

29. Thompson, D. W. 1942. *On Form and Growth.* Cambridge: Cambridge University Press.

30. Gould, S. J. 1977. *Ontogeny and Phylogeny.* Cambridge: Harvard University Press.

31. Penin, X., C. Berge, and M. Baylac. 2002. Ontogenetic study of the skull in modern humans and the common chimpanzees: Neotenic hypothesis reconsidered with a tridimensional Procrustes analysis. *Am. J. Phys. Anthropol.* 118(1):50–62.

32. Shea, B. T. 1992. Developmental perspective on size change and allometry in evolution. *Evol. Anthropol.* 4:125–134.

33. Rice, S. H. 1997. The analysis of ontogenetic trajectories: when a change in size or shape is not heterochrony. Proceedings of the National Academy of Sciences of the United States of America. 94(3)(Feb 4):907–912.

34. Dimirjian, A. 1986. Dentition. In *Human Growth, Vol. 2: Postnatal Growth,* ed. F. Falkner and J.M. Tanner. New York: Plenum.

35. Smith, B. H. 1992. Life history and the evolution of human maturation. *Evol. Anthropol.* 1:134–142.

36. Tanner, J. M. 1978. *Fetus into Man.* Cambridge: Harvard University Press.

37. Garn, S. M. 1987. The secular trend in size and maturational timing and its implications for nutritional assessment. *J. Nutr.* 117(5)(May):817–823.

38. Nishida, T., H. Takasaki, and Y. Takahata. 1990. Demography and reproductive profiles. In Nishida, T., ed. *The Chimpanzees of the Mahale Mountains: Sexual and Life History Strategies.* Tokyo: University of Tokyo Press, 63–97.

39. Watts, D. P. and A. E. Pusey. 1993. Behavior of juvenile and adolescent great apes. In: M. E. Pereira and L. A. Fairbanks, eds., *Juvenile Primates.* Oxford: Oxford University Press, 148–170.

40. Bogin, B. 1988. *Patterns of Human Growth.* Cambridge: Cambridge University Press.

41. Bogin, B. 2001. *The Growth of Humanity.* New York: Wiley-Liss.

SECTION IV

Human Adaptation and Biological Diversity

Ever since hominids left Africa, humans have expanded throughout the world, have adapted to diverse environments, and acquired specific biological and cultural traits that enabled them to survive in a given area. This section addresses the study of the biocultural processes whereby humans have adapted to their current and past environments. It includes nine chapters—chapter 15 summarizes the general principles for the study of human adaptation; chapter 16 focuses on the life cycle; chapter 17 discusses nutritional factors; chapter 18 interprets variability in body and proportion in reference to climatic factors among contemporary and past populations; chapter 19 summarizes the physiological adaptive responses of contemporary population to hot and cold climates; chapter 20 summarizes the biological adaptation to high-altitude environments; chapter 21 discusses the biocultural origins of lactose digestibility, and the evolution of skin color; chapter 22 addresses how humans are adapting to a changing world environment; and chapter 23 summarizes the factors associated with the globalization of obesity.

General Principles for the Study of Human Adaptation

Human Diversity and the Use and Misuse of the Concept of Race

To some people the term race is a four-letter word associated with negative connotations, while for others it refers to actual biologically inherited traits. Skin color is the most readily visible signifier of race, and as such is the characteristic upon which most racial classifications are based.[1,2]

Historically, the ancient Egyptians were the first to classify humans on the basis of skin color. In 1350 B.C. Egyptians classified humans into four races: red for Egyptians, yellow for people living to the east of Egypt, white for people living north of Africa, and black for Africans from the south. On the other hand, the ancient Greeks referred to all Africans as "Ethiopians." A major tenet of the biological concept of race is that the traits that identify a given race are *unchangeable* and have been *fixed* since the beginning of humankind. However, during the last century an evolutionary approach led by anthropologists and human biologists has emerged calling into question the validity of the biological concept of race.

In the early 1900s, head shape was considered an innate "racial" trait that was inherited with little environmental influence, but this concept changed with the pioneering studies of Franz Boas.[3] In 1912, Boas demonstrated that head shape (measured by the cephalic index: head width/head length) among immigrants to the United States can be changed by growing up in a different environment.[2] These and subsequent genetic studies demonstrate that the biological features distinguishing racial groups are subject to environmental influence (see chapter 3, and chapters 15 to 21) and are of recent origin. Furthermore, recent data and models from DNA suggest that common race definitions pertaining to humans have little taxonomic validity,[4] since there is no correlation between genetic markers such as blood type and markers for race such as skin color.[5]

For example, as shown in **figure 15.1** the Australian Aborigines, East and West African populations, and native populations from India have a similar dark skin color. Based on this trait they could be assigned to an "African race." However, with reference to frequencies of the B blood group and Rh blood gene C and E, the Australian Aborigines are highly different from the East or West Africans and Indian populations. In other words, there is no concordance between blood type and skin color. Likewise, the ABO blood type frequencies for natives from Taiwan and Greece are very similar (O = 45.2%, A = 32.6%, B = 18.0%, AB = 3.4%). Yet on the bases of geography and physical appearance, these two populations clearly belong to different categories. Additionally, the indigenous populations from sub-Saharan Africa, southern Europe, the Middle East, and India have similar frequencies of the sickle-cell trait (20–34 percent), yet differ in skin color. The similarity of these populations in the frequency of sickle-cell trait is related to their common adaptation to malaria,[6–9] rather than a common racial origin (see chapter 3). Similarly, lactose tolerance occurs both in European and African populations (see chapter 21), not because they have the same racial origin but because both were evolutionarily adapted to dairy products. In other words, the concept of race is both too broad and too narrow a definition of ancestry to be biologically useful.

The reason why the definitions of race lose their discriminating power for identifying race is due to the fact that humans have been constantly migrating throughout the evolutionary history and share a common origin. For example, the large-scale migrations between Africa, Europe, and its colonial expansion to Asia and the New World has resulted in the mating of individuals from different continents and the concomitant mixture of genetic traits.

For these reasons, using the biological concept of race for describing biological diversity has largely been abandoned. Nevertheless, because the risks of some diseases have a genetic basis in some populations that may have originated in a geographic region that differs from their current area, there is still great interest in understanding how genetic diversity has been structured in the human species.

Criteria for Racial Classification

In the taxonomic literature, race is any distinguishable type within a species, but among researchers the term race as a biological concept has had a variety of meanings. Some use frequency of genetic traits between and within groups, while others use geographical area as the point of reference.

Trait Frequency

Genetic studies demonstrate that about 85.4 percent of all the variation in the human species can be attributed to allelic variation within populations and that there is only a 6.3 percent difference between "races," with less than half of this value accounted for by known racial groupings.[10, 11] In other words, there is much more genetic variation *within* local groups than

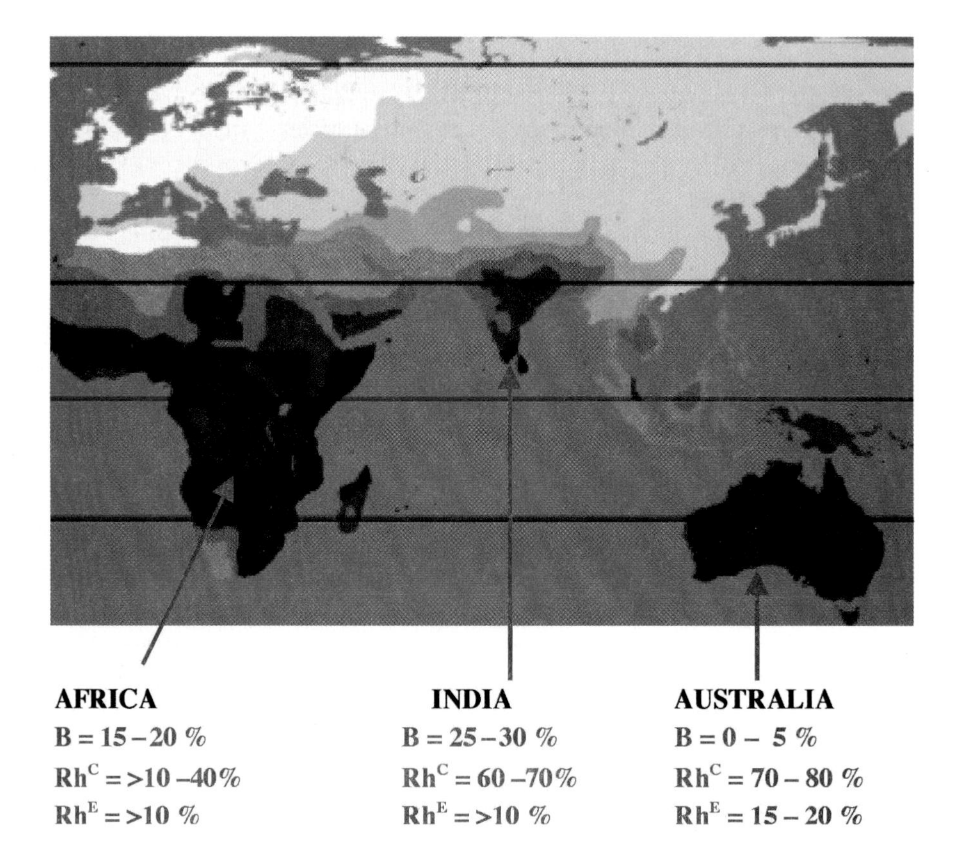

FIGURE 15.1. Distribution of skin color, B blood group, and Rh blood group Rh^C and Rh^E in native populations of the Africa, India, and Australia. Although Australian, African, and Indian populations have similar skin color, they have very different blood types. (Composite illustration made by the author.)

there is *among* local groups or among races themselves. This genetic unity means, for instance, that any local group contains on average 85 percent of the genetic variation that exists in the entire human species. As a result, between any two individuals there is about 15 percent genetic variation. Therefore, a randomly selected white European, although ostensibly far removed from black Americans in phenotype, can easily be genetically closer to an African black than to another European white. As summarized by Long and Kittles,[12] the patterns of genetic variation within and between groups are too intricate to be reduced to a single summary measure. In other words, identification of trait frequencies and statistical partitions of genetic variation do not provide accurate information for the existence of races.

Geographical Race

Because some phenotypes such as skin color, facial features, and hair form differ between native inhabitants of different regions of the world, biological anthropologists and geneticists[13–16] introduced the idea of geographical races. In this classificatory approach each geographic region such as South America, Australia, sub-Saharan Africa, East Asia or Polynesia is associated with a race. According to these authors,[13–16] geographical races refer to an aggregate of phenotypically similar populations of a species inhabiting a geographic subdivision. An underlying assumption of this approach is that in each geographical area there are clusters of genetic traits that together differentiate them from other geographic areas. Current evidences indicate that variability in the genotypic and phenotypic expression of genetic traits is affected by natural selection, migration, or genetic drift (see chapter 3). As a result of the action of these processes the genetic diversity follows a pattern characterized by gradients of allele frequencies that extend over the entire world.[17, 18] In other words, when identified, the clustering of genetic traits in a given area reflects the demographic and evolutionary history of the population rather than a racial category. Therefore, there is no reason to assume that "races" represent any units of relevance for understanding human genetic history.

In summary, and as stated by the American Association of Physical Anthropologists' statement on race:[19] (a) all human populations derive from a common ancestral group, (b) there is great genetic diversity within all human populations, and (c) the geographic pattern of variation is complex and presents no major discontinuity. In other words, race is a consequence of social history and therefore any variation is transitory. For these reasons, among biological anthropologists the biological concept of race for describing biological diversity has been largely abandoned. Yet, the issue of race with reference to IQ has recently received attention, which will be discussed next.

IQ and Race: Misuse of Scientific Information

An illustration of the dangers of misusing information on IQ and heritability is found in studies of IQ and race. Originally, the intelligence quotient (IQ) test was developed in 1904 by French psychologist Alfred Binet[20] to identify children's school reading readiness. The IQ test was intended to measure "mental age" in various categories. Binet warned that the IQ test could not be used to properly measure intelligence, "because intellectual qualities are not superposable, and therefore cannot be measured as linear surfaces are measured." Intelligence was not considered a fixed quantity, but could be increased through teaching. Yet, in the United States, tests of IQ have been used to measure general intelligence. As shown in **figure 15.2**, on

FIGURE 15.2. Schematization of the distribution of IQ in the United States (Composite made by the author.)

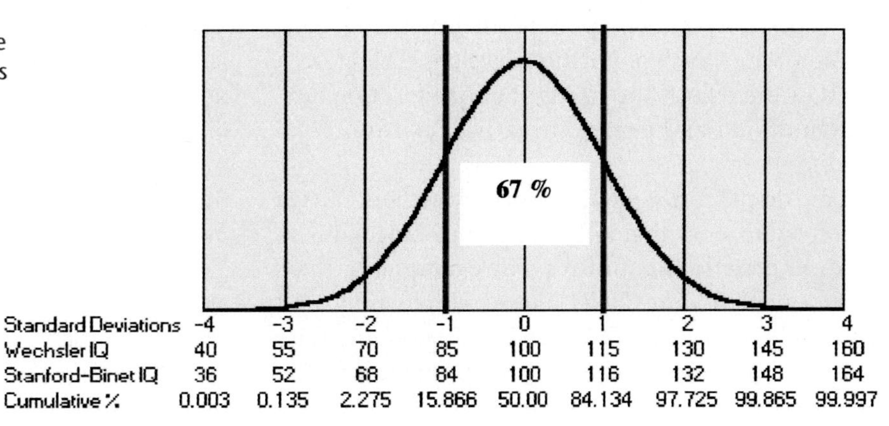

average the normal range for IQ for about 67 percent of the population falls between 85 to 115, while only 5 percent of the population attain IQ values greater than 140 and below 70.

The use of IQ as a measure of an individual's innate intelligence is not valid for two reasons. First, there are many kinds of intelligence. There are some people with outstanding memories, some with mathematical skills, some with musical talents, some good at seeing analogies, some good at synthesizing information, and some with manual and mechanical expertise. Therefore, these different kinds of intelligence cannot be subsumed with an IQ score.

Second, there is no evidence that IQ is genetically determined. It is true that about 60 percent of the variability in IQ is inherited within family lines, but the fact that it is inherited does mean that it is genetically determined. Discrete traits such as blood type that do not change through the life cycle are genetically determined and therefore have a high heritability, but continuous traits such as height, weight, or IQ are highly subject to environmental influence. Heritability is computed as the fraction of phenotypic variability due to genetic differences divided by total variability. Heritability is expressed as $h^2 = G / P = G / (G + E)$, where G is variability in genotype, E is variability in environment, and P is variability in phenotype. Depending upon whether the environmental variance (E) is large or small, the phenotypic variance (P) is either large or small, and the heritability (h^2) can be either large or small. Therefore, measures of heritability, especially of continuous traits such as intelligence, indicate the joint influence of genetic and environmental factors.[21,22] Previous twin and family studies have shown that shared environmental factors have an important effect on educational attainment.[23] Shared environmental factors such as education have a greater impact on intelligence during childhood than in adulthood.[24] In other words, heritability of intelligence—unlike genetic determination—can be very different in different populations depending upon the environmental condition in which each population develops. Therefore, a low IQ score reflects the effects of poor education during childhood and general negative environmental conditions.

Yet, despite these pitfalls, some researchers[25] have attempted to show that differences in IQ reflect differences in genetic capabilities. For example, in their book entitled *The Bell Curve,* Herrnstein and Murray[25] point to differences in IQ between white and black Americans as reflecting differences in the genetic capability of intelligence in each race. They point out that the distribution of IQ scores in black Americans is shifted to the left, so that there are higher frequencies of low IQ scores and lower frequencies of high IQ scores when compared to their white American sample. From our discussion of heritability, it is clear that this difference is more a reflection of the different educational experiences of black and white Americans. For example, in Chicago and Detroit some black children going to schools with improved learning conditions have attained scores at the 85th percentile. Similarly, in the Barclay School of Baltimore black children who used to score at the 20th percentile are now attaining scores at the 85th percentile. These findings together suggest that the lower IQ scores associated with black samples is more a function of educational experiences than of genetic determinants.

It is evident that cultural environment is an important contributor to any measures of IQ. This inference can be illustrated by several examples. First, suppose we consider two groups of 8-year-old children: (a) one from a middle-class U.S. school and (b) one from a poor rural area from Guatemala. These children are asked the following question: "Suppose you have 5 eggs and you drop 2, how many eggs do you have?" The U.S. children will likely answer that they have 3 eggs left, but the rural children may answer that they have 5 eggs. Based on that result, one may conclude that the Guatemalan rural children do not know how to add or subtract. However, if one takes into account the fact that the rural children have been raised in an environment associated with food shortage, just because an egg has been dropped does not mean it cannot be eaten. Hence, for the Guatemalan rural children there are still 5 eggs. Therefore, the answer depends on the children's past experience.

In another example, Australian Aborigine trackers and Peruvian Andean weavers are asked to identify a series of drawings that will make a complete square as fast as possible. It is quite likely that the speed for the Peruvian Andean weavers will be faster than that of Australian Aborigine trackers. This difference is related to the fact that the Australian Aborigine trackers have little or no contact with the concepts of two-dimensional geometry, while the Peruvian Andean weavers' occupation involved experience with two-dimensional geometric designs. Thus, the difference in the response reflects the individuals' or populations' past experience.

Based on the above findings IQ should be defined as a measure of an individual's sum of cultural experience rather than a measure of genetic difference. However, this does not mean that a person's genetic makeup is not a significant factor in individual intelligence in particular areas. Without the proper environment, however, this trait may not be expressed.

Use of Geographic Area and Race in Biomedical Research

It is evident that the biological concept of race is a poorly defined term that cannot be used as a surrogate for multiple environmental and genetic factors in disease causation. Recent genetic studies of DNA polymorphisms have suggested that human genetic diversity is organized in continental or geographical areas.[17, 18] This conclusion suggests that geographic area rather than race per se has a valid role in biomedical research because many medically important genes vary in frequency between populations from different regions. If, for example, there are major differences in allele frequencies between geographic areas, individuals from different origins may often be expected to respond differently to medical treatments.[26] Then the identification of the origin of geographic area of people does have justification as a proxy for differences in environmental and other factors of relevance for public health.

However, the ability to place an individual within a geographic region and range of variation does not mean that this variation is best represented by the concept of race. For example, sickle-cell disease is instead a characteristic of ancient ancestry in a geographic region where malaria was endemic such as Africa, the Mediterranean, and southern India (see chapter 3) rather than characteristic of a particular racial group. Therefore, a diagnostic approach toward sickle-cell disease must take into account the individual's geographical ancestry. Similarly, populations who throughout their evolutionary history have developed adaptive response to economize salt loss under the condition of tropic heat stress (see chapters 16 to 21), are more susceptible to developing high blood pressure than other populations with similar salt intake when living in temperate climates. In other words, in biomedical research, it is not race that is relevant, but how the forces of evolution in a geographic area have shaped the individual's genes. In summary, because an individual's genes are grounded in their genealogy, identifying all contributions to a patient's ancestry is useful in diagnosing and treating diseases with genetic influences.

In summary, the classical criteria for racial classification is based on phenotypes such as skin color, facial features, and hair form that differ between native inhabitants of different regions of the world. It is assumed that all of these defining features reflect differences in genetic determinants between races. However, recent studies of genetic diversity indicate that there is no concordance between phenotypic traits used to assign races and gene frequencies. Furthermore, genetic studies indicate there is much more genetic variation *within* local groups than there is *among* local groups or among races. The biological concept of race is a poorly defined term to be biologically useful for understanding the origin of this diversity. One productive approach is to study the biocultural processes whereby humans have adapted to their current and past environments, a topic that will be addressed in the next section and eight chapters.

General Principles for the Study of Human Adaptation

The conceptual framework of research in biological anthropology is that evolutionary selection processes have produced the human species and that the processes have produced a set of genetic characteristics, which adapted our evolving species to their environment. Recent investigations demonstrate that the phenotype measured morphologically, physiologically, or biochemically was the product of genetic plasticity operating during development. Within this framework, it is assumed that some of the biological adjustments or adaptations people made to their natural and social environments have also modified how they adjusted to subsequent environments. The adjustments we have made to improve our adaptations to a given environment have produced a new environment to which we, in turn, adapt in an ongoing process of new stress and new adaptation. The focus of this chapter is to summarize the general principles and definitions underlying the study of human adaptation.

Functional Adaptation

The term *adaptation* is used in the broad generic sense of functional adaptation, and it is applied to all levels of biological organization, from individuals to populations. A basic premise of this approach is that adaptation is a process whereby the organism has attained a beneficial adjustment to the environment.[1–59] This adjustment can be either temporary or permanent; acquired either through short-term or lifetime processes; and may involve physiological, structural, behavioral, or cultural changes aimed at improving the organism's functional performance in the face of environmental stresses. If environmental stresses are conducive to differential mortality and fertility, then adaptive changes may become established in the population through changes in genetic composition and thus attain a level of genetic adaptation. In this context, functional adaptation, along with cultural and genetic adaptation, is viewed as part of a continuum in an adaptive process that enables individuals and populations to maintain both internal and external environmental homeostasis. Therefore the concept of adaptation is applicable to all levels of biological organization from unicellular organisms to the largest mammals and from individuals to populations. This broad use of the concept of adaptation is justified not only in theory but also because it is currently applied to all areas of human endeavor so that no discipline can claim priority or exclusivity in the use of the term.[46] Functional adaptation involves changes in organ system function, histology, morphology, biochemical composition, anatomical relationships, and body composition, either independently or integrated in the organism as a whole. These changes can occur through acclimation, habituation, acclimatization, or genetic adaptation.

Acclimation

Acclimation refers to the adaptive biological changes that occur in response to a single experimentally induced stress[47, 48] rather than to multiple stresses as occurs in acclimatization. As with acclimatization, changes occurring during the process of growth may also be referred to as developmental acclimation.[49, 58]

Habituation

Habituation implies a gradual reduction of responses to, or perceptions of, repeated stimulation. [47, 48] By extension, habituation refers to the diminution of normal neural responses, for example, the decrease of sensations such as pain. Such changes can be generalized for the whole organism (general habituation) or can be specific for a given part of the organism (specific habituation). Habituation necessarily depends on learning and conditioning, which enable the organism to transfer an existing response to a new stimulus. The extent to which these physiological responses are important in maintaining homeostasis depends on the severity of environmental stress. For example, with severe cold stress or low oxygen availability, failure to respond physiologically may endanger the well-being and survival of the organism. A common confusion is that habituation does not involve a physiological process; however, one cannot get accustomed to a given stress or acquire the ability to ignore it, unless one has had a previous experience.

Acclimatization

Acclimatization refers to changes occurring within the lifetime of an organism that reduce the strain caused by stressful changes in the natural climate or by complex environmental stresses.[47, 48, 60] If the adaptive traits are acquired during the growth period of the organism, the process is referred to as either *developmental adaptation* or *developmental acclimatization.*[49, 58]

Developmental Acclimatization

The concept of developmental acclimatization (also referred to as developmental adaptation) is based upon the fact that the organism's plasticity and susceptibility to environmental influence is inversely related to developmental states of the organism, so that the younger the individual the greater is the influence of the environment and the greater the organism's plasticity.[49] Hence, variability in physiological traits can be traced to the developmental history of the individual.

Acclimatization versus Acclimation

Studies on acclimatization are done with reference to both major environmental stresses and several secondary, related stresses. For example, any difference in the physiological and structural characteristics of subjects prior to and after residence in a tropical environment are interpreted as a result of acclimatization to heat stress. In addition, because tropical climates are also associated with nutritional and disease stresses, individual or population differences in function and structure may also be related to these factors. On the other hand, in studies of acclimation any possible differences are easily attributed to the major stress to which the experimental subject has been exposed in the laboratory. For understanding the basic physiological processes of adaptation, studies on acclimation are certainly better than those of acclimatization. However, since all organisms are never exposed to a single stress but instead to multiple stresses, a more realistic approach is that of studying acclimatization responses. Thus, both studies on acclimation and acclimatization are essential for understanding the processes whereby the organism adapts to a given environmental condition. This rationale becomes even more important when the aim is to understand the mechanisms whereby humans adapt to a given climatic area, since humans in a given area are not only exposed to diverse stresses but have also modified the nature and intensity of these stresses as well as created new stresses for themselves and for generations to come.

Accommodation and Adaptation

The term **accommodation** is used to describe responses to environmental stresses that are not wholly successful because even though they favor survival of the individual they also result in significant losses in some important functions. For example, when exposed to low intake of leucine for three weeks, subjects can achieve body leucine balance at the expense of reducing protein synthesis and protein turnover. Since a low protein synthesis and protein turnover diminishes the individual's capacity to successfully withstand major stresses such as infectious diseases, under conditions of low dietary protein intake achieving body leucine balance represents only a temporary accommodation, which in the long run is not adaptive.

Individuals versus Populations

Whatever the method employed, geographical or experimental research in human adaptation is concerned with populations, not with individuals, although the research itself is based on individuals. There are two reasons for this.

The first is a practical consideration. Studying all members of a given population, unless its size is small enough, is too difficult to be attempted by any research team. Therefore, according to the objectives of the investigation, the research centers on a sample that is considered representative of the entire population. Based on these studies, the researchers present a picture of the population as a whole, with respect to the problem being investigated.

The second reason is a theoretical one. In the study of adaptation, we usually focus on populations rather than on individuals because it is the population that survives and perpetuates itself. In the investigation of biological evolution, the relevant population is the breeding population because it is a vehicle for the gene pool, which is the means for change and hence evolution. The study of an individual phenomenon is only a means to understand the process. The adaptation of any individual or individuals merely reflects the adaptation that has been achieved by the population of which he is a member.

Cultural and Technological Adaptation

Cultural adaptation refers to the nonbiological responses of the individual or population to modify or ameliorate an environmental stress. As such, cultural adaptation is an important mechanism that facilitates human biological adaptation.[57, 59] It may be said that cultural adaptation, both during contemporary times and in evolutionary perspective, represents humanity's most important tool. It is through cultural adaptation that humans have been able to survive and colonize far into the zones of extreme environmental conditions. Humans have adapted to cold environments by inventing fire and clothing, building houses, and harnessing new sources of energy. The construction of houses, use of clothing in diverse climates, certain behavioral patterns, and work habits represent biological and cultural adaptations to climatic stress. The development of medicine from its primitive manifestations to its high levels in the present era and the increase of energy production associated with agricultural and industrial revolutions are representative of human cultural adaptation to the physical environment.

Culture and technology have facilitated biological adaptation, but they have also created and continue to create new stressful conditions that require new adaptive responses. A modification of one environmental condition may result in the change of another. Such a change may eventually result in the creation of a new stressful condition. Advances in the medical sciences have successfully reduced infant and adult mortality to the extent that the world population is growing at an explosive rate, and unless world food resources are increased, the twenty-first century will witness worldwide famine. Western technology, although upgrading living standards, has also created a polluted environment that may become unfit for good health and life. If this process continues unchecked, environmental pollution will eventually become another selective force to which humans must adapt through biological or cultural processes or face extinction. Therefore, the ability to adapt to the unforeseeable threats of the future remains an indispensable condition of survival and biological success.[55] Adaptation to the world of today may be incompatible with survival in the world of tomorrow unless humans learn to adjust their cultural and biological capacities.

Purpose of Adaptation

Environmental Stress

Central to the study of adaptation is the concept of homeostasis and environmental stress. An environmental *stress* is defined as any condition that disturbs the normal functioning of the organism. Such interference eventually causes a disturbance of internal homeostasis. *Homeostasis* means the ability of the organism to maintain a stable internal environment despite diverse, disruptive, external environmental influence.[55] On a functional level, all adaptive responses of the organism or the individual are made to restore internal homeostasis. These controls operate in a hierarchy at all levels of biological organization, from a single biochemical pathway, to the mitochondria of a cell, to cells organized into tissues, tissues into organs and systems of organs, to entire organisms. For example, the lungs provide oxygen to the extracellular fluid to continually replenish the oxygen that is being used by the cells, the kidneys maintain constant ion concentrations, and the gastrointestinal system provides nutrients.

Humans living in hot or cold climates must undergo some functional adjustments to maintain thermal balance; these may comprise the rate of metabolism, avenues of heat loss, heat conservation, respiration, blood circulation, fluid and electrolyte transport, and exchange. In the same manner, persons exposed to high altitudes must adjust through physiological, chemical, and morphological mechanisms, such as an increase in ventilation, an increase in the oxygen-carrying capacity of the blood resulting from an increased concentration of red blood cells, and an increased ability of tissues to utilize oxygen at low pressures. Failure to activate the functional adaptive processes may result in failure to restore homeostasis, which in turn results in maladaptation of the organism and eventual incapacity of the individual.

Homeostasis

Therefore, homeostasis is a part and function of survival. The continued existence of a biological system implies that the system possesses mechanisms that enable it to maintain its identity, despite the endless pressures of environmental stresses.[55] These complementary concepts of homeostasis and adaptation are valid at all levels of biological organization. They apply to social groups as well as to unicellular or multicellular organisms.[55]

Homeostasis is a function of a dynamic interaction of feedback mechanisms whereby a given stimulus elicits a response aimed at restoring the original equilibrium. Several mathematical models of homeostasis have been proposed. In general, they show (as schematized in **fig. 15.3**) that when a primary stress disturbs the homeostasis that exists between the organism and the environment, to function normally the organism must resort either to biological or cultural-technological responses. For example, when faced with heat stress, the organism may simply reduce its metabolic activity so all heat-producing processes are slowed down, or may increase the activity of the heat-loss mechanisms. In either case, the organism may maintain homeostasis, but the physiological processes will occur at a different set point. The attainment of full homeostasis or full functional adaptation, depending on the nature of the stress, may require short-term responses such as those acquired during acclimation

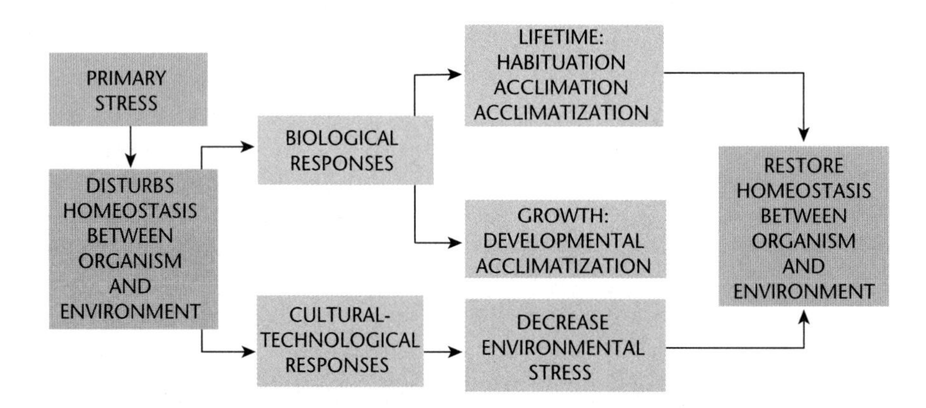

FIGURE 15.3. Schematization of adaptation process and mechanisms that enable individual or population to maintain homeostasis in the face of primary environmental disturbing stress. From *Human Adaptation and Accommodation* by A. R. Frisancho. Copyright © 1995 by University of Michigan Press. Reprinted by permission.

or acclimatization or may require exposure during the period of growth and development as in developmental acclimatization. In theory, the respective contributions of genetic and environmental factors vary with the developmental stage of the organism—the earlier the stage, the greater the influence of the environment and the greater the plasticity of the organism.[49, 55, 58] However, as will be shown in this part of this book, the principle does not apply to all biological parameters; it depends on the nature of the stress, the developmental stage of the organism, the type of organism, and the particular functional process that is affected. For example, an adult individual exposed to high-altitude hypoxia through prolonged residence may attain a level of adaptation that permits normal functioning in all daily activities, and as such, we may consider him adapted. However, when exposed to stress that requires increased energy, such as strenuous exercise, this individual may prove to be not fully adapted. On the other hand, through cultural and technological adaptation humans may actually modify and thus decrease the nature of the environmental stresses so that a new microenvironment is created to which the organism does not need to make any physiological responses. For example, cultural and technological responses permit humans to live under extreme conditions of cold stress with the result that some of the physiological processes are not altered. However, on rare occasions, humans have been able to completely avoid an environmental stress. Witness the fact that the Eskimos, despite their advanced technological adaptation to cold in their everyday hunting activities, are exposed to periods of cold stress and in response have developed biological processes that enable them to function and be adapted to their environment.

Not all responses made by the organism can be considered adaptive. Although a given response might not be adaptive per se, through its effect on another structure or function it may prove beneficial to the organism's function. Conversely, a given adaptive response may aid the organism in one function but actually have negative effects on other functions or structures. Thus, within all areas of human endeavor, a given trait is considered adaptive when its beneficial effects outweigh the negative ones. In theory this is a valid assumption, but in practice, because of the relative nature of adaptation, it is quite difficult to determine the true adaptive value of a given response. Every response must be considered in the context of the environmental conditions in which the response was measured and within the perspective of the length of time of the study and the subject population.

Genetic Adaptation and Adaptability

Genetic adaptation refers to specific heritable characteristics that favor tolerance and survival of an individual or a population in a particular total environment. A given biological trait is considered genetic when it is unique to the individual or population and when it can be shown that it is acquired through biological inheritance. A genetic adaptation becomes established through the action of natural selection.[6–16, 29–36] Natural selection refers to the mechanisms whereby the genotypes of those individuals showing the greatest adaptation or "fitness" (leaving the most descendants through reduced mortality and increased fertility) will be perpetuated, and those less adapted to the environment will contribute fewer genes to the population gene pool. Natural selection favors the features of an organism that bring it into a more efficient relationship with its environment. Those gene combinations fostering the best-adapted phenotypes will be "selected for," and inferior genotypes will be eliminated. The selective forces for humans, as for other mammals, include the sum total of factors in the natural environment. All the natural conditions, such as hot and cold climates and oxygen-poor environments, are potential selective forces. Food is a selective force by its own abundance, eliminating those susceptible to obesity and cardiac failures, or by its very scarcity, favoring smaller size and slower growth. In the same manner, disease is a powerful selective agent, favoring in each generation those with better immunity. The natural world is full of forces that make some individuals, and by inference some populations, better adapted than others because no two individuals or populations have the same capacity of adaptation. The maladapted population will tend to have lower fertility and/or higher mortality than that of the adapted population.

The capacity for adaptation (adaptability) to environmental stress varies between populations and even between individuals. The fitness of an individual or population is determined by its total adaptation to the environment—genetic, physiological, and behavioral (or cultural). Fitness in genetic terms includes more than just the ability to survive and reproduce in a given environment; it must include the capacity for future survival in future environments. The long-range fitness of a population depends on its genetic stability and variability. The greater the adaptation, the longer the individual or population will survive, and the greater the advantage in leaving progeny resembling the parents. In a fixed environment, all characteristics could be under rigid genetic control with maximum adaptation to the environment. On the other hand, in a changing environment a certain amount of variability is necessary to ensure that the population will survive environmental change. This requirement for variability can be fulfilled either genetically or phenotypically or both. In most populations a compromise exists between the production of a variety of genotypes and individual flexibility. Extinct populations are those that were unable to meet the challenges of new conditions. Thus, contemporary fitness requires both genetic uniformity and genetic variability.

Therefore, contemporary adaptation of human beings is both the result of their past and their present adaptability.[52] It is this capacity to adapt that enables them to be in a dynamic equilibrium in their biological niche. It is the nature of the living organism to be part of an ecosystem whereby it modifies the environment and, in turn, is also affected by such modification. The maintenance of this dynamic equilibrium represents homeostasis, which, in essence, reflects the ability to survive in varying environments.[46, 55] The ecosystem is the fundamental biological entity—the living individual satisfying its needs in a dynamic relation to its habitat. In Darwinian terms, the ecosystem is the setting for the struggle for existence, efficiency and survival are the measures of fitness, and natural selection is the process underlying all products.[55]

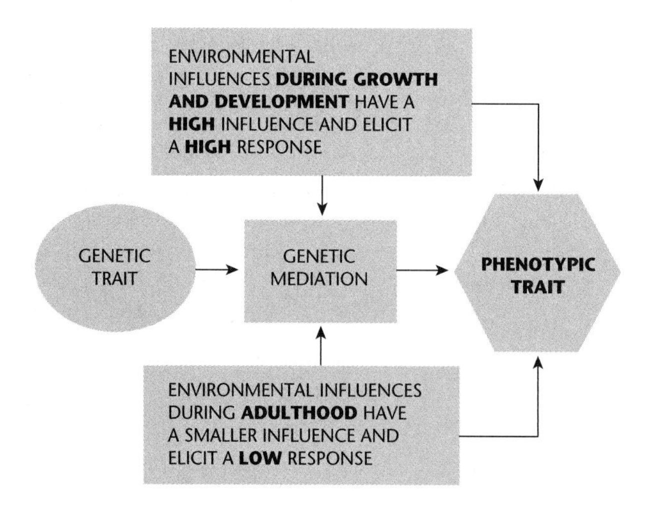

FIGURE 15.4. Schematization of interaction of genetic mediation and the environment and the phenotypic morphological outcome. Morphological and physiological diversity reflects the responses and adaptations that the organism makes to a particular environment during development. (From Frisancho, A. R., and Schechter, D. E. 1997. *Adaptation. History of Physical Anthropology: An Encyclopedia.* Garland Press, Vol. I: 6–12.)

In general, the morphological and functional features reflect the adaptability or capacity of the organism to respond and adapt to a particular environment (**fig. 15. 4**). The effect and responses to a given environmental condition is directly related to the developmental stage of the organism, so that the younger the age the greater the effect and the greater the flexibility to respond and adapt. Conversely, the later the age and, especially during adulthood, the effect of the environment is less likely to be permanent, and the capacity to respond and adapt is also diminished when compared to a developing organism. The capacity to respond positively and adapt to an environmental stress is mediated by genetic traits. A corollary of this assumption is that external features are likely to represent traits that are suitable to the environments in which they are found.

Summary

Current research on human diversity indicates that most genetic diversity is found between individuals rather than between populations. The usual criteria for classifying racial groups has been skin color, yet classification of populations based on skin color is not correlated to blood types, which is genetically inherited. Hence, there is no valid reason for classifying humans into racial groups. Current evidence indicates that using phenotypic traits such as skin color and IQ, which are subject to educational experience, is a misuse of scientific information. On the other hand, identification of an individual's genes with reference to the geographic area of ancestry is useful in diagnosing and treating diseases with genetic influences. This is because an individual's genes have been shaped by the forces of evolution in a geographic area.

The term *adaptation* encompasses the physiological, cultural, and genetic adaptations that permit individuals and populations to adjust to the environment in which they live. These adjustments are complex, and the concept of adaptation cannot be reduced to a simple rigid definition without oversimplification. The functional approach in using the adaptation concept permits its application to all levels of biological organization from unicellular to multicellular organisms, from early embryonic to adult stages, and from individuals to populations. In this context, human biological responses to environmental stress can be considered as part of a continuous process whereby past adaptations are modified and developed to permit the organism to function and maintain equilibrium within the environment to which it is exposed daily.

How the organism adapts to a given environmental condition that differs in nutritional resources and climate determines the phenotypic expression of physiology, body size, and proportion of contemporary and past populations.

Suggested Readings

Frisancho, A. R. 1993. *Human Adaptation and Accommodation.* Ann Arbor, Michigan: University of Michigan Press.

Literature Cited

1. Brown, R. A. and G. J. Armelagos. 2001. Apportionment of racial diversity: A review. *Evol. Anthropol.* 10:34–40.
2. Brace, C. L. 2005. *Race Is a Four Letter Word: The Genesis of the Concept.* New York: Oxford University Press.
3. Boas, F. 1912. *Changes in the Bodily Form of Descendants of Immigrants.* New York: Columbia University Press.
4. Lewontin, R. C. 2002. Directions in evolutionary biology. *Annu. Rev.*
5. Lieberman, L., and F. L. C. Jackson. 1995. Race and three models of human origin. *Am. Anthropol.* 97:231–242.
6. Livingstone, F. B. 1963. On the non-existence of human races. *Curr. Anthropol.* 3:279–281.
7. Haldane, J. 1949. The rate of mutation of human genes. *Proc. VIIIth Int. Congr. Genet.* 1949: 267–272.
8. Allison, A. C. 1954. Protection afforded by sickle-cell trait against subtertian malarial infection. *Br. Med. J.* 1:290–294.
9. Allison, A. C. 1956. Sickle-cells and evolution. *Sci. Amer.* 195:87–94.
10. Lewontin, R. C. 1972. The apportionment of human diversity. *Evol. Biol.* 6:381–398.
11. Barbujani, G., A. Magani, E. Minch et al. 1997. An apportionment of human DNA diversity. *Proc. Natl. Acad. Sci. (USA)* 94:4516–4519.
12. Long, J. C. and R. A. Kittles. 2003. Human genetic diversity and the nonexistence of biological races. *Hum. Biol.* 75:449–471.
13. Dobzhansky, T. 1970. *Genetics of the Evolutionary Process.* New York, NY: Columbia University Press.
14. Brues, A. M. 1977. *People and Races.* New York: Macmillan.
15. Ganr, S. M. 1961. *Human Races.* Springfield, IL: Charles C. Thomas, Publisher.
16. Mayr, E. 2002. The biology of race and the concept of equality. *Daedalus* 131:89.
17. Serre, D. and S. Pääbo. 2004. Evidence for gradients of human genetic diversity within and among continents. *Genome Res.*14:1679–1685.
18. Feldman, M. W., C. Lewontin, and M. C. King. 2004. Race: A genetic melting-pot. *Nature* 424, 374 (24 July 2003); doi:10.1038/424374.
19. American Association of Physical Anthropologists. 1996. AAPA statement on biological aspects of race. *Am. J. Phys. Anthropol.* 101:569–570.
20. Alfred Binet, A and Th. Simon. 1916. (Translated by Elizabeth S. Kite.) *The Development of Intelligence in Children (The Binet-Simon Scale).* Baltimore: Williams & Wilkins.
21. Feldman, M. W., R. C. Lewontin. 1975. The heritability hang-up. *Science* 190:1163–1168.
22. Lewontin, R. C. 1975. Genetic aspects of intelligence. *Annu. Rev. Genet.* 9:387–405.
23. Silventoinen, K., R. F. Krueger, T. J. Bouchard Jr, J. Kaprio, M. McGue. 2004. Heritability of body height and educational attainment in an international context: comparison of adult twins in Minnesota and Finland. *Am. J. Hum. Biol.*16:544–555.
24. Boomsma, D., A. Busjahn, L. Peltonen. 2002. Classical twin studies and beyond. *Nat. Rev. Genet.* 3:872–882.
25. Herrnstein, R. and C. Murray. 1994. *The Bell Curve: Intelligence and Class Structure in American Life.* New York: The Free Press.
26. Wilson, J. F., M. E. Weale, A. C. F. SmithGratrix, B. Fletcher, M. G. Thomas, N. Bradman, and D. B Goldstein. 2001. Population genetic structure of variable drug response. *Nat. Genet.* 29: 265–269.
27. Bateson, G. 1963. The role of somatic change in evolution. *Evolution* 17:529–539.
28. Bock, W. J. and G. V. Wahlert. 1965. Adaptation and the form-function complex. *Evolution* 19:269–299.
29. Dobzhansky, T. 1968. Adaptedness and fitness. R. C. Lewontin, ed. Population ecology and evolution. Proceedings of an International Symposium. Syracuse, NY: Syracuse University Press.
30. Lewontin, R. C. 1957. The adaptations of populations to varying environments. In *Cold Spring Harbor Symposia on Quantitative Biology,* vol. 22. Boston, MA: Cold Spring Harbor Laboratory of Quantitative Biology.
31. Mayr, E. 1956. Geographic character gradients and climatic adaptation. *Evolution* 10:105–108.
32. Slobodkin, L. B. 1968. Toward a predictive theory of evolution. R. C. Lewontin, ed. Population biology and evolution. Proceedings of an International symposium. Syracuse, NY: Syracuse University Press.
33. Livingstone, F. B. 1958. Anthropological implications of sickle-cell gene distribution in West Africa. *Am. Anthropol.* 60:533–562.
34. Wallace, B. and A. Sob. 1964. *Adaptation.* Englewood Cliffs, NJ : Prentice-Hall, Inc.
35. Wright, S. 1949. Adaptation and selection. In *Genetics, Paleontology and Evolution,* ed. E. L. Jepson, E. Mayr, and G. G. Simpson. Princeton, NJ: Princeton University Press.
36. Mayr, E. 1966. *Animal Species and Evolution.* Cambridge, MA: The Belknap Press of Harvard University Press.
37. Schreider, E. 1964. Ecological rules, body heat regulation, and human evolution. *Evolution* 18:1–9.

38. Roberts, D. F. 1953. Body weight, race and climate. *Am. J. Phys. Anthropol.* 11:533–558.
39. Adolf, E. F. 1972. Physiological adaptations: Hypertrophies and super functions. *Am. Sci.* 60:608–617.
40. Barcroft, J. 1932. La fixité du milieu intérieur est la condition de la vie libre. *Biol. Rev.* 7:24–87.
41. Bernard, C. 1878. *Leçons sur les phénomènes de la vie.* Baillière, Paris.
42. Baker, P. T. 1966. Human biological variation as an adaptive response to the environment. *Eugen. Quart.* 13:81–91.
43. Brauer, R. W. 1965. Irreversible effects. In *The Physiology of Human Survival,* ed. O. G. Edhol, and R. Bachrach. London: Academic Press, Inc. Ltd.
44. Cannon, W. B. 1932. *The Wisdom of the Body.* New York: W.W. Norton & Co., Inc.
45. Cracraft, J. 1966. The concept of adaptation in the study of human populations. *Eugen. Quart.* 14:299.
46. Dubos, R. 1965. *Man Adapting.* New Haven, CT: Yale University Press.
47. Eagan, C. J. 1963. Introduction and terminology. *Fed. Proc.* 22:930–932.
48. Folk, G. E. Jr. 1974. *Textbook of Environmental Physiology.* Philadelphia, PA: Lea & Febiger.
49. Frisancho, A. R. 1975. Functional adaptation to high altitude hypoxia. *Science* 187:313–319.
50. Harrison, G. A. 1966. Human adaptability with reference to IBP proposals for high altitude research. In *The Biology of Human Adaptability,* ed. P. T. Baker and J. S. Weiner. Oxford: Clarendon Press.
51. Hart, S. J. 1957. Climatic and temperature induced changes in the energetics of homeotherms. *Rev. Can. Biol.* 16:133–141.
52. Lasker, G. W. 1969. Human biological adaptability. *Science* 166:1480–1486.
53. Mazess, R. B. 1973. Biological adaptation: aptitudes and acclimatization. In *Biosocial Interrelations in Population Adaptation,* ed. E. S. Watts, F. E. Johnston, and G. W. Lasker. Paris: Mouton, the Hague.
54. McCutcheon, F. H. 1964. Organ systems in adaptation: the respiratory system. In *Handbook of Physiology, vol. 4. Adaptation to the Environment,* ed. D. B. Dill, E. F. Adolph, and C. G. Wilber. Washington, DC: American Physiological Society.
55. Proser, C. L. 1964. Perspectives of adaptation: theoretical aspects. In *Handbook of Physiology, vol. 4. Adaptation to the Environment.* D. B. Dill, E.F. Adolph, and C. G. Wilber, ed. Washington, DC: American Physiological Society.
56. Selye, H. 1950. *The Physiology and Pathology of Exposure to Stress.* Montréal: Acta.
57. Thomas, R. B. 1975. The ecology of work. In *Physiological Anthropology.* A. Damon, ed. New York: Oxford University Press, Inc.
58. Timiras, P. S. 1972. *Developmental Physiology and Aging.* New York: Macmillan, Inc.
59. Rappaport, R. A. 1976. Maladaptation in social systems. In *Evolution in Social Systems.* J. Friedman, M. Rowlands. London: Gerald Duckworth & Co. Ltd.
60. Bligh, J. and K. G. Johnson. 1973. Glossary of terms for thermal physiology. *J. Appl. Physiol.* 35:941–961.

The Human Life Cycle

The trajectory from birth to death, which forms the life cycle, is guided by the genetic information provided by each parent and the environment in which the organism grows, develops, matures, declines, and eventually dies. Nutrition is a major component of the environment and, as such, it plays an important role in the expression of the genetic characteristics of the life cycle. This chapter delineates the stages of the life cycle starting with conception, pregnancy, infancy, childhood, adolescence, adulthood, and ending with senescence. Within each section some of the environmental factors that affect the expression of the components of the life cycle are briefly presented.

Stages of the Human Life Cycle

Although growth, development, and maturation may occur simultaneously and as such are interchangeably used, each indicate different processes. Growth refers to the increase in size due to cell multiplication, while development is defined as the increase in complexity of the organism that leads it from an undifferentiated or immature state, to a highly organized, specialized, and mature state. The term *maturation* is also used to indicate the process of changing from an undifferentiated state to a fully developed organism or stable state. Thus, an organism can grow in size without becoming more mature or developed. For example, an individual may be bigger or more grown but not more mature or developed and vice versa. The life cycle for didactic purposes is divided into a sequence of functional stages that starts at the prenatal level and terminates with old age and senescence (**table 16.1**).

Prenatal Stage

Embryonic Period

The embryonic period includes the first 12 weeks of pregnancy. During the embryonic period, by the fifth day of conception the fertilized ovum (or fertilized egg cell) is **implanted**, or embedded, in the wall of the uterus. Throughout this period, which may last up to two weeks, the developing embryo is nourished by secretions of uterine milk from the uterine glands. If the implantation is successful, the next six to eight weeks

TABLE 16.1
Stages of the Human Life Cycle

Stages	Distinguishing Features
Prenatal Life	From fertilization to birth
Embryonic period	First 12 weeks of pregnancy (first trimester); formation of all the basic structures (embryogenesis)
Fetal period	From 16 weeks to birth (second and third trimester; growth of preexisting cells and tissues
Postnatal Life	From birth to death
Neonatal period	First 28 days after birth
Infancy	From second month to the end of second year; rapid growth in weight and length; appearance of deciduous (milk) teeth
Childhood	From end of second year to the eighth year. Moderate growth in weight and length; eruption of first permanent molar and incisor; cessation of brain growth
Puberty and Adolescence	From the ninth to nineteenth year; activation of central nervous system mechanism for sexual development; increased hormonal activity; adolescent growth spurt in height and weight; eruption of permanent teeth (tables 16.7 and 16.8); development of secondary sexual characteristics
Adulthood	From twentieth to fiftieth year; stability in height, but continued increase in weight and adipose tissue; childbearing age in women
Old age	From fiftieth to seventieth year; beginning of decline in height but stability in weight and body composition
Senescence	From seventieth year to death; decline in height, body composition, and physiological competence

are characterized by the formation of organs known as **organogenesis**, whereby the cells of the embryo differentiate into distinct tissues and functional units that later become organs, such as the heart, lungs, liver, and skeleton. By the end of organogenesis, all of the major organs have begun to develop. The sequence of some key developmental changes during the embryonic period is summarized in **table 16.2**.

In terms of environmental influences, the embryonic period is a particularly vulnerable time for the developing fetus. Any adverse factors, such as viruses, bacteria, or nutritional deficiencies, will directly affect and interfere with the normal development of the embryo and can cause specific congenital abnormalities. The vulnerability of the embryo is related to the fact that it does not have the protection of the placenta and, since its nourishment is derived directly from the mother's blood, any factors that may enter into the mother's circulation also enter into the embryo. Several studies indicate that exposure to viruses of varicella and small pox during the embryonic period causes irreparable damage to the development of organs of the central nervous system.

TABLE 16.2
Sequence of Developmental Changes During the Embryonic Period of Pregnancy

Week	Developmental Change
0–3	Brain development begins
4	Heart functions and liver functions
5	Eyes and limbs develop
6–7	Teeth develop
8–12	Skeleton mineralization begins

Nutrient and Embryological Development

The presence of particular nutrients at specific times is crucial for the normal development of various tissues. For example, folate (also known as folic acid) deficiency, during the critical period has been associated with improper development of the neural tube and poor skeletal formation. Various studies in humans confirm that folate is important for the prevention of neural tube defects (NTD).[1–3] In the United States each year about 2,500 to 3,000 infants are born with NTD and in the world approximately 300,000 to 400,000 infants each year are born with NTD. Although the mechanism whereby folate deficiency causes NTD is not well defined, current research has demonstrated that ample folate intake can significantly reduce the risk of this form of fetal abnormality. For this reason, the general consensus is to maintain an adequate intake of folate to satisfy the increased needs of pregnancy. Likewise, nutritional deficiencies of riboflavin, pyridoxine, and manganese have been associated with improper development of the skeletal and neuro-motor system.

Fetal Period

The fetal period ranges from the twelfth week of pregnancy to delivery. From 12 weeks onward the tissues and organs formed during the embryonic period continue to grow and mature until they are able to sustain the infant's life after birth. At the end of the embryonic period, the embryo weighs only about 6 g (0.2 oz), is about 3.0 cm long (1.20 in.), and has all the structural and functional features of a newborn infant. The fetal period is characterized by enlargement and growth of preexisting tissues. As illustrated in **figure 16.1**, growth of the fetus occurs in three phases. The first phase and until about the twentieth week of gestation is characterized by **hyperplasia**, a rapid increase in the number of cells. During this phase the fetus requires sufficient supplies of folate and vitamin B_{12}, so as to maintain the large numbers of cell divisions (cellular replication). In the second phase, cellular replication continues along with **hypertrophy**, or enlargement of preexisting cells. Once again, the maintenance of cell replication and growth requires sufficient supplies of amino acids and vitamin B_6. The third phase, which expands from the thirtieth week to delivery, is characterized by only growth and hypertrophy of preformed cells. Thus, during the first 20 weeks of gestation the fetus grows very slowly, but between weeks 28 and 37 it grows rapidly (at about 200 g per week). For example, a full-term newborn (>37 weeks of gestation) on the average weighs more than 3,292 ± 450 grams (7.37 lbs) and is more than 51.0 ± 2.5 cm (20.0 ± 1.0 in.) long. Associated with the increase in weight from 20 weeks onward, the fetus accumulates fat stores of its own and by term the proportion of fat reaches 16 percent. Conversely, the proportion of body weight that is water decreases steadily with increasing fetal weight so that by term the newborn is about 69 percent water. After birth and especially during the first 24 hours, the proportion of body water continues to decline, attaining an average of 65 percent.[4–6]

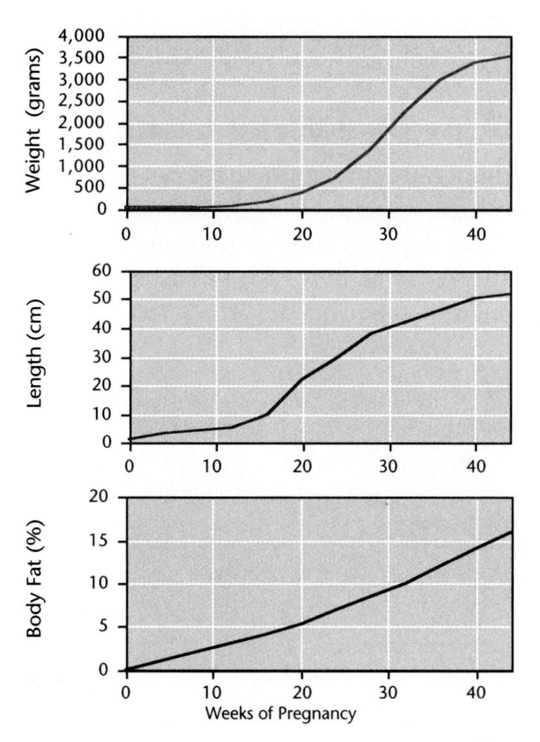

FIGURE 16.1. Prenatal growth in body weight, length, and fat. Source: Data derived from Usher and Maclean;[4] and Widdowson.[5]

The Placenta

The placenta is a highly vascularized fetal organ that functions like a digestive, respiratory, endocrine, and excretory system through which it provides oxygen, nutrients, and hormones and removes the metabolic wastes.[7–9] Although the placenta develops early in pregnancy, it begins to play its functional role in the fetal stage. On average, the placenta grows to weigh around 450 to 650 g (1.0 to 1.4 lb) at birth. The fetus and mother's circulatory system are united through the umbilical cord (**fig. 16.2**).

Energy Cost of Pregnancy

Estimates of body composition suggest that for well-nourished women[10] with a height of 165 cm (65 in.), weight gain during pregnancy averages about 11.6 kg (25.5 lb) or about 1.33 kg (2.9 lb) per month. The components of the weight gain are given in **table 16. 3.** About half of the weight gain resides in the fetus, placenta, and amniotic fluid; the remainder represents the weight of maternal reproductive tissues such as the uterus, mammary glands, blood, and extracellular

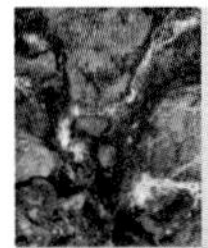

TABLE 16.3

Maternal Pregnancy Gain and Its Components

Component	Weight in kg (lb)
Fetus	3.300 (7.26)
Placenta	0.66 (1.45)
Uterus, amniotic fluid, mammary glands, blood, and extracellular fluid	5.1 (11.2)
Net maternal fat mass gain	2.5 (5.0)
Total	11.56 (25.0)

Source: Data from Raaij, Peek, and Vermaat-Miedema et al.[10]

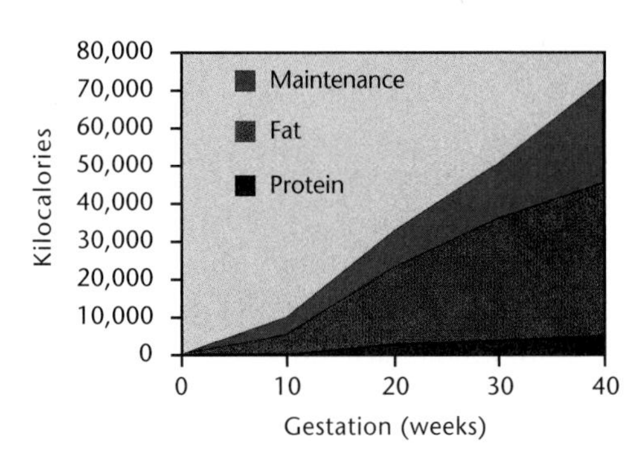

FIGURE 16.3. The Cumulative Energy Cost of Pregnancy. Source: Data from Hytten and Leitch (1971).[11]

fluid that is excreted in the urine during birth. The net maternal fat mass gain of 2.5 kg represents the maternal fat stores. However, since the fat content of adipose tissue is 80 percent, the net weight of fat gain is only 2.0 kg ($2.5 \times .80 = 2.0$ kg or 4.4 lb).

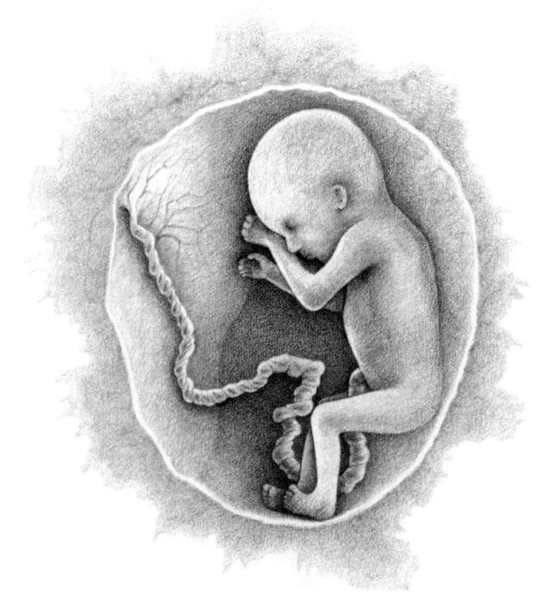

FIGURE 16.2. Illustration of the placenta-fetal relationship. © Royalty-free Corbis.

The cumulative energy cost of pregnancy and its components is about 65,700 calories (**table 16.4**). Thus, an extra 250 kcal/day must be ingested to support the increased nutritional requirements of pregnancy (**fig. 16.3**). Maternal fat gain accounts for 35 percent of the energy cost; the development of the fetus, placenta, and maternal tissue accounts for 17.5 percent of the energy cost. The cumulative effects of basal metabolic rate and an increased metabolic rate account for 48 percent of the energy cost. Since the energy needs are not uniform throughout pregnancy, the mother makes metabolic adjustments oriented at both the efficient utilization and storage of energy.

Recommended Weight Gains

The weight gain during pregnancy is quite variable and no single maternal weight gain target meets the needs of all categories of pregnant women. Individualized patterns of weight gain must be adjusted for height and pre-pregnancy weight. Current recommen-

TABLE 16.4

Energy Cost of Pregnancy Including Cost of Increased Resting Metabolic Rate (RMR) and Tissue Synthesis

Variable Tissue deposition	Scotland Energy Cost (kcal)	Netherlands Energy Cost (kcal)	Average Energy Cost (kcal)	Percent Total (%)
Maternal fat gain	25,334	21,988	23,661	34.9
Fetus	8,126	8,222	8,174	12.1
Placenta	729	741	735	1.1
Expanded maternal tissues	2,891	2,940	2,915	4.3
Cumulative cost of increased RMR	30,114	34,416	32,265	47.6
Total cost of pregnancy	67,159	68,354	67,756	

Source: Adapted from Durnin;[12] RMR = resting metabolic rate

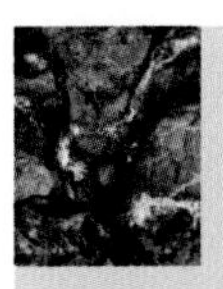

TABLE 16.5

Recommended Amount of Pregnancy Weight Gain by Category of Weight and Body Mass Index

Category	BMI* (kg/cm²)	Weight Gain (kg)	(lb)
Underweight	<19.8	12.7–18.2	28–40
Average weight	19.8–26.0	11.4–16.0	25–35
Overweight	>26.0–29.0	6.8–11.4	15–25

Source: Adapted from Subcommittee on Nutritional Status.[13]
*BMI in metric = Weight (kg) / Height (m²); BMI in English = [(Weight (lbs) / Height / Height) × 703].

dations of weight gain during pregnancy are done with reference to body mass index (BMI) (**table 16.5**).[13] In general, the recommended weight gain for a woman with a low pre-pregnancy BMI is greater than for a woman with a high pre-pregnancy BMI. Several studies have consistently found that prenatal weight gain, within the suggested range for each pregravid BMI category, is associated with more favorable outcomes than weight gain above or below the suggested range.[13–15] These outcomes include a reduction in the prevalence of low birth weight infants (<2,500 g), small for gestational age infants, large for gestational age infants, high birth weight infants (>4,500 g), cesarean deliveries,[13–15] and preterm deliveries, as well as an increase in mean birth weight. The recommended target weight gain range for women carrying twins is 16 to 20.5 kg (35 to 45 lb). Likewise, the recommended target weight gain ranges for still-growing nonadults under the age of 15 years should be increased by 1.0 kg (2.20 lb).

Birth Weight

Birth weight is the final common pathway for the expression of pregnancy and prenatal growth. Therefore, evaluation of birth weight provides retrospective information about the maturity and nutritional status of the newborn. The frequency of birth weight follows a normal (or bell-shaped) distribution[16, 17] with an extended lower tail (**fig. 16.4**). The importance of birth weight is based upon two facts. First, birth weight is an extremely powerful predictor of an individual baby's survival. The lower the weight, the higher a baby's risk of infant mortality, and, at the population level, the lower the mean birth weight, the higher the rate of infant mortality. Second, birth weight is associated with health outcomes later in life.

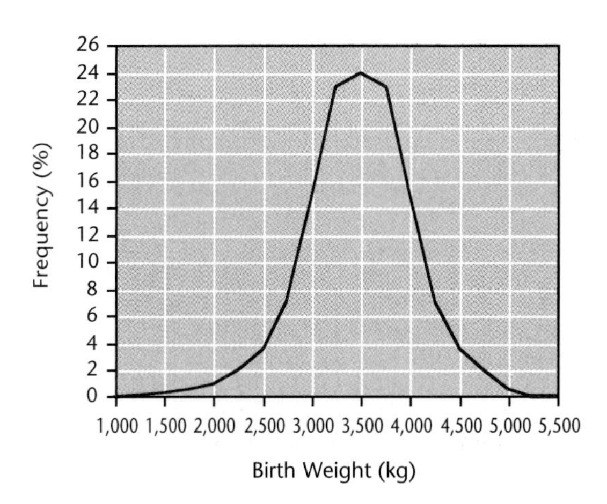

FIGURE 16.4. Distribution of birth weights. In most populations, birth weights fall in normal or bell-shaped distribution. Source: Data from Wilcox and Russell.[16]

Classification of Newborns

By Gestational Length

The average gestational length for full-term infants is 37 weeks. Based on gestational age at delivery, newborns are classified into three categories: (a) preterm or premature, infants born with gestational age of less than 37 weeks (less than 259 days); (b) term or full-term, infants born with gestational age of more than 37 weeks (259 to 293 days) but less than 42 completed weeks; and (c) postterm, infants born with gestational age of more than 42 completed weeks (≥294 days). Thus, the term *premature* is used for infants born before 37 weeks gestation. Most, but not all premature infants weigh less than 2,500 g (<5.5 lb).

By Birth Weight

Depending on the size at birth, newborns are classified into four categories: (a) low birth weight (LBW), infants born with a weight of less than 2,500 grams (<5.5 lb); (b) average birth weight (ABW), infants whose birth weight ranges from 2,501 to 3,500 grams (5.51 to 7.70 lb); (c) high birth weight (HBW), infants whose birth weight ranges between 3,500 grams and 4,540 grams (7.71 to 9.98 lb); and (d) very large birth weight (VLBW), infants whose birth weight exceeds 4,540 grams (>10 lb). In the United States, the incidence of low birth weight averages 5.9 percent of all live births.

By Gestational Age and Birth Weight

Figures 16.5 and **16.6** provide reference data weight by gestational age of newborns for the United States.[18, 19]

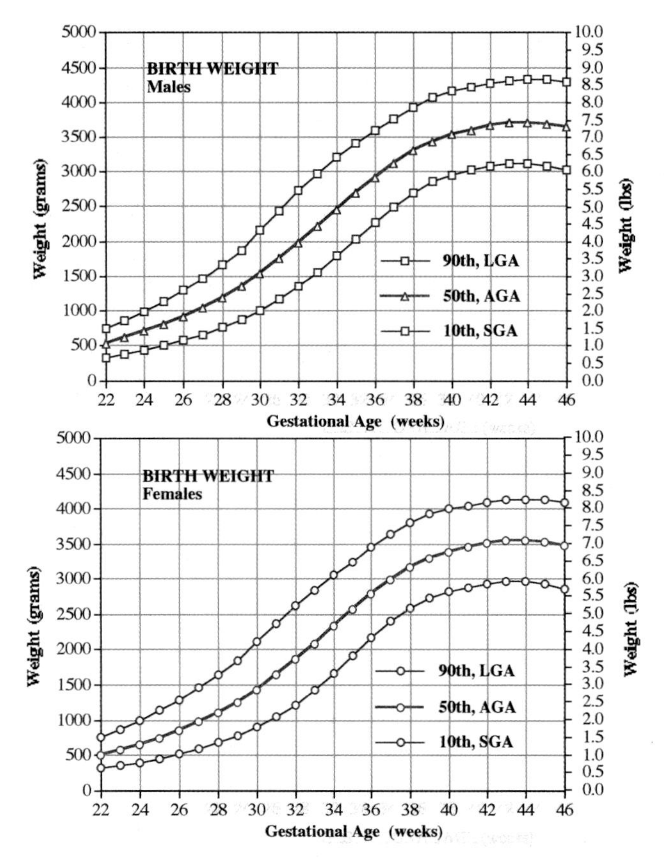

FIGURE 16.5. Percentiles of birth weight (grams) for determining prenatal growth status for white male and female Infants. Source: Data from refer. #18. Williams, R. L., Creasy, R. K., Cunningham, G. X., Hawes, W. E., Norris, F. D. 1982. Fetal growth and perinatal viability in California. *Obstet. Gynecol.* 59:624–632.

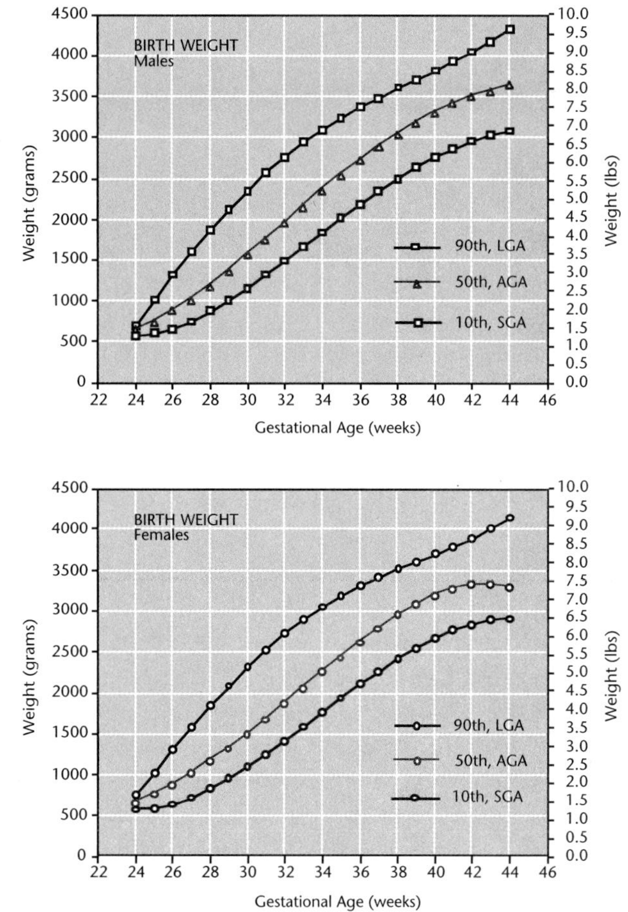

FIGURE 16.6. Percentiles of birth weight (grams) for determining prenatal growth status in black male and female infants. Source: Data from Amini, Catalano, Hirsch, and Mann.[19]

Using these references, newborns can be classified into three groups: small for gestational age (SGA), appropriate for gestational age (AGA), and large for gestational age (LGA).

Small for Gestational Age or Intrauterine Retarded. Small for gestational age (SGA) infant refers to the weight status at birth, while the term *intrauterine growth retardation* (IUGR) is sometimes applied to SGA infants. Intrauterine growth retardation refers to fetal growth that has been limited by an inadequate prenatal environment. Infants born after at least 37 weeks of gestation and weighing less than 2,500 grams (<5.5 lb) at birth are considered intrauterine growth retarded. Because of the difficulties of determining accurately the gestational age, low birth weight (<2,500 g) is often used as a proxy for IUGR.

Appropriate for Gestational Age. A newborn whose weight corresponds to its gestational age or is between the 10th and 90th percentiles for gestational age is considered appropriate for gestational age (AGA).

Large for Gestational Age. A newborn who weighs above the 90th percentile for gestational age is considered large for gestational age (LGA).

Early Pattern of Weight Gain and Birth Weight

Various studies indicate that, in both black and white women, increasing weight gain during pregnancy is associated with higher birth weight (**fig. 16.7**). These studies indicate a linear rate of about 0.66 pounds per week from 8 to 20 weeks gestation and a linear rate of 1.06 pounds per week after week 20.[20] However, a re-

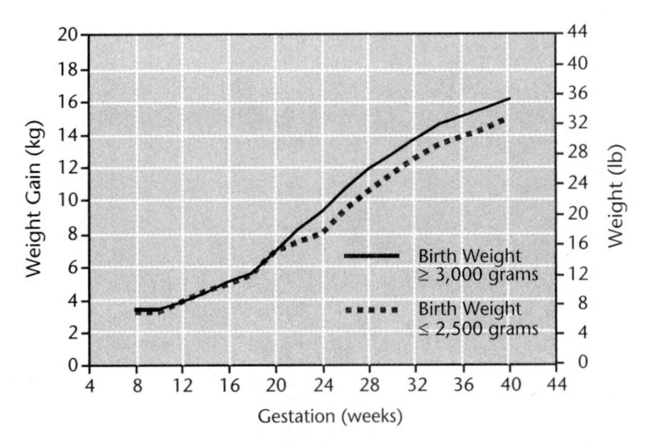

FIGURE 16.7. Weight gain by gestational age in white women who delivered singletons who weighed ≥ 3,000 grams or ≤ 2,500 grams. Source: Data from Petitti, Croughan-Minihane, and Hiatt.[20]

cent study of 389 women from Minnesota (**fig. 16.8**) indicates that maternal weight gain occurring primarily in the first and second trimesters of pregnancy is associated with higher birth weights than weight gain occurring in the third trimester.[21] From a functional and evolutionary point of view, weight gain during the first trimester of pregnancy is an opportunistic adaptation whereby energy is stored early, when the growth needs of the fetus are low, to be used during the second and third trimester when the energy needs for growth of the fetus are high. In this context, the pattern of early pregnancy weight gain would be even more advantageous under conditions of maternal undernutrition. In the developing nations, chronically undernourished women are able to reproduce, even though the cost of pregnancy (68,000 calories or 250 kcal/day) exceeds their resources. It is quite possible that women in such conditions gaining weight during the first trimester would be able to store energy that could be used to feed the growth needs of the fetus during the last trimester. This pattern of weight gain would represent an opportunistic adaptation that enables women to reproduce even under the most limited nutritional conditions. Furthermore, a pattern of weight gain during the first or second trimester, rather than the third trimester, would also have adaptive significance for well-nourished women since women who gain weight excessively after midpregnancy retained more weight after pregnancy. Therefore, gaining weight early in pregnancy may also be of benefit for promoting postpartum weight loss.

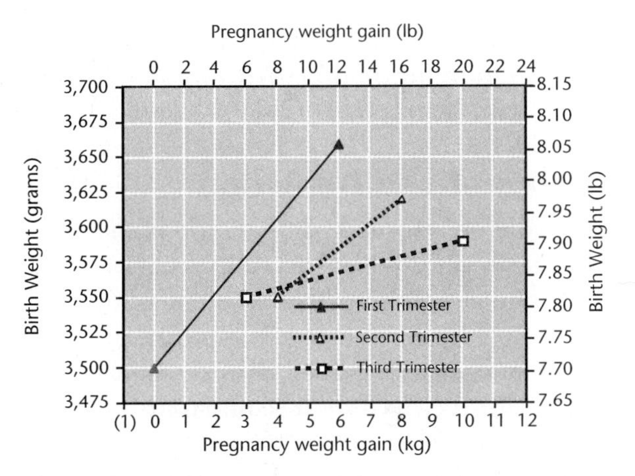

FIGURE 16.8. Early pregnancy weight gain and birth weight. Source: Data from Brown, Murtaugh, Jacobs, and Margellos.[21]

In summary, it is recommended that evaluation of nutritional status during pregnancy through measurements of pregnancy weight gain be done with reference to body mass index rather than height and weight only. This recommendation should be done taking into account that weight gain during the first trimester is conducive to higher birth weight than a uniform weight gain throughout pregnancy. Likewise, it is recommended that assessments of intrauterine growth status be done with reference to normative data specific for white and black infants.

Postnatal Stage

The neonatal period includes the first 28 days after birth, while one month to three years of age is considered the period of infancy. The neonatal period is one of the critical stages in human development, because during this stage the neonate has to adapt to a completely new environment that requires adjustments in all aspects of physiology. The neonate changes from an environment where everything from oxygen, nutrients, and thermoregulation to excretion of waste products is provided by the maternal circulatory system to an environment where everything has to be obtained by the newborn itself. It is for this reason that nearly 50 percent of infant deaths usually occur during the neonatal period.

Infancy and Body Size

The infancy stage occupies the first three years of human postnatal life and is characterized by rapid changes in body size and body composition. Weight (**fig. 16.9**) from 2 months onwards increases steadily at about 0.26 kg/month (0.57 lbs/month), so that by the end of the second year an infant weighs twice as much as at 2 months. Growth in length occurs at about 1.12 cm/month (0.45 in./month). As a consequence of the

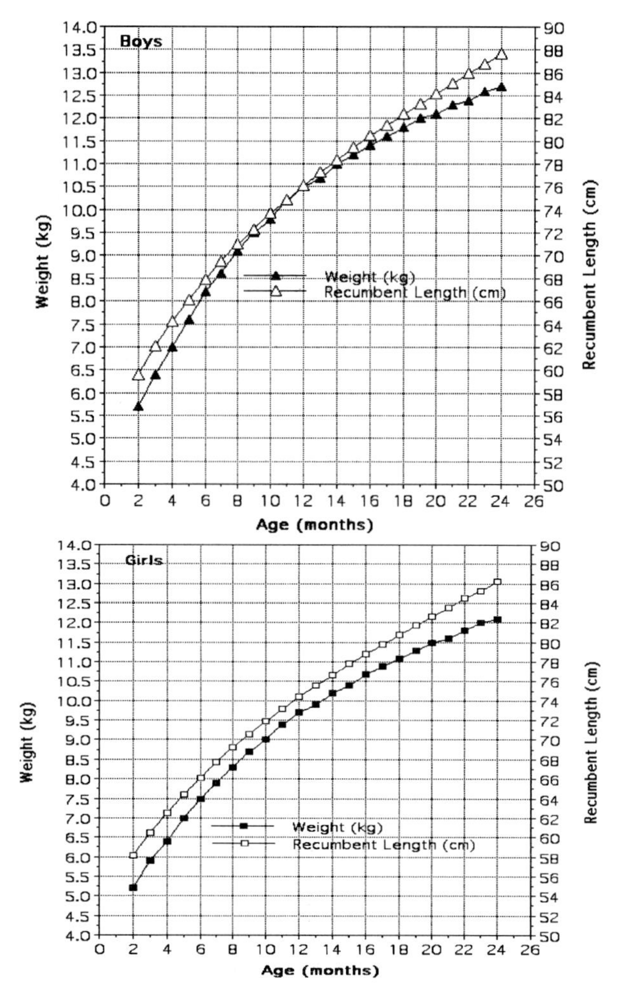

FIGURE 16.9. Infant growth in weight and recumbent length in boys and girls. Source: Adapted from Frisancho, A. R. 2005. *Nutritional Anthropometry.* Ann Arbor: University of Michigan Press.

fast growth in length the proportion of weight per unit of length as measured by the Rohrer's index (**fig. 16.10a**) declines steadily until about 16 months and remains stable thereafter. **Figure 16.10b** illustrates the age-associated changes in head circumference. From these data it is evident that head circumference continues to increase throughout infancy. The utility of measurements of head circumference during infancy lies in the fact that it is directly correlated with brain size. Recent studies indicate that a reduced head circumference at birth and a sudden increase between 1 and 12 months precedes the expression of autism in children.[22]

Infancy and Dentition

Infancy is also characterized by feeding through lactation[24] and by the eruption of the deciduous teeth.[24] As shown in **table 16.7 and figure 16.11a**, on the aver-

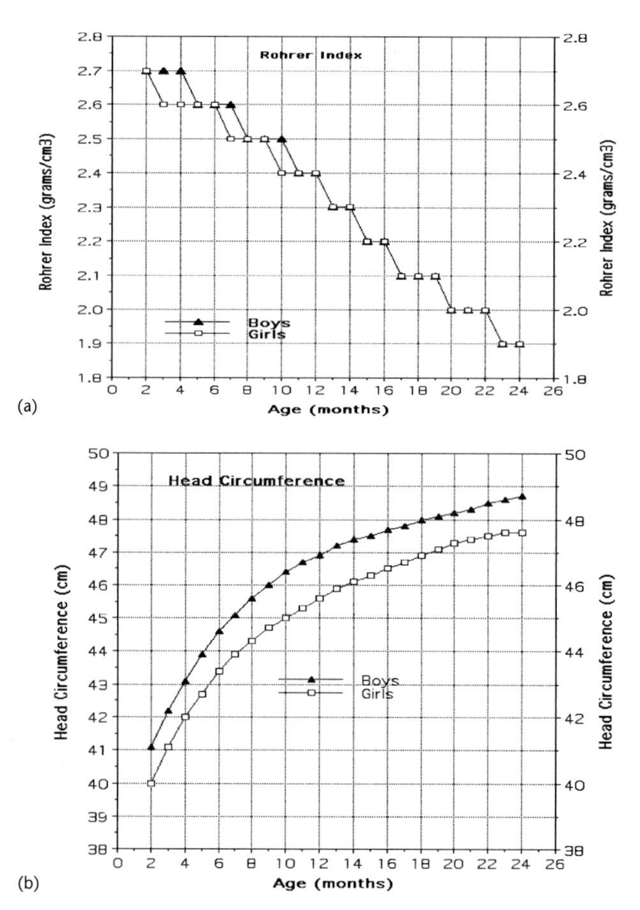

FIGURE 16.10. Infant growth in Rohrer index (a) and head circumference in (b) boys and girls. Source: Adapted from Frisancho, A. R. 2005. *Nutritional Anthropometry.* Ann Arbor: University of Michigan Press.

age, the first incisor of the mandible erupts at about 6 to 7 months but it can appear as early as 4 months. The deciduous teeth of boys emerge about one month earlier than in girls. The age of eruption of the second premolar (M_2), which occurs at between 19 and 36 months, marks the end of infancy. In general, the emergence of the deciduous teeth is associated with the switch from the infant's dependency on lactation to increased consumption of solid foods and weaning.

Childhood and Dentition

The childhood stage is a continuation of the infancy stage extending from 3 to 9 years. It is characterized by a moderate growth rate, the replacement of the deciduous teeth by the emergence of the first permanent teeth,[24] and completion of growth of the brain in weight. During childhood, compared to infancy, growth rate in body length slows and averages about 5 cm (2 in.) per year. By the end of childhood, most children have replaced the "milk" incisors with permanent ones. From **table 16.8** it is evident that the first

TABLE 16.7
Median and Range in Age of Eruption of Deciduous Teeth for French Canadian Infants

Teeth	Males Median Age (months)	Males Range (months)	Females Median Age (months)	Females Range (months)
Maxillary				
I_1= central incisor	8.49	5.70–11.84	9.42	6.19–13.33
I_2= lateral incisor	9.81	5.87–14.74	10.53	6.13–16.12
C = canine	17.56	12.44–23.56	18.31	13.28–24.15
M_1 =first premolar	15.20	11.26–19.74	15.06	11.51–19.09
M_2= second premolar	27.04	19.95–35.20	27.86	20.67–36.11
Mandibular				
I_1 = central incisor	6.86	3.90–10.63	7.31	3.96–11.68
I_2 = lateral incisor	11.48	6.95–17.13	12.54	7.70–18.57
C = canine	17.77	12.98–23.30	18.58	13.14–24.96
M_1 =first premolar	15.25	11.49–19.54	15.32	11.67–19.48
M_2 = second premolar	26.13	19.19–34.14	26.90	20.22–34.54

Source: Adapted from Dimirjian.[24]

molar (M_1) and first incisor (I_1) of the mandible erupt between the ages of 5.5 and 6.5 years and mark the end of childhood.

From Adolescence to Adulthood

Puberty refers to the period during which the reproductive system matures or the state at which a person is first capable of bearing children. On the other hand, adolescence refers to the period of growth and development from childhood to maturity. The adolescence period extends from 10 to 20 years and is associated with increased hormonal activity and changes in body size.

Hormonal Activity and Reproductive Maturation

Starting with puberty, the activity of the hypothalamic-pituitary gonadal system accelerates, which in turn increases the production of gonadotrophic and enlargement of reproductive organs.[25, 26] Reflecting the increased hormonal activity, adolescence is associated with the development of secondary sexual characteristics. In boys, these include a change in voice; an increase in the density and growth of pubic, axillary, and facial hair; an increase in the size of the penis and scrotum; and production of viable sperma-

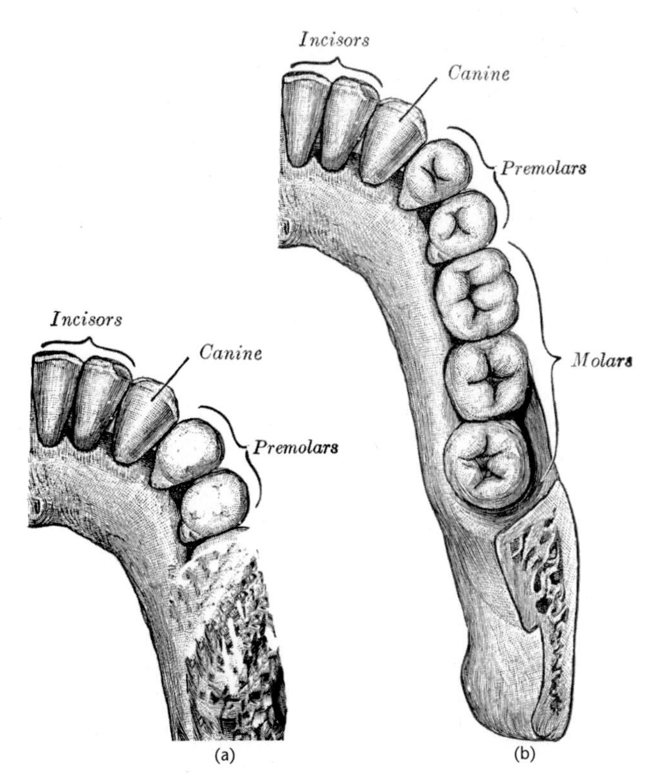

FIGURE 16.11. (a) Deciduous teeth and (b) permanent teeth. The deciduous teeth (also referred to as primary teeth or milk teeth) include 4 incisors, 2 canines, and 4 premolars, for a total of 20 teeth. In contrast, the permanent teeth consist of 32 teeth that include 4 incisors, 2 canines, 2 premolars, and 3 molars. (Composite illustration made by the author.)

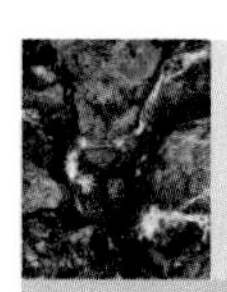

TABLE 16.8

Mean Age and Standard Deviation for Eruption of Permanent Teeth for North American Boys and Girls

	Males		Females	
Teeth	Mean (years)	SD (years)	Mean (years)	SD (years)
Maxillary				
I_1= central incisor	7.34	0.77	6.98	0.75
I_2= lateral incisor	8.39	1.01	7.97	0.91
C = canine	11.29	1.39	10.62	1.40
PM_1 = first premolar	10.64	1.41	10.17	1.38
PM_2 = second premolar	11.21	1.48	10.88	1.56
M_1 = first molar	**6.40**	**0.79**	**6.35**	**0.74**
M_2 = second molar	10.52	1.34	11.95	1.22
M_3 = third molar	20.50	-	20.50	-
Mandibular				
I_1 = central incisor	**6.30**	**0.81**	**6.18**	**0.79**
I_2 = lateral incisor	7.47	0.78	7.13	0.82
C = canine	10.52	1.14	9.78	1.26
Pm_1 = first premolar	10.70	1.37	10.17	1.28
Pm_2 = second premolar	11.43	1.61	10.97	1.50
M_1 = first molar	**6.33**	**0.79**	**6.15**	**0.76**
M_2 = second molar	12.00	1.38	11.49	1.23
M_3 = third molar	19.80	-	20.40	-

Source: Adapted from Dimirjian.[24]

tozoa and seminal emission. In girls, adolescence is associated with the growth of the breasts, appearance of pubic and axillary hair, and menarche (first menstruation). Based upon this information, references of sexual development in boys and girls have been developed. Details of these references and the use of classificatory systems are described in the Tanner puberty stage classification system.[25]

Age at Menarche in Females

Menarche, the first menstrual period, marks the onset of reproductive capacity and ovulation of adolescent females. Menarche in the United States averages 12.5 ± 1.1 years, but in 1840 it averaged 16.5 years. On average, for about 1.5 years in many females, menstrual cycles are not associated with ovulation. This period of nonovulation is referred to as the adolescence sterility. Therefore, the average girl in the 1800s was not fertile until 18 years of age (16.5 + 1.5 = 18), while at present, at 14 years (12.5 + 1.5) most girls become fertile. This change is contributing to the increased risk of teenage pregnancy among contemporary populations.

Growth in Body Size

Figure 16.12 illustrates the mean growth in height each year (also referred to as the distance curve of growth). From these data it is evident that, in both boys and girls, during the first 10 years, growth occurs steadily at about 6.3 cm/year (2.50 in./year). Thereafter, between 11 and 16 years in females and between 12 and 18 years in boys, growth velocity in height occurs at about 9 to 10 cm/year (3.5 to 4.0 in./year) for boys and 6 to 8 cm/year (2.4 to 3.2 in./year) for girls. During the adolescent period, males and females add more than 30 cm (12 in.) and 25 cm (10 in.), respectively, to the total height. The adolescent growth spurt is fol-

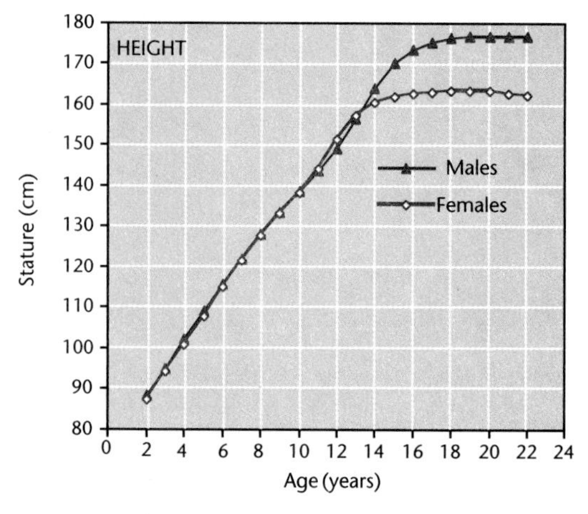

FIGURE 16.12. Changes in mean height by age. Source: Frisancho, A. R. 2005. *Nutritional Anthropometry*. Ann Arbor: University of Michigan Press.

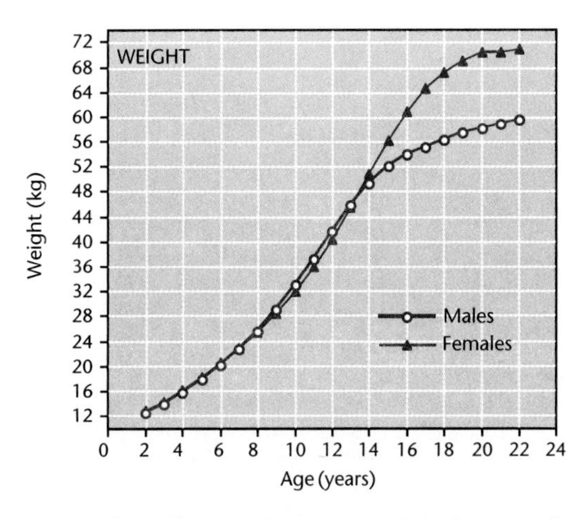

FIGURE 16.13. Changes in mean weight by age. Source: Frisancho, A. R. 2005. *Nutritional Anthropometry.* Ann Arbor: University of Michigan Press.

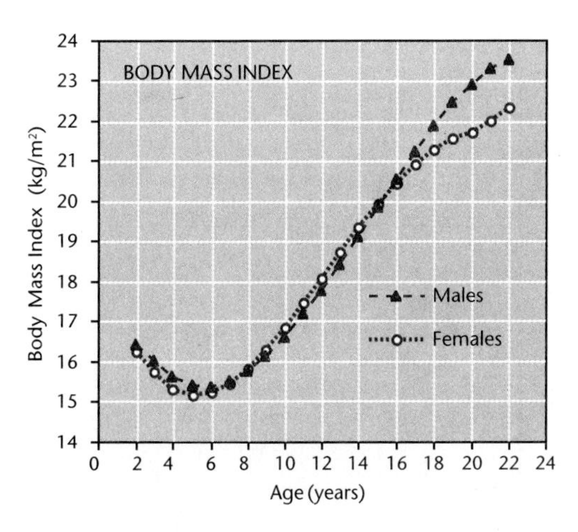

FIGURE 16.14. Changes in mean body mass index by age. Source: Frisancho, A. R. 2005. *Nutritional Anthropometry.* Ann Arbor: University of Michigan Press.

lowed by a constant decrease in growth rate that ends with the attainment of adult stature. In terms of weight (**fig. 16.13**), males and females show the same pattern of weight gain until the age of 14 years, when males exceed females by about 8 kg. The proportion of weight per unit of height, as given by the body mass index (**fig. 16.14**), declines during childhood and until the age of 6 years. Thereafter, in both males and females, it increases steadily.

Allometric Growth and Development of Body Proportions

The pattern of human growth is allometric and is characterized by a head-to-foot (cephalocaudal) gradient. During prenatal period the head is more advanced at birth than the trunk and the trunk is more advanced than the legs[25–30] and therefore children are born with disproportionately large trunks and short legs (**fig. 16.15**). Continuing this pattern throughout childhood, the growth rate of head size and trunk size is more advanced than the lower limbs. Therefore, attainment and differences of adult height are due in great part to more postnatal growth in leg length than in trunk size. This trend is demonstrated by changes in the sitting height and leg length indices (**fig. 16.16**). Growth in trunk size (expressed as percent of adult values) is completed by the age of 15 years, while growth in leg length continues until about the age of 19 years in males and 16 years in females. Since the longer the developmental period, the more the influence of the environment, growth of leg length is under greater environmental influence than the growth of trunk length.[31]

Developmental Changes in Subcutaneous Fat and Body Shape

Although subcutaneous fat can be determined by dual-energy X-ray absorptiometry, computer tomography measurements of skinfold thickness provides an effective measure of subcutaneous fat. **Figures 16.17** and **16.18** illustrate the age-associated changes in skinfold thickness at the triceps and subscapular site. From these data it is evident that fat deposition in both the arm and trunk is greater in females than in males. From **figure 16.19** it is evident that, as shown by the lower ratio of the circumference at the waist and hip, the gluteo-femoral adult female pattern of fat deposition begins during childhood and becomes well defined during adolescence.

Developmental Changes in Skeletal Muscle

There are three kinds of muscles: skeletal (or voluntary) muscles, smooth (or involuntary) muscles, and the heart muscle. Skeletal muscle is generally the main energy-consuming tissue of the body and provides the propulsive force to move about and perform physical activities. There are more than 500 skeletal muscles in the body, and each is composed of units known as fibers. A muscle fiber is a composite of about 100 to 1,000 myofibrils, which are packed together along with nuclei, mitochondria, and other cellular sarcoplasmic constituents in a sarcolem membrane. While the number of muscle fibers is established prenatally and does not change afterwards, the diameters of muscle fibers increase gradually throughout postnatal development.

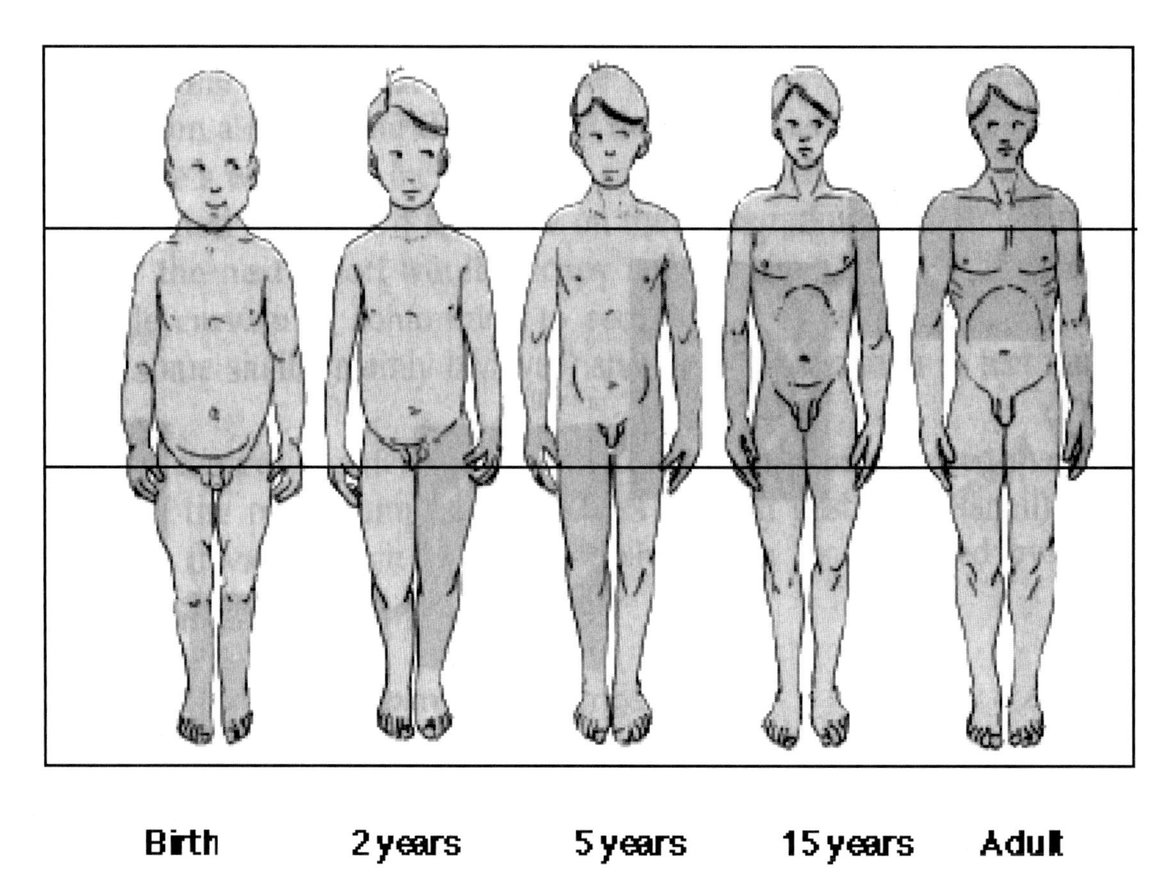

FIGURE 16.15. Change in shape of the human body from birth to adulthood. Composite made by the author based on data from Stratz.[30]

The marked postnatal increase in muscle diameter is due entirely to continued enlargement of preexisting muscle fibers (hypertrophy).

Limb Muscle

There are several indirect methods for estimating total body muscle mass. These include measurements of creatinine excretion and potassium concentration. The estimates based on creatinine excretion are grounded on the fact that creatinine is a by-product of muscle metabolism. Current estimates indicate that about 1 g of creatinine excreted in a 24-hour urine sample is derived from approximately 20 kg of muscle tissue. The estimates of muscle mass through measurements of potassium are based upon the fact that muscle tissue is rich in potassium, so that the higher the concentration of potassium, the higher the muscle mass. However, because these methods require either extensive laboratory facilities or possibly radiation risks to the subject, studies of regional development of muscle are based on anthropometric techniques that estimate limb muscle size. As shown in **figure 16.20**, in both boys and girls, muscle mass increases gradually during childhood and from 10 years onwards it increases rapidly in males but less so in females. By adulthood, males have about 20 percent more muscle mass than females.

Adult Height

Adulthood usually extends from 18 to 50 years. The attainment of adult stature, which in the United States averages about 176 cm (69 in.) for males and 164 cm (65 in.) for females, occurs between 18 and 21 years of age and is one of the hallmarks of adulthood (**fig. 16.21**). Height growth stops when the epiphysis of the long bones of the skeleton (e.g., the femur and tibia) fuses with the diaphysis of the bone. As long as there is cartilage between the epiphysis and diaphysis, growth in length continues (**fig. 16.22**). However, when the cartilage is replaced with bone or is calcified, the diaphysis and epiphysis become one single piece and lose their ability to increase in length. The fusion of epiphysis and diaphysis, that is, the skeletal maturation, is stimulated by the gonadal hormones, the androgens and estrogens. However, there are individual differences in the time of skeletal maturation such that early

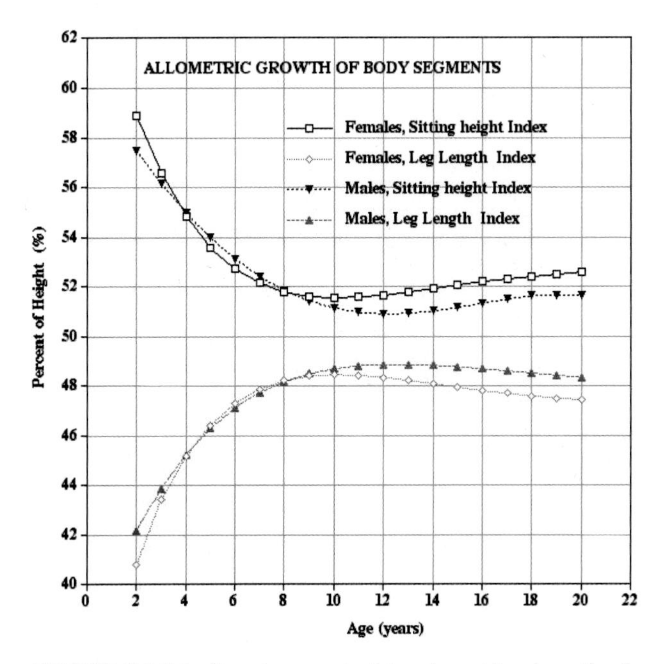

FIGURE 16.16. Development of trunk and leg length of children. Source: Frisancho, A. R. 2005. *Nutritional Anthropometry.* Ann Arbor: University of Michigan Press.

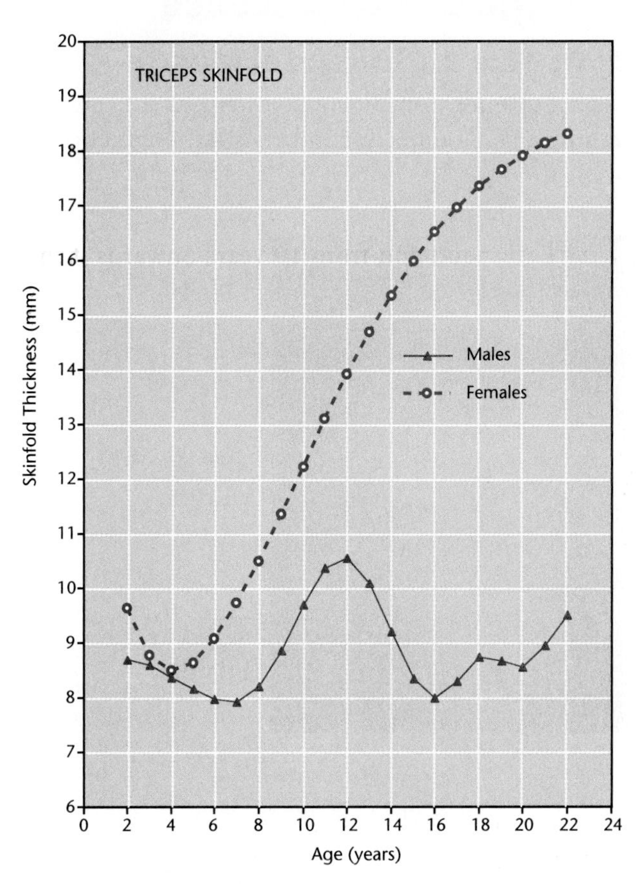

FIGURE 16.17. Age-associated changes of the triceps skinfold thickness of males and females. Source: Frisancho, A. R. 2005. *Nutritional Anthropometry.* Ann Arbor: University of Michigan Press.

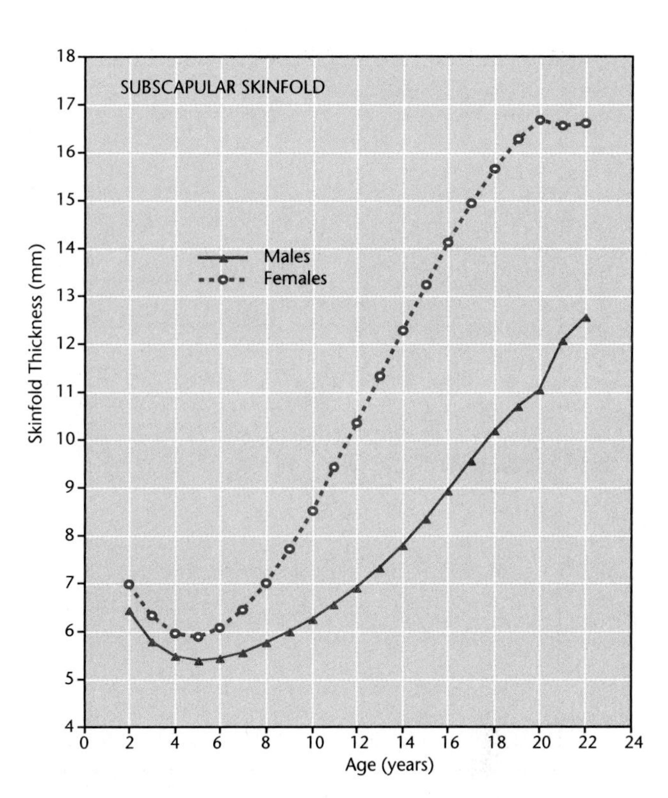

FIGURE 16.18. Age-associated changes of the subscapular skinfold thickness of males and females. Source: Frisancho, A. R. 2005. *Nutritional Anthropometry.* Ann Arbor: University of Michigan Press.

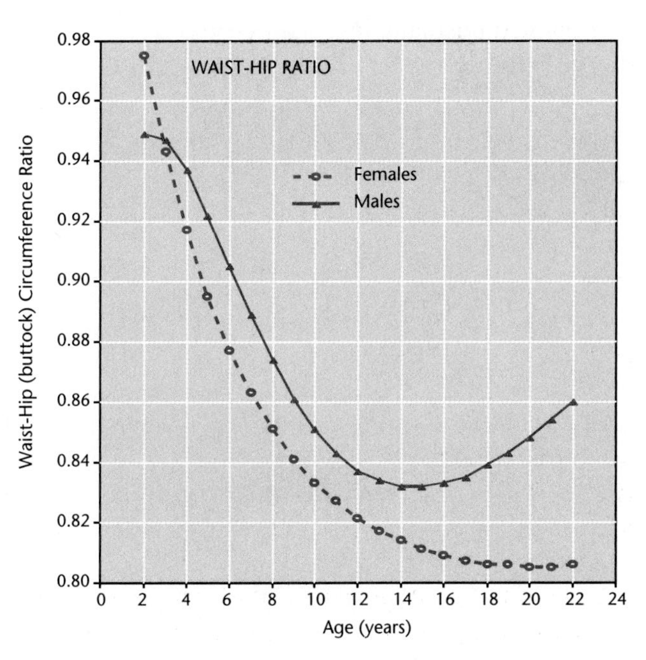

FIGURE 16.19. Age-associated changes in ratio of circumference at the waist to the circumference at hip. Source: Frisancho, A. R. 2005. *Nutritional Anthropometry.* Ann Arbor: University of Michigan Press.

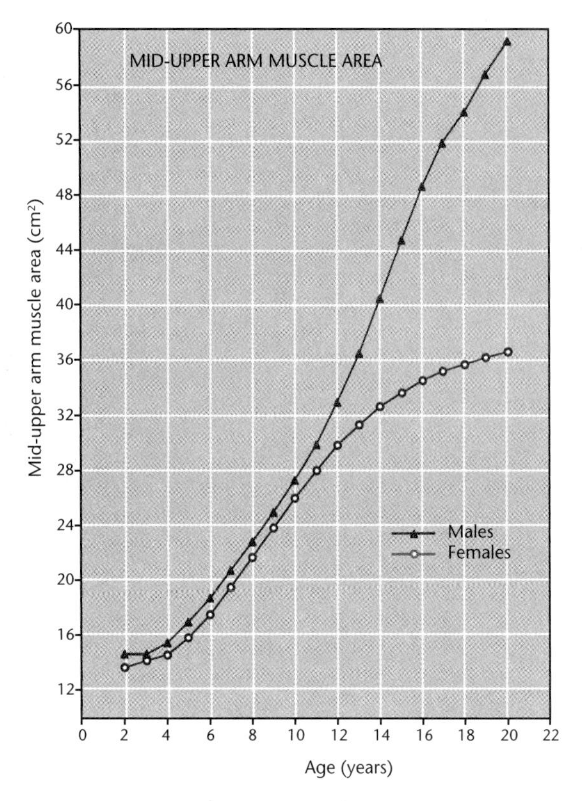

FIGURE 16.20. Growth of skeletal muscle of males and females estimated from measurements of the mid-upper arm muscle area. Source: Frisancho, A. R. 2005. *Nutritional Anthropometry.* Ann Arbor: University of Michigan Press.

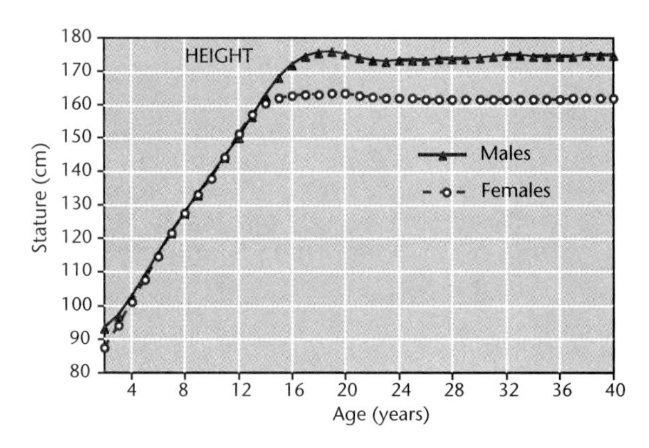

FIGURE 16.21. Attainment of adult height in males and females. Source: Frisancho, A. R. 2005. *Nutritional Anthropometry.* Ann Arbor: University of Michigan Press.

maturers will tend to terminate growth earlier than late maturers. Thus, depending on the height attained prior to adolescence, the timing of skeletal maturation can influence the final height.

Old Age and Senescence

Body Size and Skeletal Muscle

Old age extends from 50 years onwards. It is characterized by a general decline in height and reduction of

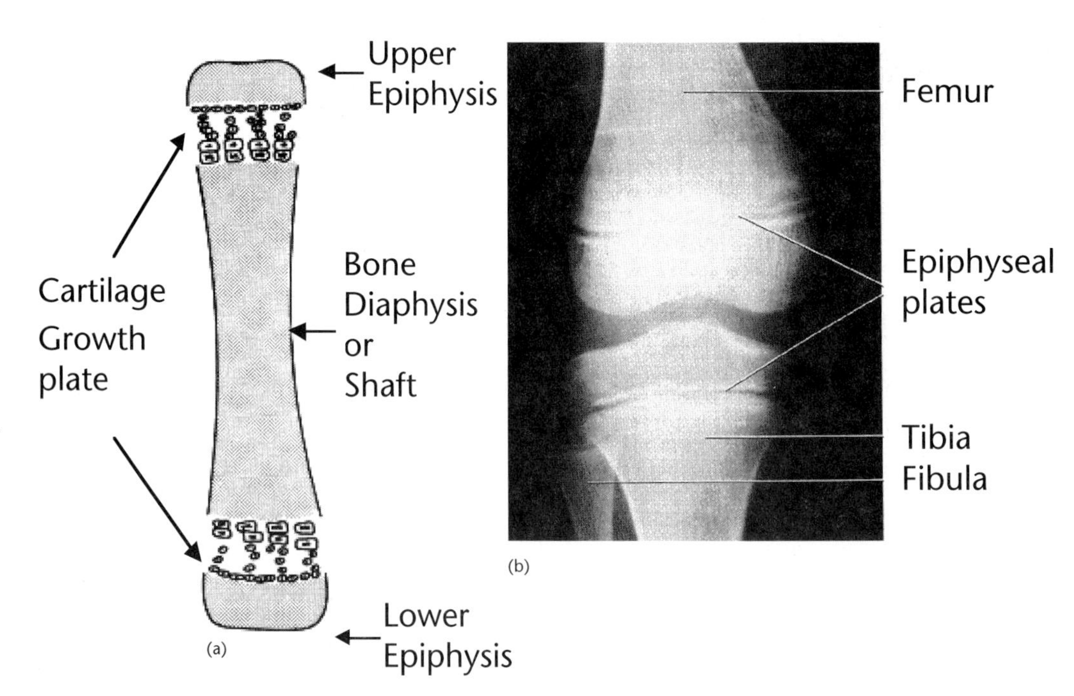

FIGURE 16.22. (a) A limb bone consists of a shaft bone, or diaphysis, and an upper and lower end called the epiphysis. Prior to adulthood the epiphysis and diaphysis are separated by cartilage. (b) Magnification of the cartilage present between the diaphysis and epiphysis. During growth and maturation, the cartilage present between the diaphysis and epiphysis is replaced by bone (calcified) and growth in length ceases thereafter. (Composite illustration made by the author.)

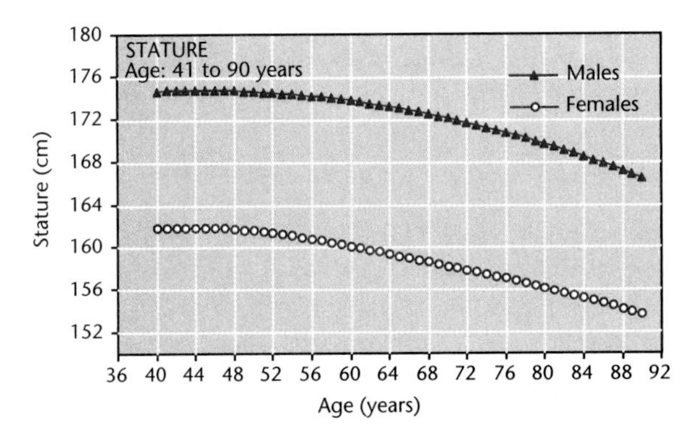

FIGURE 16.23. Decline in height associated with increased curvature of the spinal column. Source: Frisancho, A. R. 2005. *Nutritional Anthropometry.* Ann Arbor: University of Michigan Press.

skeletal muscle. **Figures 16.23** and **16.24** show that after the age of 45 years in males and 60 years in females, declines occur rapidly so that by 90 years of age, males and females have as much muscle as a 12-year-old child. Similarly, after the age of 55 years, height declines rapidly at a rate of 1.33 cm per decade in males and 1.67 cm in females. The greater age-associated decline in adult height of females is related to the greater bone loss and osteoporosis of females, which in turn reduces the intervertebral space. It is also related to greater spinal curvature. Current evidence suggests that the physiological changes that characterize old age is, to a certain extent, related to the individual's developmental experience. Studies among first-generation immigrants indicate that the occurrence of menarche is correlated with the age at menopause,[32] suggesting that the onset of maturity is related to the end of reproductive status.

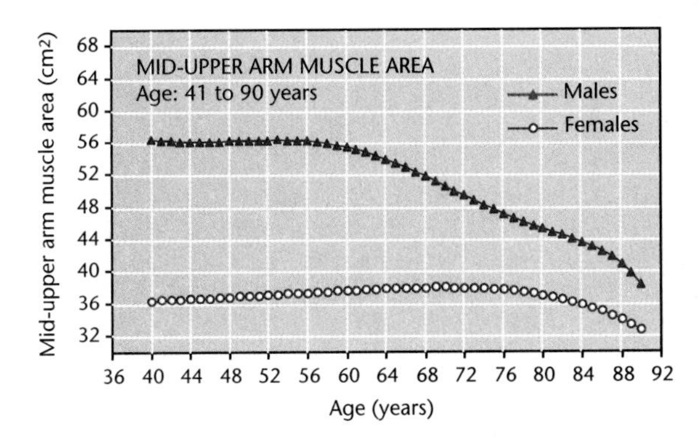

FIGURE 16.24. Age-associated decline in estimated mid-upper arm muscle area in males and females. Source: Frisancho, A. R. 2005. *Nutritional Anthropometry.* Ann Arbor: University of Michigan Press.

Summary

The fetal period is influenced by energy variability, which influences growth in body size and body composition. In the prenatal stage, the embryonic period is highly vulnerable to the effects of teratogenic agents such as drugs, bacteria and viruses, which can impair the proper development of the central nervous system. Even nutrient deficiencies, such as vitamins, can have negative effects on prenatal development. For example, a deficiency in the vitamin folate (folic acid deficiency), when deficient in the early stages of pregnancy, contributes to the development of neural tube birth defects. During pregnancy there are anatomical and physiological changes that affect almost every function of the body. These changes occur both as an adaptation to facilitate the appropriate milieu for the development of the fetus, and as a preparation of the mother for the process of labor, birth, and lactation. Owing to the increased demands required for the growth of the maternal and fetal tissues, the maternal energy needs increase drastically. It is estimated that an extra 68,000 Kcalories, or about 250 kcal/day, must be ingested to support the total cost of pregnancy. How nutritional and environmental factors modify the pattern of prenatal and postnatal growth and development is discussed in the next chapter.

Suggested Readings

Roche, A. F., and S. S. Sun. 2003. *Human Growth: Assessment and Interpretation.* Cambridge: Cambridge University Press.

Literature Cited

1. Lumley, J., L. Watson, M. Watson, and C. Bower. 2000. Periconceptional supplementation with folate and/or multivitamins for preventing neural tube defects. *Cochrane Database Syst. Rev.* 2:CD001056.
2. Feinleib, M., S. A. Beresford, B. A. Bowman, J. L. Mills, J. I. Rader, J. Selhub, and E. A. Yetley. 2001. Folate fortification for the prevention of birth defects: Case study. *Am. J. Epidemiol.* 154(12 Suppl.):S60-69.
3. Fleming, A. 2001. The role of folate in the prevention of neural tube defects: Human and animal studies. *Nutr. Rev.* 59:21–23.
4. Usher, R., and F. Maclean. 1969. Intrauterine growth of live born Caucasian infants at sea level: Standards obtained from measurements in 7 dimensions of infants born between 25 and 44 weeks of gestation. *J. Pediatr.* 74:901–910.
5. Widdowson, E. M. 1968. Growth and development of the fetus and newborn. In *Biology of Gestation: Vol 2. The Fetus and Neonate*, Assali, N. S., ed., 68–89. New York: Academic Press.

6. Stratz, C. H. 1909. Washstum und Proportionen des Menschen vor und nach der Geburt. *Arhiv für Anthropologie* 8:287–297.
7. Genuth, S. M. 1983. The endocrine system. In *Physiology,* R. M. Berne and M. N. Levy, eds. St. Louis: CV Mosby, 895–1069.
8. Handwerger, S., and M. Freemark. 2000. The roles of placental growth hormone and placental lactogen in the regulation of human fetal growth and development. *J. Pediatr. Endocrinol.* 13:343–356.
9. Hauguel-de Mouzon, S., and E. Shafrir. 2001. Carbohydrate and fat metabolism and related hormonal regulation in normal and diabetic placenta. *Placenta* 22:619–627.
10. van Raaij, J. M., M. E. Peek, S. H. Vermaat-Miedema, C. M. Schonk, and J. G. Hautvast. 1988. New equations for estimating body fat mass in pregnancy from body density or total body water. *Am. J. Clin. Nutr.* 48:24–29.
11. Hytten, F. E., and I. Leitch. 1971. *The Physiology of Human Pregnancy,* 2nd ed. Oxford: Blackwell Scientific Publications.
12. Durnin, J. V. G. A. 1987. Energy requirements of pregnancy: An integration of the longitudinal data from the five-country study. *Lancet* 2:1131–1133.
13. Subcommittee on Nutritional Status and Weight Gain During Pregnancy, Nutritional Status and Weight Gain During Pregnancy of the Food and Nutrition Board, Institute of Medicine. Part 1. 1990. Washington, DC: National Academy Press.
14. Abrams, B., and S. Selvin. 1995. Maternal weight gain pattern and birth weight. *Obstet. Gynecol.* 86:163–169.
15. To, W. W., and W. Cheung. 1998. The relationship between weight gain in pregnancy, birth-weight and postpartum weight retention. *Aust. N. Z. J. Obstet. Gynaecol.* 38:176–179.
16. Wilcox, A. J., and I. T. Russell. 1983. Birthweight and perinatal mortality: I. On the frequency distribution of birthweight. *Int. J. Epidemiol.*12:314–318.
17. Frisancho, A. R. 2000. Prenatal compared with parental origins of adolescent fatness. *Am. J. Clin. Nutr.* 72:1186–1190.
18. Williams, R. L., R. K. Creasy, G. C. Cunnigham, W. E. Hawes, F. D. Norris, and M. Tashiro. 1982. Fetal growth and perinatal viability in California. *Obstet. Gynecol.* 59:624–632.
19. Amini, S. B., P. M. Catalano, V. Hirsch, and L. I. Mann. 1994. An analysis of birth weight by gestational age using a computerized perinatal data base, 1975–1992. *Obstet. Gynecol.* 83:342–352.
20. Petitti, D. B., M. S. Croughan-Minihane, and R. A. Hiatt. 1991. Weight gain by gestational age in both black and white women delivered of normal-birth-weight and low-birth-weight infants. *Am. J. Obstet. Gynecol.* 164:801–805.
21. Brown, J. E., M. A. Murtaugh, D. R. Jacobs, Jr., and H. C. Margellos. 2002. Variation in newborn size according to pregnancy weight change by trimester. *Am. J. Clin. Nutr.* 76:205–220.
22. Courchesne, E., R. Carper, and A. N. Natacha Akshoomoff. 2003. Evidence of brain overgrowth in the first year of life in autism. *JAMA* 290:337–344.
23. Frisancho, A. R. 2005. *Nutritional Anthropometry.* Ann Arbor, MI: University of Michigan Press.
24. Dimirjian, A. 1986. Dentition. In *Human Growth, Vol. 2: Postnatal Growth,* F. Falkner and J. M. Tanner, eds. New York: Plenum, 269–298.
25. Tanner, J. M. 1978. *Fetus into Man.* Cambridge: Harvard University Press.
26. Tanner, J. M. 1981. *A History of the Study of Human Growth.* Cambridge: Cambridge University Press.
27. Scammon, R. E. 1930. The ponderal growth of the extremities of the human fetus. *Am. J. Phys. Anthropol.* 15:111–121.
28. Krogman, W. M. 1972. *Child Growth.* Ann Arbor: University of Michigan Press.
29. Bogin, B. 1988. *Patterns of Human Growth.* Cambridge: Cambridge University Press.
30. Stratz, C. H. 1909. Washstum und Proportionen des Menschen vor und nach der Geburt. *Arhiv für Anthropologie* 8:287–297.
31. Frisancho, A. R., N. Guilding, and S. Tanner. 2001. Growth of leg length is reflected in socio-economic differences. *Acta Medica Auxologica* 33:47–50.
32. Leidy, L. E. 1996. Timing of menopause in relation to body size and weight change. *Hum. Biol.* 68:967–982.

Nutrition and the Human Life Cycle

The extent to which an individual follows the normal pattern of prenatal and postnatal development in the life cycle is determined by the environmental factors that the organism encounters. If the environment provides the appropriate conditions, the organism fulfills its growth potential and reaches adult and old age following the pattern set by the genetic information. On the other hand, if the environment provides the inappropriate conditions, the preset pattern of the life cycle is disrupted and the organism does not fulfill its biological potential with consequences during adulthood. Undernutrition is one well-defined negative environmental condition that influences prenatal and postnatal growth and development. As of 2000, in the developing countries, some 30 million infants are born each year with impaired growth due to poor nutrition during fetal life. Another 182 million preschool children, or 33 percent of children under five are stunted or chronically undernourished and 27 percent of preschool children are underweight. Similarly, among adults, both under- and over-weight examples are present in many countries in the developing world. This chapter focuses on the determinants of low birth weight in industrialized and developing countries, determinants of protein-calorie malnutrition, development of immunity under conditions of chronic undernutrition in both children and adults, and consequences of chronic undernutrition on the development of body size of children and adults.

Determinants of Low Birth Weight in Industrialized Countries

Ethnic Differences

Epidemiological studies conducted in the 1960s[1, 2] have shown that the birth weight for full-term black infants (3,100 ± 450 grams or 6.82 lbs) is significantly lower than the birth weight of white infants (3,450 ± 450 grams). The difference in birth weight persisted independently of numerous social and economic risk factors. In the 1967 National Collaborative Perinatal Project, only 1 percent of the total variance in birth weight among 18,000 infants was accounted for by socioeconomic variables.[1] Epidemiological studies[2] indicate that the incidence of low birth weight (birth weight less than 2,500 grams) varied between about 4 to 5 percent for infants born to white mothers, 5.27 percent for infants born to Native American mothers, and 5.31 percent for those born to Hispanic mothers, but infants born to black mothers were much higher (11.32 percent). Likewise, comparative studies of infants born to "mixed-race" parents in the United States in the year 1983[3] found that the percentage of LBW, when the mother was black and the father was white, was higher than when the mother was white and the father was black. That is, the mother's ethnicity is a stronger predictor of low birth weight than the ethnicity of the father.

Socioeconomic Factors

The above findings have led some investigators to suggest that the lower birth weights among blacks have a genetic basis.[4–7] However, various studies indicate that foreign-born black women were less likely to have low birth weight infants than U.S.-born black women[8] when giving birth in the United States. Analysis of nearly 100,000 birth records for 1980 through 1995 from Illinois of whites and blacks born in the United States and in Africa indicates that the reduction in birth weight associated with black populations is also related to ethnic differences.[8, 9] As shown in **table 17.1**, the mean birth weight of white infants born in the United States was 3,446 grams, that of black infants born in Africa was 3,333 grams, and that of black infants born in the United States was 3,089 grams. The incidence of low birth weight (weight less than 2,500 g) infants for African-born blacks was 7.1 percent and 13.2 percent among U.S.-born blacks as compared with 4.3 percent among infants of U.S.-born white women. Furthermore, regardless of socioeconomic status, the infants of black women born in Africa weighed more than the infants of comparable black women born in the United States. Thus, it would appear that the differences in birth weight between blacks and whites in the United States reflects the influence of **both** socioeconomic and genetic factors, or ethnic history.[8, 9] Similarly, various studies indicate that although foreign-born Hispanic

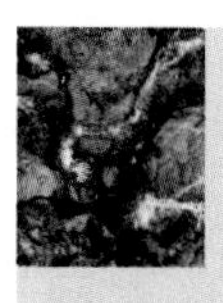

TABLE 17.1

Comparison of Birth Weight of Whites and Blacks Born in the United States and in West Africa Derived from Birth Records for 1980 through 1995 from Illinois

Variables	White Born in United States	Black Born in West Africa	Black Born in United States
No. of Births	44,046	3,165	43,322
Mean Birth Weight (grams)	3,446	3,333	3,089
Low Birth Weight (%)	4.3	7.1	13.2

Source: Adapted from David and Collins.[8]

women had less favorable maternal characteristics than U.S.-born women, they were less likely to have low birth weight infants than U.S.-born Hispanic women.[10, 11] In other words, place of residency rather than ethnicity seems to be an important factor that influences the frequency of low birth weight.

Smoking

In the industrialized countries, smoking is the most consistent and important determinant of birth weight. The reduction in birth weight associated with smoking ranges from an average of 150 g to 250 g.[10–13] The reduction in mean birth weight is inversely related to the number of cigarettes smoked and the frequency of low birth weight infants is proportional to the number of cigarettes smoked.[12–14] Comparative studies indicate that infants born to nonsmoking mothers compared to those born to continuously smoking mothers were on average 257 grams lighter, 1.0 cm shorter, and 0.5 cm smaller in head circumference.[15]

Cocaine

Clinical studies indicate that cocaine users have a higher incidence of specific complications during pregnancy including miscarriages, placental abruptions, and preterm labor, as compared to methadone users and non-drug users.[16] Furthermore, cocaine-exposed infants had significantly lower birth weights, shorter gestational periods, and smaller head circumferences than controls.[16, 17] A major deterioration of prenatal growth has been reported among infants born to cocaine and crack users. Infants of all drug abusers weighed on average 423 g less than a control,[18] amounting to a reduction of nearly 13 percent in prenatal growth. This reduction is of special significance since prenatal growth retardation is also associated with increased frequencies of congenital cardiac anomalies.[19]

Teenage Pregnancy

Pregnancy among adolescents younger than 16 years of age is associated with low birth weight.[20–24] As shown by longitudinal studies conducted by Scholl and associates,[22–24] contrary to previous assumptions, pregnant teenagers aged 12 to 15 continue growing while pregnant. These teenage gravidas gave birth to infants whose birth weights were 150 to 200 grams less than those of infants born to nongrowing gravidas.[20–24] These studies suggest that the reduction in birth weight is the result of a maternal-fetal competition for nutrients,[20, 21] whereby the pregnant teenagers compete for nutrients with their fetus, reducing birth weight (to complete their own growth). In addition, the reduction in birth weight associated with teenage pregnancy is also probably related to the small size of the birth canal that is associated with early maturation. This is because after menarche, the birth canal lags in maturity compared with height.[25, 26] Therefore, girls who mature and get pregnant earlier have smaller and less mature pelvises than do girls who mature later.

In summary, in industrialized countries, cigarette smoking and cocaine use are the most important determinants of low birth weight. In addition, early teenage pregnancy and low prepregnancy weight gain are important factors that contribute to the prevalence of low birth weight infants. Except for teenage pregnancy, the majority of LBW is due to preterm birth or prematurity rather than intrauterine growth retardation. While socioeconomic factors such as income, education, and access to prenatal care are known to influence birth weight, the effects of these factors are usually expressed through their influence on maternal biological traits. Thus, the poorer the socioeconomic status, the lower the pregnancy weight, pregnancy weight gain, and birth weight.

Determinants of Low Birth Weight in Developing Countries

Undernutrition

As of 2000, about 17 million infants were born every year with LBW, representing about 16 percent of all newborns in developing countries. About 21 percent of newborns in south central Asia and 15 percent and 11 percent, respectively, in middle and western Africa are born undernourished.[27] As shown in **figure 17.1**, the majority of LBW in developing countries is due to intrauterine growth retardation (IUGR). Intrauterine growth retardation, defined as small for gestational age, is caused by several environmental factors such as maternal undernutrition before conception and during pregnancy, malaria (where it is endemic), anemia, and acute and chronic infections. It is also associated with biological related factors such as teenage pregnancy and/or primiparity, multiple gestation, low gestational weight gain due to inadequate dietary intake, and short maternal stature due to the mother's own childhood undernutrition.

Teenage Pregnancy

The populations of the developing nations are experiencing a massive increase in migration of rural populations into the urban centers where the cultural norms that govern marriage and reproduction have been disrupted. Furthermore, in all urban populations, the age at menarche has declined by as much as 2 years when compared to rural populations. Thus, as a consequence of these sociological and biological factors, the number of reproductively capable teenagers has drastically increased. The increased incidence of teenage pregnancy has also increased the frequency of low birth weight infants. As illustrated in **figure 17.2**, the reduction of birth weight among mothers younger

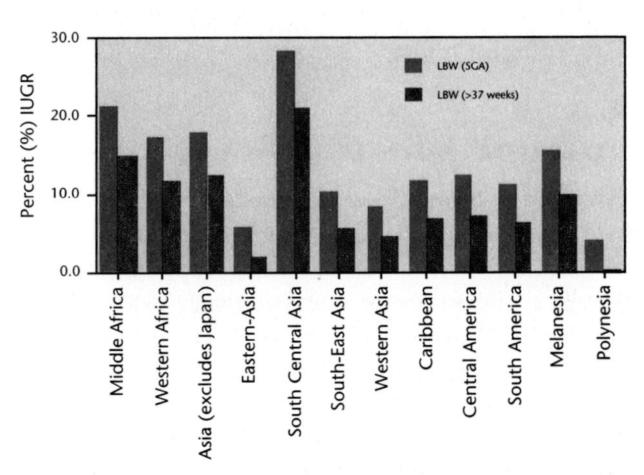

FIGURE 17.1. Incidence of intrauterine growth retardation (IUGR) associated with small for gestational age (SGA) and low birth weight at term (LBW > 37 weeks) in selected countries from Africa, Asia, and Central and South America. Source: Data derived from ACC/SCN.[27]

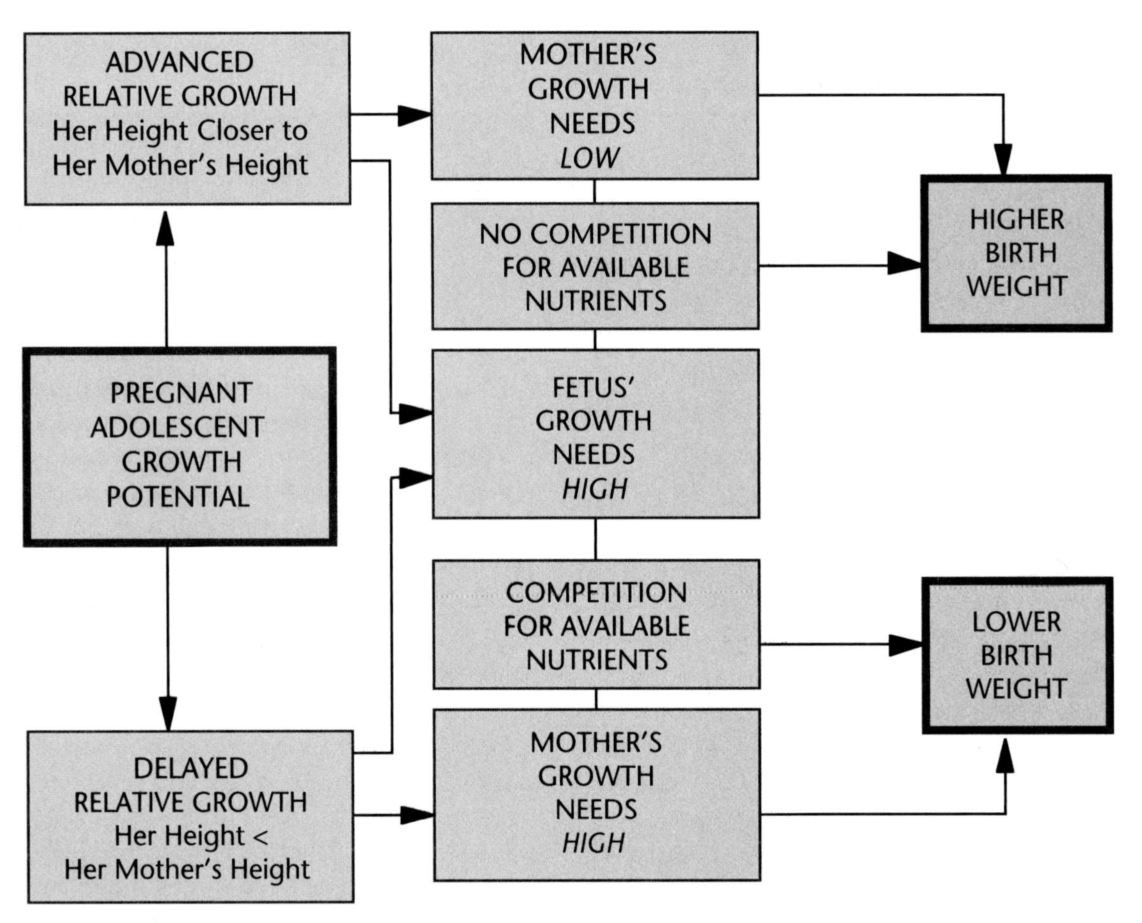

FIGURE 17.2. Schematization of the maternal-fetal growth competition. Because pregnant teenagers younger than 15 years continue growing, they compete with their fetuses for nutrients, resulting in a lower birth weight of their infants. Source: Frisancho, Matos, Leonard, and Yaroch.[21]

than 15 years of age is the result of a maternal-fetal competition for nutrients,[21] whereby the pregnant teenagers, to complete their own growth, compete for nutrients with their fetus, reducing birth weight.

Immune Function

Because the human body is constantly under attack from viruses, bacteria, and unicellular and multicellular parasites, development of an appropriate immune system is absolutely necessary for survival.[28–47] The defense of the organism includes: (a) the leukocyte-macrophage system, (b) the lymphatic system, and (c) the humoral immunity.

Leukocytes (or White Blood Cells) and Macrophages

The leukocytes, or white blood cells, are formed both in the bone marrow and the lymphoid tissue. There are about 7 million white blood cells (or leucocytes) per milliliter of blood. The leucocytes are classified into six types. These include polymorphonuclear neutrophils, polymorphonuclear eosinophils, polymorphonuclear basophils, monocytes, lymphocytes, and occasional plasma cells. The function of all leukocytes is to inactivate, digest, or kill the bacteria with their bactericidal agents.

Lymphatic System

The lymphoid tissue is located in the **lymph** nodes, the spleen, the submucosal areas of the gastrointestinal tract, and the bone marrow. The lymphoid tissues of the body together form the **reticuloendothelial system.** The lymphoid tissue in its different forms such as the tonsils and adenoids traps and digests (phagocytizes) foreign organisms such as bacteria that enter the respiratory system including the throat and pharynx (the tonsils and adenoids), the gastrointestinal mucous, and general circulation.

Humoral Immunity and Antibodies

Concurrent with protection provided by the leukocytes (or white blood cells) and lymphatic system, the organism develops immunity through two avenues: (a) **humoral immunity**, which produces circulating **antibodies**, such as globulin molecules that are capable of attacking the invading agent; and (b) **cell-mediated immunity**.

Antibodies

The antibodies are gamma globulins or proteins called immunoglobulins. Antibodies are synthesized by the plasma cells, found in the lymph nodes, spleen, and lymphoid tissues. There are five general classes of antibodies named: **IgM, IgG, IgA, IgD**, and **IgE**. Ig stands for immunoglobulin, and the other five letters simply designate the respective classes. IgG comprises about 75 percent of the antibodies of the normal person; IgE constitutes only a small percent of the antibodies; IgM also comprises a small percent of the antibodies but because they have ten binding sites they are exceedingly effective in protecting the body against invaders even though there are not many IgM antibodies. A unique feature of antibodies is that they are specific for a particular antigen. Antibodies protect the body against invading agents in one of several ways: (a) by agglutination, in which bacteria or cells are bound together into a clump; (b) by precipitation, in which the toxin and antibody becomes so large that it is rendered insoluble; (c) by neutralization, in which the antibodies cover the toxic sites of the antigenic agent; and (d) by lysis, in which the antibodies directly attack the membranes of cellular agents and thereby rupture the cell.

Cell-Mediated Immunity

The thymus produces two types of lymphocytes: **T lymphocytes** and **B lymphocytes**. The T lymphocytes are responsible for **cell-mediated immunity**. Upon exposure to the proper antigens, the T lymphocytes of the lymphoid tissue proliferate and release large numbers of activated T cells into the lymph. Like the antibodies, the T lymphocytes are specific for a particular antigen. There are several different types of T cells. These are classified into three major groups: (a) cytotoxic T cells, (b) helper T cells, and (c) suppressor T cells. All three work in concert to protect the organism from the effect of bacteria, toxins, and viruses.

In summary, all these systems protect the organism by destroying the invading agents by the process of phagocytosis and by forming antibodies and sensitized lymphocytes, which destroy or neutralize the invading bacteria, toxins, and viruses.

Immune Function, Chronic Undernutrition, and Evolution

The development of an appropriate immune system is absolutely necessary for survival. Environmental factors, such as prenatal malnutrition occurring at a critical point in the development of the immune system, compromise postnatal immunity and subsequent survival. Studies on animals and humans indicate fetal nutrient deprivation impairs immune function later in life.[28–32] In developing nations, the adverse consequences of LBW is manifested in increased rates of neonatal morbidity and mortality. Epidemiologic evidence links low birth weight to an increase in prevalence of mortality during infancy. In Brazil, for example, Victora et al.[33] found that low birth weight infants were 2.0 times more likely to die from diarrhea, 1.9 times more likely to die of respiratory infections, and 5.0 times more likely to die from other infections than were children who weighed 2,500 g at birth. Likewise, infants who weigh 2,000 to 2,499 g at birth have a fourfold higher risk of neonatal death than those who weigh 2,500 to 2,999 g, and a tenfold higher risk than those weighing 3,000 to 3,499 g.[34] Thus, the more severe the prenatal growth restriction, the higher the risk of death. Being born preterm, as well as having low birth weight carries the strongest risk of mortality.[35] Studies in the Gambia showed that mortality from infectious diseases among individuals born during or shortly after the annual rains (in rural Gambia, annual rains coincide with a hungry season when food stocks from the previous year's harvest are running out and the current year's crops have yet to be harvested) were found to be >10 times as likely to die prematurely than those born during the dry season.[36, 37]

Synergic Interaction of Malnutrition, Infections, and Mortality

In general, the immune system incorporates evolutionary processes into normal development and function.[38] Therefore, populations that live under chronic undernutrition have developed efficient immunological responses that enable them to survive. Otherwise, natural selection would have eliminated them long ago. In fact, undernourished children have an elevated production of serum immunoglobulins (IgA, IgM, IgG, IgD).[39, 40] Yet, undernourished children show an increased postnatal mortality due to infectious diseases. This paradox is related to the fact that undernutrition in children is associated with the decreased production of immunoglobulins (IgA) in nasopharyngeal salivary secretions, such as those found in the mucosa, tears, and saliva.[39, 40] There is abundant evidence indicating that prenatal and postnatal undernutrition is usually associated with many negative environmental factors such as poverty, poor hygiene, and contaminated food.[41–47] Hence, malnourished children, despite the high presence of immunoglobulins in the serum, are subject to increased infections due to deficient antibody production in the nasopharyngeal salivary secretions. This low presence of antibody facilitates the entrance of bacteria-causing infections that overcome the organism's immunity.

Protein-Energy Malnutrition During Childhood

Protein-energy malnutrition (PEM) refers to a wide spectrum of clinical manifestations ranging from pure protein deficiency to deficiencies of both protein and energy. PEM is found in children and adults and two main forms are recognized: kwashiorkor and marasmus.[49]

Kwashiorkor

Kwashiorkor is caused primarily by an acute protein deficiency occurring in the presence of relatively adequate energy intake, often from foods poor in protein but rich in carbohydrates such as starch. The name kwashiorkor is used by the Ga people of Ghana to whom the disorder is well known. This disease is usually manifested after the first year, most often between the second and fourth years of postnatal life. The height of the child with kwashiorkor during the first year may be normal. In the second year of growth, height is drastically reduced to between 60 percent and 80 percent of the expected standard. A typical kwashiorkor (**fig. 17.3a**) child is the so-called "sugar baby," with a round moon face, pitting edema, variable degrees of dermatosis with depigmentation, and hyperkeratosis. The child's hair is often depigmented, and when malnutrition alternates with periods of relatively adequate dietary intake, depigmented bands, often referred to as "flag signs," appear in the hair. In addition, hair implantation is affected so that it falls out spontaneously or can be painlessly removed. Serum albumin and protein levels are also reduced. The kwashiorkor child is apathetic, lethargic, anorexic (not hungry), withdrawn, and highly irritable. Sometimes the child becomes immobile, lying quietly in a fetal position with open, nonfixating eyes. The kwashiorkor child often maintains a monotonous whimper. Like the marasmic infant, the kwashiorkor child exhibits hypotonia and poorly developed motor skills. In some cases, the poorly developed motor skills are so severe that the child does not respond in any measurable way to standard psychomotor stimuli derived from mental developmental scales.

Marasmus

Marasmus refers to a child with exhausted protein and energy stores resulting from a chronic, symmetrical reduction of all nutrients that approaches starvation levels. It causes a drastic reduction in growth and development such that the marasmic child is short and lightweight for its age, as well as retarded in skeletal maturation. The child exhibits extreme muscular wasting and almost no subcutaneous fat (**fig. 17.3b**). These conditions reflect an attempt by the organism to use its own tissues as a source of nutriment in the face of a chronic and generalized decrease in nutrient resources. The marasmic child also has a reduced brain weight, cortical atrophy, hypotonia, reduced activity, and displays constant hunger behavior. Sometimes patients will present signs of both marasmus and kwashiorkor such as reduced muscle and subcutaneous fat and the edema of kwashiorkor, with or without its skin lesions. Biochemical markers show features of both marasmus and kwashiorkor, but the alterations of severe protein deficiency predominate.

Chronic Undernutrition During Childhood, Adolescence, and Old Age

Undernutrition in children is usually categorized as: stunting, underweight, and wasting.

Stunting

Stunting refers to short stature and is considered indicative of chronic undernutrition. It is defined with reference to height for age:

1. Z score for height = (individual's height − age − sex-specific mean height)/age-and-sex-specific SD for height.

Underweight

Underweight refers to low weight and is considered indicative of acute undernutrition of recent origin. It is defined with reference to weight for age:

FIGURE 17.3. Typical appearance of children with (a) kwashiorkor and (b) marasmus. Note the drastic reduction of subcutaneous fat and muscle wasting of the marasmic child. In contrast, the child with kwashiorkor exhibits edema and changes in skin and hair pigmentation. (a) © Omikron/Photo Researchers. (b) © Hulton-Deutsch Collection/Corbis.

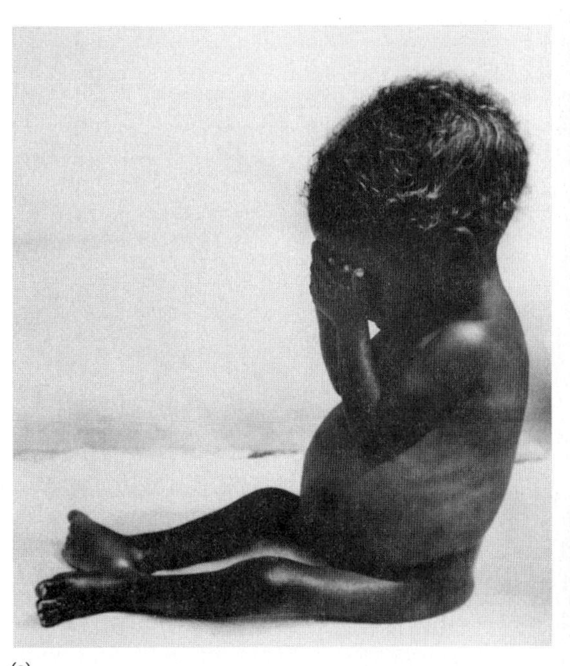

(a)

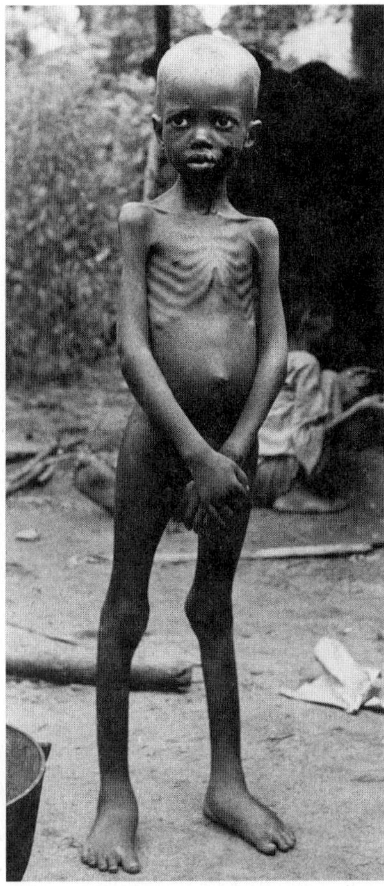

(b)

2. Z score for weight = (individual's weight – age – sex-specific mean weight)/age-and-sex-specific SD for weight.

Wasting

Wasting refers to low weight for height and is considered indicative of chronic undernutrition. It is defined with reference to weight for height for age:

3. Z score of weight for height = (individual's weight for height – age – sex-specific mean weight for height)/age- and sex-specific SD of weight for height.

In the developing countries, as of 2000,[27] 32 percent and 26.7 percent of the children under age five are stunted and underweight, respectively (**table 17.2**). Global wasting is not as common as stunting or underweight; the global prevalence is about 9.4 percent. Western Africa and south central Asia have the highest prevalence of wasting (both about 15.5 percent), followed by Southeast Asia (10.4 percent). In Central American, South American, and southern African children, the prevalence of wasting is quite low. As inferred from height censuses of children entering primary school (first grade), stunting or chronic undernutrition is much greater in nine year olds than in six year olds (**fig. 17.4**). Furthermore, stunting is more frequent in rural areas than in urban areas (**table 17.3**).

Gender Differences

In all Latin American countries a higher proportion of boys than girls are stunted (**fig. 17.5**). This difference implies that boys are either under greater nutritional stress than females, or females are more resilient to the negative effects of environmental stress than males. Various studies have documented that the long-term

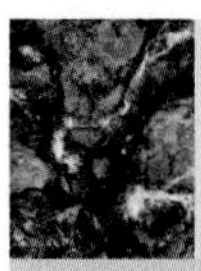

TABLE 17.2

Estimated Prevalence (%) of Stunted and Underweight Preschool Children in 2000 and 2005

	2000 Stunted (%)	2005 Stunted (%)	2000 Underweight (%)	2005 Underweight (%)
Africa	35.2	33.8	28.5	29.1
Eastern	48.1	48.5	35.9	38.7
Northern	20.2	17.0	14.9	13.2
Western	34.9	34.6	36.5	38.1
Asia	34.4	29.9	29.0	25.3
South Central	43.7	39.4	43.6	40.0
Southeast	32.8	27.9	28.9	25.3
Latin America	12.6	9.3	6.3	4.3
Caribbean	16.3	13.7	11.5	8.7
Central America	24.0	23.5	15.4	15.4
South America	9.3	5.3	3.2	2.3
All developing countries	32.5	29.0	26.7	24.3

Source: Adapted from ACC/SCN.[27]

effects of protein deprivation are more pronounced in males than in females. Experimental protein deprivation in rhesus macaques revealed significant differences in metabolic efficiency, with females being much more efficient and able to gain more weight than males when fed a low protein diet.[50] Laboratory experiments on animals[51] and famine demographics[52] show that females are better able to cope with nutritional deprivation and to survive episodes of famine than males. From an empirical perspective, females should be less vulnerable to nutritional deprivation than males, because, on the average, they have more calorie reserves in the form of body fat than males. Furthermore, in most of the countries, boys aged six to nine, in general, spend more time outside the home than girls do. Proximity to the household may allow girls better physical access to available food. Similarly, from the evolutionary perspective, females are given the demanding nature of reproductive functions such as pregnancy, lactation, and child-rearing with which the female body must cope. Thus, females should be better buffered against environmental stress than males.[53] Field studies in Africa, Nepal, and South America among nutritionally stressed children and adults lend support to the concept of significant female buffering under conditions of environmental stress.[54–61]

Polygyny and Child Mortality

Another important factor affecting child mortality is polygyny. Many preindustrial populations practice polygyny. It has been postulated that the practice of polygyny is related to a shortage of men, older age of available males, and younger age of death of males, which together produce a surplus of women on the marriage "market." According to the investigations of Strassmann,[48] among the Dogon of Mali, East Africa, male preference is the driving force that accounts for the polygyneous marriage system. Probably as a consequence of the joint effect of decrease of co-wife com-

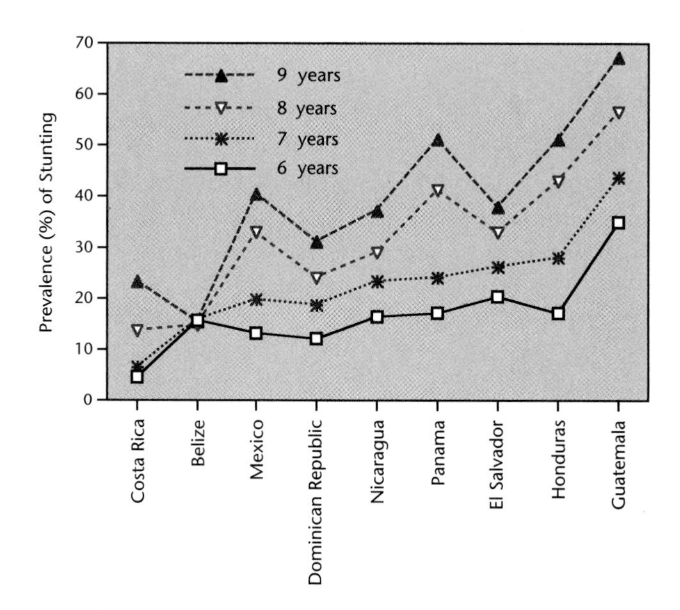

FIGURE 17.4. Prevalence of stunting among first-grade schoolchildren by age group in Latin America and the Caribbean. Source: Data derived from ACC/SCN.[27]

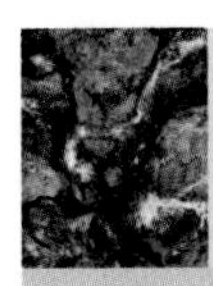

TABLE 17.3

Prevalence of Stunting among First-Grade Schoolchildren in Latin America and the Caribbean by Area of Residence

Country	Year of School Height Census	Prevalence by Area (%)		
		Rural	Urban	Total
Belize	1996	22.5	6.9	15.4
Dominican Republic	1995	23.1	13.8	19.0
Nicaragua	1986	30.1	20.1	23.9
Honduras	1997	47.6	28.2	40.6
Peru	1993	67.0	35.0	48.0

Source: Adapted from ACC/SCN.[27]

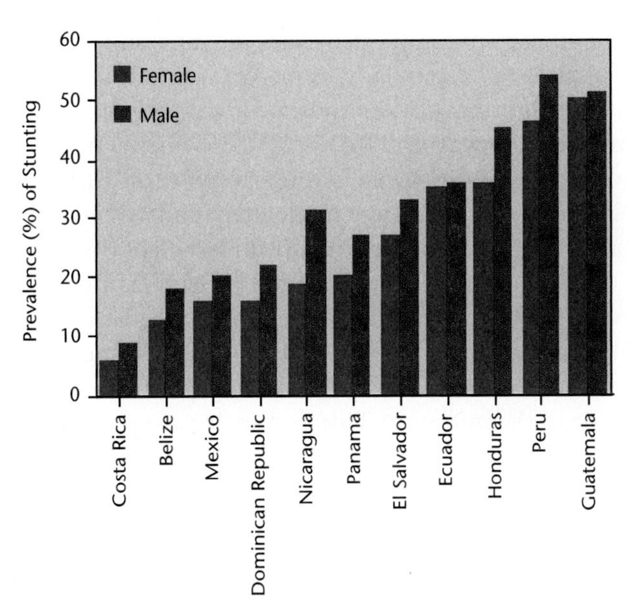

FIGURE 17.5. Gender differences in the prevalence (%) of stunting among first-grade schoolchildren in Latin America and the Caribbean from 1986 to 1997. Source: Data from ACC/SCN.[27]

petition, paternal investment, nepotistic investment, and decrease of economic resources, polygyny has a negative effect on child mortality.[48]

Stunting in Adults

Most populations of the developing nations attain adult height by about 22 to 24 years for males and 20 to 22 years for females. These values, when compared to the United States' standards, amount to an increase of 10 percent in the duration of the period of growth (**fig. 17.6**). Despite this increase in the period of growth, the adult height of undernourished populations is reduced by 10 percent. This difference is mostly accounted for by the uniform slow growth rate that characterizes undernourished populations.[54–61] Furthermore, this growth deficit is related to the growth retardation suffered in early childhood, which is not easily compensated for in adolescence. The undernourished adolescents, even though they grow for a longer period than well-nourished ones, become shorter as adults.[61] The reason why the childhood growth retardation is not easily compensated for in adolescence is related to the increased genetic control of adolescent maturation. As inferred from analyses of skeletal maturation and growth in body size of nearly 6,000 children exposed to varying environmental conditions from Central America and Panama and summarized in **figure 17.7**, the delay in skeletal maturation between the ages of 0 and 10 years averages about 20 percent, whereas the delay during adolescence does not exceed 10 percent.[61] This 10 percent delay (which reflects epiphyseal closure) coincides with the increase in the age of attainment of adult stature, which, as previously indicated, is prolonged by about 10 percent in most undernourished populations. Since the age at epiphyseal closure and termination of growth is proportionally less retarded than the delay experienced during childhood, environmentally related differences in adult stature are mostly caused by environmental influences that retard growth and maturation during childhood. A follow-up study in Guatemala found that nearly 67 percent of severely stunted and 34 percent of moderately stunted three-year-old girls later became stunted adult women.[62]

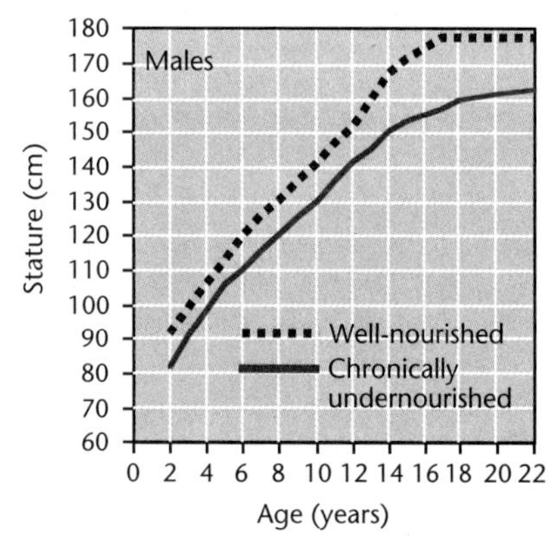

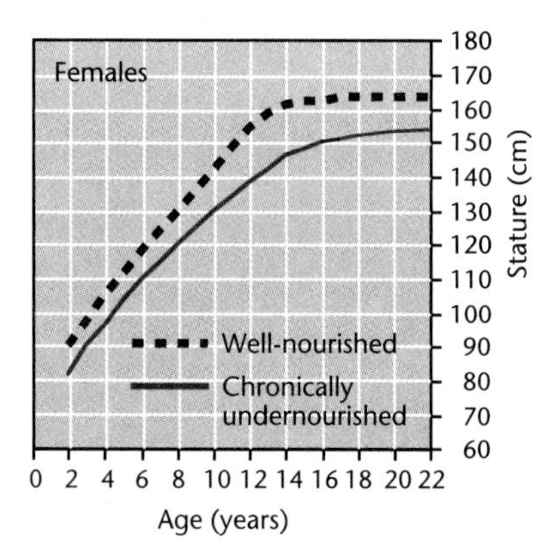

FIGURE 17.6. Comparison of patterns of growth of chronically undernourished and well-nourished populations. With chronic undernutrition, growth rate for all ages is slow and continues into the early twenties. Source: Data from Frisancho, Garn, and Ascoli.[61]

In summary, stunted children are more likely than nonstunted children to become stunted adults as long as they continue to reside in the same environment that gave rise to the stunting.

Undernutrition and Work Capacity

In men, there seems to be a continuous gradient in work capacity and productivity that is linked to chronic undernutrition. Studies in developing countries indicate that individuals with a BMI below 18.5 kg/m^2 allocate fewer days to heavy labor, are more likely to be absent from work owing to illness, and show a progressive increase in mortality rates.[63, 64] Similarly, among Nigerian men and women, the mortality rates among chronically energy deficient individuals who are severely (BMI <16.00 kg/m^2), moderately (BMI = 16.00–16.99 kg/m^2), and mildly (BMI = 17.00–18.49 kg/m^2) underweight are, respectively, 150 percent, 140 percent, and 40 percent greater than rates among non-CED individuals.[27, 63, 64] Similarly, results from a study in an urban slum in Bangladesh indicated that the mean BMI among men was 19.0 (SD = 2.9), of which 9 percent were classified as severely underweight (or grade III CED), 15.6 percent were moderately underweight (or grade II), 27 percent were mildly underweight (or grade I), and 48.2 percent were "normal." There was a significant inverse association between BMI and work-disabling morbidity. Below a BMI of 16.0, 55 percent of men had lost one or more working days in the month prior to the interview. This proportion dropped to 35 percent among those with a BMI between 16.0 and 17.0. Above a BMI "threshold" of 17.0, the percentage of men incapacitated from work was similar in each BMI category 1.[27, 63, 64]

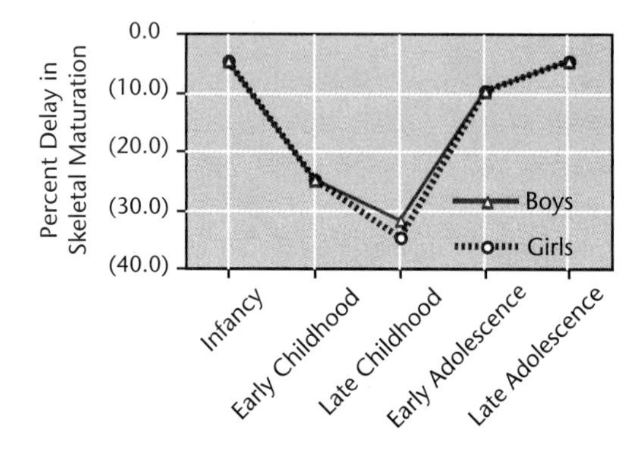

FIGURE 17.7. Skeletal maturation of chronically undernourished children expressed as percent delay with respect to U.S. references. Source: Summarized from Frisancho, Garn, and Ascoli.[61]

Undernutrition in Older People

In developing countries, the majority of poor older people enter old age after a lifetime of poverty and deprivation, poor access to health care, and a diet that is usually inadequate in quantity and quality. As shown in **table 17.4**, the prevalence of undernutrition among males and females in India and Malawi varies between 27 percent and 36 percent.

Summary

In industrialized countries, the major cause of reduction in birth weight is related to smoking and teenage pregnancy (conception under 15 years) and premature birth. Smoking during pregnancy accounts for a birth weight reduction of about 200 grams. In the developing countries smoking during pregnancy does not

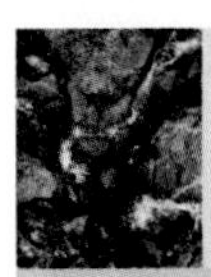

TABLE 17.4
Prevalence of Undernutrition from Two Studies of Older People

Country	Men (%)	Women (%)
India (Mumbai slums)	35.0	35.0
Malawi (rural Lilongwe)	36.1	27.0

Source: Adapted from ACC/SCN.[27]

have much influence on birth weight, but as developing nations adopt the social patterns of industrialized nations, smoking is becoming an important negative factor that is a growing contributor to the reduction in birth weight.

In the developing countries, a large proportion of low birth weight infants is born to teenage adolescents. As a result of the high migration of rural populations into urban areas, adolescents that are not culturally and nutritionally prepared to sustain the challenges of reproduction are giving birth to a high proportion of low birth weight infants. These infants, in turn, will become stunted children and adolescents that continue to give birth to intrauterine growth-retarded infants. Unless this cycle is interrupted, chronic undernutrition in the developing nations will remain a main drawback to future development. The net result of the continuous struggle with infectious diseases is that populations in the developing nations continue to suffer from chronic undernutrition. The high worldwide prevalence of low birth weight, protein-calorie malnutrition such as kwashiorkor and marasmus, stunting, and underweight indicates that, despite the success of the green revolution, shortage of food remains a major problem that constrains the growth, development, and productivity of more than a third of all children and adults in the developing world.

Suggested Readings

Barker, D. J. P. 1998. *Mothers, Babies and Health in Adult Life.* Edinburgh: Churchill Livingstone.

ACC/SCN. 2000. *Fourth Report on the World Nutrition Situation.* Geneva: ACC/SCN in collaboration with IFPRI.

ACC/SCN. 2004. *Fifth Report on the World Nutrition Situation.* Geneva: ACC/SCN in collaboration with IFPRI.

Literature Cited

1. Naylor, A. F., and N. C. Myrianthopoulos. 1967. The relation of ethnic and selected socio-economic factors to human birth-weight. *Ann. Hum. Genet.* 31:71–83.
2. Emanuel, I., W. Leisenring, M. A. Williams, C. Kimpo, S. Estee, W. O'Brien, and C. B. Hale. 1999. The Washington State intergenerational study of birth outcomes: Methodology and some comparisons of maternal birth weight and infant birth weight in four ethnic groups. *Pediatr. Perinat. Epidemiol.* 13:352–369.
3. Migone, A., I. Emanuel, B. Mueller, J. Daling, and R. E. Little. 1991. Gestational duration and birthweight in white, black and mixed-race babies. *Paediatr. Perinat. Epidemiol.* 5: 378–391.
4. Little, R. E., and C. F. Sing. 1987. Genetic and environmental influences on human birth weight. *Am. J. Hum. Genet.* 40:512–526.
5. Magnus, P. 1984. Further evidence for a significant effect of fetal genes on variation in birth weight. *Clin. Genet.* 26: 289–296.
6. Hulsey, T. C., A. H. Levkoff, and G. R. Alexander. 1991. Birth weights of infants of black and white mothers without pregnancy complications. *Am. J. Obstet. Gynecol.* 164:1299–1302.
7. Goldenberg, R. L., S. P. Cliver, G. R. Cutter, et al. 1991. Black-white differences in newborn anthropometric measurements. *Obstet. Gynecol.* 78:782–788.
8. David, R. J., J. W. Collins, Jr. 1997. Differing birth weight among infants of U.S.-born blacks, African-born blacks, and U.S.-born whites. *N. Engl. J. Med.* 337:1209–1214.
9. Steckel, R. H. 1987. Growth, depression and recovery: The remarkable case of American slaves. *Ann. Hum. Biol.* 14: 111–132.
10. Fuentes-Afflick, E., N. A. Hessol, and E. J. Pérez-Stable. 1998. Maternal birthplace, ethnicity, and low birth weight in California. *Arch. Pediatr. Adolesc. Med.* 152:1105–1112.
11. Hessol, N. A., and E. Fuentes-Afflick. 2000. The perinatal advantage of Mexican-origin Latina women. *Ann. Epidemiol.* 10:516–523.
12. Kramer, M. S. 1987. Determinants of low birth weight: Methodological assessment and meta-analysis. *Bull. World Health Organ.* 65:663–737.
13. Garn, S. M. 1985. Smoking and human biology. *Hum. Biol.* 57:505–523.
14. Frisancho, A. R., and S. L. Smith. 1990. Reduction in birth weight associated with smoking among young and older-age women. *Am. J. Hum. Biol.* 2:85–88.
15. Wang, X., I. B. Tager, H. Van Vunakis, F. E. Speizer, and J. P. Hanrahan. 1997. Maternal smoking during pregnancy, urine cotinine concentrations, and birth outcomes. A prospective cohort study. *Int. J. Epidemiol.* 26:978–988.
16. Dewan, N., B. Brabin, L. Wood, S. Dramond, and C. Cooper. 2003. The effects of smoking on birthweight-for-gestational-age curves in teenage and adult primigravidae. *Public Health* 117:31–35.
17. Chouteau, M., P. B. Namerow, and P. Leppert. 1988. The effect of cocaine abuse on birth weight and gestational age. *Obstet. Gynecol.* 72:351–354.
18. Kaye, K., L. Elkind, D. Goldberg, and A. Tytun. 1989. Birth outcomes for infants of drug abusing mothers. *N.Y. State J. Med.* 89:256–261.
19. Little B. B., L. M. Snell, V. R. Klein, and L. C. Gilstrap. 1983. Cocaine abuse during pregnancy: Maternal and fetal implications. *Obstet. Gynecol.* 73:157–160.

20. Frisancho, A. R., J. Matos, and L. A. Bollettino. 1984. Influence of growth status and placental function on birth weight of infants born to young still-growing teenagers. *Am. J. Clin. Nutr.* 40:801–807.
21. Frisancho, A. R., J. Matos, W. R. Leonard, and L. A. Yaroch. 1985. Developmental and nutritional determinants of pregnancy outcome among teenagers. *Am. J. Phys. Anthropol.* 66:247–261.
22. Scholl, T. O., M. L. Hediger, and I. G. Ances. Maternal growth during pregnancy and decreased infant birth weight. *Am. J. Clin. Nutr.* 51:790–793.
23. Scholl, T. O., M. L. Hediger, J. I. Schall, C. S. Khoo, and R. L. Fischer. 1994. Maternal growth during pregnancy and the competition for nutrients. *Am. J. Clin. Nutr.* 60:183–188.
24. Scholl, T. O., T. P. Stein, and K. S. Woollcott. 2000. Leptin and maternal growth during adolescent pregnancy. *Am. J. Clin. Nutr.* 72:1542–1547.
25. Moerman, M. 1982. Growth of the birth canal in adolescent girls. *Am. J. Obstet. Gynecol.* 143:528–532.
26. LaVelle, M. 1995. Natural selection and developmental sexual variation in the human pelvis. *Amer. J. Phys. Anthrop.* 98: 59–72.
27. ACC/SCN. 2000. *Fourth Report on the World Nutrition Situation.* Geneva: ACC/SCN in collaboration with IFPRI.
28. Andre, J. B., S. Gupta, S. Frank, and M. Tibayrenc. 2004. Evolution and immunology of infectious diseases: What's new? An E-debate. *Infect. Genet. Evol.* 4:69–75.
29. Lee, W. L., and S. Grinstein. 2004. Immunology. The tangled webs that neutrophils weave. *Science* 303:1477–1478.
30. Makala, L. H., Y. Nishikawa, N. Suzuki, and H. Nagasawa. 2004. Immunology. Antigen-presenting cells in the gut. *J. Biomed. Sci.* 11:130–141.
31. Godfrey, K. M., D. P. J. Barker, and C. Osmond. 1994. Disproportionate fetal growth and raised IgE concentration in adult life. *Clin. Exp. Allergy* 24:641–648.
32. Chandra, R. K. 1975. Antibody formation in first and second generation offspring of nutritionally deprived rats. *Science* 190:189–190.
33. Victora, C. G., P. G. Smith, and J. P. Vaughan. 1988. Influence of birth weight on mortality from infectious diseases: A case control study. *Pediatrics* 81:807–811.
34. Ashworth, A. 1998. Effects of intrauterine growth retardation on mortality and morbidity in infants and young children. *Eur. J. Clin. Nutr.* 52:S34–S42.
35. de Onis, M., M. Blossner, and J. Villar. 1998. Levels and patterns of intrauterine growth retardation in developing countries. *Eur. J. Clin. Nutr.* 52:S5–S15.
36. Moore, S. E., T. J. Cole, E. M. E. Poskitt, et al. 1997. Season of birth predicts mortality in rural Gambia. *Nature* 388:434.
37. Moore, S. E., T. J. Cole, A. C. Collinson, E. M. E. Poskitt, I. A. McGregor, and A. M. Prentice. 1999. Prenatal or early postnatal events predict infectious deaths in young adulthood in rural Africa. *Int. J. Epidemiol.* 28:1088–1095.
38. Mcdade, T. W., and C. M. Worthman. 1999. Evolutionary process and the ecology of human immune function. *Am. J. Hum. Biol.* 11:705–717.
39. McMurray, D. N., H. Rey, L. J. Casazza, and R. R. Watson. 1977. Effects of moderate malnutrition on concentrations of immunoglobulins and enzymes in tears and saliva of young Colombian children. *Am. J. Clin. Nutr.* 30:1944–1948.
40. Sirisinha, S., R. Suskind, R. Edelman, C. Asvapaka, and R. E. Olson. 1975. Secretory and serum IgA in children with protein calorie malnutrition. *Pediatrics* 55:166–170.
41. Edelman, R., R. Suskind, S. Sirisinha, and R. E. Olson. 1973. Mechanisms of defective cutaneous hypersensitivity in children with protein-calorie malnutrition. *Lancet* 1:506–508.
42. Shell-Duncan, B., and J. W. Wood. 1997. The evaluation of delayed-type hypersensitivity responsiveness and nutritional status as predictors of gastro-intestinal and acute respiratory infection: A prospective field study among traditional nomadic Kenyan children. *J. Trop. Pediatr.* 43:25–32.
43. Moore. S. E., A. C. Collinson, and A. M. Prentice. 2001. Immune function in rural Gambian children is not related to season of birth, birth size, or maternal supplementation status. *Am. J. Clin. Nutr.* 74:840–847.
44. Rowland, M. G. M., R. A. E. Barrell, and R. G. Whitehead. 1978. Bacterial contamination in traditional Gambian weaning foods. *Lancet* 1:136–138.
45. Rahaman, M. M., K. M. Aziz, Y. Patwari, and M. H. Munshi. 1979. Diarrhoeal mortality in two Bangladeshi villages with and without community-based oral rehydration therapy. *Lancet* 20:809–812.
46. Black, R. E., M. W. Merson, and K. H. Brown. 1983. Epidemiological aspects of diarrhea associated with known enteropathogen in rural Bangladesh. In *Diarrhea and Malnutrition,* L. C. Chen, and N. S. Scrimshaw, eds. New York: Plenum Press.
47. Rivera, J., and M. Martorell. 1988. Nutrition, infection, and growth Part II: Effects of malnutrition on infection and general conclusions. *Clin. Nutr.* 7:163–167.
48. Strassmann, B. I. 1997. Polygyny as risk factor for child mortality among the Dogon. *Curr. Anthropol.* 38:688–695.
49. Scrimshaw, N. S., J. B. Salomon, H. A. Bruch, and J. E. Gordon. 1966. Studies of diarrheal disease in Central America. *Am. J. Trop. Med. Hyg.* 15:625–631.
50. Riopelle, A. J. 1990. Postnatal protein deprivation in rhesus monkeys. *Am. J. Phys. Anthropol.* 83:239–252.
51. Hoyenga, K. B., and K. T. Hoyenga. 1982. Gender and energy balance: Sex differences in adaptations for feast and famine. *Physiol. Behav.* 28:545–563.
52. Ali, M. 1984. Women in famine: The paradox of status in India. In: *Famine as a Geographical Phenomenon.* B. Currey, and G. Hugo, eds. Dordrecht: D Reidel, 113–133.
53. Stini, W. A. 1969. Nutritional stress and growth: Sex difference in adaptive response. *Am. J. Phys. Anthropol.* 31:417–426.
54. Stinson, S. 1985. Sex differences in environmental sensitivity during growth and development. *Yrbk. Phys. Anthropol.* 28:123–147.
55. Dettwyler, K. A. 1992. The biocultural approach in nutritional anthropology: Case studies of malnutrition in Mali. *Med. Anthropol.* 15:17–39.
56. Leonard, W. R. 1991. Age and sex differences in the impact of seasonal energy stress among Andean agri-culturalists. *Hum. Ecol.* 19:351–368.

57. Sellen, D. W. 1999. Growth patterns among seminomadic pastoralists. Datoga. of Tanzania. *Am. J. Phys. Anthropol.* 109:187–209.
58. Guatelli-Steinberg, D., and J. R. Lukacs. 1999. Interpreting sex differences in enamel hypoplasia in human and non-human primates: Developmental, environmental, and cultural considerations. *Ybk. Phys. Anthropol.* 42:73–126.
59. Panter-Brick, C. 1996. Proximate determinants of birth seasonality and conception failure in Nepal. *Popul. Stud. Camb.* 50:203–220.
60. Panter-Brick, C., P. G. Lunn, R. Baker, and A. Todd. 2001. Elevated acute-phase protein in stunted Nepali children reporting low morbidity: Different rural and urban profiles. *Br. J. Nutr.* 85:125–131.
61. Frisancho, A. R., S. M. Garn, and W. Ascoli. 1970. Childhood retardation resulting in reduction of adult body size due to lesser adolescent skeletal delay. *Am. J. Phys. Anthropol.* 33:325–336.
62. Martorell, R., U. Ramakrishnan, D. G. Schroeder, P. Melgar, and L. Neufeld. 1998. Intrauterine growth retardation, body size, body composition and physical performance in adolescence. *Eur. J. Clin. Nutr.* 52: S43–S53.
63. Shetty, P. S., and W. P. James. 1994. Body mass index: A measure of chronic energy deficiency in adults. *FAO Food and Nutrition Paper* 56:1–57.
64. Pryer, J. A. 1993. Body mass index and work-disabling morbidity: Results from a Bangladeshi case study. *Eur. J. Clin. Nutr.* 47:653–657.

Biocultural Adaptation to Hot and Cold Climates

As tropical primates, humans have developed a unique capacity to adapt to hot climates. However, since they left Africa, humans have been exposed to climates that differ in temperature and humidity. As a consequence, some populations have developed physiological responses that are well suited for hot and humid climates, but not for hot and dry environments. Successful tolerance of heat stress requires the development of synchronized responses that permit the organism to lose heat in an efficient manner and maintain homeostasis. On the other hand, biological responses to cold stress involve mechanisms of heat production and conservation. These adaptive mechanisms to cold stress are more complex than those of heat adaptation. Successful responses to cold stress require the synchronization of cardiovascular and circulatory systems and, most importantly, the activation of the metabolic process. In this chapter, we will focus on the immediate physiological responses to heat stress, the individual factors that affect heat tolerance, and the process of acclimation. In addition, this chapter will focus on the basic physiological responses with which organisms counteract cold stress, the individual factors that modify and affect these responses, and the process of acclimation and acclimatization of native and nonnative populations.

Heat Balance and Heat Exchange

Humans, like most mammals, are *homeothermic*, which means they can maintain a relatively constant internal body temperature independent of environmental temperature. The internal body temperature is in dynamic equilibrium between the factors that add or produce heat and the factors that facilitate heat loss (**fig. 18.1**). This balance is achieved through the integration of mechanisms that produce heat and regulate heat transfer from the internal organs to the shell or periphery, and mechanisms that facilitate the gain or loss of heat to the environment. The factors that produce heat include the basal metabolic rate (BMR), which is the minimum

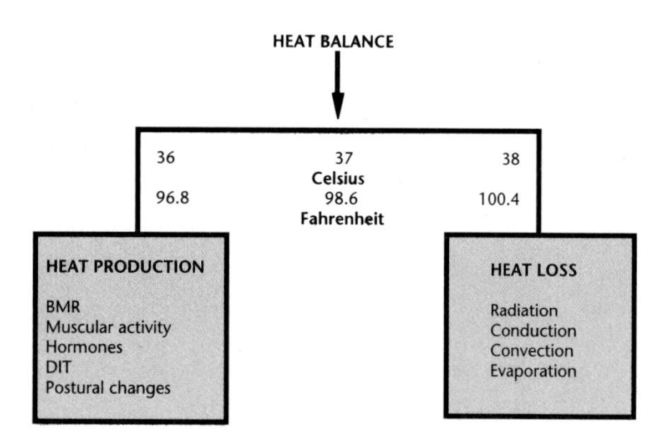

FIGURE 18.1. Heat balance. It is attained throughout the action of mechanisms of heat production and heat loss. From *Human Adaptation and Accommodation* by A. R. Frisancho. Copyright © 1995 by University of Michigan Press. Reprinted by permission.

level of energy required to sustain the body's vital functions in the waking state (see Chapter 13) and the energy to maintain daily activities.

Avenues of Heat Exchange

The avenues of heat exchange include radiation, convection, conduction, and evaporation.

Radiation (R)

Radiation refers to the heat transfer that occurs when particular electromagnetic waves are emitted by one object and absorbed by another. The organism radiates heat to other objects and receives heat from other warmer objects (thus, *R* is + or –). Skin color does not have an effect on heat radiation. However, the color of cloth does influence heat radiation. In general, white clothing under cold conditions radiates less heat than dark color. Heat radiation also depends on the posture of the organism. Upright posture decreases the effective radiating surface area by as much as 75 percent when compared to quadrupedal posture. Therefore, when a person faces the sun rising on the horizon, 24.7 percent of his body surface receives direct radiation, but when the sun is directly overhead, a standing person receives direct radiation on only 4.4 percent of the body surface.[1, 2] In other words, vertical posture increases the area available for warming when the sun is low in the morning or evening and minimizes heat load during the middle of the day. Variations in body surface area have a potential influence on heat exchange by radiation. As indicated below, body surface area influences heat exchange in two directions. Under conditions in which the ambient temperature is lower than that of the skin, a large surface area per unit of body weight can be advantageous for facilitating heat loss. However, when the ambient temperature is higher, as is true in the desert, such a ratio can be a disadvantage because it increases the amount of heat that can be gained from the larger surface area exposed.

Convection (C)

Convection refers to the transport of heat by a stream of molecules from a warm object toward a cooler object. The most common exchange of heat by convection begins with heat conduction from a warm body to surrounding air molecules. The heated air expands, becomes less dense, and rises, taking the heat with it, to be replaced by cooler, denser air. If the ambient air temperature is lower than that of the body, the skin will heat the air with which it comes in contact, but if it is higher than that of the body, the hot air will warm the skin (thus, *C* is + or –). Heat exchange by convection depends on (a) the difference in temperature between the body surface and the air, which in turn determines the amount of heat gained or lost by a unit mass of air contacting the skin; and (b) the air movements present, which in turn determine the mass of air that will come in contact with the skin surface. In general there are two different forms of convection: natural convection (rising of warm air) and forced convection, caused by the actions of an outside force (e.g., a fan). Natural convection depends on the natural buoyancy of heated material, and forced convection depends on the force of the air current and the movement of heated material within a medium.

Conduction (Cd)

Conduction refers to the transfer of heat by direct physical contact. Heat exchange by conduction takes place within and outside the body. Internal conduction occurs from tissue to tissue, especially in the blood, and represents an important force in the distribution of body heat. External conduction occurs through physical contact by the skin with external objects. Conductive heat-loss experiments have usually demonstrated that external conduction represents only a small percentage of total heat exchange. Thus, in studies of adaptation to heat stress, conduction is not usually considered separately, but is discussed with radiation and convection. Furthermore, the areas of skin in contact with surrounding objects are usually small, and direct contact with highly conductive materials is avoided. However, in a clothed individual, conduction of metabolic heat from skin to clothing does take place. When heat reaches the clothing, it is dissipated from the outer clothing surfaces by evaporation and/or convection and radiation, depending on the amount of air movement, the ambient temperature, and the vapor pressure gradients between the clothing and the environment.

Evaporation (E)

Evaporation refers to the conversion of water into vapor. Heat is lost by evaporation because heat is required in the endothermic conversion of water to vapor. This heat is called the *latent heat of evaporation.* It is estimated that when water evaporates from the body surface, 580 kilocalories (kcal) of heat are lost for each liter of water that evaporates. Therefore, evaporative heat loss is determined by the rate of water evaporation multiplied by the latent heat of water evaporation (580 kcal). Evaporation of water from the skin (in the form of sweat) and respiratory passages will always result in heat loss and never in heat gain.

Thermal sweating occurs through the *eccrine* glands that are distributed throughout the skin on the head, face, trunk, arms, and legs. Humans have between 2 and 5 million sweat glands, with an average distribution of 150 to 340 per square centimeter. Although there are no ethnic differences in the number of sweat glands or in the gradient of distribution over the body, the extent to which the sweat glands become active or dormant appears to depend on the exposure to heat stress during the development period.[3] Individuals exposed to heat stress since childhood tend to have a greater number of active sweat glands than those exposed only during adulthood.

When the heat lost by radiation, convection, and conduction is not able to maintain thermal homeostasis, the only means by which the body can rid itself of heat is by evaporation. Therefore, water output from sweat glands increases in proportion to the need for evaporative cooling. In general, the sweat rate increases 20 ml for every 1°C rise in air temperature. Once sweating begins, the skin blood flow must continue to increase as the heat load increases to transfer more heat from the internal body core toward the periphery, where it will be dissipated to the environment by the evaporation of sweat. When evaporation is incomplete because of increased humidity, sweat production is a far less efficient means of heat dissipation. For these reasons, water lost through the kidneys, skin, and respiratory tract may range from 10 to 12 L/24 hr for individuals working in a hot environment.

TABLE 18.1

Mechanisms of Heat Loss from the Body (Nude) at Different Room Temperatures (at Constant Low Air Movement)

Room temperature	Radiation (%)	Convection (%)	Evaporation (%)
Comfortable = 25°C (77°F)	67	10	23
Warm = 30°C (86°F)	41	33	26
Hot = 35°C (95°F)	4	6	90

Source: Folk.[6]

In summary, as shown in **table 18.1** and **figure 18.2**, the relative role played by each of the four avenues of heat loss depends on the interaction of ambient temperature and humidity. Thus, in a comfortable climate with a temperature of 25°C (77°F), an unclothed seated person loses metabolic heat mostly through radiation (67%), and very little heat is lost through evaporation (23%) and convection (10%). At a warm temperature of 30°C (86°F), heat is lost in about equal proportions by radiation (41%), convection (33%), and evaporation (26%). At temperatures higher than 35°C (95°F), 90 percent of the heat lost is by evaporation, and very little (4% and 6%, respectively) is lost by radiation and convection.[9]

In addition, evaporation of water from the body occurs in the form of insensible perspiration. Insensible perspiration represents the constant diffusion of water from the body. This moisture diffuses through the pores of the sweat glands at a rate of approximately 900 ml/24 hr. Similarly, diffusion of water from the lungs occurs at about 400 ml/24 hour. Thus, on average in temperate zones, insensible perspiration results in a heat loss of 522 kcal from the sweat glands (900 ml × 0.58 kcal/ml = 522 kcal) and 232 kcal from the lungs (400 ml × 0.58 kcal/ml = 232 kcal). This water loss through the skin and lungs may vary with altitude and the general quantity of moisture in the air—the higher the altitude, the greater the insensible moisture loss through the skin. Emotional evaporation occurs through the skin on the forehead, the eccrine glands (soles of the feet and palms of the hands), and the apocrine glands (axillary and pubic). In general, the contributions of insensible water loss and emotional perspiration to thermoregulation of heat stress are minimal.

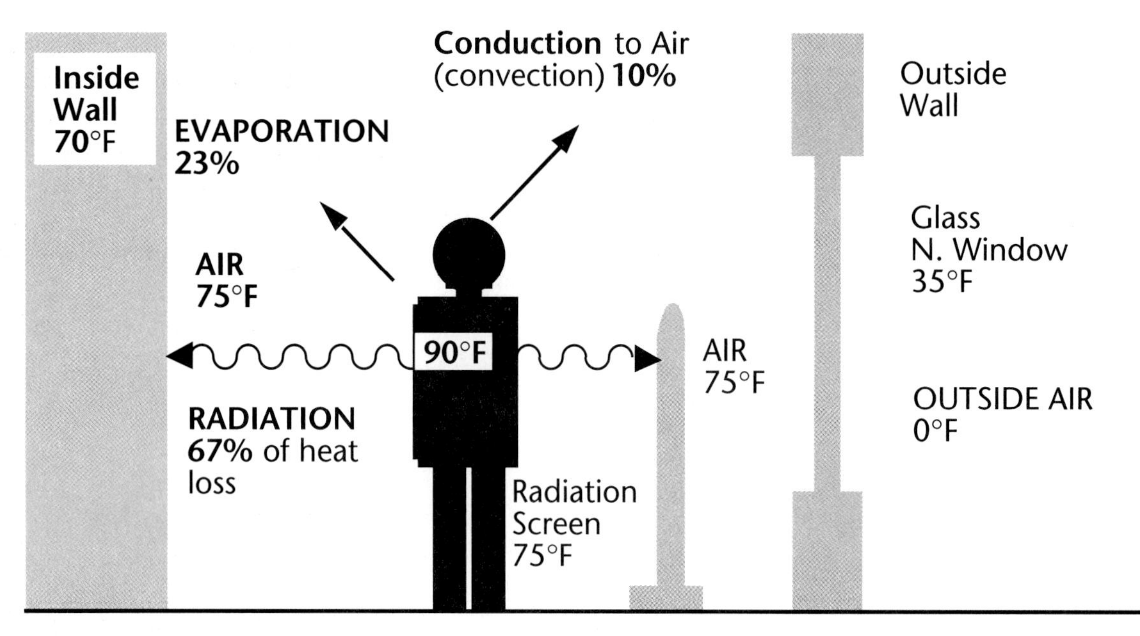

FIGURE 18.2. Partitioning of heat exchange avenues from man. At air temperature of 24°C (75.2°F), most heat is lost by radiation and the rest by conduction and evaporation. From *Human Adaptation and Accommodation* by A. R. Frisancho. Copyright © 1995 by University of Michigan Press. Reprinted by permission.

Sweating and Sodium Adjustments

Stimulation of the sweat glands produces secretion of the precursor fluid that contains high concentrations of sodium (about 142 mEq per liter), chloride (about 104 mEq per liter), and iron. Normally, the sodium and chloride are reabsorbed into the blood and the concentration of these ions in the sweat is decreased to as low as 5 mEq per liter. However, when sweat glands are strongly stimulated, as occurs during heat stress, the reabsorption of ions is not so great that the concentration of sodium and chloride in the sweat may be as high as 65 mEq per liter. Acclimation to heat stress leads to decreased concentration of sodium in the sweat. For example, in a nonheat acclimated person the sweat contains high concentrations of sodium that may range from 40 to 65 mEq per liter. After a 2-week period of acclimation, the salt concentration in sweat decreases to between 10 and 20 mEq per liter. The importance of this adjustment attains critical importance for populations chronically exposed to heat stress (see below).

Adaptation to Hot Environments

The fundamental problem for humans and all other homeotherms exposed to heat stress is heat dissipation. Therefore, most of the physiological responses to heat stress are aimed at facilitating heat loss. Successful tolerance of heat stress requires the development of synchronized responses that permit the organism to lose heat in an efficient manner and maintain homeostasis. Humans encounter heat stress not only in tropical equatorial areas, but also during the summer in many of the vast land areas of the temperate zones. In general, hot climates are classified as either hot-dry or hot-wet.

Hot-Wet

The hot-wet or hot-humid climates are typical of the tropical rain forests usually located within the latitudes of 10° to 20° above or below the equator. Hot-humid climates have the following characteristics: (a) the air temperature does not exceed 35°C (95°F), usually ranging between 26.7°C and 32.2°C (80° to 90°F); (b) the average relative humidity exceeds 50 percent, usually reaching as high as 95 percent; and (c) there is marked seasonal precipitation. As a result of the high precipitation and hot climate, vegetation is quite abundant and provides ample shade. Because of the combination of moisture and vegetation, much of the solar energy is used to convert liquid water to vapor, which exits in the atmosphere as insensible heat. There is little day-night or seasonal variation in temperature and dew point.

Hot-Dry

The hot-dry climates are usually found in desert regions such as those in the southwestern United States, the Kalahari, the Sahara of Africa, and other areas of the world. A hot-dry climate is characterized by (a) high air temperatures, which during the day range from 32.2° to 51.7°C (90° to 125°F); (b) low humidity, usually from 0 percent to 10 percent; (c) intense solar radiation; (d) very little precipitation; (e) little or no vegetation; and (f) marked day-night variation in temperature, often exceeding 50°C (122°F). Because of the lack of vegetation, the ground absorbs considerable solar energy and may heat to 32.2°C (90°F). When the ground temperature exceeds that of the surrounding environment, the soil acts as a radiator for long-wave (infrared) radiation. Moreover, the terrain may reflect up to 30 percent of the incident sunlight. The ambient day air temperature is almost always higher than the skin and clothing; therefore, this hotter air heats the individual's body instead of cooling it. In other words, in the hot-dry climate, because of the low moisture content, the solar energy either directly or indirectly heats surfaces as well as the ambient air. In this manner, solar energy exists as a sensible heat in contrast to that of the hot-wet climate. Finally, pervasive desert winds, while facilitating evaporative cooling, may, at the same time, increase the body's heat load by boosting the rate of heat exchange between the hot air and the cooler skin.

In summary, it is evident that the difference between dry and humid heat stress is based on the physiological principle that heat conductance is directly proportional to the amount of humidity in the air. Therefore, the higher the degree of humidity, the lesser the heat dissipation, and the greater the heat stress suffered by the organism.

Acclimation to Heat Stress

Acclimation to humid heat (28° to 30°C wet bulbs) or dry heat (50°C dry bulb) stress is attained through adjustments of the cardiovascular system, which enables the organism to attain thermal homeostasis (**fig. 18.3**).

Initial Phase

In the initial phase of acclimation, which usually lasts one week (or about 4 days of 60 to 120 minutes of daily physical exercise on a treadmill or ergometric bicycle), the organism responds by increasing the peripheral heat conductance through an increase in

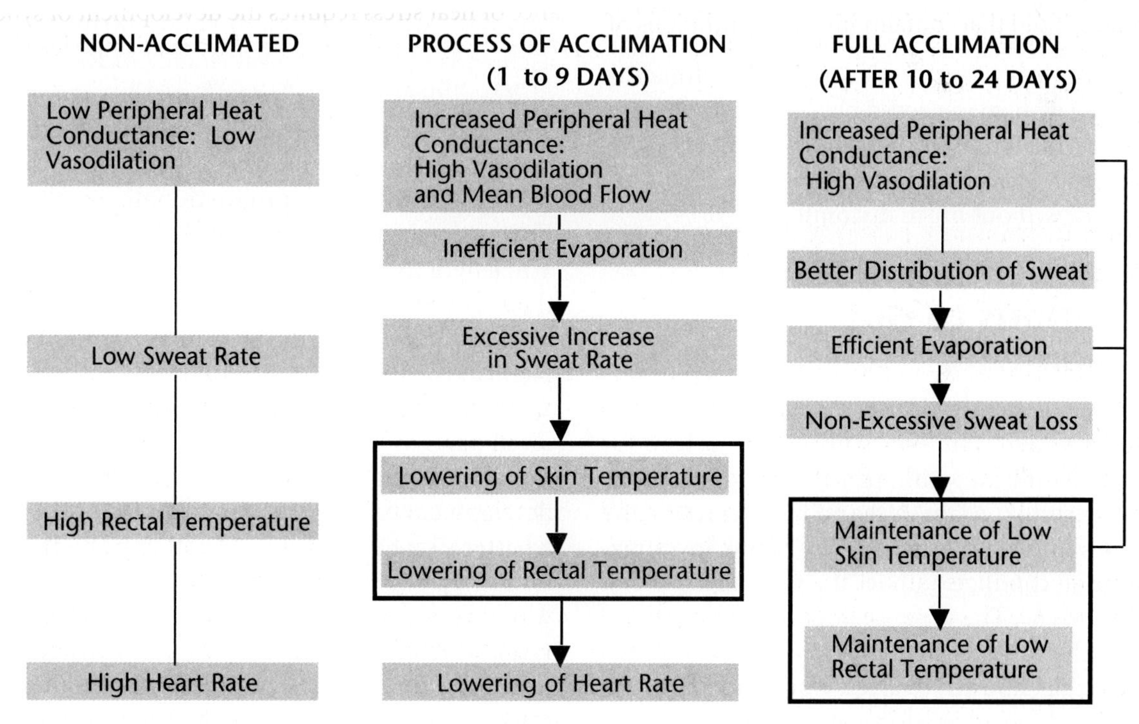

FIGURE 18.3. Schematization of Mechanism of Acclimation to Heat Stress. During initial acclimation, thermal homeostasis is attained through an increased peripheral conductance, but evaporation is inefficient so that sweat rate becomes excessive. In full acclimation, thermal homeostasis is achieved through an increased peripheral heat conductance, better distribution of sweat, and efficient evaporation. From *Human Adaptation and Accommodation* by A. R. Frisancho. Copyright © 1995 by University of Michigan Press. Reprinted by permission.

blood flow from the internal core to the shell. Along with this change, there is an excessive increase in sweat rate and sodium loss, which may amount to more than 50 percent of preacclimation levels. Only a small proportion of this sweat is evaporated; a 10 percent increase in evaporation is accompanied by a 200 percent increase in nonevaporated sweat.[4, 5] This means that during acclimation to heat stress t, homeostasis is attained through a wasteful overproduction of sweat.

Attainment of Full Acclimation

Attainment of full acclimation is characterized by an efficient evaporation of sweat and low sweat output. Repeated exposure to heat stress and after about one week, peripheral heat conductance is increased by a more complete and even distribution of sweat over the skin.[3–10] As a result, evaporation of sweat is more efficient and the sweat output is less than during the first week of acclimation. Furthermore, the skin temperature threshold for the onset of sweating is decreased, which means that an equivalent rate of sweating is achieved at a lower skin temperature.[4–10] As a result of the decrease in skin temperature, the rectal temperature decreases simultaneously and circulatory stability is maintained. During acclimation the renal sodium output is drastically reduced and sweat-sodium concentration is also decreased. However, when considering the total greater sweat loss, even after full acclimation the total loss of sodium exceeds preacclimation levels. Along with an improved physiological response, the disagreeable sensations associated with heat exposure are progressively reduced until individuals are able to work without much discomfort.[6]

Acclimatization of Native Populations to Hot Environments

Prominent among investigations of acclimatization to hot environments are the studies of indigenous samples of Bantus in South Africa,[11] Bushmen in the Kalahari Desert,[12] Arabs in the Sahara Desert,[13] and Aborigines in central Australia.[14,15] These investigations were all conducted under the same thermal and work conditions. The experimental routine consisted of having the subjects step up and down on a bench with the height adjusted to give a work load of 216 kg-m/min (1,560 foot-pounds/min) at a rate of 12 steps per minute. The external work was set to require an oxygen consumption of about 1 L/min. The experiment lasted for 4 hours and was conducted at 33.9°C (93°F) dry bulb and 32.2°C (90°F) wet bulb temperatures, with 24 m/min (80 feet/min) air movement.

Each of the samples were studied in their respective habitats and, as illustrated in **figure 18.4**, indicate that the indigenous populations, while working in heat stress, are able to maintain thermal homeostasis with a generally lower sweat rate than acclimated Europeans. In other words, native or acclimatized populations have been able to maintain thermal homeostasis in heat stress with a lower body fluid loss. Therefore, native populations of tropical climates have developed other mechanisms that are more efficient than those of populations that are not constrained by water availability. This trait is clearly exemplified by the thermoregulatory responses to exercise of the Australian Aborigines. As illustrated in **figure 18.5**, throughout the heat stress tests, the rectal temperatures and heart rates of the Australian Aborigines were higher than those of the acclimated whites, but their sweat rates were significantly lower than the acclimated whites. The same trend of economizing sweat loss is evident among the Arabs from the Chaamba tribe who live in the Sahara Desert.[13–15] These studies found that the Chaamba Arabs, even though their rectal temperatures and heart rates were higher, had lower sweat rates than the Europeans.

Role of Growth and Development in Acquiring Tolerance to Heat Stress

American white children of good nutritional status, acclimatized to the tropical climate of Rio de Janeiro of Brazil develop smaller calf girths as compared to non-heat-stressed controls.[71] Similarly, Quechua children acclimatized to the tropical climate of the Peruvian lowlands have a proportionally smaller trunk (low sitting height: stature ratio) and relatively greater arm length (high arm length: sitting height ratio) than their Quechua counterparts of the same genetic composition living in the cold climate of the Peruvian Highlands.[72] Since the extremities have a greater density of functional sweat glands than the trunk and the distal parts of the limbs have greater density of functional sweat glands than the proximal parts,[73, 74] the development of leaner extremities would contribute to a better sweat evaporation. Furthermore, the reduction in trunk size and increase in relative arm length would result in an increase in the surface area:weight ratio, which would facilitate radiative and convective heat loss. In addition, populations who have been raised in tropical climates do have a significantly greater number of sweat glands than their counterparts, raised in temperate climates.[75] Therefore, native populations acclimatized to hot climates have a more efficient sweat evaporation and do not have to sweat as excessively as those non-heat-acclimatized

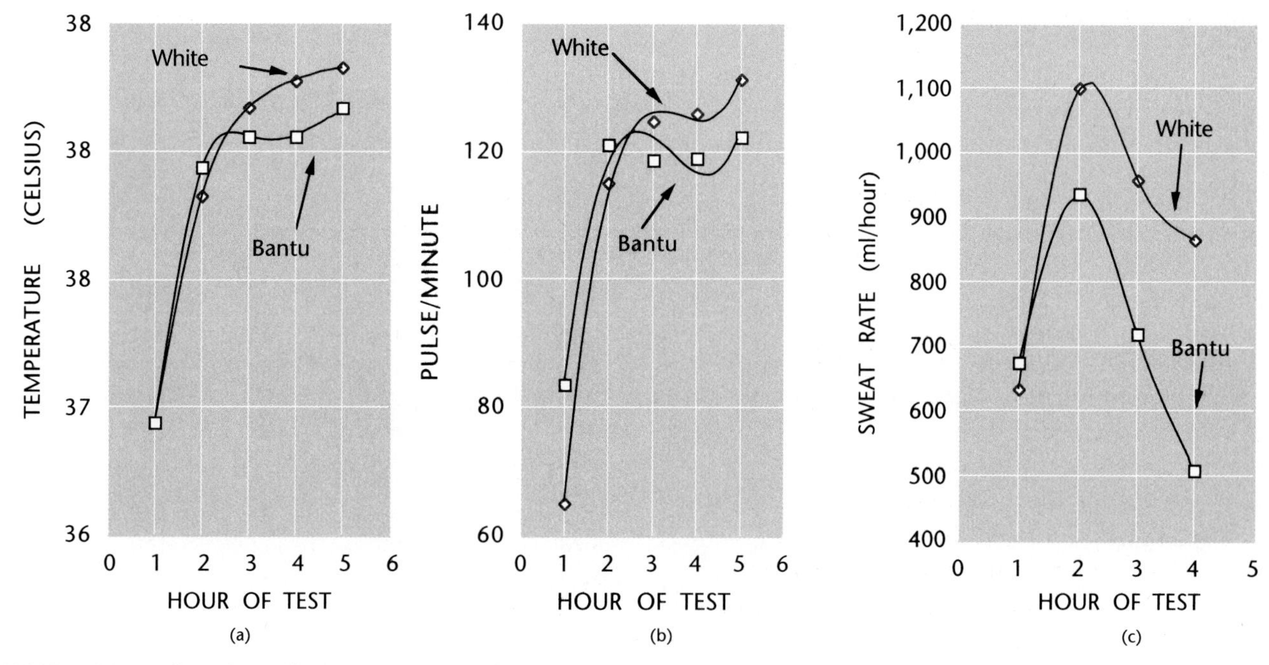

FIGURE 18.4. Physiological responses to standardized heat stress test of acclimatized Bantus and acclimated whites. The Bantus are able to maintain thermal homeostasis with a lower sweat rate and a lower heart rate than the acclimated whites. Source: Based on data from Wyndham.[15]

populations. As a result, tropical native populations have an enhanced tolerance to heat stress and a low sweat rate.

Viewed in this context, population differences in acclimatization to heat stress is to a certain extent the result of morphological adaptation acquired during growth and development (**fig. 18.6**).

Adaptation to Cold Environments

Biological responses to cold environments are more complex than those of heat adaptation. Successful responses to cold environments require the synchroniza-

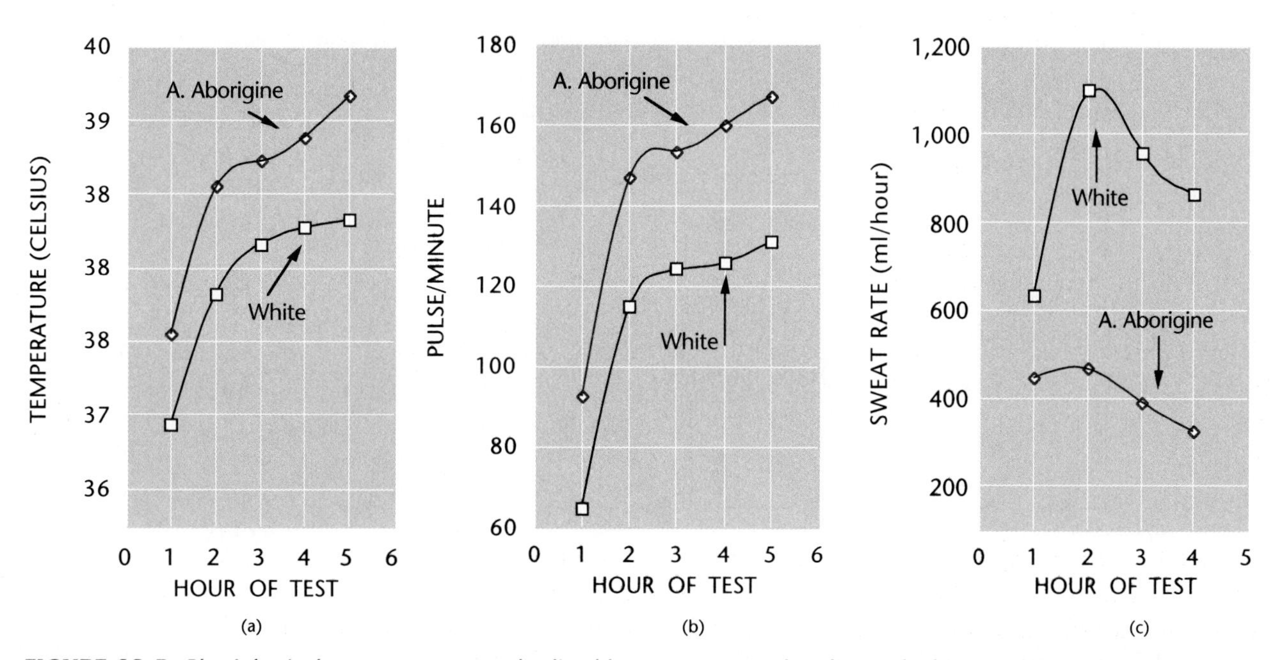

FIGURE 18.5. Physiological responses to standardized heat stress test of acclimated whites and Australian Aborigines. The Australian Aborigines show a higher increase in rectal temperatures and heart rates than acclimated whites but have lower sweat rates. Source: Based on data from Wyndham.[15]

TABLE 18.2

Wind Chill Effect: Equivalent Effective Temperature

Wind speed (mph)	Actual air temperature (°F) 50	40	30	20	10	0	−10	−20	−30	−40	−50	−60
	Equivalent temperature (°F)											
5	48	36	27	17	−5	−5	−15	−25	−35	−46	−56	−66
10	40	29	18	5	−8	−20	−30	−43	−55	−68	−80	−93
15	35	23	10	−5	−18	−29	−42	−55	−70	−83	−97	−112
20	32	18	4	−10	−23	−34	−50	−64	−79	−94	−108	−121
25	30	15	−1	−15	−28	−38	−55	−72	−88	−105	−118	−130
30	28	13	−5	−18	−33	−44	−60	−76	−92	−109	−124	−134
35	27	11	−6	−20	−35	−48	−65	−80	−96	−113	−130	−137
40	26	10	−7	−21	−37	−52	−68	−83	−100	−117	−135	−140
45	25	9	−8	−22	−39	−54	−70	−86	−103	−120	−139	−143
50	25	8	−9	−23	−40	−55	−72	−88	−105	−123	−142	−145

Source: Data from Ward, M. 1975. *Mountain Medicine: A Clinical Study of Cold and High Altitude.* London: Crosby Lockwood Staples Ltd.

tion of cardiovascular and circulatory systems and, most importantly, the activation of the metabolic process. Besides low temperatures, there are several other environmental factors that must be taken into account when considering human responses to cold stress. Among these, the most important are wind velocity, humidity, and duration of exposure to cold. In general, it is assumed that a low temperature, usually around 0°C (32°F), with high humidity results in greater cold sensation than low humidity. Heat loss is strongly affected by wind velocity; a given temperature and a rapid wind result in greater cold stress than the same temperature with slow wind. The interaction of wind velocity and temperature is given by the wind chill index (**table 18.2**). For example, if the local temperature is 0°C (32°F) and wind velocity is 40 mph, the person loses heat as if the temperature were −38.9°C (−38°F). Therefore, the chilling power of the wind can produce an almost supercooling effect on exposed skin.

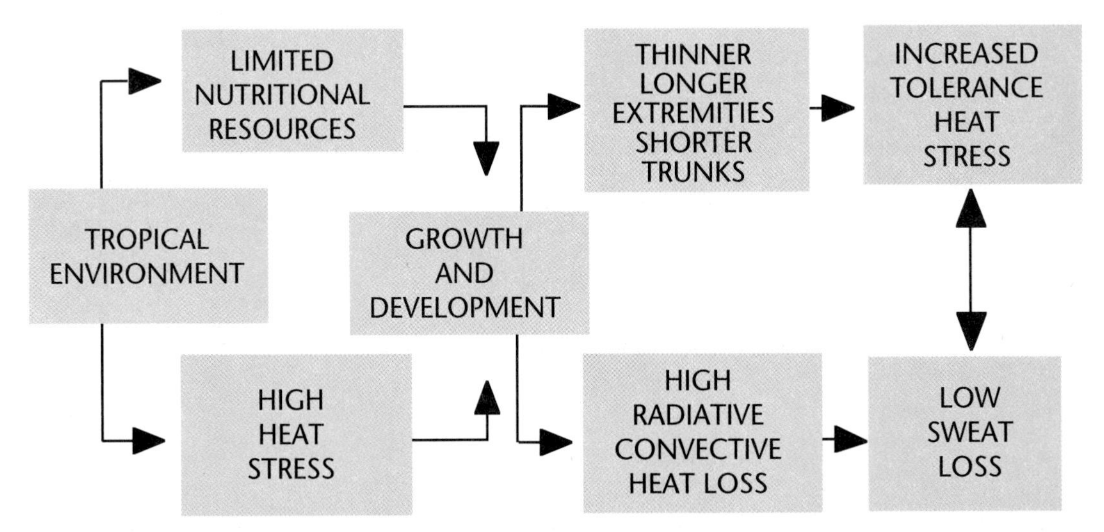

FIGURE 18.6. Schematization of interaction of tropical environmental stresses influencing adaptation to heat stress. Joint influences of limited nutritional stress, heat stress, and growth and development in tropical environment results in development of adaptive morphology and increase in heat tolerance capacity. From *Human Adaptation and Accommodation* by A. R. Frisancho. Copyright © 1995 by University of Michigan Press. Reprinted by permission.

Types of Cold Stress and the Value of Clothing Insulation

In general, the degree of cold stress to which a person is exposed is classified as either *acute* or *chronic. Acute cold stress* refers to severe cold stress for short periods of time. *Chronic cold stress* refers to moderate cold stress experienced for prolonged periods of time, either seasonally or throughout the year. Obviously, the degree of cold stress depends on the amount of insulation. The thermal insulation value of clothing is measured in terms of "Clo" units. One Clo of thermal insulation will keep a resting sitting person with a metabolic rate of 50 kcal/m^2/hr comfortable in an environment of 21°C (70°F) with relative humidity less than 50 percent and air movement of 6 m/min (20 feet/min). In these basal conditions, 1 Clo is equivalent to a business suit or 0.64 cm (1/4 inch) of clothing. A heavy article of arctic clothing provides about 5 Clo units. In general, at the same level of thickness, several layers of insulation has better thermal insulation than one thick layer.

Biological Responses to Cold Stress

Thermoneutral temperature range is between 25° to 27°C (77° to 80°F) and when a nude individual is exposed below this range, the body responds immediately through mechanisms that permit both the conservation of heat and an increase in heat production. The major mechanisms concerned with heat conservation are vasoconstriction, alternated with vasodilation and synchronized with the countercurrent system; the major mechanism concerned with heat production is shivering.

Heat Conservation through Vasoconstriction

On exposure to cold stress, such as a temperature of 0°C (32°F) or even 15°C (60°F), the subcutaneous blood vessels constrict (vasoconstriction), which limits the flow of warm blood from the internal body (core) to the shell (skin). The result of this lower blood flow is a decrease in skin temperature, a reduction of the temperature gradient between the skin surface and the environment, and, consequently, a reduction in the rate of heat loss (**fig. 18.7**). As a result, during full vasoconstriction, heat conductivity of the blood is reduced by as much as eight times. Concomitant with decreased blood flow to the skin, the blood flow to the viscera and internal organs is augmented. This shift in blood flow is probably responsible for increased blood pressure and heart rate under severe cold stress.

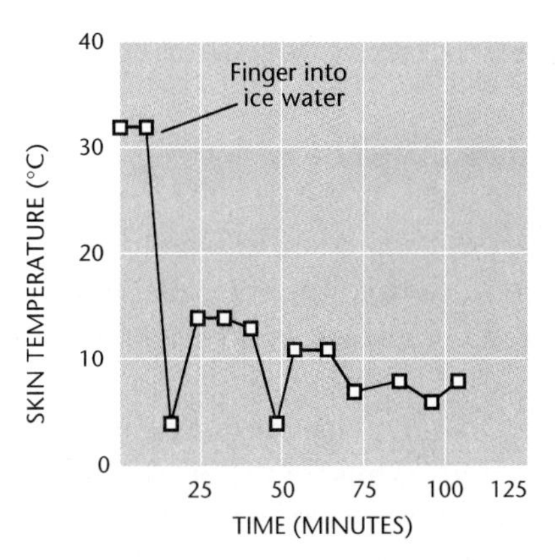

FIGURE 18.7. Alternation of vasodilation and vasoconstriction (Lewis's hunting phenomenon) in response to immersion of finger in crushed ice. Curve shows large, prolonged temperature oscillations finally giving way to smaller, more rapid ones. Source: Based on data from Lewis.[16]

Heat Conservation through Both Vasoconstriction and Cold-Induced Dilation

Hunting Response

A typical response to immersion of the finger in cold water is the spontaneous, semirhythmic changes in temperature. As shown in **figure 18.7**, on immersion, the digital thermometer drops for about 15 minutes until the temperature is about 2.5°C above the ice water temperature. After about 16 minutes, the digital temperature rises by some 6° to 8°C and fluctuates thereafter between 4° and 6°C. These fluctuations in temperature are caused by fluctuations in the blood flow and have been termed the Lewis's "hunting" phenomenon.[16] Similar responses to cold were noted in the skin of the ear, cheek, nose, chin, and toes. The adaptive function of these fluctuations in temperature is to protect exposed parts of the skin from excessive cooling and injury. Another effect of cold stress is that pain sensation increases with falling skin temperatures; with rewarming of the skin by cold-induced vasodilation (CIVD), this sensation disappears, giving the sensation that the hand is being immersed in lukewarm water. Recurrence of vasoconstriction coincides with an elevation in the pain sensation. However, pain is observed not only when skin temperature cools, but also when the skin warms up after cooling. Indeed, frequent rewarming of the hand or face at room temperature produces a marked pain sensation. This increase in pain occurs when skin temperature increases rapidly.

Heat Production: Shivering

When the vasoregulatory mechanisms of heat conservation are not sufficient to counteract heat loss, the organism adjusts by increasing the rate of heat production. The most rapid and efficient way to increase heat production is by voluntary exercise, such as running, which may increase the metabolic rate from a basal value of 1.17 cal/min to 37.94 cal/min. However, such high rates of activity cannot be maintained for prolonged periods. Thus, in the absence of voluntary exercise, shivering of the skeletal muscle is the main source of increased heat production. The major function of shivering is to increase the rate of heat production. As a result of shivering, the metabolic rate may be increased two to three times the basal value. This increase in heat production is progressive throughout the cold stress. For example, during intense cold, heat production increases from a basal level at pre-exposure of 35.4 kcal/m^2/hr to 54 kcal/m^2/hr by the end of the first hour. By the end of the second, third, and fourth hours, production had risen to 72, 92, and 96 kcal/m^2/hr, respectively.[17]

Along with shivering, subcutaneous blood vessels dilate (vasodilation) to keep the skin warm and prevent tissue injury from frostbite. Thus, adjustment to cold stress is an interplay between mechanisms to conserve heat and mechanisms to produce and dissipate heat. In terms of energy expenditure, defense against cold is achieved more economically by increasing body heat conservation than by increasing heat production. However, the extent to which these mechanisms are operative in humans depends on the degree of cold stress experienced and on technological and cultural adaptation, as will be shown in later chapters.

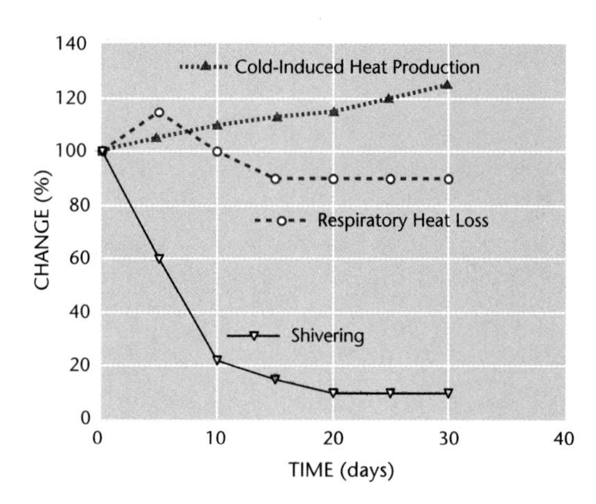

FIGURE 18.8. Human acclimation within 20 days. After 20 days (except nights) of exposure to cold chamber (12°C) shivering declines whereas heat production continues to increase. Source: Data from Davis.[20]

Acclimation to Cold Stress

Studies of acclimation to cold stress have been conducted with animals and human subjects.

Animal Studies

When rats are continuously exposed to temperatures of 5°C (41°F) for 2 or 3 weeks, they are able to maintain a high metabolism (elevated about 80%) and normal temperature, without resorting to shivering. The increased heat production associated with continuous cold exposure is *not* the result of shivering.[17, 18] These studies found that the increased production results from an increased sensitivity to noradrenaline and thyroxine.[19] For this reason, noradrenaline (or norepinephrine) and thyroxine are considered to be the hormones of cold acclimation. The energy for the maintenance of nonshivering is derived from increased total daily food consumption and rapid utili-

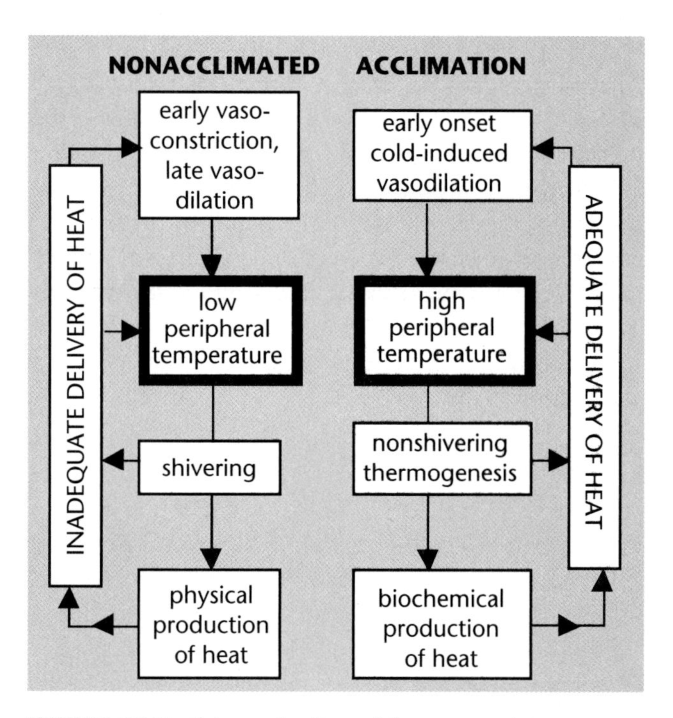

FIGURE 18.9. Schematization of thermoregulatory responses to cold before and after acclimation. Before the onset of acclimation, cold-induced vasodilation is delayed and an increase in heat production is attained through shivering, but delivery of heat is still inadequate, resulting in low peripheral temperature. After acclimation, cold-induced vasodilation occurs earlier, heat production is increased through nonshivering thermogenesis, and peripheral temperature is higher. From *Human Adaptation and Accommodation* by A. R. Frisancho. Copyright © 1995 by University of Michigan Press. Reprinted by permission.

zation of fat stores. The derivation of energy activation appears to be related to the activation and sensitivity of noradrenaline present in brown adipose fat.

Humans

Experimental studies of humans included exposing nude subjects for 8 to 24 hours a day to a temperature of 5° to 15°C for 2 to 4 weeks. These studies indicate that by the second week, shivering decreases while the metabolic rate increases through the action of cold-induced nonshivering thermogenesis[20, 21] (**fig. 18.8**). As schematized in **figure 18.9**, when exposure to cold stress is severe, before acclimation the organism increases its rate of heat production through shivering. As a result of shivering, the metabolic rate increases to two to three times the normal rate; thus, the rate of heat production is increased. However, because of inefficient delivery of heat, the peripheral temperature is cold. On the other hand, acclimation to cold stress appears linked to the maintenance of warm skin temperatures, made possible by an increased ability to produce heat. Increased heat production is achieved without shivering. The source of energy for nonshivering thermogenesis appears to be white and brown adipose tissue deposits converted into free fatty acids. Lipid sources of energy appear to be mediated through the increased activity of calorigenic hormones such as noradrenaline (norepinephrine) and thyroxine. Tolerance to cold stress appears to be influenced by the individual's age and physical fitness, the degree of insulation derived from subcutaneous fat, and the ratio of surface area to weight. Females, despite their higher amount of subcutaneous fat, are less tolerant to cold stress than males and this difference appears to be related to their high surface area and low weight, which facilitates heat loss and limits heat production.

Acclimatization to Cold Environments: Native Populations

Investigators concerned with human acclimatization and adaptation to cold have centered their investigations on the study of indigenous populations who live and work in the cold. In general, the methodology employed in these studies can be classified into three major types: (a) the night-long cold-bag technique, which has been used to study the thermoregulatory characteristics of Australian Aborigines, Kalahari Bushmen, Alacaluf Indians, Norwegian Lapps, Eskimos, Athapascan Indians, and Andean Quechuas; (b) the short-period laboratory whole-body-cooling technique, which has been used to evaluate the thermoregulatory characteristics of Eskimos, Asiatic divers, Andean Quechuas, and European whites; (c) the short-period laboratory extremity-cooling technique, which complements the whole-body-cooling technique. These approaches, although their results are not strictly comparable and they have produced contradictory results within the same population, provide valuable information about human adaptation.

In this section, the thermoregulatory responses to cold of the Australian Aborigines, Eskimos, and Andean Quechuas will be summarized to illustrate the adaptation to moderate night and chronic severe cold stress. The aim is not to determine population differences in cold adaptation, but to ascertain the mechanisms that enable a given population to overcome cold stress.

Australian Aborigines

The thermoregulatory characteristics of the Aborigines from central and northern Australia have been studied both in winter and summer.[22, 23] The test consisted of the subjects sleeping over night (8 hours) in an air temperature ranging from 3° to 5°C (37.4° to 41°F), with an insulation ranging from 2.9 to 3.4 Clo units (1 Clo = 38 cal/m²/hr). As shown in **figure 18.10**, the Aborigines compared to white control subjects studied under the same conditions tolerated a greater lowering of the skin and rectal temperatures, resulting

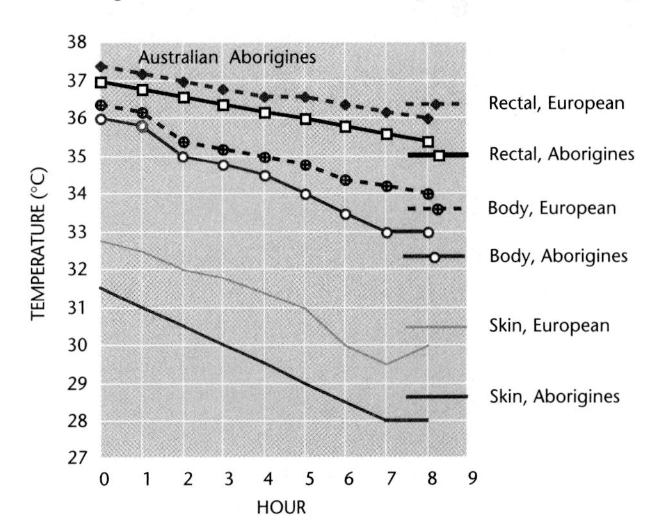

FIGURE 18.10. Thermal and metabolic responses of six central Australian Aborigines and four control whites during a night of moderate cold exposure in winter. For the Australian Aborigines, adaptation to the moderate night cold temperatures (about 3°C) involves a decrease in skin temperature and metabolic rate. Source: Data from Scholander, Hammel, Hart, LeMessurier, and Steen.[22]

in a 30 percent reduction in heat conductance from core to shell. Hence, metabolic heat production was also lower. The Aborigines, despite this cold stress, were able to sleep comfortably without shivering, whereas the white controls studied under the same conditions shivered continuously without being able to sleep. In other words, the Aborigines respond to cold stress by both increasing insulation of the body shell through vasoconstriction and by tolerating moderate hypothermia without metabolic compensation.

Eskimos: Inuits

Cultural Adaptation

The world's population of Eskimos, also referred to as Inuits, numbers between 50,000 and 60,000 and occupies the northwestern coast of America and across the Bering Strait into Asia. They live in one of the most inhospitable environments, where the average winter temperature ranges from –46°C (–50°F) to –31°C (–35°F) and the summer temperature rarely goes above 46°F. Equally important is the drastic change in sunlight hours that range from almost continuous sunshine in the summer and fall to complete absence of sun in midwinter. To overcome this inhospitable environment, the Eskimos have developed an efficient technological and physiological adaptation.

The native Eskimo housing such as the "igloo" type is well insulated and designed to maintain an excellent microclimate that averages between 0° and 21°C, despite subzero temperatures outside. In addition, the Eskimo clothing is made of caribou fur (1½ to 3 in. thick), snowshoes and short skin mittens that provide insulation equivalent to 7 to 12 Clo units,[24]

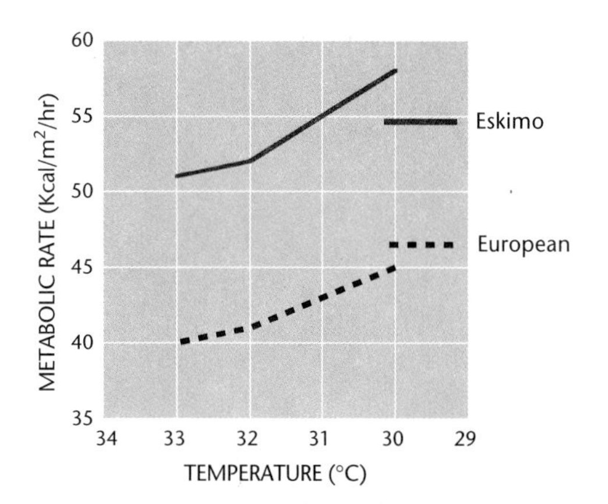

FIGURE 18.11. Relationship between skin temperature and metabolic rate during cold exposure of Eskimos and whites. At any given temperature, Eskimos respond with a greater metabolic rate. Source: Data from Adams and Covino.[36]

which results in an insulative efficiency that ranges from 266 to 456 cal/m²/hr. However, despite this cultural adaptation, the subsistence economy based upon fishing and hunting causes continuous exposure of the Eskimos' hands and feet to cold stress. For example, according to ethnographic accounts,[25–27] during hunting and fishing the Eskimos continually dip their hands in cold water and expose their feet to severe cold and experience whole-body chilling while waiting motionless at breathing holes of seals for up to 72 hours at a time in subzero temperatures.[28] Thus, despite their efficient technological adaptations due to the nature of their subsistence activities, the Eskimos are not always "tropical men in arctic clothing." In response to this cold stress, the Eskimos have developed specialized physiological adaptations.

Physiological Adaptations

Metabolic Rate

In general, the metabolic rate of Eskimos and the indigenous populations living in cold climates, whether during warm or cold conditions, is between 13 percent and 45 percent higher than that of white controls or predicted from standards (**fig. 18.11**).[29–32] To maintain this high heat production, the diet includes a high proportion of protein and fat derived from marine animals such as seals and whales. Earlier investigations in East Greenland described the Eskimos as "the most exquisitely carnivorous people on earth."[29]

Peripheral Temperature

As shown in **figure 18.12**, the minimum finger temperature of Eskimo children and women, when exposed to air temperatures between –3° and –7°C, despite their smaller hand volumes, is as high or higher than those of white men accustomed to outdoor cold.[42] Thus, the Eskimos' ability to maintain high extremity temperatures and their ability to tolerate cold appear to be acquired during growth. That is, the thermoregulatory characteristics of Eskimos reflect the influence of developmental adaptation (or developmental acclimatization).

In summary, despite the outstanding cultural and nutritional adaptations to overcome the severe climatic conditions in which Eskimos live, because of their subsistence economy, the Eskimos are still exposed to prolonged extremity, and at times, whole-body chilling. Hence, the Eskimos' increased metabolic rates, high peripheral temperatures, and remarkable tolerance to cold exposure of the extremities form an integral part of human adaptation to cold stress.

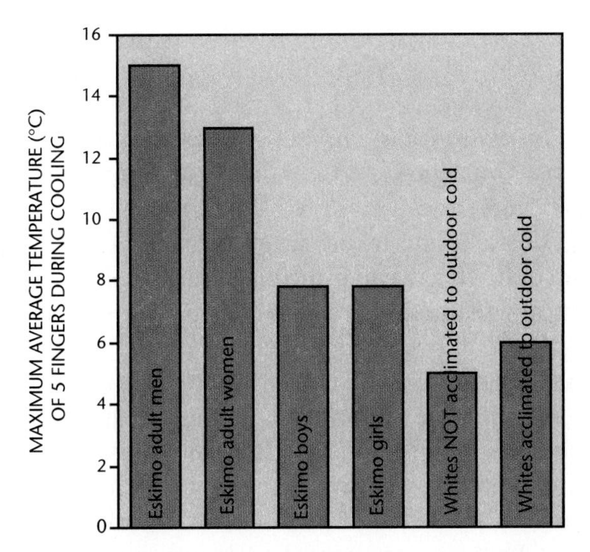

FIGURE 18.12. Peripheral temperature of the left middle finger of Eskimo men, women, boys, girls, outdoor whites, and indoor whites during cold air exposure (–3° to –7°C). Eskimo men maintain higher skin temperatures than both outdoor and indoor whites, and Eskimo children, in spite of smaller hand volumes, maintain temperatures equal to adult indoor whites. Source: Data from Miller and Irving.[42]

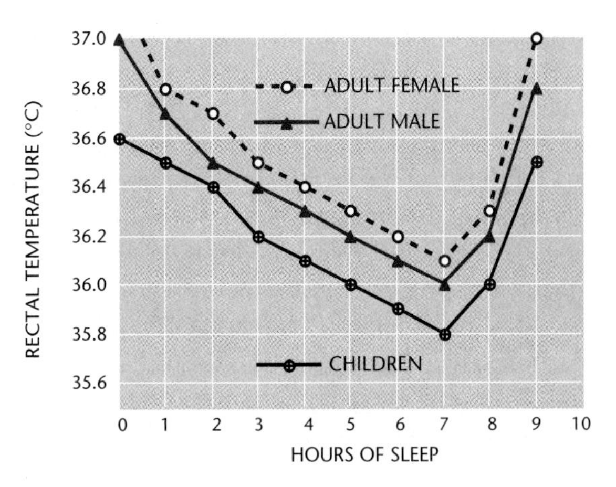

FIGURE 18.13. Rectal temperature of Quechua females, males, and children while sleeping in their own houses. Rectal temperatures did not decrease to initial low levels even though interior house temperatures were only 3.6° C. The children show a rapid fall in rectal temperature for the first hour and then attain a certain equilibrium for remainder of the night. Source: Data Hanna, J. M. 1976. Natural exposure to cold. In *Man in the Andes: A Multidisciplinary Study of High-Altitude Quechua,* ed. P. T. Baker, and M. A. Little, 315–330. Stroudsburg, Pa.: Dowden, Hutchinson & Ross, Inc.

Andean Quechuas

The highland Quechua populations from the Andes and other mountain areas of South America are exposed to a variety of stresses including hypoxia, cold, low humidity, and high levels of solar radiation. These climatic stresses interact with other stresses such as limited energy, food resources, and disease. This discussion is based on the multidisciplinary research centered in the altiplano population of the district of Nuñoa situated at a mean altitude of 4,150 m (13,695 feet) in southern Peru. [43]

Environment

By most standards the altiplano of the Nuñoa region is cold. The mean annual temperature is 8.3°C (47°F) (**fig. 18.14**), which is well below the thermoneutral zone for humans. The mean monthly temperature ranges from 10°C in the warmest month of November to 5.5°C in the coldest month of June. Daily temperatures throughout the year are lower than those considered comfortable for most populations. Seasonal variation in temperature in the highlands is primarily the result of the monsoon pattern in the Pacific, which creates dry and wet seasons. The economic focus of the indigenous population is influenced by altitudinal factors. Between 3,000 and 4,000 meters of altitude the economy is based both on cultivation of corn, potatoes, barley, wheat, and native chenopodium (canihua and quinua) and the herding of sheep and llamas; above 4,000 m, due to the severe cold and frost, the economy is based only on the herding of sheep, llamas, and alpacas. The sheep, llama, and alpaca skins are used for bedding, the sheep and alpaca wool is used for clothing, the dung is used for fuel and fertilizer, and the meat is used for private consumption as well as for trade. The hides and animals are also often traded for cash, although the greater part of the family income is derived from the sale of wool. With this income, cereal foods, additional clothing, yarn dyes, cocoa leaves, alcohol, and other small luxuries can be purchased. All the meat and the wool for basic clothing is provided by herding. Every Indian owns some of these domestic animals, and many own highland ponies as well. Llamas and horses are used as pack animals, whereas sheep and alpacas are raised solely for their wool and meat.

Cultural Adaptation

Evaluations of the thermoregulatory responses of individuals in their natural state were obtained during both the dry and wet seasons.[44] The greatest cold stress in the dry season is at nighttime.[44–46] As shown in **figure 18.13**, even though the interior house temperatures averaged only 3.6°C (38.5°F), the subjects were not greatly cold stressed. Even though the severity of the ambient cold stress is ameliorated by the fact that the subjects usually sleep clothed and in groups of two to four, and as shown in **figure 18.14**, the children are the most cold stressed. The greater cold stress of children is related to the fact that traditionally they do not wear trousers until about the age of 6 years, and until then wear poorly insulated clothing. Furthermore, young children assist in the herding of alpacas and llamas, an activity that leads to considerable cold stress, especially in the afternoons. Thus, the everyday activities of Quechua children provide stimulus for the development of adaptive physiological responses to cold stress. The success of the Quechua populations in preventing severe body cold stress reflects the effectiveness of their technological adaptations, the most important of which are housing, bedding, and clothing.

Housing

The housing of the highland natives differs with variations in altitude and subsistence patterns. A distinct advantage of the more sedentary, mixed economy populations living below 4,000 m is individual or community ownership of land. In higher regions of the altiplano, the large herds make individual ownership of pasture land impractical. Personal ownership provides for a greater economic investment in the land itself. Thus, at elevations of 4,000 m, the houses are built of adobe and are permanent. These adobe houses seem quite effective in protecting against cold stress in that they maintain the indoor temperatures at more than 10°C above outdoor temperatures.[58] On the other hand, the housing at those elevations above 4,300 m, because of a pastoral economy requiring high mobility, is more temporary. It usually consists of two or three circular dwellings, about 9 m^2 (30 square feet) in area, constructed of piled stones and roofed with straw. These houses have the advantage of minimal economic investment and may be abandoned without great economic loss each time the family moves on to new grazing lands (**fig. 18.14**).

In summary, housing at around 3,500 to 4,000 m provides adequate protection against cold stress; however, at higher elevations, because the housing is very

FIGURE 18.14. Natural cold exposure and circular house of pastoral Quechuas living above 4,500 m in the district of Nuñoa in southern Peruvian highlands. Source: Frisancho, A. R. 1993. *Human Adaptation and Accommodation.* Ann Arbor: University of Michigan Press.

temporary, it has only a minimal effect on the severity of cold stress. Nevertheless, it does provide protection against wind and rain and consequently helps reduce heat loss through conduction and radiation.

Clothing

The effectiveness of the Nuñoa Quechua clothing has been studied by comparing the thermoregulatory responses of men and women to a standard cold stress of 10°C for 2 hours with and without clothing.[47] This study indicated that the insulative value of the men's clothing without poncho and hat equaled 1.21 Clo units and 1.43 Clo units for women without shawls and hats. However, the use of gloves and socks for protection of the hands and feet of adults and children is not normally practiced.

Physiological Adaptations

Whole-Body Cold Exposure

Studies of whole-body cooling, including nude exposures of 2 hours at 10°C, were conducted in a laboratory built especially for this purpose at 150 m in the town of Nuñoa.[43] As illustrated in **figure 18.15**, the highland Nuñoa Quechua maintained higher foot temperatures and hence higher mean-weighted skin temperatures than the sea level United States whites. Associated with increased peripheral extremity temperatures, the Quechuas had higher rectal temperatures. Furthermore, although metabolic heat production during the first hour of the test was higher for the Quechuas than for sea level whites, the rest of the test results were comparable for the two groups. In other words, the central highland Quechuas, when compared to sea level controls, showed a metabolic compensation that enabled them to maintain warmer skin temperatures.[48]

Extremity Cooling

Tests of extremity cooling were done under a variety of conditions, including exposure of the hand and foot to air temperatures of 0°C; exposure of the foot in water at 4°, 10°, and 15°C; and exposure of the hand in water at 4°C.[49, 50] These studies indicate that in each test the highland Nuñoa Quechuas displayed warmer skin temperatures and hence greater blood flow to the surface of the extremities than white subjects who were tested at sea level and after a residency of 14 months at high latitude (**fig. 18.16**). It must be noted that the greatest highland Quechua–white differences

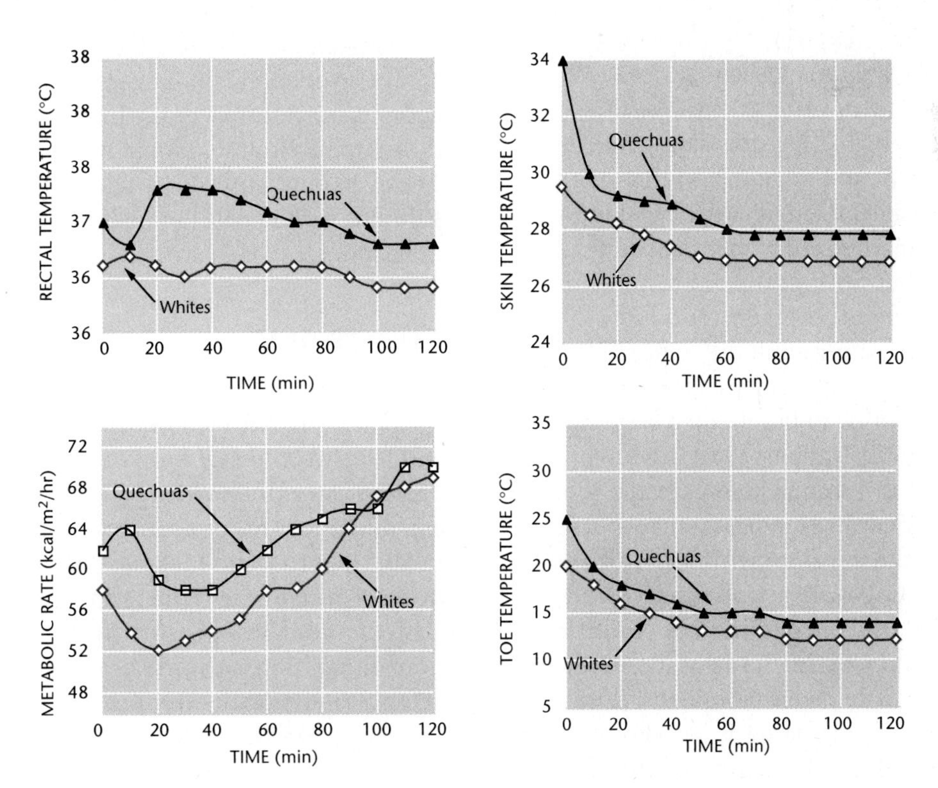

FIGURE 18.15. Thermal and metabolic responses of highland Quechua Indians from Nuñoa and control whites during 2-hour exposure at 10°C at 4,150 m altitude. Highland Quechuas maintain higher metabolic rate, rectal and peripheral temperatures, especially of the toe, than white controls. Source: Modified from Hanna.[44]

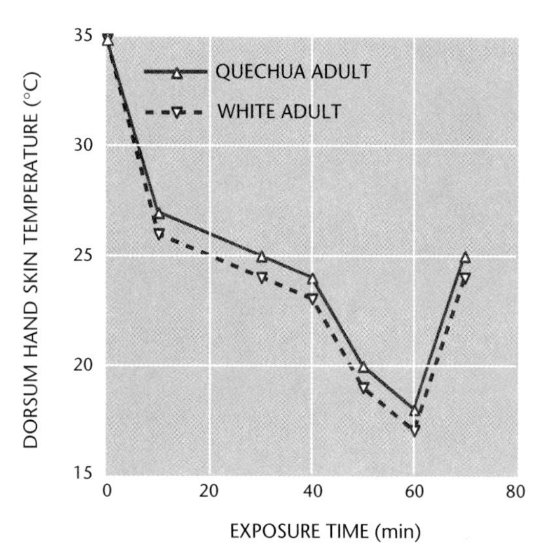

FIGURE 18.16. Skin temperature of the hand of Nuñoa highland Quechuas and white controls during 1-hour exposure to 0°C air and recuperation at 4,150 m. The highland Nuñoa Quechuas had warmer peripheral temperatures than white controls who resided 14 months at high altitude. Source: Data from Little.[49]

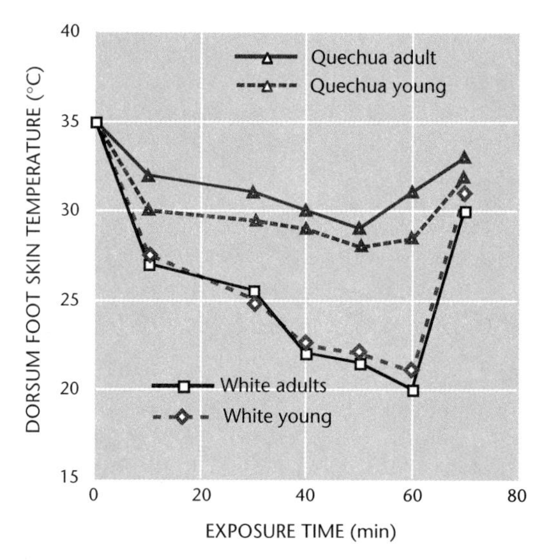

FIGURE 18.17. Skin temperatures of the dorsum of the foot of adult Nuñoa highland Quechuas, young highland Quechuas, adult lowland whites, and young lowland whites during 1-hour exposure to 0°C air tested and recuperation at 4,150 m. Both adult and young highland Quechuas had a higher peripheral temperature than white controls, suggesting that developmental acclimatization influences high peripheral temperatures of Andean Indians. Source: Modified from Little.[49]

occurred at a water temperature of 10° to 15°C. This finding suggests that the peripheral vasomotor system of the highland Quechuas operates more effectively in moderate cold stress.[49, 50] **Figure 18.17** compares the responses of 1-hour exposure of the foot to 0°C air temperatures of 30 Nuñoa Quechua adults, 29 young Quechuas aged 7 to 19 years, 26 adult whites, and 28 young whites aged 7 to 18 years. These data demonstrate that the adult and nonadult whites maintained the same low peripheral temperatures. In contrast, the adult Quechuas maintained higher temperatures than the nonadults, but both groups had systematically warmer foot temperatures than the whites. The fact that there were no young adult differences among the whites suggests that developmental acclimatization is one factor contributing to the elevated extremity temperatures of Andean Indians.[49] In other words, the existence of a relationship between age and foot temperatures of the Quechuas indicates that cold stress is present during the developmental period and that some acclimatization to this stress has taken place.

In summary, the environmental conditions in which the highland Quechua Indians live provide potential cold stress. The Quechua Indians, through the use of clothing, sleeping patterns, and housing, have successfully modified and ameliorated the severity of cold stress. Evaluations of the thermoregulatory characteristics in natural and laboratory settings demonstrate that the Quechua Indians respond to cold stress with metabolic compensation and great heat flow to the extremities. The fact that maintenance of high peripheral temperatures characterizes both children and adult Quechua Indians suggests that the thermoregulatory characteristics of the Indian native are acquired through developmental acclimatization to cold.[49, 50] The everyday activities of Quechua children provide stimulus for the development of adaptive physiological responses to cold stress.

Summary

The major population differences in adaptation to heat stress are reflected in high tolerance to heat stress and low sweat loss of indigenous populations. At the present stage of knowledge, these differences can be explained as a result of interaction of limited nutritional resources and high heat stress associated with a tropical environment operating during the period of growth and development (**fig. 18.6**). Thus, populations inhabiting tropical climates during growth develop proportionally thinner, longer extremities and shorter trunks than those raised in temperate climates. In turn, this development affects the rate of radiative and convective heat loss. As a result, tropical native populations demonstrate an enhanced tolerance to heat stress and a low sweat rate.

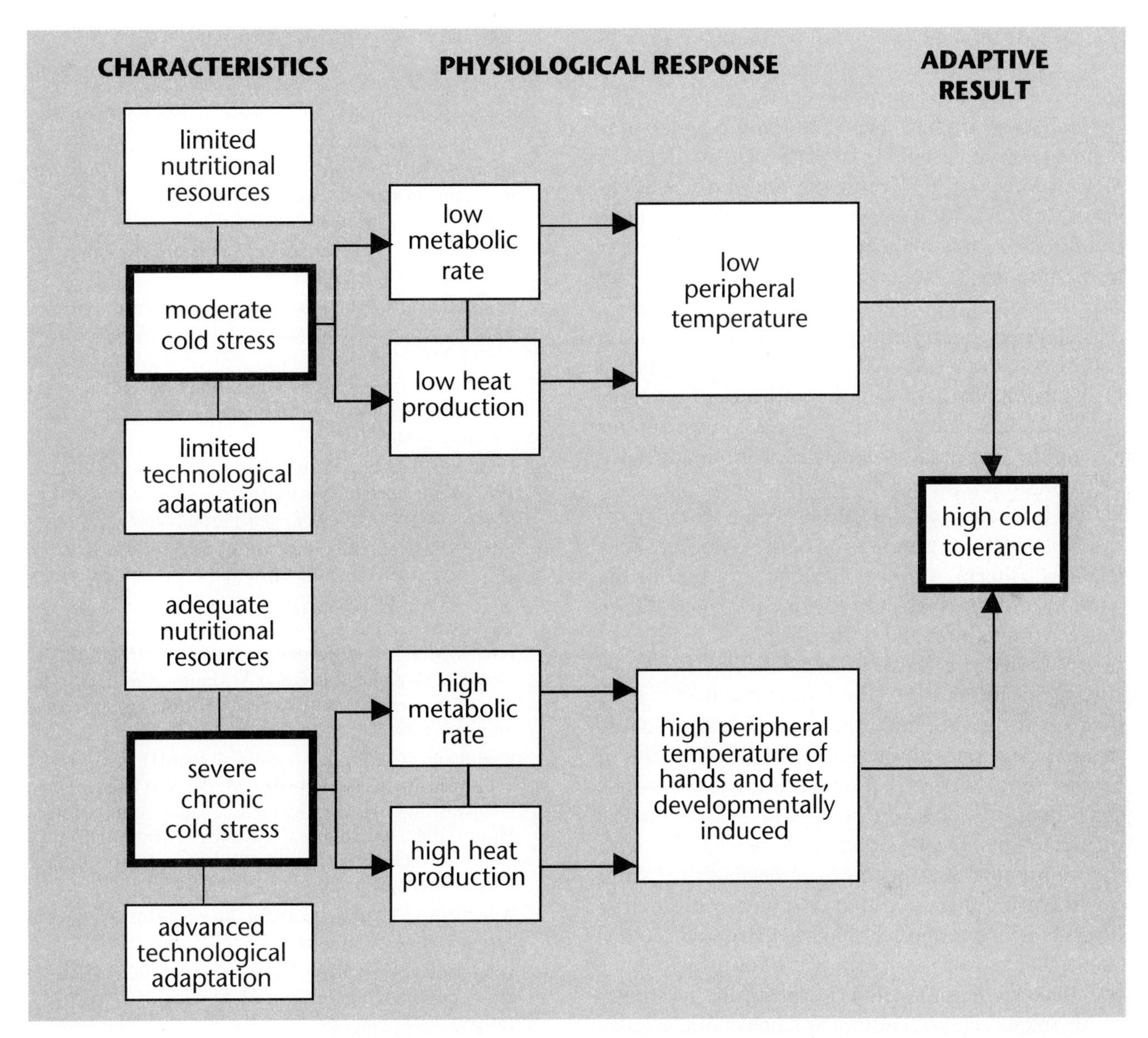

FIGURE 18.18. Schematization of adaptive responses to cold of indigenous and nonindigenous populations acclimatized to environments of moderate and severe cold stress. Adaptation to cold stress has resulted in the development of specialized thermoregulatory responses, which are intimately related to the severity of cold stress and degree of access to nutritional resources and technological adaptation. From *Human Adaptation and Accommodation* by A. R. Frisancho. Copyright © 1995 by University of Michigan Press. Reprinted by permission.

It is quite obvious that humans have developed a great capacity to tolerate varying degrees of cold stress. In general, populations inhabiting regions of moderate cold stress, such as those of the desert, have developed physiological mechanisms oriented toward conserving body heat. This kind of adaptation is attained through continuous exposure leading to habituation, whereby the organism becomes accustomed to sustaining some degree of hypothermia. As indicated by LeBlanc,[51] "If *life is not endangered, the responses of the body seem oriented to retaining individual identity and preserving homeostasis, avoiding unnecessary challenges.*" In a sense, one can say that the thermostat is lowered to a more economical level. On the other hand, populations living in the regions of severe cold stress, such as the Arctic and Andes, have developed efficient technological and cultural adaptations that ameliorate, but do not eliminate, the cold stress, in response to which they have developed physiological responses. The various adaptive responses that enable populations to overcome cold stress are schematized in **figure 18.18**.

At least three physiological mechanisms of cold adaptation set the Eskimos and Quechuas apart from white controls or tropical populations, such as the Australian Aborigines. First, they have a high metabolic rate, which is not explainable solely on the basis of nutritional factors. Second, a test of whole-body exposure as well as local extremity exposure to cold indicates that they maintain high levels of blood flow to the extremities, resulting in warm hands and feet and a correspondingly greater heat loss than whites. Third, the high peripheral temperatures of the extremities and high tolerance to cold of Eskimos and highland Quechuas appear to reflect the influence of developmental acclimatization. Among the Eskimos, but not among the Quechuas, these characteristics appear to be based on specific genetic contributions.

As learned from the studies of the vascular response to cold water immersion of the Gaspé and British Isles fishermen,[52, 53] continued exposure of the hand to cold stress results in the maintenance of high finger temperatures and decreased pain sensation. In other words, the Gaspé fishermen, like the Quechuas,[43–50] Eskimos,[29–42] and Athapaskans[54–56] have developed thermoregulatory responses that enable them to carry on with their daily activities. Studies of Korean Ama divers[57–64] provide conclusive evidence that chronic exposure to cold stress increases the metabolic rate. On the other hand, the fact that with the adoption of wet suits the high metabolic rates and low shivering threshold that characterized the Ama divers is no longer present indicates that acclimatization to cold stress is reversible.[64] Successful adaptation to cold environments as learned from the studies of Algonkian natives' cultural and behavioral adaptations[65–70] is as important as physiological responses.

Suggested Readings

Folk, G. E., Jr. 1974. *Textbook of Environmental Physiology.* Philadelphia: Lea & Febiger.

Frisancho, A. R. 1993. *Human Adaptation and Accommodation.* Ann Arbor: University of Michigan Press.

Hammel, H. T. 1964. Terrestrial animals in cold: Recent studies of primitive man. In *Handbook of Physiology, Vol. 4. Adaptation to the Environment,* D. B. Dill, E. F. Adolph, and C. G. Wilber, eds. Washington, D.C.: American Physiological Society.

Literature Cited

1. Underwood, C. R. and E. J. Ward. 1966. The solar radiation area of man. *Ergonomics* 9:155–168.
2. Hardy, J. D. 1961. The physiology of temperature regulation. *Physiol. Rev.* 41:521–606.
3. Kuno, Y. 1956. *Human Perspiration.* Springfield: C.C. Thomas.
4. Mitchell. D., L. C. Senay, C. H. Wyndham, A. J. Van Rensburg, G. G. Rogers, and N. B. Strydom. 1976. Acclimatization in a hot, humid environment: Energy exchange, body temperature, and sweating. *J. Appl. Physiol.* 40:768–778.
5. Senay, L. C., D. Mitchell, and C. H. Wyndham. 1976. Acclimatization in a hot, humid environment: Body fluid adjustments. *J. Appl. Physiol.* 40:786–796.
6. Folk, G. E., Jr. 1974. *Textbook of Environmental Physiology.* Philadelphia: Lea & Febiger.
7. Gisolfi, C. and S. Robinson. 1969. Relations between physical training, acclimatization and heat tolerance. *J. Appl. Physiol.* 26:530–534.
8. Colin, J. and Y. Houdas. 1965. Initiation of sweating in man after abrupt rise in environmental temperature. *J. Appl. Physiol.* 20:984–990.
9. Houdas, Y., J. Colin, J. Timbal, C. Bontelier, and J. D. Guien. 1972. Skin temperatures in warm environments and the control of sweat evaporation. *J. Appl. Physiol.* 33:99–104.
10. Wyndham, C. H., G. G. Rogers, L. C. Senay, and D. Mitchell. 1976. Acclimatization in a hot, humid environment: Cardiovascular adjustments. *J. Appl. Physiol.* 40:779–785.
11. Wyndham, C. H., J. F. Morrison, C. G. Williams, G. A. G. Bredell, M. J. E. Von Raliden, L. D. Holdsworth, C. H. Van Graan, A. J. Van Rensburg, and A. Munro. 1964. Heat reactions of Caucasians and Bantu in South Africa. *J. Appl. Physiol.* 19:598–606.
12. Wyndham, C. H., N. B. Strydom, S. Ward, J. F. Morrison, C. G. Williams, G. A. G. Bredell, M. J. E. Von Raliden, L. D. Holdsworth, C. H. Van Graan, A. J. Van Rensburg, and A. Munro. 1964. Physiological reactions to heat of Bushmen and of unacclimatized and acclimatized Bantu. *J. Appl. Physiol.* 19:885–888.
13. Wyndham, C. H., B. Metz, and A. Muno. 1964. Reactions to heat of Arabs and Caucasians. *J. Appl. Physiol.* 19:1951–1954.
14. Wyndham, C. H., R. K. McPherson, and A. Munro. 1964. Reactions to heat of Aborigines and Caucasians. *J. Appl. Physiol.* 19:1055–1058.
15. Wyndham, C. 1966. Southern African ethnic adaptation to temperature and exercise. In *The Biology of Human Adaptability,* P. T. Baker and J. S. Weiner, eds. Oxford: Clarendon Press.
16. Lewis, T. 1930. Vasodilation in response to strong cooling. *Heart* 15:177–181.
17. Glickman N., H. Mitchell, R. Keeton, and E. Lambert. 1967. Shivering and heat production in men exposed to intense cold. *J. Appl. Physiol.* 22:1–8.
18. LeBlanc, J. 1975. *Man in the Cold.* Springfield, IL: Charles C. Thomas.
19. Hsieh, A. C. L. and L. D. Carlson. 1957. Role of the thyroid in metabolic response to low temperature. *Am. J. Physiol.* 188:40–44.
20. Davis, T. R. A. 1961. Chamber cold acclimatization in man. *J. Appl. Physiol.* 16:1011–1015.
21. Bittel, J. H. M. 1987. Heat debt as an index for cold adaptation in men. *J. Appl. Physiol.* 62:1627–1634.
22. Scholander, P. F., H. T. Hammel, J. S. Hart, D. H. LeMessurier, and J. Steen. 1958. Cold adaptation in Australian Aborigines. *J. Appl. Physiol.* 13:211–218.

23. Hammel, H. T., R. W. Elsner, D. H. LeMessurier, H. T. Anderson, and F. A. Milan. 1959. Thermal and metabolic responses of the Australian Aborigine exposed to moderate cold in summer. *J. Appl. Physiol.* 14:605–615.
24. Scholander, P. F., V. Walters, R. Hock, and L. Irving. 1950. Body insulation of some arctic and tropical mammals and birds. *Biol. Bull.* 99:225–236.
25. Murdock, G. P. 1964. *Our Primitive Contemporaries.* New York: Macmillan.
26. Arctic Aeromedical Laboratory. 1966. *Literature review of Eskimo knowledge of the sea ice environment.* By R. K. Nelson. Tech. Rep. AAL-TR-65-7. Fort Wainwright, Alaska.
27. Forde, C. D. 1963. *Habitat, Economy and Society.* New York: E. P. Dutton.
28. Perry, R. 1966. *The World of the Polar Bear.* Seattle: University of Washington Press.
29. Krough, A. and M. Krogh. 1913. *A Study of the Diet and Metabolism of Eskimos.* Copenhagen: Bianco Lund.
30. Heinbecker, P. 1928. Studies on the metabolism of Eskimos. *J. Biol. Chem.* 80:461–475.
31. Rabinowitch, I. M. and F. C. Smith. 1936. Metabolic studies of Eskimos in Canadian Eastern Arctic. *J. Nutr.* 12:337–356.
32. Hoygarrd, A. 1941. Studies on the nutrition and physiopathology of Eskimo. Skrifter 9. Norske Videnskaps Akademi, Oslo.
33. Brown, G. M. and J. Page. 1953. The effect of chronic exposure to cold on temperature and blood flow of the hand. *J. Appl. Physiol.* 5:221–227.
34. Brown, G., J. Malcom, J. D. Hatcher, and J. Page. 1953. Temperature and blood flow in the forearm of the Eskimo. *J. Appl. Physiol.* 5:410–420.
35. Brown, G. M., G. S. Bird, L. M. Boag, D. J. Delahaye, J. E. Green, J. D. Hatcher, and J. Page. 1954. Blood volume and basal metabolic rate of Eskimos. *Metabolism* 3:247–254.
36. Adams, T. and B. G. Covino. 1958. Racial variations to a standardized cold stress. *J. Appl. Physiol.* 12:9–12.
37. Rennie, D. W. and T. Adams. 1957. Comparative thermoregulatory responses of Negroes and white persons to acute cold stress. *J. Appl. Physiol.* 11:201–204.
38. Rennie, D. W., B. G. Covino, M. R. Blair, and K. Rodahl. 1962. Physical regulation of temperature in Eskimos. *J. Appl. Physiol.* 17:326–332.
39. Hart, J. S., H. B. Sabean, J. A. Hildes, F. Depocas, H. T. Hammel, K. L. Andersen, L. Irving, and G. Foy. 1962. Thermal and metabolic responses of coastal Eskimos during a cold night. *J. Appl. Physiol.* 17:953–960.
40. Arctic Aeromedical Laboratory. 1966. *Oxygen consumption and body temperature of Eskimos during sleep.* By F. W. Milan and E. Evonuk. Tech. Rep. AAL-TR-66–10. Fort Wainwright, Alaska.
41. Arctic Aeromedical Laboratory. 1965. Nutritional Requirements in the Cold. Paper presented at Symposia on Arctic Biology and Medicine. Fort Wainwright, Alaska.
42. Miller, L. and L. Irving. 1962. Local reactions to air cooling in an Eskimo population. *J. Appl. Physiol.* 17:449–455.
43. Baker, P. T. and M. A. Little, eds. 1976. *Man in the Andes: A Multidisciplinary Study of High-Altitude Quechua.* Stroudsburg, PA.: Dowden, Hutchinson & Ross.
44. Hanna, J. M. 1976. Natural exposure to cold. In *Man in the Andes: A Multidisciplinary Study of High-Altitude Quechua,* P. T. Baker and M. A. Little, eds. 315–330. Stroudsburg, PA: Dowden, Hutchinson & Ross.
45. Mazess, R. B. and R. Larsen. 1972. Responses of Andean highlanders to night cold. *Int. J. Biometeorol.* 16:181–192.
46. Baker, P. T. 1966. Micro-environment cold in a high altitude Peruvian population. In *Human Adaptability and Its Methodology.* H. Yoshimura and J. S. Weiner, eds. Tokyo: Japanese Society for the Promotion of Sciences.
47. Hanna, J. M. 1970. A comparison of laboratory and field studies of cold response. *Am. J. Phys. Anthropol.* 32:227–232.
48. Blatteis, C. M. and L. O. Lutherer. 1976. Effect of altitude exposure on thermal regulatory response of man to cold. *J. Appl. Physiol.* 41:848–858.
49. Little, M. A. 1976. Physiological responses to cold. In *Man in the Andes: A Multidisciplinary Study of High-Altitude Quechua,* P. T. Baker and M. A. Little, eds. Stroudsburg, PA: Dowden, Hutchinson & Ross.
50. Little, M. A. and J. M. Hanna. 1978. The responses of high-altitude populations to cold and other stresses. In *The Biology of High-Altitude Peoples,* ed. P. T. Baker, 332–362. New York: Cambridge University Press.
51. LeBlanc, J. 1975. *Man in the Cold.* Springfield, Ill: Charles C. Thomas.
52. LeBlanc, J., J. A. Hildes, and O. Heroux. 1960. Tolerance of Gaspé fishermen to cold water. *J. Appl. Physiol.* 15:1031–1034.
53. LeBlanc, J. 1962. Local adaptation to cold of Gaspé fishermen. *J. Appl. Physiol.* 17:950–952.
54. Meehan, J. P. 1955. Individual and racial variations in a vascular response to a cold stimulus. *Milit. Med.* 116:330–334.
55. Elsner, R. W., J. D. Nelms, and L. Irving. 1960. Circulation of heat to the hands of Arctic Indians. *J. Appl. Physiol.* 15:662–666.
56. Irving, L., K. L. Andersen, A. Bolstad, R. W. Elsner, J. A. Hildes, Y. Loyning, J. D. Nelms, L. P. Peyton, and R. D. Whaley. 1960. Metabolism and temperature of Arctic Indian men during a cold night. *J. Appl. Physiol.* 15:635–644.
57. Hong, S. K. 1963. Comparison of diving and non-diving women of Korea. *Fed. Proc.* 22:831–833.
58. Kita, H. 1965. Review of activities: Harvest, seasons, and diving patterns. In *Physiology of Breath-Hold Diving and the Ama of Japan,* ed. H. Rahn and T. Yokoyama. National Research Council Publication 1341. Washington, DC: National Research Council.
59. Rennie, D. W., B. G. Covino, B. J. Howell, S. H. Song, B. S. Kang, and S. K. Hong. 1962. Physical insulation of Korean diving women. *J. Appl. Physiol.* 17:961–966.
60. Rennie, D. W. 1965. Thermal insulation of Korean diving women and nondivers in water. *Physiology of Breath-Hold Diving and the Ama of Japan,* ed. H. Rahn and T. Yokoyama. National Research Council Publication 1341. Washington, DC: National Research Council.

61. Sasaki, T. 1966. Relation of basal metabolism to changes in food composition and body composition. *Fed. Proc.* 25:1163–1168.
62. Paik, K. S., B. S. Kang, D. S. Han, D. W. Rennie, and S. K. Hong. 1972. Vascular responses of Korean Ama to hand immersion in cold water. *J. Appl. Physiol.* 32:446–450.
63. Kang, D. H., Y. S. Park, Y. D. Park, I. S. Lee, D. S. Yeon, S. H. Lee, S. Y. Hong, D. W. Rennie, and S. K. Hong. 1983. Energetics of wet-suit diving in Korean women divers. *J. Appl. Physiol.: Respirat. Environ. Exercise Physiol.* 54:1702–1707.
64. Park, Y. S., D. W. Rennie, I. S. Lee, Y. D. Park, K. S. Paik, D. H. Kang, D. J. Suh, S. H. Lee, S. Y. Hong, and S. K. Hong. 1983. Time course of deacclimatization to cold water immersion in Korean women divers. *J. Appl. Physiol.: Respirat. Environ. Exercise Physiol.* 54:1708–1716.
65. Steegman, A. T. 1977. Finger temperatures during work in natural cold: The Northern Ojibwa. *Hum. Biol.* 49:349-374.
66. Hurlich, M. D. and A. T. Steegmann, Jr. 1979. Contrasting laboratory response to cold in two sub-arctic Algonkian villages: An admixture effect? *Hum. Biol.* 51:255–278.
67. Winterhalder, B. 1983. History and ecology of the boreal zone in Ontario. In *Boreal Forest Adaptations: The Northern Algonkians,* ed. A. T. Steegmann, Jr. New York: Plenum Press.
68. Marano, L. 1983. Boreal forest hazards and adaptations: The present. In *Boreal Forest Adaptations: The Northern Algonkians,* ed. A. T. Steegmann, Jr. New York: Plenum Press.
69. Steegmann, A. T., Jr, M. G. Hurlich, and B. Winterhalder. 1983. Coping with cold and other challenges of the Boreal forest: An overview. In *Boreal Forest Adaptations: The Northern Algonkians,* ed. A. T. Steegmann, Jr. New York: Plenum Press.
70. So, J. K. 1975. Genetic acclimatizational and anthropometric factors in hand cooling among North and South Chinese. *Am. J. Phys. Anthropol.* 43:31–38.
71. Eveleth, P. B. 1965. The effects of climate on growth. *Ann. N.Y. Acad. Sci.* 134:750–755.
72. Stinson, and A. R. Frisancho. 1978. Body proportions of highland and lowland Peruvian Quechua children. *Hum. Biol.* 50:57–68.
73. Hofler, W. 1968. Changes in the regional distribution of sweating during acclimatization to heat. *J. Appl. Physiol.* 25:503–506.
74. Ogawa, T. 1972. Local determinants of sweat gland activity. In S. Itoh, K. Ogata, and H. Yoshimura, eds., *Advances in Climate Physiology.* Heidelberg, NY: Springer-Verlag New York, Inc.
75. Kuno, Y. 1956. *Human Perspiration.* Springfield, IL: C. C Thomas.

Climate and Evolution of Variability in Body Size, Body Proportions, and Hairlessness

There is considerable difference in human body size and proportion among populations living today. Similar differences also occurred among the early and recent hominids. Understanding the source of these differences is a major goal of the study of human adaptation. This chapter focuses on the role of the environment on variability in height, weight, head shape, and nose shape among contemporary and early hominids. In addition, this chapter focuses on the evolution of hairlessness, which played an important role in the evolution of thermoregulatory effectiveness of humans.

Body Size and Proportions

Thermoregulation and Body Shape: Bergmann and Allen's Rules

Body and proportion in animals varies in association with climate. The association of body size and climate is described by Bergmann and Allen's rule. The Bergmann rule states that within a polytypic warm-blooded species, the body size of the subspecies usually increases with the decreasing mean temperature of its habitat.[2] That is, species that live in cold places tend to be heavier than those living in warmer climates. Allen's rule states that "in warm-blooded species, the relative size of exposed portions of the body decreases with decrease of mean temperature."[3] That is, animals that live in cold climates tend to have shorter limbs than those animals that live in warmer climates. Recent analysis of an extensive database of mammals and birds confirms that there is a relationship between Bergmann's rule and body size.[4, 5]

The explanation for these associations lies in principles of thermoregulation and body size. First, heat dissipation is directly related to the amount of body surface area. Therefore, homeothermic animals, in order to conserve heat in colder regions, minimize their surface area/weight ratios (SA/W). On the other hand, the loss of excess heat in hot environments may be facilitated by a relatively high surface area/volume ratio. Second, heat production is a function of the total mass of a mammal, such that the greater the volume of a mammal, the greater the heat production. Third, a change in body weight is not proportional to a change in surface area. In general the change in body mass is greater than the change in surface area; increases in body size (Bergmann's rule) tend to reduce this ratio. Therefore, per unit of weight, heavier individuals have a lower surface area than lighter weight persons. Thus, heavier individuals produce more heat and are comparatively less efficient at radiating off their body heat into the surrounding environment than lighter weight individuals. Conversely, lighter weight individuals produce less heat and are comparatively more efficient at radiating off their body heat into the surrounding environment than heavier individuals. Hence, the larger animal will lose heat less rapidly and will therefore be better adapted to colder climates. Conversely, the smaller animal will lose heat more rapidly and will therefore be better adapted to warmer climates, where the ability to lose heat is advantageous. Fourth, the reduction in limb length associated with Allen's rule alters relative surface area. Since the trunk has a lower SA/W ratio than do the limbs, varying the amount of mass in the trunk versus the limbs affects the relative surface area of the entire organism. Therefore, an animal with relatively long limbs will tend to have a higher surface area/weight ratio than a shorter limbed conspecific.

Body Weight and Climate

Contemporary Populations

Most natives of sub-Sahara Africa and especially the East African pastoralists such as the Massai, Samburu, Turkana, and other population living in tropical climates are characterized by slim bodies and relative long limbs (**figs. 19.1a** and **b**). In contrast, natives of the Arctic such as Greenland, Alaska, and Syberia, and Andean cold regions (**figs. 19.2a** and **b**) are character-

(a)

(b)

FIGURE 19.1. Body proportion and body shape among inhabitants of hot and cold climates. The (a) lowland Quechua (Lamas) women who have lived for more 500 years in the tropical lowlands of Peru are characterized by a slim body and relatively long limbs. In contrast, the (b) high altitude Andean Quechuas who live in a cold climate are characterized by a compact body shape and relatively short limbs. (Composite pictures made by the author.)

(a)

(b)

FIGURE 19.2. Body proportion and body shape among inhabitants of hot and cold climates. The Turkana pastoralists of northern Kenya (a) are characterized by a slim body and relatively long limbs. On the other hand, the Evenki reindeer herders of Syberia (b) are characterized by a compact body shape and relatively short limbs. Source: (a) Courtesy of Bettina Shell-Duncan, Department of Anthropology, University of Washington, Seattle, WA.; (b) courtesy of William R. Leonard, Department of Anthropology, Northwestern University.

ized by a short and stocky body shape. Similar differences occur among world populations. For example, the mean stature for populations of adults varies from about 145 cm (4.7 ft) for men and 136 cm (4.5 ft) for women of the Efe Pygmies of Africa to 184 cm (6 ft) for Dutch males and 171 cm (5.7 ft) for females of Europe.[6] Along with these differences there are also differences in weight and proportions.

Statistical Correlates of Climate

Evaluation of a large database of native human populations indicates that body size and body proportions do conform to the Bergmann and Allen's rules such that those individuals inhabiting colder regions are heavier and have shorter relative limb lengths, resulting in a decreased ratio of surface area to body mass.[7–11] As shown in **figure 19.3**, on a worldwide scale, body size is correlated with mean annual temperature. That is, the lower the mean annual temperature, the higher the mean body weight.

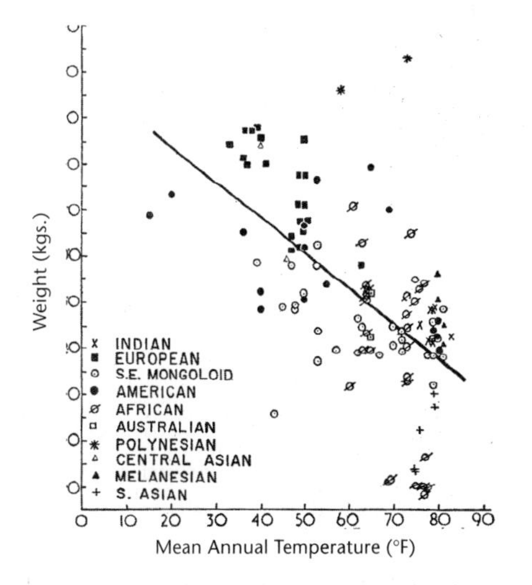

FIGURE 19.3. Relationship between body weight (kg) and mean annual temperature (°F). Source: Adapted from Roberts.[8]

Ratio of Surface Area to Weight and Climate

There are several approaches for measuring surface area of the body. The direct approach involves measuring the amount of material needed to cover the whole body from head to toe. The indirect approach involves predicting surface area from equations based on weight and height:[12, 13]

$$\text{Surface area (m}^2\text{)} = (\text{Weight in kg} \times \text{Height in m}) / 35.37)^{0.5}$$

The ratio of surface area to weight (SA/W × 100) gives the relative amount of surface area available for heat dissipation.[1] As shown in **figure 19.4**, in both men and women there is a significant relationship between the ratio of body surface area to body weight and mean annual temperature. These findings suggest that climate plays an important role in shaping variation in body mass.

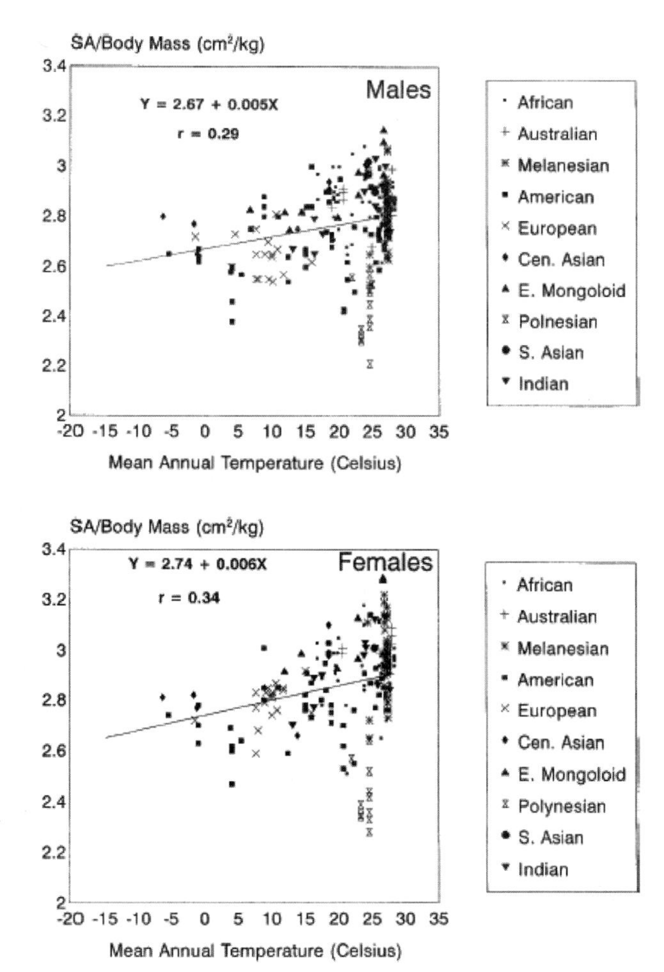

FIGURE 19.4. Relationship of SA/body mass ratio and mean annual temperature among worldwide samples. Source: Adapted from Katzmarzyk.[14]

Relative Sitting Height

Relative sitting height, or sitting height index, is a ratio of sitting height divided by total height multiplied by 100. This index is inversely associated with mean annual temperature, such as that the lower the mean annual temperature, the higher the proportion of trunk size (**fig. 19.5**). Likewise, the relative arm span is directly related to mean annual temperature, such as that the lower the temperature, the shorter the arms. Conversely, the higher the temperature, the longer the arms.

Modification of Body Proportions

The pattern of human growth is characterized by a head-to-foot (cephalocaudal) gradient, whereby the head is more advanced at birth than the trunk and the trunk is more advanced than the legs.[15–19] This trend occurs throughout the prenatal stage and, as such, children are born with disproportionately large trunks and short legs (see **fig. 17.15**). Therefore, during the postnatal stage, the leg length changes more than the trunk length. For this reason, the growth of leg length is under greater environmental influence than the growth of trunk length.[20]

Relative leg length (i.e., leg length/height × 100) and sitting height index are measures of body proportions. These body proportions during development are highly modifiable by environmental influences. First, the secular trend of growth in height that occurred during the last twenty years in the industrialized countries is associated with a greater change in leg length than in trunk length.[36] For example, Japanese born and raised in the United States are taller than Japanese raised in Japan, mostly due to a large increase in leg length.[21, 22] Second, evaluations of changes in long bone length and proportion among ethnically black and white men and women from the years 1800 to 1970 found that the lower limb bones (femur, tibia, and fibula) increased significantly more in length than the upper bones (humerus). Furthermore, the distal long bones (tibia and fibula) increased in length much more than the femur.[23] These changes were associated with improvements in the nutritional and health environment in which the people lived.[23] Third, comparative studies of Mayan children raised in Guatemala and in the United States indicate that in Mayan children raised in the United States, the leg length was relatively longer (averaging 7.02 cm longer) than in their counterparts raised in Guatemala.[24] Further-

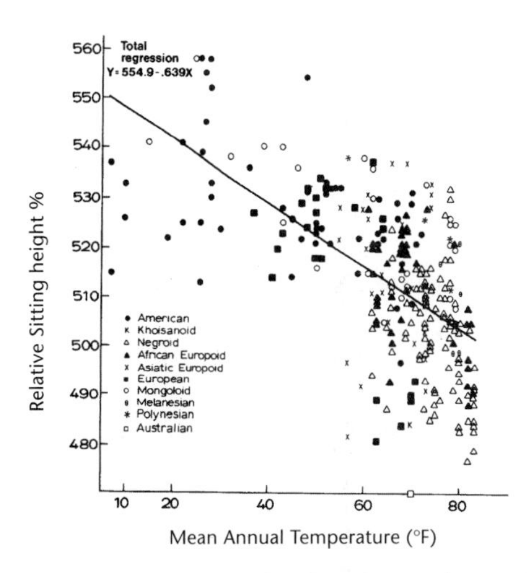

FIGURE 19.5. Relationship between relative sitting height (%) and mean annual temperature. Source: Adapted from Roberts.[8]

more, the Mayan children raised in the United States have body proportions more like those of white children in the United States than like Mayan children in Guatemala. These differences were related to the better nutritional and health conditions experienced by Mayan children raised in the United States. Fourth, differences in socioeconomic status among U.S. and Mexican American children are associated with differences in leg length but not in sitting height.[20] In other words, growth under better socioeconomic conditions is associated with relatively longer legs (**fig. 19.6**).

Moreover, Mexican American children derived from the second Nutrition and Health Examination Survey (NHANES II) of 1982–1984 have a relatively shorter leg length than Mexican American children included in the third Nutrition and Health Examination Survey (NHANES III) of 1988–1994, but do not differ in trunk length (**fig. 19.7**). These differences were related to the better nutritional and health conditions experienced in the United States by Mexican American children during the intervening years.

Weight and Surface Area

Recent statistical analysis of body size among native world populations[14] indicate that the correlation of body surface area and mean annual temperature have decreased by about 50 percent (**table 19.1**) when compared to the values found in the 1950s.[8] The decrease in the correlation coefficients (r) were related to the changes in body weight that the populations in question experienced during the last 50 years.

In summary, it is quite evident that on a worldwide scale, due to the changes in the nutritional environment, the relationship of body size and related indices and mean annual temperature have been modified. Furthermore, during the years of growth and development, leg length and, hence, the resulting body proportion are highly plastic and modifiable by environmental conditions.

Environmental Adaptation of Body Build and Proportions of Neanderthals

Body Build

The body morphology in earlier hominids follows geographic trends that are similar to those of living and recent past populations. For example, *Homo erectus* (such as the Narikotome boy: KNM-WT 15000) who lived in a very hot and dry environment is characterized by a tall but thin physique that emphasizes a high surface-area-to-weight ratio. In contrast, the European Neanderthals who lived in a very cold environment are characterized by a short, stocky body build[25–29] and relatively short lower limbs (**fig. 19.8**). As shown in **table 19.2**, the stocky body build of Neanderthals is associated with a surface-area-to-weight ratio that is smaller than populations that live in warm climates, but a similar ratio to the populations that live in cold climates. The distinctive Neanderthal body shape associated with small surface area has been interpreted to reflect the biological adaptation to the cold environment where the Neanderthals lived.[25–29] Specifically, it has been postulated that the European Neanderthals, living as they did in glacial Europe, gained via "natural selection, the physical features that gave them an advantage for survival in the cold."[42, 43]

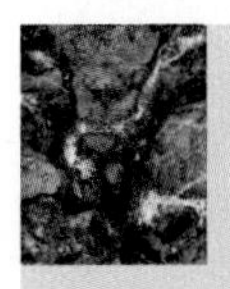

TABLE 19.1

Relationships of Weight and SA/Weight and Mean Annual Temperature among Robert's (1953) and Katzmarzyk and Leonard's (1998) Samples

Study Reported	Roberts 1953 (n = 116) r	Katzmarzyk and Leonard 1998 (n = 223) r
Weight	–0.59	–0.27
Surface/Weight	–0.59	–0.29

Source: Roberts,[8] Katzmarzyk and Leonard.[14]

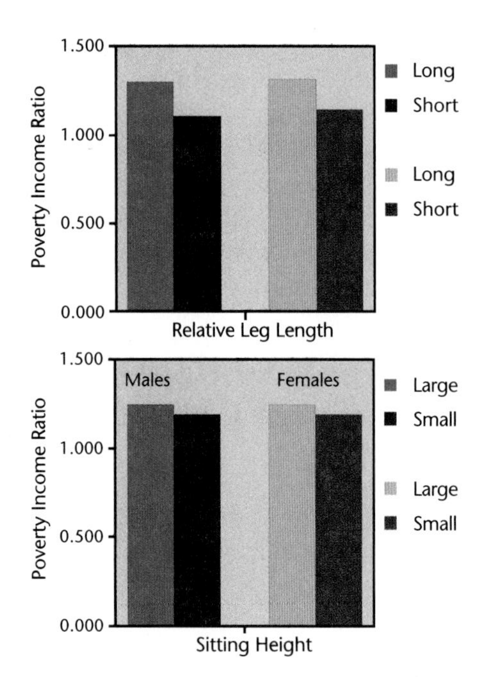

FIGURE 19.6. Comparison of poverty income ratio (pir) to leg length and sitting height. Differences in relative leg length are associated with a significant difference in PIR. In contrast, differences in sitting height are not reflected in differences in PIR. Source: Frisancho, Guilding, and Tanner.[20]

It has been argued that the Neanderthal features are related to ecogeographical rules of Bergmann[2] and Allen.[3] As previously discussed, a low surface area to body weight ratio (SA/W) is associated with more efficient heat retention than a high SA/W ratio.

Limb Proportions

From **figure 19.8** it is evident that Neanderthal, European Late Upper Paleolithic, European Mesolithic, and European Medieval samples have relatively short legs in proportion to trunk height.[25–29] In contrast, the European Early Upper Paleolithic and sub-Saharan recent samples have relatively long legs in proportion to trunk height. As shown in **table 19.2**, the Neanderthals have a smaller surface area than do populations that live in warm climates, but a similar surface area to the populations that live in cold climates. The relatively short legs and low body surface area of Neanderthals has been interpreted to reflect biological adaptations to the cold environment where the Neanderthals lived.[25–30] Analysis of the skeletal dimensions of adult samples from the Southern African Later Stone Age who lived between 10,000 and 2,000 years ago and

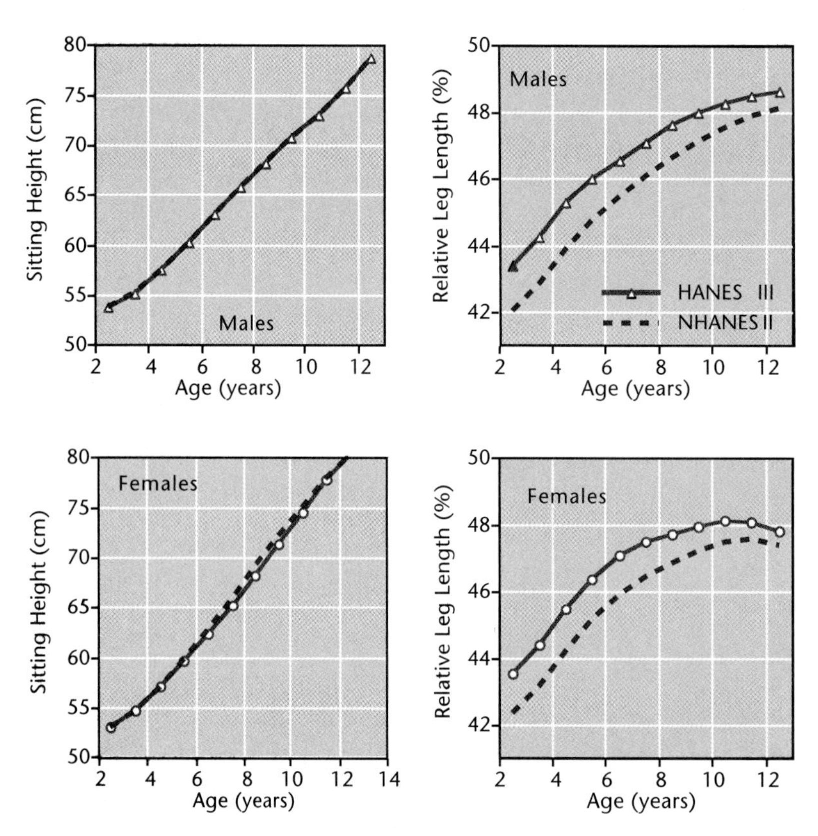

FIGURE 19.7. Comparison of growth in trunk and leg length of mexican americans included in the first and second National Health and Nutrition Examination Surveys (NHANES II and III) conducted during 1982–1984 and 1988–1994.

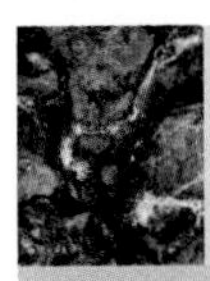

TABLE 19.2

Body Size of Neanderthals Compared to *Homo erectus* and Modern *Homo sapiens*

	Weight (kg)	Height (m)	Body Surface Area* (m^2)	Ratio Surface Area to Weight (%)
Homo erectus	65	1.80	1.82	2.80
Neanderthals	70	1.58	1.77	2.53
Eskimo	70	1.58	1.77	2.53
Modern *Homo sapiens*	70	1.75	1.86	2.66

*Body surface area (m^2) = [(weight, kg × (height, m,/35.37] $^{0.5}$
Source: Wang and Hihara.[13]

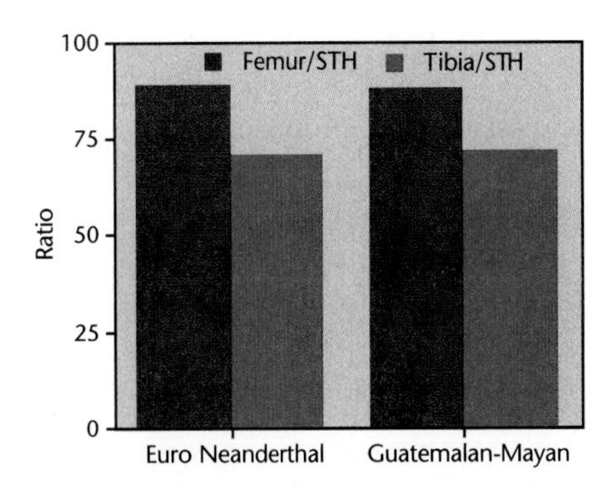

FIGURE 19.9. Ratio of femur length/skeletal trunk height and tibia length/skeletal trunk height for hominids. Source: Data from Bogin.[24]

nineteenth-century indigenous Andaman Islanders[31] found that the postcranial robustness is correlated with overall limb function, and the more proximal elements within the limb are more responsive to mechanical loading. Experimental studies with animals indicate that tibia length is reduced in animals raised in a cold environment relative to warm-environment controls.[32, 33]

Researchers have postulated that the earliest modern *H. sapiens* in Europe are likely to be derived from African ancestors.[25–27] According to this hypothesis, because both Africans and Europeans have relatively long legs, the heat-adapted, relatively long-legged *Homo sapiens* from Africa replaced the cold-adapted, relatively short-legged Neanderthals of the Levant and Europe.[25–27] This hypothesis assumes that body proportions have a genetic basis and hence are not easily modifiable by environmental factors. As discussed above, leg length and, hence, body proportion during development are highly modifiable by environmental factors such as undernutrition. First, studies conducted by Bogin and Rios[24] demonstrate that the undernourished Mayan adults from rural Guatemala have body proportions similar to Neanderthals (**fig. 19.9**). Second, the frequencies of enamel hypoplasias in Neanderthals is as high or higher than that found in undernourished contemporary populations.[35] Therefore, since the higher the frequency of enamel hypoplasia, the greater the growth reduction of leg length, it is quite possible that the reduced leg length of Neanderthals may be related to the effect of undernutrition.[36] By all accounts, the diet of Neanderthals included a high proportion of animal protein,[37] but given the stressful cold climate, it probably was not sufficient to overcome the energy cost of living in a cold climate with limited cultural adaptation. It is estimated that Neanderthals, to support the strenuous winter foraging and cold resistance costs, required a very high caloric intake that ranged from 3,360 to 4,480 kcal per day.[38]

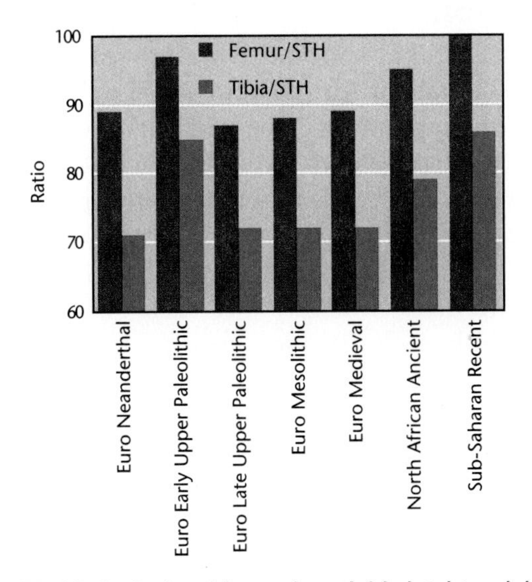

FIGURE 19.8. Ratio of femur length/skeletal trunk height (STH) and tibia length/skeletal trunk height for hominids. Source: Data from Holliday.[26]

In summary, these findings together suggest that the Neanderthal body proportions may be related to both adaptation to cold climate and to the growth-retarding effects of high energy requirements associated with a stressful cold environment. However, despite the nutritional limitations, the Neanderthals attained a skeletal structure and bone mineral content that was much greater than any other contemporary cold-adapted populations.[39, 40] The cortical bones' thickness was probably related to the lifestyle of Neanderthals that required a great deal of strength.[41] In other words, skeletal body breadth is a much more evolutionarily conservative morphological feature than stature, limb length, or the proportion of distal to proximal limb elements. Viewed in this context, populations would need to inhabit a climatic zone for a very prolonged period of time (perhaps tens of thousands of years) for it to show any significant effects on body breadth. Hence, the development of the strong and stocky body morphology of Neanderthals was influenced by their lifestyle that required physical strength, while the reduction of limb length was influenced by the high energy requirements that resulted from the extreme cold, which required high energy expenditure.

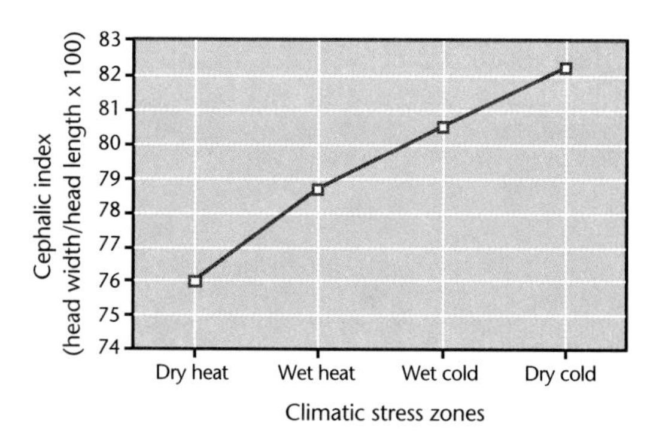

FIGURE 19.10. Relationship of cephalic index and climatic stress zones. As the temperature decreases, the head shape tends to become rounder (high cephalic index). Source: Data from Beals.[45]

Head Shape and Climate

Contemporary Populations

The head is one of the most vascularized body segments and, as such, is an important avenue of heat exchange. The thermoregulatory function of the head is influenced by the amount of surface area that is exposed to the environment. The surface area of the head is related to the shape of the head: a long head has a greater surface area than a round one. Hence, under conditions of cold stress, heat loss would be lesser when the head shape is round rather than long. Conversely, under conditions of heat stress, a long head would be associated with greater heat dissipation than a round one.

The cephalic index (which is a ratio of head width divided by the head length × 100) gives a measure of head shape. The higher the value, the rounder the head shape, while the lower the value, the longer the head shape. In 1912, Franz Boas published a study demonstrating the plastic nature of the head shape in response to changes in the environment.[44] Specifically, in this pioneering study he demonstrated that the offspring of immigrants born in the United States showed a "significant" difference from their immigrant parents in cephalic index. This finding was perhaps one of the most instrumental in demonstrating that the environment had an important role in the expression of such traits. Following this adaptive conceptual framework, Beals,[45,46] based upon a large database of skull measurements representing 362 samples of 82 contemporary ethnic groups, found an inverse relationship between the mean cephalic index and mean annual temperature (**fig. 19.10**). That is, the colder the temperature, the rounder the head shape, while the warmer the temperature, the longer the head shape. This association agrees with the thermodynamics of heat regulation. The closer a structure approaches a spherical shape, the lower the surface area to volume ratio, while the longer the shape, the greater the surface area to volume ratio. Hence, rounded heads dissipate heat slowly and are, therefore, at an advantage in cold climates. On the other hand, long or narrow heads lose heat more quickly and are therefore at an advantage in hot climates. Since up to 80 percent of body heat can be lost through the head, variations in cephalic index of modern populations are, in part, a consequence of adaptation to cold climate during the Holocene.

Change of Head Shape throughout Evolution

It is a truism that hominid evolution started in the tropical climate of Africa about 4 to 5 million years ago and the appearance in the cold climate of Asia and Europe occurred about 1 million years ago. Beals[46] has analyzed the trends of cranial index from *Homo erectus* to Neanderthals to early modern *Homo sapiens.* Within each species, the cephalic index has been classified into three climatic zones. As shown in **figure 19.11**, within each climatic zone there is a geometric trend toward round head shape. The species that live in tropical climates have a lower cephalic index than those living in the glacial climates. The trend toward high cephalic index (round head) is drastically magnified between early modern *H. sapiens* and the ethnographic present *H. sapiens.* For example, Neanderthals found in the tropical and temperate regions have a lower cephalic index (70.9 and 73.9, respectively) or

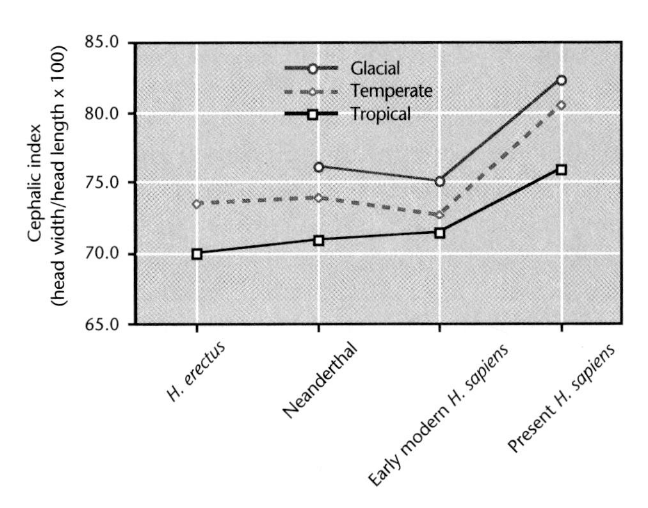

FIGURE 19.11. Cephalic index for hominids from tropical, temperate, and glacial climatic zones, and present ethnographic *H. sapiens*. Source: Data from Beals;[45] Beals, Smith, and Dodd.[46]

narrower head shape than the Neanderthals found in the cold climate of Europe (76.2). A similar trend toward round head shape with cold climate is evident in the early modern *H. sapiens*.

In summary, these findings together suggest that the trend toward round head shape occurred throughout the Pleistocene. The change in head form associated with the climatic zones among the Neanderthals supports the inference of cold adaptation. However, there exists a trend toward brachicephalization (round-headed model) even among the tropical populations supposedly unexposed to cold stress. Therefore, climatic adaptation is not the sole factor that accounts for the trend toward the round head that characterizes *H. sapiens*.

Nose Shape and Climate

Nasal Anatomy and Function

The nasal airway provides heat and fluid exchange for warming and humidifying inspired air; alters the nasal airway resistance by congestion and decongestion of nasal mucosa blood vessels; cleans and filters inspired air by encountering the mucus-coated surface; and senses the environment with specialized (olfactory) and general sensory (trigeminal) nerves.[47] In other words, the nose functions as an air-conditioning unit, whereby the inspired air is warmed to the level of body temperature.[48, 49] Therefore, the humidifying capacity of the nose is directly related to the total surface area of the internal structure of the nose, which, in turn, is directly related to the size and shape of the nose. A high narrow nasal opening with a large internal cavity can warm and moisten air more efficiently than a short and wide nasal opening. In addition to the humidifying function, the nose also plays an important role in temperature regulation of the brain. This is because the venous blood from the scalp, midface, and nasal mucous membrane communicate with the cavernous sinus intracaneally via the ophthalmic veins, pterygoid plexus, and maxillary vein arterial blood vessels (**fig. 19.13**).

Nasal Configuration of Contemporary Native Populations

Recently Hall[74] conducted a comprehensive evaluation of nasal configuration in a sample of 355 individuals representing 14 native population samples that live in diverse climates. The study (**fig. 19.14**) confirms that nose shape, especially at the extremes of temperature and humidity, varies with climate. For example, populations that live in cold climates tend to have the least projection of the nose and tall nasal height.

Nasal Configuration of Neanderthals

The nasal configuration of Neanderthals is characterized by an unusually depressed, sloping internal nasal

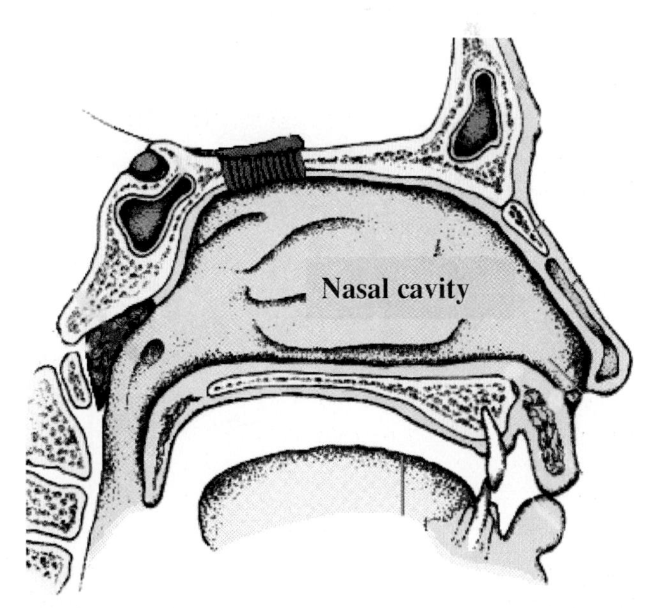

FIGURE 19.13. The interior shape of the human nose. The nasal cavity moistens and warms incoming air, while small hairs and mucus in the nose filter out harmful particles and microorganisms before it gets into the lungs. (Composite picture made by the author.) See also www.sun.ac.za.[50]

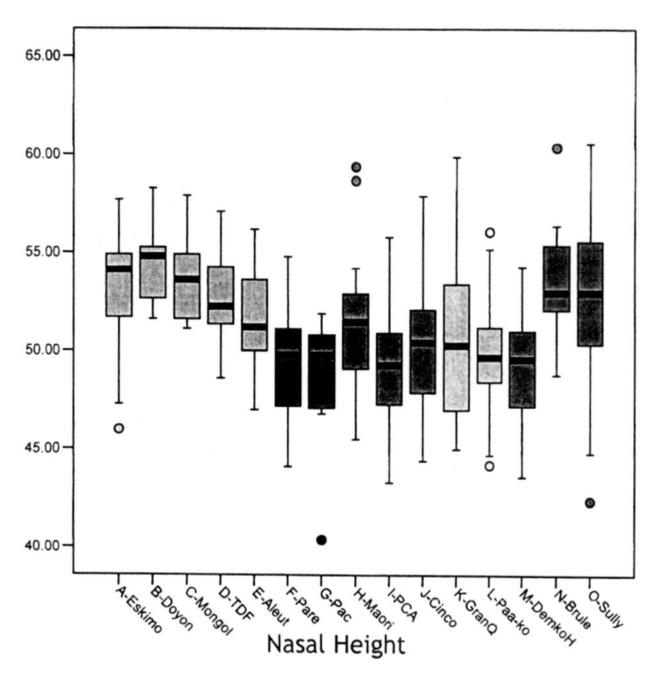

FIGURE 19.14. Median values of nasal height among 14 native population samples. The tallest nasal measures are associated with the cold-climate samples from such as the Eskimos, Doyon-Ingalik, and the Sully Arikara and the Brule Sioux of North Dakota. (Courtesy of Dr. Roberta Hall, Department of Anthropology, Oregon State University).

floor (**fig. 19.15**). About 80 percent of Neanderthals show a high frequency of the depressed nasal configuration. On the other hand, only 15 to 50 percent of Middle Pleistocene African, Late Pleistocene non-Neanderthal (Skhul, Qafzeh), and European Later Upper Paleolithic samples have the depressed nasal configuration.[51] Similarly, only 10 to 20 percent of recent human samples have the depressed nasal configuration.[51] Several hypotheses have been advanced to account for the large nasal volume of Neanderthals' nasal configuration.

First, the depressed nasal floor in Neanderthals has been seen as one reflection of an unusually large internal nasal fossa or chamber that is capacious in all its dimensions[52] and functions as a nasal radiator.[42, 51–54] Coon[42] postulated that the characteristic mid-facial projection of Neanderthals, especially those from western Europe, was entirely due to the necessity of distancing the nasal chamber from the neurocranium. Statistical analysis of the morphological features of human skeletal nasal protrusion found a significant cline of increasing nose protrusion with decreasing absolute humidity and with increasing latitude.[53] Cold climatic variables appeared to be of greater importance than warm measures.

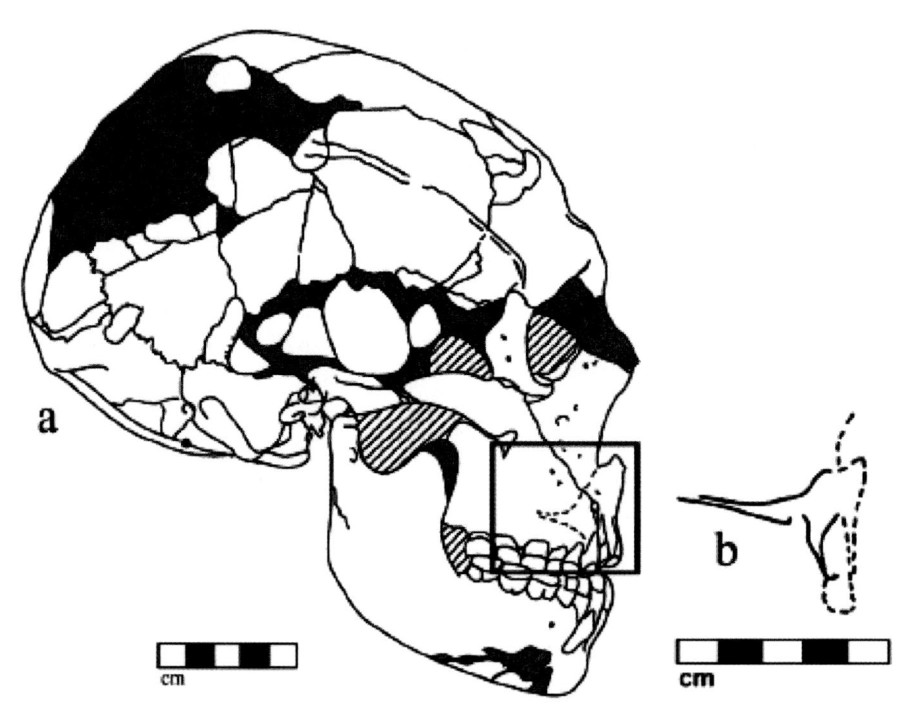

FIGURE 19.15. Internal nasal floor depression in Neanderthals. (a) Shanidar 1. (b) Shanidar 2 maxilla. Source: Adapted from Franciscus.[51]

Second, the large nasal cavities and nasal mucosa area have also been hypothesized to be involved in brain thermoregulation in terms of cooling rather than warming under conditions of heightened activity.[55] As previously indicated, because of the communication of blood vessels with arterial blood that supplies the brain, the nose influences brain thermoregulation. This factor becomes important when the body temperature increases to high levels due to strenuous exercise. For example, at an ambient temperature of 18.8°C (65.8°F), the trunk core temperature of a marathon runner can rise to 41.9°C (107.4°F) with no clinical signs of heat stress.[56] If the brain temperature increased at the same rate, it would rapidly impair cerebral integrating function. Therefore, the influence of ambient temperature and its effect on nasal respiratory mucosa is potentially important in cooling cerebral temperature.

Third, the large and depressed sloping internal nasal floor is a biomechanical response to the unusual use of the anterior teeth as tools.[57, 58] At present, there is no evidence that variation in nasal floor configuration or other aspects of nasal fossa anatomy are functionally tied to masticatory biomechanics in *Homo*.[59]

Fourth, it has also been suggested that the lowered, sloping nasal floor in Neanderthals acted to enhance turbulent airflow during inspiration, as opposed to laminar airflow that is induced in a smooth or unilevel internal fossa configuration.[51, 60] Turbulent airflow results in a larger portion of air coming into direct contact with the nasal mucosa during both inspiratory and expiratory cycles, and, therefore, directly affects the efficiency of temperature and moisture exchange in the nasal cavity.[51]

A comprehensive evaluation of nasal configuration of Neanderthals and other early archaic *H. sapiens* populations indicates that cold-climate and activity-related thermal adaptation as an explanation for the high frequency of pronounced nasal floor depression in Neanderthals is not supported.[51] Instead, it is postulated that variation in internal nasal floor configuration is more likely related to stochastically derived population differences in fetal nasofacial growth patterns.[51] In this context, the Neanderthal craniofacial evolution is due primarily to increased reproductive isolation and largely stochastic differentiation among Neanderthals during a series of glacial advances over the past 450,000 years. These glacial advances resulted in population-specific patterns of very early fetal growth that emerged and crystallized during the increasingly colder and more isolating phases of the last Ice Ages.

Evolution of Hairlessness

Among primates and mammals, humans are unique in that they lack a dense layer of hair covering their bodies. Likewise, females are less hirsute than males. Several hypotheses have been proposed to account for human hairlessness (refers to a lack of a dense layer of thick fur): bipedal locomotion, aquatic evolution, reduction of parasite load, sexual selection, and axillary and pubic hair.

Bipedal Locomotion

This hypothesis postulates that hairlessness evolved along with bipedal locomotion to promote cooling of the body[61–63] (see chapter 9). The premise of this proposition is that the combination of an upright posture and lack of hair made it easier to radiate heat back into the environment or to lose heat by convective cooling from the wind. This proposition is based upon two principles. First, hairlessness promotes heat loss, including evaporative heat loss by increasing airflow directly on the skin. Second, upright posture greatly reduces the negative effects of increased heat absorption with hairlessness by exposing less of the body surface to the sun, especially near midday in the tropics where heat exposure is the most critical. In addition, upright posture raises more of the body surface area above the ground and surface vegetation, increasing air movement over the skin and also aiding in evaporative and convective cooling.

However, because naked skin increases the rates of heat absorption during periods of too much heat and increases rates of heat loss during periods of too little heat, naked skin is not an appropriate adaptation.[64] For this reason, hairlessness most likely evolved very early in hominid evolution when hominids were limited to forested environments where naked skin could have aided in heat radiation without increasing direct heat absorption.

Investigations in molecular genetics found that the gene referred to as melanocortin 1 receptor (MC1R) makes a protein that affects the color of skin and hair. These studies found that sun-resistant MC1R alleles became present when human skin was first regularly exposed, hairless, to strong sunlight and occurred more than 1 million years ago.[65] These investigations indicate that hairlessness in humans might have evolved in the African savanna at least 1.2 million years ago.

Aquatic Evolution

Proponents of this hypothesis maintain that the hominid ancestors evolved between 6 and 8 million years ago in an aquatic environment and the aquatic adaptations were retained as ancestral characteristics throughout at least 5 million years of subsequent hominid evolution in predominantly terrestrial habitats.[66] According to this hypothesis, hairlessness and high levels of body fat evolved in these ancestors because fur is not an effective thermal layer underwater.[66] This hypothesis fails to explain why features supposedly adaptive to an aquatic lifestyle should have been retained. Likewise, fossil and genetic evidence for an aquatic phase of proto-hominid existence has not been found.

In summary, it appears that the aquatic adaptation hypothesis does not account for the evolution of hairlessness.

Reduction of Parasite Load

This hypothesis maintains that hairlessness evolved as an adaptation to reduce ectoparasite loads such as ticks, lice, fleas, mosquitoes, and other parasites and has been maintained by its naturally selected advantages in reducing disease. In furry or feathered animals, hair retains ectoparasites such as fleas and ticks that can affect animals directly by biting and causing local irritation, and indirectly by carrying a variety of infections.[67] Animals have specialized muscles for twitching their skin, long tails to swat at flies and many other antiparasites, morphological, and behavioral adaptations.[68] In primates, grooming is an important social practice, primarily to remove ectoparasites.[69] In humans, ectoparasites are largely confined to the head and pubic hair probably because it provides a warm and humid environment favorable to ectoparasites.

Sexual Selection

Darwin[70] observed that in all parts of the world, women are less hairy than men and suggested that hairlessness has developed through sexual selection, but did not provide any mechanism for its prevalence. Building on Darwin's hypothesis, Pagel and Bodmer[71] propose that hairlessness is maintained by sexually selected effects arising from mate choice for hairless partners. Since sexual selection typically relies upon a trait having a naturally selected advantage to begin the process of its exaggeration,[72] it is quite possible that hairlessness in humans, by virtue of advertising reduced ectoparasite loads, became a desirable trait in a mate.[71] Coincidentally, reflecting sexual preference of hairlessness, use of depilatory agents is more common in females than in males.

Axillary and Pubic Hair

The retention of pubic and axillary hair cannot be accounted for by the hypothesis of ectoparasite reduction because both the axilla and pubic area provide a warm and humid environment favorable to ectoparasites. In addition, these areas have a density of sweat glands[73] that favor the development of odors perceptible to the sense of smell. Therefore, it is quite possible that axillary and pubic hair may be vestigial traits that functioned in our primate ancestors for pheromonal signaling between the sexes.

Summary

Variability in body size and body proportions reflects adaptations to the environmental conditions occurring during development. Low body surface area (SA) associated with high body weight (W) tends to occur among populations living in cold climates, while a high body surface area (SA) associated with low body weight (W) occurs more frequently in populations living in warm climates. This geographic patterning of body morphology reflects thermoregulatory adaptation to climatic conditions associated with heat and cold, whereby a high SA/W ratio facilitates heat dissipation and low SA/W ratio conserves heat. However, variability in SA/W ratio is profoundly influenced by environmental factors that influence the growth of limbs. Growth under positive nutritional conditions leads to long leg length and high SA/W ratio. In contrast, growth under negative nutritional conditions results in relatively short legs and low SA/W ratio. The influence of these allometric growth trends are evident in the reduction of the association of SA/W ratio and ambient temperature among contemporary living populations. It is quite possible that the variability in body size and surface area of early hominids were also affected by nutritional factors.

Earlier hominids also follow geographic trends that are similar to those of living and recent past populations. For example, the *Homo erectus* from Africa (such as the Narikotome body referred to as

KNM-WT 15000) that lived in a very hot and dry climate is characterized by high body surface and low body weight. In contrast, the European Neanderthals who lived in a very cold climate are characterized by low surface area and high body weight. Viewed in this context, the origin of variability in body size and body shape of earlier hominids was the result of adaptation to both climatic and nutritional conditions.

Current evidence suggests that along with sexual selection, human hairlessness might have become an adaptation to reduce ectoparasite loads. As hominids moved from areas of high solar radiation toward the northern climates, the protective role of body hair was probably compensated for by cultural adaptations, such as clothing and control of fire.

Suggested Readings

Frisancho, A. R. 1993. *Human Adaptation and Accommodation.* Ann Arbor: University of Michigan Press.

Stinson, S. B., B. Bogin, R. Huss-Ashmore, and D. O'Rourke. 2000. *Human Biology: An Evolutionary and Biocultural Perspective.* New York: John Wiley & Sons.

Literature Cited

1. Frisancho, A. R. 1993. *Human Adaptation and Accommodation.* Ann Arbor: University of Michigan Press.
2. Bergmann, C. 1847. Über die verlhältnisse der Wärmeökonomie der Thiere zu ihrer Grösse. *Göttinger Studien* 3:595–708.
3. Allen, J. A. 1877. The influence of physical conditions on the genesis of species. *Radical Rev.* 1:108–140.
4. Freckleton, R. P., P. H. Harvey, and M. Pagel. 2003. Bergmann's rule and body size in mammals. *Am. Nat.* 161:821–5. Epub 2003.
5. Graves, G. R. 1991. Bergmann's rule near the equator: Latitudinal clines in body size of an Andean passerine bird. *Proc. Natl. Acad. Sci.* 15:2322–2325.
6. Eveleth, P. B. and J. M. Tanner. 1976. *Worldwide Variation in Human Growth.* Cambridge: Cambridge University Press.
7. Schreider, E. 1950. Geographical distribution of the body-weight/body-surface ratio. *Nature* 165:286.
8. Roberts, D. F. 1953. Body weight, race and climate. *Am. J. Phys. Anthropol.* 11:533–558.
9. Roberts, D. F. 1973. *Climate and Human Variability.* Addison-Wesley Module in Anthropology, No. 34. Reading, MA: Addison-Wesley.
10. Roberts, D. F. 1978. *Climate and Human Variability,* 2d ed. Menlo Park, CA: Cummings.
11. Crognier, E. 1981. Climate and anthropometric variations in Europe and the Mediterranean area. *Ann. Hum. Biol.* 8: 99–107.
12. DuBois, D. F., and E. F. DuBois. 1916. A formula to estimate the approximate surface area if height and weight be known. *Arch. Intern. Med.* 17:863–871.
13. Wang, J., and E. Hihara. 2003. Human body surface area: A theoretical approach. *Eur. J. Appl. Physiol.* 22:22–25.
14. Katzmarzyk, P. T. and W. T. Leonard. 1998. Climatic influences on human body size and proportions: Ecological adaptations and secular trends. *Am. J. Phys. Anthropol.* 106: 483–503.
15. Tanner, J. M. 1978. *Fetus into Man.* Cambridge: Harvard University Press.
16. Scammon, R. E. 1930. The ponderal growth of the extremities of the human fetus. *Am. J. Phys. Anthropol.* 15:111–121.
17. Krogman, W. M. 1972. *Child Growth.* Ann Arbor: University of Michigan Press.
18. Bogin, B. 1988. *Patterns of Human Growth.* Cambridge: Cambridge University Press.
19. Stratz, C. H. 1909. Washstum und Proportionen des Menschen vor und nach der Geburt. *Arhiv für Anthropologie* 8:287–297.
20. Frisancho, A. R., N. Guilding and S. Tanner. 2001.Growth of leg length is reflected in socio-economic differences. *Acta Medica Auxologica* 33:47–50.
21. Greulich, W. W. 1957. A comparison of the physical growth and development of American-born and native Japanese children. *J. Phys. Anthropol.* 15:489–515.
22. Greulich, W. W. 1976. Some secular changes in the growth of American-born and native Japanese children. *Am. J. Phys. Anthropol.* 45:553–568.
23. Jantz, L. M. and R. L. Jantz. 1999. Secular change in long bone length and proportion in the United States, from 1800 to 1970. *Am. J. Phys. Anthropol.* 110:57–67.
24. Bogin, B. and L. Rios. 2003. Rapid morphological change in humans: Implications for modern human origins. *Comp. Biochem. Physiol. A. Mol. Integr. Physiol.* 136:71–84.
25. Holliday, T. W. 1995. Body size and proportions in the Late Pleistocene Old World and the origins of modern humans. Ph.D. diss., University of Michigan.
26. Holliday, T. W. 1997a. Body proportions in Late Pleistocene Europe and modern human origins. *J. Hum. Evol.* 32: 423–447.
27. Holliday, T. W. 1997b. Postcranial evidence for cold adaptation in European Neandertals. *Am. J. Phys. Anthropol.* 104:245–258.
28. Ruff, C. B. 1994. Morphological adaptation to climate in modern and fossil hominids. *Yrbk. Phys. Anthropol.* 37: 65–107.
29. Ruff, C. B. and A. Walker. 1993. Body size and body shape. In A. Walker and R.E. Leakey, ed., *The Nariokotome Homo erectus skeleton.* Cambridge: Harvard University Press, 234–265.

30. Ruff, C. B. 2002. Variation in human body size and shape. *Annu. Rev. Anthropol.* 31:211–232.

31. Stock, J., and S. Pfeiffer. 2001. Linking structural variability in long bone diaphyses to habitual behaviors: foragers from the Southern African Later Stone Age and the Andaman Islands. *Am. J. Phys. Anthropol.* 115:337–348.

32. Harrison, G. A. 1963. Temperature adaptation as evidenced by growth of mice. *Federation Proc.* 22:691–698.

33. Chevillard, L., R. Portet, and M. Cadot. 1963. Growth rate of rats born and reared at 5° and 30° C. *Fed. Proc.* 22:699–703.

34. Hahn, P., O. Koldovsky, J. Krecek, J. Martinek, and Z. Vacek. 1963. Temperature adaptation during postnatal development. *Fed. Proc.* 22:824–827.

35. Ogilvie, M. D., B. K. Curran and E. Trinkaus. 1989. Incidence and patterning of dental enamel hypoplasia among the Neandertals. *Am. J. Phys. Anthropol.* 79:25–41.

36. Stinson, S. 2000. Growth variation: biological and cultural factors. In *Human Biology: An Evolutionary and Biocultural Perspective,* eds., S. Stinson, B. Bogin, R. Huss-Ashmore and D. O'Rourke. New York: Wiley-Liss, 425–463.

37. Bocherens, H., D. Billiou, A. Mariotti, M. Toussaint, M. Patou-Mathis, D. Bonjean, and M. Otte. 2001. New isotopic evidence for dietary habits of Neandertals from Belgium. *J. Hum. Evol.* 40:497–505.

38. Steegman, A. T., Jr., F. J. Cerny, and T. W. Holiday. 2002. Neanderthal cold adaptation: Physiological and energetic factors. *Am. J. Hum. Biol.* 14:566–583.

39. Harper, A. B., W. S. Laughlin, R. B. Mazess. 1984. Bone mineral content in St. Lawrence Island Eskimos. *Hum. Biol.* 56:63–78.

40. Mazess, R. B., H. S. Barden, C. Christiansen, A. B. Harper, and W. S. Laughlin. 1985. Bone mineral and vitamin D in Aleutian Islanders. *Am. J. Clin. Nutr.* 42:143–146.

41. Trinkaus, E. and P. Shipman. 1993. *The Neandertals: The Changing Image of Mankind.* New York: Knopf.

42. Coon, C. S. 1962. *The Origin of Races.* New York: Alfred A. Knopf.

43. Coon, C. S., S. M. Garn, and J. B. Birdsell. 1950. *Races: A Study of the Problems of Race Formation in Man.* Springfield, IL: Charles C. Thomas.

44. Boas, F. 1912. *Changes in the Bodily Form of Descendants of Immigrants.* New York: Columbia University Press.

45. Beals, K. L. 1972. Head form and climatic stress. *Am. J. Phys. Anthropol.* 37:85–92.

46. Beals, K. L., C. L. Smith, and S. M. Dodd. 1983. Climate and the evolution of brachycephalization. *Am. J. Phys. Anthropo.* 62:425–437.

47. Li, A. Q., Y. G. Sun, G. H. Wang, Z. K. Zhong, and C. Cutting. 2002. Anatomy of the nasal cartilages of the unilateral complete cleft lip nose. *Reconstr. Surg.* 109:1835–1838.

48. Weiner, J. S. 1954. Nose shape and climate. *Am. J. Phys. Anthropol.* 12:615–618.

49. Morris, I. R. 1988. Functional anatomy of the upper airway. *Emerg. Med. Clin. North Am.* 6:639–669.

50. www.sun.ac.za/internet/academic/health/schools/medicine/orl/dept/ppt/nosesinus

51. Franciscus, R. G. 2003. Internal nasal floor configuration in *Homo* with special reference to the evolution of Neandertal facial form. *J. Hum. Evol.* 44:701–729.

52. Stringer, C. B. and E. Trinkaus. 1981. The Shanidar Neandertal crania. In *Aspects of Human Evolution,* ed. C. B. Stringer, London: Taylor & Francis, 129–165.

53. Carey, J. W., and A. T. Steegmann. 1981. Human nasal protrusion, latitude, and climate. *Am. J. Phys. Anthropol.* 56:313–319.

54. Wolpoff, M. H. 1999. *Paleoanthropology,* 2nd ed. Boston: McGraw-Hill.

55. Dean, M. C. 1988. Another look at the nose and the functional significance of the face and nasal mucous membrane for cooling the brain in fossil hominids. *J. Hum. Evol.* 17: 715–718.

56. Maron, M. B., J. A. Wagner, S. M. Horvath. 1977. Thermoregulatory responses during competitive marathon running. *J. Appl. Physiol.* 42:909–914.

57. Smith, F. H. 1991. The Neanderthals: Evolutionary dead ends or ancestors of modern people? *J. Anthropol. Res.* 47: 219–238.

58. Spencer, M. A., and B. Demes. 1993. Biomechanical analysis of masticatory system configuration in Neandertals and Inuits. *Am. J. Phys. Anthropol.* 91:1–20.

59. Glanville, E. W. 1969. Nasal shape, prognathism and adaptation in man. *Am. J. Phys. Anthropol.* 30:29–38.

60. Franciscus, R. G., and E. Trinkaus. 1988. The Neandertal nose. *Am. J. Phys. Anthropol.* 75:209–210.

61. Wheeler, P. 1984. The evolution of bipedality and loss of functional body hair in humans. *J. Hum. Evol.* 13:91–98.

62. Wheeler, P. 1991. The influence of bipedalism on the energy and water budgets of early hominids. *J. Hum. Evol.* 21: 117–136.

63. Wheeler, P. 1992. The influence of the loss of functional body hair on hominid energy and water budgets. *J. Hum. Evol.* 23:379–388.

64. Amaral, L. Q. 1996. Loss of body hair, bipedality and thermoregulation. *J. Hum. Evol.* 30:357–366.

65. Rogers, A. R., D. Iltis, and S. Wooding. 2004. Genetic variation at the MC1R locus and the time since loss of human body hair. *Curr. Anthropol.* 45:105–108.

66. Morgan, E. 1997. *The Aquatic Ape Hypothesis.* London: Souvenir Press.

67. Lehmann, T. 1993. Ectoparasites—direct impact on host fitness. *Parasitol. Today* 9:8.

68. Hart, B. L. 1997. Behavioural defense. In *Host–Parasite Evolution: General Principles and Avian Models,* eds. D. H. Clayton, and J. Moore. Oxford: Oxford University Press, 59–77.

69. Dunbar, R. 1991. Functional significance of social grooming in primates. *Folia Primatol.* 57:121.

70. Darwin, C. 1888. *The Descent of Man and Selection in Relation to Sex,* 2d ed. London: John Murray, 600.

71. Pagel, M., and W. Bodmer. 2003. A naked ape would have fewer parasites. *Proc. R. Soc. Lond. B. Biol. Sci.* 270 (Suppl. 1):S117–S119.

72. Fisher, R. A. 1930. *The Genetical Theory of Natural Selection.* Oxford: Clarendon Press of Oxford University Press.

73. Stoddart, D. M. 1990. *The Scented Ape: The Biology and Culture of Human Odour.* Cambridge: Cambridge University Press.

74. Hall, R. I. 2005. Energetics of nose and mouth breathing, body size, body composition, and nose volume in young adult males and females. *Am. J. Hum. Biol.* 17:321–330.

Human Adaptation to High-Altitude Environments

The Andean and Tibetan plateaus represent one of the few regions of the world where human adaptation by sea-level inhabitants is not easily overcome. This difficulty is related to the fact that these regions are characterized by hypoxia, which is a severe physiological stress caused by lowered barometric pressure. High-altitude hypoxia is a pervasive and ever-present stress that is not easily modified by cultural or behavioral responses. In addition, high plateaus are cold, dry, and generally harsh and forbidding places. Yet, large numbers of people reside in both regions today. It is estimated that at least six or seven million people reside on the Andean altiplano, and some two million occupy the Tibetan plateau. In these regions developed the Inca and Tibetan empires that in prehistoric times ruled much of South America and central Asia, respectively. An important question is how these populations became adapted and became so successful. This chapter discusses the effects and initial physiological responses of sea-level natives to high-altitude hypoxia, and the similarities and differences in the adaptive strategies of the high-altitude Andean and Tibetan natives.

Nature of High-Altitude Stress

High-Altitude Areas of the World

As shown in **figure 20.1**, the two largest high-altitude regions of the world are the Tibetan plateau in Asia

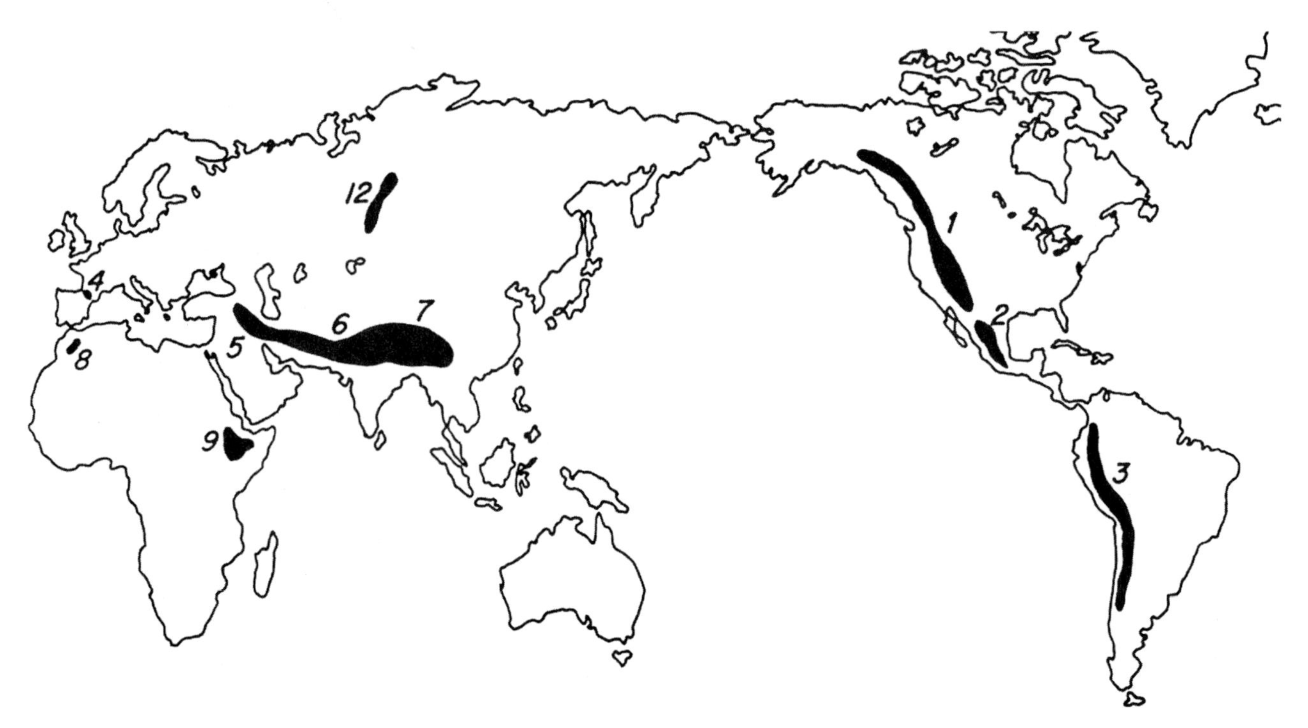

FIGURE 20.1. Some of the areas of the world where people live above 2500 meters of altitude. (1) Rocky Mountains; (2) Sierra Madre; (3) Andes; (4) Pyrenes; (5) mountain ranges of eastern Turkey, Persia, Afghanistan, and Pakistan; (6) Himalayas; (7) Tibetan Plateau and southern China; (8) Atlas Mountains; (9) high plains of Ethiopia; and (12) Tien Shan Mountains. From *Human Adaptation and Accommodation* by A. R. Frisancho. Copyright © 1995 by University of Michigan Press. Reprinted by permission.

and the Andean region of South America, which each contain vast areas higher than 2,500 meters in elevation. In addition, some regions in Africa such as the high plains of Ethiopia, the Tien Shan Mountains of Russia, and the Rocky Mountains of the United States and Canada are inhabited by populations that live above 2000 meters of altitude. It is estimated that about 140 million people worldwide reside in altitudes above 2500 m, or 8000 ft.[1]

Characteristics of High-Altitude Hypoxia

Hypoxia refers to the low availability of oxygen in the inspired air or in the body. High-altitude hypoxia results from a decrease in barometric pressure with increasing altitude (**fig. 20.2**). Although the atmosphere always contains 20.95 percent oxygen, due to the decrease in *barometric pressure* the partial pressure of oxygen also decreases. The decrease in the partial pressure of oxygen in inspired air leads to less oxygen in the alveoli of the lungs to diffuse to the pulmonary bloodstream for transport. This, in turn, reduces the amount of oxygen available to the tissues, where a constant supply of oxygen is required for all the metabolic activities of the organism. For example, at 4000 m (13,200 ft) a liter of air has about 37 percent less pressure than at sea level; therefore, the air has just 63 percent of the number of oxygen molecules compared with sea level. The initial physiological responses to hypoxia include changes in pulmonary ventilation, heart rate, and red blood cell concentration. In general, the effects of high-altitude hypoxia depend on the level of physical activity and age. Under resting conditions young adults do not notice a marked effect at altitudes below 2500 meters (8250 ft), but for older individuals under rest or during physical activity, the effects are clearly manifested above 2500 m. For didactic purposes altitudes between 2500 and 4500 meters are considered high altitude and altitudes beyond 4500 meters are classified as very high altitude.

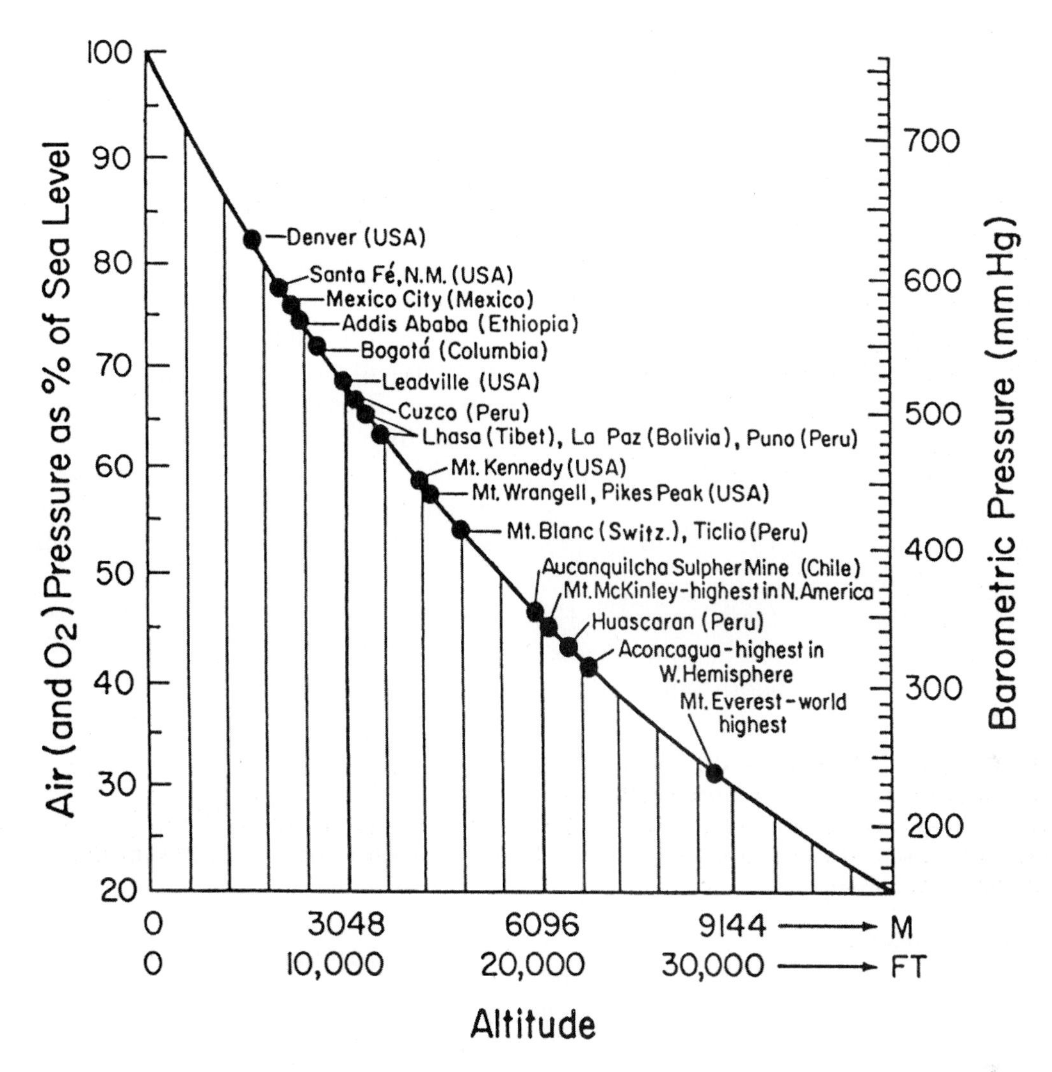

FIGURE 20.2. Barometric pressure and oxygen pressure at high altitudes. With an increase in altitude there is a percentage decrease in the air and oxygen pressure. From *Human Adaptation and Accommodation* by A. R. Frisancho. Copyright © 1995 by University of Michigan Press. Reprinted by permission.

Effects and Initial Physiological Responses to High-Altitude Hypoxia

Pulmonary Ventilation

An immediate response to hypoxia in inspired air is an increase in pulmonary ventilation. On exposure to high-altitude hypoxia, sea-level natives show, both at rest and during exercise, a progressive increase in pulmonary ventilation that may reach as much as 100 percent of sea-level values. For example, the breath rate changes from an average of twelve breaths per minute to as much as twenty breaths per minute. The resting ventilation of sea-level young men in Pike's Peak, Colorado, at 4300 m increased from 6.8 L/min at sea level to 8.7 L/min on the first day and to 10.5 L/min by the eighth day.[3] The adaptive significance of the increase in pulmonary ventilation is that it moves more air and oxygen through the lungs in a given time to offset the hypoxia in the air; however, the increased ventilation also has a negative effect, in that it decreases the concentration of carbon dioxide (CO_2) in the alveoli of the lung, which in turn, changes the pH of the blood from a normal (pH = 7.4) to an alkaline state (pH > 7.4), resulting in respiratory alkalosis. The effects of alkalosis are reflected in high frequency of headaches and hiccups. Concomitant with the increase in pulmonary ventilation, the resting heart rate increases from an average of 70 beats/min at sea level to as many as 105 beats/min at 4500 m.

Pulmonary Vasoconstriction

Sea-level adults exposed to 3700 m (12210 ft) altitude respond to hypoxia by narrowing the blood vessels in the lungs, where oxygen diffuses into the bloodstream; this is referred to as pulmonary vasoconstriction. Pulmonary vasoconstriction also occurs in Andean and Tibetan high-altitude natives.[3–9] However, the Andeans differ from the Tibetans in the "muscularization" of the pulmonary artery.[10–11] The muscularization of the pulmonary arteries of the Andeans has been associated with the high incidence of patent ductus arteriosus (a fetal blood vessel joining the aorta and the pulmonary artery that usually closes functionally and anatomically after birth when the systemic blood pressure far exceeds the pulmonary artery pressure).[11] It has been postulated that hypoxic pulmonary vasoconstriction is a general trait that evolved at sea level as a local homeostatic mechanism to redistribute blood from temporarily poorly ventilated and oxygenated to better ventilated and oxygenated portions of the lungs.[12] Others hypothesize that hypoxic pulmonary vasoconstriction is a vestige of prenatal adaptation and has no adaptive function but remains because there has been no selection against it.

Red Blood Count

The blood volume averages about 5 liters in males and 4.5 in females. The blood is composed of a yellowish liquid called plasma, billions of red cells, and several types of infection-fighting white cells and tiny cell fragments called platelets, which are essential for clotting. The plasma is a pale fluid that is mixture of water, proteins, and salts. The plasma accounts for 55 percent of the total blood volume. The major function of plasma is to act as a carrier for blood cells, nutrients, enzymes, and hormones.

The red blood cell (RBC) count per cubic milliliter of blood averages 5 million in males and 4.5 million in females. The red blood cells, all types of white blood cells, and platelets are made in the red bone marrow located in the epiphysis (head) of the long bones, the vertebrae, and the flat bones such as the skull, the sternum, and the pelvis. The life span of red blood cells is 120 days, after which they deteriorate and are absorbed into the spleen where they are digested.

The blood is red in color because it contains the red-colored protein hemoglobin. The major purpose of hemoglobin is to transport oxygen from areas of high concentration, such as the lungs, and deliver oxygen to sites where they can be used, such as the body tissues. The blood cells also transport carbon dioxide from the areas of high concentration, such as the body tissues, and deliver to the lungs where they are exhaled (**fig. 20.3**).

Hemoglobin is made from two similar proteins (alpha and beta) that "stick together." Hemoglobin contains four molecules of iron (Fe++), which are attached to the porphyrin portions of the hemoglobin molecule. A unique property of iron is that it attracts oxygen and carbon dioxide. Hence, there are four binding sites for oxygen and carbon dioxide per hemoglobin molecule. Males on the average have 14.5±1.5 gram/dl of blood and 12.5 ±1.5 grams/100 ml for females. Values below 12 gr/dL and 10 gr/dL, respectively, are considered indicative of anemia. The synthesis of hemoglobin requires an adequate intake of dietary iron. In general, the daily absorption of iron from the diet is about 1–2 mg (5–10 percent) and about 25 mg/d of iron is needed to maintain red cell production. Hence, anemia can also occur in the presence of adequate amounts of hemoglobin associated with deficiency of dietary iron, which is referred to as iron deficiency anemia (IDA). Iron deficiency anemia is a condition where one has insufficient amounts of iron to meet body demands, such as during periods of rapid growth and pregnancy. IDA is usually due to a diet insufficient in iron or from blood loss. Blood loss can be acute, as in hemorrhage or trauma, or long-term, as in heavy menstruation.

Within three days of exposure to chronic high-altitude (4,540 m) hypoxia, the bone marrow is stimulated by an erythropoietin factor to increase production of red blood cells. Likewise, in 1 to 2 weeks, the average iron turnover rate increases on the average from 0.37 to as high as 0.91 mg/day/kg body weight.[13, 14] During this period, erythropoietin activity is about three times that at sea level. The degree of erythropoietin stimulation has been found to be proportional to severity of the hypoxia; it is high during initial hours of exposure to high altitude and falls in a few days to a value intermediate between the initial peak response and that at sea level.[13, 14] This secondary decrease in erythropoietin activity is related to the fact that after the first day several acclimatization processes operate and thus decrease the hypoxic stimuli. Furthermore, erythropoietin activity is not linearly related to altitude exposure because above 6000 m the rate of erythrocytic and hemoglobin formation is decreased.[9]

As a result of the increased erythropoietin activity, the hemoglobin concentration gradually increases. On the average, young sea-level natives exposed to altitudes more than 4300 m, the mean red cell count during ac-

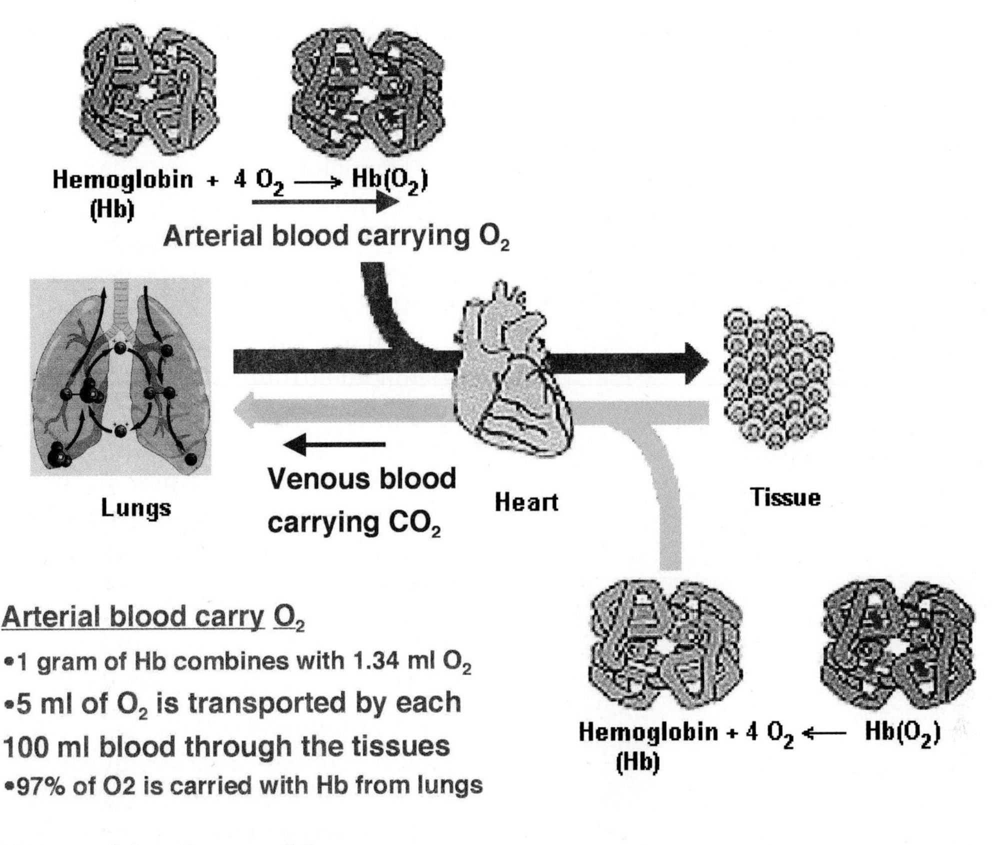

FIGURE 20.3. Hemoglobin and iron function in oxygen and carbon dioxide transport. (Adapted from Brown J. E. Adapted from figure in: Sizer, F., Rolfes, S., DeBruyne, L., and Beerman, K. *Nutrition Interactive.* Version 2. CD–ROM. Wadsworth-Thomson Learning).

climatization changes from about 5 million to 6 to 7 million/cu mm. Along with the increased red blood cell count, hemoglobin concentration also increases from 14 to 18.0 gm/dL in males and from 12 to 16 gm/dL in females. The highest increase in hemoglobin takes place seven to fourteen days after exposure, and during this time erythropoietin activity is about three times that observed at sea level. This process may continue for as long as 6 months, but the hemoglobin level stabilizes between 18 and 20 g/100 ml.[13–17]

The pronounced polycythemia with exposure to high altitude results in an increased red cell volume. As shown in **figure 20.4**, the red cell volume, after a residence at 4540 m for 1 to 3 weeks, increases from 40 ml/kg to 50 ml/kg at high altitude,[14] but the plasma volume decreases. Consequently, the total blood volume increases minimally from 5,000 ml at sea level to 5,220 ml at high altitude, while the hematocrit increases from an average of 43 percent at sea level to about 60 percent after 2 weeks' exposure to high altitude. In other words, the viscosity (thickness) of the blood flow at high altitudes is high. This increased viscosity results in greater strain on the heart; at a high altitude the heart must compensate for increased blood viscosity by doing more work. This explains the increased heart rate observed during initial exposure to high altitude.[18]

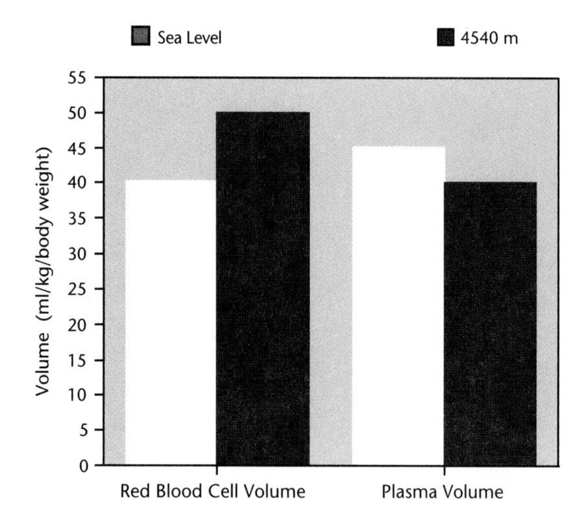

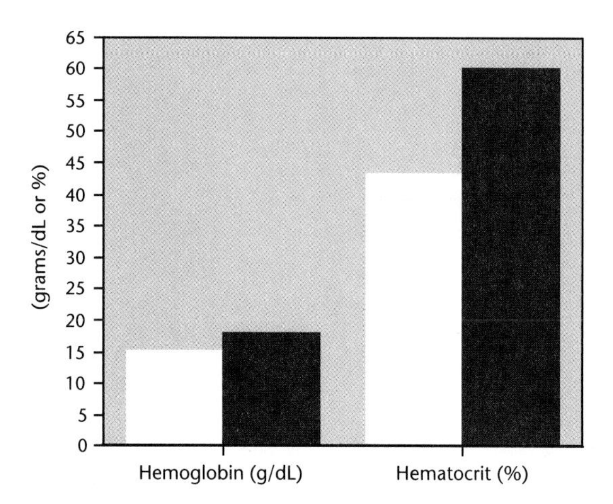

FIGURE 20.4. Changes in red blood cell volume, plasma volume, hemoglobin, and hematocrit among sea-level subjects after a residence for 1 to 3 weeks at high altitude. Source: Merino, C. F. 1950. Blood formation and blood destruction in polycythemia of high altitude. *Blood* 5:1–31.

Oxygen Saturation of Arterial Hemoglobin

The percentage of all the available binding sites of the arterial hemoglobin occupied by oxygen molecules is referred to as the oxygen saturation hemoglobin of arterial blood (SaO_2). The percent of arterial hemoglobin that is saturated with oxygen (SaO_2) is a measure of the capability of the cardiovascular and respiratory systems to deliver oxygen for diffusion into the bloodstream. The percent of oxygen saturation of arterial hemoglobin is usually measured under resting conditions with either an ear or finger pulse oximeter (**fig. 20.5**). The procedure includes attaching a noninvasive sensor to the ear lobe or the index finger for 5 to 15 min. while the participant sits quietly and a total of six measurements are recorded every 10 sec. As shown in **table 20.1**, if the partial pressure of oxygen in ambient air is 159 mm Hg and 104 mm Hg in the alveoli, as it is at sea level, the hemoglobin in arterial blood is 98 percent saturated with oxygen. On the other hand, if the partial pressure of oxygen in ambient air is 110 mm Hg and 67 mm Hg in the alveoli, as it is at an altitude of 3048 m (10,000 feet), the hemoglobin in arterial blood is only 90 percent saturated with oxygen. This means that at an altitude of 3000 m, there is a 10 percent decrease in oxygen for each unit of blood that leaves the lungs. Between 4000 and 5000 m, this decrease might be as high as 30 percent. In other words, the arterial oxygen saturation is directly related to the partial pressure of oxygen in the air. When this pressure is high, as occurs at sea level, a greater proportion of hemoglobin in the arterial blood is saturated with oxygen (greater formation of oxyhemoglobin). On the other hand, when the partial pressure of oxygen is low, as occurs at high altitude, a lower proportion of hemoglobin is combined with oxygen (less formation of oxyhemoglobin and, hence, less arterial oxygen saturation).

Net Effect of the Initial Physiological Responses

Despite the respiratory, circulatory, and hematological responses, sea-level natives sojourning to high altitudes are not able to overcome the high-altitude hypoxic stress. For example, as shown in **table 20.2**, at sea level the arterial hemoglobin is 98 percent saturated with oxygen, the hemoglobin concentration averages 15.0 gr/dL, and the blood volume averages 5,000 ml. Since on the average, each gram of hemoglobin has a carrying capacity of 1.34 ml of oxygen, the total quantity of oxygen in blood equals 98,490 ml ($0.98 \times 1.34 \times 15.0 \times 5{,}000 = 98{,}490$). On the other hand, when a sea-level individual is exposed to high altitude (4545 m) for more than 3 days, the arterial hemoglobin is 81 percent saturated with oxygen, the hemoglobin increases to about 17.0 gr/dl, and the blood volume increases to 5,220 ml. As a result of these changes at high altitude the total quantity of oxygen in blood equals 96,318 ml blood ($0.81 \times 1.34 \times 17.0 \times 5{,}220 = 96{,}318$). It is evident that despite these changes, sea-level natives sojourning to high altitude still have a deficit in their blood of 2,172 ml of oxygen ($98{,}490 - 94{,}318 = 2{,}172$). The resulting hypoxemia causes the oxygen-dependent metabolic processes throughout the organism a great deal of stress.

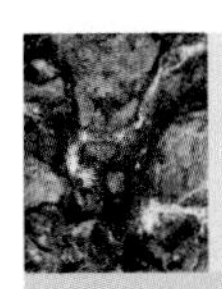

TABLE 20.1

Effects of Altitude and Barometric Pressure on Partial Pressure of Oxygen in the Air, Inspired (Tracheal) Air, and Alveoli and Arterial Oxygen Saturation

Altitude (m)	Altitude (ft)	Barometric pressure (mm Hg)	Partial Pressure of Oxygen (PO_2)† Air (mm Hg)	Trachea (mm Hg)	Alveoli (mm Hg)	Arterial blood (mm Hg)	Oxygen Saturation (HbO_2) Arterial blood (%)
0	0	760	159	149	96	90	97
3,048	10,000	523	110	100	67	62	90
3,151	10,398	515	108	98	65	60	87
3,735	12,325	493	103	93	60	50	81
4,340	14,322	458	96	86	50	46	75
4,540	14,982	446	93	83	47	44	73
5,791	19,110	364	79	69	45	40	60
6,096	20,000	349	73	63	40	35	70
9,144	30,000	226	47	37	30	25	20
12,192	40,000	141	29	19	10	8	5
15,240	50,000	87	18	8	5	3	1

†The partial pressure of oxygen (P_{O2}) in the trachea is reduced because as inspired air enters the airways, it is warmed to almost body temperature (37° C) and saturated with water vapor, which exerts a vapor pressure of 47 mm Hg, regardless of barometric pressure. Therefore the P_{O2} of tracheal air at sea level = (760 – 47) × 0.2095 = 149 mm Hg, and at 3048 m = (523 – 47) × 0.2095 = 100 mm Hg.
Source: Frisancho, A. R. 1993. *Human Adaptation and Accommodation.* Ann Arbor: University of Michigan Press.

Due to this incomplete functional adaptation, sea-level natives exposed to high altitudes do experience disturbances of the sensory nervous system, adrenal activity, reproductive function, and reduced work capacity. Along with physiological disturbances, high-altitude hypoxia also affects appetite, producing anorexia, which in turn results in weight loss. The weight loss is also caused by dehydration or loss of body water, which occurs because of inhibition of fluid intake and excessive fluid loss through the increased altitude-induced ventilation and low humidity of the high altitudes. Eventually the sea-level native, although able to achieve full acclimatization to high altitude, is usually incomplete in the sense that sea-level baseline values of arterial oxygen saturation are not achieved. On the other hand, high-altitude indigenous populations attain a functional capacity that is comparable to that attained at sea level. This is achieved through the long-term process of acclimatization discussed below.

TABLE 20.2

Changes in Oxygen-Carrying Capacity of the Blood with Altitude

	Sea Level	High Altitude (4545 m)	High Altitude (5950 m)
%Hb Saturation (%)	98.0	81.0	65.3
Hb (gr/dl)	15.0	17.0	20.7
Arterial (ml O_2/100 of blood)	19.7	18.5	18.1
Total O_2 in blood (ml)	98,490	96,318	94,549

Adaptive Strategies to High Altitude of Sea-Level, Andean, and Tibetan Natives

Both Andean and Tibetan high-altitude natives (**fig. 20.6**) are descended from sea-level ancestors who migrated to high altitudes at a comparable time. The Tibetans are estimated to have migrated about 7,000 years ago,[19] whereas the Andean regions might have been occupied between 11,500 and 11,000 years ago.[20] Yet, the Andean and Tibetan high-altitude natives show distinctly different physiological adaptive strategies to high-altitude hypoxia, which will be summarized in this section.

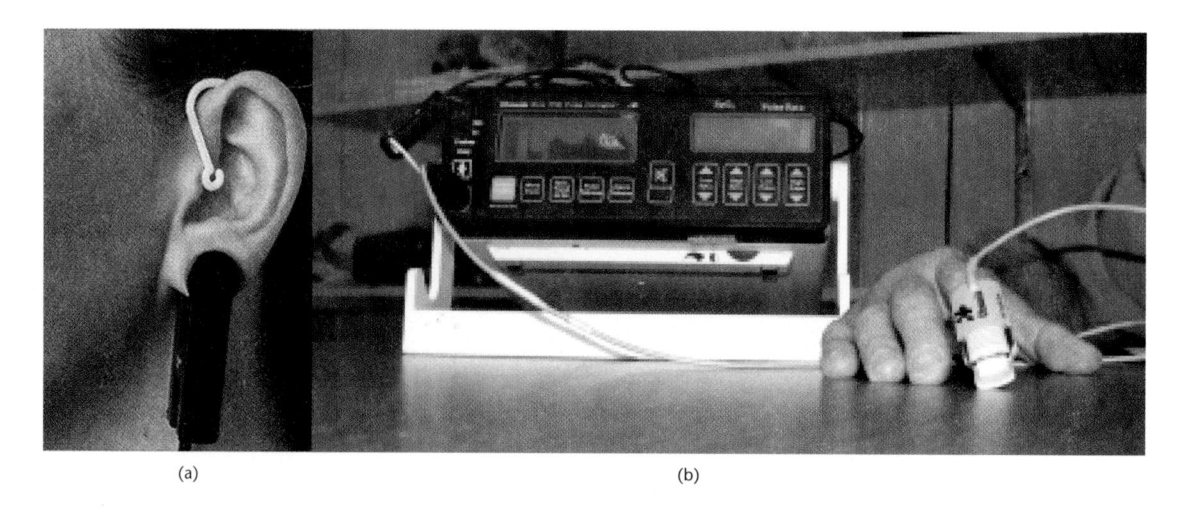

(a) (b)

FIGURE 20.5. Measurement of oxygen saturation of arterial hemoglobin using an ear (a) and finger (b) probe pulse oximeter. (Picture composed by the author.)

Pulmonary Ventilation

Andean Natives versus Tibetan Natives

As shown in **figure 20.7**, Andean natives have lower resting ventilation than Tibetan natives. An analysis of 28 samples studied above 3000 m indicates that, at a mean altitude of 3896 m, the estimated mean ventilation of Tibetan samples was 15.0 L/min as compared with 10.5 L/min among Andean samples and 11.7 L/min among acclimatized lowlanders.[3, 21] That is, Tibetans are more similar to sea-level natives acutely exposed to hypoxia.

Measurements of the age-associated changes of resting ventilation of high-altitude and low-altitude Andean natives tested at sea level and at 4,200 m of altitude (**fig. 20.8**) confirm that Andean high-altitude

(a) (b)

FIGURE 20.6. Andean (a, left side) and Tibetan high-altitude natives (b, right side) living above 4,350 m (14,355 ft). These populations have adapted to high-altitude hypoxia through different biological adaptations.

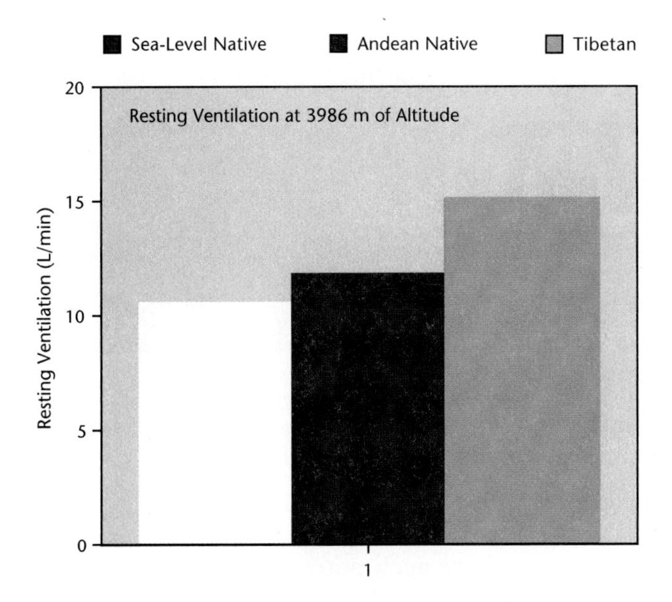

FIGURE 20.7. Comparison of resting ventilation rate of sea-level natives, Andean high-altitude natives and Tibetans. Source: Beall, C. M., Brittenham, G. M., Strohl, K. P., Blangero, J., Williams-Blangero, S., et al. 1997. Ventilation and hypoxic ventilatory response of Tibetan and Aymara high altitude natives. *Am. J. Phys. Anthropol.* 104:427–447.

natives have a lower resting ventilation than high-altitude Tibetans.[22] This study also shows that the low resting ventilation that characterizes the Andean high-altitude native is developmentally expressed in the same manner as it is at low altitude.[22]

Lung Volume

Lung volume is usually divided into two compartments: forced vital capacity (FVC) and residual lung volume (RV) (**fig. 20.9**).

Forced Vital Capacity

The forced vital capacity (FVC) measures the volume of air in the lung during a forced maximal expiration following maximal inspiration of air. At sea level, it usually averages about 3 to 4 liters in women and 4 to 5 liters in men. The volume varies with height and age.

Residual Lung Volume

Residual lung volume (RV) represents the amount of air in the alveolar areas of the lung and is measured with helium gas. The RV averages about 0.8 to 1 liter for women and 1 to 1.5 liters for men. The volume varies with height and age, level of physical activity during childhood, and exposure to high altitude during childhood. The alveolar area that the residual volume measures consists of about 480 million alveoli. The number of alveoli varies with the size of the RV, and it can range from 274 to 790 million. The alveoli have a thin membrane that provide the vital surface for gas exchange between the lungs and the blood. The alveolar area is highly vascularized, where millions of short, thin-walled capillaries lie side by side with air moving on one side and blood on the other. Diffusion of oxygen and carbon dioxide occurs through these

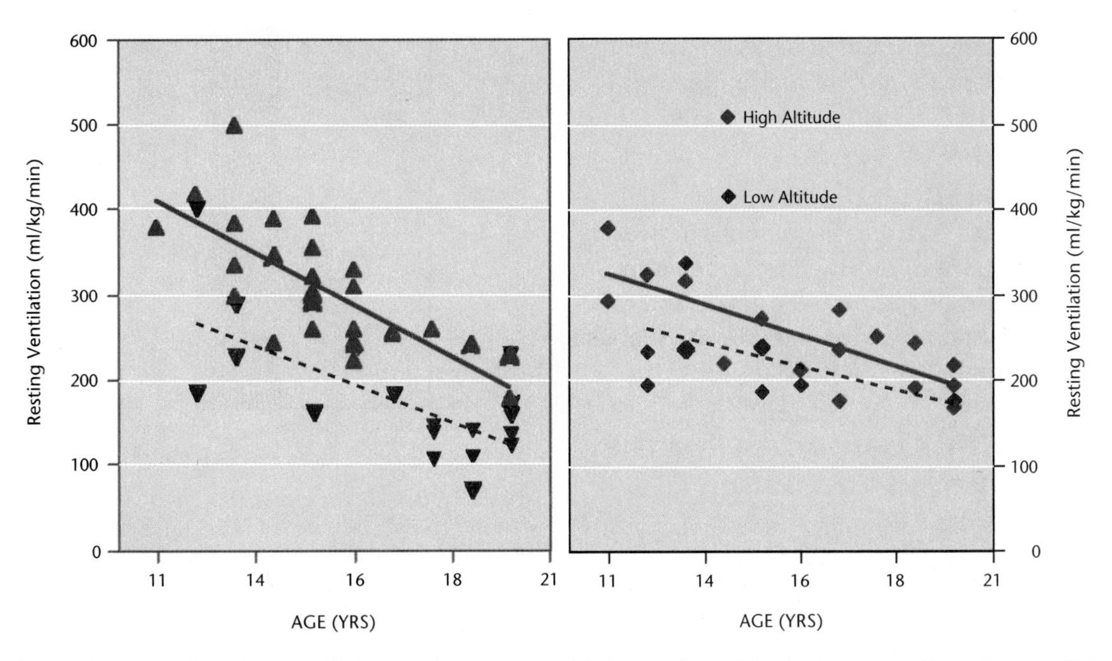

FIGURE 20.8. Relation of resting ventilation and age among high- and low-altitude Andean natives. Source: Frisancho, A. R. 1999. Developmental components of resting ventilation among high and low altitude Andean children and adults. *Am. J. Phys. Anthropol.* 109:295–301.

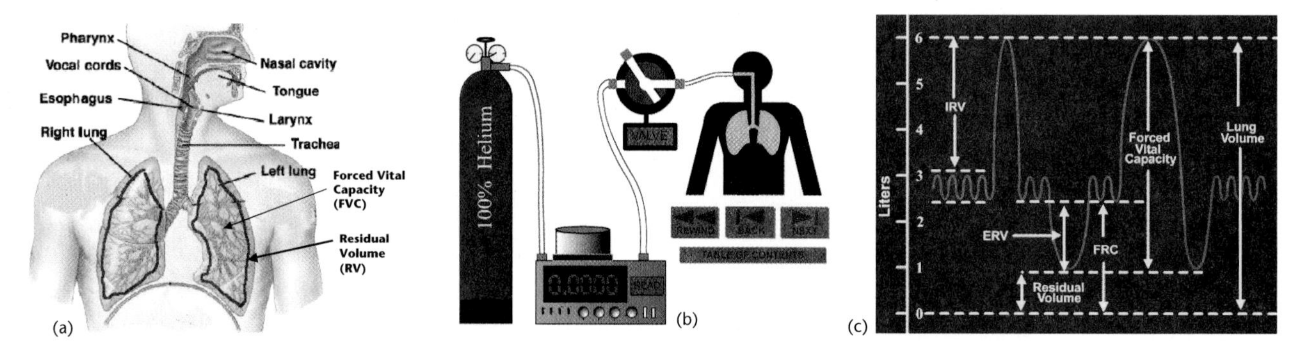

FIGURE 20.9. (a) The respiratory system showing the areas of the lung measured with forced vital capacity (FVC) and residual volume (RV) of the lung. (Picture composed by the author.) (b, c) Measurement of forced vital capacity (FVC) and residual volume (RV) of the lung. (Adapted from http://nutrition.uvm.edu/bodycomp/uww/lung-vol.html.)

capillaries. During each minute at rest, approximately 250 ml of oxygen leave the alveoli and enter the blood, and about 200 ml of carbon dioxide diffuse from the blood into the alveoli to be exhaled. Thus, the alveolar area, and hence the residual volume of the lung, plays an important function in supply of oxygen and getting rid of carbon dioxide.

Sea-Level and Andean Natives

Andean highland natives have larger lung volumes and residual volumes than sea-level natives when adjustments are made for differences in body size.[23, 24] These findings have recently been confirmed by studies of (**fig. 20.10**) subjects born and raised at sea level and measured at high altitude. They have significantly smaller lung volumes (measured by FVC) than subjects who were born and raised at high altitude.[25]

Andean and Tibetan Natives

As shown in **figure 20.11**, both Andean and Tibetan highland natives differ from the sea-level natives in having large lung volumes. However, when adjustments are made for differences in body size and total lung capacity, the Andean natives have relatively larger residual lung volumes than the Tibetans.[24, 26, 27]

Developmental Component for the Enlargement of Lung Volume

Comparison of the lung volumes of sea-level subjects of foreign ancestry acclimatized to high altitude—La Paz, Bolivia situated at mean altitude of 3750 m (12,375 ft)—indicate that, compared to that expected at sea level, those acclimatized since birth or during growth showed a similar increase in residual lung volume as the high altitude (urban) natives (**fig. 20.12**).[28] In contrast, those acclimatized to high altitude as adults exhibited smaller increases in residual lung volume. These findings suggest that acclimatization to high altitude during development (that is, since birth or during growth) is an important component for the increased residual lung volume that characterizes high-altitude natives.[23, 29] The inference that the enlarged lung volume of high-altitude natives results

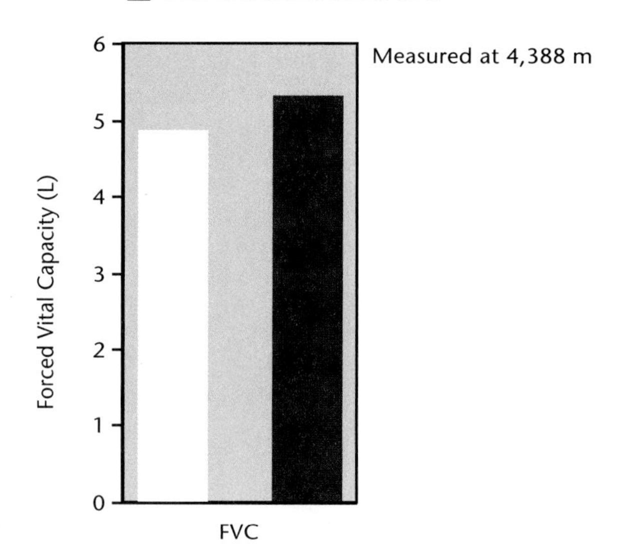

FIGURE 20.10. Comparison of lung volumes (measured by forced vital capacity) of male subjects born and raised at sea level and those born and raised at high altitude and measured in Peru. Source: Brutsaert, T. D., Parra, E., Shriver, M., Gamboa, A., Palacios, J. A., Rivera, M., Rodriguez, I., and Leon-Velarde, F. 2004. Effects of birthplace and individual genetic admixture on lung volume and exercise phenotypes of Peruvian Quechua. *Am. J. Phys. Anthropol.* 123:390–398.

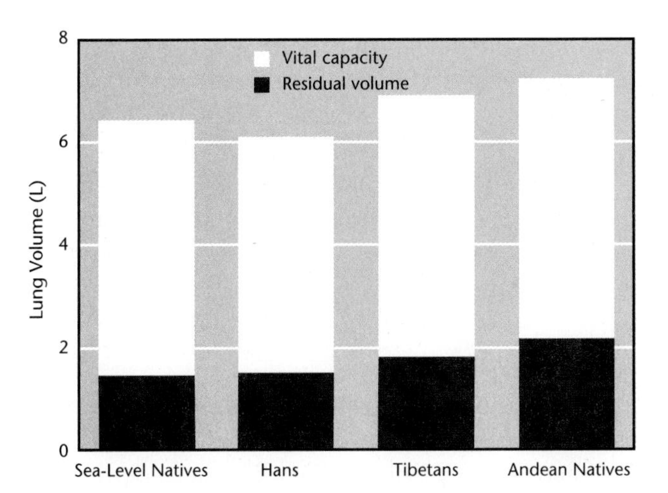

FIGURE 20.11. Comparison of lung volumes of sea-level natives (Peruvian), sea-level natives (Hans), high-altitude Tibetan natives, and Andean high-altitude natives. Source: Hurtado, A. 1964. Animals in high altitudes: resident man. In: D. B. Dill, E. F. Adolph, and C. G. Wilber. *Handbook of Physiology, Vol. 4. Adaptation to the Environment.* Washington, DC: American Physiological Society. Droma, T., McCullough, R. S., McCullough, R. E., Zhuan, J., Cymerman, A., et al. 1991. Increased vital and total lung capacities in Tibetan compared to Han residents of Lhasa (3,658 m). *Am. J. Phys. Anthropol.* 86:341–351.

from developmental adaptation is supported by experimental studies on animals. Various studies[30–33] have demonstrated that young rats, after prolonged exposure to high-altitude hypoxia (3450 m), exhibited an accelerated proliferation of alveolar units and accelerated growth in alveolar surface area and lung volume. In contrast, adult rats, after prolonged exposure to high-altitude hypoxia, did not show changes in quantity of alveoli and lung volume.[32, 33] Recent studies conducted among Peruvian subjects born and raised at sea level and born and raised at high altitude demonstrates that the attainment of lung volume (measured by FVC) is strongly influenced by birthplace, emphasizing the importance of developmental adaptation to high altitude.[25]

In summary, even though both Tibetans and Andean natives have large lung volumes, when adjustments are made for differences in body size and total lung volume, the Andean natives have a relatively larger residual lung volume than the Tibetans. Comparative studies of samples residing at high altitudes indicate that the attainment of a large residual lung volume at high altitude is in part mediated by developmental factors.

Hemoglobin Concentration

Sea-Level and Andean Natives

The hematological response to hypoxia of sea-level natives is similar to that of Andean highlanders in that hemoglobin concentration increases in proportion to the increase in altitude. For example, the hemoglobin concentration of sea-level and Andean populations living at above 3700 m averages about 18.8 g/dl.[5, 23, 34] However, Andean natives who live in rural, nonmining regions of southern Peru have a lower hemoglobin concentration than those who reside near mining centers.[35–37]

Tibetans

Tibetans differ from both sea-level subjects acclimatized to high altitude and Andean high-altitude natives, in that they do not maintain a large elevation of hemoglobin concentration. As documented by the studies of Beall and colleagues,[38–40] the hemoglobin concentration among Himalayan natives is between 1 and 2 g/dl lower than those found among Andean populations living at comparable or lower altitudes. For example, the mean hemoglobin concentration for Tibetans living between 4850 and 5450 m is about

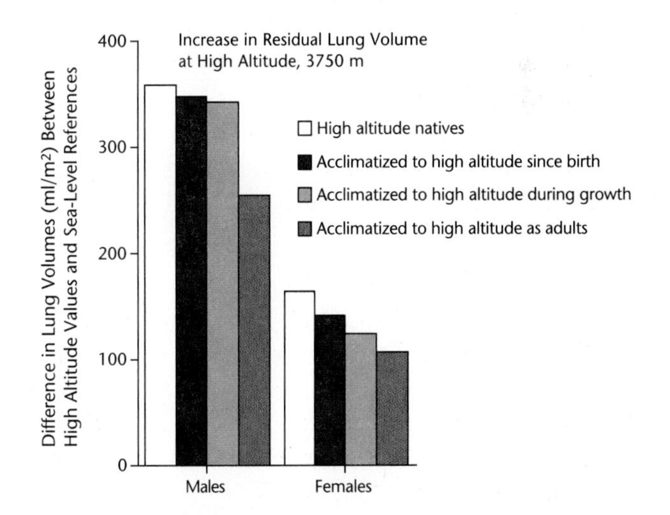

FIGURE 20.12. Comparison of altitude-associated increase in residual lung volume in La Paz, Bolivia, situated at mean altitude of 3750 m (12,375 ft). Among high-altitude natives, sea-level subjects of foreign ancestry acclimatized to high altitude since birth, during growth and during adulthood. Source: Frisancho, A. R., Frisancho, H. G., Milotich, M., Albalak, R., Spielvogel, H., Villena, M., Vargas, E., and Soria, R. 1997. Developmental, genetic and environmental components of lung volumes at high altitude. *Amer. J. Hum. Biol.* 9:191–204.

18.2 g/dl,[38–40] while for Bolivians living at about 3700 m, it is about 18.8 g/dl.[38–40] Thus, for Tibetans, the increase in hemoglobin concentration is not proportional to altitude.

In summary, Andean and Tibetans differ in the hematological response to high altitude. Tibetans increase hemoglobin concentration minimally, while acclimatized lowlanders and Andean highlanders maintain substantially elevated hemoglobin concentrations proportional to altitude.

Physical Work Capacity

The maximum oxygen intake during maximal work is a measure of the individual's work capacity because it reflects the capacity of the working muscles to use oxygen and the ability of the cardiovascular system to transport and deliver oxygen to the tissues. The rate of oxygen consumption increases linearly with the magnitude of work. As an exercising subject approaches the point of exhaustion or fatigue, his oxygen consumption will reach a maximum and remain at that level even with a further increase in work. This peak value is referred to as the individual's *maximal oxygen* consumption (VO_2max). It is usually measured while the subject is performing a maximal or exhaustive work task for short time periods (9 to 15 minutes), either on a stationary bicycle ergometer or a treadmill.

Andean Natives

Acclimatization to high altitude leads to improvement of aerobic capacity. However, even after 3 months of acclimatization, the physical work capacity declines irrespective of physical training. On the other hand, as shown by measurements of oxygen consumption during maximal exercise, the aerobic capacity of high-altitude natives is comparable to that attained by lowland natives tested at sea level (see **table 20.3**).[23, 41–53] In general, the decrease in physical work capacity aver-

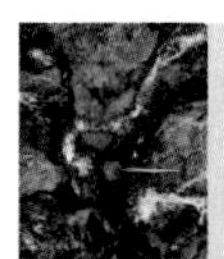

TABLE 20.3

Comparison of Maximum Aerobic Power (VO_2 max: ml/kg/min) Attained by Samples Tested at High and Low Altitudes (LA)*

			VO_2 max (ml/kg/min)			
Sample	N	Tested at altitude	Time at altitude	Attained at altitude	Attained or expected at sea level	D (%)
			Sedentary or Untrained			
Sedentary–U.S.	12	4000	4 weeks	38.1	50.4	24.4
Urban–Bolivia	28	3600	Life	46.0	40–50	0.0
Rural–Chile	37	3650	Life	46.4	40–50	0.0
Rural–Peru	8	4000	Life	51.8	40–50	0.0
Rural–Peru	5	4000	10 years	49.2	40–50	0.0
Urban–Peru	20	3400	2–15 years	46.3	40–50	0.0
Urban–Peru	8	4350	2 weeks	49.0	53.6	8.6
Urban–Peru	5	4540	23 weeks LA	50.0	50.2	0.4
			Active			
Urban–Bolivia	29	3700	Life	37.7	50–60	0.0
Urban–Peru	28	4500	Life	51.2	50–60	0.0
Urban–Peru	20	3750	Life	50.9	50–60	0.0
Rural–Tibet†	16	3658	Life	51.0	50–60	0.0+
Rural–Tibet†	20	3658	8–14 years	46.0	50–60	10.0+
			Trained or Athletes			
Athletes–U.S.	6	4000	7 weeks	49.0	63.0	22.2
Athletes–U.S.	6	3090	2 weeks	59.3	72.0	17.6
Athletes–U.S.	5	3100	4 weeks LA	45.5	61.7	26.3
Athletes–Peru	10	3700	6 months	55.0	70.0	21.4

*Frisancho, A. R. 1993. *Human Adaptation and Accommodation.* Ann Arbor: University of Michigan Press.

† Sun, S. F., Droma, T. S., Zhuang, J. G., Tao, J. X., Huang, S. Y., McCullough, R. G., McCullough, R. E., Reeves, C. S., Reeves, J. T. and Moore, L. G. 1990. Greater maximal O_2 uptakes and vital capacities in Tibetan than Han residents of Lhasa. *Respiration Physiology* 79:151–162.

ages between 20 and 30 percent when compared to sea-level values (**figure 20.13**). For example, the maximal physical work capacity at 3700 m was 20 percent higher than that of Europeans born and raised at high altitude.[54] This decline is also evident among sea-level natives with highland ancestry sojourning to high altitudes (>4,000 m).[55] Thus, high-altitude natives as judged by their ability to attain an aerobic capacity similar to that at sea level have acquired a full (complete) functional adaptation to high altitude.

Tibetan Natives

The maximal work capacity of Tibetans studied at 3658 m was 11 percent higher than that of Han Chinese residents at the same altitude.[56] Thus, it is quite evident that as measured by maximal physical work capacity, both Andean and Tibetan highlanders have higher functionality than acclimatized lowlanders.

Development Component for the Attainment of Normal Work Capacity

Comparison of the work capacity of sea-level subjects of foreign ancestry acclimatized to high altitude—La Paz, Bolivia situated at mean altitude of 3750 m (12,375 ft)—indicate those acclimatized since birth or during growth attained similar values of maximal oxygen consumption as the high-altitude (urban) natives (**fig. 20.14**).[47, 54] In contrast, those acclimatized to high altitude as adults exhibited lower values of maximum oxygen consumption.[54] These findings suggest that acclimatization to high altitude during development (that is, since birth or during growth) is an important component for the attainment of normal maximal aerobic capacity that characterizes high-altitude natives.[54] Recent studies conducted among Peruvian subjects born and raised at sea level and born and raised at high altitudes demonstrate that the attainment of normal VO_2max during exercise is strongly influenced by birthplace, emphasizing the importance of developmental adaptation to high altitude.[55]

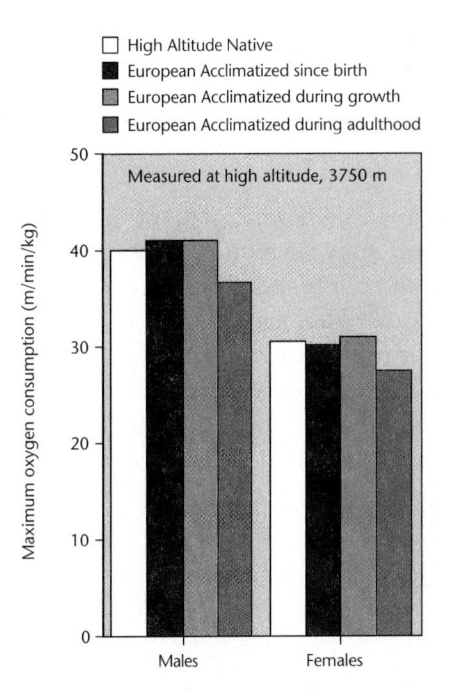

FIGURE 20.14. Comparison of maximal oxygen consumption among high-altitude natives, sea-level subjects of foreign ancestry acclimatized to high altitude since birth, during growth and during adulthood measured in La Paz, Bolivia, situated at mean altitude of 3750 m (12,375 ft). Source: Frisancho, A. R., Frisancho, H. G., Milotich, M., Albalak, R., Brutsaert, T., Spielvogel, H., Villena, M., and Vargas, E. 1995. Developmental, genetic and environmental components of aerobic capacity at high altitude. *Amer. J. Phys. Anthrop.* 96:431–442.

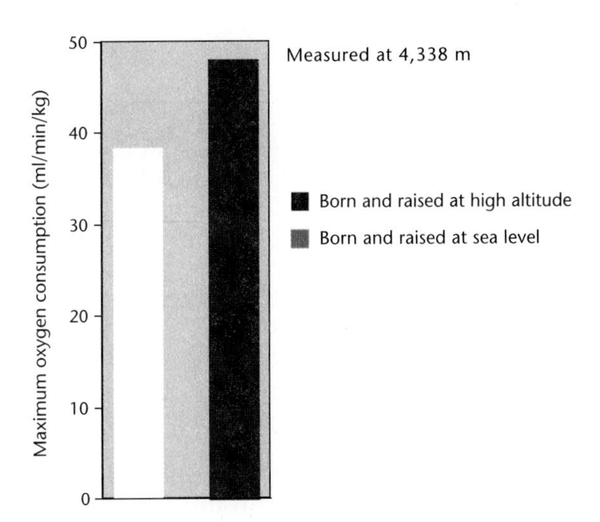

FIGURE 20.13. Comparison of maximal oxygen consumption of Peruvian sea-level male subjects who were born and raised at sea level and at high altitude and measured in Peru at 4338 m. Source: Brutsaert ,T. D., Parra, E., Shriver, M., Gamboa, A., Palacios, J. A., Rivera, M., Rodriguez, I., and Leon-Velarde, F. 2004. Effects of birthplace and individual genetic admixture on lung volume and exercise phenotypes of Peruvian Quechua. *Am. J. Phys. Anthropol.* 123:390–398.

Summary of Similarities and Differences of Adaptive Strategies between Tibetans and Andeans

Irrespective of the level of physical fitness, all sea-level sojourners to high altitudes experience a reduction in maximal aerobic capacity, which is not recuperated by increased length of residence. As schematized in **figure 20.15**, it is evident that the Andean and Tibetan high-altitude natives show distinctly different physiological adaptive strategies to high-altitude hypoxia:

1. Andean natives compensate for the low availability of oxygen with high concentrations of hemoglobin in their blood, large residual lung volume, and increased diffusing of oxygen into the cell that together restores the organism's functional adaptation. As such, they do not need to maintain a high pulmonary ventilation under both resting conditions or when challenged with additional hypoxic stress, referred to as blunted ventilatory response.
2. Tibetans, even though they developed large residual lung volume (like the Andeans), cope with the low levels of atmospheric oxygen with high pulmonary ventilation under both resting conditions and when challenged with additional hypoxic stress referred to as hypoxic ventilatory response. This response is necessary because the Tibetans do not increase the hemoglobin concentration, which would have increased the oxygen-carrying capacity of blood.

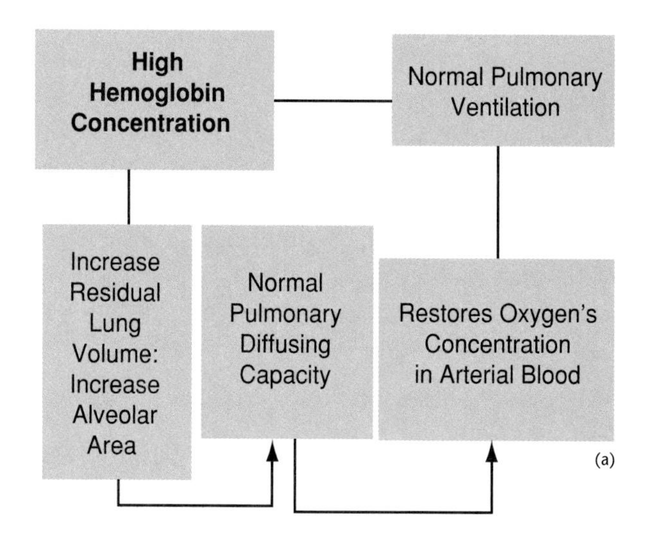

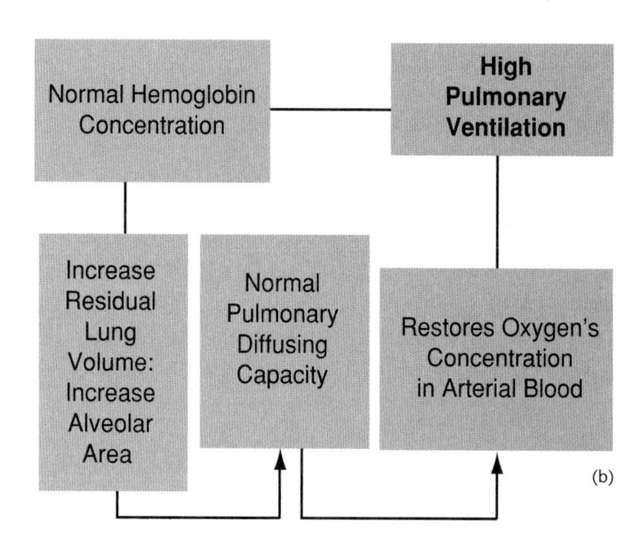

20.15. Andean (a) and Tibetan (b) natives show distinctly different physiological adaptations to high-altitude hypoxia. Andeans compensate for the low availability of oxygen with high concentrations of hemoglobin in their blood and a large residual lung volume that together restore the organism's functional adaptation. In contrast, Tibetans cope with low levels of atmospheric oxygen by maintaining high rates of pulmonary ventilation.

Arterial Oxygen Saturation and Genetic Adaptation

As noted before, the percent of arterial hemoglobin that is saturated with oxygen (SaO_2) is a measure of the capability of the cardiovascular and respiratory systems to deliver oxygen for diffusion into the bloodstream. Therefore, under hypoxic conditions, a high oxygen saturation of hemoglobin implies less physiological stress than a low saturation. Comparative studies indicate that Tibetan natives are more hypoxemic than the Andean natives. For example, among Tibetans living at an altitude of 3,900–4,000 m, the percent oxygen saturation of arterial hemoglobin averages 89.4 percent, with a range of 84–97 percent.[39] On the other hand, the SaO_2 of the Andean natives living at the same altitude was about 2.6 percent, averaging 92.0 percent, with a range of 84–99 percent .[39, 54, 92] In other words, the percent of arterial hemoglobin that is saturated with oxygen in Tibetans departs further from sea level than Andean means. Furthermore, the variance in the percent of oxygen saturation of arterial hemoglobin is greater among the Tibetan than the Andean natives. Quantitative heritability analyses indicate that among the Tibetans, between 21 percent and 39 percent of the phenotypic variance in the percent arterial hemoglobin saturation is accounted by heritable factors.[93, 94] In contrast, among Andean natives no significant variance in percent arterial hemoglobin saturation attributable to heritable factors were found.[92] This finding is expected in view of the high variance in the percent arterial hemoglobin saturation of Tibetans, since heritability to a certain extent is influenced by high variance. Furthermore, Beall and associates[95] report that the infant mortality (infant deaths/1000 live births) of children born to women with high oxygen saturation genotype was five times lower (0.48 deaths) than their counterparts who were born to mothers with low oxygen saturation (2.53 deaths). Beall et al.[39, 40] concludes that "high-altitude hypoxia is acting as an agent of natural selection on the heritable trait of oxygen saturation of hemoglobin." These findings together suggest that the Tibetans are still in the process of adapting to high-altitude hypoxia while the Andean natives have already attained a genetic adaptation. This difference in time of genetic adaptation occurs despite that fact that both Andean and Tibetan high-altitude natives migrated to high altitudes at a comparable time. The extent to which this hypothesis is validated remains to be determined by future research.

Prenatal Growth and Development at High Altitude

Birth Weight

Birth weight is the final common pathway for the expression of pregnancy and prenatal growth. Therefore, evaluation of birth weight provides retrospective information about the maturity and nutritional status of the newborn. The importance of birth weight is based upon two facts. First, birth weight is an extremely powerful predictor of an individual baby's survival: the lower the weight, the higher a baby's risk of infant mortality, and, at the population level, the lower the mean birth weight, the higher infant mortality. Second, birth weight is associated with health outcomes later in life.

Sea-Level Natives

In all populations, birth weight declines with increasing altitude, averaging a 100 g reduction per 1000 m altitude increase or nearly a quarter of a pound per 3000 ft. [57–59] The reduction is reflected in the entire distribution of birth weights, so that the proportion of low-birth-weight infants (<2500 g) is nearly twice those at low altitudes.[58] As inferred from comparisons of birth weights of premature newborns, the fetal growth retardation is due to a slowing growth in the third trimester.[60, 61]

Andean and Tibetan Natives

Meta-analysis showed that the altitude-associated decline in birth weight varied, depending on the population ancestry and length of residence at the high altitude:[62] Tibetans and Andeans who have lived at high altitudes for more years show less reduction in birth weight than sea-level European or Chinese (Han) populations who have resided at high altitudes for fewer years. As shown in **figure 20.16**, sea-level Chinese (Han) and Europeans residing at high altitude show the highest altitude-associated birth weight reduction. In the United States, infant birth weight starting at 2000 m or 6,600 ft decreases with increasing altitude, averaging about 100 g per 1000 m of increase in altitude.[57, 58] Among Tibetans the birth weight is less reduced than those of European or Han (Chinese) populations residing at high altitude. For example, babies born at 2700–3000 m to Tibetan mothers weighed 310 g more than babies born to Han mothers residing in the Tibet Autonomous Region of southwestern China. Similarly, infants born at 3000–3800 m to Tibetan mothers weighed 530 g more than babies born to Han mothers residing in the Tibet Autonomous Region of southwestern China. The reduction in birth weight for Tibetans living at 2700–4800 m averages 15 g per 1000 m altitude increase. In contrast, for Han (Chinese) residing at 2700–3800 m, the birth weight is

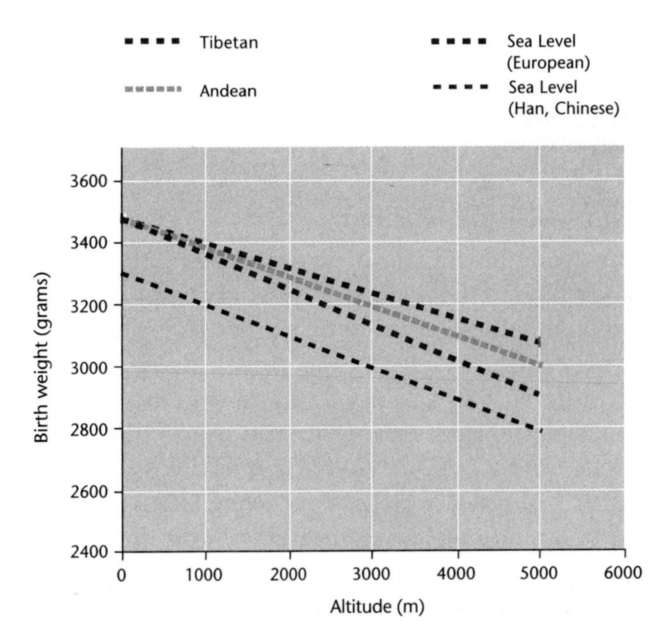

FIGURE 20.16. Relationship of altitude and birth weight among sea-level Chinese (Han), sea-level Europeans residing at high altitude, and Andean and Tibetan high-altitude natives. Best fit regression lines are based on average values for 4 million births occurring in the populations and altitudes represented. Source: Moore, L. G., Shriver, M., Bemis, L., Hickler, B., Wilson, M., Brutsaert, T., Parra, E., and Vargas, E. 2004. Maternal adaptation to high-altitude pregnancy: an experiment of nature—a review. *Placenta Suppl*:S60–71.

reduced by about 45 g/1000 m of altitude gain.[63] In other words, indigenous Andean and Tibetan high-altitude native women give birth to heavier babies than do sea-level women residing at high altitude whether European or Chinese.[64, 65]

Placental Function

Sea-Level Natives Residing at High Altitude

Extensive studies indicate that the reduction of birth weight among infants born to sea-level European or Chinese women residing at high altitudes is due to a reduction of the utero-placental circulation and blood flow in the main uterine artery (UA).[66–69] Under normoxic or sea-level conditions, pregnancy is associated with an increase in ventilation, and a decrease in hemoglobin concentration, due to expansion of plasma fluid. As a consequence, the arterial O_2 is reduced, and to compensate for this reduction, the blood flow is increased, which, in turn, increases the O_2 delivery in the utero-placental circulation. Concomitant with the increase in utero-placental circulation, the cardiac output rises. In contrast, under conditions of chronic high-altitude hypoxia, the utero-placental circulation and blood flow in the main uterine artery (UA) is reduced and the expansion of blood volume that, at low altitude, usually occurs in the second trimester, occurs in the last trimester at high altitude.[68]

High-Altitude Natives

The reduction in utero-placental circulation and blood flow in the main uterine artery (UA) is much less in Andean populations than in sea-level natives residing at high altitude.[66] According to Moore et al.,[66] the difference in UA blood flow between European and Andean high-altitude natives is related to the differences in the action of hypoxia-inducible factor one (HIF-1)-targeted or -regulatory genes. HIF-1 initiates the transcription of genes that underlie the increase in breathing and activates the transcription of the erythropoietin gene, which underlies the increase in hemoglobin concentration.[70, 71] As shown in **figure 20.17**, during pregnancy, Andeans have a lower concentration of the vasoconstrictor ET-1 than women of European ancestry residing at the same altitude.[66] Based on this evidence, Moore and associates propose that the long-term residents of high altitudes, such as Andean and Tibetan natives, may be protected from the adverse effects of chronic hypoxia via the actions of HIF-targeted genes.[66]

Summary of Prenatal Growth at High Altitude

Despite differences in sampling and methods of study, the general consensus is that the fetus at high altitude is subjected to the double insult of hypoxia due to lowered maternal arterial PO_2, decreased uterine blood flow, and reduced placental nutrient transporter densities.[69] As a result of this additive and synergistic effect, birth weight at high altitude is reduced by an average of 200 grams when compared to sea-level values. However, not all infants born to residents of high altitudes are small. The birth weight of Andean and Tibetan high-altitude natives is less reduced than those sea-level natives acclimatized and residing at a high altitude. There is the possibility that such differences might reflect genetic adaptations that control capillarization and thinning of the villous membranes of the placenta, which increases oxygen diffusion capacity that underlie oxygen delivery at the prenatal level.

Postnatal Growth in Body Size at High Altitude

Body Size

More than thirty years ago, various investigators reported that the growth in height of Andean high-altitude natives was delayed when compared to low-altitude populations (**fig. 20.18a**).[72–75] Similar differences in growth were reported from Ethiopian[76] and Indian

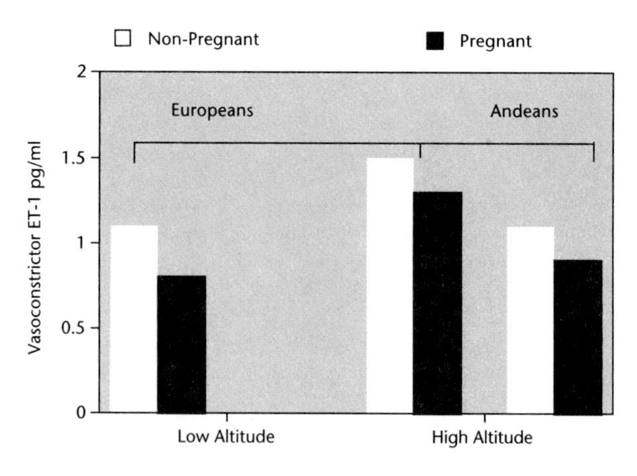

FIGURE 20.17. During pregnancy, Andeans have a lower concentration of the vasoconstrictor ET-1 than women of European ancestry residing at the same altitude (3600 m). Source: Moore, L. G., Shriver, M., Bemis, L., Hickler, B., Wilson, M., Brutsaert ,T., Parra, E., and Vargas, E. 2004. Maternal adaptation to high-altitude pregnancy: An experiment of nature—a review. *Placenta Suppl*:S60–71.

samples.[77] Frisancho and colleagues claimed that the observed delay in physical growth was related to high altitude.[72] However, new research of Peruvian and Bolivian high-altitude children indicates that nutrition may be a more significant cause of reduced growth at high altitude than hypoxia.[78, 79] Likewise, Pawson and associates[80] found that the highland Peruvian children that lived in a community with improved socioeconomic conditions were taller than their counterparts who lived in another village at the same altitude but with poorer socioeconomic conditions. Similarly, studies of Tibetan high-altitude children found significant improvement in physical growth among previously growth-delayed children.[81]

In summary, these findings together suggest that the growth delay of the high-altitude native is due to the joint effects of high-altitude hypoxia and the poor socioeconomic conditions in which the children developed.

Chest Dimensions and Lung Volumes

Several studies reported that Andean high-altitude children had, relative to stature, bigger chest dimensions than Tibetan and Sherpa children.[82–85] However, studies by Weitz and colleagues,[81] after measuring chest dimensions among the Tibetan male and female children to be at 3,200 m and 3,800 m, found only minor differences in thorax dimensions. However, Tibetan males at 4,300 m possess slightly narrower and deeper chests (during and after adolescence) than males at 3,200 m and 3,800m.[81] The origin of this difference has not been defined.

Among Andean children, chest dimensions are similar among highland and lowland Quechua, despite differences in stature,[82, 83, 86] indicating that genetic factors may regulate to thorax growth in this population.[82, 83, 86] In Chile and Bolivia, increased chest depth among individuals at higher elevations has been linked to Aymara ancestry.[83, 87, 88] The enlarged chest dimensions of high-altitude natives may also be related to differences in the degree of exposure to high-altitude hypoxia-related factors. Among Aymara highlanders, chest shape (depth/width) was positively correlated with lung volume (measured by forced vital capacity and forced expiratory volume).[83] These findings indicate that Andean and Tibetan high-altitude natives exhibit differences in chest dimensions and shapes. The functional significance and origins of these differences remain to be determined.

Forced Vital Capacity. A unique trait of high-altitude Andean natives is that adjusted for body size they have a larger lung volume than sea-level adults. They characteristicly appear to have developmental origin as shown by the fact that high-altitude boys from 8 years have a greater forced vital capacity than low-altitude children (**fig. 20.18b**).

Summary of Postnatal Growth in Body Size at High Altitude

There is strong evidence that a delayed growth in height and weight that characterize both Andean and Tibetan high-altitude natives is due to both the effects of high-altitude hypoxia and poor socioeconomic conditions in which the children developed. However, the large volume of Andean natives is probably explained as developmental adaptive response to high-altitude

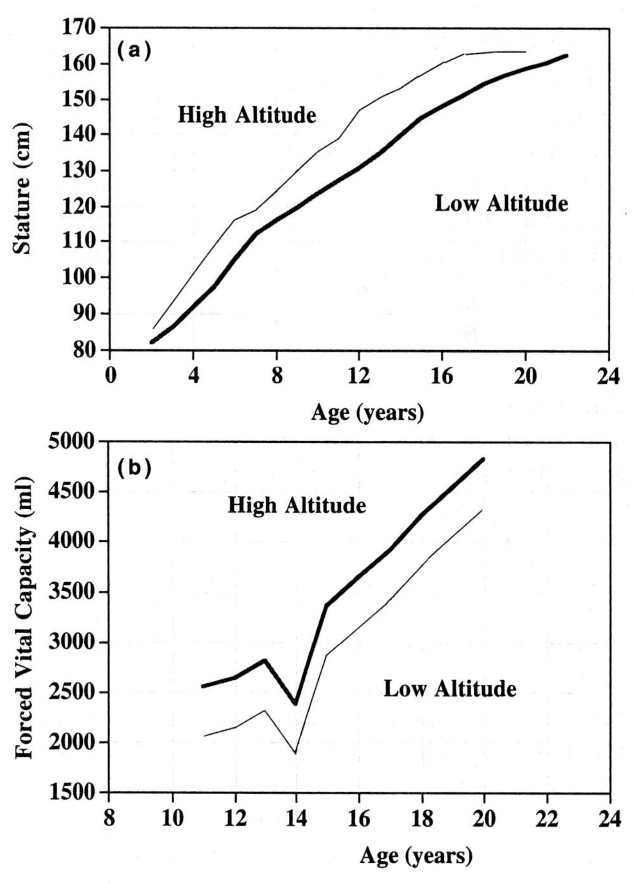

FIGURE 20.18. Distance curves for Growth in height (a) and forced vital capacity, (b) for high-altitude and low-altitude Peruvian native boys. (Based on data from reference 72. Frisancho, A. R., Baker, P. T. 1970. Altitude and growth: a study of the patterns of physical growth of a high altitude Peruvian Quechua population. *Am. J. Phys. Anthropol.* 32:279–292.)

hypoxia. On the other hand, the larger chest dimensions of Andean natives appears to be a population trait.

Problems of Acclimatization to High Altitude

Three clinical entities specific to altitude have been identified: acute mountain sickness, pulmonary edema, and chronic mountain sickness.

Acute Mountain Sickness

Acute mountain sickness (AMS) has been clearly associated with hypoxic stress and is characterized by headaches, malaise, dizziness, shortness of breath, sleep difficulties, and an upset stomach.[89] AMS occurs in lowlanders as well as in highlanders who are unable to restore functional homeostasis when exposed to hypoxia. Several factors precipitate the onset of acute mountain sickness, among which rapid ascent, dehydration, and excessive activity are the most important. Usually symptoms of acute mountain sickness disappear without treatment. Some individuals benefit from the use of carbonic anhydrase inhibitors, such as acetazolamide (*Dimox*), which has been used with some success[90] to decrease the frequency of occurrence of AMS.

Pulmonary Edema

Pulmonary edema (PE) is a severely debilitating disease and can be fatal if the patient is not treated rapidly.[89] The disease is associated with excessive pulmonary hypertension, impaired pulmonary oxygen exchange, and histological alterations of the pulmonary vessels. As a result, the lowlander or highlander with pulmonary edema suffers a greater loss of functional homeostasis than that individual with acute mountain sickness. Furthermore, there are multiple causes of the inability to restore functional homeostasis with pulmonary edema. For this reason treatment includes oxygen administration, corticosteroids, and transports to lower altitudes. Preventive measures include initial avoidance of excessive physical activity when ascending to altitude for the first time or when returning from sea level.

Chronic Mountain Sickness

Chronic mountain sickness (CMS), or Monge's disease,[91] is a severely debilitating disease that occurs mostly among Andean highland natives with prolonged residence at a high altitude and is associated with a loss of functional adaptation. CMS is interpreted as a loss of adaptation in a previously adapted individual, and is characterized by very low oxygen saturation, hypoventilation, very high hemoglobin concentration, and high pulmonary artery pressure.[90] Treatment includes transfer to a lower altitude and administration of corticosteroids.

Summary of Problems of Acclimatization to High Altitude

Andeans and Tibetans also differ in the prevalence of altitude-associated diseases. In both regions, acute mountain sickness and pulmonary edema occur, but chronic mountain sickness happens only among Andean populations.

Summary

High-altitude hypoxia is a pervasive, chronic, and unavoidable physiological stress caused by lowered barometric pressure that is not modifiable by cultural adaptation. As such, high-altitude hypoxia offers natural experimental settings for investigating biological adaptations both in a contemporary and evolutionary perspective. Sea-level natives sojourning to high altitude usually respond with increased in pulmonary ventilation shown by the increased breath rate, pulmonary vasoconstriction, and increased red blood cell count and hemoglobin concentration. These responses are adaptive in that they increase the oxygen availability at the level of the lung and increase the oxygen-carrying capacity of the blood. However, these responses are not sufficient to overcome the hypoxic stress, so those sea-level subjects residing at high altitude attain only 80 percent of the sea-level physiological work capacity.

Andean and Tibetan natives have attained a full functional adaptation that has overcome the hypoxic stress. At the physiological level, the Andean and Tibetan natives differ in their adaptive strategies. The Andean natives rely more on expansion of the lung volume and increasing the red blood cell count, hemoglobin concentration, and oxygen saturation of arterial hemoglobin, while Tibetans depend mostly on increasing the pulmonary ventilation and cardiac work shown by the increased pulse rate. These different adaptive strategies are rooted in genetic adaptation that in Tibetans are of recent origin while among Andeans are of ancient origin.

Prenatal growth as measured by birth weight is negatively influenced by high-altitude hypoxia. The effects of hypoxia on Andean and Tibetan natives are less than those experienced by sea-level natives acclimatized to high altitude. Postnatal growth in body size is not affected by high-altitude hypoxia and the observed growth delay that characterize high-altitude natives is related to the poor socioeconomic conditions of a high-altitude environment.

The prevalence of altitude-associated diseases such as acute mountain sickness (also known as mountain sickness) and pulmonary edema is similar in both the Andes and Tibet. However, chronic mountain sickness (also known as Monge's disease) has been reported only among Andean natives.

Suggested Readings

Frisancho, A. R. 1993. *Human Adaptation and Accommodation.* Ann Arbor: University of Michigan Press.

Hultgren, Herbert N. 1997. *High Altitude Medicine.* Stanford, CA.: Hultgren Publications.

Pollard, Andrew J. 2003. *The High-Altitude Medicine.* Abingdon, UK: Radcliffe Medical Press.

Literature Cited

1. Niermeyer, S., Zamudio, S., and Moore, L. G. 2001. The people. In *Adaptations to Hypoxia,* ed. T. Hornbein and R. B. Schoene. New York: Marcel Dekker and Co., p. 43–100.
2. Frisancho, A. R. 1993. *Human Adaptation and Accommodation.* Ann Arbor: University of Michigan Press.
3. Moore, L. G., Cymerman, A., Huang S, McCullough, R. E., McCullough, R. G., et al. 1987. Propranolol blocks metabolic rate increase but not ventilatory acclimatization to 4300 m. *Respir. Physiol.* 70:195–204.
4. Grover, R. F., Vogel, J. H. K., Averill, K. H., and S. G. Blount. 1963. Pulmonary hypertension. Individual and species variability relative to vascular reactivity. *Am. Heart. J.* 66:1–3.
5. Penaloza, D., Sime, F., Banchero, N., Gamboa, R., Cruz, J., and E. Marticorena. 1963. Pulmonary hypertension in healthy men born and living at high altitudes. *Am. J. Cardiol.* 11:150–157.
6. Sime, F., Penaloza, D., Ruiz, L., Gonzales, N., Covarrubias, E., and R. Postigo. 1974. Hypoxemia, pulmonary hypertension, and low cardiac output in newcomers at low altitude. *J. Appl. Physiol.* 36:561–565.
7. Yang, J. S., He, Z. Q., Zhai, H. Y., Yang, Z., Zhang, H. M., et al. 1987. A study of the changes in pulmonary arterial pressure in healthy people in the plains and at high altitude under exercise load. *Chin. J. Cardiol.* 15:39–41.
8. Sui, G. J., Liu, Y. H., Cheng, X. S., Anand, I. S., Harris, E., et al. 1988. Subacute infantile mountain sickness. *J. Pathol.* 155:161–170.
9. Gupta, M. L., Rao, K. S., Anand, I. S., Banerjee, A. K., Boparai, M. S. 1992. Lack of smooth muscle in the small pulmonary arteries of native Ladakhi: Is the Himalayan highlander adapted? *Am. J. Respir. Crit. Care Med.* 145:1201–1204.
10. Sime, F., Banchero, N, Penaloza, D., Gamboa, R., Cruz, J., and E. Marticorena. 1963. Pulmonary hypertension in children born and living at high altitudes. *Am. J. Cardiol.* 11:143–149.
11. Miao, C. Y., Zuberbuhler, J. S., and J. R. Zuberbuhler. 1988. Prevalence of congenital cardiac anomalies at high altitude. *JACC* 12:224–228.
12. Harris, P. 1986. Evolution, hypoxia and high altitude. In *Aspects of Hypoxia,* ed. D. Heath. Liverpool, UK: Liverpool Univ. Press, 207–216.
13. Reynafarje, C. 1959. Bone-marrow studies in the newborn infant at high altitudes. *J. Pediatr.* 54:152–167.
14. Merino, C. F. 1950. Blood formation and blood destruction in polycythemia of high altitude. *Blood* 5:1–31.
15. Forster, H. V., Dempsey, J. A., Birnbaum, M. L., Reddan, W. G., Thoden, J., Grover, R. F., and J. Rankin. 1971. Effect of chronic exposure to hypoxia on ventilatory response to CO_2 and hypoxia. *J. Appl. Physiol.* 31:586–592.
16. Pugh, L. G. C. E. 1964. Blood volume and hemoglobin concentration at altitudes above 18,000 ft. (5500 m.) *J. Physiol.* 170:344–354.
17. Tufts, D. A., Haas, J. D., Beard, J. L., and H. Spielvogel. 1985. Distribution of hemoglobin and functional consequences of anemia in adult males at high altitude. *Am. J. Clin. Nutr.* 42:1–11.
18. Moret, P., Covarrubias, E., Coudert, J., and Duchosal, F. 1972. Cardiocirculation adaptation to chronic hypoxia: III. Comparative study of cardiac output, pulmonary and systematic circulation between sea-level and high-altitude residents. *Extr. Acta. Cardiol.* 27:596–619.
19. Chang, K. C. 1992. China. In *Chronologies in Old World Archaeology,* 3rd ed., ed. R.W. Ehrich. Chicago/London: Univ. Chicago Press.
20. Aldenderfer, M. 1999. The Pleistocene/Holocene transition in Peru and its effects upon human use of the landscape. *Quat. Int.* 53/54:11–19
21. Beall, C. M., Brittenham, G. M., Strohl, K. P., Blangero, J., Williams-Blangero, S., et al. 1997. Ventilation and hypoxic ventilatory response of Tibetan and Aymara high-altitude natives. *Am. J. Phys. Anthropol.* 104:427–447.
22. Frisancho, A. R. 1999. Developmental components of resting ventilation among high- and low-altitude Andean children and adults. *Am. J. Phys. Anthropol.* 109:295–301.
23. Hurtado, A. 1964. Animals in high altitudes: resident man. In: *Handbook of Physiology, Vol. 4. Adaptation to the Environment,* ed. D. B. Dill, E. F. Adolph, and C. G. Wilber. Washington, DC: American Physiological Society.
24. Frisancho, A. R., Velásquez, T. and J. Sanchez 1973. Influences of developmental adaptation on lung function at high altitude. *Hum. Biol.* 45:583–594.
25. Brutsaert, T. D., Parra, E. Shriver, M., Gamboa, A., Palacios, J. A., Rivera, M., Rodriguez, I., and F. Leon-Velarde. 2004. Effects of birthplace and individual genetic admixture on lung volume and exercise phenotypes of Peruvian Quechua. *Am. J. Phys. Anthropol.* 123:390–398.

26. Moore, L. G., Niermeyer, S., and S. Zamudio. 1998. Human adaptation to high altitude: regional and life cycle perspectives. *Yearb. Phys. Anthropol.* 41:25–64.

27. Droma, T., McCullough, R. S., McCullough, R. E., Zhuan, J., Cymerman, A., et al. 1991. Increased vital and total lung capacities in Tibetan compared to Han residents of Lhasa (3,658 m). *Am. J. Phys. Anthropol.* 86:341–351.

28. Frisancho, A. R., Frisancho, H. G., Milotich, M., Albalak, R., Spielvogel, H., Villena, M., Vargas, E., and R. Soria. 1997. Developmental, genetic and environmental components of lung volumes at high altitude. *Amer. J. Hum. Biol.* 9:191–204.

29. Brody, J. S., Lahiri, S., Simpser, M., Motoyama, E. K., and T. Velásquez. 1977. Lung elasticity and airway dynamics in Peruvian natives to high altitude. *J. Appl. Physiol.* 42:245–251.

30. Bartlett, D. 1972. Postnatal development of the mammalian lung. In *Regulation of Organ and Tissue Growth,* ed. R. Goss. New York: Academic Press, Inc.

31. Burri, P. H., and E. R. Weibel. 1971. Morphometric evaluation of changes in lung structure due to high altitudes. In *High Altitude Physiology. Cardiac and Respiratory Aspects,* ed. R. Porter and J. Knight. Edinburgh: Churchill Livingstone.

32. Cunningham, E. L., Brody, J. S., and B. P. Jain. 1974. Lung growth induced by hypoxia. *J. Appl. Physiol.* 37:362–366.

33. Bartlett, D. Jr, and Remmers, J. E. 1971. Effects of high-altitude exposure on the lungs of young rats. *Respir. Physiol.* 13:116–125.

34. Tufts, D. A., Haas, J. D., Beard, J. L., and H. Spielvogel. 1985. Distribution of hemoglobin and functional consequences of anemia in adult males at high altitude. *Am. J. Clin. Nutr.* 42:1–11.

35. Garruto, R. M. 1976. Hematology. In *Man in the Andes. A Multidisciplinary Study of High-Altitude Quechua,* ed. P. T. Baker and M. A. Little. Stroudsburg, Pa: Dowden, Hutchinson & Ross, 261–282.

36. Garruto, R. M., and J. S. Dutt. 1983. Lack of prominent compensatory polycythemia in traditional native Andeans living at 4,200 meters. *Am. J. Phys. Anthro.* 61:355–366.

37. Frisancho, A. R. 1988. Origins of differences in hemoglobin concentration between Himalayan and Andean populations. *Resp. Physiol.* 72:13–18.

38. Beall, C. M., Reichsman, A. B. 1984. Hemoglobin levels in a Himalayan high-altitude population. *Am. J. Phys. Anthropol.* 63:301–306.

39. Beall, C. M., and M. C. Goldstein. 1990. Hemoglobin concentration, percent oxygen saturation and arterial oxygen content of Tibetan nomads at 4,850 to 5,450 m. In *Hypoxia: The Adaptations,* ed. J. R. Sutton, G. Coates, and J. E. Remmers. Philadelphia: B. C. Decker Inc.

40. Beall, C. M., Brittenham, G. M., Macuaga, F., Barragan, M. 1990. Variation in hemoglobin concentration among samples of high-altitude natives in the Andes and the Himalayas. *Am. J. Hum. Biol.* 2:639–651.

41. Grover, R. F., Reeves, J. T., Grover, E. B. and J. S. Leathers 1967. Muscular exercise in young men native to 3,100 m altitude. *J. Appl. Physiol.* 22:555–564.

42. Buskirk, E. R. 1976. Work performance of newcomers to the Peruvian highlands. In *Man in the Andes: A Multidisciplinary Study of High-Altitude Quechua Natives,* eds. P. T. Barker and M. A. Little. Stroudsburg, PA: Dowden, Hutchinson, & Ross, Inc.

43. Kollias, J., Buskirk, E. R. Akers, R. F., Prokop, E. K., Baker, P. T. and E. Piconreategui. 1968. Work capacity of long-time residents and newcomers to altitude. *J. Appl. Physiol.* 24:792–799.

44. Mazess, R. B. 1969. Exercise performance of indian and white high altitude residents. *Hum. Biol.* 41:494–518.

45. Lahiri, S., Milledge, J. S., Chattopadhyay, H. P., Bhattacharyya, A. K. and A. K. Sinha. 1967. Respiration and heart rate of Sherpa highlanders during exercise. *J. Appl. Physiol.* 23:545–554.

46. Mazess, R. B. 1969. Exercise performance at high altitude (4000 meters) in Peru. *Fed. Proc.* 28:1301–1306.

47. Frisancho, A. R., Martinez, C., Velásquez, T., Sanchez, J. and H. Montoye. 1973. Influence of developmental adaptation on aerobic capacity at high altitude. *J. Appl. Physiol.* 34:176–180.

48. Velásquez, T. 1966. Acquired acclimatization to sea level. In *Life at High Altitudes.* Pub. 140. Washington, DC: Pan American Health Organization

49. Velásquez, T. 1970. Aspects of physical activity in high altitude natives. *Am. J. Phys. Anthropol.* 32:251–258.

50. Velásquez, T., and B. Reynafarje. 1966. Metabolic and physiological aspects of exercise at high altitude. Part 2. Response of natives to different levels of workload breathing air and various oxygen mixtures. *Fed. Proc.* 25:1400–1402.

51. Elsner, R. W., Bolstad, A. and C. Forno. 1964. Maximum oxygen consumption of Peruvian Indians native to high altitude. In *The Physiological Effects of High Altitude,* ed. W. H. Weihe. Oxford: Pergamon Press Ltd.

52. Baker, P. T. 1976. Work performance of highland natives. In *Man in the Andes: A Multidisciplinary Study of High-Altitude Quechua Natives,* eds. P. T. Baker and M. A. Little. Stroudsburg, PA: Dowden, Hutchinson & Ross, Inc.

53. Way, A. B. 1976. Exercise capacity of high-altitude Peruvian Quechua Indians migrant to low altitude. *Hum. Biol.* 48:175–191.

54. Frisancho, A. R., Frisancho, H. G., Milotich, M., Brutsaert, T., Albalak R., Brutsaert, T., Spielvogel, H., Villena, M., and E. Vargas. 1995. Developmental, genetic, and environmental components of aerobic capacity at high altitude. *Am. J. Phys. Anthropol.* 96:431–442.

55. Brutsaert, T. D., Parra, E., Shriver, M., Gamboa, A., Palacios, J. A., Rivera, M., Rodriguez, I., and F. Leon-Velarde. 2004. Effects of birthplace and individual genetic admixture on lung volume and exercise phenotypes of Peruvian Quechua. *Am. J. Phys. Anthropol.* 123:390–398.

56. Sun, S. F., Droma, T. S., Zhuang, J. G., Tao, J. X., Huang, S. Y., McCullough, R. G., McCullough, R. E. Reeves, C. S. Reeves, J. T. and L. G. Moore. 1990. Greater maximal O_2 uptakes and vital capacities in Tibetan than Han residents of Lhasa. *Respir. Physiol.* 79:151–162.

57. Yip, R. 1987. Altitude and birth weight. *J. Pediatr.* 111:869–876.

58. Jensen, G. M., and L. G. Moore. 1997. The effect of high altitude and other risk factors on birthweight: independent or interactive effects? *Am. J. Public Health* 87:1003–1007.

59. Giussani, D. A., Seamus, P., Anstee, S., and D. J. P. Barker. 2001. Effects of altitude versus economic status on birth weight and body shape at birth. *Pediatr. Res.* 49:490–494.

60. Unger, C., Weiser, J. K., McCullough, R. E., Keefer, S., Moore, L. G. 1988. Altitude, low birth weight, and infant mortality in Colorado. *J. Am. Med. Assoc.* 259:3427–3432.

61. Krampl, E., C. Lees, J. M. Bland, J. E. Dorado, M. Gonzalo, and S. Campbell. 2000. Fetal biometry at 4300 m compared to sea level in Peru. *Ultrasound Obstet. Gynecol.* 16:9–18.

62. Moore, L. G. 2001a. Human genetic adaptation to high altitude. *High Alt. Med. Biol.* 2:257–279.

63. Moore, L. G., Young, D. Y., McCullough, R. E., Droma, T. S., and S. Zamudio. 2001a. Tibetan protection from intrauterine growth restriction (IUGR) and reproductive loss at high altitude. *Am. J. Hum. Biol.* 13:635–644.

64. Zamudio, S., Droma, T., Norkyel, K. Y., Acharya, G., Zamudio, J. A., et al. 1993. Protection from intrauterine growth retardation in Tibetans at high altitude. *Am. J. Hum. Biol.* 91:215–224.

65. Haas, J. D. 1980. Maternal adaptation and fetal growth at high altitude in Bolivia. In *Social and Biological Predictors of Nutritional Status, Physical Growth, and Neurological Development*, ed. L. S. Greene, and F. E. Johnston. New York/London: Academic, 257–290.

66. Moore, L. G., Shriver, M., Bemis, L., Hickler, B., Wilson, M., Brutsaert, T., Parra, E., and E. Vargas. 2004. Maternal adaptation to high-altitude pregnancy: An experiment of nature-a review. *Placenta Suppl.* S60–71.

67. Zamudio, S., Palmer, S. K., Droma, T., Stamm, E., Coffin, C., et al. 1995. Effect of altitude on uterine artery blood flow during normal pregnancy. *J. Appl .Physiol.* 79:7–14.

68. Zamudio, S., Palmer, S. K., Dahms, T. E., Berman, J. C., McCullough, R. G., R. E. McCullough, et al. 1993. Blood volume expansion, preeclampsia, and infant birth weight at high altitude. *J. Appl. Physiol.* 75:1566–1573.

69. Zamudio, S. 2003. The placenta at high altitude. *High Alt. Med. Biol.* 4:171–191.

70. Semenza, G. L. 2000. HIF-I and human disease: one highly involved factor. *Genes Dev.* 14:1983–1991.

71. Guillemin, K., and M. A. Krasnow. 1997. The hypoxic response: huffing and HIFing. *Cell* 89:9–12, Guleria, J. S., Pande, J. N., Sethi, P. K., Roy, S. B. 1971. Pulmonary diffusing capacity at high altitude. *J. Appl. Physiol.* 31:536–543.

72. Frisancho, A. R., and P. T. Baker. 1970. Altitude and growth: A study of the patterns of physical growth of a high altitude Peruvian Quechua population. *Am. J. Phys. Anthropol.* 32:279–292.

73. Hoff, C., and W. R. Blackburn. 1981. A multivariate discriminant method for comparing growth in a study group with age-matched means and standard deviations from a control group. *Hum. Biol.* 53:513–520.

74. Greksa, L. P., Spielvogel, H., Paredes-Fernandez, L., Paz-Zamora, M., and E. Caceres. 1984. The physical growth of urban children at high altitude. *Am. J. Phys. Anthropol.* 65:315–322.

75. Mueller, W. H., Murillo, F., Palamino, H., Badzioch, M., Chakraborty, R., Fuerst, P., et al. 1980. The Aymara of western Bolivia: V. growth and development in an hypoxic environment. *Hum. Biol.* 52:529–546.

76. Clegg, E. J., Pawson, I. G., Ashton, E. H., and R. M. Flinn. 1972. The growth of children at different altitudes in Ethiopia. *Phil. Trans. R. Soc. Lond. Series B* 264:403–437.

77. Malik, S. L., I. P. Singh. 1978. Growth trends among male Bods of Ladakh—a high altitude population. *Am. J. Phys. Anthropol.* 48:171–176.

78. Leonard, W. R. 1989. Nutritional determinants of high altitude growth in Nuñoa, Perú. *Am. J. Phys. Anthropol.* 80:341–352.

79. Obert, P., Fellman, N., Falgairette, G., Bedu, M., Van Praagh, E., Kemper, H., Post, B., Spielvogel, H., Tellez, V., and A. Qintela. 1994. The importance of socioeconomic and nutritional conditions rather than altitude on the physical growth of prepubertal Andean highland boys. *Ann. Hum. Biol.* 21:145–154.

80. Pawson, I. G., Huicho, L., Muro, M., and A. Pacheco. 2001. Growth of children in two economically diverse Peruvian high-altitude communities. *Am. J. Hum. Biol.* 13:323–340.

81. Weitz, C. A., Garruto, R. M., Chin, C. T., Liu, J. C., Liu, R. L., and X. He. 2000. Growth of Qinghai Tibetans living at three different high altitudes. *Am. J. Phys. Anthropol.* 111:69–88.

82. Beall, C. M. 1982. A comparison of chest morphology in high altitude Asian and Andean populations. *Hum. Biol.* 54:145–163.

83. Greksa, L. P., and C. M. Beall. 1989. Development of chest size and lung function at high altitude. In *Human Population Biology: A Transdisciplinary Science,* ed. M. A. Little and J. D. Haas, New York: Oxford University Press, 222–238.

84. Pawson, I. G. 1977. The effects of high altitudes on child growth and development. *Int. J. Biometeor.* 21:171–178.

85. Malik, S. L., and I. P. Singh. 1978. Growth trends among male Bods of Ladakh—a high altitude population. *Am. J. Phys. Anthropol.* 48:171–176.

86. Hoff, C. 1974. Altitudinal variations in the physical growth and development of Peruvian Quechua. *Homo.* 24:87–99.

87. Palomino, H., Mueller, W. H., and W. J. Schull. 1979. Altitude, heredity and body proportions in northern Chile. *Am. J. Phys. Anthropol.* 50:39–50.

88. Mueller, W. H., Schull, V. N., Schull, W. J., Soto, P., and R. Rothhammer. 1978. A multinational Andean genetic and health program: Growth and development in an hypoxic environment. *Ann. Hum. Biol.* 5:329–352.

89. Rodway, G. W., Hoffman, L. A., and M. H. Sanders. 2003. High-altitude-related disorders—Part I: Pathophysiology, differential diagnosis, and treatment. *Heart Lung* 32:353–359.

90. Beeckman, D., E. R. Buskirk, 1988. Drug use at high terrestrial altitudes and in cold climates: a brief review. *Hum. Biol.* 60: 663–677.

91. Winslow, R. M., and C. Monge. 1987. *Hypoxia, Polycythemia, and Chronic Mountain Sickness.* Baltimore, MD: Johns Hopkins Univ. Press.

92. Beall, C. M., Almasy, L. A., Blangero, J., Williams-Blangero, S., Brittenham, G. M., Strohl, K. P., Decker, M. J., Vargas, E., Villena, M., Soria, R., Alarcon, A. M., and C. Gonzales. 1999. Percent of oxygen saturation of arterial hemoglobin among Bolivian Aymara at 3,900–4,000 m. *Am. J. Phys. Anthropol.* 108: 41–51.

93. Beall, C. M., Blangero, J., Williams-Blangero, S. and M. C. Goldstein. 1994. A major gene for percent of oxygen saturation of arterial hemoglobin in Tibetan highlanders. *Am. J. Phys. Anthropol.* 95:271–276.
94. Beall, C. M., Strohl, K. P., Blangero, J., Williams-Blangero, J., Brittenham, G. M., and M. C. Goldstein. 1997. Quantitative genetic analysis of arterial oxygen saturation in Tibetan highlanders. *Hum. Biol.* 69:597–604.
95. Beall, C. M., Song, K., Elston, R. C., and M. C. Goldstein. 2004. Higher offspring survival among Tibetan women with high oxygen saturation genotypes residing at 4,000 m. *PNAS* 102:1–5.

Biocultural Origins of Lactose Intolerance and the Evolution of Skin Color

Milk, besides protein and other nutrients, contains lactose and calcium, which are necessary for the growth and development of the skeletal system. Lactose is a sugar that is synthesized exclusively in the mammary gland of placental mammals. Therefore, all mammals, with the exception of the sea lion, have the enzyme lactase to digest the lactose present in milk. Yet, in humans there are individual and population differences in the ability to digest milk. Since the dietary environment is composed of both cultural and natural factors, the utilization of milk is similarly composed of biological and cultural adaptations. On the other hand, there is evidence that the absorption of calcium is dependent on the availability of vitamin D, which is synthesized in the body under the influence of solar radiation. On the other hand, the synthesis of vitamin D is influenced by both the intensity of solar radiation and the amount of melanin present in the skin, which varies between individuals and populations. Therefore, in this chapter, we will examine the biocultural adaptations for the utilization of lactose and its interaction with the evolution of skin color of humans.

Biology of Lactase Activity

Lactose in Milk

Milk contains proteins, lipids, fat, vitamins, minerals, trace elements, and carbohydrates, such as lactose. The lactose content of human milk is about 7 percent. The lactose content of milk in other species varies considerably, and among those animals used in dairying, lactose concentration ranges from 2.5 percent (reindeer milk) to 6.1 percent (cow's milk). In all cases, milk consumption and lactose digestion are necessary to the development and survival of mammalian young.

Lactose is a disaccharide composed of two monosaccharides, such as glucose and galactose, which are bonded by enzymes into a single larger molecule. In order to be absorbed, the disaccharide lactose must be broken down into its component monosaccharides: glucose and galactose. This separation takes place through the action of enzymes referred to as **lactases**. These enzymes are synthesized within the mucosal cells that line the many small absorptive protrusions of the duodenal wall of the small intestine.[1, 2] There, lactases cause the enzymatic hydrolysis of lactose back into its constituent monosaccharides, glucose and galactose. These are then readily absorbed across the cell membranes of the intestinal wall and may be transported on to the bloodstream.

Variability in Lactase Activity

In humans, lactase activity starts in the fetus at the beginning of the second trimester and is at its highest level shortly after birth.[1, 2] However, during childhood the production of the enzyme normally begins a dramatic decline, resulting in low levels of lactase activity. Although this pattern appears to hold true for all, there is considerable variation in the timing of the decline in lactase activity. A low level or absence of intestinal lactase activity will result in an inability to hydrolyze or digest lactose in the small intestine, which causes a number of physiological consequences. First, the unabsorbed lactose produces osmotic pressure in the small intestine, which results in an inflow of water and electrolytes and the dilation of the small intestine. Second, the unabsorbed carbohydrates (including lactose), as result of bacterial metabolism in the colon, produce short-chain fatty acids from which H_2 and CO_2 are formed.[1, 2] This then increases the production of gases in the colon. The joint effects of these actions cause flatulence, intestinal cramps, "colicky abdominal pain," diarrhea, nausea, and even vomiting. In general, differences in lactose digestibility occur in four types: (a) congenital or infantile lactase deficiency, (b) secondary lactase deficiency, (c) primary or adult lactase inactivity, and (d) persistent lactase activity.

Congenital Lactase Deficiency

Congenital lactase deficiency in humans is an extremely rare disorder in which neonates have negligible or no brush border lactase activity at all; it is inherited as an autosomal recessive gene.[3] If undiagnosed and untreated, the infant's response to lactose in the diet will result in severe diarrhea, dehydration, acidosis, and death. At present in most industrialized countries, treatment consists of either removing lactose from the diet or by adding three drops of lactase to each 200 ml bottle of human or cow's milk.

Secondary Lactase Deficiency

Secondary lactase deficiency occurs as a result of damage to the small intestinal mucosa, which affects the brush border disaccharides. This damage can be the result of celiac disease, tropical and nontropical sprue, enteritis, giardiasis, and other diseases affecting the gastrointestinal tract.[4] In addition, certain medications can affect the production of brush border disaccharidases. These include colchicine (for gout) and such antibiotics as neomycin and kanamycin.[4] Secondary lactase deficiency can be a temporary disorder, as in the case of successfully treated cases of celiac disease.

Primary Lactase Deficiency

Primary lactase deficiency occurs in the majority of the world's adult population. It is characterized by a **normal** level of lactase activity during infancy, which declines right after weaning. The age of onset varies in decline of lactase activity and varies between populations from around 3–4 years among African Americans to 2–4 years among Thais, and 7–13 years among Greek children.[4] Current investigations indicate that primary lactase deficiency during adulthood is the normal pattern. Individuals with primary lactose deficiency are "lactose malabsorbers," and are said to have the phenotype of lactose malabsorption (LM). Some lactose malabsorbers also suffer from "lactose intolerance" (LI), which, as described before, is characterized by a generalized indigestion associated with flatulence, intestinal cramps, diarrhea, nausea, and even vomiting.

Persistent Lactase Activity

Persistent lactase activity (PLA) refers to those individuals who retain lactase activity throughout life. Current investigations indicate that persistent lactase activity during adulthood occurs in the minority of the world populations. Individuals with persistent lactase activity are referred to as "lactose absorbers" (LA). In such individuals, lactase activity continues at physiologically active concentrations straight through into adulthood. A lactose absorber can consume milk and dairy products without any discomfort.

In the current literature,[1–4] individuals with persistent lactase activity, or lactose absorbers, are said to have a high lactose digestion capacity (high LDC) while lactose malabsorbers have a low lactose digestion capacity (low LDC). This chapter will use this nomenclature interchangeably, namely low LDC for lactose malabsorbers and high LDC for persistent lactase activity or lactose absorbers.

Genetic Inheritance of Lactase Activity

The persistence of lactase activity during adulthood is a genetic trait that is inherited in a Mendelian form.[1–3, 6, 7, 20, 21] The persistence of lactase activity is controlled by a dominant (D) allele, and decreased lactase activity is controlled by a recessive allele (d). Hence, homozygous dominant (D) and heterozygous individuals (Dd) express high lactose digestion, while homozygous recessive individuals are characterized by low lactase activity.

Evolution of the Gene for Lactase Persistence

At present, there is consensus that low lactose digestion capacity (low LDC) during adulthood is the normal trait, yet some individuals in some populations retain the capacity to digest lactose or have persistent lactase activity (PLA). For example, more than 80 percent of adults of Northern European ancestry have high lactose activity, while only 25 percent of African Black, Asian, and Native American ancestry have high lactose activity (**table 21.1**).[8] The gene for the lactase persistence first appeared in Afro-Arabia with the origin of herding and dairying, between 10,000 and 6000 B.C.[12–14] Then, between 4000 and 2000 B.C., the lactase persistence gene was brought into Northern Europe, along with the introduction of farming.[12–14] It has been estimated that the gene for lactase persistence gene frequency of 0.70 percent observed today in northern European populations evolved between 6000 and 9000 years ago.[23–25] This time scale agrees with the time of first animal domestication in Africa and the Middle East.[26, 27] The hypothesis that the gene for lactase persistence first appeared in Afro-Arabia is supported by the fact about 91 percent of the milk-dependent pastoralists from Africa and milk-dependent populations of northern Europe are lactose absorbers (**table 21.2**).

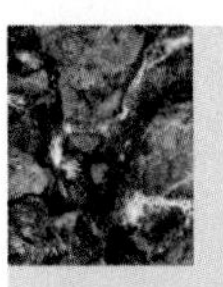

TABLE 21.1

Distribution of Lactose Phenotypes in Selected Populations in the United States*

Population	Low Lactase Activity (%)	High Lactase Activity (%)
Northern European	7	93
White	22	78
Black	65	35
American Indian	95	5
Vietnamese	100	0

Sourcec: Data from Flatz, G. 1987. Genetics of lactose digestion in humans. In *Advances in Human Genetics*. Harris, H. and Hirschhorn, K. (eds). New York: Plenum, pp. 1–77. Adapted from Flatz and from Simoons, F.J. 1989. The geographic hypothesis and lactose malabsorption. *Dg. Dis. Sci.* 23: 963–980.

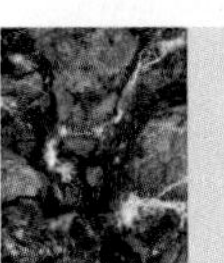

TABLE 21.2

Frequency of Persistent Lactase Activity (Lactose Absorbers) among the Milk-Dependent Pastoralists from Africa and Dairying Peoples of Northern Europe

Populations Economy	N	Lactose Absorbers (%)	Milk Consumption (L/p/y)
Milk-Dependent Pastoralists from Africa			
Arabs of Saudi Arabia	22	86.4	
Hima pastoralists	11	90.9	
Tussi, in Uganda	17	88.2	
Tussi, in Congo	15	100.0	
Tussi, in Rwanda	27	92.6	
Total N	92	91.3	
Dairying Peoples of Northern Europe			
Danes	761	97.5	1032.8
Swedes	491	97.8	393.1
Finns	578	85.1	677.9
Northwest Europeans	158	87.3	283.3
French	14	92.9	580.3
Germans from Central Europe	55	85.5	390.2
Dutch (living in Surinam)	14	85.7	828.1
Poles (living in Canada)	21	71.4	516.6
Czechs (living in Canada)	17	82.4	391.5
Czechs (Bohemia, Moravia)	20	100.0	391.5
Spaniards	265	85.3	171.2
North Italians (Ligurians)	40	70.0	210.9
Total N	2,434	91.5	

Source: Data from Simoons, F. J. 1989. The geographic hypothesis and lactose malabsorption. *Dg. Dis. Sci.* 23: 963–980.

Hypothesis to Account for Population Differences in Persistent Lactase Activity

Earlier studies maintained that differences in LDC are not primary, but are secondary to disease. According to this hypothesis, some groups that have high prevalence of intestinal diseases, such as dysentery or protein-calorie malnutrition (that damage the intestinal mucosa), have a low LDC. Likewise, according to the hypothesis of dietary inhibition, differences in LDC are related to differences in the consumption of some food or drugs (like spicy foods or betel nut) that inhibit lactase activity. These hypotheses have been disproved by the fact that many LDCs are healthy individuals with no unusual medical history. They also had no evidence of duodenal abnormality, and no difficulty in absorbing other disaccharides and no consistent food sources or set of food sources that would cause an LDC to have been identified. Other investigators maintained that lactase activity levels were the result of milk consumption, and that continuation of milk ingestion would maintain lactase enzyme at newborn levels.[9] While various investigations have demonstrated that individuals with LDC can be induced to drink, consume probiotics, and digest milk in small quantities, this digestion appears to be a function of the action of colonic bacteria rather than of lactase activity.[4, 10, 11] Besides these possibilities to account for the population differences in lactose digestion capacity in adults, two hypotheses have been postulated: (1) culture historic lactose dependence and (2) calcium absorption, which will be discussed next.

Culture-Historical Milk Dependence Hypothesis

The premise of the culture-historical milk dependence hypothesis is that the lactase persistence in adulthood is an adaptation to millennia of pastoralism and milk consumption.[12–15] According to this hypothesis, the capacity to digest lactose has a selective advantage in adults in pastoralist populations. Contingent on this hypothesis is that an advantage would have occurred where milk was an especially critical part of the diet and where the group was under dietary stress. In such an environment, individuals with a low digestive capacity would be at a selective disadvantage.[14] Individuals with a high LDC were able to derive a nutritional benefit from the lactose in milk, which is not available to individuals with low LDC. Individuals with low LDC may also suffer from symptoms of lactose intolerance when they consume fresh milk, including abdominal discomfort, flatulence, and diarrhea. In this scenario, because of these symptoms, milk could be nutritionally detrimental to individuals with low LDC. As a result, their general nutritional states were impaired and were less likely to survive and reproduce than those who could digest milk, and gradually over time the gene for persistence of lactase activity became common in the population as a whole. In this context, the cultural environment of the herding and milking of livestock became the selection force that determined the genetic capacity to digest milk. The process whereby populations adapted to the need to digest milk involved biological and cultural solutions.

Cultural Solution

It is quite evident that not all populations whose subsistence depended on milk developed the capacity to digest lactose. As shown in **tables 21.3** and **21.4**

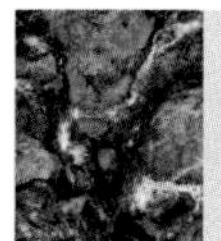

TABLE 21.3

Frequency of Persistent Lactase Activity (Lactose Absorbers) Among Recently Dairying Agriculturalists and Non-Dairying Agriculturalists, and Hunter-Gatherers: Traditionally Lacking Dairy Animals

Populations Economy	N	Lactose Absorbers (%)	Milk Consumption (L/p/y)
Recently Dairying Agriculturalists			
Kenyans (mainly Bantu)	71	26.8	67.5
Bantu of Zambia	26	0.0	8.8
Bantu of South Africa	31	9.7	92.1
Shi, Bantu of Lake Kivu area	28	3.6	7.3
Ganda, other Bantu of Uganda	70	5.7	29.8
Total N	226	11.9	
Non-Dairying Agriculturalists			
Yoruba	10	9.0	4.5
Ibo	15	20.0	4.5
Bantu of Zaire	52	1.9	0.2
Hausa	17	23.5	4.5
Total N	94	9.5	
Hunter-Gatherers: Traditionally Lacking Dairy Animals			
Eskimos of Greenland	119	15.1	
Twa Pygmies of Rwanda	22	22.7	
!Kung Bushmen of S.W. Africa	40	2.5	
!Kun Bushmen of Botswana	25	8.0	
Total N	206	12.1	

Source: Data from Simoons, F. J. 1989. The geographic hypothesis and lactose malabsorption. *Dg. Dis. Sci.* 23: 963–980.

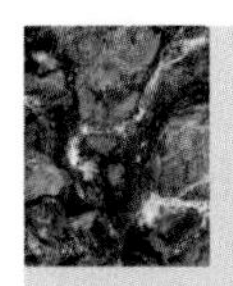

TABLE 21.4

Frequency of Persistent Lactase Activity (Lactose Absorbers) Dairying of North Africa

Populations Economy	N	Lactose Absorbers (%)	Cheese Production (%)
Jews in Israel	201	40.8	40.1
Ashkenazic Jews	53	20.8	40.1
N. African Sephardim	32	37.5	10.7
Other Sephardim	36	27.8	18.0
Iraqi Jews	38	15.8	38.4
Other Oriental Jews	20	15.0	33.8
Arab villagers in Israel	67	19.4	40.1
Syrian Arabs	40	5.0	36.3
Jordanian Arabs	56	23.2	54.3
Arabs (Jordan, Syria, etc.)	19	0.0	54.3
Other Arabs	26	19.2	36.3
Egyptian Fellahin	14	7.1	54.9
Greeks (mostly mainland)	730	52.1	48.8
Greek Cretans	50	44.0	48.8
Greek Cypriots	67	28.4	56.4
Ethiopians/Eritreans	58	10.3	5.4
Total N	1,507	38.8	

Source: Data from Simoons, F. J. 1989. The geographic hypothesis and lactose malabsorption. *Dg. Dis. Sci.* 23: 963–980.

among the recently dairying agriculturists, non-dairying agriculturalists, and hunter and gatherers traditionally lacking dairy animals less than 15 percent are lactose absorbers. According to the culture-historical milk dependence hypothesis some populations adopted a cultural solution that circumvented the need to digest lactose.[13, 14] Among these populations, fresh milk is processed into one or more of the "soured" and fermented milk products such as kefir, maconi, kurt, yogurt, dahi, chal (from camel's milk), or kourniss (from mare's milk), and aged cheese. Through these techniques lactose is either: (i) *externally predigested*, by lactic acid bacteria, *(Lactobacillis bulgaricus)* and yeast, (ii) *autodigested*, by lactase from the bacteria within the processed milk, effectively substituting for intestinal lactase; (iii) *physically drained away*, with the separation of cheese curds from lactose-rich whey. The result of these techniques is that the problem of lactose digestion is minimized or even eliminated. In other words, milk processing evolved culturally as an alternative solution to the problem of lactose digestion.[13, 14]

Milk-processing techniques originated about 6000 B.C. in southwestern Asia, from which it spread as far out as northern Poland by 4300 B.C.[16, 17] As illustrated in **figure 21.1**, the percentage of cheese production is significantly ($r = -0.36$ $p < 0.05$) correlated with the percent of lactose absorbers so that the higher the proportion of milk that is processed into cheese, the lower the percent of individuals who have maintained a high lactose digestive capacity. These data together support the hypothesis of a *cultural* compensation to the lactose problem.[17] That is, cheese-processing techniques, and hence, cultural technology, circumvents the selection pressure of some populations.[22] Thus, as pointed out by Kretchmer,[18, 19] the advent of dairying did not automatically create a selective pressure in favor of lactose digesters as opposed to nondigesters.

Calcium Absorption Hypothesis

This hypothesis postulates that the adult lactose digestive capacity evolved in northern Europe as an adaptation to facilitate absorption of calcium in areas of low solar radiation and low synthesis of vitamin D.[18] The rationale for this hypothesis is that low solar radiation is associated with low synthesis of vitamin D.[28, 29] In such conditions, the presence of lactose in milk might be advantageous for increasing calcium absorption and preventing rickets.[30, 31] In other words, in an environment where solar radiation was reduced by the high latitudes and frequent cloud cover, cutaneous vitamin D synthesis may have been insufficient; however, having the ability to digest lactose might compensate for this deficiency.

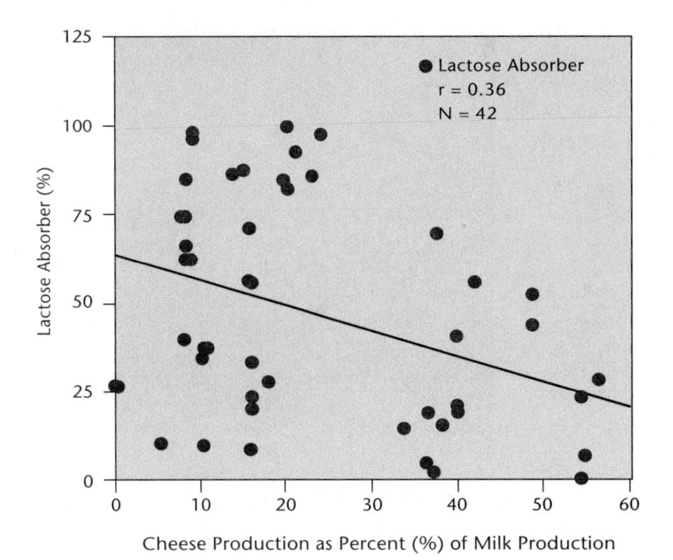

FIGURE 21.1 Relationship between cheese production and lactose absorption among pastoralists. It is evident that the amount of cheese production is inversely related to lactose persistence. Source: Adapted from Durham.[18]

Animal studies produced strong evidence that lactose has beneficial effects on intestinal calcium absorption[32–37] and on calcium retention in bone.[37] Studies in postmenopausal women indicate that the addition of lactose enhances the bioavailability of calcium in milk,[38] but the addition of lactose to lactose-tolerant men did not increase calcium bioavailability.[39] These findings suggest that the beneficial effects of lactose on calcium absorption are more evident in females than in males. This different response to lactose makes sense in an evolutionary perspective because of women's enhanced needs for calcium during pregnancy.

In summary, the absorption of calcium is enhanced by vitamin D, and the synthesis of vitamin D in the skin is directly related to the amount of solar radiation. Then, under conditions of low solar radiation and to compensate for the low synthesis of vitamin D, the development of a gene that digests lactose might have been favored because it enhanced the absorption of calcium.

Nutrient Needs and Evolution of Skin Color

Geographic Association and Skin Color

Variability in human skin color is one of the most conspicuous traits of humans. In general, population differences in skin color are associated with solar radiation and latitude.[40] Native peoples of the tropics, as measured by reflectance spectrometry on the upper inner arm, tend to be darker than those living far from the equator. Likewise, in the Old World, skin color correlates strongly with latitude and with the intensity of ultraviolet (UV) radiation reaching the earth's surface.[41] Furthermore, at equivalent latitudes, skin color tends to be darker in the Southern Hemisphere (below the equator) than in the Northern Hemisphere[41, 42] (**fig. 21.2**). Pedigree analysis indicates that skin color is controlled by three to five genes acting additively[43, 44] and population differences in skin color are related to differences in gene frequency.[45] Given the regular pattern of geographical variation in skin color that clearly has a genetic basis, several hypotheses have been postulated to explain population differences in skin color. This chapter will focus on the evolution of skin color of humans. Accordingly, first we will focus on the biology of skin color.

Biology of the Skin and Skin Color

Skin Structure

Human skin consists of the dermis and epidermis, which are connected by a basement membrane and an intricate system of cells and blood vessels that enable the skin to function as one of the important factors in

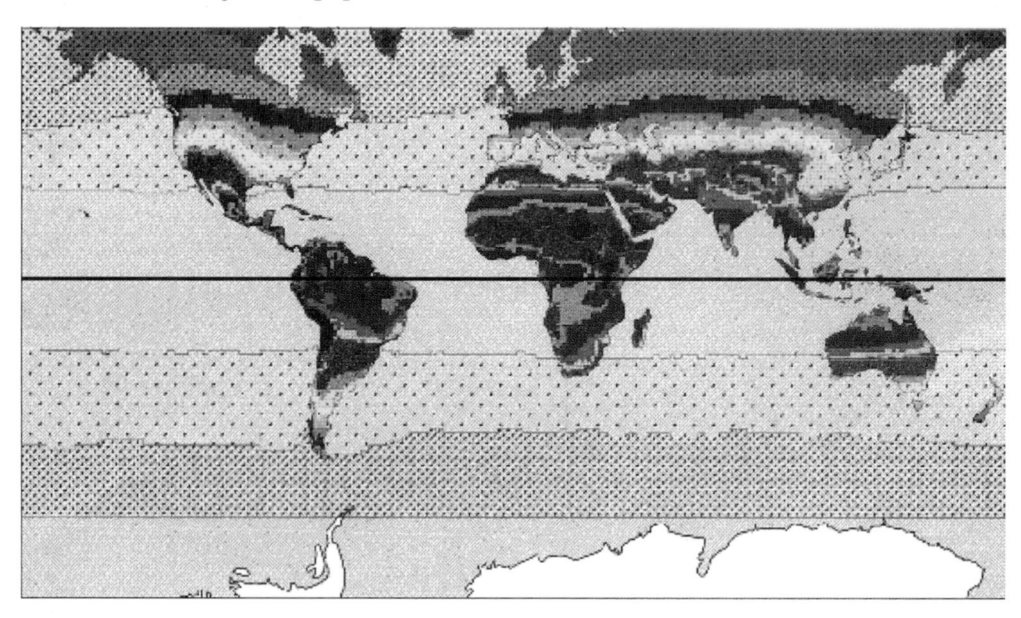

FIGURE 21.2. Distribution of human skin color according to latitude. Dark and light shades indicate levels of pigmentation. The closer to the equator, the darker the skin pigmentation and the further from the equator, the lighter the pigmentation. Source: Adapted from Jablonski and Chaplin.[41] Reprinted from *Journal of Human Evolution,* Vol. 39, N. G. Jablonski & G. Chaplin, "The Evolution of Human Skin Coloration," pages 57–106. Copyright 2000, with permission from Elsevier.

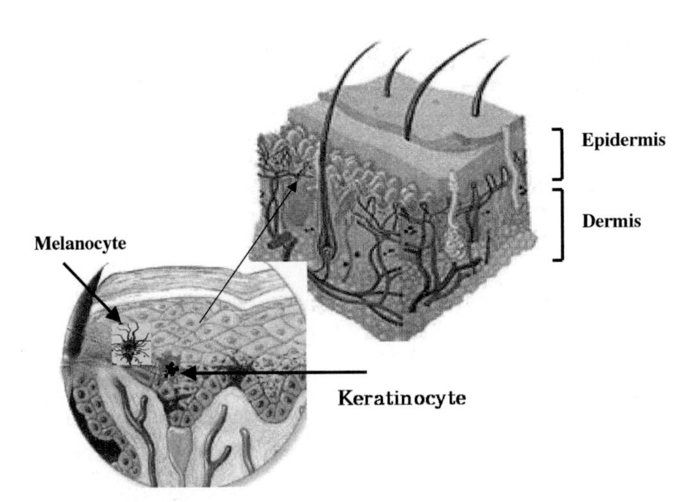

FIGURE 21.3. Diagram of the structure of the human skin. (Composite picture made by the author.)

human thermoregulation and as protection against the deleterious effects of solar radiation. As shown in **figure 21.3**, the outer surface of the epidermis has a stratum corneum (horny layer). This layer is extremely tough, chemically resistant, and almost impenetrable. Its average thickness is about 15 μm. It consists of many flat cells that have lost most of their cytoplasm and nuclei, but contains many filaments of keratin in a matrix formation. These cells are held together by an extremely strong cement substance of unknown composition. The stratum corneum layer is made up of thick cells called keratinocytes, and because of its appearance, is known as the prickle cell. These prickle cells migrate outward during cell duplication and division. As they approach the stratum corneum, or horny layer, they accumulate granules, becoming the granular layer located directly below the stratum corneum. Below the prickle cells are located the basal cells and the Langerhans' cells; interspersed among these are the melanocytes, or melanoblasts. The melanocytes synthesize the pigment melanin in a process that begins with oxidation of the amino acid tyrosine, which is then acted on by the enzyme tyrosinase. Melanin is deposited along with protein in specific subcellular organelles called melanosomes. These melanosomes are then transferred to keratinocytes of the skin and hair through the dendritic processes.[46] It has been estimated that each melanocyte "services" about 36 keratinocytes. In other words, there is functional as well as structural integration of a melanocyte with its associated keratinocytes.

Skin Color and Its Components

Skin color is a function of the pigment melanin, hemoglobin, and carotene. Melanin is a brown pigment produced by specialized dendritic cells called melanocytes, which are located at the bottom of the epidermis. Once the melanocytes synthesize melanin, it is transferred to the keratinocytes of the epidermis. The mature melanosome, or melanin granule, is dispersed into smaller melanin particles as the cells move outward. These cells range in color from brown to black. The melanin granules are about 1 mm in size and scatter themselves to form a cap over the nucleus of the keratinocyte, thus protecting the nuclear deoxyribonucleic acid (DNA) of the epithelium. As shown in recent studies, individual and population differences in skin color appear related to differences in both the amount of melanin synthesis and in the number of melanocytes.[46, 47] Among blacks or dark-skinned populations, such as the Australian Aborigines and Solomon Island natives, there are greater number of melanocyte cells and the melanin granules are larger and more dispersed than those found in whites or Asians[46, 47] (**fig. 21.4**). Hemoglobin influences the skin color only of light-skinned people because they have little melanin near the surface of the skin and so the red color shows through.

Measurement of Skin Color

The most useful method of evaluating skin color is by measurement of skin reflectance with either a photoelectric reflectometer or a reflectance spectrophotometer (**fig. 21.5**). Both of these instruments work on the same principle of reflectance and have similar parts.

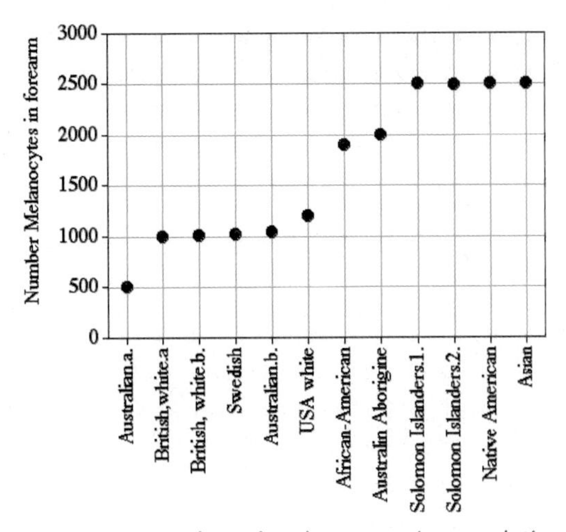

FIGURE 21.4. Number of melanocytes in populations indigenous to high and low latitudes. Populations 1 and 5 are Australian white, 2 and 3 are British white, 4 is Swedish white, 6 is American white, 7 is African American, 8 is Australian Aborigine, 9 and 10 are Solomon Islanders, 11 is Native American, and 12 is Asian. Source: Adapted from Stinson, Bogin, Huss-Ashmore, and O'Rourke.[48]

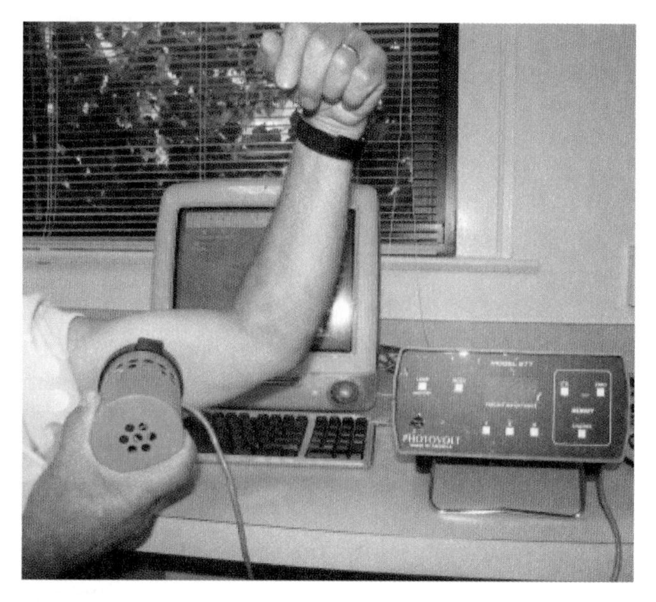

FIGURE 21.5. Measurement of skin reflectance. The percent skin reflectance gives an indication of skin color. The higher the reflectance, the lighter the skin color. (Photo by author.)

Therefore, with appropriate statistical adjustments, their values are comparable.[49] The main unit includes a galvanometer, a constant voltage transformer, a search unit consisting of a lamp, and a photocell. Light from the lamp passes through exchangeable glass filters of known wavelength to the surface being measured. The light that is diffusely reflected from this surface acts on the photocell. Thus, the amount of light reflected from a skin surface, as compared with reflectance by a standard white magnesium surface (giving 100 percent reflectance), is measured by the photocell and recorded on the galvanometer.[44, 45, 49–53] Reflectance readings are usually taken with glass filters identified as blue, green, tri-amber, tri-blue, tri-green, and red. These filters have transmission peaks that range from 420 to 670 nanometers (nm). The highest correlation between skin reflectance and UV levels occurs at 545 to 600 nm. The filters and their transmission peaks and absorption capacities as used with the reflectometer are given in **table 21.5**.

The most common body sites where skin reflectance is measured include: (a) the inner upper arm distal to the axillary region of the arm, (b) the subscapular area, and (c) the forehead. Readings of the inner upper arm are used to evaluate the biological characteristics of skin color prior to tanning, while the forehead and subscapula, if usually uncovered, are used to evaluate the effects of tanning on skin color.

Age and Sex Differences

As measured by skin reflectance of the inner arm, from infancy to the beginning of adolescence there is rapid darkening in skin color, followed by a gradual darkening through adulthood. Furthermore, females through all ages are consistently lighter than males.[2–4, 13–15]

Genetics

Pedigree analysis indicates that skin reflectances are moderately heritable in a polygenic fashion.[4, 5] As shown by the studies of a mixture of white Europeans and South American blacks,[6] population differences in skin color are related to three or four genes. However, identification of the actual genes has been more complex than simple Mendelian inheritance. One of the genes that accounts for the variance in normal skin color (and hair), referred to as melanocortin 1 receptor (MC1R), has been identified.[56–62]

Index of Erythemal Threshold and Tanning

Studies of the effects of UV radiation on human skin have utilized the minimum-erythemal dose (UVMED) as a standard, which is the quantity of ultraviolet (UV) radiation required to produce a barely perceptible reddening of lightly pigmented skin. The tanning response has two distinct phases:[63] immediate and delayed.

Immediate Tanning

Immediate pigment darkening begins after being exposed to the sun. This initial tan is due to oxidation by the UVA rays of the sun of the melanin pigments already in the skin and a redistribution of melanosomes, which are transported from the bottom of the epider-

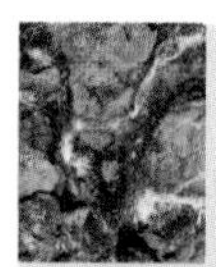

TABLE 21.5
Glass Filter Wavelength and the Absorption of Components of Skin Color Measured by Skin Reflectometry

Glass Filter	Wavelength (nm)	Components Absorbed
Blue	420	Melanin and oxyhemoglobin
Green	525	Hemoglobin and carotene
Tri-amber	600	Melanin
Tri-blue	450	Melanin
Tri-green	550	Oxyhemoglobin and melanin
Red	670	Melanin

Source: From Frisancho, Wainwright, and Way.[43]

mis to the dendrites of epidermal melanocytes. The color change is subtle and disappears after a few hours, depending on the individual complexion. The tanned area fades almost to nonexposure levels and offers no measurable photoprotection. The rate of depigmentation appears to be related to length of exposure to ultraviolet light; after prolonged exposure (90 to 120 minutes), residual hyperpigmentation may be visible for as long as 24 to 36 hours.

Delayed Tanning

Delayed tanning appears after 48 hours, resulting from the synthesis of melanin. It is detectable 48 hours after irradiation of the skin and its action spectrum and threshold dose are the same as those that induce sunburn. Delayed tanning is associated with an increase in the number of functional melanocytes, resulting from a proliferation of melanocytes and possible activation of dormant or resting melanocytes. This increase in pigmentation is photoprotective and can increase the threshold at which a sunburn occurs.

Negative and Positive Effects of Solar Radiation

Sunburn and Skin Cancer

When solar radiation is intense, 290 to 315 nm, and the skin is not tanned, an individual develops a sunburn. The general sequence of events includes a latent period of several hours, followed by blood vessel dilation manifested in erythema, which reaches its maximum between 8 and 24 hours after exposure. As a result of a sunburn, there is general discomfort, a reduction in the pain threshold, and severe blistering. From a sunburn the skin may also develop a secondary infection or a suppression of sweating. Susceptible individuals, such as those with low melanization, may develop desquamation and peeling of the sunburned area with continued exposure; this in turn reduces the possibility of maintaining adequate melanization. Therefore, repeated sunburns may lead to degenerative changes in both the dermis and epidermis, and skin cancer is one of the possible consequences.[64–66]

Pigmentary traits, such as red hair, fair skin, lack of tanning ability, and the propensity to freckle, have been identified as genetic risk factors for skin cancers when combined with the environmental risk factor of high ultraviolet light exposure. Heterozygote carriers for the propensity to freckle are more prone to skin cancer risk than those who are not carriers. These studies indicate that the capacity for melanin synthesis rather than skin color is the best indicator for skin cancer risk.[61–63]

Photolysis of Folate

Photolysis refers to the degradation of vitamins by the effects of ultraviolet radiation. Although several vitamins, such as vitamin A and E, are light sensitive and can be degraded by exposure to UV radiation, the folate, or folic acid, is the most sensitive to the effects of ultraviolet light. Intense or prolonged exposure to UV light leads to photolysis of folate and subsequent folate depletion.[67]

Synthesis of Vitamin D in Human Skin

Fish and other marine animals obtain vitamin D from algae, fungi, and plankton.[68] Humans, on the other hand, synthesize vitamin D in the skin under the influence of solar radiation[69–71] (**fig. 21.6**). Vitamin D

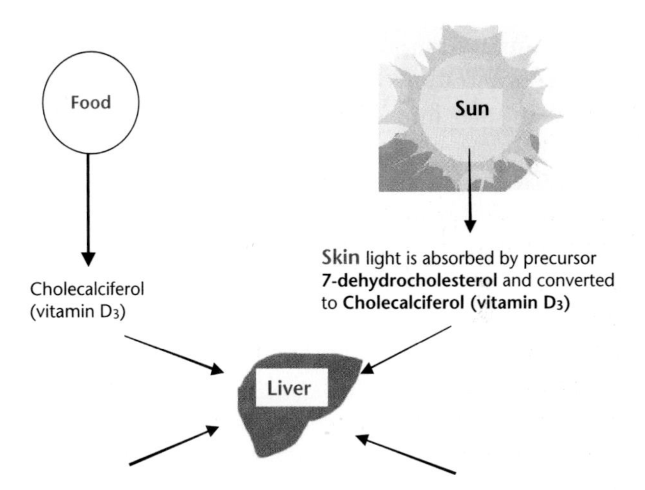

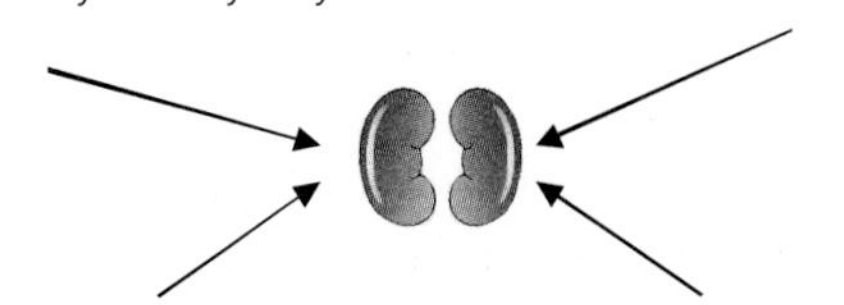

Within the kidney, 25-vitamin D is converted to the biologically active form of vitamin D (1,25-dihydroxycholecalciferol), which is used by the body.

FIGURE 21.6. Vitamin D synthesis. Vitamin D is produced in the body from the photosynthetic reaction that occurs when 7-dehydrocholesterol in skin cells is exposed to ultraviolet light from the sun. This inactive precursor travels to the liver, where it is changed to 25-hydroxyvitamin D_3, and the kidneys, in turn, convert this intermediate form of the vitamin to the active circulating form of the vitamin. Humans can get a precursor (inactive) form of vitamin D from food. (Composite illustration made by the author.)

exists in two forms: cholecalciferol (vitamin D_3) and ergocalciferol (vitamin D_2). Upon exposure to the sun, the solar radiation is absorbed in the skin by a compound called provitamin D, which in interaction with cholesterol in the blood is converted into vitamin D_3.[69–74] Thus, on average, more than 90 percent of the vitamin D circulating in the blood is synthesized in the epidermis of the skin.[69–74] Vitamin D_2 (ergocalciferol) is found in animal products[75] such as saltwater fish (herring, salmon, and sardines) (**table 21.6**).

During the winter, the solar zenith angle of the sun becomes more oblique. This configuration causes the ultraviolet-B photons to be absorbed more efficiently by the stratospheric ozone layer, thereby decreasing the total number of photons that reach the earth's surface. As a result, cutaneous synthesis of previtamin D_3 is less during the winter than in the summer (**fig. 21.7**). Reflecting this difference, exposure to sunlight in Boston (42° N) and in Edmonton (52° N) Canada from March through October promoted a higher cutaneous photosynthesis of previtamin D_3 than from November through February. In the same manner, in London, variations in blood 25-OH vitamin D concentrations are associated with seasonal differences in solar intensity.[74] Comparative studies indicate that the concentrations of 25-hydroxyvitamin D_3 were substantially and significantly lower in black than in white women in both the winter (February–March) and summer (June–July) months.[74] As shown in **table 21.7**, the seasonal skin color–associated difference of vitamin D concentration is confirmed by the epidemiological data of the third National Health and Nutritional Examination Survey of the U.S.A. conducted from 1988 to 1994.[76]

Likewise, in Swedish adults, intestinal calcium absorption is more efficient in the summer, declining progressively during the winter.[74] In contrast, in Puerto Rico, where there are minimal seasonal differences in solar intensity, there are no seasonal variations in vitamin D levels.[74] In Scotland, the incidence of infantile hypocalcemia and defective teeth is much higher among children who developed in utero during the winter.[70] It is evident, then, that the limiting factor in cutaneous vitamin D production is the amount of absorbed solar radiation. On the other hand, exposure to prolonged intense sunlight does not result in an overproduction of vitamin D_3 in the skin.[73] This is because any excess of vitamin D_3 is photodegraded to a biologically inert form. Thus, once vitamin D_3 is made from previtamin D_3, it is rapidly photodegraded during exposure to sunlight. Therefore, even lifeguards on southern beaches are safe from vitamin D toxicity from the sun.

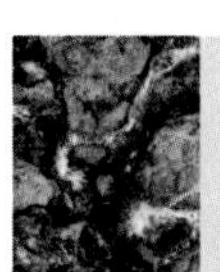

TABLE 21.6

Selected Food Sources of Vitamin D*

Food	International Units (IU)	Percent DV
Cod liver oil, 1 Tablespoon	1,360	340
Salmon, cooked, 3½ ounces	360	90
Mackerel, cooked, 3½ ounces	345	90
Tuna fish, canned in oil, 3 ounces	200	50
Sardines, canned in oil, drained, 1¾ ounces	250	70
Milk, nonfat, reduced fat, and whole, vitamin D fortified, 1 cup	98	25
Margarine, fortified, 1 Tablespoon	60	15
Pudding, prepared from mix and made with vitamin D fortified milk, ½ cup	50	10
Egg, 1 whole (vitamin D is found in egg yolk)	20	6
Liver, beef, cooked, 3½ ounces	15	4
Cheese, Swiss, 1 ounce	12	4

*DV = Daily Value. The recommended daily vitamin D intake varies with age and health status: Birth to age 50 years = 200 IU, 51 to 70 years = 400 IU, older than 71 years = 600 IU, and for homebound or institutionalized elderly = 800 IU.
Source: U.S. Department of Agriculture's Nutrient Database Web site: *http://www.nal.usda.gov/fnic/cgi-bin/nut_search.pl.*

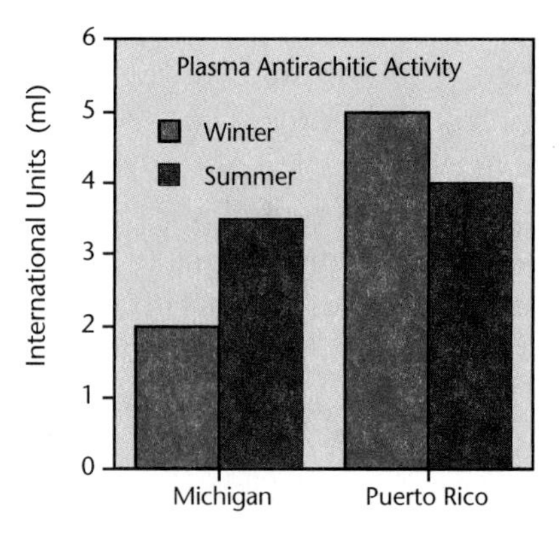

FIGURE 21.7. Seasonal differences in plasma antirachitic activity. In Michigan, despite the availability and consumption of vitamin D–fortified dairy products, differences in vitamin D levels parallel seasonal differences in solar radiation between winter and summer, but in Puerto Rico no such differences exists. From *Human Adaptation and Accommodation* by A. R. Frisancho. Copyright © 1995 by University of Michigan Press. Reprinted by permission.

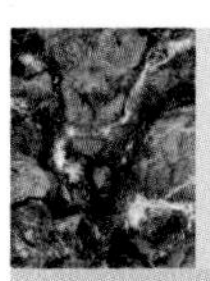

TABLE 21.7

Mean Serum 25-hydroxyvitamin D Concentrations (nmol/L) by Season and Residence for African American and White Women Aged 15–49 Years in NHANES III

	African American Black			White		
	(N)	(nmol/L)	(%)	(N)	(nmol/L)	(%)
Spring	263	38.8 ± 2.62	45.7	377	66.0 ± 2.02	4.0
Summer	412	43.3 ± 1.40	28.6	531	82.9 ± 2.50	2.3
Fall	523	49.6 ± 0.82	45.7	315	90.9 ± 2.02	3.4
Winter	348	43.7 ± 1.72	52.3	203	78.8 ± 1.82	10.9
Urban	908	43.1 ± 1.30	46.6	552	78.3 ± 1.52	4.9
Rural	638	45.8 ± 1.60	36.4	874	85.9 ± 2.40	3.6

Source: Data from Nesby-O'Dell, Scanlon, Cogswell, et al.[76]

Function and Deficiency of Vitamin D

The most important function of vitamin D is to facilitate the absorption of dietary calcium and phosphorus and thus maintain calcium homeostasis and facilitate bone formation. The maintenance of adequate calcium levels is important for bone formation and the proper function of nerves and muscles.

Rickets

When vitamin D is deficient, the serum calcium level drops and the calcium stores in bone are mobilized. Deficiency in vitamin D leads to rickets in children, and osteomalacia and osteoporosis in adults. Rickets is a bone disorder of children in which growth of the cartilage fails to mature and mineralize adequately. Growth is stunted and various deformities about the epiphyseal abnormalities occur. It is characterized by bowlegs, knock-knees (**fig. 21.8**), curvature of the spine, and pelvic and thoracic deformities resulting from the application of normal mechanical stress to demineralized bone. Clinical studies clearly indicate that childhood rickets is readily cured with 0.05 to 0.1 mg (2,000 to 4,000 I.U.) of vitamin D daily for 6 to 12 weeks, prolonged exposure to winter sunlight, or exposure to artificial ultraviolet radiation equivalent to strong summer sunlight at 36° latitude.[69–72]

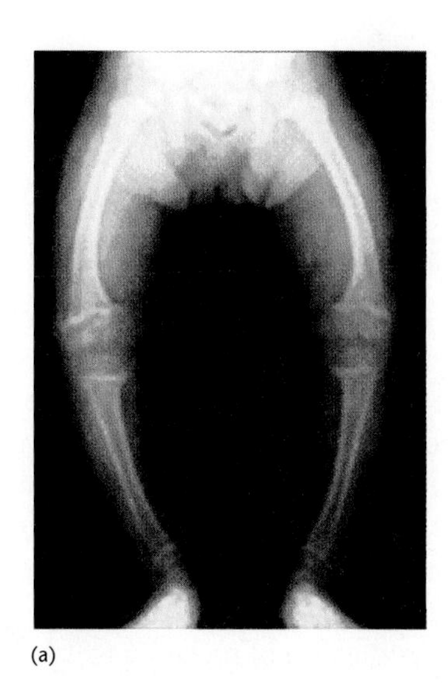

(a)

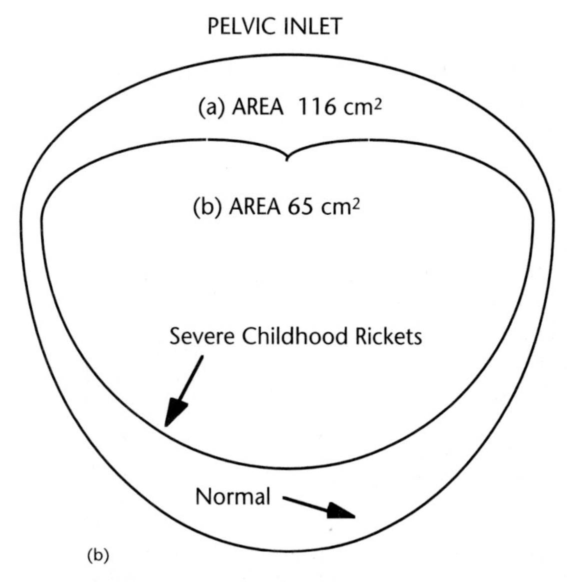

(b)

FIGURE 21.8. Rickets. Vitamin D deficiency during childhood can lead to adult rickets, which is characterized by bowed legs, associated with poor calcium absorption due to vitamin D deficiency. Childhood rickets can also lead to reduction of the pelvis during adulthood. From *Human Adaptation and Accommodation* by A. R. Frisancho. Copyright © 1995 by University of Michigan Press. Reprinted by permission.

Osteomalacia and Osteoporosis

Osteomalacia

Osteomalacia can occur in children and adults. Because bones are in a constant state of turnover of calcium deposition and resorption, severe vitamin D deficiency leads to **osteomalacia.** In osteomalacia, the collagenous bone matrix is preserved but bone mineral is progressively lost, resulting in low density of bone and bone pain (**fig. 21.9**). The demineralization can become very extensive and the bones may lose their stiffness and become severely deformed.

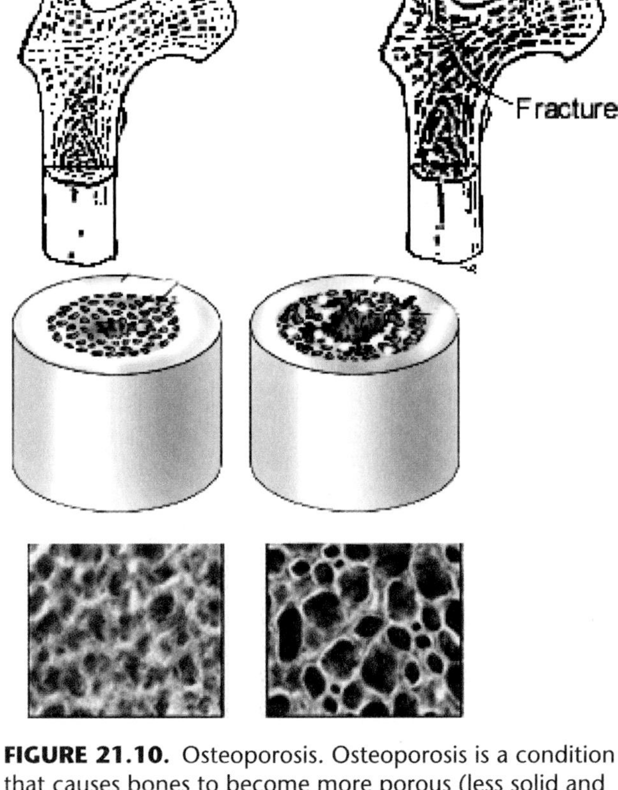

FIGURE 21.10. Osteoporosis. Osteoporosis is a condition that causes bones to become more porous (less solid and less dense) than a normal bone. The porosity gradually diminish the strength of the bone and makes it brittle. (Composite illustration made by the author.)

Osteoporosis

Osteoporosis is a disorder of the skeleton of adults. It is characterized by a reduction in bone mass (i.e., both the size and the bone mineral). Furthermore, the amount of bone per unit of area to that of the external size or frame of the bone is low and the bone looks very porous. Osteoporosis is defined as an age-related reduction in the quantity of bone mass (**fig. 21.10**). Osteoporosis occurs when bone resorption exceeds bone formation. It is characterized by a loss in bone density and bone mass. Osteoporosis is a disease that typically reflects an imbalance in skeletal turnover so that bone reabsorption exceeds bone formation. Its major symptom is an increased vulnerability to bone fractures. The fragility of the bones is most severe in the pelvic bone, which may become so brittle that it breaks when the person is simply walking. Osteoporosis is also associated with increased curvature of the spinal column and collapse of the vertebrae, resulting in reduction of stature.

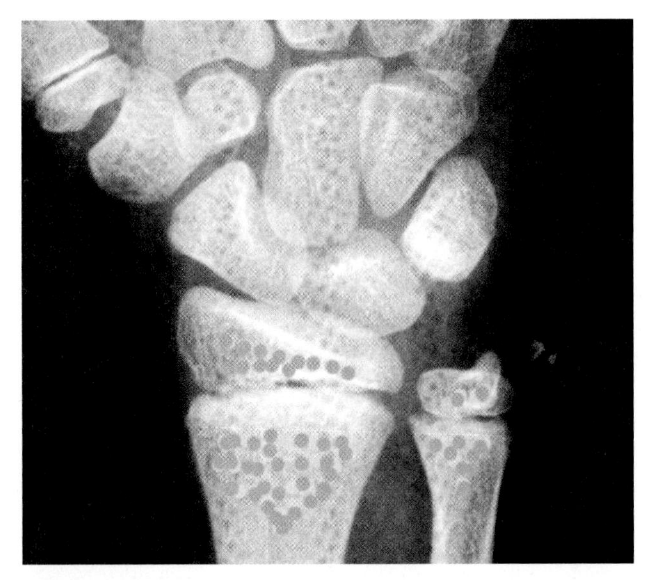

FIGURE 21.9. Anteroposterior radiograph of the wrist in a child with osteomalacia. The dark spots indicate the areas of low bone density. (Picture from the author's files.)

Hypotheses to Explain the Evolution of Skin Color

It is evident that, depending on the degree of acclimatization, the effects of solar radiation on the health and well-being of the organism can be both positive and negative. Intense exposure of the skin to ultraviolet and visible solar radiation can result in sunburn and related cellular injuries and, possibly, in skin cancer, and in the degradation of essential nutrients, such as the folate vitamin. The organism's capacity to protect against such injuries is related to the production of melanin, which blocks the solar radiation from penetrating in to the blood vessels. Conversely, solar radiation plays an important role in the synthesis of vitamin D. Because vitamin D plays an important role in the metabolism of calcium, variations in solar radiation intensity are reflected in marked seasonal differences in the incidence of rickets. Therefore, individual and population differences in melanin-producing capacity have profound influences on the well-being, survival, and, hence, evolution of the organism. Several hypotheses have been proposed to explain population differences in skin color, skin cancer, degradation of folate, vitamin D synthesis, and sexual selection.

Skin Cancer and the Evolution of Dark Skin Color

In general, the greater the intensity of ultraviolet radiation, the greater the risk for skin cancer. Experimental evidence[79–82] indicates that a low production of melanin is an important risk factor for sunburn and cancer. Epidemiological evidence indicates that in places where the intensity of ultraviolet radiation is high, skin cancer is more frequent among light-skinned individuals.[82] In the United States alone, about a million new cases occur annually, rivaling the incidence of all other types of cancer combined.[66, 79] Despite the advances of medicine, the incidence of cancer and mortality in the U.S. population is inversely related to latitude[81, 82] so that the closer to the equator, the higher the rate, and the further to the north, the lower the rate of skin cancer.

Since skin cancer affects primarily individuals past their reproductive years, some argue that mortality from cancer has no evolutionary consequences.[83] This argument is not valid for several reasons. First, the damage caused by skin cancer begins decades earlier.[66, 79] Sunburn interferes with sweating, leading to hyperthermia, especially in infants, and often leads to secondary infection.[84, 85] Second, natural selection, prior to modern medicine, would have favored high melanin in areas of high solar radiation. As shown in epidemiological data, in Australia, for whites living in a region of high solar radiation approximately 669 deaths per 100,000 between the ages of 10 and 49 years were recorded to be caused by malignant melanoma, and about 8 deaths per 100,000 in the same age range were caused by other skin cancers.[86–88] In the United States and Australia, the survival rate for individuals suffering from melanoma under the age of 50 years is approximately 50 percent because of improved medical care and is approximately 90 percent for skin cancers.[86–88] One would expect that among populations without access to adequate medical care, such as those of early man, the actual mortality rate from melanoma would equal about 1,338 deaths per 100,000 and about 89 deaths per 100,000 for other skin cancers. The total combined mortality for these two diseases without modern medical care would equal 1,427 per 100,000, or approximately 142/10,000. In view of the lower incidence of malignant melanoma and skin cancer in dark-skinned populations, the selection against light skin color would be high in areas of intense solar radiation. Given the nearly 3 million years of human evolution, it is conceivable that this level of selection could have had a strong influence on the evolution of dark skin color. Through computer simulation, it has been estimated that with optimizing selection and a 6 percent maximum difference in fitness, the evolution of the range of human skin color differences would have taken about 800 generations with no dominance and about 1,500 generations with 80 percent dominance.[89] This would suggest that changes in skin color could have taken place within a range of 24,000 to 45,000 years ago.

In summary, although mortality from skin cancer occurs at old age, the etiology that leads to death from skin cancer starts decades earlier and thus could have been a selective factor. Furthermore, prior to modern medicine, natural selection would have favored high melanin-producing genotypes in regions of high ultraviolet stress.

Protection of Folate Levels and the Evolution of Dark Skin Color

Clinical studies conducted by Branda and Eaton[90] found that exposure of plasma in vitro to high levels of UV resulted in a greater loss of folate among light-skinned rather than dark-skinned individuals. In this study, ten light-skinned patients who had undergone at least 3 months prophylactic UV therapy for various dermatological disorders had abnormally low serum folate levels. This suggested to them that photolysis of folate may also occur in vitro, and that the original function of darkly pigmented skin was protection against extensive photodegradation of folate and other light-sensitive nutrients.[90] It has also been reported that three women who had spent time in tanning salons during the first weeks of pregnancy gave birth to infants with neural tube defects.[91] Based on these findings, Jablonski and Chaplin[41, 92] propose that dark skin color evolved to protect the vitamin folate in regions of high solar radiation.

Epidemiological studies[93–99] indicate that there is a direct connection between folate deficiency during the first weeks of pregnancy and neural tube defects

(NTDs), including spina bifida (**fig. 21.11**). Under normal conditions in the first month of prenatal development the neural groove fuses together to form a tube (the neural tube), which develops into the spinal cord and brain. The spinal vertebrae (backbones) then begin to form around the tube. However, if part of the neural tube does not close, the spine also does not close, and spina bifida occurs. There are three types of spina bifida: oculta, meningocele, and myelomeningocele. In spina-oculta there is no opening on the back, but the outer part of some of the vertebrae are not completely closed. In spina meningocele the outer part of some of the vertebrae are split and the meninges are damaged and pushed out through the opening, appearing as a sac. In myelomeningocele the outer part of some of the vertebrae are split, and the spinal cord and meninges are damaged and pushed out through the opening. Although there is no cure for spina bifida, it can be operated to close the opening in the back. On a worldwide basis, as many as 10 to 20 infants per 10,000 births are affected with some type of NTD.[93] Statistical analysis[95] has also suggested that UV light might be involved in causing NTDs. Epidemiologic data from Africa and the United States has suggested that the prevalence of these particular birth defects is higher in "whites" than in "blacks".[97–98] A recent review also indicated that epidemiologic studies have suggested that environmental and genetic factors have a joint role in the causation of NTDs, but no mention was made of an association between UVL and NTDs.[99] Furthermore, there is evidence that melanin protects dark skins against the photodegradation of DNA.[100] Likewise, there is evidence that folate deficiency might lead to diminished spermatogenesis.[101, 102]

In summary, although there are no controlled experimental studies,[103] in view of the direct connection between folate and individual reproductive success, the epidemiological evidence supports the hypothesis that protection against nutrient photolysis, and specifically the photolysis of folate, was a prime selective agent that brought about the evolution of deeply pigmented skins among people living under regimes of high UVB radiation throughout most of the year. In other words, the primary reason for the evolution of human skin color is the balance between the need for darker skin to protect against photodestruction of folate and the need for lighter skin to facilitate vitamin D synthesis. At any given latitude, the optimal skin color reflects the balance between these risks. Near the equator, the primary risk is folate deficiency, leading to darker skin for protection.

Vitamin D Synthesis and the Evolution of Light Skin Color

This hypothesis proposes that light skin evolved to facilitate vitamin D synthesis in regions of low solar radiation.[104, 105] Specifically, according to this hypothesis, depigmentation of the skin was a necessary adaptation for humans attempting to inhabit regions outside the tropics, especially those north of 40°N receiving low average amounts of UV radiation throughout the year. Several studies indicate that when specimens of white and black skin were exposed to simulated sunlight under the same conditions, the exposure times for the formation of previtamin D increased exponentially with the degree of pigmentation.[69–74]

The hypothesis that white skin is adaptive under conditions of low solar radiation is supported by several findings. First, dark-skinned individuals are more susceptible to rickets than whites in northern latitudes.[106]

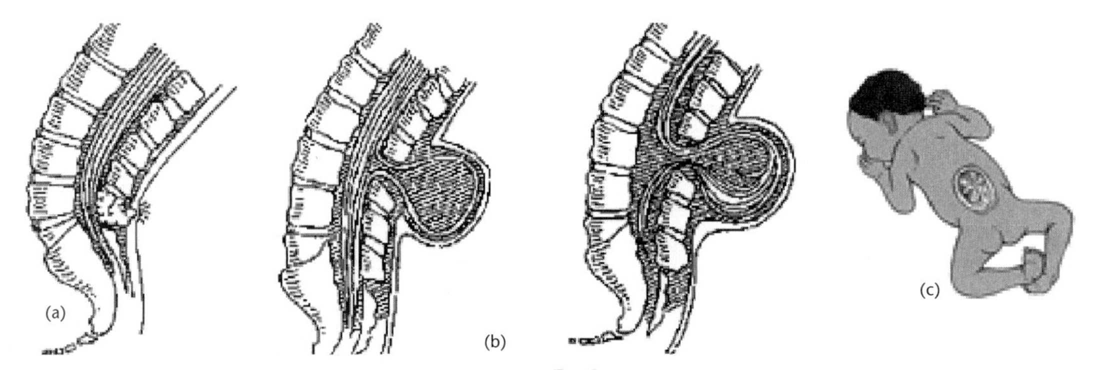

FIGURE 21.11. Folate deficiency and spina bifida. If folate is inadequate during neural tube closure, the normal closure (a) of the spine does not occur, which can result in neural tube defects, such as spina bifida (b and c). (Composite illustration made by the author.)

Before the widespread use of dietary vitamin D supplements that started in the 1930s, the incidence of deformed pelvises among black women studied in the 1950s was significantly greater (15%) than that of white women (2%).[107] The size reduction in the pelvic inlet and the absence of cesarean operations in any population can have profound effects on survival and fertility. Second, recently there has been an increased frequency of rickets among the children of dark-skinned mothers in England and Canada who live in an area of low solar radiation.[108–113] Together these findings suggest that under the condition of low solar radiation, dark-skinned people, and especially children who are breast-fed, are at risk of low vitamin D synthesis.

Since melanin is a natural sunscreen that decreases the speed at which ultraviolet radiation enters into the epidermis where vitamin D_3 is synthesized, dark-skinned people require longer sunlight exposure than light-skinned people. In England and Canada, most of the people affected with vitamin D deficiency were either immigrants or natives whose dark skin pigmentation or cultural rules dictating women be heavily veiled when outside in public prevented sufficient ultraviolet light from penetrating their skin.[112] Furthermore, even though Indian and Pakistani immigrants have the same capacity as Caucasians to produce vitamin D,[114] Asians do need longer exposure to sunlight than do Caucasians to produce vitamin D.[112, 113, 115]

In summary, the available evidence supports the hypothesis that depigmentation or light skin color evolved to accommodate the nutritional needs of vitamin D in regions of low solar radiation. On the other hand, Brace[116] proposes that light skin evolved from its original dark color by the relaxation of the selective forces and the accumulation of random mutations. However, the molecular genetic data of Rogers and associates[117] show that light skin has been selected for at high latitudes and might have evolved more than 1 million years ago.

Sexual Selection and the Evolution of Light Skin Color

Following Darwin, this hypothesis maintains that sexual preference for lighter than average skin color contributed to the evolution of light skin.[118, 119] This hypothesis is based upon both ethnographic and quantitative information. As reviewed by van den Berghe and Frost,[120] most of the ethnically diverse societies present in the Human Relations Area Files (see **table 21.8**) state a preference for the lighter skin color. According to historical record, the Romans, during the first century B.C., had a predisposition to find light-skinned females attractive even when they were members of a conquered group.[121] Similarly, in Japan since recorded time, "white" skin has been considered an essential characteristic of feminine beauty.[122]

TABLE 21.8

Number of Societies in Which a Preference for Lighter Skin Is Expressed, Classified by Area and Sex

Area	Males prefer	Females prefer	Number of societies with information
Sub-Saharan Africa	11	3	12
Europe and Soviet Union	2	1	2
Asia	9	4	10
Middle East and Muslim Africa	2	0	2
North America	4	2	5
South America and Caribbean	7	3	9
Insular Pacific	9	4	11

Source: Adapted from van den Berghe and Frost.[120]

The preference for lighter skin has also been documented in the United States.[123] A survey among a sample of 1,039 white undergraduate students at the University of Wyoming indicates that both sexes prefer a lighter rather than median skin color (**table 21.9**). On the other hand, most males dislike "black," whereas the great majority of females dislike the two lightest shades.

In summary, the historical, ethnographic, and anthropometric information[2, 3, 4, 13–15] indicates that males are darker than females. Since in each society a lighter than average skin color is preferred in a sexual partner, it is quite possible that such a preference would generate sexual selection for light skin that counteracts natural selection for dark skin. However, there is no evidence of differential reproduction or mortality associated with skin color preference and therefore the hypothesis of sexual selection is speculative.

Summary

Even though all human infants have the capacity to digest the lactose present in milk, there are individual and population differences in the ability to digest milk during adulthood. The persistence of lactase activity

TABLE 21.9

Percentage of Males and Females that Like or Dislike Eight Verbally Described Skin Colors in the Opposite Sex

Skin Color	Males prefer	Females prefer	Males dislike	Females dislike
Black	1	2	30	10
Brown	2	4	4	2
Red-brown	4	5	5	2
Dark white	23	27	3	1
Medium white that tans to gold	51	42	1	1
Medium white that tans to red-brown	13	15	1	2
Very light that freckles	3	3	32	42
Very light that does not freckle	3	2	24	40

Source: Data from Feinman and Gill.[123]

during adulthood is a genetic trait that is inherited in a Mendelian form. Two hypotheses have been presented for the evolution of the lactase persistence gene. The ancestral pastoralism—cultural historical hypothesis maintains that the ability to digest lactose developed as a selective advantage in certain dairying populations. The other hypothesis states that the adult lactose digestive capacity evolved as an adaptation to facilitate absorption of calcium in areas of low solar radiation and low synthesis of vitamin D. In such conditions, the capacity to digest lactose in milk might be advantageous by increasing calcium absorption. Thus, the evolution of persistent lactose digestive capacity during adulthood is a clear example of co-evolution.

The distribution of human skin color in the world today suggests past evolutionary events relating to natural selection. Current evidence suggests that the primary reason for the evolution of human skin color is the balance between the need for darker skin to protect against skin cancer, protection from the destructive effects of high solar radiation of essential nutrients such as folate, and the need for lighter skin to facilitate synthesis of vitamin D. In this context, dark skin color might have evolved as an adaptation against skin cancer and folate deficiency in areas of high solar radiation. In contrast, depigmentation or light skin color might have evolved as an adaptation to accommodate the nutritional needs of low synthesis of vitamin D in regions of low solar radiation.

Suggested Readings

Simoons, F. J. 2001. Persistence of lactase activity among northern Europeans: A weighing of evidence for the calcium absorption hypothesis. *Ecol. Food Nutr.* 40:397–469.

Frisancho, A. R. 1993. *Human Adaptation and Accommodation.* Ann Arbor: University of Michigan Press.

Stinson, S. B., B. Bogin, R. Huss-Ashmore, and D. O'Rourke. 2000. *Human Biology: An Evolutionary and Biocultural Perspective.* New York: John Wiley & Sons.

Literature Cited

1. Flatz, G. 1987. Genetics of lactose digestion in humans. In *Advances in Human Genetics,* ed. H. Harris and K. Hirschhorn. New York: Plenum, 1–77.
2. Montgomery, K., H. A. Buller, E. H. H. M. Rings, and R. J. Grand. 1991. Lactose intolerance and the genetic regulation of intestinal lactase-phlorizin hydrolase. *FASEB J* 5:2824–2832.
3. Savilathi, E., K. Launiala, P. Kuitunen. 1983. Congenital lactase deficiency: a clinical study on 16 patients. *Arch. Dis. Child.* 58:246–252.
4. Scrimshaw, N. S. and E. B. Murray. 1988. The acceptability of milk and milk products in populations with a high prevalence of lactose intolerance. *Am. J. Clin. Nutr.* 48:1083–1159.
5. Saltzman, J. R., R. M. Russell, B. Golner, S. Barakat, G. E. Dallal, and B. R. Goldin. 1999. A randomized trial of Lactobacillus acidophilus BG2FO4 to treat lactose intolerance. *Am. J. Clin. Nutr.* 69:140–146.
6. Johnson, J. D., et al. 1977. Lactose malabsorption among the Pima Indians of Arizona. *Gastroenterology* 73:1299–1304.
7. Sahi, T. and K. Launiala. 1977. More evidence for the recessive inheritance of selective adult type lactose malabsorption. *Gastroenterology* 73:231–232.

8. Kretchmer, N. 1978. Genetic variability and lactose tolerance. *Progress in Human Nutrition* 197–205.
9. Mendel, L. B. and P. H. Mitchell. 1907. Clinical studies on growth: The enverting enzymes of the alimentary tract, especially in the embryo. *Am. J. Physiol.* 20:81–96.
10. Vrese, M., A. Stegelmann, B. Richter, S. Fenselau, C. H. Laue, and J. Schrezenmei. 2001. Probiotics—compensation for lactase insufficiency. *Am. J. Clin. Nutr.* 73:421S–429S.
11. Suarez, F. L., D. Savaiano, P. Arbisi, and M. D. Levitt. 1997. Tolerance to the daily ingestion of two cups of milk by individuals claiming lactose intolerance. *Am. J. Clin. Nutr.* 65:1502–1506.
12. Simoons, F. J. 1970. Primary adult lactose intolerance and the milking habit: A problem in biological and cultural Interrelation: II. A culture historical hypothesis. *Am. J. Dig. Dis.* 15:695–710.
13. Simoons, F. J. 1989. The geographic hypothesis and lactose malabsorption. *Dig. Dis. Sci.* 23:963–980.
14. McCracken, R. D. 1971. Lactase deficiency: An example of dietary evolution. *Curr. Anthropol.* 12:479–517.
15. Johnson, J. D., N. Kretchmer, and F. J. Simoons. 1974. Lactose malabsorption: Its biology and history. In *Advances in Pedriatics*, I. Schulman, ed. Chicago: Year Book Medical Publishers, 21:197–237.
16. Kosikowski, F. V. 1985. Chesse. *Sci. Am.* 252 (no. 5):88–99.
17. Bogucki, P. 1986. The antiquity of dairying in temperate Europe. *Expedition* 28:2;51–58.
18. Durham, W. 1991. *Coevolution.* Stanford, CA: Stanford University Press.
19. Kretchmer, N. 1977. The geograph and biology of lactose digestion and malabsorption. *Postgraduate Medical J* supp.2; 53:65–72.
20. Flatz, G. and H. W. Rotthauwe. 1977. The human lactase polymorphism: Physiology and genetics of lactose absorption and malabsorption. In *Progress in Medical Genetics,* A. G. Steinberg, A. G. Bearn, A. G. Motulsky, and B. Childs (eds). Philadelphia: W. B. Saunders, 2:205–249.
21. Sahi, T. 1994. Genetics and epidemiology of adult-type hypolactasia. *Scand. J. Gastroenterol.* 29 suppl 202:7–20.
22. Holden, C. and R. Mace. 1997. Phylogenetic analysis of the evolution of lactose digestion in adults. *Hum. Biol.* 69:605–628.
23. Aoki, K. 1986. A stochastic model of gene-culture coevolution suggested by the "culture historical hypothesis" for the evolution of adult lactose absorption in humans. *Proc. Nad. Acad. Sci. USA* 83:2929–2933.
24. Feldman, M. W. and L. L. Cavalli-Sforza. 1989. On the theory of evolution under genetic and cultural transmission with application to the lactose absorption problem. In *Mathematical Evolutionary Theory,* ed. M. W. Feldman. Princeton: Princeton University Press, 145–172.
25. Bodmer, W. F. and L. L. Cavalli-Sforza. 1976. *Genetics, Evolution, and Man.* San Francisco: W.H. Freeman.
26. Pringle, H. 1998. The slow birth of agriculture. *Science* 282: 1446–1450.
27. Clutton-Brock, J. 1987. *A Natural History of Domesticated Mammals.* London: British Museum (Natural History) and Cambridge University Press.
28. Pettifor, J. M., G.P. Moodley, F. S. Hough, H. Koch, T. Chen, Z. Lu, and M. F. Holick. 1996. The effect of season and latitude on in vitro vitamin D formation by sunlight in South Africa. *S. Ajr. Med. J.* 86:1270–1272.
29. Holick, M. F. 1996. The role of sunlight in providing vitamin D for bone health. In *Biologic Effects of Light 1995,* ed. M. F. Holick, E. G. Jung Berlin. New York: W. de Gruyter, 4–12.
30. Gibbs, D. 1994. Rickets and the crippled child: An historical perspective. *J. Roy. Soc. Med.* 87:729–732.
31. Drummond, J. C. and A. Wilbraham. 1958. *The Englishman's Food: A History of Five Centuries of English Diet,* 2nd ed. London. First published in 1939.
32. Vaughan, O. W. and L. J. Filer, Jr. 1960. The enhancing action of certain carbohydrates on the intestinal absorption of calcium in the rat. *J. Nutr.* 71:10–14.
33. Leichter, J. and A. F. Tolensky. 1975. Effect of dietary lactose on the absorption of protein, fat and calcium in the postweaning rat. *Am. J. Clin. Nutr.* 28:238–241.
34. Armbrecht, H. J., and R. Wassermann 1979. Enhancement of Ca 2+ uptake by lactose in the rat small intestine. *J. Nutr.* 106:1265–1271.
35. Schaafsma, G. and R. Visser. 1980. Nutritional interrelationships between calcium, phosphorus and lactose in rats. *J. Nutr.* 110:1101–1111.
36. Schaafsma, G., W. J. Visser, P. R. Dekker, and M. van Schaik. 1988. Effect of dietary calcium supplementation with lactose on bone in vitamin D-deficient rats. *Bone* 8:357–362.
37. Gregor, J. L., C. M. Gutkowski, and R. R. Khazen. 1989. Interactions of lactose with calcium, magnesium and zinc in rats. *J. Nutr.* 119:1691–1697.
38. Schuette, S. A., N. J. Yasillo, and C. M. Thompson. 1991. The effect of carbohydrates in milk on the absorption of calcium by postmenopausal women. *J. Am. Coll. Nutr.* Apr;10(2):132–139.
39. Zittermann, A., P. Bock, C. Drummer, K. Scheld, M. Heer, and P. Stehle. 2000. Lactose does not enhance calcium bioavailability in lactose-tolerant, healthy adults. *Am. J. Clin. Nutr.* 71:931–936.
40. Roberts, D. F., and D. P. S. Kahlon. 1976. Environmental correlations of skin colour. *Annals of Human Biology* 3:11–22.
41. Jablonski, N. G., and G. Chaplin. 2000. The evolution of human skin color variation. *J. Hum. Evol.* 39:57–106.
42. Relethford, J. H. 1997. Hemispheric differences in human skin color. *Am. J. Phys. Anthrop.* 104:449–458.
43. Frisancho, A. R., R. Wainwright, and A. Way. 1981. Heritability and components of phenotypic expression in skin reflectance of mestizos from the Peruvian low lands. *Am. J. Phys. Anthropol.* 55:203–208.
44. Williams-Blangero, S., and J. Blangero. 1992. Quantitative genetic analysis of skin reflectance: A multivariate approach. *Human Biology* 64:35–49.
45. Harrison, G. A., and J. J. T. Owen. 1964. Studies on the inheritance of human skin colour. *Ann. Hum. Genet.* London 28:27–37.
46. Garcia, R. I., R. E. Mitchell, J. Bloom, and G. Szabo. 1977. Number of epidermal melanocytes, hair follicles, and sweat ducts in the skin of Solomon Islanders. *Am. J. Phys. Anthropol.* 47:427–434.

47. Szabo, G., A. B. Gerald, M. A. Pathak, and T. B. Fitzpatrick. 1972. The ultrastructure of racial color differences in man. In *Pigmentation: Its Genesis and Biologic Control,* ed. V. Riley. New York: Appleton-Century-Crofts, 123–127.

48. Stinson, S., B. Bogin, R. Huss-Ashmore, and D. O'Rourke eds. 2000. *Human Biology: An Evolutionary and Biocultural Perspective.* New York: Wiley-Liss.

49. Garrad, C., G. A. Harrison, and J. J. T Owen. 1967. Comparative spectrophotometry of skin color with EEL and photovolt instruments. *Am. J. Phys. Anthropol.* 27:389–396.

50. Lasker, G. W. 1954. Seasonal changes in skin color. *Am. J. Phys. Anthropol.* 12:553–558.

51. Weiner, J. S., G. A. Harrison, R. Singer, A. Harris, and W. Japp. 1964. Skin color in Southern Africa. *Hum. Biol.* 36:294–307.

52. Hulse, F. S. 1967. Selection for skin color among Japanese. *Am. J. Phys. Anthropol.* 27:143–156.

53. Conway, D., and P. I. Baker. 1972. Skin reflectance of Quechua Indians: The effects of genetic admixture, sex and age. *Am. J. Phys. Anthropol.* 36:267–282.

54. Garn, S. M., and N. Y. French. 1963. Postpartum and age changes in areolar pigmentation. *Am. J. Obst. and Gyn.* 85:873–875.

55. Kubota, R., Y. Yang, S. Minoshima, J. Kudoh, Y. Mashima, Y. Oguchi, and N. Shimizu. 1995. Mapping of the human gene for a melanocyte protein Pmel 17 (D12S53E) to chromosome 12q13-q14. *Genomics* 26:430–431.

56. Mountjoy, K. G., L. S. Robbins, M. T. Mortrud, and R. D. Cone. 1992. The cloning of a family of genes that encode melanocortin receptors. *Science* 257:1248–1251.

57. Gantz, I., Y. Konda, T. Tashiro, Y. Shimoto, H. Miwa, G. Munzert, S. J. Watson, J. Delvalle, and T. Yamada. 1993. Molecular-cloning of a novel melanocortin receptor. *J. Biol. Chem.* 268:8246–8250.

58. Valverde, P., E. Healy, I. Jackson, J. L. Rees, and A. J. Thody. 1995. Variants of the melanocyte-stimulating hormone receptor gene are associated with red hair and fair skin in humans. *Nat. Genet.* 11:328–330.

59. Cone, R. D., D. Lu, S. Koppula, D. I. Vage, H. Klungland, B. Boston, W. Chen, D. N. Orth, C. Pouton, and R. A. Kesterson. 1996. The melanocortin receptors: Agonists, antagonists, and the hormonal control of pigmentation. *Recent Prog. Horm. Res.* 51:287–318.

60. Box, N. F., J. R. Wyeth, L. E. O' Gorman, N. G. Martin, and R. A. Sturm. 1997. Characterization of melanocyte stimulating hormone receptor variant alleles in twins with red hair. *Hum. Mol. Gen.* 6:1891–1897.

61. Rana, B. K., D. Hewett-Emmett, L. Jin, B. H. J. Chang, N. Sambuughin, M. Lin, S. Watkins, M. Bamshad, L. B. Jorde, M. Ramsay, T. Jenkins, and W. H. Li. 1999. High polymorphism at the human melanocortin 1 receptor locus. *Genetics* 151:1547–1557.

62. Harding, R. M., E. Healy, A. J. Ray, N. S. Ellis, N. Flanagan, C. Todd, C. Dixon, A. Sajantila, I. J. Jackson, M. A. Birch-Machin, and J. L. Rees. 2000. Evidence of variable selective pressures at MC1R. *Am. J. Hum. Gen.* 66:1351–1361.

63. Gilchrest, B. A., H. Y. Park, M. S. Eller, and M. Yaar. 1996. Mechanisms of ultraviolet light-induced pigmentation. *Photochem. Photobiol.* 63:1–10.

64. Sturm, R. A., N. F. Box, and M. O. Ramsay. 1998. Human pigmentation genetics: The difference is only skin deep. *Bioessays.* 20:712–721.

65. Sturm, R. A. 2002. Skin colour and skin cancer—MC1R, the genetic link. *Melanoma Res.* 12(5):405–416.

66. Leffell, D. J., and D. E. Brash. 1996. Sunlight and skin and cancer. *Sci. Amer.* 275:52–59.

67. Akhtar, M. J., M. A. Khan, and I. Ahmad. 1999. Photodegradation of folic acid in aqueous solution. *J. Pharm. Biomed. Anal.* 19:269–275.

68. Bjorn, L. O., and T. Wang. 2000. Vitamin D in an ecological context. *Int. J. Circumpolar Health* 59:26–32.

69. Holick, M. F., J. A. MacLaughlin, and S. H. Doppelt. 1981. Regulation of cutaneous previtamin D photosynthesis in man: Skin pigment is not an essential regulator. *Science* 211:590–593.

70. Holick, M. F. McCollum Award Lecture, 1994: Vitamin D—new horizons for the 21st century. *Am. J. Clin. Nutr.* 60(4): 619–630.

71. Holick, M. F. 2000. Sunlight and vitamin D: The bone and cancer connections. *Radiation Protection Dosimetry* 91: 65–71.

72. DeLuca, H. E. 1974. Vitamin D: The vitamin and the hormone. *Fed Proc.* 33:2211–2219.

73. Webb, A. R., and M. F . Holick. 1988. The role of sunlight in the cutaneous production of vitamin D_3. *Ann. Rev. Nutr.* 8:375–399.

74. Webb, A. R., L. Kline, and M. F. Holick. 1988. Influence of season and latitude on the cutaneous synthesis of vitamin D_3: exposure to winter sunlight in Boston and Edmonton will not promote vitamin D_3 synthesis in human skin. *J. Clin. Endocrinol. Metab.* 67(2):373–378.

75. Neer, R. M. 1975. The evolutionary significance of vitamin D, skin pigmentation, and ultraviolet light. *Am. J. Phys. Anthropol.* 43:409–416.

76. Nesby-O'Dell, S., K. S. Scanlon, M. E. Cogswell, C. Gillespie, B. W. Hollis, A. C. Looker, C. Allen, C. Doughertly, E. W. Gunter, and B. A. Bowman. 2002. Hypovitaminosis D prevalence and determinants among African American and white women of reproductive age: Third National Health and Nutrition Examination Survey, 1988–1994. *Am. J. Clin. Nutr.* 76:187–189.

77. Frisancho, A. R. 1993. *Human Adaptation and Accommodation.* Ann Arbor, Michigan: University of Michigan Press.

78. Gilchrest, B. A., H. Y. Park, M. S. Eller, and M. Yaar. 1996. Mechanisms of ultraviolet light-induced pigmentation. *Photochem. Photobiol.* 63:1–10.

79. Freedman, D. M., M. Dosemeci, and K. McGlynn. 2002. Sunlight and mortality from breast, ovarian, colon, prostate, and non-melanoma skin cancer: A composite death certificate based case-control study. *Occup. Environ. Med.* 59:257–262.

80. Decarli, A., and C. La Vecchia. 1986. Environmental factors and cancer mortality in Italy: Correlational exercise. *Oncology* 43:116–126.

81. Jemal, A., S. S. Devesa, T. R. Fears, and P. Hartge. 2000. Cancer surveillance series: Changing patterns of cutaneous malignant melanoma mortality rates among whites in the United States. *J. Natl. Cancer Inst.* 92(10):811–818.

82. Scotto, J., and T. R. Fears. 1987. The association of solar ultraviolet and skin melanoma incidence among Caucasians in the United States. *Cancer Invest.* 5(4):275–283.

83. Robins, A. H. 1991. *Biological Perspectives on Human Pigmentation.* Cambridge: Cambridge University Press.

84. Thomson, M. L. 1951. The cause of changes in sweating rate after ultraviolet radiation. *J. Physiol.* 112:31–42.

85. Thomson, M. L. 1955. Relative efficiency of pigment and horny layer thickness in protecting the skin of Europeans and Africans against solar ultraviolet radiation. *Am. J. Physiol.* 127:236–246.

86. Gellin, G. A., A. W. Kopf, and L. Garfinkel. 1969. Malignant melanoma: A controlled study of possibly associated factors. *Arch. Dermatol.* 99:43–48.

87. Lee, J. A. H. 1972. Sunlight and the etiology of malignant melanoma. In *Melanoma and Skin Cancer.* W. H. McCarthy, ed. Sydney, Australia: Government Printer.

88. Beardmore, G. L. 1972. The epidemiology of malignant melanoma in Australia. In *Melanoma and Skin Cancer.* W. H. McCarthy, ed. Sydney, Australia: Government Printer.

89. Livingstone, F. 1969. Polygenic models for the evolution of human skin color differences. *Hum. Biol.* 41:480–493.

90. Branda, R. F., and J. W. Eaton. 1978. Skin color and nutrient photolysis: An evolutionary hypothesis. *Science* 201:625–626.

91. Lapunzina, P. 1996. Ultraviolet light-related neural tube defects? *Am. J. Med. Genet.* 67:106.

92. Jablonski, N. G. 1999. A possible link between neural tube defects and ultraviolet light exposure. *Med. Hypotheses* 52: 581–582.

93. Medical Research Council Vitamin Study Research Group. 1991. Prevention of neural tube defects: Results of the MRC vitamin study. *Lancet* 338:131–137.

94. Frey, L., and W. A. Hauser. 2003. Epidemiology of neural tube defects. *Epilepsia* 44 (Suppl.) 3:4–13.

95. Van Rootselaar, F. J. 1993. The epidemiology of neural tube defects: A mathematical model. *Med. Hypotheses* 41:78–82.

96. Grace, H. J. 1981. Prenatal screening for neural tube defects in South Africa: An assessment. *S. Afr. Med. J.* 60:324–329.

97. Khoury, M. J., J. D. Erickson, and L. M. James. 1982. Etiologic heterogeneity of neural tube defects: Clues from epidemiology. *Am. J. Epidemiol.* 115:538–548.

98. Stevenson, R. E., W. R. Allen, P. Shashiclhar, R. Best, L. H. Seaver, J. Dean, et al. 2000. Decline in prevalence of neural tube defects in a high-risk region of the United States. *Pediatrics* 106:677–683.

99. Botto, L. D., C. A. Moore, M. J. Khoury, and J. D. Erickson. 1999. Neural-tube defects. *N. Engl. J. Med.* 341:1509–1519.

100. Barker, D., K. Dixon, E. Medrano, et al. 1995. Comparison of the responses of human melanocytes with different melanin contents to ultraviolet B irradiation. *Cancer Res.* 55:4041–4046.

101. Mathur, U., S. L. Datta, and B. B. Mathur. 1977. The effect of aminopterin-induced folic acid deficiency on spermatogenesis. *Fertil. Steril.* 28:1356–1360.

102. Cosentino, M. J., R. E. Pakyz, and J. Fried. 1990. Pyrimethamine: An approach to the development of a male contraceptive. *Proc. Natl. Acad. Sci.* USA 87:1431–1435.

103. Cohn, B. A. 2002. Sunlight, skin color, and folic acid. *J. Am. Acad. Dermatol.* 46:317–318.

104. Murray, F. G. 1934. Pigmentation, sunlight and nutritional disease. *Am. Anthropol.* 36:438–445.

105. Loomis, W. F. 1967. Skin-pigment regulation of vitamin-D biosynthesis in man. *Science* 157:501–506.

106. Specker, B. L., R. C. Tsang, and B. W. Hollis. 1985. Effect of race and diet on human-milk vitamin D and 25-hydroxyvitamin D. *Am. J. Dis. Child.* 139:1134–1137.

107. Eastman, N. J. 1956. Obstetrics, 11th ed. New York: Appleton-Century-Crofts.

108. Holmes, A. M., B. A. Enoch, J. L. Taylor, and M. E. Jones. 1973. Occult rickets and osteomalacia amongst the Asian immigrant population. *Q. J. Med., New Series* 42:125–149.

109. Stephens, W. P., P. S. Klimiuk, S. Warrington, J. L. Taylor, J. L. Berry, and E. B. Mawer. 1982. Observations on the natural history of vitamin D deficiency amongst Asian immigrants. *Q. J. Med.* 51:171–188.

110. Pal, B. R., and N. J. Shaw. 2001. Rickets resurgence in the United Kingdom: Improving antenatal management in Asians. [Letter] *J. Pediatr.* 139(2):337–338.

111. Haworth, J. C., and L. A. Dilling. 1986. Vitamin-D-deficient rickets in Manitoba, 1972–84. *Can. Med. Assoc. J.* 134(3): 237–241.

112. Binet, A., and S. W. Kooh. 1996. Persistence of vitamin D-deficiency rickets in Toronto in the 1990s. *Can. J. Public Health* 87:227–230.

113. Welch, T. R., W. H. Bergstrom, and R. C. Tsang. 2000. Vitamin D-deficient rickets: The reemergence of a once-conquered disease. *J. Pediatr.* 137:143–145.

114. Lo, W., P. W. Paris, and M. F. Holick. 1986. Indian and Pakistani immigrants have the same capacity as Caucasians to produce vitamin D in response to ultraviolet radiation. *Am. J. Clin. Nutr.* 44:683–685.

115. Norman, A. W. 1998. Sunlight, season, skin pigmentation, vitamin D, and 25-hydroxyvitamin D: Integral components of the vitamin D endocrine system. *Am. J. Clin. Nutr.* Norman. 67:1108–1110.

116. Brace, C. L. 1973. A nonracial approach towards the understanding of human diversity. In C. L. Brace, and J. Metress, eds. *Man in Evolutionary Perspective*. New York: Wiley, 341–363.

117. Rogers, A. R., D. Iltis, and S. Wooding. 2004. Genetic variation at the MC1R locus and the time since loss of human body hair. *Curr. Anthropol.* 45:105–108.

118. Aoki, K., M. W. Feldman, and B. Kerr. 2001. Models of sexual selection on a quantitative genetic trait when preference is acquired by sexual imprinting. *Evolution* 55:25–32.

119. Aoki, K. 2002. Sexual selection as a cause of human skin colour variation: Darwin's hypothesis revisited. *Ann. Hum. Biol.* 29:589–608.

120. van den Berghe, P. L., and P. Frost. 1986. Skin color preference, sexual dimorphism and sexual selection: A case of gene culture coevolution? *Ethn. Racial Stud.* 9:87–113.

121. Thompson, L. A. 1989. *Romans and Blacks.* Norman: University of Okla homa Press.

122. Wagatsuma, H. 1967. The social perception of skin color in Japan. *Daedalus* 96:407–443.

123. Feinman, S., and G. W. Gill. 1978. Sex differences in physical attractiveness preferences. *J. Soc. Psychol.* 105:43–52.

Adapting to a Changing World

One of the traits that distinguishes humans from other primates is our ability to modify the environment to fit to our needs. During the first five to eight million years of evolutionary history, the foods available and consumed by hominids were probably similar to that of extant hominoids. Then, about two million years ago, when *Homo erectus* harnessed and controlled fire, the nutritional resources and environment changed. As shown by the charcoal remains of burned animals found in South Africa and Asia, fire was probably first used for keeping warm, keeping predators away, and, perhaps, for roasting meat and vegetables. It is not clear for which purpose fire was used first; what is clear is that by using and controlling fire, humans changed the environment in which they initially evolved, and thus set a chain of events in motion. The shift from hunting and gathering to agriculture changed the human dietary source from a pattern that included both animal foods and plant foods hunted and gathered with high energy expenditure to a pattern where plant foods and animals were produced rather than hunted or gathered. Eventually, agriculture led to the industrial revolution, agribusiness, and modern food-processing techniques, all of which have changed the human nutritional and cultural environment to which they have had to adapt. This chapter outlines some of the effects and the ongoing adaptations that humans are making in this changing and evolving world.

Changing Health Conditions

Blood Pressure

The circulation of the blood through the body is a continuous circuit, in that if a given amount of blood is pumped by the heart, the same amount also flows through each subdivision of the circulation. The circulatory system is divided into the systemic circulation and the pulmonary circulation (**fig. 22.1**). The pulmonary circulation supplies blood from the heart to the lungs, while systemic circulation supplies blood to all the tissues of the body except the lungs. The systemic circulation is also called the peripheral circulation or the greater circulation, because the greater amount of the blood in the circulatory system is contained in the systemic veins.

Measurement of Blood Pressure

Blood pressure refers to the force exerted by the blood against the wall of the blood vessel. It is measured in millimeters of mercury (mm Hg). Hence, a blood pressure of 120 mm Hg means that the force exerted upon the wall of a blood vessel is sufficient to push a column of mercury up to a level of 100 mm high. Un-

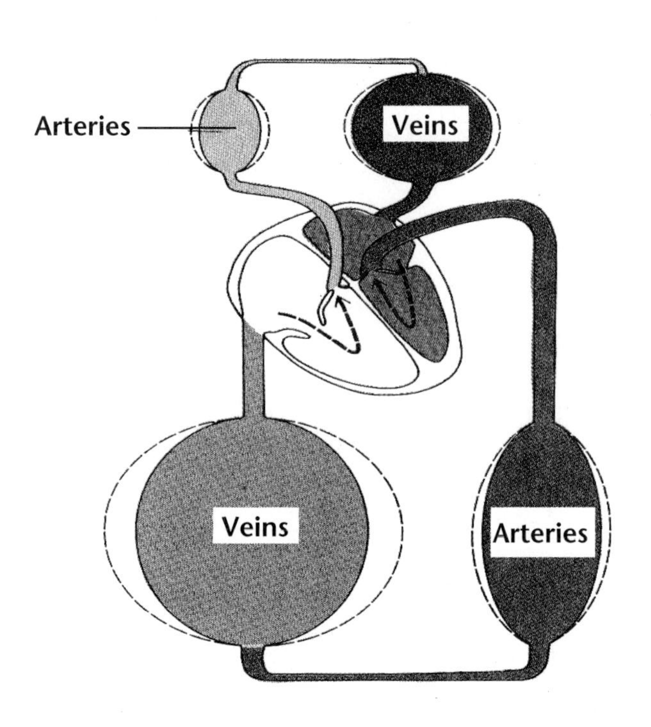

FIGURE 22.1. The pulmonary and systemic circulatory system of the blood. (Composite illustration made by the author.)

der experimental conditions, a cannula or catheter is inserted into an artery, a vein, or even the heart, and the pressure from the cannula or catheter is transmitted to the left-hand side of the manometer where it pushes the mercury down while raising the right-hand mercury column. The difference between the two levels of mercury is approximately equal to the pressure in the circulation in terms of millimeters of mercury.

Blood pressures can be measured by invasive and noninvasive methods. The invasive technique assesses blood pressure using a needle sensor (either fluid-filled or solid-state) inserted directly into the blood flow of the brachial artery at the end of a catheter connected to high-speed electrical pressure transducers. Because of impracticality, the invasive technique is not commonly done. Instead, measurements of blood pressure are usually obtained by indirect means such as the auscultatory method (**fig. 22.2**). It is determined by occluding the artery with an air-inflated bladder inside a cuff (called a sphygmomanometer), which is attached to a mercury column or gauge. A listener, using a stethoscope placed over the antecubital artery, slowly releases the pressure inside the bladder. The blood pressure of the arterial includes two pressures: systolic and diastolic blood pressure.

Systolic Pressure. Systolic blood pressure refers to the maximum pressure exerted against the arterial wall at the peak of ventricular contraction. To obtain the systolic pressure, a blood pressure cuff is inflated so its pressure exceeds 170 mm Hg or the highest pressure within the artery. As long as this pressure is high, the brachial artery remains collapsed and no blood whatsoever flows into the lower artery during any part of the pressure cycle. But then the cuff pressure is gradually reduced; as soon as the pressure in the cuff falls below systolic pressure, blood slips through the artery beneath the cuff during the peak of systolic pressure. One begins to hear distinct tapping sounds in the antecubital artery in synchrony with the heartbeat. The appearance of the first sounds (called Korotkoff phase I) is the systolic pressure.

Diastolic Pressure. Diastolic pressure is the minimum pressure exerted when the heart is at rest. The pressure in the cuff is lowered further, blood slips through the artery beneath the cuff during the peak of systolic pressure, and one begins to hear distinct tapping

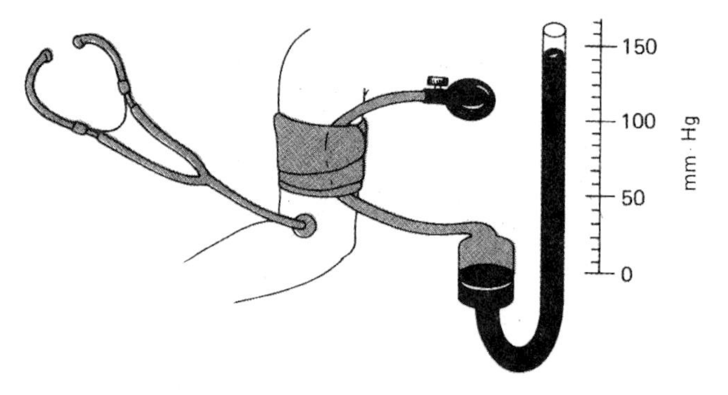

FIGURE 22.2. Manual measurement of arterial blood pressures. Blood pressure is measured by occluding the brachial artery, usually in the left arm, with an air-inflated bladder inside a cuff (called a sphygmomanometer), which is attached to a mercury column or gauge. The cuff is inflated to about 185 mm Hg. A listener, using a stethoscope placed over the antecubital artery, slowly releases the pressure inside the bladder. The first sound that is heard following the release of pressure is the systolic blood pressure (Korotkoff phase I) and the sound disappearing corresponds to the diastolic blood pressure (Korotokoff phase V). (Composite illustration by the author.)

sounds in the antecubital artery in synchrony with the heartbeat. Then, finally, as the pressure is decreased, the artery no longer closes and the sounds suddenly change to a muffled quality. When the sounds disappear (called Korotokoff phase V) the diastolic pressure can be determined.

Although it is very easy to take a blood pressure measurement, a single determination does not accurately reflect the continuing adaptation of blood pressure to changing conditions. For example, it has been documented that blood pressures measured in a physician's office under quiet conditions are not reliable indicators of blood pressure during the rest of the day or of the average level of pressure of the individual. For this reason, health researchers use ambulatory blood pressure monitors[1] that take measurements every 15 min during waking hours and every 30 min during sleep for a 24-hour period on a normal workday.

Normal blood pressures ranges from less than 120 mm Hg systolic and less than 80 mm Hg diastolic (**table 22.1**).

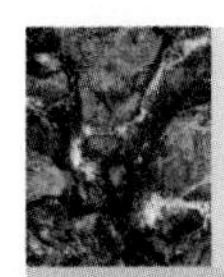

TABLE 22.1

Categories for Blood Pressure Levels in Adults

Blood Pressure Level (mm Hg) Category	Systolic	Diastolic
Normal	< 120	< 80
Prehypertension	120–139	80–89
Stage 1 Hypertension	140–159	90–99
Stage 2 Hypertension	≥ 160	≥ 100

Source: http://www.nhlbi.nih.gov.

Blood Pressure and Stress

There is considerable evidence indicating that high blood pressure is linked to persistent stress and the way in which people cope.[2] Chronic psychological stress is associated with increased circulating levels of adrenaline and noradrenaline resulting from the increased activation of the sympathetic-adrenomedullary axis. Chronically elevated adrenaline levels have been implicated in the increase in blood pressure and the development of hypertension.[3, 4] In contrast, conditions that reduce stress and sympathetic arousal have been shown to lower blood pressure and prevent the development of hypertension.[4–6] For this reason, measurements of blood pressures are used as an indicator of stress to which the individual or population is exposed. Stress, as defined in Chapter 15, refers to any condition that disrupts the homeostatic equilibrium of the organism. Stressors can include work, commuting, noise, pollution, crowding, alcohol, poor nutrition, poor sanitation, and psychological distress. The organism responds to the stressors through adaptive processes oriented at restoring the homeostatic equilibrium, but not all individuals respond in the same fashion. Some cope very well, while others are not able to restore homeostasis. One criteria for measuring the success of the adaptive process on the effects of environmental stressors is obtained from evaluations of blood pressure of individuals in different environments.

Blood Pressure and Stress of Adoption of Modern Western Lifestyle

Studies of Brown and colleagues have demonstrated that Filipino immigrants who had lived longer in the United States had elevated norepinephrine levels in the work and home setting, higher diastolic blood pressure, and lower dips in blood pressure during sleep.[8] A study of female teachers working in public schools located in Hilo, Hawaii, found a greater increase in blood pressure with age in Japanese Americans than those of Caucasian ethnicity (**fig. 22.3**).[9] Similarly, 24-hour evaluations of ambulatory blood pressures found that Japanese American women had a significantly higher mean diastolic blood pressure during sleep and a lower blood pressure dip between waking and sleeping than those of Caucasian ethnicity.[9]

In summary, the above studies,[6–9] and other studies,[10–15] suggest that adoption of modern Western lifestyles is associated with increases in the blood pressure of many populations. In other words, the Western industrialized world—while it has job opportunities for many—has also created stressful conditions that have profound negative effects on the health condition of many populations not accustomed to such conditions. Hence, industry must develop more appropriate conditions that will lead to better health. Since healthy workers are more productive, the benefits of such a change to humankind are obvious.

Sodium Concentration and Blood Pressure

The concentration of sodium in the blood is maintained within a narrow range (80 to 100 g) through homeostatic processes. In general, when the amount of sodium intake is high, the blood pressure is elevated. For example, observations in normotensive subjects

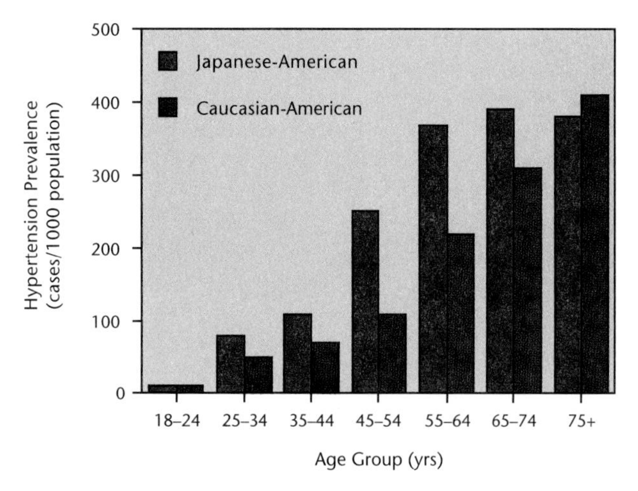

FIGURE 22.3. Prevalence of hypertension by age, in Hawaii, of Japanese American and Caucasian ethnic groups. Source: data derived from Brown, D. E., Leidy Sievert, L., Aki, S. L., Mills, P. S., Etrata, M. B., Paopao, R. N. K., and G. D. James. 2001. Effects of age, ethnicity and menopause on ambulatory blood pressure: Japanese American and Caucasian school teachers in Hawaii. *Am. J. Hum. Biol.* 13:486–493.

who were subjected to an incremental increase of sodium intake from 10 to 1500 mmol/day showed that the blood pressure rose significantly from the lowest to the highest salt intake.[17] The reason for the salt-associated increase in blood pressure is related to two factors: (a) change in osmolarity and (b) increased secretory activity of the hypothalamic-posterior pituitary gland.

Change of Osmolarity. When the amount of salt in the body is high, the osmolarity of the body fluids increases; this, in turn, stimulates the brain's thirst center, making the person drink extra amounts of water to dilute the extracellular salt to a normal concentration, which also increases the extracellular fluid volume.

Activity of the Hypothalamic-Posterior Pituitary Gland. The increase in osmolarity in the extracellular fluid also increases the secretion of large quantities of antidiuretic hormone by the hypothalamic-posterior pituitary gland secretory mechanism. As a result of the increased secretion of the antidiuretic hormone, the kidneys reabsorb large amounts of water from the urine before it is excreted, thus diminishing the volume of urine while increasing the extracellular fluid volume.

Thus, for these two important reasons, when the amount of salt in the body is high, the organism responds by enhancing fluid intake and fluid retention, which, in turn, leads to an elevation of the arterial pressure.

High Blood Pressure, Sociocultural Stress, and Past Adaptation to Hot Climates

In general, African Americans, Puerto Ricans, and Cubans[18, 19] have a predisposition to hypertension. It has been indicated that the increased risk of high blood pressure among African Americans is related to sociocultural stress resulting from the struggle of dark skin individuals to establish and maintain a middle-class lifestyle with individuals of light skin color.[20–25] Others propose that the tendency to maintain higher blood pressure in African Americans is a by-product of an adaptation to the hot tropical climate of Africa and/or as a result of selection during the transport of slaves from Africa to America.[18, 26, 27] This hypothesis is based upon several studies.

First, a common denominator of African Americans is that they originally descended from the hot tropical climate of Africa. Acclimatization to high temperature is associated with more than a 50 percent reduction in salt excretion in sweat for any specific sweating rate.[28–29] Hence, the retention of salt is adaptive in an environment with a high sweat rate, because it ensures that an appropriate amount of salt is present in the blood. Therefore, it is quite possible that populations, such as African blacks, adapted to a hot, tropical climate and have developed thermoregulatory traits oriented at economizing sodium loss during sweating. This adaptation probably became established in the genetic profile of populations adapted to hot tropical climates.

Second, experimental and clinical studies have shown that blacks in the United States have an enhanced sodium retention and a greater increase in blood pressure in response to greater Na+ intake than whites.[19, 30-32] Similarly, epidemiological studies indicate that African Americans ingest similar amounts of dietary salt as non-African Americans.[33]

To test these hypotheses, the author and associates[34] conducted an anthropological study of blacks of Bolivia. This population lives in a semitropical area of the eastern lowland of Bolivia and has better socioeconomic conditions than the indigenous Bolivian population that lives in the same area. In this study we found that the frequency of hypertension (defined as systolic pressure >140 mm Hg, and diastolic pressure > 99 mm Hg) of blacks living in the lowlands of Bolivia is only 4.5 percent, while in African Americans in the United States exceeds 30 percent. However, as

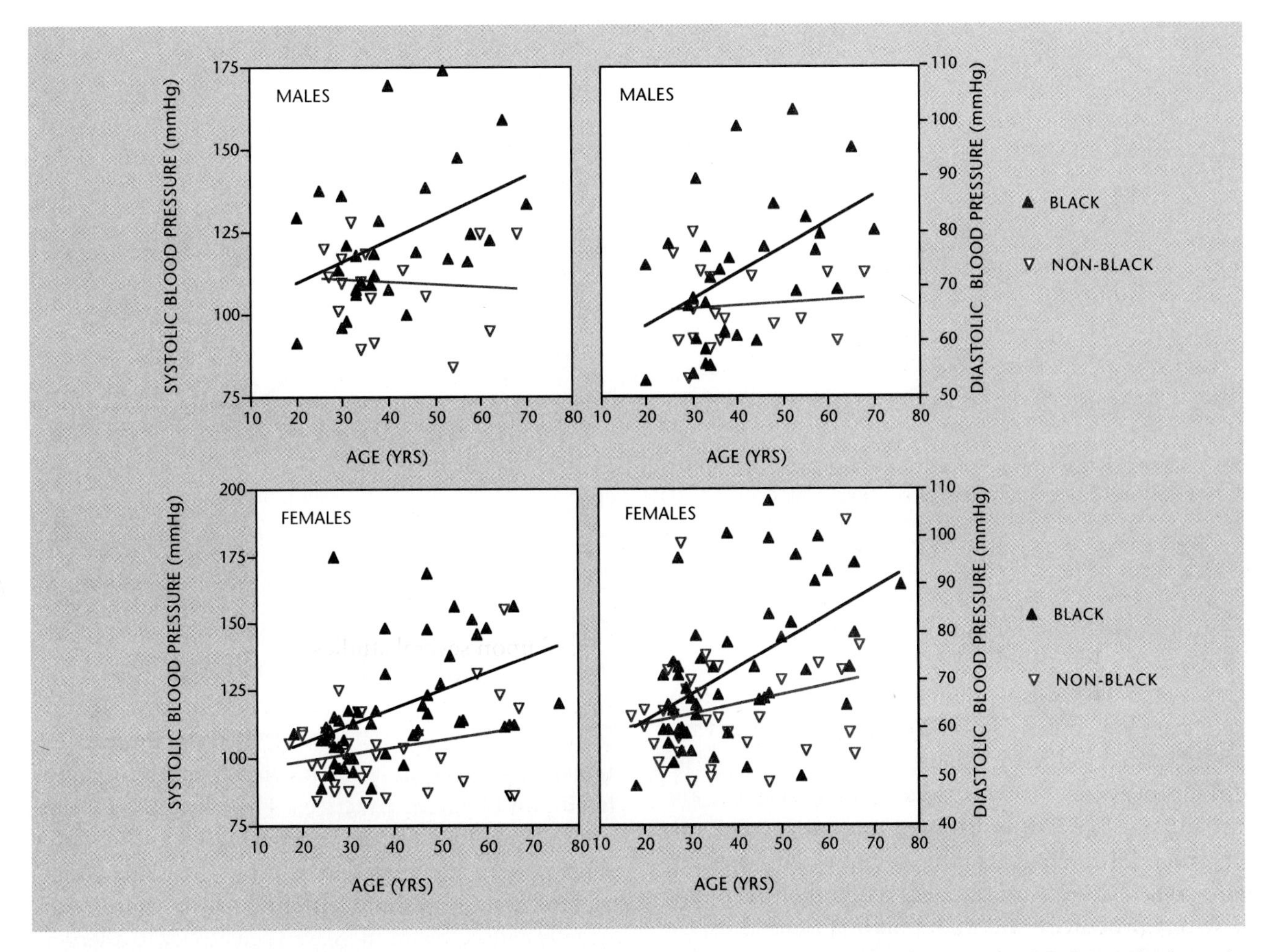

FIGURE 22.4. Comparison of the age-associated increase in systolic and diastolic blood pressures of Bolivian Blacks and non-Blacks living in the same village of Chicaloma, Bolivia. [Data derived from Frisancho AR, Farrow S, Friedenzohn I, Johnson T, Kapp B, Miranda CH, Perez M, Rauchle I, Sanchez N, Swaninger K, G. Wheatcroft G, Woodill L, Ayllon I, Soria R, Rodriguez A, Machicao J, Villena M, Vargas E. 1999. Role of Genetic and Environmental Factors in the Increased Blood Pressures of Bolivian Blacks. Amer. J. Hum. Biol. 11: 489-498).]

shown in **figure 22.4**, the age-associated increase in blood pressure is higher among blacks living under better socioeconomic conditions, than non-blacks, who live in the same village of the lowlands of Bolivia.[34] The salt excretion was lower in blacks than in non-blacks despite the fact that both samples had the same daily dietary intake of salt. These findings suggest that the tendency toward high blood pressure of African Americans is related to their high retention of salt,[34] which probably is a by-product of a past adaptation to a tropical, hot climate. In other words, in an environment where the risk of dehydration is low, the ability to retain body salt along with normal dietary intake of salt increases the concentration of salt, resulting in increased extracellular fluid retention and water intake. As result of the increased amount of extracellular fluid volume, the blood pressure increases.

These findings together support the hypothesis that the increased risk of high blood pressure of African American blacks is due to the **compound** effects of genetic susceptibility to retain salt[31] derived from past adaptations to a tropical climate and to the sociocultural stress to which dark-skinned populations are exposed.[22–25]

Dental Malocclusion and Food Processing

Orthodontists usually classify occlusion patterns in three categories: edge-to-edge bite, overbite, and underbite (**fig. 22.5**). Edge-to-edge bite is the normal pattern and refers to the maximum contact of the teeth, while overbite and underbite are considered malocclusion. There is considerable evidence indicat-

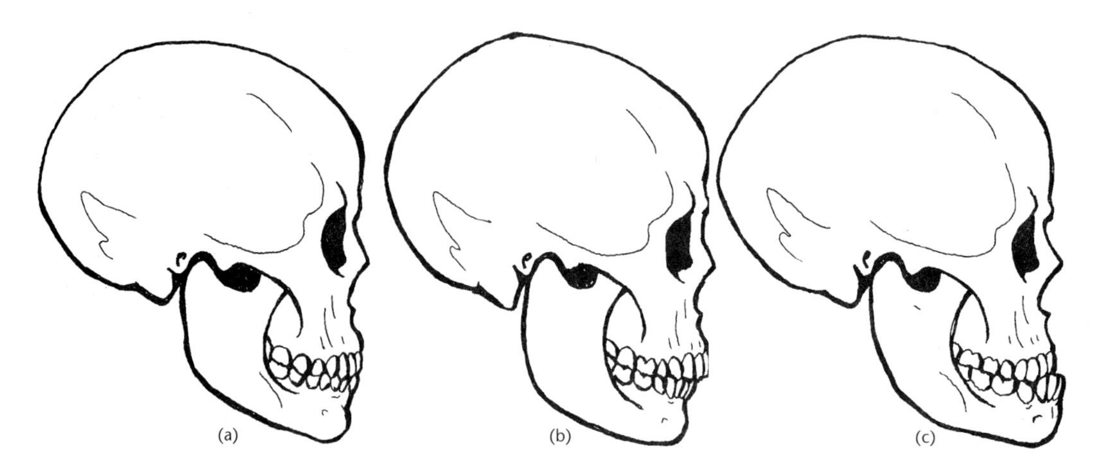

FIGURE 22.5. Occlusion patterns in humans. The centric or edge-to-edge bite (a) is considered a normal occlusion, while (b) overbite or (c) underbite reflect malocclusion. (Drawing made for the author by Paul J. Bohensky.)

ing that, with the exception of prehistoric Native Americans from Ohio,[35, 36] hunter-gatherers had larger teeth (especially premolars and molars) than industrialized peoples. The reduction of teeth size is said to be related to the increasing role of technology for food processing, brought about with agriculture and sedentism. As humans acquired the ability to modify their food through pounding, grinding, and cooking, they relied more on tools and less on their teeth and jaws for procuring and manipulating food; thus, the need for large teeth decreased.[37] The changes in teeth and jaw size were not always congruent, resulting in a high frequency of malocclusion.[38, 39] For example, Inuit populations, such as Eskimos, prior to contact with industrialized societies, had a low frequency of malocclusion.[40] However, it rose to 50 percent or more after contact with industrialized societies.[40] Likewise, among hunter-gatherers and nonintensive agriculturists, the frequency of an overbite averages less than 20 percent compared to more than 30 percent for advanced agricultural and industrialized people.[41] Likewise, the increased frequency of malocclusion among Japanese populations was associated with the use of cooking and eating utensils.[42]

In summary, it appears that with modernization humans acquired the ability to modify their food through pounding, grinding, and cooking. Consequently, they relied more on tools and less on their teeth and jaws for procuring and manipulating food. Therefore, the need for large teeth decreased; causing disproportion in the relationship of the teeth, mandible, and jaws, which, in turn, is reflected in the high frequency of malocclusion that characterize industrialized populations.

Allergies and Parasites: A Paradigm of Immunology

In recent years, the prevalence of allergies (also referred to as atopy) has risen in industrialized countries, concomitant with improvements in public health and hygiene practices. The causes of allergies are multifactorial and are reflected by an increased sensitivity to allergens that result in the high production of histamines and inflammation. Allergies are usually caused by an increased sensitivity to allergens that result in the high production of histamines and inflammation. In general, the stimulatory signals for allergies are supposed to occur via the action of helminthic parasites, which selectively stimulate the immunological system. Yet, the prevalence of allergies and sensitivity to allergens is much less in populations where the prevalence of helminths is high. Several studies indicate that children living in tropical and nontropical rural areas have decreased allergy reactions than children with a low parasite load.[43–46] An important question is, "why, then, is allergic disease less frequent in these populations?"

It has been suggested that the association of parasitism and allergy lies in the interaction of parasite infestation and the production of serum immunoglobin IgE (see Chapter 17). Infestation with helminth parasitism (measured by determining the presence of eggs in stool samples, or the presence of serologic markers of chronic infection) elicits the production of immunoglobin IgE, which causes an elevated allergic response (measured by means of allergen skin test reactivity). However, helminth parasitism also triggers the production of anti-inflammatory cytokine interleukin-10,[43] which suppress the effect of IgE and limits the responses to allergens. Viewed in this context, childhood parasitic infections protect against the expression of allergies.[43–45] Consequently, de-infestation of childhood parasites also removes the anti-inflammatory mechanisms, increasing the severity of allergic disease in individuals free of parasites. In other words, children in developed countries free of parasites—when exposed to aeroallergens—respond with high hypersensitivity and inflammatory reactions, precipitating clinical allergy. Therefore, some investigators[46] postulate that the high frequency of allergy in industrialized populations may be the price paid for being relatively free of helminths. In view of the multifactorial nature of allergies, the relationship between parasite infestation and atopy probably represents only one aspect of this major paradigm of immunology.

Life Expectancy

The human life span (which is the maximum number of years that a human can live) has remained unchanged for the past 100,000 years at about 100 years, with few exceptions. On the other hand, life expectancy at birth (which is the average total number of years that a human expects to live) has increased in the United States and other developed countries from about 49 years in 1900 to about 76 years in 1999.[47] The increase in life span is due, in large part, to the very low mortality among the young and to the excellent treatments for cardiovascular diseases and cancer among older individuals. It should be noted, however, that the actual "years of healthy life" are much less than the life span. It has been calculated that "years of healthy life" lost due to disability represent 8 percent of the total life expectancy.[48] Based on this estimate, the "healthy life expectancy" in the United States is 70.0 years rather than 76 years. It is quite evident that the gains in life expectancy will not be worth it, unless the quality of life is also enhanced, so that older age individuals can continue being self-sufficient members of society. Perhaps, if humans change their habits from being passive consumers to active, selective lifestyles that include eating less, eating more plant foods, being much more physically active, and restricting non-dietary risk factors such as smoking, living longer will be worth the effort.

Summary

Changing Health Conditions

As a species we continue to evolve both biologically and culturally, and will continue to do so in the future. It was only 12,000 to 14,000 years ago that we subsisted on hunting and gathering. The transition to agriculture and animal domestication was a major change in human adaptation. It increased humans' ability to feed themselves and provided the opportunity to maintain much larger populations, which in turn allowed for the development of complex state-level societies, but it also created new problems and dilemmas for human populations.

We live in an environment where populations migrating and settling in new environments are faced with new stresses. As shown by the measurement of blood pressure, some population differences in the prevalence of hypertension can be traced to past adaptations to tropical environments, while others are related to different levels of stress. Likewise, the change in food processing brought about by industrialization has changed human dietary habits, and in doing so, it has also changed the size of teeth and their relationship to the size of the mandible. Similarly, the success of modern medicine in getting rid of debilitating parasites might have also removed the anti-inflammatory traits that prevent respiratory allergies.

Suggested Readings

Frisancho, A. R. 1993. *Human Adaptation and Accommodation.* Ann Arbor, MI: University of Michigan Press.

Literature Cited

1. James, G. D., and P. T. Baker. 1995. Human population biology and blood pressure: Evolutionary and ecological considerations and interpretations of population studies. In: *Hypertension: Pathophysiology, Diagnosis, and Management.* J. H. Laragh and B. M. Brenner, ed. New York: Raven Press, 115–126.
2. Shapiro, A. P. 1996. *Hypertension and Stress: A Unified Concept.* Mahwah, NJ: Lawrence Erlbaum Associates.
3. Rumantir, M. S., G. L. Jennings, G. W. Lambert, D. M. Kaye, D. R. Seals, and M. D. Esler. 2000. The "adrenaline hypothesis" of hypertension revisited: Evidence for adrenaline release from the heart of patients with essential hypertension. *J. Hypertension* 18:717–723.
4. Julius, S., N. Schork, and A. Schork. 1988. Sympathetic hyperactivity in early stages of hypertension: The Ann Arbor data set. *J. Cardiovasc. Pharmacol.* 12 (Suppl 3):S121–S129.

5. McCraty, R., M. Atkinson, and D. Tomasino. 2003. Impact of a workplace stress reduction program on blood pressure and emotional health in hypertensive employees. *J. Altern. Complement. Med.* 9:355–369.
6. Brown, D. E., G. D. James, L. Nordloh, and A. A. Jones. 2003. Job strain and physiological stress responses in nurses and nurse's aides: predictors of daily blood pressure variability. *Blood Press. Monit.* 8:237–242.
7. Brown, D. E., L. Leidy Sievert, S. L. Aki, P. S. Mills, M. B. Etrata, R. N. K. Paopao, and G. D. James. 2001. Effects of age, ethnicity and menopause on ambulatory blood pressure: Japanese-American and Caucasian school teachers in Hawaii. *Am. J. Hum. Biol.* 13:486–493.
8. Brown, D. E., and G. D. James. 2000. Physiological stress responses in Filipino-American immigrant nurses: the effects of residence time, lifestyle, and job strain. *Psychosom. Med.* 62:394–400.
9. Brown, D. E., L. L. Sievert, S. L. Aki, P. S. Mills, M. B. Etrata, R. N. Paopao, and G. D. James. 2001. Effects of age, ethnicity and menopause on ambulatory blood pressure Japanese-American and Caucasian school teachers in Hawaii. *Am. J. Human Biol.* 13:486–493.
10. Labarthe, D., D. Reed, J. Brody, and R. Stallones. 1973. Health effects of modernization in Palau. *Am. J. Epidemiol.* 98:161–174.
11. Marmot, M. G., and S. L. Syme. 1976. Acculturation and coronary heart disease in Japanese-Americans. *Am. J. Epidemiol.* 104:225–247.
12. Cruz-Coke, R. 1987. Correlation between prevalence of hypertension and degree of acculturation. *J. Hypertens.* 5:47–50.
13. Dressler, W. W., A. Mata, A. Chavez, and F. E. Viteri. 1987. Arterial blood pressure and individual modernization in a Mexican community. *Soc. Sci. Med.* 24:679–687.
14. McGarvey, S. T. 1992. Biocultural predictors of age increases in adult blood pressure among Samoans. *Am. J. Hum. Biol.* 4:27–35.
15. Schall, J. I. 1995. Sex differences in the response of blood pressure to modernization. *Am. J. Hum. Biol.* 7:159–172.
16. Murphy, J. K., and S. T. McGarvey. 1994. Modernization in the Samoas and children's reactivity: A pilot study. *Psychosom. Med.* 56:395–400.
17. Weinberger, M. H. 1996. Salt sensitivity of blood pressure in humans. *Hypertension* 27:481–490.
18. Grim, C. E., F. C. Luft, J. Z. Miller, G. R. Meneely, H. D. Batarbee, C. G. Hames, and L. K. Dahl. 1980. Racial differences in blood pressure in Evans County, Georgia: Relationship to sodium and potassium intake and plasma renin activity. *J. Chronic Dis.* 33:87–94.
19. Aviv, A., N. K. Hollenberg, and A. Weder. 2004. Urinary potassium excretion and sodium sensitivity in blacks. *Hypertension* 43:707–713.
20. Dressler, W. W. 1990. Lifestyles, stress, and blood pressure in a southern black community. *Psychosom. Med.* 52:182–198.
21. Dressler, W. W. 1991. Social class, skin color, and arterial blood pressure in two societies. *Ethn. Dis.* 1:60–77.
22. Dressler, W. W., J. E. Dos Santos, P. N., Gallagher, Jr., and F. E. Viteri. 1987. Arterial blood pressure and modernization in Brazil. *Am. Anthropol.* 89:398–408.
23. Calhoun, D. A. 1992. Hypertension in blacks: Socioeconomic stress and sympathetic nervous system activity. *Am. J. Med. Sci.* 304:306–311.
24. Moorman, P. G., C. G. Hames, and H. A. Tyroler. 1991. Socioeconomic status and morbidity and mortality in hypertensive blacks. *Cardiovasc. Clin.* 21:179–194.
25. Williams, D. R. 1992. Black-White differences in blood pressure: The role of social factors. *Ethn. Dis.* 2:126–141.
26. Helmer, O. M. 1967. Hormonal and biochemical factors controlling pressure. In *Les Concepts de Claude Bernard Sur le Milieu Interieur.* Paris: Masson, 115–128.
27. Gleibermann, L. 1973. Blood pressure and dietary salt in human populations. *Ecol. Food Nutr.* 2:143–156.
28. Sawka, M. N., J. Scott, and S. J. Montain. 2000. Fluid and electrolyte supplementation for exercise heat stress. *Am. J. Clin. Nutr.* 72:564S–572S.
29. Allan, J. R., and C. G. Wilson. 1971. Influence of acclimatization on sweat sodium concentration. *J. Appl. Physiol.* 30:708–712.
30. Jackson, F. L. 1991. An evolutionary perspective on salt, hypertension, and human genetic variability. *Hypertension.* 17 (1 Suppl):I129–I132.
31. Luft, F. C., R. Bloch, A. Weyman, et al. 1978. Cardiovascular responses to extremes of salt intake in man. *Clin. Res.* 26: 365A.
32. Hall, W. D. 1990. Pathophysiology of hypertension in blacks. *Am. J. Hyperten.* 3:366S–371S.
33. Frisancho, A. R., W. R. Leonard, and L. A. Bollettino. 1984. Blood pressure in blacks and whites and its relationship to dietary sodium and potassium intake. *J. Chron. Dis.* 34:515–519.
34. Frisancho, A. R., S. Farrow, I. Friedenzohn, T. Johnson, B. Kapp, C. H. Miranda, M. Perez, I. et al. 1999. Role of genetic and environmental factors in the increased blood pressures of Bolivian blacks. *Amer. J. Hum. Biol.* 11:489–498.
35. Sciulli, P. W. 2001. Evolution of dentition in prehistoric Ohio Valley Native Americans III. Metrics of deciduous dentition. *Am. J. Phys. Anthropol.* 116:140–153.
36. Sciulli, P. W., and M. C. Mahaney. 1991. Phenotypic evolution in prehistoric Ohio Amerindians: Natural selection versus random genetic drift in tooth size reduction. *Hum. Biol.* 63(4):499–511.
37. Brace, C., S. Xiang-qing, and Z. Zhen-biao. 1984. *Prehistoric and Modern Tooth Size in China. The Origins of Modern Humans: A World Survey of the Fossil Evidence.* New York: Alan R. Liss, Inc., 485–516.
38. Calcagno, J. 1989. *Mechanisms of Human Dental Reduction.* University of Kansas Publications in Anthropology, 18.
39. Carlson, D. S., and D. P. Van Gerven. 1977. Masticatory function and post-Pleistocene evolution in Nubia. *Am. J. Phys. Anthropol.* 46:495-506.
40. Liu, K. L. 1977. Dental condition of two tribes of Taiwan aborigines: Ami and Atayal. *J. Dental Research* 56:117–127.

41. Corruccini, R. 1991. Anthropological aspects of orofacial and occlusal variations and anomalies. In *Advances in Dental Anthropology,* Kelley, M. A., and C. S. Larson, eds. New York: Wiley-Liss, Inc., 295–323.

42. Seguchi, N. 2000. *Secular Change in the Japanese Occlusion: The Frequency of the Overbite, and Its Association with Food Preparation Techniques and Eating Habits.* Ph.D. dissertation in Anthroplogy. Ann Arbor: The University of Michigan.

43. van den Biggelaar, A. H. J., R. van Ree, L. C. Rodrigues, et al. 2000. Role for parasite-induced interleukin-10 in children infected by Schistosoma haematobium. *Lancet* 356:

44. Holt, P.G. 2000. Parasites, atopy, and the hygiene hypothesis: resolution of a paradox? *Lancet* 356:1699–1701.

45. Cooper, P. J., M. E. Chico, L. C. Rodrigues, M. Ordonez, D. Strachan, G. E. Griffin, and T. B. Nutman. 2003. Reduced risk of atopy among school-age children infected with geohelminth parasites in a rural area of the tropics. *J. Allergy Clin. Immunol.* 111:995–1000.

46. Gerrard, J. W., C. A. Geddes, P. L. Reggin, C. D. Gerrard, and S. Horne. 1976. Serum IgE levels in white and Metis communities in Saskatchewan. *Ann. Allergy* 37:91–100.

47. Mathers, C. D., R. Sadana, J. A. Salomon, C. J. L. Murray, and A. D. Lopez. 2001. Healthy life expectancy in 191 countries, 1999. *Lancet* 357:1685–1700.

48. Anderson, R. N. 1999. United States Life Tables (1997 National Vital Statistics Reports Vol. 47, No. 28) Centers for Disease Control and Prevention. National Center for Health Statistics.

CHAPTER 23

Globalization of Obesity

Globalization of Obesity: The Price of Industrialization

Obesity, however defined, represents an excess accumulation of calories, and occurs when the caloric intake exceeds energy expenditure. The frequency of obesity in the developed and developing nations is reaching epidemic proportions. It is estimated that in 2010 more than 50 percent of all deaths worldwide will be related to obesity, cardiovascular disease, and diabetes. A similar trend is occurring in the developing nations. The origins of obesity are associated with four factors such as attitudes toward body image, the energy density of foods, amount of physical activity, and past experience with undernutrition, each of which play a different role in the developed and developing nations.[1] This chapter summarizes current understanding about the etiology of obesity.

Obesity in the Industrialized Countries

Cultural Attitudes about Body Image

Until recently, and throughout the Paleolithic, heavy weight in women and children was considered attractive. Prior to the eradication of debilitating diseases, such as tuberculosis, being heavy was considered an indicator of good health in the United States and fat babies were usually considered healthy. In some cultures, such as those of Polynesia, fatness indicated wealth or high economic status. In fact, the term "fat cat," which is associated with wealth and power, is said to be a vestige of the idea that fatness is associated with success.[2] Similarly, among South American populations, the term "gordo" or "gorda" is used in a friendly fashion, and studies of Puerto Ricans in the United States indicate that being fat does not have the social stigma of obesity given in other populations.[3] For this reason, Mexican Americans use the term of "gordura mala" (bad fatness) when describing fatness that is associated with a negative health risk.[4] The San Antonio Heart study[5, 6] and analyses of the anthropometric data of the Hispanic Health and Nutritional Examination Survey[7] indicate that among Mexican Americans, the maintenance of traditional cultural values was associated with an increase in the incidence of obesity. This finding suggests that levels of fatness in Mexican Americans, especially women, may reflect cultural preferences toward fatness. However, recent concerns about the negative health consequences of obesity are changing the cultural attitudes about body image.

Decrease in the Level of Physical Activity

Search for food and the concomitant high level of activity has been a major component that ensures the survival of humans. Contemporary hunter-gatherers are characterized by periods of fairly intense physical activity lasting between three and four days per week dedicated to hunting combined with periods of light activity dedicated to gathering, tool making, preparing clothing, carrying firewood and water, butchering, and moving to new campsites.[8, 9] Assuming that the lifestyle of extant hunter-gatherers was similar to that of Paleolithic populations (see Chapter 11) a daily high level of physical activity was an integral and constituted an important component of food procurement through human evolution. In other words, the genes of our ancestors were not selected for sedentary existence. According to the archeological evidence, the overall genetic makeup of *Homo sapiens* has not changed during the past 15,000 years and therefore contemporary humans have inherited genes that evolved to support the physically active lifestyle of the hunter-gatherer. However, during the past 100 years, there has been a dramatic decline in physical activity and increase in food availability and thus modern society is remarkably sedentary. For example, many individuals, especially in the industrialized countries, no longer have to use manual labor to procure food. Likewise, in the United States about 26 percent of 8- to 16-year-old children watch television for at least 4 hours a day and 67 percent watched television for at least 2 hours a day. Furthermore, only about half of adolescent and young adults participate in vigorous activity and one-fourth reported no physical activity.[8–10] In other words, industrialized populations with the genetic makeup that evolved to maintain an active lifestyle and survival in the late Paleolithic era are now living in a sedentary and food-abundant society. Consequently, our present sedentary lifestyle and the increased food availability have led to incongruity in gene-environmental interactions; thus, in modern *Homo sapiens* the calories used to maintain an active lifestyle are now stored as body fat manifested in the increased risk of obesity. Regardless of the measurement used, there is considerable evidence indicating an inverse association between physical activity and weight gain.[11–15] Thus, in general, as activity level declines, the risk of obesity increases. In some studies, the increase in body mass index occurred despite the fact that the proportion of persons reporting leisure-time physical activity had increased, suggesting that the increase in leisure-time physical activity has not been enough to compensate for the decrease in overall physical activity.

In summary, a decrease in the level of physical activity and the sedentarism that characterizes industrialized populations are strong factors for the increased risk of becoming overweight and obese.

Inclusion of Plant Foods

The dietary pattern of contemporary populations has changed drastically when compared to the diet of the Paleolithic hunter-gatherers. Isotopic analysis of human remains indicate that Paleolithic populations had a wide dietary breadth that included a significant amount of aquatic (fish, mollusks, and/or birds) foods and terrestrial herbivores.[16] This broad-spectrum subsistence economy based on animal food was maintained through the upper Paleolithic. Continuing this pattern among contemporary hunter-gatherer populations,[17, 19] with the exception of the !Kung San Bushmen from the Kalahari desert of South Africa,[18] animal foods provide the major food energy (50 percent) (**see table 23.1**). It should be noted, however, that the fat content of animal foods was intrinsically low when fat content was compared to animal foods in Western diets. Along with the high reliance on lean animal foods, the hunter-gatherers' diet also included high intakes of antioxidants, fiber, vitamins, and phytochemicals.[17] With the development of agriculture about 12,000 to 14,000 years ago, human dietary patterns changed from an emphasis on animal foods to a dietary pattern where plant foods became the dominant dietary energy source. In general, among the ma-

TABLE 23.1

Dietary Characteristics of Contemporary Hunter and Gatherers

Population	Location	Plant Food (%)	Animal Food (%)
Aborigines (Arhem Land)	Australia	77	23
Ache	Paraguay	78	22
Anbarra	Australia	75	25
Efe	Africa	44	56
Eskimo	Greenland	96	4
Gwi	Africa	26	74
Hadza	Africa	8	52
Hiwi	Venezuela	75	25
!Kung	Africa	33	67
!Kung	Africa	68	32
Nukak	Columbia	41	59
Nunamiut	Alaska	99	1
Onge	Andaman Islands	79	21

Source: Data from Kaplan, H., Hill, K., Lancaster, J., and M. Hurtado. 2000. A theory of human life history evolution: Diet, intelligence, and longevity. *Evol. Anthropol.* 9, 156–185.

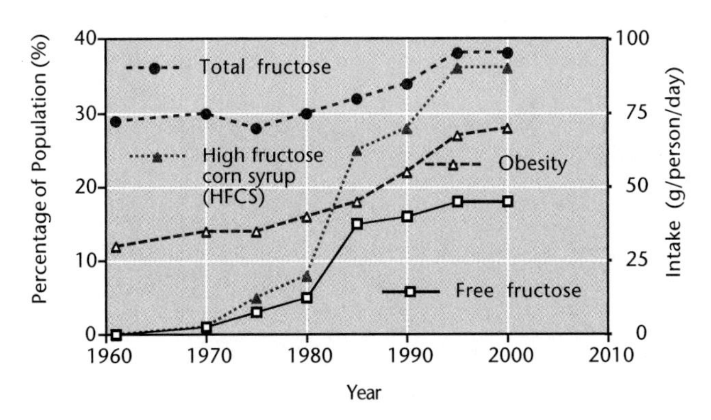

FIGURE 23.1. Changes in the intake of total fructose and high-fructose corn syrup (HFCS) and trends in the prevalence of obesity (BMI > 30 kg/m^2) in the United States. Source: Data derived from Gray, G. A., Nielsen, S. J., and B. M. Popkin. 2004. Consumption of high-fructose corn syrup in beverages may play a role in the epidemic of obesity. *Am. J. Clin. Nutr.* 79:537–543.

jority of the agricultural populations, about 25 to 35 percent of the daily energy is derived from animal foods and the rest from plant foods.

Shift from Unrefined to Refined Carbohydrates

With the development of agriculture came increased development of food technology that resulted in the production of refined carbohydrates. Since 1960, the consumption of refined carbohydrates in the form of sweeteners and sucrose increased dramatically while the consumption of whole grains declined. As illustrated in **figure 23.1**, since 1909 the consumption of refined carbohydrates has increased dramatically. For example, the consumption of high fructose corn syrup (HFCS) between 1970 and 1990 increased, far exceeding the changes in intake of any other food or food group (**fig. 23.1**). During the same interval, the percent of the obese (body mass greater than 30 kg/m^2) in the United States also increased. According to various estimates, HFCS now represents more than 40 percent of caloric sweeteners added to foods and beverages and is the sole caloric sweetener in soft drinks in the United States.[20]

In summary, it is quite evident that a shift in dietary source from unrefined to refined carbohydrates, as exemplified by the consumption of high fructose corn syrup, which is the major ingredient of soft drinks, mirrors the rapid increase in obesity in the United States. Therefore, the overconsumption of HFCS in calorically sweetened beverages may play a role in the increased trend toward excess weight and obesity that is becoming epidemic throughout the world.

Poverty and Obesity

In the United States, there is an inverse relationship between obesity and socioeconomic status (**fig. 23.2**), such that the highest rates of obesity occur among

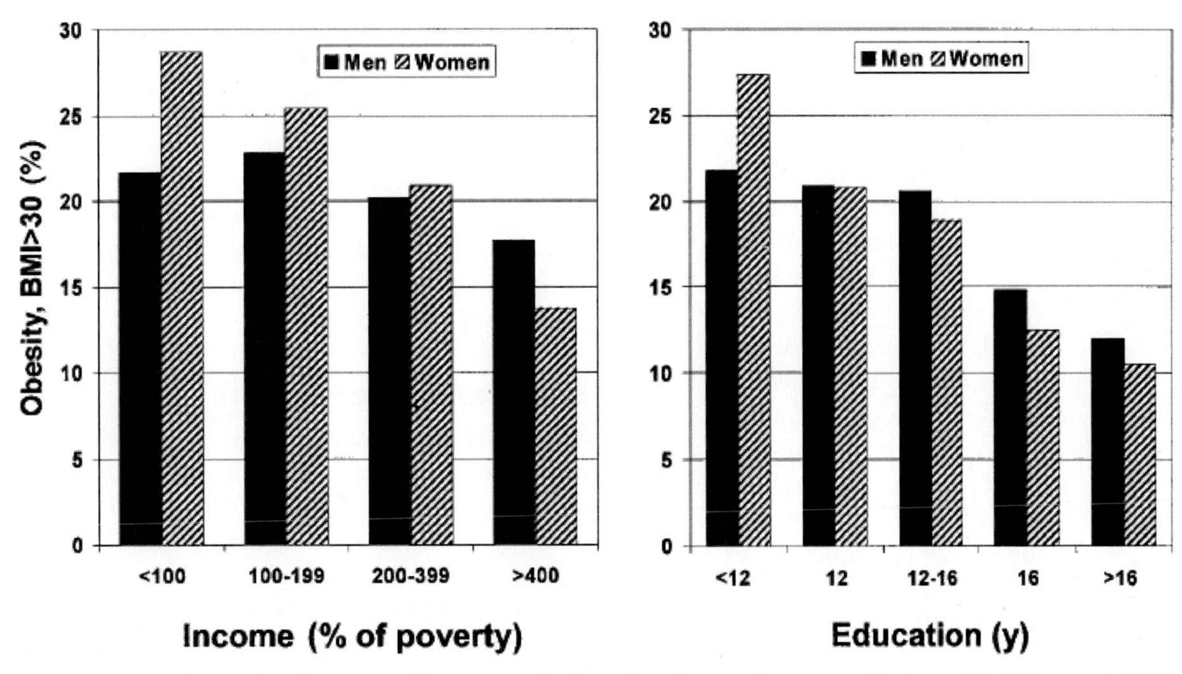

FIGURE 23.2. Obesity as a function of income and education. Source: Drewnowski, A., and S. E. Specter. 2004. Poverty and obesity: The role of energy density and energy costs. *Am. J. Clin. Nutr.* 79:6–16.

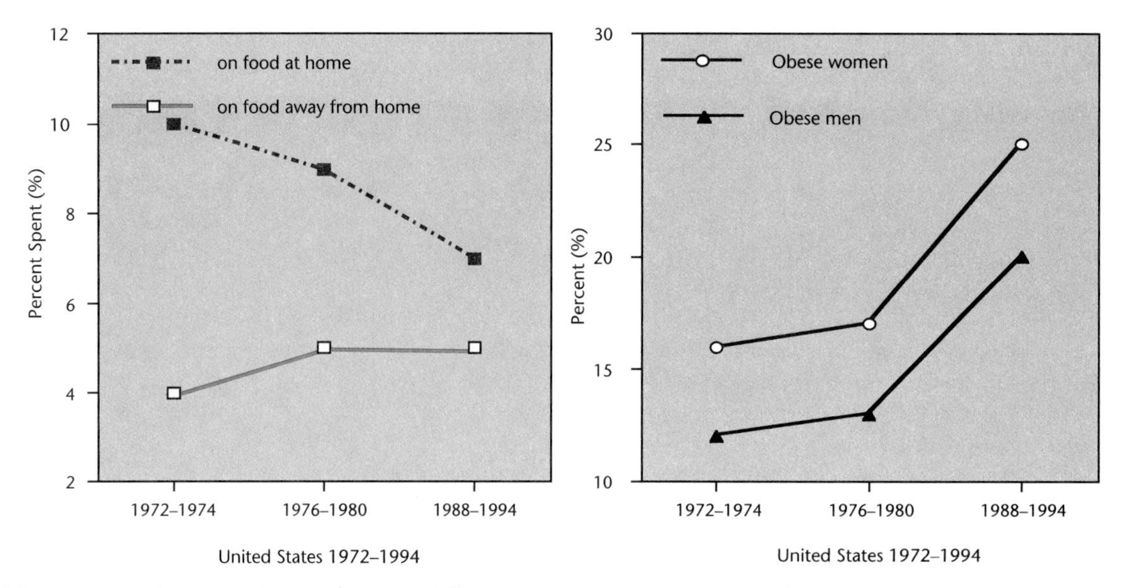

FIGURE 23.3. Change in the percentage of disposable income spent on food at home and away from home, and the percentage of the population that was obese in the United States from 1972 to 1994. Source: Drewnowski, A. 2003. Fat and sugar: An economic analysis. *J. Nutr.* 133:838S–840S.

population groups with the highest poverty rates and the least education.[21–26] Analysis of incomes and family size in the United States indicate that when household income declines, energy intakes are maintained at a lower cost than when the income increases.[27] Specifically, this study demonstrates that when household members are faced with diminishing incomes, the consumption of less-expensive foods increases.[27] For example, in the United States during the last twenty years the percentage of disposable income that is spent on food has decreased dramatically (**fig. 23.3**). Associated with the decline in the percentage of disposable income that is spent on food, obesity rates also increased in both males and females. As shown in **figure 23.4**, there is an inverse relationship between the energy density (MJ/kg) of selected foods and their energy cost (cents/10 MJ). In other words, energy-dense foods provide dietary energy at a lower cost than low energy density foods. It follows from this that the association between poverty and obesity is mediated by the low cost of energy-dense foods, which may, in turn, lead to overconsumption of inexpensive foods.

In summary, current evidence suggests that energy-dense foods containing refined grains, added sugars, and fat provide energy at a much lower cost than do fresh vegetables and fruit. Based on this information, it has been proposed that the high energy intakes associated with obesity are driven by the very low cost of energy-dense foods.[28]

Obesity in the Developing Nations

As shown in **figure 23.5**, the prevalence of being overweight (defined as body mass index > 30.0 kg/m^2) is increasing throughout the developing nations. Although the prevalence of being overweight in developing nations is less than those of the industrialized countries, by the year 2010 it will reach the same frequency. The risk of obesity in the developing countries is associated with decreased level of physical activity and increased consumption of foods rich in energy. For example, in Brazil and China the prevalence of overweight people is greater in urban areas than in rural areas due to the increased dietary intake derived from nutrient-dense and energy-dense food, increased use of public transportation, and closer proximity of residence to work. There has been a decrease in energy expenditure and an increase in inactive leisure time.[20–27]

Income and Overweight

The responses to increased economic development have followed a different pattern than those in the industrialized countries. As shown in **figure 23.6**, in the developing countries such as Brazil and China, the prevalence of overweight people is greater in the high-income, rather than in the low-income groups, while in United States it is the opposite.[31]

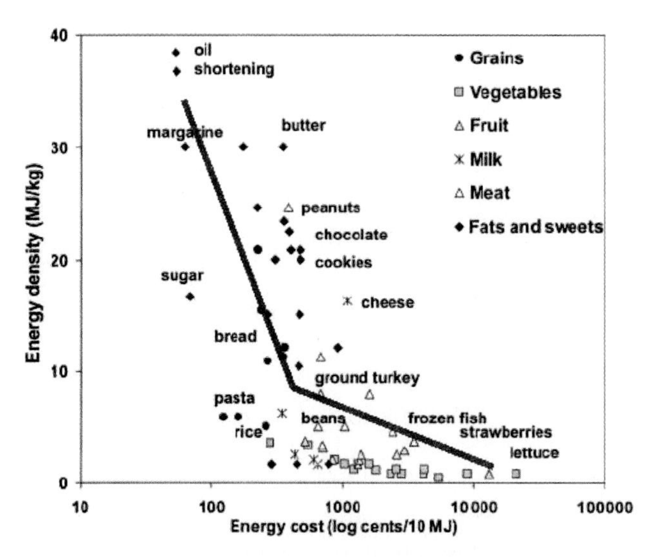

FIGURE 23.4. Relationship between the energy density of selected foods and energy cost ($/MJ). Source: Drewnowski, A. and S. E. Specter. 2004. Poverty and obesity: The role of energy density and energy costs. *Am. J. Clin. Nutr.* 79:6–16.

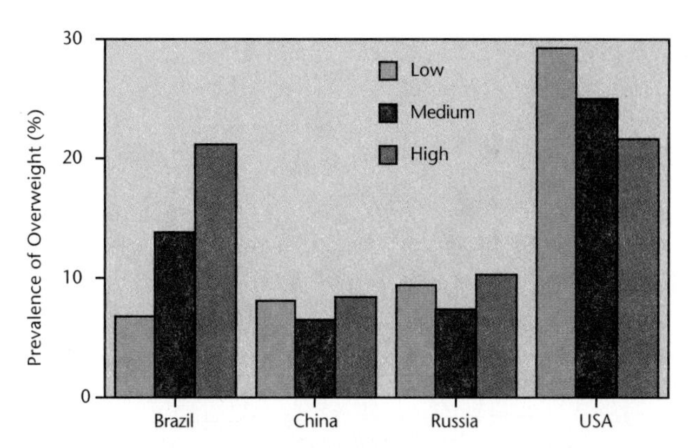

FIGURE 23.6. Relation between per capita household income and the prevalence of overweight people. Source: Data from Wang, Y., Monteiro, C., and B. M. Popkin. 2002. Trends of obesity and underweight in older children and adolescents in the United States, Brazil, China, and Russia. *Am. J. Clin. Nutr.* 75:971–977.

Coexistence of Underweight and Overweight

Traditionally, there are three indices for determining childhood nutrition status of preschool children and these include: weight-for-age, height-for-age, and weight-for-height. Underweight is defined as <- 2 standard deviations of the weight-for-age median value of NCHS/WHO international reference data; stunting is defined as <- 2 standard deviations of the height-for-age median value of NCHS/WHO international reference data; and wasting is defined as <- 2 standard deviations of the weight-for-height median value of NCHS/WHO international reference data. As shown in **figure 23.7**, it is evident that among preschool (0 to 5 years) children the frequency of childhood undernutrition and overnutrition occurs simultaneously in each country. Likewise, **figure 23.8** shows the prevalence of stunting is related to the frequency of overweight among children.

In summary, from these data it is evident that in the developing nations both undernutrition and overnutrition coexist. An important question is why underweight and overweight co-exist in the developing nations. The answer lies in the developmental response to the joint effects of the "thrifty gene" and undernutrition.

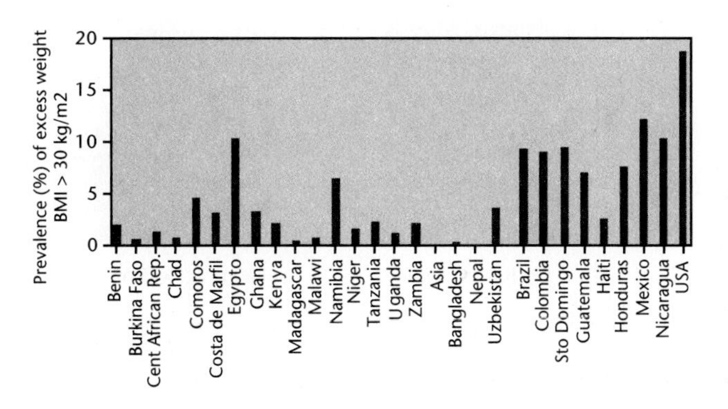

FIGURE 23.5. Prevalence of excess weight among women from the developing nations. Source: ACC/SCN. 2000. *Fourth Report on the World Nutrition Situation.* Geneva: ACC/SCN in collaboration with IFPRI.

The "Thrifty Gene" and Undernutrition

In 1962 Neel[33] postulated the existence of the "thrifty gene." It proposed that during periods of feast the genes that were exceptionally efficient in the intake and/or conversion of food into glucose and eventually storage in the adipose tissue were selected into the human genome because of their selective advantage over the less "thrifty" ones (i.e., less economical). Subsequently, those with the thrifty phenotype during periods of food shortage and famine would have a selective advantage because they relied on larger, previously stored energy to maintain homeostasis, whereas those without "thrifty" genotypes would be at a disadvantage and less likely to survive. Given the fact that the human species went through periods of shortage and abundance of food, the "thrifty gene" forms part of the human species as a whole rather than a particular

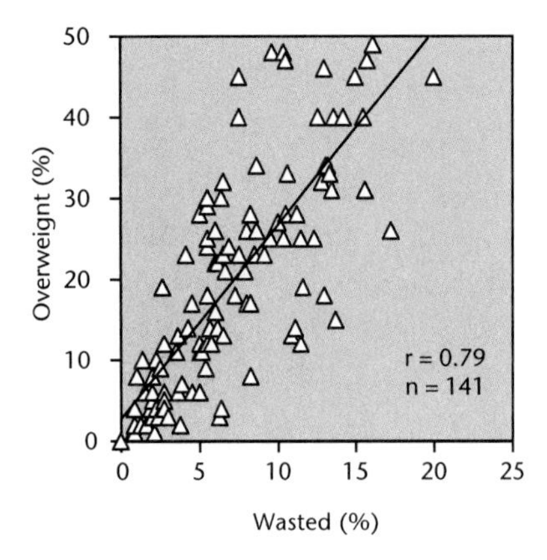

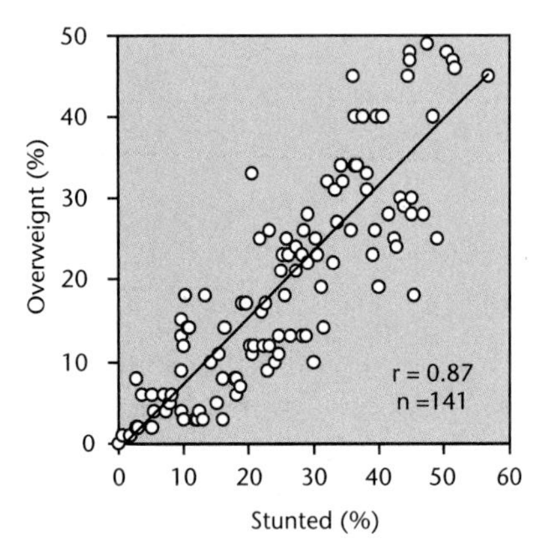

FIGURE 23.7. Relationship of prevalence of (a) wasting and overweight, and (b) stunting and overweight in children from the developing nations. Source: ACC/SCN. 2004. *Fifth Report on the World Nutrition Situation.* Geneva: ACC/SCN in collaboration with IFPRI.

population. However, the extent to which the "thrifty gene" is expressed depends on the nutritional challenge to which the population is exposed. I propose that populations with a history of undernutrition would be more likely to express the "thrifty gene" than those without nutritional restrictions. As learned from experimental studies of animal and human exposure to undernutrition in utero or during postnatal development, a reduction in energy intake below an acceptable level of requirement results in a series of adaptive physiological, biochemical, and behavioral responses oriented at economizing energy expenditure and maximizing nutrient absorption.(34) This increased efficiency of nutrient absorption results in enhanced fat accumulation during development and adulthood.

Development

Experimental studies with rats indicate that male offspring of rats, food restricted during the first two weeks of pregnancy and then allowed to eat freely,(35) became obese by five weeks of age. Studies of pigs who were growth retarded and protein-deficient during weaning became quite fat later in life.(36) Similar investigations in rats who were food restricted showed an elevated efficiency of energy utilization and a preferential accumulation of fat.(37) Supplementation studies conducted on growth-retarded children (with an average age of 9.0 years) from Bundi, Papua, New Guinea,(38) indicate that protein and calorie supplementation for a thirteen-week period did not have a major effect on growth in body length (**fig. 23.8a**). In contrast, supplementation did have a significant effect on increasing subscapular skinfold thickness (**fig. 23.8b**). These data would suggest that both the protein and energy provided some extra energy that was used for fat deposition rather than growth.

This conclusion is supported by recent studies conducted among Jamaican children.(39) In these studies, nutritional supplementation (calories and protein) of growth-retarded children (with an average age of 18.5 months) for one year did not have a major effect on growth in length or mid-upper arm circumference,(39) which is an indicator of body muscle mass. In contrast, supplementation did have a significant effect on increasing skinfold thickness. Studies conducted in the shantytown population in the city of São Paulo(45, 46) found that in children the prevalence of undernutrition (low weight-for-age and/or low height-for-age) was 30 percent. In contrast, during adolescence 21 percent of girls and 8.8 percent of boys were overweight

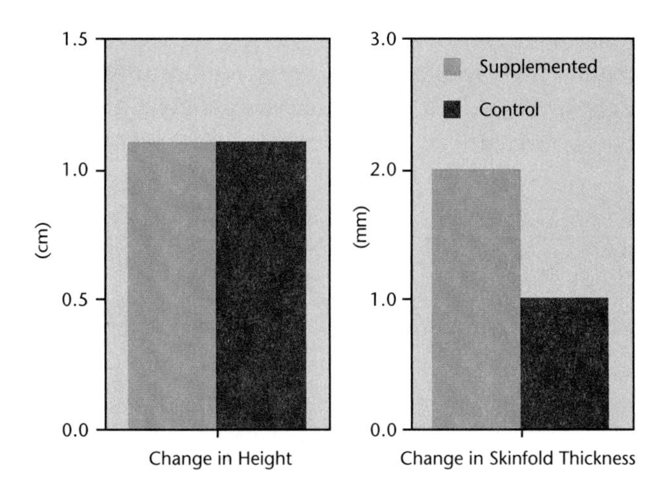

FIGURE 23.8. Supplementation effects on height (a) and skinfold thickness (b) of school children from Bundi, New Guinea. Source: Malcomb, L. A. 1970. *Growth and Development in New Guinea—A Study of the Bundi People of the Madang District.* Institute of Human Biology; Papua, New Guinea. Monograph Series No. 1. Australia: Surrey Beatty and Sons.

and 14.6 percent were overweight during adulthood (high weight-for-height and BMI). Likewise, a study of a sample from the city of Alagoas, one of the poorest states in Brazil[41] found the prevalence of stunting was 22.6 percent, but 30 percent of the stunted subjects were overweight or obese, compared with 23 percent for the nonstunted individuals. Evaluation of the dietary intake showed that the increased body fatness was not related to difference in the proportion of macronutrients and total energy.

Adulthood

Analyses of survivors of the Dutch famine of World War II found that the Dutch army draftees whose mothers had been deprived of food in the first two trimesters of pregnancy had a greater incidence of obesity than did the general population, even when adjusted by occupation.[40] The Minnesota experimental studies of semistarvation showed that six months and one year of severe energy restriction in thirty-two lean men led to a marked reduction in energy expenditure (EE). This was due to a reduction in both physical activity and in the resting metabolic rate (RMR), which decreased not only in absolute terms (39 percent) but also when expressed per kilogram of metabolically active tissue (16 percent). Analysis of subjects who were refed after the starvation indicate that weight regain was due to increased fat accumulation.[34] Reanalysis of data from the Minnesota experiment revealed that the reduction in the resting metabolic rate (SMR) in response to food deprivation is related to the activation of the autoregulatory feedback system in which signals from the depleted fat stores suppress thermogenesis, which results in decreased RMR and a specific metabolic component favoring fat storage.[34, 41, 42]

Investigations of nonobese cyclical dieters (having dieted for ≥ 7–10 days four times/year) indicated that cyclical dieters had a significantly lower energy expenditure (per unit of body weight) in treadmill exercise compared to controls with a similar calorie intake.[43] In other words, nonobese chronic dieters with similar lean body weights (but more body fat) use less energy to perform their daily activities than do non-dieters.[43] Likewise, individuals confined inside Biosphere 2[44] for two years with a markedly restricted food supply, during most of the time exhibited marked weight loss, and an increase in fat mass when their body weight reverted to pre-entry values.[44]

In summary, these findings together suggest that humans exposed to life-threatening undernutrition (such as after several months of semistarvation) and moderate energy restriction that can be sustained over years adapt through a reduction in EE and preferential accumulation of body fat. This form of energy conservation is a biologically meaningful mechanism in the face of dangerously and moderately low energy supplies and has been referred to as metabolic adaptation and "post-starvation obesity." [34] In other words, in the face of food energy deficits, the organism increases metabolic efficiency by suppressing thermogenesis and hence reducing the rate at which the body's tissues are being depleted. The survival value of such an energy-sparing regulatory process that limits tissue depletion during food scarcity is obvious. Therefore, exposure to undernutrition even during adulthood accelerates the replenishment of fat stores during refeeding and thereby increasing the risk for obesity.

Undernutrition and Reduced Fat Oxidation

One productive way of determining how the organism uses the storage of nutrient substrates such as carbohydrates, fats, or protein is to measure the ratio of carbon dioxide production to oxygen consumption known as respiratory quotient (RQ). In general, under steady fasting and resting conditions an RQ that is close to 1.0 indicates that the material being used is chiefly carbohydrate; an RQ around 0.70 indicates that it is mainly fat; while an RQ around 0.08 indicates that the material being used is chiefly protein (which occurs only under starvation). That is, the higher the RQ the higher is the metabolism of carbohydrate. Conversely, the lower the RQ the higher is the metabolism of fat. Metabolic studies conducted in the shantytown population in the city of São Paulo[45, 46] found that the chronic undernourished children (i.e., stunted: height-for-age ≤ –1.50 Z scores) compared to control children had significantly higher respiratory quotient (RQ), which indicated higher oxidation of carbohydrates (**fig. 23.9**). Evaluations of dietary intake showed that the increased metabolic utilization of carbohydrates was not related to differences in the proportion of macronutrients and total energy.

In summary, it appears that under conditions of nutrition restriction, the organism prefers to metabolize carbohydrates rather than fats. Metabolic studies indicate that chronically undernourished adults have higher rates of carbohydrate oxidation and lower rates of fat oxidation (indicated by high RQs) than well-nourished controls.[48, 49] The reason for the enhanced

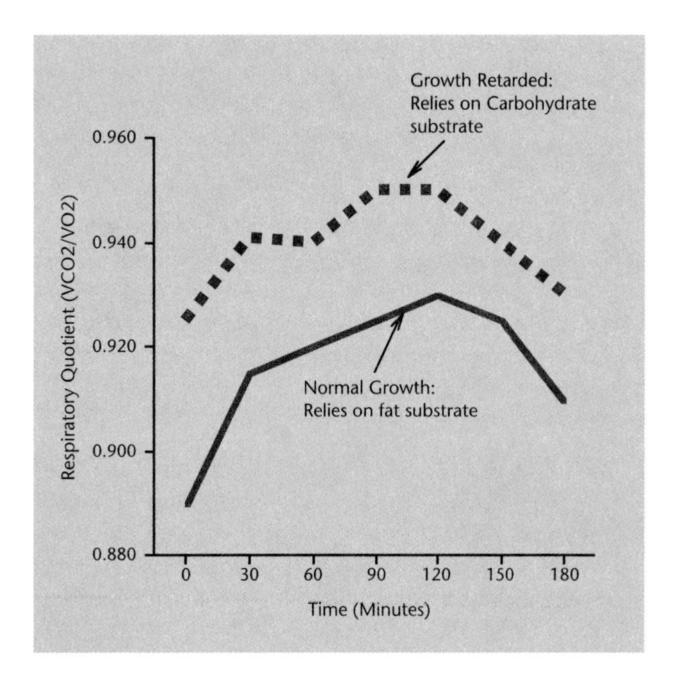

FIGURE 23.9. Increased reliance on carbohydrates, rather than fat substrate in chronically growth-retarded children from Brazil. Source: Hoffman, D. J., Sawaya, A. L., Verreschi, I., Tucker, K. L., and S. B. Roberts. 2000. Why are nutritionally stunted children at increased risk of obesity? Studies of metabolic rate and fat oxidation in shantytown children from São Paulo, Brazil. *Am. J. Clin. Nutr.* 72:702–707).

reliance on carbohydrate to maintain resting metabolic rate is related to the fact that the metabolic utilization of *stored* carbohydrate requires fewer biochemical steps than the utilization of *stored* fat.[48, 49] In other words, in the same way that dietary carbohydrates are faster to absorb than fats, carbohydrate substrates stored in the blood are also more rapidly metabolized than stored fatty acids. The net effect of the preferential metabolism of carbohydrates is that fats are stored, increasing the risk of obesity under condition of undernutrition.

In summary, the available experimental and epidemiological evidence suggests that exposure to undernutrition, especially during development, induces the activation of energy-conserving mechanisms and improvement in the efficiency of energy utilization. This response is responsible for the increased tendency to gain weight and risk obesity among individuals exposed to acute and chronic undernutrition. Thus, applying this inference to the impoverished populations of the developing nations it would appear that the same adaptive response that enabled them to survive the effects of undernutrition is now the main cause of obesity. Furthermore, in the developing nations as in the industrialized countries, energy-dense foods containing refined grains, added sugars, and fat provide energy at a much lower cost than do fresh vegetables and fruit. Earlier studies suggested that the export of refined carbohydrates in the form of soft drinks is an important component of undernutrition among rural populations of Latin America.[24] Thus, it would appear that both in the industrialized and developing nations the high energy per unit of currency along with the undernutrition induced response is an important factor that is contributing to the globalization of obesity.

Summary

Changing Health Conditions

As a species we continue to evolve both biologically and culturally, and will continue to do so in the future. It was only 12,000 to 14,000 years ago that we subsisted on hunting and gathering. The transition to agriculture and animal domestication was a major change in human adaptation. It increased humans' ability to feed themselves and provided the opportunity to maintain much larger populations, which in turn allowed for the development of complex state-level societies, but it also created new problems and dilemmas for human populations.

We live in an environment where populations migrating and settling in new environments are faced with new stresses. As shown by the measurement of blood pressure, some population differences in the prevalence of hypertension can be traced to past adaptations to tropical environments, while others are related to different levels of stress. Likewise, the change in food processing brought about by industrialization has changed human dietary habits, and in doing so, it has also changed the size of teeth and their relationship to the size of the mandible. Similarly, the success of modern medicine in getting rid of debilitating parasites might have also removed the anti-inflammatory traits that prevent respiratory allergies.

Obesity

We live in an environment where public and private transportation has decreased the amount of physical activity to the minimum and changed conspicuously the food supply and consumption. These changes in environment and lifestyle have resulted in an epidemic-like increase in the prevalence of obesity in both the industrialized and developing nations (**fig. 23.10**). The available evidence suggests that the prevalence of obe-

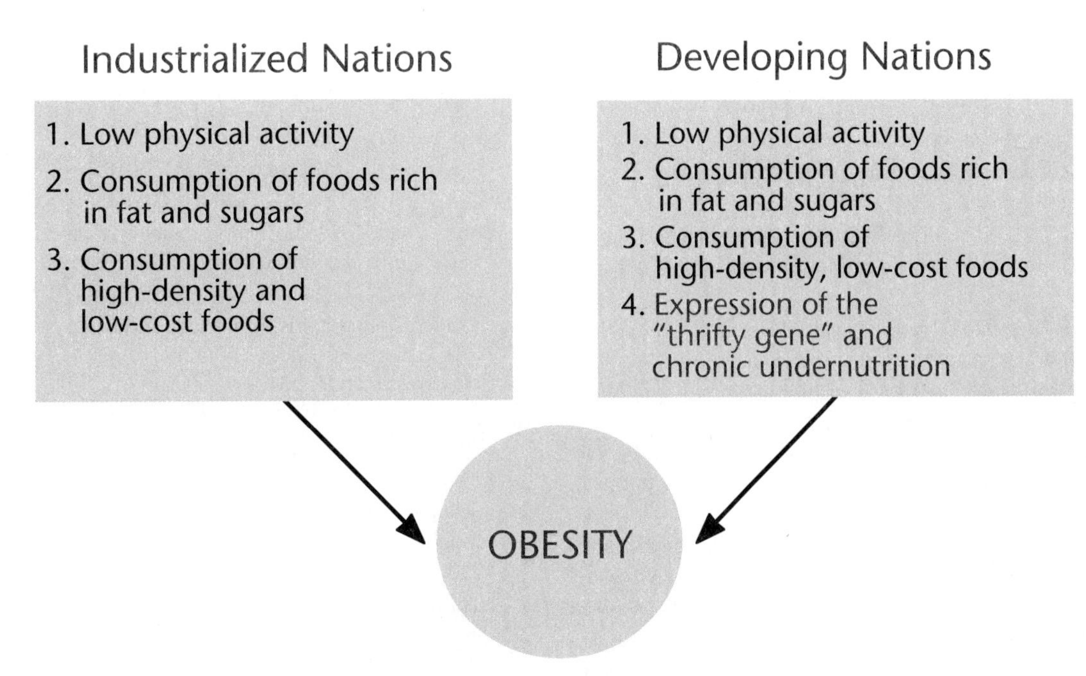

FIGURE 23.10. Schematization of factors associated with the increased risk of obesity in the industrialized and developing nations. (Illustration made by the author.)

sity is associated with a high intake of energy-dense foods, with emphasis on refined carbohydrates and decreased physical activity. Energy-dense foods provide energy at a much lower cost than do fresh vegetables and fruit, which are perceived as luxury items and are not always easily accessible. Hence, due to the low cost of energy-dense foods, the consumption of low-cost and energy-dense foods is directly related to income. As a result of this interaction, the association between poverty and obesity, is to a large extent, mediated by economic factors that determine the purchasing of foods that differ in energy density.

In the developing nations, it is predicted that the same relationship between energy density and the energy cost is contributing to the rapid increase in the prevalence of obesity. The fact that the trend towards low levels of physical activity is occurring throughout the world suggests that low levels of physical activity are global in nature. In addition, the chronic exposure to undernutrition during the period of development that stimulates the activation of energy-saving mechanisms, resulting in the selective metabolism of carbohydrates and storage of fat, is an important contributing factor for the prevalence of obesity and its coexistence of childhood undernutrition.

The fact that diabetes type 2, due to the activation of the "thrifty gene," is more likely to be expressed among chronically undernourished populations[50–53] suggests that the negative consequences of industrialization that occurred during the twentieth century will have even greater effects on the developing nations. Overcoming the negative effects of industrialization on general health is the major challenge confronting human biologists studying past and present human evolution. We certainly will continue evolving and adapting to the old and new environment we created, for it is the human ability to change and adapt that made us the most successful mammalian species.

Suggested Readings

Frisancho, A. R. 1993. *Human Adaptation and Accommodation.* Ann Arbor, MI: University of Michigan Press.

American Diabetes Association. 1998. Economic consequences of diabetes mellitus in the U.S. in 1997. *Diabetes Care* 21, 296–309.

Literature Cited

1. Frisancho, A. R. 2003. Reduced rate of fat oxidation: A metabolic pathway to obesity in the developing nations. *Am. J. Hum. Biol.* 15:35–52.

2. Schwartz, H. 1986. *Never Satisfied: A Cultural History of Diets, Fantasies, and Fat.* New York: Free Press.
3. Massara, E. B. 1989. *Que gordita!: A Study of Weight among Women in a Puerto Rican Community.* New York: AMS Press.
4. Ritenbaugh, C. 1982. Obesity as a culture-bound syndrome. *Cult. Med. Psychiatry* 6:347–361.
5. Haffner, S. M., M. P. Stern, B. D. Mitchell, and H. P. Hazuda. 1991. Predictors of obesity in Mexicans. *Am. J. Clin. Nutr.* 53:1571–1576S.
6. Hazuda, H. P., S. M. Haffner, M. P. Stern, and C. W. Eifler. 1988. Effects of acculturation and socioeconomic status on obesity and diabetes in Mexican Americans. *Am. J. of Epidem.* 128:1289–1301.
7. Pawson, G., R. Martorell, and F. E. Mendoza. 1991. Prevalence of overweight and obesity in U.S. Hispanic populations. *Am. J. Clin. Nutr.* 53:1522S–1528S.
8. Andersen, R. E., C. J. Crespo, S. J. Bartlett, L. J. Cheskin, and M. Pratt. 1998. Relationship of physical activity and television watching with body weight and level of fatness among children. *JAMA* 279:938–942.
9. Crespo, C. J., E. Smit, R. P. Troiano, S. J. Bartlett, C. A. Macera, and R.E. Andersen. 2001. Television watching, energy intake, and obesity in US children: results from the third National Health and Nutrition Examination Survey, 1988–1994. *Arch. Pediatr. Adolesc. Med.* 55:360–365.
10. Crespo, C. J., E. Smit, R. E. Andersen, O. Carter-Pokras, and B. E. Ainsworth. 2000. Race/ethnicity, social class and their relation to physical inactivity during leisure time: Results from the Third National Health and Nutrition Examination Survey, 1988–1994. *Am. J. Prev. Med.* 8:46–53.
11. Lahti-Koski, M., P. Pietinen, M. Heliövaara, and E. Vartiainen. 2002. Associations of body mass index and obesity with physical activity, food choices, alcohol intake, and smoking in the 1982–1997 FINRISK Studies. *Am. J. Clin. Nutr.* 75:809–817.
12. Lahti-Koski, M., E. Vartiainen, S. Männistö, and P. Pietinen. 2000. Age education and occupation as determinants of trends in body mass index in Finland from 1982 to 1997. *Int. J. Obes. Relat. Metab. Disord.* 24:1669–1976.
13. Telama, R., and X. Yang. 2000. Decline of physical activity from youth to young adulthood in Finland. *Med. Sci. Sports Exerc.* 32:1617–1622.
14. Bautista-Castano, I., J. Molina-Cabrillana, J. A. Montoya-Alonso, and L. Serra-Majem. 2004. Variables predictive of adherence to diet and physical activity recommendations in the treatment of obesity and overweight, in a group of Spanish subjects. *Int. J. Obes. Relat. Metab. Disord.* 5:697–705.
15. Tammelin, T., J. Laitinen, and S. Näyhä. 2004. Change in the level of physical activity from adolescence into adulthood and obesity at the age of 31 years. *Int. J. Obes.* 23:1–9.
16. Richards, M. P., P. B. Pettitt, M. C. Stiner, and E. Trinkaus. 2001. Stable isotope evidence for increasing dietary breadth in the European mid-Upper Paleolithic. *Proc. Natl. Acad. Sci. USA* 98:6528–6532.
17. Cordain, L., J. B. Miller, and S. B. Eaton, N. Mann, S. H. Holt, J. D. Speth. 2000. Plant-animal subsistence ratios and macronutrient energy estimations in worldwide hunter-gatherer diets. *Am. J. Clin. Nutr.* 71:682–692.
18. Lee, R. B. 1979. *The !Kung San: Men, Women, and Work in a Foraging Society.* Cambridge: Cambridge University Press.
19. Kaplan, H., K. Hill, J. Lancaster, and M. Hurtado. 2000. A theory of human life history evolution: Diet, intelligence, and longevity. *Evol. Anthropol.* 9:156–185.
20. Gray, G. A., S. J. Nielsen, and B. M. Popkin. 2004. Consumption of high-fructose corn syrup in beverages may play a role in the epidemic of obesity. *Am. J. Clin. Nutr.* 79:537–543.
21. Garn, S. M., S. M. Bailey, and I. T. T. Higgins. 1980. Effects of socioeconomic status, family line, and living together on fatness and obesity. *Childhood Prevention of Atherosclerosis and Hypertension,* 187–204.
22. Flegal, K. M., M. D. Carroll, C. L. Ogden, and C. I. Johnson. 2002. Prevalence and trends in obesity among US adults 1999–2000. *JAMA* 288:1723–1727.
23. Schoenborn, C. A., P. F. Adams, and P. M. Barnes. 2002. Body weight status of adults: United States, 1997–98. *Adv Data.* 330:1–15.
24. Thomas, R. B., and T. L. Leatherman. 1990. Household coping strategies and contradictions in response to seasonal food shortage. *Eur. J. Clin. Nutr.* 44 Suppl 1:103–111.
25. Wardle, J. 2002. Sex differences in association with SES and obesity. *Am. J. Public Health.* 92:1299–1304.
26. Paeratakul, S., J. C. Lovejoy, D. H. Ryan, and G. A. Bray. 2002. The relation of gender, race, and socioeconomic status to obesity and obesity comorbidities in a sample of U.S. adults. *Int. J. Obes. Relat. Metab. Disord.* 26:1205–1210.
27. Basiotis, P. P. 1992. Validity of the self-reported food sufficiency status item in the U.S. In *American Council on Consumer Interests 38th Annual Conference, Columbia, Mo.,* ed. V. A. Haldeman. Washington, DC: US Department of Agriculture.
28. Drewnowski, A., and S. E. Specter. 2004. Poverty and obesity: The role of energy density and energy costs. *Am. J. Clin. Nutr.* 79:6–16.
29. Drewnowski, A. 2003. Fat and sugar: an economic analysis. *J. Nutr.* 133:838S–840S.
30. ACC/SCN. 2000. *Fourth Report on the World Nutrition Situation.* Geneva: ACC/SCN in collaboration with IFPRI.
31. Wang, Y., C. Monteiro, B. M. Popkin. 2002. Trends of obesity and underweight in older children and adolescents in the United States, Brazil, China, and Russia. *Am. J. Clin. Nutr.* 75:971–977.
32. ACC/SCN. 2004. *Fifth Report on the World Nutrition Situation.* Geneva: ACC/SCN in collaboration with IFPRI.
33. Neel, J. V. 1962. Diabetes mellitus: a "thrifty" genotype rendered detrimental by "progress." *Am. J. Hum. Genet.* 14:353–362.
34. Keys, A., J. Brozek, A. Henschel, O. Mickelsen, and H. L. Taylor. 1950. *The Biology of Human Starvation.* Minneapolis: University of Minnesota Press.
35. Jones, A. P., S. A. Assimon, and M. I. Friedman. 1986. The effect of diet on food intake and adiposity in rats made obese by gestational undernutrition. *Physiol. Behav.* 37:381–386.
36. McCance, R. A., and E. M. Widdowson. 1974. The determinants of growth and form. *Proc. R. Soc. Lond.* 185:1-17.

37. Dulloo, A. G., L. Girardier. 1990. Adaptive changes in energy expenditure during refeeding following low-calorie intake: Evidence for a specific metabolic component favoring fat storage. *Am. J. Clin. Nutr.* 52:415–420.

38. Malcolm, L. A. 1970. *Growth and Development in New Guinea—A Study of the Bundi People of the Madang District.* Institute of Human Biology; Papua, New Guinea. Monograph Series No. 1. Australia: Surrey Beatty & Sons.

39. Walker, S. P., C. A. Powell, S. M. Grantham-McGregor, J. H. Himes, and S. M. Chang. 1991. Nutritional supplementation, psychosocial stimulation, and growth of stunted children: The Jamaican study. *Am. J. Clin. Nutr.* 54:642–648.

40. Ravelli, G., Z. A. Stein, and M. W. Susser. 1976. Obesity in young men after famine exposure in utero and early infancy. *N. Eng. J. Med.* 295:350–353.

41. Dulloo, A. G. 1997. Regulation of body composition during weight recovery: Integrating the control of energy partitioning and thermogenesis. *Amer. J. Clin. Nutr.* 16 (suppl):25–35.

42. Dulloo, A. G., and J. Jacquet. 1998. Adaptive reduction in basal metabolic rate in response to food deprivation in humans: A role for feedback signals from fat stores. *Am. J. Clin. Nutr.* 68:599–606.

43. Manore, M. M., T. E. Berry, J. S. Skinner, and S. S. Carroll. 1991. Energy expenditure at rest and during exercise in nonobese female cyclical dieters and in nondieting control subjects. *Am. J. Clin. Nutr.* 54:41-46.

44. Weyer, C. H., R. L. Walford, I. T. Harper, M. Milner, T. MacCallum, P. A. Tataranni, and E. Ravussin. 2000. Energy metabolism after 2 y of energy restriction: the Biosphere 2 experiment. *Am. J. Clin. Nutr.* 72:946–953.

45. Sawaya, A. L., L. P. Grillo, I. Verreschi, A. Carlos da Silva, and S. B. Roberts. 1998. Mild stunting is associated with higher susceptibility to the effects of high-fat diets: Studies in a shantytown population in Sao Paulo, Brazil. *J. Nutr.* 128: 415S–420S.

46. Hoffman, D. J., A. L. Sawaya, I. Verreschi, K. L. Tucker, and S. B. Roberts. 2000. Why are nutritionally stunted children at increased risk of obesity? Studies of metabolic rate and fat oxidation in shantytown children from São Paulo, Brazil. *Am. J. Clin. Nutr.* 72:702–707.

47. Florenciol, T. T., H. S. Ferreira, J.C. Cavalcante, S. M. Luciano, and A. L. Sawaya. 2003. Food consumed does not account for the higher prevalence of obesity among stunted adults in a very-low-income population in the Northeast of Brazil (Maceió, Alagoas). *Eur. J. Clin. Nutr.* 57:1437–1446.

48. Shetty, P. S. 1992. Respiratory quotients and substrate oxidation rates in the fasted and fed state in chronic energy deficiency. In: *Protein-Energy Interactions,* ed. N. S. Scrimshaw and B. Schurch, 139–150. Lausanne: IDEGG.

49. Zurlo, F., S. Lillioja, A. Esposito-Del Puente, B. L. Nyomba, I. Raz, M. F. Saad, B. A. Swinburn, W. C. Knowler, C. Bogardus, and E. Rayussin. 1990. Low ratio of fat to carbohydrate oxidation as predictor of weight gain: Study of 24h RQ. *Am. J. Physiol.*, 259:650–657.

50. Barker, D. 1994. *Mothers, Babies, and Disease in Later Life.* London, UK: BMJ Publishing.

51. Bateson, P., D. Barker, T. Clutton-Brock, D. Deb, B. D'Udine, R. A. Foley, P. Gluckman, K. Godfrey, T. Kirkwood, M. M. Lahr, J. McNamara, N. B. Metcalfe, P. Monaghan, H. G. Spencer, and S. E. Sultan. 2004. Developmental plasticity and human health. *Nature* 430:419–421.

52. Wolf, G. 2003. Adult type 2 diabetes induced by intrauterine retardation. *Nutr. Rev.* 61:176–179.

53. Johnston, F. E., and G. D. Foster. 2001. *Obesity, Growth and Development.* London: Smith-Gordon-Nishimura.

54. Lampl, M., and P. Jeanty. 2004. Exposure to maternal diabetes is associated with altered fetal growth patterns: A hypothesis regarding metabolic allocation to growth under hyperglycemic-hypoxemic conditions. *Am. J. Human Biol.* 16:237–263.

Appendix A

CONVERSION FACTORS

TO CONVERT	INTO	MULTIPLY BY
Acre	hectare	0.4047
Acres	sq feet	43,560.0
Acres	sq meters	4,047.0
Acres	sq miles	1.562×10^{-3}
Alcohol (gram)	kilocalories	7.0
Bushels	cu ft	1.2445
Bushels	cu in.	2,150.4
Bushels	cu meters	0.03524
Bushels	liters	35.24
Calories	joules	4.186
Calories	megajoules	0.4186
Carbohydrate (gram)	kilocalories	4.0
Centigrade	fahrenheit	$(C° \times 1.8) + 32$
Centimeters	feet	3.281×10^{-2}
Centimeters	inches	0.3937
Centimeters	kilometers	10^{-5}
Centimeters	meters	0.01
Centimeters	miles	6.214×10^{-6}
Centimeters	millimeters	10.0
Centimeters	mils	393.7
Centimeters	yards	1.094×10^{-2}
Cubic centimeters	cubic inches	0.06102
Cubic inches	cubic centimeters	16.39
Deciliters	liters	0.1
Fat (gram)	kilocalories	9.0
Fahrenheit	centigrade	$(F° - 32) / 1.8$
Feet	centimeters	30.48
Feet	kilometers	3.048×10^{-4}
Feet	meters	0.3048
Gallons	liters	3.785
Grams	kilograms	0.001
Grams	milligrams	1,000
Grams	ounces (troy)	0.03215
Grams	pounds	0.2204
Inches	centimeters	2.540
Inches	meters	0.254
Inches	miles	1,000.0
Inches	yards	2.778×10^{-2}
Joules	kg-calories	2.389×10^{-4}
Kcalories	Kilo Joules	4.186
Kilograms	grams	1,000.0
Kilogram-Calories	joules	4.186
Kilogram-Calories	kilojoules	4.186
Kilometers	feet	3,281
Kilometers	inches	3.937×10^{4}

TO CONVERT	INTO	MULTIPLY BY
Kilometers	meters	1,000.0
Kilometers	miles	0.6214
Kilometers	millimeters	10^6
Kilometers	yards	1,094
Mega Joules	kcalories	2.389
Meters	centimeters	100.0
Meters	feet	3.281
Meters	inches	39.37
Meters	kilometers	0.001
Micrograms	grams	10^{-6}
Ounces	pounds	0.0625
Oxygen (L/min)	kilocalories	4.83
Pounds	grams	453.5924
Pounds	kilograms	0.4536
Pounds	ounces	16.0
Protein (gram)	kilocalories	4.0
Yards	centimeters	91.44
Yards	kilometers	9.144×10^{-4}
Yards	meters	0.9144
Yards	millimeters	914.4

Appendix B

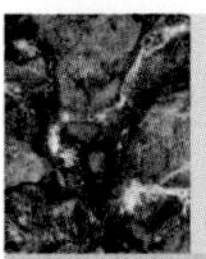

CONVERSION FACTORS FOR MEASUREMENTS USED IN THE BIOLOGICAL SCIENCES*

Blood Chemistries	Present Symbol	Conversion Factor	SI Unit Symbol	Significant Digits	Suggested Minimum Increment
alpha-fetoprotein (S)	ng/mL	1.00	mg/L	XX	1 mg/L
alpha-fetoprotein (Amf)	mg/dL	10	mg/L	XX	1 mg/L
chloride (S)	mEq/L	1.00	mmol/L	XXX	1 mmol/L
cholesterol (P)	mg/dL	0.02586	mmol/L	X.XX	0.05 mmol/L
ethanol (P)	mg/dL	0.2171	mmol/L	XX	1 mmol/L
ferritin (S)	ng/mL	1.00	mg/L	XXO	10 mg/L
folate (S)	ng/mL	2.266	nmol/L	XX	2 nmol/L
glucagon (S)	pg/mL	1	ng/L	XXO	10 ng/L
glucose (P,S)	mg/dL	0.05551	mmol/L	XX.X	0.1 mmol/L
iron (S)	mg/L	0.1791	mmol/L	XX	1 mmol/L
lead (B,U)	mg/dL	48.26	mmol/L	X.XX	0.05 mmol/L
lipoproteins (P)		(10)	(g/L)		
[LDL][HDL]	mg/dL	0.02586	mmol/L	X.XX	0.05 mmol/L
Protoporphyrin (Ere)	mg/dL	0.0177	mmol/L	X.XX	0.02 mmol/L
sodium (S,U)	mEq/L	1.00	nmol/L	XXX	1 mmol/L
thiocyanate (P)	mg/dL	0.1722	mmol/L	X.XX	0.1 mmol/L
triglycerides (as triolein)	mg/dL	0.1129	mmol/L	X.XX	0.02 mmol/L
vitamin A (retinol) (P,S)	mg/dL	0.03491	mmol/L	X.XX	0.05 mmol/L
vitamin B_1 (thiamine hydrochloride) (U)	mg/24 h	0.002965	mmol/d	X.XX	0.01 mmol/L
vitamin B_2 (S) (riboflavin)	mg/dL	26.57	nmol/L	XXX	5 nmol/L
vitamin B_6 (B) (pyridoxal)	ng/mL	5.982	nmol/L	XXX	5 nmol/L
vitamin B_{12} (P,S) (cyanocobalamin)	ng/dL	7.378	pmol/L	XXO	l0 pmol/L
vitamin C (B,P,S) (ascorbate)	mg/dL	56.78	mmol/L	XO	10 mmol/L
vitamin D_3 (P) (cholecalciferol)	mg/mL	2.599	nmol/L	XXX	5 nmol/L
vitamin E (P,S) Symbols (alpha-tocopherol)	mg/dL	23.22	mmol/L	XX	1 mmol/L
Hematology erythrocyte count (B)	10^6/mm^3	1	10^{12}/L	X.X	0.1 10^{12}/L
hematocrit	0/0	0.01	(1)	O.XX	0.01
hemoglobin (B) (mass concentration)	g/dL	10	g/L	XXX	1 g/L

Symbols

(1)	number one	Ere	Erythrocyte	n	nano (10^{-9})
Amf	amniotic fluid	g	gram	p	pico (10^{-12})
B	Blood	L	Liter	P	plasma
d	deci (10^{-1})	m	milli (10^{-3})	S	serum
		mol	mole	u	micro (10^{-6})

*Source: Yankauer, A. 1987. Back to the future with SI Units. *American Journal of Public Health* 77: 1398–1399.

Appendix C

Equations for Estimating Resting Energy Expenditure

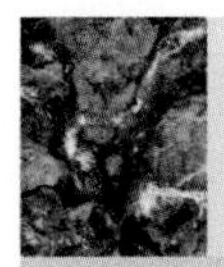

Table 1. Equations for Estimating Resting Energy Expenditure (REE) for Subjects Ranging in Age from 16 to 78

Harris and Benedict Equation[1]
Males: BMR (Kcalories/day) = (13.397 × weight, kg) + (4.799 × height, m) – (5.677 × age, years) + 88.362
Females: BMR (Kcalories/day) = (9.247 × weight, kg) + (3.098 × height, m) – (4.330 × age, years) + 447.593

World Health Organization Equations [2]
Males: BMR (Kcalories/day) = (11.6 × weight, kg) + 879
Females: BMR (Kcalories/day) = (8.7 × weight, kg) + 829

Equations by Owen et al. [3]
Males: BMR (Kcalories/day) = (10.2 × weight, kg) + 879
Females: BMR (Kcalories/day) = (7.18 × weight, kg) + 795

Equations by Mifflin et al. [4]
Males: BMR (Kcalories/day) = (10 × weight, kg) + (6.25 × height, cm) – (5 × age, years) + 5
Females: BMR (Kcalories/day) = (10 × weight, kg) + (6.25 × height, cm) – (5 × age, years) – 161

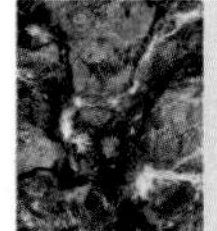

Table 2. Equations for Estimating Resting Energy Expenditure (RMR) from Weight (kg) and Height (m). RMR is given in Mega joules (MJ) /24h (5).

Age Group (Years)	Males Mega Joules (Mj) /24h	Females Mega Joules (Mj) /24h
0.0–3.0	BMR = 0.0007 Wt + 6.349 Ht – 2.584	BMR = 0.068 Wt + 4.281 Ht – 1.730
3.0–10.0	BMR = 0.082 Wt + 0.545 Ht + 1.736	BMR = 0.071 Wt + 0.677 Ht + 1.553
10.0–18.0	BMR = 0.068 Wt + 0.574 Ht + 2.157	BMR = 0.035 Wt + 1.948 Ht + 0.837
19.0–30.0	BMR = 0.063 Wt – 0.042 Ht + 2.953	BMR = 0.057 Wt + 1.184 Ht + 0.411
30.0–60.0	BMR = 0.048 Wt – 0.011 Ht + 3.670	BMR = 0.034 Wt + 0.006 Ht + 3.530
60.0 +	BMR = 0.038 Wt – 4.068 Ht + 3.491	BMR = 0.033 Wt + 1.917 Ht + 0.074

Mega Joules to K calories = MJ × 2.389; Kcalories to Kilo Joules = Kcal × 4.186

1. Roza, A. M., and Shizgal, H. M. 1984. The Harris Benedict equation reevaluated: Resting energy requirements and the body cell mass. *Am. J. Clin. Nutr.* 40(1):168–182.
2. Report of a joint FAO/WHO/UNU expert consultation. *Energy and Protein Requirements.* Technical Report Series No 724. Geneva, Switzerland: World Health Organization, 1985.
3. Owen, O. E., Holup, J. L., D'Alessio, D.A., Craig, E. S., Polansky, M., Smalley, K. J., Kavle, E. C., Bushman, M. C., Owen, L. R., and Mozzoli, M. A. 1987. A reappraisal of the caloric requirements of men. *Am. J. Clin. Nutr.* 46:875–885.
4. Mifflin, M. D., S. T. St Jeor, L. A. Hill, B. J. Scott, S. A. Daugherty, and O. K. Young. 1990. A new predictive equation for resting energy expenditure in healthy individuals. *Am. J. Clin. Nutr.* 1:241–247.
5. Schofield, W. N. 1985. Predicting basal metabolic rate: New standards and a review of previous work. *Human Nutrition*, 39C, Suppl. 1: 5–41.

Appendix D

Approximate calculation of body mass index (kg/m²)

	Body Mass Index (kg/m²)																
	19	20	21	22	23	24	25	26	27	28	29	30	31	32	33	34	35
Height	Weight (in pounds)																
4'10" (58")	91	96	100	105	110	115	119	124	129	134	138	143	148	153	158	162	167
4'11" (59")	94	99	104	109	114	119	124	128	133	138	143	148	153	158	163	168	173
5'0" (60")	97	102	107	112	118	123	128	133	138	143	148	153	158	163	168	174	179
5'1" (61")	100	106	111	116	122	127	132	137	143	148	153	158	164	169	174	180	185
5'2" (62")	104	109	115	120	126	131	136	142	147	153	158	164	169	175	180	186	191
5'3" (63")	107	113	118	124	130	135	141	146	152	158	163	169	175	180	186	191	197
5'4" (64")	110	116	122	128	134	140	145	151	157	163	169	174	180	186	192	197	204
5'5" (65")	114	120	126	132	138	144	150	156	162	168	174	180	186	192	198	204	210
5'6" (66")	118	124	130	136	142	148	155	161	167	173	179	186	192	198	204	210	216
5'7" (67")	121	127	134	140	146	153	159	166	172	178	185	191	198	204	211	217	223
5'8" (68")	125	131	138	144	151	158	164	171	177	184	190	197	203	210	216	223	230
5'9" (69")	128	135	142	149	155	162	169	176	182	189	196	203	209	216	223	230	236
5'10" (70")	132	139	146	153	160	167	174	181	188	195	202	209	216	222	229	236	243
5'11" (71")	136	143	150	157	165	172	179	186	193	200	208	215	222	229	236	243	250
6'0" (72")	140	147	154	162	169	177	184	191	199	206	213	221	228	235	242	250	258
6'1" (73")	144	151	159	166	174	182	189	197	204	212	219	227	235	242	250	257	265
6'2" (74")	148	155	163	171	179	186	194	202	210	218	225	233	241	249	256	264	272
6'3" (75")	152	160	168	176	184	192	200	208	216	224	232	240	248	256	264	272	279

Body Mass Index (BMI) = [(Weight, kg / (Height, m²)
Body Mass Index (BMI) = [(Weight, lb / Height in., / Height in.) × 703]

Glossary

Internet Sources

http://www.anth.ucsb.edu/glossary/glossary.html

http://www.glossarist.com/glossaries/humanities-socialsciences/anthropology.asp

http://www.alphadictionary.com/directory/Specialty_Dictionaries/Anthropology/

http://www.webref.org/anthropology/anthropology.htm

A

ABO blood type system Pertaining to the genetic system for one of the proteins located on the surface of red blood cells. Composed of one gene with three alleles: A, B, and O.

aboriginal Native, indigenous.

absolute dating Age estimation in calendar years before the present; also known as numerical dating, chronometric dating.

acclimatization Biological features acquired under natural, noncontrolled environmental conditions.

acclimation Biological features acquired under laboratory-controlled, experimentally induced, environmental conditions.

acetabulum A cup-shaped depression on the side surface (ilium) of the pelvis into which the head of the femur fits; the socket in the ball-in-socket hip joint. The three bones of the os coxa, which include the ilium, ischium, and pubis, meet in the acetabulum.

Acheulian tradition Pertaining to a stone tool industry of the Lower and Middle Pleistocene characterized by a large proportion of bifacial tools (flaked on both sides). Acheulian tool kits are very common in Africa, southwest Asia, and western Europe, but are less common elsewhere. (Also spelled "Acheulean.")

acrocentric chromosome A chromosome with a centromere near the end such that it has one long arm plus a stalk and a satellite.

activity budget The pattern of waking, eating, moving, socializing, and sleeping that all nonhuman primates perform every day.

adapoids Family consisting mainly of Eocene primates, perhaps ancestral to all strepsirhines.

adaptability An individual organism's capability to make positive anatomical or physiological changes after short- or long-term exposure to stressful environmental conditions.

adaptation Biological or behavioral responses of organisms or population to the environment. Genetic adaptation is a result of natural selection.

adaptive niche The place and environment where an organism lives successfully.

adaptive radiation The expansion and diversification of organisms or species into new ecological niches or adaptive zones.

additive genetic variance (VA) Genetic variance that arises from the additive effects of genes on the phenotype.

adduction Movement of a body limb towards the midline, or movement of a digit toward the long axis of a limb.

adenine (A) A purine base found in DNA and RNA; in double-stranded DNA adenine couples with the pyrimidine thymine.

affiliative Pertaining to friendly associations between individuals, as in the case of grooming, which reinforces social cohesion and bonds among primates.

affinity Relationship by marriage. May include the relationship between corporate groups linked by marriage between their members.

African replacement model The hypothesis that modern humans evolved as a new species in Africa between 150,000 and 200,000 years ago and then spread throughout the Old World, replacing preexisting Neanderthal and other archaic human populations.

age at menarche The age at first menstrual period in human females.

allele One, two, or more alternative forms of a single gene.

allele frequency The percentage of an allele at a particular locus relative to all the different alleles at the locus, usually calculated for a population.

Allen's Rule The tendency of warm-blooded animals to have longer and leaner limbs in warmer climates than their relatives of the same species in colder regions.

allometry (allometric scaling) The differential proportion among anatomical structures. For instance, the size of the brain does not change in proportion to body size. Opposed to isometry, which refers to equal proportion among anatomical structures. For example, tall individuals have proportionally bigger lungs than short individuals.

alloparenting Care and interaction of adults with the infants and offspring of other adults.

allopatric speciation Species formation when there is geographical isolation (cf. sympatric speciation).

allopatry Geographic ranges that do not overlap.

altricial Young being born in an immature and helpless condition.

altruism Behavior that benefits another individual at a cost to the individual performing the altruistic behavior, where cost and benefit are defined in terms of reproductive success.

alveolar prognathism Forward projection of the portions of the jaws that hold the teeth and their roots.

alveolus Socket, as in tooth socket.

amino acids The building blocks of polypeptides; there are 20 different amino acids.

amino-acid racemization A method used to date material up to 100,000 years old.

amniocentesis Procedure by which a sample of amniotic fluid is withdrawn from the amniotic sac of a developing fetus and cells are cultured and examined for chromosomal abnormalities.

anagenesis An anatomical change in a single lineage over time that is sufficient to name a new species.

analogies Commonalities between organisms based strictly on similar function with no assumed common evolutionary descent.

analysis of variance (ANOVA) Statistical procedure used to examine differences in means and to partition variance.

anaphase The stage in mitosis or meiosis during which the sister chromatids or homologous chromosomes separate and then migrate toward the opposite poles of the cell.

anaphase II The second stage of meiosis during which the chromatids (and therefore the centromeres) are pulled to the opposite poles of the spindle. The separated chromatids are now referred to as chromosomes in their own right.

ancestral (primitive) Pertaining to characters inherited by a group of organisms from a remote ancestor that diverged after the character first appeared.

aneuploidy A condition in which one or more whole chromosomes of a normal set of chromosomes either are missing or are present in more than the usual number of copies; abnormal condition in which a part or parts of a chromosome or chromosomes are duplicated or deleted.

anterior Frontal portion.

anterior iliac spines In hominids, two bony projections (superior and inferior) for muscles to attach that extend the leg, protruding from the front edge of the ilium. *Sartorius* attaches on the anterior iliac spine and *rectus femoris* (one of the quadriceps) attaches on the anterior inferior spine.

anterior pillars (of the maxilla) Two vertical columns of bone, on either side of the nasal aperture, extending above and lengthening the pilaster of bone surrounding the canine roots.

anthropocentric Viewing nonhuman organisms with reference to human experience and capabilities; stressing the importance of humans over everything else.

anthropoids Members of a suborder of Primates, the *Anthropoidea*, which includes monkeys, apes, and humans.

anthropology The study of humanity—our physical characteristics as animals, and our non-biological characteristics collectively referred to as culture. Traditionally, the subject is broken down into four subdisciplines biological (physical) anthropology, ethnology (cultural or social), archaeology, and linguistics.

anthropometry The measurement of human body parts.

antibody A protein molecule that recognizes and binds to a foreign substance that has been introduced into the organism.

antigen Any large molecule that stimulates the production of specific antibodies or that binds specifically to an antibody.

apes The large hominoid arboreal primates, based on similarity and not common descent (apes are not a monophyletic group).

appendicular skeleton The bones that make up the limbs and the shoulder girdle.

arboreal Living mainly in trees; for instance, arboreal quadrupeds are animals that use all four limbs in walking and running on tree limbs.

Arboreal theory A theory that states that many of the cranial and postcranial adaptations found in primates are actually adaptations for life in the trees.

archaeology A subdiscipline of anthropology that involves the study of the human past through its material remains.

archaic Ancient.

Ardipithecus ramidus An early primitive hominid species from Africa who lived between 5.8 and 4.4 million years ago.

argon-argon (Ar-40/Ar-39) Radiometric technique modified from K-Ar that gauges K-40 by proxy using Ar-39. Provides for the measurement of smaller samples with less error.

artifacts Objects that have been modified by humans.

artificial selection Selection made by humans about which individuals will survive and reproduce. If the selected traits have a genetic basis, they will change and evolve.

ascending ramus Vertical portion of the mandible that extends from the corpus to the condyle.

asexual reproduction Reproduction in which a new individual develops either from a single cell or from a group of cells; occurs in the absence of any sexual process.

assemblage All the artifacts, fauna, and other debris found in a single layer or excavation unit at a site.

association area Refers to the cerebral cortex of the frontal, parietal, and temporal regions of the brain. These areas are involved in complex associations between these modalities.

atlas The most superior of the cervical vertebrae, supporting the cranial articulation.

atomic absorption spectrometry (AAS) A method that measures energy in the form of visible light waves. AAS is capable of measuring up to 40 different elements with an accuracy to approximately 1 percent.

auricular point (au) A point located a few millimeters vertically above the center of the auditory meatus and porion, on the root or base of the zygomatic arches.

Aurignacian Relating to an Upper Paleolithic stone tool industry in Europe beginning at about 40,000 years ago.

Australopithecus The collective name for the genus of fossil hominid that lived between 2 million and 5 years ago in East Africa and is characterized by bipedal locomotion but small brain size.

Australopithecus aethiopicus Member of the robust species of australopithecines that lived 2.5 million years ago in East Africa. It is characterized by derived features seen in other robust australopithecines combined with primitive features seen in *A. afarensis.*

Australopithecus afarensis Member of the gracile species of australopithecines that lived between 2.9 million and 4 million years ago in East Africa. The teeth and postcranial skeleton show a number of primitive and apelike features.

Australopithecus africanus Member of the gracile species of australopithecines that lived between 2.3 million and 3.3 million years ago in South Africa.

Australopithecus anamensis A hominid species that lived in East Africa between 3.9 million and 4.2 million years ago. May be the ancestor of the australopithecines.

Australopithecus boisei Member of the robust species of australopithecines that lived between 1.2 million and 2 million years ago and found in East Africa.

Australopithecus garhi A species of australopithecine that lived about 1.5 million years ago in East Africa.

Australopithecus robustus Member of the robust species of australopithecines that lived between 1.5 million and 2 million years ago and found in South Africa.

autoimmune diseases Occur when the immune system of the body attacks its own tissues.

autonomic Relating to physiological responses not under voluntary control.

autosomal recessive disease Disease resulting from a recessive allele; one copy of the allele needs to be inherited from each parent in order for the disease to develop.

autosome A chromosome other than a sex chromosome.

auxology A science concerning human growth and development.

axial skeleton The part of the skeleton along the central axis of the body including the vertebral column, pelvis, and thorax.

axillary Of or pertaining to the armpit.

axion The nerve fiber that extends away from the nerve cell body to meet the axion of another nerve cell.

B

B.P. Before present.

bacteria Spherical, rod-shaped, or spiral-shaped, single-cellular or multicellular, filamentous prokaryotic organisms.

balanced polymorphism A system of genes in which two alleles are maintained in stable equilibrium because the heterozygote is selected for while selection works against the extremes in either of the homozygotes.

band A simple form of human social organization, composed of one or more families.

Barr body A highly condensed mass of chromatin found in the nuclei of normal females, but not in the nuclei of normal male cells; represents a cytologically condensed and inactivated X chromosome.

base A variable aspect of the nucleotides that form the nucleic acids DNA and RNA. The bases for DNA include adenine, guanine, thymine, and cytosine. For RNA, thymine is replaced by uracil.

base-pair substitution mutation A change in a gene such that one base pair is replaced by another base pair; for instance, a TA pair is replaced by a CG pair.

basicranium Base of the skull, formed mainly by the occipital, petrosal (of the temporal), and sphenoid bones.

basion (ba) The point at which the anterior border of the foramen magnum crosses the midline.

behavioral evolutionary ecology The study of behavior, emphasizing the role of ecological factors as agents of natural selection. Behaviors and behavioral patterns have been selected for because they increase the reproductive fitness of individuals (i.e., they are adaptive) in specific environmental contexts.

Bergmann's rule The tendency of warm-blooded animals of the same species to be heavier in cold climates than animals that lived in warm climates.

Beringia The continent, or wide landbridge, spanning the Bering Strait between Alaska and Siberia at times of low sea level, including contiguous parts of both.

bicondylar angle The angle (from vertical) that the shaft of the femur makes when the bone is stood upright on its condyles.

bicuspid A premolar.

biface Stone tool that has been flaked on two opposing sides or faces, thus forming a cutting edge between the two flake scars.

bilateral symmetry Anatomical features for which the right and left sides are close to being mirror images of one another.

bilophodont Teeth with two crests, a type of molar construction in which there are two parallel enamel ridges on the occlusal surface, running from side to side and connecting the cusps.

binocular vision Overlapping fields of vision in which both eyes can focus on a distant object to produce a stereoscopic (three-dimensional) image.

binomial distribution The theoretical frequency distribution of events that have two possible outcomes.

binomial nomenclature (refers to "two names") Used in taxonomy, the convention established by Carolus Linnaeus whereby genus and species names are used to refer to species. For example, *Homo sapiens* refers to human beings.

biocultural evolution The interactive evolution of human biology and culture; the concept that both biology and culture evolve interdependently.

biological anthropology The study of humans as biological organisms in the context of an evolutionary framework; also referred to as physical anthropology.

biological species Defines a species as a reproductively isolated aggregate of populations that can actually or potentially interbreed and produce fertile offspring. This concept stresses the importance of reproductive isolation and the anatomical and behavioral mechanisms that create it.

biomass The total combined weights of all living things in a particular area.

biomedical anthropology A subfield of biological anthropology that deals with issues of health and illness.

biostratigraphy A relative dating technique that compares fossils from different stratigraphic sequences in order to estimate the age of the layers.

bipedal Two-legged.

bipolar technique The removal of flakes from a core resting on a hard surface, which gives the flakes different thickness characteristics and the core a unique form due to a shockwave from percussion at both ends. A technique of resting the objective piece on an anvil and striking it with a hammer to split or remove a detached piece.

blade tool An artifact formed on a parallel-sided stone flake that is usually removed from a carefully prepared core but can be defined as any flake whose length is more than double its breadth.

bonobo A species of chimpanzee, *Pan paniscus*, that live in forested habitats of Central Africa south of the Zaire River; the closest living relative of the common chimpanzee, *Pan troglodytes*.

bovid A member of the family *Bovidae*, cloven-hoofed ungulates, including bison, antelopes, deer, goats, sheep, etc. Most bones found at African sites are bovid.

Brachial Index Ratio of the length of the forearm divided by the length of the upper arm (radius/humerus × 100).

brachiation Arm-over-arm arboreal locomotion in which the animal progresses below branches by swinging its body between forelimb supports.

brachycephalic Broad-headed, possessing a Cephalic Index (cranial breadth/cranial length × 100) over 80.

brain stem The portion of the brain controlling basal metabolic rates, respiration, pulse, and other basic body functions.

bregma (b) The point at the top of the head where the coronal and sagittal sutures of the skull meet, or the point at which the two parietal bones meet with the frontal.

bridewealth Marriage payments from the husband and his kin to the bride's kin. Characteristically these payments balance a transfer of rights over the wife's sexuality, work services, residence, fertility, and so on.

Broca's area A cortical region of the human brain located on the side of the frontal lobe, just above the temporal lobe (directly beneath a finger placed at the temple). This area is important to speech production and therefore, injury to this area will result in aphasia (language dysfunction).

browridge A thickened ridge or shelf of bone above the orbits at the base of the forehead, continuously, although not necessarily evenly, developed from the middle of the cranium to each side.

buccal The cheek-facing side of a postcanine tooth.

burin A chisel-like stone tool for engraving bone, wood, horn, or soft stone.

butchering site A location at which archaeological evidence of butchering of carcasses by hominids exists. The evidence typically includes tool cut marks on fossilized animal bones or the presence of the stone tools themselves.

C

calcaneus Heel bone.

calibrated relative dating techniques Techniques that can be correlated to an absolute chronology.

calvaria (calvarium) The bones of the cranium without the face or mandible.

cancellous (trabecular, spongy) bone Internal bone tissue that is porous and lightweight.

canine A conical or spade-like tooth (depending on species) located between the incisors and premolars.

canine fossa A vertical furrow on the maxilla under the infraorbital foramen, extending toward the base of the zygomatic process of the maxilla and to the side of the nose.

captive study A study of primate behavior performed in an enclosed environment, such as a zoo or laboratory.

carnivore (1) An animal that eats primarily the flesh of other animals; or (2) members of the mammalian order *Carnivora*, which include cats, dogs, skunks, raccoons, and bears.

carpal bones Small bones of the wrist; includes the scaphoid, lunate, triquetram, pisiform, hamate, capitate, trapezoid, and trapezium bones in humans.

carpal tunnel A tunnel on the palm side of the hand created between the arch formed by the carpal bones and the fibrous band that draws the ends of the arch together. The long flexor tendons for the fingers pass through this tunnel, which holds them in and prevents them from bowing when the wrist is moved.

carrying capacity Given a particular subsistence adaptation, the number of individuals that can be optimally supported.

cartilage A flexible connective tissue; important part of most of the skeleton that calcifies at various stages of growth.

Catarrhina The infraorder of the order Primates that is composed of Old World monkeys, apes, and hominids.

catastrophism The view promoted by Cuvier, especially in opposition to Lamarck, that states that the earth's geological landscape is the result of violent cataclysmic events.

cell division A process whereby one cell divides to produce two cells.

cementum A soft, bone-like tissue covering tooth roots that anchors them to the ligament that covers the alveolar bone.

centromere A location of the chromosome that is condensed and constricted and where sister chromatids attach to one another during the processes of mitosis and meiosis.

Cephalic Index The ratio of the skull's breadth to its length.

cercopithecine Pertaining to members of the monkey subfamily Cercopithecinae.

cerebellar fossa Two broad depressions on the internal surface of the occiput, which hold the cerebellar lobes of the brain.

cerebellar lobe One of the lobes of the cerebellum.

cerebellum Area of the brain lying below and behind the cerebrum. Functions in proprioception, replaying feedback from muscle activity and motions back to the cortex for finer adjustments and coordination of movement.

cerebral cortex Referring to the cerebrum, the front and upper portion of the brain, including the frontal, parietal, temporal, and occipital lobes.

cerebral fossae Two broad depressions on the internal surface of the occiput, holding the posterior part of the occipital lobes of the cerebral cortex.

cerebral lobe A division of the cerebral cortex at the rear of the brain.

cerebrum The major part of the brain, which occupies the upper part of the cranium and is comprised of the two cerebral hemispheres connected by the corpus callosum.

cervical Referring to the neck, as in the seven neck vertebrae.

Chatelperronian Relating to an Upper Paleolithic industry found in France and Spain, containing blade tools and associated with Neanderthals.

cheek teeth Premolars and molars.

chiefdom A term used to describe a society that operates on the principle of ranking (i.e., differential social status). Different lineages are graded on a scale of prestige, calculated by how closely related one is to the chief. The chiefdom generally has a permanent ritual and ceremonial center, in addition to being characterized by local specialization in crafts.

chopper A stone made by taking a few flakes off a pebble or rock fragment in order to produce a sharp cutting edge.

chiasma A cross-shaped structure formed during crossing-over and visible during meiosis in the diplonema stage.

chi-square (X^2) test A statistical procedure that determines what constitutes a significant difference between observed results and results expected on the basis of a particular hypothesis; a test to determine goodness-of-fit.

Chordata The phylum of the animal kingdom including vertebrates.

chromatid One of the two visibly distinct longitudinal subunits of all replicated chromosomes that becomes visible during mitosis between early prophase and metaphase.

chromatin The piece of DNA-protein complex that is studied and analyzed. Each chromatin fragment reflects the general features of chromosomes but does not reflect the specifics of any individual chromosome.

chromosome The genetic material of a cell, complexed with protein and organized into a number of linear structures. It literally means "colored body," due to the fact that the threadlike structures are visible under the microscope only after they are stained with dyes.

chronometric dating (*chrono*, meaning "time," and *metric*, meaning "measure") A dating technique that gives an estimate in actual numbers of years.

cingulum A shelf of enamel running partially or completely around the base of a tooth crown.

clade A group composed of all the species descended from a single common ancestor; a monophyletic group.

cladistic homology Homologies based on comparing features in different clades or lineages.

cladogenesis Species formation, the branching of a single lineage to create two lineages.

cladogram A branching diagram, or dendrogram, based on genealogy and used to represent phyletic relationships (in a cladogram, *rates* of evolutionary divergence are ignored).

clan A unilineal descent group or category whose numbers trace patrilineal descent (patriclan) or matrilineal descent (matriclan) from an apicle ancestor/ancestress, but are unaware of the genealogical links that connect them to this apical ancestor.

clavicle Collarbone; the bone connecting the sternum (breastbone) to the scapula (shoulder blade).

cleaver A kind of Acheulean bifacial tool, typically oblong with a broad cutting edge on one end.

clinal variation Continuous, gradual variation of a trait. Clinal variation is in the context of geographic patterns of a trait's variation.

cline A gradual change in the frequency of genotypes and phenotypes from one geographical region to another geographical region.

clone An organism that is genetically identical to another organism; may also be used in reference to genetically identical DNA segments, molecules, and cells.

cloning The regeneration of many copies of a DNA molecule (e.g., a recombinant DNA molecule) by replication in a suitable host.

coadapted genes Sets of genes whose frequencies in a population reflect a compromise from different magnitudes and directions of selection acting on their various effects.

coalescence (of genes) The genealogical relations between genes show that all of the extant varieties of a gene must have originated in a single gene. If to look from the present to the past, we could say that they descended from a single form to which they coalesce.

coccyx Tailbone; consists of the caudal vertebrae, it is most often fused into a single unit.

coding sequence The part of an mRNA molecule that specifies the amino acid sequence of a polypeptide during translation.

codominance The situation in which the heterozygote exhibits the phenotypes of both homozygotes.

codon A group of three adjacent nucleotides in an mRNA molecule that specifies either one amino acid in a polypeptide chain or the termination of polypeptide synthesis.

coefficient of variation (CV) A statistical measure defined as the ratio of standard deviation to the mean multiplied by 100.

cognition The process of acquiring knowledge, which includes perception, intuition, and reasoning.

cognitive universals Cognitive phenomena like sensory processing, basic emotions, consciousness, motor control, memory, and attention, which are conveyed by all normal individuals.

collagen A fibrous protein that is the chief constituent part of connective tissue in bone.

collateral flaking The process of removing expanding flakes removed from the lateral margins of an objective piece at right angles to the longitudinal axis.

colobine Pertaining to members of the monkey subfamily Colobinae.

colonizing species A species with a high rate of reproduction, readily able to take advantage of new habitats because of its genetic variation and internal subdivisions.

communication Any action that conveys information, in the form of a message, to another individual. Often, the result of communication is a change in the behavior of the recipient. Communication may not be deliberate but rather may be the result of involuntary processes or a secondary consequence of an intentional action.

compact (cortical) bone The dense bone tissue found on the outside of bones or the walls of long bone shafts.

competitive exclusion The concept that no two species can coexist in the same locality if they rely on the same limiting resources.

complete dominance The case in which one allele is dominant to the other such that the phenotypic level of the heterozygote is essentially indistinguishable from the homozygous dominant.

complete recessiveness A situation in which an allele is phenotypically expressed only when it is homozygous.

computed axial tomography (CAT or CT) The method of examining body organs by scanning them with X-rays and using a computer to construct a series of cross-sectional scans along a single axis. The body is passed into the machine and images of cross-sectional "slices" through the body are created.

condyle A smooth, rounded articular surface found in pairs.

conjugation A process having an unidirectional transfer of genetic information through direct cellular contact between a donor ("male") and a recipient ("female") bacterial cell.

consanguineal A relative by birth, as distinguished from in-laws and step-relatives.

conspecific Belonging to the same species.

continental drift The movement of continents on sliding plates of the earth's surface. As an effect of continental drift, the positions of large landmasses have shifted dramatically during the earth's history.

convergence The independent evolution of the same (or very similar) features in two or more species from different features in their last common ancestor.

convolutions Wrinkles on the brain's surface.

coprolites Fossilized feces containing food residues that can be used to reconstruct diet.

core A nucleus or mass of rock that shows signs of detached piece removal. A core is often considered an objective piece that functions primarily as a source for detached pieces.

coronal suture The suture between the frontal bone and the parietal bones behind it.

coronoid process A hooked or curved projection. (1) The frontal part of the ascending ramus of the mandible, forming a pointed projection at its top, where part of the *temporalis* muscle attaches. (2) The lower projecting part of the ulna's trochlear notch, below the olecranon process.

corpus Body; the principal part of the bone.

corpus callosum The bundle of neurons connecting the left and right cerebral hemispheres.

correlation coefficient A statistic measuring the strength of the association between two variables. It can vary from –1 (perfect negative correlation) through 0 (no correlation) to +1 (perfect positive correlation).

covariance A statistic used in calculating the correlation coefficient between two variables.

cranial base angle The angle between the basiocciput and the body of the sphenoid, erroneously thought to be related to speech capacity.

cranial crests Muscles attach to these bony ridges located on the skull.

cranium The skull, without the mandible, which is made up of 28 bones.

creation science Attempt formulated by creationists to refute the evidence in favor of evolution.

crest A ridge with a sharp edge caused by muscle pull. A **simple crest** is created by the unidirectional pull of a single muscle; a **compound crest** is created by the opposing pulls of two muscles.

cross-cultural universals Behavioral phenomena like singing, dancing, and mental illness, which can be found in almost all human cultures, but are not necessarily displayed by every member of a particular cultural group.

cross-fertilization A term used to refer to the fusion of male and female gametes from different individuals; the bringing together of genetic material from different individuals for the purpose of genetic recombination.

crossing-over A term introduced by Morgan and E. Cattell, in 1912, to detail the process of reciprocal chromosomal interchange by which recombinants arise.

cross-sectional studies Examination of changes through the study of many individuals of different ages at a single time.

Crural Index The ratio of the length of the lower leg (tibia), divided by the length of the upper leg (femur), multiplied by 100.

cultural anthropology A subdiscipline of anthropology concerning with the nonbiological, behavioral aspects of society (i.e., the social, linguistic, and technological components underlying human behavior). Two important branches of cultural anthropology include ethnography (the study of living cultures) and ethnology (which attempts to compare cultures using ethnographic evidence). In Europe, it is referred to as social anthropology.

cultural ecology A term used to account for the dynamic relationship between human society and its environment, in which culture is viewed as the primary adaptive mechanism; this term was devised by Julian Steward.

culture A term used by anthropologists in reference to the nonbiological characteristics unique to a particular society.

cusp An elevated part on the occlusial surface of an unworn tooth.

cytoplasm The portion of the cell contained within the cell membrane, but not including the nucleus. Rather, consisting of a semifluid material and containing numerous structures involved with cell function.

cytosine (C) A pyrimidine base found in RNA and DNA. In double-stranded DNA, cytosine pairs with the purine guanine.

D

Darwinian fitness The relative reproductive ability of a genotype.

data (*sing.*, datum) Facts from which conclusions can be drawn; scientific information.

decay In physics, the breaking apart of nuclei into smaller constituent nuclei, releasing energetic particles in the process as there is less total energy.

deciduous (milk) teeth The first set of teeth in a mammalian jaw, which are in turn replaced by the permanent dentition.

deduction A process of reasoning by which more specific consequences are inferred by rigorous argument from more general propositions.

deleted Pertains to when a chromosome breaks off spontaneously and is lost.

deletion mutation A change in a gene's base sequence resulting from the loss of one or more base pairs in the DNA.

deltoid muscle A complex shoulder muscle with humerus, clavicle, and scapula attachments that brings the arm up, flexing or extending the humerus depending on which part of the muscle is being used.

deme A local population of a species; the community of potentially interbreeding individuals at a given locality; a population or race sampled over time.

demography The study of a population's main life-history parameters, including its growth, size, composition, and the age-specific rates of births and deaths.

dendrochronology The study of tree-ring patterns; annual variations in climatic conditions producing differential growth can be used both as a measure of environmental change, and as the basis for a chronology.

dendrogram A branching diagram that begins with a single source.

dental arcade The tooth row.

dental caries A pathological process, with destruction of tooth enamel and dentine, that can lead to infection and tooth loss.

dental comb A primate feature in which the incisors and canines of the lower jaw are of similar size and form, short peg-like teeth that are set horizontally across the front of the mouth.

dental eruption A continuous process by which teeth emerge out of their crypts (alveolar eruption), through the gums (gingival eruption), and into occlusion with the opposing teeth of the opposite jaw (occlusal eruption), finally ending when the tooth is lost because it erupts out of the mouth.

dental formula A shorthand notation of the number of incisors, canines, premolars, and molars on one side of the upper and lower dentition of a species (denoted by a single formula when upper and lower quadrants are the same). For example, the normal adult human dental formula is 2/1/2/3 or 2/1/2/2.

dental hypoplasia Defects in the enamel of teeth created by interuptions in enamel development, usually because of stresses; for instance, poor nutrition or infection.

dental microwear Scratches, pits, and gouges on the occlusal surface of teeth that are so small that an optical or **scanning electron microscope** is needed to study them.

dentin Internal tissue in a tooth crown and tissue of the root, surrounding the pulp cavity and surrounded by the crown enamel. This bone-like substance is softer than the enamel.

deoxyribonuclease (DNase) The enzyme that catalyzes the degradation of DNA to nucleotides.

deoxyribonucleic acid (DNA) A polymeric molecule composed of deoxyribonucleotide building blocks that in a double-stranded, double-helical form is the genetic material of most organisms.

deoxyribonucleotide The basic building blocks of DNA, composed of a sugar (deoxyribose), a base, and a phosphate.

deoxyribose The pentose (five-carbon) sugar found in DNA.

development The process of regulated growth that is a result of the interaction of the genome with cytoplasm and the environment. It involves a programmed sequence of phenotypic events that are typically irreversible.

diagnosis In taxonomy, a statement of the characters that uniquely differentiate a taxon from other taxa.

diagnostic Differentiating or uniquely characteristic.

diaphysis In a long bone, the shaft portion.

diastema A space or gap between adjacent teeth in a tooth row, often existing to accommodate a projecting canine from the opposing jaw; in the mandible the diastema is between the canine and anterior premolar, in the maxilla between the lateral incisor and canine.

differential reproduction Differences in successful reproduction, some individuals have greater numbers of surviving offspring than others.

differential tooth wear Differences in the amount of wear between adjacent teeth.

digits Fingers or toes.

digital flexors Muscles that create the ability to bend the fingers or toes.

digitigrade A kind of quadrupedal locomotion in which animals support their body weight on their phalanges.

dimorphism The existence of differences of polymorphic character such as coloration, size, or shape between the males and the females of the same species.

diploid (2N) A eukaryotic cell that has two sets of chromosomes.

directional selection A type of natural selection that perpetuates evolutionary change via selection for greater or lesser frequency of a particular trait in a population.

discontinuous trait A heritable trait in which the mutant phenotype is sharply different from the alternative, wild-type phenotype.

disjunction The process in anaphase during which sister chromatid pairs separate.

displays Sequences of repetitious behaviors that serve as a means of communicating emotional states. Nonhuman primate displays are most frequently associated with reproductive or agonistic behavior.

distal When applied to the appendicular skeleton, meaning away from the midline of the body, or, when applied to teeth, meaning further from the center of the jaw as directed along the tooth row.

diurnal Usually active during hours of daylight.

division of labor Different cooperative strategies, usually for both males and females.

DNA (deoxyribonucleic acid) The molecule that carries the genetic information (genes) in all organisms except the RNA viruses. It is composed of two long polysugar-phosphate strands connected by base pairs connecting congruent bases out of a set of four, and twisted in a double helix.

DNA fingerprinting The use of restriction fragment length polymorphisms; the analysis of DNA to identify an individual; also referred to as DNA typing.

dolichocephalic Long headed, possessing a Cephalic Index of less than 75.

dominance hierarchies Systems of social organization in which individuals within a group are ranked relative to one another. Higher-ranking individuals possess greater access to preferred food items and mating partners than lower-ranking individuals. Dominance hierarchies are sometimes referred to as "pecking orders."

dominant A phenotype or allele that is expressed in either the heterozygous or the homozygous state.

dominant lethal allele Allele that will exhibit a lethal phenotype when existing in the heterozygous condition.

double crossover Two crossovers happening in a particular region of a chromosome in meiosis.

Down syndrome *See* **trisomy-21**.

dowry Goods sent with a woman at her marriage, either as payment to the husband's kin or as payment of the woman's share of her family estate.

drift (genetic drift) Changes in gene frequencies that are a result of random or stochastic variation and not the result of selection, mutation, or genic exchanges. Drift changes are most prominent in small populations.

dryopithecine Member of the subfamily Dryopithecinae.

Duffy blood group A red blood cell system that has proved useful in studying admixture between African- and European-derived populations.

duplication A chromosomal mutation that causes the doubling of a segment of a chromosome.

E

Early Stone Age (or Lower Paleolithic) Referred to as the ESA in Africa and the Lower Paleolithic outside Africa, the earliest stone tool industries, which include the Oldowan and Acheulean industries.

ecological niches The positions of species within their physical and biological environments, together making up the *ecosystem.* A species' ecological niche is described in terms of its components, such as diet, terrain, vegetation, type of predators, relationships with other species, and activity patterns, and each niche is unique to a given species.

ecological species concept The idea that a species is a group of organisms exploiting a single niche. This view stresses the role of natural selection in separating species from one another.

ecology The interrelationships between populations or organisms and their respective environment (including such aspects as temperature, predators, nonpredators, vegetation, availability of food and water, types of food, disease organisms, parasites, etc.).

edentulous Toothless; jaws without any remaining or preserved teeth.

elasticity A property of stone that allows it to return to its former state after being depressed by the application of force.

embryo An organism just after conception; for example, the first eight weeks of in utero development in humans.

empirical Depending on experiment or observation; from the Latin *empiricus,* meaning "experienced."

enamel The prismatically structured, very hard outer surface of a tooth crown.

enamel wrinkling Secondary folding of the enamel at the occlusal surface of a tooth.

encephalization An evolutionary trend characterized by a relative increase in brain size.

encephalization quotient (EQ) The ratio of a species between the actual brain size to its expected brain size as related to a statistical regression of brain size to body size based on a large number of species.

endemic Continuously existing in a population.

endocast A cast made of the mold formed by the impression the brain makes on the inside of the neurocranium; to provide a replica of the brain with the grosser details of its outer surface. Also referred to as endocranial cast.

endocranium The neurocranium's inside.

endogamy The mating or marriage within a cultural or social unit.

endoplasmic reticulum (ER) A type of organelle in the cytoplasm that consists of a folded membrane.

endothermic (*endo,* meaning "within" or "internal") Capable of maintaining internal body temperature through the production of energy by means of metabolic processes within cells; characteristic of mammals, birds, and perhaps some dinosaurs.

environmental sex determination The process by which the environment plays a major role in determining the sex of an organism is possible.

environmental variance (V_E) Any nongenetic source of phenotypic variation between individuals.

enzymes Specialized proteins produced by living organisms that act to initiate and direct chemical reactions in the body.

Eocene epoch The second epoch of the Cenozoic era, dating roughly between 34 million and 54 million years ago. The first primates, early prosimians, appeared during this epoch as well as the first anthropoid.

epidemiology The quantitative study of disease's cause and occurrence in populations.

epiphysis A secondary center of ossification (bone formation) usually located at the ends of long bones, separated from the primary center (diaphysis, the shaft) by cartilage growth plates that fuse at the completion of bone elongation.

epistasis A form of gene interaction in which one gene interferes with the phenotypic expression of another nonallelic gene such that the phenotype is governed by the former gene and not by the latter when both genes are present in the genotype.

epoch A particular subdivision of a particular geological period (e.g., the Pliocene is an epoch in the Tertiary Period).

era The longest division of geological time (e.g., the Cenozoic era).

erythroblasts Precursors of red blood cells.

estrus Occurring at and around the time of ovulation, a time of increased female sexual activity, often accompanied by enhancement of visual or olfactory sexual signals.

ethnic group A human group defined by sociological, cultural, and linguistic qualities.

ethnobiology The study of how things and organisms in the natural world are classified by traditional cultures.

ethnocentric Viewing other cultures from the inherently biased perspective of one's own culture. Other cultures being perceived as inferior to one's own is often a result of ethnocentrism.

ethnography A subset of cultural anthropology pertaining to the study of contemporary cultures through firsthand observations.

ethnology A subset of cultural anthropology pertaining to the comparative study of contemporary cultures, with a view to deriving general principles about human society.

eugenics The philosophy of "race improvement" via the forced sterilization of members of some groups and increased reproduction among others; an overly simplified, often racist view that is now discredited.

eukaryote A term literally meaning "true nucleus." Eukaryotes are organisms that have cells in which the genetic material is located in a membrane-bound nucleus. Eukaryotes can be unicellular or multicellular.

euploid The condition in which an organism or cell has one complete set of chromosomes, or an exact multiple of complete sets.

eutherian Placental mammal.

Eve theory Sometimes referred to as "Out of Africa" or "African replacement model, " this theory postulates that all modern humans have a common recent origin in an African population that became a new species between 150,000 and 200,000 years ago and then spread throughout the Old World, replacing preexisting Neanderthal and other archaic human populations.

evolution Genetic change; differentiation in a population's gene pool from generation to generation (Darwin's descent with modification).

evolutionary psychology An approach to comprehending human behavioral evolution that stresses the selection of specific behavioral patterns in relation to the environment of evolutionary adaptedness.

exogamy Social rules that prescribe mating or marriage outside a social or cultural unit.

extant Living, in comparison to extinct.

extended family A domestic group or composite of domestic groups composed of two or more nuclear families linked together through parent and child or through siblings.

external auditory meatus The ring of bone surrounding the outer opening for the ear canal.

F

F_1 generation The first filial generation, which is produced by crossing two parental strains.

F_2 generation The second filial generation, which is produced by selfing the F_1.

factor analysis A multivariate statistical technique assessing the degree of variation between artifact types, and based on a matrix of correlation coefficients measuring the relative association between any two variables.

falsifiable Capable of being tested (verified or falsified) by experiment or observation.

familial trait A trait that is shared by members of a family.

family The major group within an order; a monophyletic group of genera separated from similar groups by distance of common ancestry.

fauna Animals.

faunal assemblage A group of living or fossil animals found in a particular geographic or geological context and thought to sample part of a naturally occurring community.

faunal correlation A method of relative dating; a determination of the relative ages of different geological strata made by comparing the fossils within the strata and assigning similar ages to strata with similar fossils.

faunal dating A method of relative dating based on the observation of evolutionary changes in a particular species of mammals so as to form a rough chronological sequence.

faunivore An animal that primarily eats other animals (includes insectivores and carnivores).

fecundity Number of offspring a female can or does give birth to over the span of her lifetime.

female philopatry A type of social system for primates in which the females remain and breed in the same group into which they were born, while the males emigrate from the group.

femoral condyles The femur's enlarged inferior end, which forms the top of the knee joint.

femur (femora) Long bone of the upper leg or thigh.

fertility Over a given interval or lifetime, the number of offspring produced.

fetus Human organism from eight weeks of development until birth.

fibula This is the more slender, lateral one of the two long bones of the lower leg.

fission-fusion A type of mating system displayed by chimpanzees, bonobos, and a few other primates in which no stable cohesive groups exist; rather, there are temporary subgroups.

fission track dating A radiometric technique used for dating noncrystalline materials using Ur-238 and counting the tracks produced by this particular fission. Also approximates the age of the sediments in which fossils are found.

fitness A measure of fertility and survivorship that reflects genetic variation or Darwinian fitness.

fixity of species The concept that species, once created, can never change; an idea diametrically opposed to theories of biological evolution.

flake A stone fragment that has been struck from a core and hypothesized to have been the primary tools of the Oldowan.

flake tool A flake that has been subsequently modified via either intentional retouch and/or wear resulting from use.

flexor A type of muscle whose action decreases the angle between the bones in a joint.

flintknapper One who forms stone implements through the controlled fracture of the objective piece.

flora Plants.

folivore (folivorous, folivory) Leaf-eating; folivores constitute those animals whose primary source of food is foliage.

fontanelle A region between skull bones that, at birth, is unossified.

foraging strategy Behaviors leading to the acquisition of food.

foramen (foramina) Hole or tube-like passageway into or through bone.

foramen magnum The large opening in the occipital bone on the base of the skull through which the spinal cord passes in order to join the base of the brain.

foramen ovale A passageway with an oval opening, through the sphenoid bone on the base of the skull, which transmits meningeal arteries and mandibular nerves.

forebrain The brain's most anterior part.

forensic anthropology An applied anthropological approach that deals with legal matters. Physical anthropologists work with coroners and others to identify and analyze human remains.

fossil Preserved remains of once-living plants or animals in which the replacement of organic or inorganic materials by soil minerals has begun. Naturally occurring casts are also referred to as fossils.

founder effect A phenomenon occurring when the isolate effect is exhibited by a small breeding unit that has formed by migration of a small number of individuals from a large population.

Frankfurt plane A widely agreed upon plane for orienting crania to facilitate valid comparisons, and to approximate the position of the head during life. In this plane a line between the top of the external auditory meatus and the lowest point on the orbit is made horizontal.

free-ranging In reference to noncaptive animals living in their natural habitat. Ideally, the behavior of wild study groups would be free of human interference.

frontal bone The cranial bone that forms the forehead, the top of the orbits, and nose.

frontal sinus An open space present in the frontal bone where the squama meets the top of the orbits.

frugivore An animal that feeds primarily on fruit.

G

gamete Mature reproductive cells that are specialized for sexual fusion. Each gamete is haploid and fuses with a cell of similar origin but of opposite sex to produce a diploid zygote.

gametogenesis The formation of male and female gametes via meiosis.

gene (Medelian factor) A unit of inheritance carried on a chromosome, which is transmitted from generation to generation by the gametes and controls some aspects of the development of an individual.

gene flow A type of genic exchange in which genetic material is transferred between populations because of interbreeding or mate exchange between them.

gene frequency The percentage or proportion of a specified allele in a sample.

gene mutation A heritable alteration of the genetic material, usually occurring from one allelic form to another.

gene pool The total genetic information encoded in the total genes in a breeding population existing at a given time.

generalized (1) Primitive or plesiomorphic, similar to the ancestral condition; (2) adapted to a wide range of resources.

genetic bottleneck A form of genetic drift that occurs when a population is drastically reduced in size. As a result of chance, some genes may be lost from the gene pool.

genetic code The base-pair information specifying the amino acid sequence of a polypeptide.

genetic drift A mechanism for evolutionary change that occurs as a result of the random fluctuations of gene frequencies from one generation to the next, or from any form of random sampling from a gene pool. Any change in gene frequency due to chance in a population.

genetic marker (gene marker) Any genetically controlled phenotypic difference used in genetic analysis, especially in the detection of genetic recombination events.

genetics The science of heredity involving the structure and function of genes and the way genes pass from one generation to the next.

genetic variance (V_G) Genetic sources of phenotypic variation among individuals of a population; including dominance genetic variance, additive genetic variance, and epistatic genetic variance.

genome The entire DNA component of a cell, a structured array that consists of genes and their parts, units of DNA replication, and nonfunctioning regions.

genotype The genetic makeup of an organism; the organism's total genetic material.

genotypic frequencies The frequencies or percentages of different genotypes found in a population.

genus (genera) A group of closely related species, a monophyletic category for the taxon above the species level including one or more species.

geological time scale (GTS) The organization of earth history into eras, periods, and epochs; commonly utilized by geologists and paleoanthropologists.

geology The science concerned with the study of the earth.

geomagnetic reversal A component of archaeomagnetism relevant to the dating of the Lower Paleolithic, involving complete reversals in the earth's magnetic field.

gestation The period beginning at fertilization and ending at birth.

glabella A location on the midline of the frontal bone between the browridges, superciliary arches, or upper orbital borders.

glaciation Ice Age, the nonpermanent enlargement of continental and momentum glaciers associated with worldwide climate changes.

gluteus maximus Large muscle that extends from the lateral and rear surface of the ilium to the gluteal tuberosity of the femur and causes extension and rotation at the top.

gluteus medius Muscle that extends from the lateral surface of the ilium to the greater trochanter of the femur and causes abduction at the hip.

gluteus minimus Large muscle that extends from the lateral surface of the ilium to the greater trochanter of the femur and causes abduction and rotation at the hip.

gracile Slender, delicately built, weak muscle attachments or bony butresses. A relative condition, meaning that it is referenced to another condition.

grade A grouping that is characterized by a general level of organization (or sharing a suite of features). Grades composed of independent lineages that may or may not be monophyletic.

gradualism A theory proposing that evolution progresses by the extension of microevolutionary processes over long periods of time; the gradual modification of populations. This concept does not imply continuous evolution, nor does it imply evolution at a constant rate.

great apes The four large living apes: bonobos, chimpanzees, gorillas, and orangutans. Great apes do not constitute a monophyletic group.

greater trochanter A very large process on the lateral and proximal end of the femur shaft, for the attachment of muscles stabilizing the hip during one-legged balance (whether standing or in bipedal locomotion).

grooming Cleaning the surface of the body via licking, biting, picking with fingers or claws, or other kinds of manipulation.

group selection Evolutionary process that involves differential survival and reproduction of competing groups.

guanine (G) The purine base present in RNA and DNA. In double-stranded DNA, guanine pairs with the pyrimidine cytosine.

gyri (*sing.*, gyrus) Ridges on the brain's surface that are formed by sulci.

H

habiline Pertaining to the *Homo* species of *Homo habilis* and *Homo rudolfensis.*

habitat The normal environment or home of a group.

half-life The amount of time required for half of the original amount of an unstable isotope, an element to decay into a more stable form.

hallux The first digit on the foot; the big toe.

hammerstone A stone that is used to strike cores in order to produce flakes or bones, thereby exposing marrow.

hamstrings A group of muscles including biceps femoris, semimembranosus, and semitendinosus; it extends from the ischial tuberosity to the back of the femur and top of the tibia, mainly acting to flex the hip joint.

hand axe A teardrop-shaped or pear-shaped bifacially flaked stone implement.

haploid Possessing only a single set of chromosomes, half the number in a normal somatic cell. Gametes are normally haploid.

haplorhine (Haplorhini) An infraorder of the order Primates, which includes anthropoids and the tarsier.

haplotype (N) Sets of genes present at more that one locus.

hard object feeding To chew tough, hard-to-break food items like nuts or fibrous vegetation.

Hardy-Weinberg law (Hardy-Weinberg equilibrium, Hardy-Weinberg law of genetic equilibrium) An extension of Mendel's laws of inheritance describing the expected relationship between gene frequencies in natural populations and the frequencies of individuals of various genotypes in the same populations.

hearth A circle of stones that encloses a camp fire to focus, contain, and sustain its heat.

hemizygous The condition of X-linked genes in males. Males who have an X-chromosome with an allele for a particular gene but do not have another allele of that gene present in the gene complement are considered hemizygous.

hemoglobin A protein molecule occurring in red blood cells and binding to oxygen molecules.

herbivore (herbivorous) Plant eater.

heritability A measure of the extent to which a feature is inherited; the proportion of variation of a trait in a population that is caused by the variation of genotypes.

hermaphroditic In animals, the species that have both testes and ovaries (e.g., nematode); in plants, the species that have both stamens and pistils on the same flower.

heterodontic Teeth of the same type (e.g., incisors, premolars, etc.) but that differ in size or form.

heterosis The phenomenon in which the heterozygous genotypes with respect to one or more characters are superior in terms of growth, survival, phenotypic expression, and fertility, in comparison with the corresponding homozygous genotypes.

heterozygosity The proportion of individuals heterozygous at a locus; the state of being heterozygous. See also **heterozygous**.

heterozygote A type of a polymorphism controlled by different alleles at a locus.

heterozygous A term to describe a diploid organism that has different alleles of one or more genes and therefore produces gametes of different genotypes.

heterozygous advantage Pertaining to a specific genetic system, a situation in which heterozygotes possess a selective advantage over homozygotes (e.g., sickle-cell disease); a mechanism used to maintain a balanced polymorphism.

Holocene The most recent epoch of the Cenozoic. The Holocene followed the Pleistocene, and it is estimated to have begun 10,000 years ago.

home base A specific place at which individuals can expect to meet each other and engage in social and other activities.

home range The area within which a group of primates usually moves during the course of their yearly cycle.

homeobox A 180-bp consenus sequence present in the protein-coding sequences of genes that regulate development.

homeostasis A condition of balance, or stability, within a biological system maintained by the interaction of physiological mechanisms compensating for changes (both external and internal).

Hominidae Humans belong to this taxonomic family; also includes other, now extinct, bipedal relatives.

hominid Extant humans and their unique ancestors and collateral relatives extending back in time until the split with the line leading to chimpanzees (the closest living human relative).

Hominoidea The formal designation for the superfamily of anthropoids including apes and humans.

hominoid Member of Hominoidea, the superfamily that includes apes and their unique ancestors. In fossils refers to members of the apes.

homodont Possessing teeth of uniform shape, form, and function.

Homo habilis A species of early *Homo*, well known from East Africa but possibly also found in other regions.

homologous Pertaining to members of chromosome pairs. Homologous chromosomes carry loci that govern the same traits. During meiosis, homologous chromosomes pair and exchange segments of DNA. They are alike with regard to size, position of centromere, and banding patterns.

homologous chromosomes The members of a chromosome pair that are identical in the arrangement of genes they possess and in their visible structure.

homology A feature in two or more species that is the same as a result of descent; it evolved from the same feature in the last common ancestor of the species.

homozygosity The occurrence of two identical alleles at a single locus.

homozygote A feature controlled by a locus where the two alleles are the same.

homozygous A term to describe a diploid organism having the same alleles at one or more genes and therefore producing gametes of identical genotypes.

homozygous dominant A diploid organism possessing the same dominant allele for a given gene locus on both members of a homologous pair of chromosomes.

homozygous recessive A diploid organism possessing the same recessive allele for a given gene locus on both members of a homologous pair of chromosomes.

hormones Substances (usually proteins) that are created by specialized cells and that travel to other parts of the body, where they influence chemical reactions and regulate various cellular functions.

human biology A particular subfield of biological anthropology concerning human growth and development, adaptation to environmental extremes, and human genetics.

human evolutionary ecology The study of ecological and demographic aspects significant in determining individual reproductive success and fitness in a cultural context.

hunter-gatherers Populations that live via hunting (and often scavenging dead) animals, gathering plant foods, insects, and other small and relatively sedentary animals, and then sharing the results of these planned economic activities.

hylobatid (Hylobatidae) A member of the gibbon or lesser ape family.

hyoid bone Located in the front part of the throat and kept in place by muscles and ligaments, a small "floating bone."

hypervitaminosis A condition that results from a dietary excess of the vitamin concerned.

hypoplasia Interrupted enamel formation that leaves transverse lines, pits, or grooves visible on the enamel surface.

hypothesis (pl., hypotheses) A provisional explanation of a phenomenon. Hypotheses necessitate verification or falsification through testing.

hypoxia Low oxygen availability due to lowered barometric pressure on the air.

I

ilium (iliac blades) The side, or broad and flat blade of the innominate, forming its upper portion.

immunoglobulins Specialized proteins (antibodies) secreted by B cells, which circulate in the blood and lymph and which are responsible for humoral immune responses.

inbreeding Preferential mating that occurs between close relatives.

incest A violation of cultural rules that regulate mating behaviors.

incidence rate The total number of new occurrences of a disease over a particular time span, divided by population size.

incisor Broad tooth at the front-most part of the jaw.

inclusive fitness The enhanced transmission to the next generation of genetic material of members of the group that are relatives.

incomplete (partial) dominance The condition that results when one allele is not completely dominant to another allele such that the heterozygote has a phenotype between that shown in individuals homozygous for either individual allele involved. An example of partial dominance is the frizzle chicken.

infanticide Killing of infants.

innominate bones (os coxae) The couple of bones that are comprised of the lateral parts of the pelvis; every innominate is composed of three bones which fuse during adolescence.

intelligence The ability to learn, reason, or comprehend and interpret information, facts, relationships, and meanings based on the sum of cultural experiences.

interglacial A warm period occurring between two major periods of multiple glaciations.

interorbital Between orbits.

interphase The portion of a cell's cycle during which metabolic processes and other cellular activities occur. Chromosomes are not visible as discrete structures at this time. DNA replication occurs during this portion of the cell's cycle.

interspecific Between species; pertains to variation beyond that seen within the same species to include additional aspects seen between two different species.

intraspecies clade A group of ancestral-descendant populations that share common descent (although not *unique* common descent) within a species.

intraspecific Within species; pertains to variation seen within the same species.

ischial callosities Areas of tough, hard skin on the buttocks of Old World monkeys and chimpanzees.

ischial tuberosity The roughened area at the base of the ischium for the attachment of hamstrings.

ischium The lower rear bone of the innominate.

isolating mechanisms Individual biological or behavioral characteristics that prevent sympatric groups from interbreeding.

isometry Change in overall size that allows the same relative proportional shape to be maintained.

isotope Chemically identical but anatomically different forms of an element (the number of neutrons are different so the atomic weight is different).

isotopic analysis An important source of information on the reconstruction of prehistoric diets; this technique analyzes the ratios of the principal isotopes preserved in human bone; in effect, the method reads the chemical signatures remaining in the body by different foods. Additionally, isotopic analysis is used in characterization studies.

J

Java Man A hominid fossil found in Java now classified as *Homo erectus.*

Jurassic period The period that encompasses the rise of the dinosaurs and the first birds, and the beginning of Pangaea.

Juvenile The period in an individual's life cycle that lasts from the eruption of the first to the eruption of the last permanent teeth.

K

K-selected A reproductive strategy whereby individuals produce relatively few offspring, in whom they invest increased parental care. Although only a few infants are born, chances of survival are increased for each individual due to parental investments in time and energy. It characterizes most mammals and specially primates, but birds also use the K-selected strategy.

karyotype A full set of all the metaphase chromatid pairs in a cell (literally, "nucleus type").

kin selection Differential aid or favoritism toward relatives who promote the inclusive fitness of shared genes.

kinship Relationships between people that are based on real or imagined descent or, at times, marriage.

Klinefelter syndrome A human clinical syndrome that results from disomy for the X chromosome in a male, which results in a 47,XXX male. Many of the affected males are mentally deficient, possess underdeveloped testes, and are taller than average.

knuckle-walking A form of quadrupedal walking used by chimpanzees or gorillas in which the forearms rest on the dorsal surface of the middle phalanges of the hands.

L

labial In a direction toward the lips; on the anterior teeth toward the outside.

labiolingual The breadth dimension of an incisor or canine, extending from the lip side to the tongue side.

lacrimal bone In the skull, a small bone that forms part of the medial orbit wall.

lacrimal duct Tear duct; connects the orbit with the nasal cavity.

lacrimal foramen Opening of the tear duct.

lactation The secretion of milk from mammary glands.

lactose intolerance The inability to digest fresh milk products, which is a result of the low activity of lactase, the enzyme that breaks down lactose, or milk sugar.

lambda A point on the back of the skull at which the juncture of the occiput and the parietal bones, where the sagittal and lambdoidal sutures meet.

lambdoidal flattening A flattened surface at lambda, always extending anteriorly onto the parietal bones and in some instances extending posteriorly onto the occiput when a true occipital bun is formed.

lambdoidal suture The transverse suture at the back of the cranium at which the parietal and occipital bones join.

large-bodied hominoids Those hominoids that include the great apes (orangutans, chimpanzees, gorillas) and hominids, as well as all ancestral forms back to the time of divergence from small-bodied hominoids (i.e., the gibbon lineage).

larynx The uppermost portion of the windpipe, the sphincter guarding the entrance to the trachea and serving as the sound-producing organ of the throat.

lateralization The transfer of a function to one side of a bilaterally symmetric structure or body.

lateralized In relation to lateralization, the functional specialization of the hemispheres of the brain for specific activities.

lesser trochanter Large blunt process on the posterior face of the femoral shaft, just below the neck, for attachment of muscles that flex the thigh.

lethal allele An allele that causes the death of an organism.

Levallois technique A method for flake production in which a stone core is shaped like a tortoise shell and a single flake with a preformed shape is struck from it.

life history From birth to death, the stages of life an organism passes through.

limbic system A complex part of the brain composed of deep nuclei and fiber tracts pertaining to the control and expression of the emotions.

lineage A group of ancestral-descendant species, which are reproductively isolated from other lineages, a line of common descent.

lingual Toward the tongue, the side of a tooth that faces the tongue.

linkage A term that describes genes found on the same chromosome.

linked genes Genes that are found on the same chromosome.

lithic Of or relating to stone.

locus A particular location (1) on a chromosome, matched to the corresponding position on the other chromosome of a pair, the site of the maternal and paternal alleles that are often considered together as a gene; (2) defined within a paleontological or archaeological site.

longitudinal study A type of study that examines changes in individuals over a given time span.

lumbar Related to the lower back, the vertebrae that lie between the thoracic vertebrae and the sacrum.

M

macaques Group of Old World monkeys composed of several species, including rhesus monkeys.

macroevolution Evolution above the species level; the evolution of higher taxa and the processes that are caused by differences in species survivorship or rates of speciation.

Magdalenian Relating to the final phase of the Upper Paleolithic stone tool industry in Europe.

magma Molten rock that cools and solidifies below the earth's surface.

magnetic reversal Alteration of the earth's magnetic field, which results in a reversed polarity.

mandible Lower jaw.

mandibular corpus (body) The tooth-bearing or horizontal portion of the mandible.

mandibular (glenoid) fossa Joint for the mandibular articulation with the skull, a depression on the base of the temporal bone, just in front of the ear opening, into which the mandibular condyles fit.

mandibular groove A groove that extends down from the lower rim of the mandibular foramen or from just below it.

mandibular ramus *See* **ascending ramus.**

mandibular symphysis The midline joining plane, which connects the two sides of the mandible; fused in adult Anthropoidea.

mandibular (transverse) torus Shelf-like thickening of bone on the inside of the mandibular symphysis; **superior** and **inferior** transverse tori can be evident but there can also only be one. The inferior transverse torus is a simian shelf if it is thin and projects so far to the rear that its lowest point is also its most posterior.

mandibular (symphyseal) trigone An upward facing triangle-shaped form at the base on the symphysis.

manuport An unmodified piece of rock known to have been carried to a locality by a hominid because it could not have gotten there naturally.

masseter muscle A short, quadrangular muscle located between the zygomatic arch and the lower edge of the jaw, along its outside, supplying bite power.

mastication To chew.

mastoid notch The notch at the bottom-rear of the parietal bone, found over the mastoid process.

mastoid process A pyramid-shaped prominence of cancelous bone located on the temporal bone behind the external auditory meatus. Muscles that extend and turn the head attach on it.

mastoid tubercle A distinct bump on the lateral face of the mastoid process, located just behind the external auditory meatus. It is typically treated as a nonmetric trait.

mate recognition system The system of signals (chemical, olfactory, vocal, visual) that bring potential breeding partners together.

material culture The buildings, tools, and other artifacts that constitute the material remains of former societies.

maternal effect The phenotype in an individual, which is established by the maternal nuclear genome, as the result of mRNA and/or proteins that are deposited in the oocyte prior to fertilization. These inclusions direct early development in the embryo.

maternal inheritance A phenomenon occurring when the mother's phenotype is expressed exclusively.

matrifocal Family or other group that is headed by a female.

matrilineal Descent reckoned through the female line.

maxilla Paired bone of the upper jaw, encompassing the nose and the inner and lower rims of the eye and holding the upper teeth.

meiosis Two successive nuclear divisions of a diploid nucleus resulting in the formation of haploid gametes or of meiospores having one-half the genetic material of the original cell.

meiosis I The first meiotic division resulting in the reduction of the number of chromosomes. This division consists of four stages prophase I, metaphase I, anaphase I, and telophase I.

meiosis II The second meiotic division, which results in the separation of chromatids.

menarche Onset of menstruation, signaled by the first menstrual period.

Mendelian population An interbreeding group of individuals who share a common gene pool; the basic unit of study in population genetics.

Mendelian traits Characteristics that are influenced by alleles at only one genetic locus. Instances include many blood types, such as ABO. Many genetic disorders, such as sickle-cell anemia and Tay-Sachs disease, are also Mendelian traits.

menopause The end of menstruation in women; usually this occurs at around age 50.

mental foramen A large, sometimes multiple, foramen located on the lateral anterior surface of the mandibular corpus, for the mantal nerve and vessels.

mesial The side of the tooth closest to the midline of the jaw, as directed along the tooth row.

mesiodistal A forward to back direction, as taken along the tooth row arch, the tooth's length.

Mesolithic An Old World chronological period, which began about 10,000 years ago, situated between the Paleolithic and Neolithic, and associated with the rise to dominance of microliths.

messenger RNA (mRNA) The RNA molecule that contains the coded information for the amino acid sequence of a protein.

metabolism The internal processes that make energy available.

metacarpals Five parallel bones of the hand that connect the phalanges (fingerbones) with the carpals (bones of the wrist).

metaphase A stage in mitosis or meiosis during which chromosomes become aligned along the equatorial plane of the spindle.

metaphase II The second stage of meiosis in which the centromeres line up on the equator of the second-division spindles (in each of two daughter cells that formed from meiosis I).

metastasis The spreading of malignant tumor cells throughout the body in such a way that tumors develop at new sites.

metatarsals Five parallel bones of the foot that connect to the phalanges (bones of the toe) with the tarsal bones (bones of the arch).

Metazoa Multicellular animals; one of the major divisions of the animal kingdom.

microblade A bladelet or small blade. This term is usually referenced with bladelets found in the Arctic areas of North America and northeastern Asia.

microevolution Evolution of populations over short periods of time, as a result of observable causes.

microwear The traces of wear on stone tools that are not visible unless magnified. Such wear may be in the form of a retouch or polish.

Middle Pleistocene The portion of the Pleistocene epoch that began 780,000 years ago and ended 125,000 years ago.

midface The central portion of the face, composed mainly of the cheeks and nose.

midline An anatomical term pertaining to a hypothetical line that divides the body into right and left halves.

migration (genic) The movements of genes that occur as a result of individuals moving, including new individuals entering (immigration) or leaving (emigration) a population, introducing or removing genetic material and thereby changing allele frequencies.

mitochondria The small extra-nuclear organelles (bodies) within a cell's cytoplasm that control the production of energy from food via the production of ATP (adenosine triphosphate).

mitochondrial DNA (mtDNA) The single (double-stranded) DNA molecule that controls the development and functioning of the mitochondrion containing it. Due to the fact that reproduction is by cloning, mtDNA is usually passed along female lines, as part of the egg's cytoplasm.

mitosis The process of nuclear division in haploid or diploid cells that produces daughter nuclei that contain identical chromosome complements and are genetically identical to one another and to the parent nucleus from which they arose.

moiety A type of division of a society into two social categories or groups, characteristically divided by a rule of patrilineal descent (patrimoiety) or matrilineal descent (matrimoiety).

molecular clock A mean to determine dates of evolutionary divergences using genetic similarities between extant species and assuming that molecular evolution proceeds at a constant rate.

molecular genetics A subdivision of the science of genetics that involves how genetic information is encoded within the DNA and how biochemical processes of the cell translate the genetic information into the phenotype.

molecules Structures composed of two or more atoms. Molecules can combine with other molecules to form more complex structures.

monogamy A social system that is based on mated pairs and their offspring.

mosaic evolution Evolution that proceeds at different rates for different features.

motor area (of the brain) The posterior region of the frontal cortex that controls motor movements.

Mousterian Relating to the stone tool industry associated with Neanderthals and some modern *H. sapiens* groups; also called Middle Paleolithic. This industry is characterized by a larger proportion of flake tools than is found in Acheulian tool kits.

MSA Middle Stone Age.

multifactorial trait A trait that is influenced by multiple genes and environmental factors.

Multiregional evolution The hypothesis that the anatomically modern humans evolved from *A. Homo erectus, Archaic Homo sapiens* into an interconnected polytypic species in Africa, Asia, and Europe.

multivariate statistics Statistical procedures that are designed to simultaneously treat (and assess relationships among) many variables per object.

mutation An error in replication or other alteration of the nucleotide base sequence causing a change in the sequence of base pairs on a DNA molecule. Mutations in the germ cells can cause heritable changes in the offspring. Mutation in the somatic cells can lead to various diseases that are not heritable such as cancers.

mutation frequency The number of events of a particular kind of mutation in a population of cells or individuals.

mutation rate The probability of a specific kind of mutation as a function of time.

N

nasion The location on the midline where the two nasal bones and the frontal come together.

natal group A type of group in which animals are born and raised. (*Natal* pertains to birth.)

natural selection Difference in reproductive success and/or survivorship of individuals that results in the unequal contribution of genotypes to the gene pool of the next generation.

negative assortative mating A type of mating that occurs more often between dissimilar individuals than it does between randomly chosen individuals.

neocortex The cortex, or outer surface, of the brain's cerebral hemispheres.

Neolithic An Old World chronological period characterized by the development of agriculture and, thus, an increased stress on sedentism.

Neolithic Revolution A term developed by V. G. Childe in 1941 used to describe the origin and consequences of farming (i.e., the development of stock raising and agriculture), providing for the widespread development of settled village life.

neonate A newborn infant.

neural canal The large opening through the vertebrae, which encloses the spinal cord; also known as the vertebral canal.

neural tube In early embryonic development, the anatomical structure that develops to form the brain and spinal cord.

neuron A nerve cell.

New World primate A primate from North or South America.

niche The limited portions of the environment, in terms of space, resources, etc., that a species fits and/or that it needs for purposes of survival and reproductive success.

nitrogenous base A base that contains nitrogen, which, along with a pentose sugar and a phosphate, is one of the three parts of a nucleotide, the building block of RNA and DNA.

nocturnal Primarily active at night.

nondisjunction (primary disjunction) At anaphase, the failure of homologous chromosomes or sister chromatids to separate.

nonhomologous chromosomes The chromosome containing dissimilar genes that do not pair during meiosis.

nonrandom mating Patterns of mating in a population by which individuals choose mates according to preference.

normal distribution In statistics, distribution in which an equal proportion of individuals are above and below the mean; a probability distribution that is graphically displayed as a bell-shaped curve.

nuchal crest A raised bony ridge located on the back of the skull, which is caused by the attachment of neck muscles.

nuchal torus A thickened bony prominence that extends transversely across some or all of the back of the head, on the occipital bone; it reflects the pattern of muscle use as it separates the nuchal plane below from the occipital plane above.

nuclear family A family unit composed of parents and their dependent children.

nucleotide A monomeric molecule of RNA and DNA that consists of three distinct parts—a pentose (ribose in RNA, deoxyribose in DNA), a nitrogenous base, and a phosphate group.

nucleus A discrete structure within the cell that contains most of the cell's genetic material; it is bounded by a nuclear membrane.

O

obsidian A volcanic rock that is formed into natural glass. Typically, this rock is black but may be found in greenish and reddish colors or banded.

occiptal bone The bone that forms the vault posterior and much of the basicranium.

occipital bun A backward extension of the cranial rear taking the form of a protuberance bounded by the nuchal plane below, a shaft vertical face for the occiput behind, and a flat surface above (lambdoidal flattening).

occipital condyles Raised, elongated-oval, convex articular eminences on both sides of the foramen magnum of the occipital.

occipital lobes The back part of the brain's cerebral hemispheres.

occlusal The surfaces of the opposing teeth that make contact for chewing; in an occlusal view the crowns of the teeth are shown.

occlusion The position of teeth when the jaws are closed and their biting surfaces touch.

Oldowan industry One of the earliest toolkits, composed of flake and pebble tools, used by hominids in the Oldovai Gorge, East Africa.

Old World primate Any primate from Africa or Eurasia.

olecranon fossa A depression at the posterior side of the distal humerus, at the elbow, for accommodating the olecranon process of the ulna when the elbow is extended.

olecranon process A beak-like projection on the ulna's proximal end, at the elbow, for articulation with the humerus and attachment of the triceps muscles.

olfaction Sense of smell.

omnivore An organism that eats a variety of food types, including animals and plants.

ontogeny From egg to adult, the developmental history of an individual.

opposability The capability to touch the tip of the thumb to the fingertips of the same hand.

orbit The eye's bony socket.

order A monophyletic higher-level taxon (comprised of suborders, superfamilies, etc.) whose members typically share a basic structural pattern.

organelles Structures contained within cells, surrounded by a membrane. Many different types exist, and each performs specific functions.

origin A specific location on a chromosome at which the double helix denatures into single strands and continues to unwind as the replication fork(s) migrates.

orthognathous Possessing a relatively vertical, nonprotruding face.

Orrorin tugenensis An early primitive, possibly hominoid-hominid, species from Africa that lived about 6 million years ago during the late Miocene.

ossicle Very small bone; for instance, the ear ossicles or the finger joint sesamoid bones.

ossification The process by which new bone is formed.

Osteodontokeratic Artifacts comprised of bone, tooth, or horn, as in the Osteodontokeratic "culture" of the Makapansgat australopithecines.

osteology The study of bones and the bone's variation.

outbreeding Preferential mating between nonrelated individuals.

overspecialized Adapted to a specific niche so particularly that the genetic variation necessary to meet changing conditions has been lost.

ovulation The release of an unfertilized gamete (egg) from the ovary.

ovum A mature egg cell. In the second meiotic division, the secondary oocyte creates two haploid cells; the large cell quickly matures into the ovum.

P

pair bonding Developing an intense social connection between monogamous mates.

palate The mouth's bony roof.

paleoanthropology The interdisciplinary approach to the study of earlier hominids; includes such aspects as their chronology, physical structure, archaeological remains, habitats, etc.

paleoecologist (*paleo*, meaning "old," and *ecology*, meaning "environmental setting") A scientist who studies ancient environments.

paleoentomology The study of insects from archaeological contexts. The survival of insect exoskeletons, which have proved quite resistant to decomposition, is a crucial source of evidence in the reconstruction of paleoenvironments.

paleoethnobotany The recovery and identification of plant remains from archaeological contexts, significant to the reconstruction of past environments and economies.

Paleolithic Literally the Old Stone Age, the period during which humans relied on a stone technology to sustain a scavenging/hunting/gathering adaptation. The archaeological period before 10,000 B.C.

paleomagnetic stratigraphy The arrangement of geological strata based upon the alternating direction of residual magnetism, compared with the world geomagnetic polarity column by other facets of the sequence such as preserved fauna.

paleopathology The branch of osteology that studies the evidence of disease and injury in human skeletal (or, on occasion, mummified) remains.

parietal Wall. One of the flat paired bones that form part of the skull's lateral sides.

parietal association area A portion of the parietal association complex that is posterior to the sensory region of the parietal lobe. The integration of sensory, motor, and cognitive actions, or cross-modal transfer, occurs in this area.

parsimony In an explanation or theory, the use of as few assumptions as possible. "Occam's razor" is an example.

patella Kneecap; large sesamoid bone at the knee.

pathogens Substances or microorganisms, such as bacteria, fungi, or viruses, that cause disease.

patrilineal Based on relationship via the father's side.

patrilineal descent Descent traced via a line of ancestors in the male line.

patrilocal Residence after marriage in association with the husband's father's relatives.

pebble tools Simple artifacts made on stone cores. Sometimes this term is also applied to the cores themselves even when they are not used as tools per se.

pedigree analysis A family tree investigation that involves the careful compilation of phenotypic records of the family over many generations.

pedigree chart A diagram displaying family relationships in order to trace the hereditary pattern of particular genetic (usually Mendelian) traits.

pelvis The bony structure composed of the sacrum and three paired bones, the ischium, ilium, and pubis, which fuse together in adults as paired innominates.

pentose sugar A 5-carbon sugar that, along with a nitrogenous base and a phosphate group, is one of the three parts of a nucleotide, the building block of RNA and DNA.

peptide bond A covalent bond in a polypeptide chain that joins the carboxyl group of one amino acid to the amino group of the adjacent amino acid.

perikymata Elevations between the grooves encircling tooth crowns that are a result of growth-related segmentation of enamel crystals.

period A division of an era in geology, such as the Quaternary period of the Cenozoic era.

permafrost Subsoil that is permanently frozen.

phalanx (phalange) Finger bone or toe digit bone.

phenotype Appearance of an individual; the observed set of traits, the result of the interaction between genotype and environment.

phenotypic variance (V_p) A measure of the variability of a trait.

phosphate group A component, along with a pentose sugar and a nitrogenous base, of a nucleotide, the building block of RNA and DNA. Due to the fact that phosphate groups are acidic in nature, DNA and RNA are called nucleic acids.

phylogenetic tree A chart displaying evolutionary relationships as determined by phylogenetic systematics. It contains a time component and implies ancestor-descendant relationships.

phylogeny A hypothesis concerning how fossils and living species are related in a genealogical framework.

physical anthropology A subdiscipline of anthropology dealing with the study of human biological or physical traits and their evolution. (Also referred to as biological anthropology.)

pistil The female reproductive organ in a flowering plant that typically is composed of the stigma, the style, and the ovary.

plantigrady A stance or locomotion in which the body is positioned in such a way that the palms and soles point downward.

plaque A round, clear area in a lawn of bacteria on solid medium, which results from the lysis of cells by repeated cycles of phage lytic growth.

plasma membrane Lipid bilayer that surrounds the cytoplasm of both animal and plant cells.

plasmid An extrachromosomal genetic element composed of double-stranded DNA that replicates autonomously from the host chromosome.

plasticity The ability to change. In a behavioral context, the ability of animals to modify behaviors in response to differing circumstances.

platyrrhine An infraorder of New World monkey.

pleiotropic Genes that influence the expression of more than one trait.

Pleiocene The epoch of the Cenozoic from 1.8 million to 10,000 years ago. Often referred to as the Ice Age, this epoch is associated with continental glaciations in northern latitudes.

plesiomorphic A characteristic whose form is similar to the ancestral condition.

Plio-Pleistocene Shorthand term literally referring to the Pliocene and Pleistocene together, but usually referring to the Pliocene and Early Pleistocene.

pluvial An unusually wet period occurring continent-wide.

pollex The thumb.

polyandry The marriage of a woman to more than one man.

polygene (multiple-gene) hypothesis for quantitative inheritance The hypothesis that states that quantitative traits are controlled by many genes.

polygenic A character, controlled by numerous genes.

polygenic traits Characteristics encoded by many loci.

polygyny Any form of social organization in which one male mates with more than one female.

polymerase chain reaction (PCR) A method performed in order to replicate defined DNA sequences selectively and repeatedly from a DNA mixture.

polymorphic Displaying a variety of forms; a feature with alternative character states.

polymorphisms Loci with more than one allele. Polymorphisms can be expressed in the phenotype as the result of gene action (as in ABO), or they can exist only at the DNA level within noncoding regions.

polypeptide chain A sequence of amino acids that may act alone or in combination with others as a functional protein.

polypeptide A polymetric, covalently bonded linear arrangement of amino acids that have been joined by peptide bonds.

polyploidy The condition of a cell or organism that has more than its normal number of sets of chromosomes.

polytypic A variable taxon that contains more than one taxon of the next lower category; for example, a species with several subspecies or races.

pongid Belonging to the family Pongidae, humans and the great apes (chimpanzees, gorillas, and orangutans) and their unique ancestors and collaterals.

population A community of potentially breeding individuals, usually at a specified locality or within a limited geographic region.

population genetics A branch of genetics that describes in mathematical terms the consequences of Mendelian inheritance on the population level.

positive assortative mating A type of mating that occurs more frequently between individuals who are phenotypically similar than it does among randomly chosen individuals.

postcranium (postcranial skeleton) All of the elements of the skeleton below the skull.

postcranial (*post*, meaning "after") In a quadruped, pertaining to that portion of the body *behind* the head; in a biped, pertaining to all parts of the body *beneath* the head (i.e., the neck down).

postorbital bar Bony ridge that surrounds the lateral side of the orbit in some primates and many other mammals.

potassium-argon dating A method used to date rocks up to billions of years old, although it is limited to volcanic material no more recent than approximately 100,000 years old; one of the most widely used methods in the dating of early hominid sites in Africa.

preadaptation The idea that species, or their features, can be predesigned to meet the requirements of future adaptations.

precocial Early or advanced in development.

preferential Pertaining to a marriage pattern (e.g., marriage with a cross-cousin, a brother's widow, etc.), socially valued and desirable, but not conjoined.

prehensile Ability to grasp.

prehistory The period of human history prior to the advent of writing.

premolar Tooth found between the canine and molars, typically smaller than the molars and generally flat except for the most anterior lower tooth in species with a canine cutting complex.

pressure flaker A tool used to press a detached flake from an objective piece. This tool is often pointed and made of antler, wood, or bone.

pressure flaking The removal of a detached piece from an objective piece by pressing rather than by means of percussion.

primates Members of the order of mammals, including prosimians, monkeys, apes, and humans.

primatology The study of the behavior and biology of nonhuman primates (including prosimians, monkeys, and apes).

principle of independent assortment (second law) The law that states the factors (genes) for different traits assort independently of one another. In other words, genes on different chromosomes behave independently in the production of gametes.

principle of segregation (first law) The law that states two members of a gene pair (alleles) segregate (separate) from each other during the formation of gametes. Consequently, one-half of the gametes carry one allele and the other half carry the other allele.

probability The ratio of the number of times a particular event occurs to the number of trials during which the event could have occurred.

proband In human genetics, the affected person with whom the study of a character in a family begins.

prognathous A forward protrusion in the facial region, as a whole or in part (*See* alveolar prognathism).

projectile point A biface that contains a shaft area and is used as a projectile tip. These are frequently identified as arrow points, dart points, and spear points.

prokaryote A cellular organism whose genetic material is not found within a membrane-bound nucleus.

pronation Rotation of the forearm such that the palm faces downward; the reverse movement from supination.

prophase The first stage in mitosis or meiosis; during this stage the chromosomes (already replicated) condense and become visible under a microscope.

prophase I The first stage of meiosis. The several stages of prophase I include leptonema, zygonema, pachynema, diplonema, and diakinesis.

prophase II The second stage of meiosis. Chromosome contraction occurs during prophase II.

propliopithecine Any member of the subfamily Propliopithecinae.

proportion of polymorphic loci A ratio that can be calculated by determining the number of polymorphic loci and then dividing by the total number of loci examined.

prosimians Members of a suborder of Primates, the *Prosimii* (pronounced "pro-sim'-ee-eye"). Lemurs, lorises, and tarsiers are traditionally included in this suborder.

protein One of a group of nitrogen-containing organic compounds of complex shape and composition and of high-molecular weight.

protein synthesis The assembly of chains of amino acids into functional protein molecules. DNA directs this process.

proximal Closer to the midline of the body, refers to the appendicular skeleton.

pterygoid muscle A two part muscle that extends from the lateral pterygoid plate to the medial ramus and gonial angle (medial part, closing the jaw and generating occlusal force) and the top of the ramus (lateral part, opening the jaw and moving it from side to side).

pubis The front of the pelvis formed by the parts of the innominate that meet at the midline.

pulp (cavity) The vascularized and innervated tissue that is enclosed in the center of a tooth.

Punctuated Equilibrium Theory A model of evolution in which changes occur when new species are formed and only rarely are changes slowly and gradually accumulated during the stable periods between speciations.

Punnett square A matrix describing all the possible genetic fusions that will give rise to the zygotes that will produce the next generation.

purine A kind of nitrogenous base. In DNA and RNA the purines include adenine and guanine.

pyrimidine A kind of nitrogenous base. Cytosine is a pyrimidine in DNA and RNA, thymine is a pyrimidine in DNA, and uracil is a pyrimidine in RNA.

Q

quadriceps femoris A package of four muscles joining in a large tendon surrounding the patella and attaching to the anterior tuberosity of the tibia. The largest of these, rectus femoris, has an ilium attachment on the anterior inferior iliac spine and brim of the acetabulum. The bundle flexes the hip and extends the knee.

quadrupedalism Posture and locomotion on four feet.

quartz A mineral comprised of the elements silicon and oxygen (silicon dioxide), which occurs in multiple forms.

R

r-selected Referring to an adaptive strategy that stresses relatively large numbers of offspring and reduced parental care (compared to k-selected species). *K-selection* and *r-selection* are relative terms; for example, mice are r-selected compared to primates but k-selected compared to many fish species.

race A set of individual traits that are associated with some populations and/or geographical regions.

radioactive decay The regular process via which radioactive isotopes break down into their decay products with a half-life that is specific to the isotope in question.

radiocarbon dating An absolute dating method that measures the amount of decay of the radioactive isotope of carbon (^{14}C) in organic material.

radioimmunoassay A method of protein analysis whereby it is possible to identify protein molecules that survived in fossils that are thousands, and even millions, of years old.

radiometric A method based on nuclear decay, as in radiometric dating.

radius Located on the thumb side, one of two long bones of the forearm, which rotates against the ulna so that its lower end, the hand, can be turned.

ramus The area of bone at an angle to the body, as in ascending ramus (mandible) or pubic ramus (innominate).

random mating Matings between genotypes that occur in proportion to the frequencies of the genotypes in the population.

recessive An allele or phenotype that is expressed only in the homozygous state.

reciprocal altruism An exchange of favors occurring between two individuals in which one individual temporarily sacrifices potential fitness in expectation of a return.

recombinant chromosome A chromosome that emerges from meiosis with a combination of genes different from a parental combination of genes.

recombinant DNA molecule A new type of DNA sequence that has been constructed or engineered in the test tube from two or more distinct DNA sequences.

regional continuity The observation that a sequence of anatomical features exists, often found together, that span the time from earlier to later populations in a geographic region, and that seem to reflect some degree of ancestral-descendant relationship.

relative dating A determination of the ordered sequence of sites, artifacts, or fossils.

reproductive strategies The complex of behavioral patterns that contributes to individual reproductive success. It is not necessary that the behaviors are deliberate, and they often vary considerably between males and females.

reproductive success The number of offspring that an individual produces and rears to reproductive age; the genetic contribution of an individual to the next generation.

retrovirus Single-stranded DNA virus that replicates via double-stranded DNA intermediates. The DNA integrates into the host's chromosome where it can be transcribed.

rhinarium A hairless patch of skin between the nose and upper lip, which is kept moist as a means of enhancing the sense of smell.

ribonucleic acid (RNA) A usually single-stranded polymeric molecule that consists of ribonucleotide building blocks. RNA is chemically very like DNA. The three major types of RNA in cells include ribosomal RNA (rRNA), transfer RNA (tRNA), and messenger RNA (mRNA), each of which acts in an essential role in protein synthesis (translation). In some viruses, RNA is the genetic material.

ribonucleotide The basic building block of RNA that consists of a sugar (ribose), a base, and a phosphate.

ribose The pentose sugar portion of the nucleotide building block of RNA.

ribosomal DNA (rDNA) The regions of the DNA that house the genes for the rRNAs in prokaryotes and eukaryotes.

ribosome A complex cellular particle comprised of ribosomal protein and rRNA molecules that is the location of amino acid polymerization during protein synthesis.

robust A large or heavily constructed body/ body part.

Robusticity Index An index that can be found by expressing a diameter (or circumference) of a bone in terms of its length.

S

sacculated Subdivided; pertaining to stomach of ruminants.

sacral vertebrae The fused vertebrae that constitute the sacrum, at the back of the pelvis.

sacroiliac joint The joint located at the back of the pelvis between the sacrum and the ilium.

sagittal (plane) A vertical plane on the midline that separates the body into a right and left half.

sagittal crest A compound crest of bone that runs along the midline of the skull for attachment of enlarged temporalis muscles meeting along the midline.

Sahelanthropus tchadensis An early possible hominoid-hominid species from Africa that lived between 7 million and 8 million years ago that has a number of hominid dental traits and may have been bipedal.

sample The subset used to provide information about a population. In order to provide accurate information about the population, it is necessary that it be of reasonable size and be a random subset of the larger group.

savanna A plain characterized by coarse grasses and scattered trees, usually with seasonal rainfall.

scanning electron microscope (SEM) An instrument used in the analysis of the surfaces of tiny structures via a focused beam of electrons, which produce an enlarged image.

scapula The flat, triangle-shaped bone located at the back of the shoulder.

scientific method A method of research in which a problem is identified, a hypothesis (or hypothetical explanation) is stated, and that hypothesis is tested through the collection and analysis of data. The hypothesis becomes a theory once verified.

scraper A generalized word that is used to describe a flake tool with a retouched edge angle of approximately 60° to 90°.

seasonality Components of the environment, or adaptations to it, that vary from one time of year to another.

secondary sexual characteristics Gender-related traits that are not directly involved in reproduction and develop during or following puberty.

sedentary To settle permanently in one location.

sedimentary rock A rock made up of the by-products of other rocks that have eroded or dissolved. Instances of sedimentary rocks include sandstone, mudstone, halite, and chert.

selection coefficient The measure of the relative intensity of selection against a genotype.

selective pressures Forces existing in the environment that influence individual reproductive success.

self-fertilization (selfing) The joining of male and female gametes from one individual.

senescence The process of physiological decline in body function that occurs naturally as a result of aging.

sensory area One of the three areas of the cerebral cortex focused on the reception of information from the body's senses.

sesamoid A bone that forms within a tendon.

settlement pattern Occurs when semipermanent or permanent human habitations are distributed on the landscape and within archaeological communities.

sex chromosome A chromosome in eukaryotic organisms that is represented differently in the two sexes. In many organisms, one sex contains a pair of visibly different chromosomes. One is an X chromosome, and the other is a Y chromosome. Often, the XX sex is female and the XY is male.

sex-influenced traits The characteristics that appear in both sexes but either the frequency of occurrence in the two sexes varies or a different relationship between genotype and phenotype exists.

sex-limited trait A trait that is genetically controlled and is phenotypically exhibited in only one of the two sexes.

sex-linked Pertaining to genes found on the X chromosome.

sex-linked character Feature whose expression is controlled by genes found on the sex chromosomes.

sex-linked dominant trait A trait caused by a dominant mutant gene carried on the X chromosome.

sexual dimorphism A polymorphic trait in which males and females of a species vary in some aspect of their anatomy not directly related to reproduction or birth.

sexual reproduction The type of reproduction that involves the fusion of haploid gametes produced by meiosis.

sexual selection The increased reproductive success of males or females due to characters that either enhance their ability to compete with members of the same sex, or their attractiveness to members of the opposite sex.

shaft Created from the diaphyses; the shaft is the long part of long bone.

shovel-shaped incisors Incisors with a scooped out lingual surface due to lingual marginal ridges, crown curvatures, or a basal tubercle alone or in combination.

sickle-cell anemia A severe inherited hemoglobin disorder resulting from the inheritance of two copies of a mutant allele. This allele is a result of a single base substitution in the DNA.

simian In reference to any member of Anthropoidae (monkeys, humans, apes).

simple random sampling A type of probabilistic sampling in which areas to be sampled are chosen according to a table of random numbers.

sinus A pocket or cavity within a cranial bone; can also be applied to describe the grooved pathways for blood vessels on the endocranial surface.

Sivapithecus Possible ancestor of modern orangutans that lived in Asia between 7 million and 14 million years ago.

slash-and-burn agriculture The traditional practice of clearing lands in which trees and vegetation are cut and burned. In several areas, fields are abandoned after a few years and clearing occurs elsewhere.

sociobiology A type of biology that focuses on the biological basis of all social behavior.

somatic cell At the most basic level, this refers to all the cells in the body minus those involved with reproduction.

speciation The process through which species multiply; the acquisition of reproductive isolation between populations, splitting one species into two.

species In living animals, a group of populations (**biological species**) that can actually or potentially interbreed and produce fertile offspring and are reproductively isolated from other species.

spermatogenesis The development of the male animal germ cell occurring within the male gonad.

sperm cells (spermatozoa) The male gametes; the testes in male animals produce the spermatozoa.

sphenoid bone Irregularly shaped bone that forms part of the base and sides of the skull and the back of the orbit.

spina bifida A condition in which the arch of one or more vertebrae is incapable of fusing and forming a protective barrier around the spinal cord.

spine A projection that is sharp or a short ridge.

stable carbon isotopes Isotopes of carbon that are produced in plants in varying proportions, depending on environmental conditions. Through the analysis of the proportions of the isotopes contained in fossil remains of animals (who ate the plants), it is possible to reconstruct aspects of ancient environments (especially temperature and aridity).

stasis Little to no evolutionary change occurring over an extended period of time; *See* also **Punctuated Equilibrium Theory**

state Pertaining to a social formation that is defined by distinct territorial boundedness, and characterized by a strong central government in which the operation of political power is sanctioned by legitimate force. In cultural evolutionist models, it ranks second only to the empire as the most complex societal development state.

stereoscopic vision A condition in which visual images are, to differing degrees, superimposed on one another. This allows for depth perception, or the perception of the external environment in three dimensions. Stereoscopic vision is partly a function of brain structures.

sternum The breastbone.

stochastic Random.

stone tool An artifact that has been intentionally modified by retouch or unintentionally modified via usewear. Instances of stone tools include projectile points, unifaces, scrapers, and microliths. Debitage would not be considered tools, but rather would be considered artifacts.

strategies Behaviors or behavioral complexes that have been favored by natural selection to increase the reproductive fitness of an individual.

stratification The laying down or depositing of strata or layers (also called deposits) one above the other. A succession of layers should allow a relative chronological sequence, with the earliest at the bottom and the latest at the top.

stratigraphy The location or position of fossil or other deposits in relation to other buried layers or features.

stress In a physiological context, any aspect that acts to disrupt homeostasis; more precisely, the body's response to any aspect that threatens its ability to maintain homeostasis.

striking platform The surface area of an objective piece that receives the force to detach a piece of material. This surface is frequently removed with the detached piece so that the detached piece will contain the striking platform at the point of applied force.

subadults An age category; young individuals that include infants, children, and juveniles.

supination Rotation of the forearm such that the palm faces upward; the opposite movement from pronation.

supramastoid The pneumatized region on the temporal bone located just above the mastoid process base. It serves as a marker of the backward extension of the root of the zygomatic process of the temporal.

supraorbital torus Browridge; a thickened ridge or shelf of bone located above the orbits at the base of the forehead, continuously, although not necessarily evenly, developed from the middle of the cranium to either side.

suspensory Hanging, locomotor and postural habits in which the body is below or among branches.

suture A joint at which two bones interdigitate and are separated by fibrous tissue. The joints between most of the bones of the skull would be considered sutures. As individuals grow older, most sutures join and the bones eventually fuse together.

sympatric speciation A type of speciation without geographic isolation with isolating mechanisms developing within populations.

sympatric Living in the same area; referring to two or more species whose habitats partly or largely overlap.

symphyseal angle An angle formed by the mandibular symphysis face and the lower border of the body, such as the mandibular symphysis and pubic symphysis.

symphysis A flexible fibrocartilaginous joint located on the middle of the body, such as the mandibular symphysis and the pubic symphysis.

synapsis The intimate association of homologous chromosomes caused by the formation of a zipperlike structure along the length of the chromatids referred to as the synaptonemal complex.

synostosis In human skeletons, the joining of separate pieces of bone; the precise timing of such processes is a significant indicator of age.

systematics A science dealing with the diversity of organisms and their relationships and classification.

T

talus (astragalus) The anklebone.

taphonomy Study of the processes that affect the remains of organisms, beginning with the organism's death and continuing through to its fossilization.

tarsals The small bones of the ankle and foot. In humans, these bones include the talus, calcaneus, navicular, cuboid, and three cuneiforms.

taurodont Teeth characterized as possessing enlarged pulp cavities in their roots.

taxon A monophyletic group of organisms that is recognized as a formal unit, at any level of a hierarchic classification.

taxonomy The theory and practice by which organisms are classified.

technology The techniques used in producing artifacts.

tectonic movements Displacements in the plates that constitute the crust of the earth, frequently responsible for the occurrence of raised beaches.

telocentric chromosome A chromosome with the centromere more or less at one end.

telomere-associated sequences Repeated, complex DNA sequences that extend from the molecular gene of chromosomal DNA; suspected to mediate several of the telomere-specific interactions.

telophase A stage during which completion of the migration of the daughter chromosomes to the two poles occurs.

telophase II The final stage of meiosis II during which a nuclear membrane forms around each set of chromosomes, and cytokinesis occurs.

temporal bone Complex bone on the side and base of the cranium, which includes the ear, mandibular joint, and a portion of the side of the braincase.

temporal fossa Space enclosed by the side of the skull and the zygomatic arch, which is occupied by the temporalis muscle as it passes from its attachment on the mandible to its attachment on the cranium.

temporal line The line that is due to the edge of the temporalis muscle where it attaches along the cranial vault. Two lines are present, an inferior line from the deep part of the muscle and a superior line from the superficial part.

temporalis muscle A fan-shaped muscle that moves the jaw in mastication and causes force between the teeth, joining the inside of the mandibular ramus and the skull's side and rear.

temporonuchal crest A compound crest on the back of the skull that is formed by convergence of the temporal line of the nuchal crest.

tendon A strong, inelastic cord of connective tissue that joins muscle to bone.

terrestrial quadruped Type of animal that lives on the ground and moves about primarily on all four limbs.

territory The portion of a home range that is exclusive to a group of animals and is actively defended from other groups of the same species.

tetrasomy The aberrant, aneuploid state in a normally diploid cell or organism in which an extra chromosome pair causes the presence of four copies of one chromosome type and two copies of every other chromosome type.

tetratype (T) One of the three types of tetrads possible when two genes segregate in a cross. The T tetrad possesses two parental and two recombinant nuclei, one of each parental type and one of each recombinant type.

theory A broad statement of scientific relationships or underlying principles, which has been verified at least in part.

thermoluminescence (TL) A dating technique that is indirectly reliant on radioactive decay, overlapping with radiocarbon in the time period for which it is useful, but also has the potential for dating earlier periods. It is similar to electron spin resonance (ESR).

thoracic Referring to the thorax (chest), particularly the rib-bearing vertebrae below the cervical and above the lumbar vertebrae.

thymine (T) A pyrimidine base existing in DNA but not in RNA. In double-stranded DNA, thymine and adenine pair together.

tibia The lower leg's long bones, found between the knee and the foot.

tomography A CAT scan is a radiographic technique that has the capability to show "slices" taken through bones or skulls that will display the shape and extent of internal cavities.

toothpick grooves Elongated grooves located between adjacent teeth, typically on the roots just below the crown level and marked on their facing sides.

trace element analysis The use of chemical techniques, such as neutron activation analysis, or X-ray fluorescence spectrometry, to determine the incidence of trace elements in rocks. These methods are widely used in the identification of raw material sources for stone tool production.

trajectory In systems thinking, this pertains to the series of successive states through which the system proceeds over time. It is possible to say it represents the long-term behavior of the system.

transcription Transfer of information from a double-stranded DNA molecule to a single-stranded RNA molecule; also referred to as *RNA synthesis.*

trapezius muscle A muscle that extends from the nuchal plane to the clavicle and scapula, which stabilizes the shoulder and brings the scapula upwards.

tribes Pertaining to a social grouping generally larger than a band, but rarely outnumbering more than a few thousand; unlike bands, tribes are typically settled farmers, though they also include nomadic pastoral groups whose economy is based on exploitation of livestock. Individual communities tend to be integrated into the larger society via kinship ties.

trisomy An aberrant, aneuploid state in a normally diploid cell or organism in which three copies of a particular chromosome exist, rather than two.

trisomy-21 A human clinical condition that is characterized by various abnormalities. It is the result of the presence of an extra copy of chromosome 21.

trochanter A greater process, for muscle attachment.

trochlea Any smooth, saddle-shaped bony surface that forms part of a joint.

trochlear (sigmoid) notch The notch located within the hook-like proximal end of the ulna, which slides in and out of the olecranon fossa of the humerus.

tubercle An eminence that is small.

Turner syndrome A human clinical syndrome that is a result of monosomy for the X chromosome in the female, which gives a 45,X female. These females fail to develop secondary sexual characteristics, tend to be short, possess weblike necks and poorly developed breasts, are usually infertile, and display mental deficiencies.

tympanic bone The portion of the temporal bone that encloses the inner ear.

tympanic membrane The eardrum.

typology A scheme to order multiple types in a manner that is relational. A typical typology orders types in a manner that is hierarchical.

U

ulna A lower arm bone that hinges at the elbow with the humerus.

uniformitarianism A concept that originated in geology, it is the precept that the processes observable in the present can also be used to explain the past.

Upper Paleolithic A cultural period typically associated with modern humans (but also found with some Neanderthals) and distinguished by technological innovation in many stone tool industries. Best known from western Europe, similar industries are additionally known from central and eastern Europe and Africa.

Upper Pleistocene The portion of the Pleistocene epoch that began 125,000 years ago and ended approximately 10,000 years ago.

uracil (U) A pyrimidine base present in RNA but not in DNA. It pairs with adenine when in double-stranded RNA.

uranium series dating A dating technique based on the radioactive decay of isotopes of uranium. It has proved especially useful for the period before 50,000 years ago, which lies outside the time range of radiocarbon dating.

V

vascularized Possessing blood vessels.

vasoconstriction Narrowing of blood vessels as a means of reducing blood flow to the skin. Vasoconstriction is an involuntary response to cold and results in a reduction of heat loss at the surface of the skin.

vasodilation Expansion of blood vessels, allowing for increased blood flow to the skin. Vasodilation permits warming of the skin and also allows for the radiation of warmth as a means of cooling. An involuntary response, vasodilation is triggered by warm temperatures, various drugs, and even emotional states (blushing).

ventral An animal's belly side; the opposite side of dorsal.

ventral surface of flake The smooth surface of a detached piece, which contains no previous flake removals, minus an occasional eraillure flake scar on the bulb of force.

vertebra One of the vertebral column's bony segments.

vertebral column A structure composed of the vertebrae, from the cervical to the thoracic and lumbar.

vertebral spine A blade of bone that projects dorsally from a vertebra, serves as an attachment site for several ligaments and muscles, also referred to as a spinous process.

vertebrates Animals that possess segmented bony spinal columns; includes fishes, amphibians, reptiles, birds, and mammals.

visual cortex The outer portion of the brain that is responsible for visual input and association; located at the rear of the cerebrum.

viviparous To give birth to live young offspring.

W

weight The force that gravity places on a body of mass.

woodland A vegetation type marked by discontinuous stands of relatively short trees separated by grassland.

world view General cultural orientation or perspective members of a society have in common.

Wormian bones Small bones that form within sutures from isolated centers of ossification between major components of the skull vault; typically located between the occipital and parietal bones.

X

X chromosome A sex chromosome occurring in two copies in the homozygous sex and in one copy on the heterozygous sex.

X chromosome-autosome balance system A system used for genotypic sex determination. The main factor in sex determination concerns the ratio between the numbers of X chromosomes and autosomes. Sex is determined at the time of fertilization, and sex differences are assumed to be a result of the action during development of two sets of genes located in the X chromosomes and in the autosomes.

X chromosome nondisjunction An event that occurs when the two X chromosomes fail to separate in meiosis so that eggs are produced either with two X chromosomes or with no X chromosomes, rather than the usual one X chromosome.

X-linked Pertaining to genes found on the X chromosome.

X-linked dominant trait A trait caused by a dominant mutant gene carried on the X chomosome.

X-linked recessive trait A trait caused by a recessive mutant gene carried on the X chromosome.

X-ray fluorescence spectrometry (XRF) A method used to analyze artifact composition, in which the sample is irradiated with a beam of X-rays that excite electrons associated with atoms on the surface.

Y

Y-5 A cusp pattern existing in lower molars in which there exists five main cusps separated by grooves, and the mesiolingual and distobuccal cusps touch.

Y chromosome A sex chromosome, which when present is located in one copy in the heterogametic sex, along with an X chromosome, and is not present in the homogametic sex. Not all organisms with sex chromosomes possess a Y chromosome.

Y-intercept In a regression analysis, when x is zero, the value of y.

Z

zygomatic arch Bony arch located on the lateral part of the cheek formed by projections of the zygomatic bone and the temporal bone enclosing the temporalis muscle's fibers and for attachment of the masseter muscle.

zygomatic base Location at which the lower border of the cheek merges with the outer wall of the palate.

zygomatic bone The facial bone that comprises the cheek corner and outer orbital pillar, and encloses the front part of the temporal fossa.

zygomatic process Portion of a bone that extends towards and meets the zygomatic bone; there are zygomatic processes of the frontal, temporal, and maxilla.

zygomatic root The rearward base of the zygomatic process of the temporal, as it extends over the opening of the outer ear.

zygomaxillary region The cheek, composed of the anterior face of the maxilla and zygomatic bones.

zygomaxillary ridge A low ridge that extends along the suture between the zygomatic and the maxillary bones.

zygonema The stage during meiosis in prophase I during which homologous chromosomes begin to pair in a highly specific way (similar to a zipper).

zygote The fertilized egg that results from the union of two gametes.

Index

Page numbers in *italic* indicate figures.
Page numbers followed by "t" indicate tables.

A

B

C

Q

R

S